W9-DFQ-391

TX
051
H27

FOOD PREPARATION

Third Edition

an **aTP** publication

Robert G. Haines
R. T. Miller

LIBRARY
WAUKESHA COUNTY TECHNICAL COLLEGE
WITHDRAWN
800 MAIN STREET
PEWAUKEE, WI 53072

Food Preparation contains procedures commonly practiced in the food service industry. Specific procedures vary with each task and must be performed by a qualified person. For maximum safety, always refer to specific manufacturer recommendations, insurance regulations, specific facility procedures, applicable federal, state, and local regulations, and any authority having jurisdiction. The material contained is intended to be an educational resource for the user. American Technical Publishers, Inc. assumes no responsibility or liability in connection with this material or its use by any individual or organization.

American Technical Publishers, Inc., Editorial Staff

Editor in Chief:
    Jonathan F. Gosse
Production Manager:
    Peter A. Zurlis
Art Manager:
    Jim M. Clarke
Technical Editor:
    Aimée M. Brucks
Copy Editor:
    Richard S. Stein
Cover Design:
    Carl R. Hansen
Illustration/Layout:
    Sarah E. Kaducak
    Peter J. Jurek
    So Yun Kim
    Nicole D. Bigos
    Catherine A. Mini
    Gianna C. Butterfield

Drifond is a registered trademark of Ingredient Technology Corporation. Liederkranz is a registered trademark of Beatrice Cheese, Inc.

© 2006 by American Technical Publishers, Inc.
All rights reserved

3 4 5 6 7 8 9 – 06 – 9 8 7 6 5 4 3 2 1

Printed in the United States of America

ISBN 0-8269-4250-4

Acknowledgments

The author and publisher are grateful for the technical information and assistance provided by the following companies, organizations, and individuals.

- Alaska Seafood Marketing Institute
- American Egg Board
- American METALCRAFT, Inc.
- Anchor Food Products, Inc.
- Ansul Fire Protection
- Bunn-O-Matic Corporation
- Burgers' Smokehouse
- Carlisle FoodService Products
- Cress-Cor
- Detecto, A Division of Cardinal Scale Manufacturing Co.
- Florida Department of Citrus
- Florida Tomato Commission
- Frymaster
- Henry Penny Corporation
- Hobart Corporation
- Idaho Potato Commission
- International Banana Association
- International Ice Cream Association
- Joliet Junior College

- Mississippi Department of Marine Resources
- National Broiler Council
- National Cattlemen's Beef Association
- National Chicken Council
- National Cherry Growers and Industries Foundation
- National Fisheries Institute
- National Onion Association
- National Pasta Association
- National Pork Producers Council
- National Potato Promotion Board
- National Turkey Federation
- North American Meat Processors Association
- Poultry and Egg National Board
- Procter and Gamble Co.
- Russell Harrington Cutlery, Inc.
- Southern Pride Catfish
- USA Rice Federation
- Washington Apple Commission
- Wisconsin Milk Marketing Board

- Brigitta McGreal, FACS
  Bolingbrook High School
  Bolingbrook, IL

- Michael McGreal, CEC, CCE, CHE, FMP
  Joliet Junior College
  Joliet, IL

- Mark Muszynski, CEPC
  Joliet Junior College
  Joliet, IL

- Eric Olsen, CSC
  Joliet Junior College
  Joliet, IL

- Kyle Richardson, CEC, CCE, CHE
  Joliet Junior College
  Joliet, IL

- Linda Trakselis, FACS
  Food Service Consultant
  Aurora, IL

- Keith Vonhoff, CEPC, CCE, CHE, FMP
  Joliet Junior College
  Joliet, IL

Contents

Introduction

Food Preparation, 3rd Edition, is a proven introductory text for the food service industry, focusing primarily on large-scale food production. This edition has been thoroughly updated to cover the fundamentals of food service. Basic food items and cooking methods and techniques are presented in addition to food service careers, math, nutrition, sanitation, use of equipment, and safety. More than 860 recipes offer a comprehensive variety of menu items using standard quantities of food service establishments typically serving 25 or more. Detailed illustrations provide a visual reference for cooking procedures and techniques throughout the text.

Both new and expanded coverage in this edition address contemporary technology pertinent to the food service industry, such as computerized food service management systems and commercial kitchen equipment. Safety topics have been updated to address OSHA concerns in the workplace as well as to provide detailed first-aid and fire safety procedures related to the professional kitchen. Diet trends and nutrition, the latest USDA Dietary Guidelines for Americans, and the revised food pyramid are introduced in the new Chapter 4—Nutrition. The latest FDA standards for cooking and storage temperatures are included as well as OSHA sanitation requirements for the prevention of food-borne illnesses.

Math applications in the food service industry are incorporated throughout the text. The Appendix covers basic math concepts, such as weight and measure conversion, recipe conversion, and cost control techniques, that are helpful for new instruction or as a review. The Food Preparation Workbook reinforces these math concepts and other key topics presented in each chapter of the text.

The Glossary contains over 550 food service terms commonly used in the industry and is an essential reference for the aspiring chef. The comprehensive Index may be used to quickly locate particular procedures, techniques, topics, or recipes in the text.

Food Preparation, 3rd Edition, is one many high-quality instructional products from American Technical Publishers, Inc. Additional information about related products can be accessed at www.go2atp.com.

The Publisher

Food Service Careers

The food service and hospitality industry employs millions of people. It is an industry that grows with increases in population. Growth in the industry has also been affected by the changes in lifestyles. More people are eating out more often. The National Restaurant Association estimates that Americans spend more than 475 billion dollars in restaurants each year.

Food service establishments vary in size and in products and services offered. Careers in food service are also varied. Opportunities exist in management, production, sales and service, and sanitation areas from entry-level positions to ownership. The level of knowledge and skills acquired depend upon the desire of the individual to improve and move ahead.

Training for a career in food service was once only accomplished through on-the-job training. Today, food service and culinary training programs offer training for all areas of food service. New advances have been made in food products, food preparation techniques, equipment, and management methods. A career in food service could offer exciting challenges, but requires constant updating to stay current in the field. The food service industry will change, but the need for skilled professionals will continue.

CAREERS IN FOOD SERVICE

Careers in food service are increasing as people are eating out more than ever before. Many new workers are needed each year to fill new jobs in the food service industry. Growth in the food service industry is expected to continue. Successful completion of an accredited culinary or food service program will be a valuable aid to obtaining a desirable job in the industry.

Food Service Training Programs

All food service occupations require training. In the past, training was usually completed on the job. Persons seeking a career in food service would start at an entry-level position and work their way up by hard work and a desire to advance. Today, with higher operating costs, on-the-job training is less common. Persons starting a career in food service usually acquire basic skills and knowledge in a food service or culinary training program. The employer then provides the opportunity for more advanced training while the worker is productive on the job.

Food service and culinary training programs are commonly offered at vocational or career and technical high schools, junior colleges, and some private schools. These programs range in length from two to four years and offer training in the areas of management, production, sales and service, sanitation, and nutrition. In addition, the American Culinary Federation (ACF), in conjunction with selected junior colleges, has initiated an apprenticeship program that requires a minimum of 6000 hours of training and a college level certification as a certified culinarian upon successful completion of an accredited program.

FOOD SERVICE OCCUPATIONS

There are many food service occupations from which one may choose. The best approach when seeking a career in food service is to acquire basic training in all areas of food service before specializing in one particular area. This permits a person to acquire a variety of knowledge and skills to sell to a prospective employer. A person that has a broad base of knowledge and skills may advance in position within a food service establishment from employee to employer. With the proper skills, a person may eventually open and own a food service establishment. According to the National Restaurant Association, approximately one-half of the food service establishments in the United States are sole proprietorships or partnerships.

Food service training programs offer training in the four main food service areas of management, production, sales and services, and sanitation.

Management

The six job classifications in the management area of food service are owner, manager, assistant manager, executive chef, sous chef, and supervisor. The *owner*, or *proprietor*, is the person who has the legal title to or sole possession of the food service establishment. The owner may have an active role in running the operation by assuming the duties of the manager or may give the responsibilities of the operation to the manager. The *manager* conducts and directs all affairs of the operation and oversees food preparation and service. In some cases, the manager orders equipment and determines the budget.

The *assistant manager* helps the manager carry out all the affairs of the operation. The assistant manager is responsible for inventory, scheduling, ordering, personnel, determining food and labor cost, and other related duties.

The *chef* may have the title of executive chef, head chef, working chef, sous chef, or chef steward. The chef is the person of authority in the kitchen and has complete charge of all food preparation and serving of food. The ACF is in charge of certifying and re-

certifying persons such as certified cook, certified sous chef, certified working chef, certified executive chef, certified master chef, certified baker, certified pastry chef, certified executive pastry chef, certified master pastry chef, and certified culinary educator.

The *supervisor* oversees the operation of a food service establishment and confers with and directs the manager of each operation. The title "supervisor" is used most often in franchised establishments with the supervisor in charge of two or more franchised establishments.

Production

The production area includes all personnel required to prepare food as directed by the executive chef. Specific job titles in food service may vary depending on the size and type of the establishment. Job titles classified in the production area include:

- chef, executive chef, or head chef
- chef steward
- working chef
- sous chef
- chef saucier
- night chef
- banquet chef
- pastry chef
- assistant pastry chef
- swing cook
- fry cook
- roast cook
- broiler cook
- soup cook
- garde manger
- breakfast cook
- butcher
- baker
- assistant baker
- night second cook
- night fry cook
- assistant garde manger
- vegetable cook
- cook's helper
- pantry person

The *chef, executive chef*, or *head chef* is in charge of the kitchen in a large operation with a full staff. The chef is responsible for menu planning, supervision of the kitchen staff, attending conferences with management, sous chef, chef steward, and maître d'hotel.

The *chef steward* is a position created by medium-sized hotels. Besides the regular duties of the chef, the chef steward also purchases food supplies. When the chef steward is absent while purchasing food supplies, the sous chef or chef saucier is placed in charge. Preparation and serving of meals for luncheon, dinner, and banquets is supervised by the chef steward.

The *working chef* is a position created for smaller food service establishments as an economy factor. In small hotels, restaurants, and cafeterias, food production is on a smaller scale and requires a smaller crew. Under these conditions, the working chef, in addition to regular duties, assists in production by working where needed. During the hours of service, the working chef works as well as supervises.

The *sous chef* is the chef's first assistant or second in command. *Sous* is a French word for "under." The sous chef carries out the chef's orders for each day, instructs the personnel in the preparation of some foods, and assists the chef in directing all kitchen production and service.

The *chef saucier* or *second cook* is an all-around experienced worker and the lead person of the production crew. The chef saucier follows the sous chef in order of authority. The chef saucier is responsible for preparing boiled, stewed, braised, sautéed and combination cream dishes, as well as taking care of all special a la carte and chafing dish preparations.

The *night chef* is in charge of the kitchen when the executive and sous chefs have left for the day. The night chef supervises the preparation and service of the night menu.

The *banquet chef* is in charge of all parties. The banquet chef supervises the preparation of all party foods and is under the direct supervision of the chef.

The *pastry chef* supervises the pastry department, prepares dessert menus, schedules work performed in the pastry department, and, on many occasions, decorates cakes and special pastries. The pastry chef is under the direct supervision of the executive chef.

The *assistant pastry chef* is under the direct supervision of the pastry chef. The assistant pastry chef participates in the production of cakes, tarts, cookies, and other pastry items.

The *swing cook* relieves cooks at their stations on their day off. The swing cook must cover a different job each day of the week. The swing cook possesses a variety of skills and is able to adapt to irregular working hours

and different tasks. A swing cook has an opportunity to be involved in different areas of food preparation.

The *fry cook* is responsible for work performed around the range and deep fat fryer. Responsibilities include the preparation of eggs, fritters, omelets, crepes (pancakes), potatoes, and other fried items that appear on the menu. The fry cook is also in charge of all vegetable preparation and directs the work of the vegetable person if one is on the staff.

The fry cook's responsibilities include the preparation of omelets.

The *roast cook* is responsible for work performed around the ovens and range. Roast cooks are responsible for the preparation of all roasts and gravies that may accompany the roast.

In smaller food service establishments, the roast cook and *broiler cook* may be a combined job. This requires thorough training in both areas. The broiler cook is responsible for preparing all broiled foods, such as steaks, fish, and chicken.

The *soup cook* is responsible for the preparation of all soup stocks, consommés, hot and cold soups, etc. Soup cooks boil chicken, turkey, and ham in preparation for later use in food production. In some food service establishments, this job title may be combined with the duties of the chef saucier.

The *garde manger*, a French term meaning "one who keeps food," is responsible for the cold meat department. The garde manger oversees the preparation of sandwiches, salad dressings and other cold sauces, seafood and meat salads, the breading of meat and fish, cold appetizers and canapes, and other preparations featuring cold meats. The garde manger must also be experienced in decorating foods for buffets and smorgasbords.

The *breakfast cook* is responsible for preparing all breakfast orders, such as eggs, bacon, ham, sausage, potatoes, toast, biscuits, pancakes, waffles, and hot cereals. The breakfast cook sets up the fry cook station for the luncheon business, and, in many cases, will perform as a fry cook after the breakfast business has been completed.

The *butcher*, or *meat cutter*, is responsible for boning, cutting, and preparing all beef, pork, veal, and their by-products for cooking. In many food service establishments today, butchers are also responsible for cleaning, cutting, and preparing fish and poultry for cooking.

The *head baker* is responsible for the operation of the bakery, but usually is under the supervision of the pastry chef. The head baker prepares all bread, rolls, and quickbreads.

The *assistant baker* assists the head baker in most preparations and also performs the task of keeping the bakery clean and orderly.

The *night second cook* helps serve lunch and sets up the second cook station for the dinner business.

The *night fry cook* helps serve lunch and sets up the fry cook station for the dinner business.

The *assistant garde manger* prepares meat and seafood salads and sandwiches for the luncheon business, assists the garde manger in all types or work, and sets up for the dinner business.

The *vegetable cook* cleans all vegetables and, in some cases, cooks them under the direction of the fry cook.

The *cook's helper* assists all cooks in preparing and serving food. Duties of the cook's helper include cleaning shrimp and removing meat from cooked poultry, cleaning and preparing fruits and vegetables, setting up relish trays, helping serve food at parties, and straining soups and stocks. The cook's helper has an excellent opportunity to learn how many different foods are prepared and served.

The *pantry person* is responsible for the preparation of all side salads and beverages and helps serve desserts.

Sales and Service

The sales and service area includes personnel required to provide service to the customer. The quality of service may determine how the prepared food is received. Good food with poor service still results in a bad impression. Job titles classified in the sales and service area include:

- maître d'
- host or hostess
- server
- cashier
- expediter
- bus person

The *maître d'* is the head of the dining room service. Responsibilities of the maître d' include overseeing the dining room, assigning stations, and directing the hostess, waitress, and waiters.

The *host* or *hostess* is responsible for seating customers in the food service establishment. Particular attention must be paid to seating patrons in their choice of smoking or non-smoking areas. Many states now require that restaurants have separate smoking and non-smoking areas, with some restaurants banning smoking altogether. The host or hostess assigns waiters or waitresses and acts as a representative of the restaurant in the event of a customer concern.

The *server* is responsible for the service of all food and beverage to the guests. No other employee is in a position to have as much influence on the guests as the server. They must take the order, fill the order, serve the order, and keep track of all food and beverages they serve. In addition, they must also enter costs for food on the guest check. Computerized order entry and guest check preparation have had a large impact on this area.

The *cashier* controls the cash. The cashier receives payment of the sales checks, makes change, and is responsible for filling out the cashier's daily worksheet. This is another area in which computerized operations are becoming more frequent.

The *expediter* (food checker) is responsible for all food that leaves the kitchen. The expediter must organize the food orders received, check all the trays that leave the kitchen, and make sure that only the foods ordered are on the tray. The expediter also keeps track of the number of each item served so a check can always be made to see which items are selling.

The *bus person* removes dirty dishes from the table, resets the table, and takes dirty dishes to the dishwashing area. Bus persons assist the waiter or waitress by carrying trays of food to the dining room, filling glasses of water, and assuming other related duties.

Sanitation

The sanitation area includes all personnel required to maintain a sanitary environment. Job titles classified in the sanitation area include:

- steward
- dishwasher
- general kitchen help
- stock clerk

The *steward* is in charge of the dishwashing area and also purchases silverware, china, and glassware. In many large food service establishments, this is the first step up the management ladder.

The *dishwasher* operates the dishwashing machine, which washes all china, glassware, and silverware, and keeps breakage to a minimum. Equipment for washing dishes is also the responsibility of the dishwasher.

General kitchen help is an unskilled job that requires very little training. It is a job that can be a starting point toward a food service career. General kitchen help is responsible for duties that vary from mopping and general clean up to minor preparation work.

The *stock clerk* receives, stores, and organizes food items delivered to the food service establishment. The stock clerk also performs other duties as assigned.

FOOD SERVICE AS A PROFESSION

Food service is a challenging and rewarding profession. Hard work and a desire to learn can result in advancement with better pay, more responsibility, and benefits. A food service employee must adjust to an often-changing work tempo. The work tempo will change from fast to slow, depending on the meal, time, and number of guests. The difference in work tempo makes for an interesting change in the day's activities.

In addition to monetary rewards, compliments received from the clientele or management when the service or food is of excellent quality are extremely rewarding to food service employees. Also, in food service, a person has an opportunity to practice artistic or creative ability. Many people have become well-known for original ideas introduced while working as a chef. Most people consider the outstanding chef as an artist, creating beauty and eye appeal in the preparation and presentation of foods. A job in food service is seldom routine because the menu changes on a daily or monthly basis.

Career opportunities in the food service industry are expected to increase.

The qualities required for a successful food service career include:

- *Attitude*: Workers must be willing to follow instructions and accept constructive criticism.
- *Dependability*: Workers must report to the job on time every day without fail.
- *Ability to work with others*: In some cases, kitchens require working at close quarters. Teamwork is an important part of food preparation.
- *Willingness to work*: When directed to do a task, do it quickly and do it well. Hard work is a way of showing desire.
- *Initiative*: Workers must be willing to do a task without being told.
- *Cleanliness*: Cleanliness is absolutely necessary for good health and sanitation in the kitchen.
- *Interest*: Interest in food service as a career will make the job more enjoyable.

- *Artistic ability*: This ability is developed through practice.
- *Health*: Good health is necessary for optimum performance on the job.
- *Persistence*: Workers must have the ability to follow a job through.

FOOD SERVICE ESTABLISHMENTS

There are many different types of food service establishments in the United States where the food service worker may seek employment. All provide food for their customers, but they serve or dispense food in different ways. Categories in which food service establishments are classified include restaurant, fast-food restaurant, cafeteria, specialty house, coffee shop, institutional unit, caterer, in-plant food service, buffet, and smorgasbord.

Restaurants

A *restaurant* is a public eating house where meals and refreshments are served. A restaurant offers a menu with a variety of items from which to choose. A waiter or waitress takes and serves the orders. The restaurant may be part of another establishment, such as a hotel, motel, convention center, stadium, or department store. Its dining atmosphere and cuisine may vary from simple to deluxe.

Fast-food Restaurants

Fast-food restaurants are designed for fast service and rapid customer turnover. The menu is usually limited, featuring items that can be prepared quickly. The menu is often listed on a large board posted above the counter showing the products and giving prices. These establishment are generally self-serve, or have a limited waiter/waitress service. Fast-food restaurants are commonly found in locations such as airports, shopping malls, and along interstate roads.

Cafeterias

Cafeterias are self-service food service establishments that feature a variety of food preparations. Food is displayed and organized according to courses: salads and juices, vegetables, hot entrees, breads, desserts, and beverages. Prices are low to moderate, with high volume necessary for a successful operation.

Specialty House

This type of food service establishment specializes in one type of food. The speciality may be hamburgers, fish, fried chicken, pancakes, or pizza, among others. Food can be eaten in the establishment or taken out. Low prices and fast service make these operations attractive. A large sales volume is essential for a successful operation.

Coffee Shops

These are usually found in areas where people are on the move. Coffee shops feature a limited menu and specialize in breakfast and luncheon service. Counter and table service are provided, as is coffee, a popular menu item.

Institutional Units

Institutional units provide food service in schools, nursing homes, hospitals, and other institutions. They may be operated by the institution, by a charitable organization, or by a catering company. Institutional units serve large groups of people three meals a day with little or no menu choices. The price of each meal is closely controlled to regulate the operating cost of the institutions.

Caterers

Caterers serve at locations specified by the customer. The food is usually prepared in the caterer's own kitchen, sent to the location designated by the customer, and set up and served as requested. Service varies from a sit-down meal to a buffet or box lunch. Catering is very popular for weddings, business meetings, conventions, office and club parties, and dinner dances.

The Tufts University Nutrition Navigator (www.navigator.tufts.edu) has over 140 links to general nutrition information.

In-Plant

In-plant food service has grown quite rapidly in the recent past. The service is usually run by a catering company specializing in plant service, with few companies operating their own in-plant food service. The food service may be a snack bar, vending machine, or cafeteria. Most in-plant food services offer the worker a limited choice of food and low prices.

Buffets

Buffets offer food set out on a table for ready access and informal service. A buffet can include hot foods, cold foods, or both. Food is usually served on a long table with the cold foods, such as appetizers and salads, first, followed by hot foods, such as vegetables and entrees, and ending with desserts. This type of food service establishment must attract a large volume of business to become and remain a success.

Smorgasbord

The smorgasbord is of Swedish origin and very similar to the buffet. The difference between the smorgasbord and buffet is the variety of food on display and how it is consumed by the guest. A buffet is the way in which food is served. A smorgasbord allows guests to serve themselves a great variety of hot and cold foods. A smorgasbord arranges hot and cold foods separately and directs the guest to eat the cold foods first, followed by the hot foods. Separate tables or service areas may be offered for salads and desserts. Many restaurants will use the term buffet for both buffet and smorgasbord.

COMPUTERIZED FOOD SERVICE MANAGEMENT SYSTEMS

Food service management requires careful control of all facets of the food service establishment, including inventory, sales, labor, and other related concerns. Food service management can be enhanced by using computer technology to provide data to the food service manager.

Computerized restaurant management systems use a personal computer (PC) and specially designed software programs to provide a comprehensive analysis of a food service establishment operation. These systems are commonly referred to as POS (point of sale) systems because data is input into the PC at the point of sale. The data is then processed by the PC and is used to generate reports to be used by the food service manager. The reports can be compiled on location or sent to a remote location using a modem. Reports available using the computerized restaurant management system include:

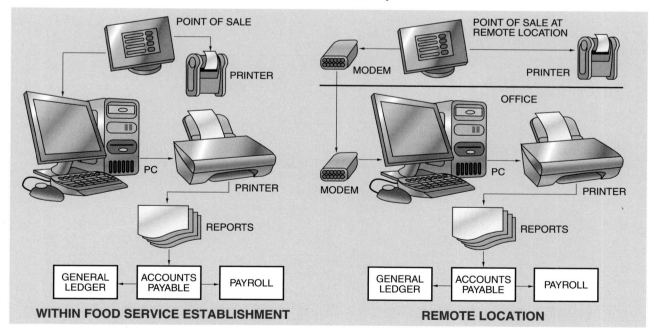

Food service establishment operation reports within the establishment are printed out at the establishment. Reports can also be sent over telephone wires using a modem to a remote location. This allows monitoring of several food service establishments from one location.

- *Daily summary*: cost of sales, labor, and cash control
- *Daily activity*: transaction count, average ticket, labor hours, labor dollars, and labor percentage
- *Sales analysis*: groups menu items by category, dollar volume sold, percentages of each category sold
- *Inventory report*: quantity on hand, price per unit, dollar value of inventory, number of units received
- *Food cost report*: current and period-to-date theoretical and actual dollar usage and efficiency percentage

In addition, other software programs are available for labor scheduling, automatic ordering, accounting, and food bar reports. A computerized restaurant management systems helps increase food service efficiency by providing more useful data to the food service manager.

Fast-food restaurants use computerized food service management systems to improve food preparation efficiency.

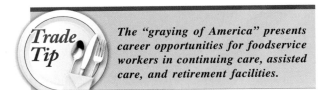

Trade Tip

The "graying of America" presents career opportunities for foodservice workers in continuing care, assisted care, and retirement facilities.

```
P.O.S. DATE  : 09/18   02 (FRI)      ProfitMAX System Report        STORE: 001       PAGE NO. 1
PROCESS DATE : 09/19   11:28                     PIER 66
                                          * * * DAILY REPORT * * *

                              ENTREE  TICK  LABOR  LABOR   LABOR
                              COUNT   AVRG  HRS    DOLLARS  %                    THEORETICAL FOOD COST
        DAY-PART SALES                                                 ------------------<----CUR----> <----PTD----->
                                                                      ITM#  ITEM       CU   VAR   EFF%   VAR   EFF%
PREP     10-11     0.00     0.0% :   0 : 0.00 :  5.0 :   24.50 : 100.00 : -----------------------------------------
LUNCH    11-16 1,506.62 :  21.8% : 191 : 7.89 : 70.5 :  339.70 :  22.55 :
HAPPY HR 16-19 1,076.83 :  15.6% : 157 : 6.87 : 54.0 :  254.65 :  23.62 :  304 GROUPER FIL  LB  -0.39  99.02  -9.1  97.56
DINNER   19-23 4,042.29 :  58.4% : 236 : 17.13: 71.5 :  394.15 :   9.75 :  311 SALMON       LB  -2.04  94.58  -7.3  98.97
EVENING  23-02   293.29 :   4.2% :  33 : 8.89 : 23.0 :  109.68 :  37.4  :  326 OYSTERS RAW  BX  -0.28  99.35  -1.9  98.84
...      ...       0.00 :   0.0% :   0 : 0.00 :  0.0 :    0.00 :   0.0  :  327 OYSTERS GAL  GL  -0.53  99.73  -3.2  99.14
...      ...       0.00 :   0.0% :   0 : 0.00 :  0.0 :    0.00 :   0.0  :  332 WH LOBSTER   EA  -0.00 100.00  -4.0  99.23
...      ...       0.00 :   0.0% :   0 : 0.00 :  0.0 :    0.00 :   0.0  :  337 SHRIMP 31-35 LB -51.62  63.45 -76.3  89.98
...      ...       0.00 :   0.0% :   0 : 0.00 :  0.0 :    0.00 :   0.0  :  338 SHRIMP 91-110LB    .84 101.23 -18.4  93.67
TOTAL         6,921.03 : 100.0% : 619 : 11.18:224.0 : 1122.88 :  16.22 :  341 KING CRAB    LB  -0.58  98.99  -8.3  99.15
                                                                         343 LOB TAIL     LB  -0.23  99.36  -5.1  99.07
COST      THEO $  % SLS          ACTUAL $              ACTUAL $          454 RIBEYE       LB  -1.23  96.52 -12.7  97.83
------------------------ -------------------- -------------------------  455 STRIP LOIN   LB  -0.79  97.24  13.4  96.31
FOOD : 2095.97 : 30.28 :  CASH +/- :  -5.37 :  DEPOSIT :  7261.71 :      458 PRIME RIB    LB  -0.31  99.73 -38.7  88.20
WASTE :  93.82 :  1.36 :  TAX : 346.05 : GROSS PROFIT : 4819.69 :


                          VARIANCE  PROJ           VAR     PROJ                   VARIANCE
              PROJECTED  NET  PROJ/ACT ENTREE ENTREE ENTREE  TICK  LABOR  OPTIMUM  ACT  OPTIMUM/ACT
        HR    SALES    SALES  SALES   COUNT  COUNT  COUNT   AVRG   HRS    HRS     HRS     HRS     LABOR$  LAB %
        ------------------------------------------------------------------------------------------------------
        3 AM  0.00     0.00   0.00    0      0      0       0.00   0.0    0.0     0.0     0.0     0.00    0.00
        4 AM  0.00     0.00   0.00    0      0      0       0.00   0.0    0.0     0.0     0.0     0.00    0.00
        5 AM  0.00     0.00   0.00    0      0      0       0.00   0.0    0.0     0.0     0.0     0.00    0.00
              0.00     0.00   0.00    0      0      0       0.00   0.0    0.0     0.0     0.0     0.00    0.00
```

Reports from the computerized food service management system provide a detailed summary of the food service operation.

Sanitation

The reputation of any food service establishment is earned by the quality of food and service provided. The good reputation of a food service establishment can be quickly ruined by inadequate and improper personal hygiene and sanitation habits. In addition, sickness can result from poor sanitation procedures. Food must be carefully inspected and properly stored. If there is any doubt regarding the quality of a food item, the food item must be thrown out immediately.

Cleanliness is an absolute requirement in the commercial kitchen. Personal hygiene and sanitation procedures determine the cleanliness of a food service establishment. Foodborne illnesses and bacteria can occur as a result of improper food handling, sanitation, and personal hygiene procedures. Management is responsible for maintaining a clean food service establishment for the customer. The food service worker is responsible for maintaining personal cleanliness and a neat appearance. Managers must be familiar with local health codes and certification requirements.

FOOD SAFETY

The Council for Agricultural Science and Technology has estimated that 33,000,000 Americans suffer from food poisoning each year, with 9000 dying from food poisoning. Food safety is a concern for all people involved with the food service industry. Practicing proper food safety guidelines helps eliminate the risk of food poisoning and the spread of foodborne illness.

Many organizations and government agencies are concerned with establishing food safety standards:
- The U.S. Food and Drug Administration (FDA) ensures the safety of all food except meat, poultry, and egg products. The FDA is also responsible for revising and publishing the *Food Code*. The *Food Code* establishes standards that assist food control jurisdictions at the national, state, and local levels in regulating the food service industry. It also provides guidelines to the food service industry for prevention of foodborne illness.
- The Food Safety and Inspection Service (FSIS) ensures the safety of meat, poultry, and egg products.
- The Animal and Plant Health Inspection Service (APHIS) ensures protection of animals and plants from diseases or pests.
- The Environmental Protection Agency (EPA) establishes levels of pesticide residues that can be tolerated by humans.
- The Centers for Disease Control (CDC) investigates foodborne illness.
- The Occupational Safety & Health Administration (OSHA) maintains workplace safety standards including standards regarding foodborne illness.

Potentially Hazardous Foods

A *potentially hazardous food* is any food that requires temperature control because it is in a form that is capable of supporting the rapid and progressive growth of infectious or toxigenic microorganisms. This includes production of *Clostridium botulinum* or, in raw shell eggs, the growth of *Salmonella enteritidis*. These microorganisms may cause foodborne illnesses.

Refrigerated potentially hazardous foods must be maintained at a temperature at or below 41°F per FDA *Food Code* § 3-202.11, *Specifications for Receiving—Temperature*. Exceptions to this include milk and shellfish, which may have alternate minimum temperatures regulated under local laws. Additionally, eggs are allowed to be maintained at a minimum of 45°F. Potentially hazardous foods that have been cooked must be maintained at temperatures at or above 135°F.

Highly Susceptible Populations

Some people are more likely to experience foodborne illness than others. These people are referred to as *highly susceptible populations*. Highly susceptible populations include infants and pre-school age children as well as senior adults. People may also be considered highly susceptible if they are receiving food from a facility that provides custodial care, health care, or assisted living. This includes adult day care centers, kidney dialysis centers, hospitals, nursing homes, and senior centers.

Particular care must be taken when serving highly susceptible populations. For example, raw or undercooked eggs, fish, or meat should not be served to people who are considered highly susceptible and therefore are more likely to experience foodborne illness. If food contains items that should not be served to highly susceptible populations, it should be noted on the menu. Substitute ingredients such as powdered egg whites or pasteurized eggs can be used to avoid health-hazardous situations.

PERSONAL HYGIENE

Personal hygiene is the physical care maintained by an individual. It encompasses important areas of health, cleanliness, attitude, and outward appearance. Good personal hygiene is essential to all food service workers. Not only does maintaining good personal hygiene aid in reducing the risk of contamination of food and work surfaces, but it also helps to instill a sense of personal and professional pride in the workplace. People are generally judged by first impressions until others are better acquainted with them.

A poor first impression may unfortunately be the end of a relationship. First impressions are made primarily by appearance. If personal appearance is neat, clean, and in good taste, the first impression will be a good one.

Personal hygiene is important to consider on a job interview, because a prospective employer may view personal appearance as a reflection of the quality of work a person will produce. On the job, a good physical appearance and personality do much determine a person's future in a food service career. Most importantly, food service workers must maintain good personal hygiene to eliminate the spread of bacteria and disease in the commercial kitchen. Personal hygiene requires the following basic rules regarding personal grooming:

- Keep hands and fingernails clean at all times. Use soap and water to clean hands, and wash vigorously. Rinse hands well and dry them with a clean paper towel.
- Wash hands thoroughly with soap and water after using the restroom, touching anything that may contain bacteria including hair and face, and eating.
- Handle food only as required and with clean hands. Avoid touching clean utensils with hands after touching food.
- Never work with open cuts or sores around food. Cuts or sores should be bandaged. Gloves should be worn to cover injured hands.
- Do not cough, spit, or sneeze near food or in food preparation areas. Always cover a cough with your arm or sneeze into a handkerchief. Wash hands immediately after using a handkerchief or coughing into your hand.
- Notify a supervisor and stay home when sick or if experiencing diarrhea.
- Wear clean, proper clothing for the job.
- Control hair by keeping it washed, neatly trimmed, combed, and covered as required.
- Keep body clean by taking a shower or bath daily.
- Keep facial hair clean and trimmed.
- Do not chew gum or smoke while on the job.
- Do not use nail polish, as it may chip off into food.
- Do not wear artificial fingernails unless wearing gloves, as they may break off into food.
- Remove jewelry which may drop into food or cause a safety hazard.
- Do not allow dirty utensils or equipment to touch food.
- Always take the grooming time required to project the best appearance possible and to maintain sanitary conditions.

Hand Washing

All food service employees must keep hands and exposed portions of their arms clean using approved hand washing procedures. The FDA *Food Code* § 2-301.12, *Personal Cleanliness—Hands and Arms*, specifies the following hand washing procedure for food service employees:

1. Wet hands and arms with hot water.
2. Work up a lather of soap on fingers, fingertips, areas between the fingers, hands, and arms.
3. Scrub the lathered areas vigorously for at least 15 seconds.
4. Clean under fingernails with a nail brush, if necessary.
5. Rinse hands and arms thoroughly with clean, running warm water.
6. Dry hands with either individual disposable paper towels or a heated-air hand drying device per FDA *Food Code* § 6-30.12, *Numbers and Capacities—Handwashing Facilities*.

All handwashing facilities mush be equipped with soap and an approved means for drying hands. All washrooms used by food service employees must display signs or posters notifying food service employees that they must wash their hands before returning to work. Sinks used for food preparation or utensil washing do not need to be equipped with handwashing aids such as soap or hand drying equipment. All hand sanitizers must be FDA-approved for contact with food and food preparation utensils.

Food service employees must wash hands and exposed portions of arms immediately before beginning any food preparation task involving food, clean equipment, and utensils. In addition, hands must be washed under the following circumstances:

- After touching bare human body parts other than clean hands or arms
- After using the washroom
- After coughing, sneezing, or using a handkerchief or disposable tissue
- After using tobacco

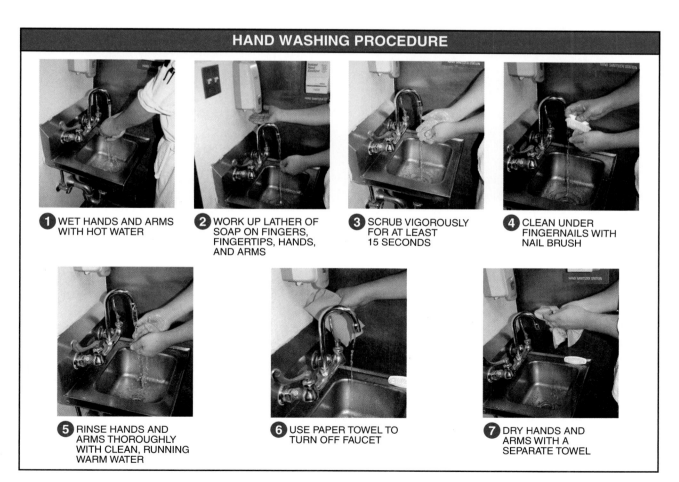

HAND WASHING PROCEDURE

1 WET HANDS AND ARMS WITH HOT WATER

2 WORK UP LATHER OF SOAP ON FINGERS, FINGERTIPS, HANDS, AND ARMS

3 SCRUB VIGOROUSLY FOR AT LEAST 15 SECONDS

4 CLEAN UNDER FINGERNAILS WITH NAIL BRUSH

5 RINSE HANDS AND ARMS THOROUGHLY WITH CLEAN, RUNNING WARM WATER

6 USE PAPER TOWEL TO TURN OFF FAUCET

7 DRY HANDS AND ARMS WITH A SEPARATE TOWEL

The hand washing procedure specified in the FDA Food Code requires hands to be vigorously scrubbed for at least 15 sec.

- After eating
- After drinking
- After handling soiled equipment or utensils
- During food preparation to remove soil and contamination and to prevent cross-contamination
- When switching between working with raw food and ready-to-eat food
- Before donning gloves for working with food
- After engaging in activities that contaminate the hands

BACTERIA

All foods contain bacteria. *Bacteria* are single-celled microorganisms that live in soil, water, organic matter, or the bodies of plants and animals and receive their nourishment by supplying their own food, absorbing dissolved organic matter, or obtaining food from their host, which they usually injure. Bacteria must be controlled using proper sanitation procedures. Foodborne illness is the result of eating food that has been contaminated by harmful bacteria or their toxins (poisons). Persons eating contaminated foods can become extremely ill and, in some cases, die. Bacteria cannot move about on their own and must be transmitted by some vehicle. *Cross-contamination* is the transfer of microorganisms or bacteria from one location to another by means of a vehicle. The most common vehicle transmitting bacteria is the hands, which is why proper hand washing is critical to food safety.

The three main causes of foodborne illness are yeast contamination, bacterial growth, and mold. Some particular types of food will be contaminated more easily than other types of food. Acidic foods (foods containing acid), such as orange juice and tomato mixtures, resist bacteria but are subject to yeast contamination and mold growth. Meat resists yeast contamination, but is subject to bacterial or mold growth unless an acidic mixture such as vinegar is applied. Bread is extremely subject to mold growth.

Bacteria of any type grow rapidly under favorable conditions. Bacteria divide once every 20 minutes, and in a matter of hours can multiply into the millions. Favorable environments for bacterial growth include warm, moist areas with an available food supply. The food service worker must make sure these conditions do not exist when storing foods.

Bacteria grow rapidly in moisture. If all moisture is extracted from the food, the food can be stored with little chance of bacterial growth. This is the reason for the extended shelf life of powdered eggs, dry milk, and other popular dehydrated foods.

Because bacteria and moisture are present in all foods, the most effective method used to control bacterial growth is temperature control. Bacteria grow very slowly at temperatures below 41°F. Bacterial growth is stopped completely at 0°F and below; however, bacteria will not be killed at 0°. Bacterial growth is minimal at 135°F. Bacteria are typically destroyed at temperatures of 180°F or above. The food service worker must heat foods above 135°F or quickly cool them below 41°F to control the growth of harmful bacteria. Bacteria in refrigerated or frozen food are only retarded, not stopped. At room temperature, the bacteria will resume its growth, beginning from whatever stage of development was underway prior to freezing. The *temperature danger zone* is between 41°F and 135°F. Food left in the danger zone longer than 4 hours must be discarded. Local health authorities may have stricter regulations regarding the temperature danger zone.

Heating Foods

Heating foods to their minimum required internal cooking temperature ensures that bacteria are destroyed. The FDA *Food Code* establishes minimum internal cooking temperature standards. Food must be heated to the specific minimum temperature, and that temperature must be maintained for the required time to effectively destroy bacteria. The higher the internal temperature, the shorter the required time period. At some internal temperatures bacteria destruction is instant, so the temperature does not have to be maintained for a minimum time. For example, when ground beef is heated to 168°F, the bacteria are instantly killed. At an internal temperature of 145°F, however, the temperature must be maintained for 3 min. to kill all bacteria that may be present.

Beef may be served raw or undercooked if it is served to persons not included in a highly susceptible population. However, the raw or undercooked beef must be extremely fresh. The beef must meet FDA code requirements for "whole-muscle, intact beef," and all external surfaces must be cooked to a surface temperature of 145°F and achieve a color change. In addition, raw fish or egg products may also be served to persons not included in a highly susceptible population under specific guidelines listed in FDA *Food Code* § 3-401.11(D). Items unacceptable for consumption by highly susceptible populations should be noted on the menu.

Care should be taken when thawing potentially hazardous foods to prevent growth of bacteria. Foods must be thawed either in a refrigerated environment of 41°F or less, or completely submerged under running water per FDA *Food Code* § 3-501.13, *Temperature and Time Control—Thawing*. Frozen food may also be thawed directly as part of the cooking process or thawed in a microwave and immediately cooked.

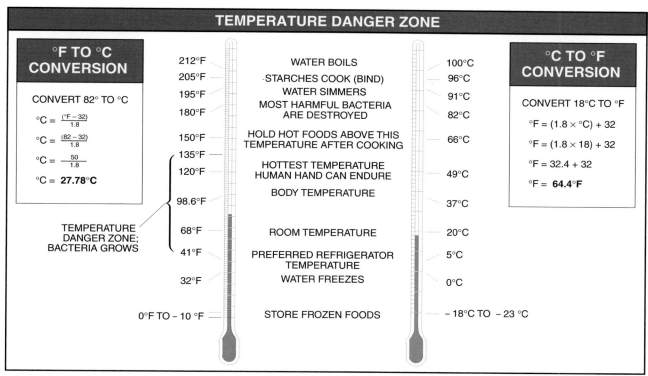

TEMPERATURE DANGER ZONE

°F TO °C CONVERSION

CONVERT 82° TO °C

$$°C = \frac{(°F - 32)}{1.8}$$

$$°C = \frac{(82 - 32)}{1.8}$$

$$°C = \frac{50}{1.8}$$

$$°C = \mathbf{27.78°C}$$

TEMPERATURE DANGER ZONE; BACTERIA GROWS

°C TO °F CONVERSION

CONVERT 18°C TO °F

$$°F = (1.8 \times °C) + 32$$

$$°F = (1.8 \times 18) + 32$$

$$°F = 32.4 + 32$$

$$°F = \mathbf{64.4°F}$$

212°F	WATER BOILS	100°C
205°F	STARCHES COOK (BIND)	96°C
195°F	WATER SIMMERS	91°C
180°F	MOST HARMFUL BACTERIA ARE DESTROYED	82°C
150°F	HOLD HOT FOODS ABOVE THIS TEMPERATURE AFTER COOKING	66°C
135°F		
120°F	HOTTEST TEMPERATURE HUMAN HAND CAN ENDURE	49°C
98.6°F	BODY TEMPERATURE	37°C
68°F	ROOM TEMPERATURE	20°C
41°F	PREFERRED REFRIGERATOR TEMPERATURE	5°C
32°F	WATER FREEZES	0°C
0°F TO – 10 °F	STORE FROZEN FOODS	– 18°C TO – 23 °C

The temperature danger zone is between 41°F and 135°F.

FDA MINIMUM INTERNAL COOKING TEMPERATURES

Food	Temperature*	Time†	Food	Temperature*	Time†
Eggs	145	0.15	Roasts (beef, pork)	130	112
Fish	145	0.15		131	89
stuffed	165	0.15	Minimum oven temp (conventional) 350˚F if less than 10 lb	133	56
Pork, beef, lamb, or veal	145	0.15		135	36
				136	28
	145	3		138	18
injected/ground	150	1		140	12
	168	instant	Minimum oven temp (convection) 325˚F if less than 10 lb	142	8
stuffed	165	0.15		144	5
				145	4
Poultry	165	0.15		147	2.25
Stuffed pasta	165	0.15		149	1.5
Stuffing containing fish, meat, or poultry	165	0.15	Minimum oven temp (conventional or convection) 250˚F if greater than 10 lb	151	1
				153	0.5
Fruits and vegetables	140	instant		155	0.25
Reheated fully-cooked potentially hazardous food	165	0.15		151	0.14
				158	instant

* in ˚F
† in min

Foods must be heated to minimum internal cooking temperature standards as established by the FDA *Food Code*.

Cooling Foods

Cooling foods for storage also helps prevent bacteria growth. Potentially hazardous food that has been cooked must be cooled to 70°F within 2 hr following cooking. Food must then be cooled to 41°F or less for storage within the next 4 additional hr. If the food was prepared with ingredients held at room temperature, it must be cooled to 41°F within 4 hr. Raw eggs must be immediately cooled in refrigerated equipment maintaining a temperature of 45°F or less. Local health requirements may vary. Always follow local health department requirements. The following methods may be used to decrease cooling times:
- Place the food items in shallow pans.
- Separate the food items into small, thin portions.
- Use rapid cooling equipment.
- Place the food in a container held in an ice water bath.
- Add ice to the food.
- Use storage containers that maximize heat transfer, such as aluminum pans.

Carlisle FoodService, Inc.
Sneeze guards protect food from germs. Ice keeps food cold.

Sanitizers, Disinfectants, and Antiseptics

While temperature is the best method used to destroy bacteria, it is not always practical to use. Bacteria can also be destroyed using chemical agents. *Sanitizers* are chemical agents used to reduce the level of microorganisms to a safe level. Sanitizers include chlorine, iodine, and quaternary ammonia compounds. The number of bacteria destroyed depends on the strength of the chemical used. If the chemical is dissolved in a solution and used to kill disease-producing organisms only, it is known as a *disinfectant*. A milder chemical solution used to treat a wound and inhibit the growth of disease organisms is known as an *antiseptic*. A chemical used in foods to retard the growth of bacteria that cause spoilage is known as a *preservative*. Preservatives are commonly used in many foods to extend shelf life.

TRANSPORTING FOOD

Some food service operations may require transporting foods from one area to another. Caterers and larger food service operations transport prepared or partially prepared foods from a central kitchen. In transporting foods, steps must be taken to ensure the food is not contaminated or bacteria do not spoil the food by multiplying while in transit. Certain precautions should be followed when transporting foods:
- The transport containers used must be clean, tightly sealed, and designed for efficient cleaning.
- The transport containers must have the required refrigeration or heating elements to maintain the proper temperature. Cold foods require temperatures of 41°F or below. Hot food should be above 135°F.
- Use the shortest route possible to the area where the food will be served. Minimize loading and unloading time.
- Foods on display for a buffet or salad bar or smorgasbord foods should not be at room temperature for more than 1 hr. Cold food should be kept at a temperature of 41°F or less and hot foods at 135°F or more. Cold foods are kept cold using ice or a refrigeration unit. Hot foods are kept hot using steam tables or chafing dishes.
- Displayed foods must be protected with a sneeze guard (shield that protects food from being contaminated). Leftover foods that are not individually packaged should be discarded at the end of the meal.

FOODBORNE ILLNESS

Foodborne illness is often caused by bacteria or viruses that may be present in foods or beverages. Poor habits such as improper hot and cold holding temperatures, poor personal hygiene habits by food service workers, storing food in punctured or opened cans, or allowing food to be contaminated by rodents or insects can all contribute to foodborne illness. Harmful bacteria and viruses causing foodborne illnesses include Campylobacter, Salmonella, E. coli O157:H7, Calicivirus, Shigella, Hepatitis A virus, Staphylococcus, Listeria monocytogenes, Clostridium botulinum, and Streptococcus.

ILLNESS-CAUSING FOOD CONTAMINANTS		
Illness	**Symptoms**	**Sources**
Campylobacter	fever, diarrhea, abdominal cramps	poultry
Salmonella	fever, diarrhea, abdominal cramps	poultry, eggs, dairy products
E. coli O157:H7	severe diarrhea	contamination-cow feces
Calicivirus (Norwalk-like virus)	gastrointestinal illness, vomiting, diarrhea	water, fruit, vegetables, shellfish
Shigella	chronic disease of large intestines	flies, unwashed hands
Hepatitis A virus	liver inflammation	contaminated foods, infected people
Staphylococcus	can cause meningitis, pneumonia	contagious bacteria
Listeria	fever, muscle aches, gastrointestinal symptoms	contaminated foods
Clostridium botulinum	constipation, diarrhea, vomiting- can be fatal	contaminated foods-extremely rare
Streptococcus	sore throat, fever, strawberry tongue	contaminated foods, infected people, highly contagious

Foodborne illness is often caused by bacteria or viruses that may be present in foods or beverages.

- *Campylobacter*: Bacteria causing fever, diarrhea, or abdominal cramps. Bacteria live in the intestines of birds. Food poisoning caused by contaminated poultry and poultry products such as eggs. The most commonly diagnosed cause of diarrheal illness worldwide.
- *Salmonella*: Bacteria in poultry, mammals, and reptiles causing food poisoning. Symptoms include fever, diarrhea, and abdominal cramps. It can invade the bloodstream and cause life-threatening infections.
- *E. coli O157:H7*: Bacterial pathogen found in cattle and similar species. Usually contracted through contamination by microscopic amounts of cow feces. Symptoms include severe and sometimes bloody diarrhea. Hemolytic uremic syndrome (HUS) is a rare and severe complication of E. coli.
- *Calicivirus (Norwalk-like virus)*: Virus believed to be spread through infected humans. Common cause of foodborne illness, though rarely diagnosed. Symptoms include gastrointestinal illness, vomiting, and diarrhea. Normal duration of two days.
- *Shigella*: Bacteria that cause shigellosis, an acute or chronic disease of the large intestine of humans, coming from defective plumbing, water contaminated at its source, and food contaminated by flies or unwashed hands after using the washroom.
- *Hepatitis A virus*: A virus causing hepatitis A, an acute viral infection causing inflammation of the liver. Transmitted by contaminated food, improperly sterilized hypodermic needles, or infected people.
- *Staphylococcus aureus (staph)*: Very contagious, round, parasitic bacteria showing up as boils, carbuncles, infection, etc. Can cause meningitis, pneumonia, etc.; passed through blood and mucous membranes.
- *Listeria monocytogenes*: Bacteria causing infection *(listeriosis)* primarily in pregnant women, newborns, and adults with weakened immune systems. Symptoms include fever, muscle aches, and some gastrointestinal symptoms. Infections in pregnant women could cause stillbirth, miscarriage, premature delivery, or infection of the newborn.
- *Clostridium botulinum*: A poisonous bacterium causing a type of food poisoning *(botulism)*. Improper canning techniques can allow *Clostridium botulinum* to grow and cause botulism.
- *Streptococcus (strep)*: Spherical, aerobic bacteria which cause strep throat, scarlet fever, etc.

Food contamination can also be caused by insect and rodent infestation, improperly used or stored cleaning supplies, and physical hazards such as hair, bone, or broken glass. Most rodents and insects live in colonies. If one is spotted, it is likely there are more on the premises. To eliminate these pests, ensure that all openings in doors, windows, air vent screens, and other openings are sealed.

Check incoming supplies for any indication of pests. Clean all garbage areas before garbage accumulates. An integrated pest managment program is essential in a food service facility. Retain the services of a professional exterminator. Pest control poisons and insecticides are dangerous if improperly used.

The quality of food, and controls used to prevent foodborne illnesses, are primarily regulated by the FDA, the CDC, and local public health authorities. In addition, OSHA maintains workplace standards related to foodborne illnesses. OSHA 29 CFR 1910.141, *General Environmental Controls,* lists specific guidelines in regard to waste disposal, potable (drinkable) water sources, and washing. In regard to food handling, OSHA requires all food service facilities and operations to conform to sound hygienic principles. Food dispensed shall be wholesome and free from spoilage, and shall be processed, prepared, handled, and stored in such a manner as to be protected against contamination.

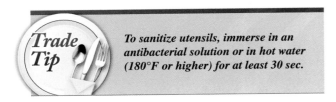

Trade Tip

To sanitize utensils, immerse in an antibacterial solution or in hot water (180°F or higher) for at least 30 sec.

WASHING AND SANITIZING

Many products are available to management to help ensure a good sanitation program. However, the food service worker is the most important part of any sanitation program. It has been stated that any cleaning job requires 95% human effort and only 5% mechanical effort. Sanitation efforts are becoming increasingly more mechanized, but the food service worker is still responsible for following and maintaining the necessary sanitation standards, regardless of the equipment used.

Many problems relating to sanitation can be traced to improper cleaning and sanitizing of dishes, silverware, glassware, pots, pans, and utensils. Pretreating these items by soaking, rinsing, and scraping is necessary in order to remove any buildup of dirt, grease, or grime. If they are placed in the wash water without pretreatment, the effectiveness of the detergent is greatly diminished.

Once ready for washing, these items may be washed utilizing automatic or manual techniques. If utilizing the automatic method, follow the codes of your local health department regarding water temperatures for washing, rinsing, and sanitizing. Proper spacing of the items in an automatic dishwashing machine is also an important factor in whether the wash water reaches all areas of the items being washed. If utilizing a manual method, hot water is necessary for washing, rinsing, and sanitizing. Regardless of method, equipment, or cleansing and sanitizing products used, the temperature of wash water should exceed 160°F. The temperature of rinse water should exceed 180°F. In most cities, wash and rinse water temperature requirements are specified by the local health department. This ensures that all bacteria are destroyed in the washing and rinsing process. Glassware should be washed in clear water using a compound recommended for glassware.

During the wash stage, proper and effective scrubbing of all surface areas helps to ensure that the sanitizing solution will effectively contact the item. Hot water, friction, and detergent are necessary for proper results when pot washing. Of the three, friction is generally the most neglected. Pots should be scraped clean, placed in hot water containing a good detergent, and scrubbed thoroughly. Care must be taken when using steel wool and other types of scouring pads because pieces will come off as they are used.

Proper rinsing is essential to remove any detergent residue as it may weaken the strength of the sanitizer. Some health departments require that approved sanitizing chemicals such as iodine or chlorine be added to the rinse water. Follow the codes of your local health department regarding allowed chemical sanitizers and chemical sanitizer concentration. The items being washed are passed through hot rinse water, then submerged in an approved sanitized solution before being allowed to air dry.

After proper washing procedures, dishes, silverware, glassware, and pots must be carefully handled and safely stored to prevent contamination before the next use. Regardless of the cleaning method used, all products must be allowed to air dry prior to storage. Drying with a towel may recontaminate the item.

Preparation and cooking tools are also a source of bacterial growth. For example, food particles will cling to the blade of a knife when cutting certain foods. The unclean knife may be stored and used the following day. When it is used again, cross-contamination occurs when the knife transfers bacteria from the blade to the item cut. This also can occur with forks and other tools. Whether the tools are used by the entire kitchen staff or are personal tools of the chef or cook, they must be thoroughly washed and sanitized.

Strict sanitation standards must also be followed in cleaning stationary equipment. All stationary equipment and attachments must be cleansed after use, usually with soap and water. Hard rubber and plastic cutting surfaces should be cleaned as recommended by the manufacturer.

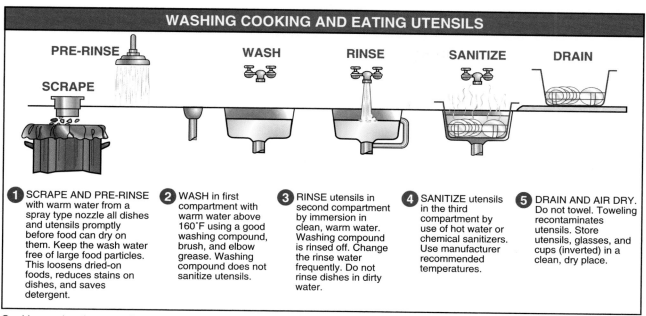

WASHING COOKING AND EATING UTENSILS

PRE-RINSE   **WASH**   **RINSE**   **SANITIZE**   **DRAIN**

SCRAPE

1 SCRAPE AND PRE-RINSE with warm water from a spray type nozzle all dishes and utensils promptly before food can dry on them. Keep the wash water free of large food particles. This loosens dried-on foods, reduces stains on dishes, and saves detergent.

2 WASH in first compartment with warm water above 160°F using a good washing compound, brush, and elbow grease. Washing compound does not sanitize utensils.

3 RINSE utensils in second compartment by immersion in clean, warm water. Washing compound is rinsed off. Change the rinse water frequently. Do not rinse dishes in dirty water.

4 SANITIZE utensils in the third compartment by use of hot water or chemical sanitizers. Use manufacturer recommended temperatures.

5 DRAIN AND AIR DRY. Do not towel. Toweling recontaminates utensils. Store utensils, glasses, and cups (inverted) in a clean, dry place.

Cooking and eating utensils must be properly washed and sanitized to prevent contamination from bacteria.

SANITATION GUIDELINES

The following guidelines for sanitation should be practiced:

- Get hot foods hot quickly, and keep hot at 135°F or above.
- Get cold foods cold quickly, and keep cold at 41°F or below.
- Always wash hands with soap and water before starting work, after visiting the restroom, after working with raw meat, poultry, or seafood, any time they are dirty, and as required at least once every four hours or as directed by local codes.
- Always use sanitized cooking utensils. Always clean and sanitize work areas after they are used.
- Keep all foods sealed and/or covered as much as possible.
- Use professional food handling tools for cutting, cooking, and serving.
- Purchase USDA inspected meat only.
- Thoroughly rinse all fruits and vegetables before using.
- Only use pasteurized milk.
- Do not prepare too much food in advance.
- Do not allow food to remain between 41°F to 135°F (temperature danger zone) for more than 4 hr.
- Do not refreeze thawed meat, fish, or vegetables. Thawing and refreezing causes cellular breakdown and increases susceptibility to decay.
- Do not store opened canned foods in the opened can.
- Use only pasteurized egg products in items that will not be fully-cooked.
- Make sure cans have not been damaged.
- Inspect all foods thoroughly. Contaminated foods do not always have an unusual odor, taste, or appearance.
- Frozen foods should be thawed under controlled conditions, such as in a refrigerator with the temperature set between 34°F and 38°F to prevent cross-contamination.
- Properly dispose of all garbage and rubbish promptly.
- The refrigerator and freezer are the most important pieces of equipment for controlling bacteria. Check their temperature on a regular basis each day.
- Check all fish, shellfish, and other orders for freshness when delivered.
- Exercise caution with leftovers. Always refrigerate as soon as possible. All leftovers must be reheated quickly to an internal temperature of 165°F.
- Avoid handling food more than necessary. Use disposable plastic gloves when possible and dispose of after contact.
- Properly cook all foods to their minimum internal cooking temperature or higher.
- Food should not be handled by any person having an acute illness, diarrhea, or open or infected sores.

Strict sanitation standards must be followed in the commercial kitchen.

- Immediately clean and sanitize all equipment that has been used on potentially hazardous foods. Equipment such as slicing machines, food shredders and grinders, cutting boards, can openers, and knives are particularly susceptible to contamination.
- When washing dishes, the wash water temperature should be 160°F, and the rinse water should have an approved concentration of sanitizer.
- Store glasses, cups, pots, and bowls with their bottoms up.
- Keep dirty dishes, utensils, and towels away from food.
- If ever in doubt about any food, throw it away.

Many states require food service managers to be certified in the area of sanitation. Certification is obtained by taking certain courses required by the state board of health. These courses usually pertain to the classification and types of bacteria, foodborne illnesses, and other related sanitation subjects.

Trade Tip

To sanitize cutting boards and counter surfaces, use an FDA-registered antibacterial product (follow label directions carefully) or mix 3 tsp of chlorine bleach in 1 gal. of water. Wipe surfaces clean and allow to air dry.

Tools and Equipment

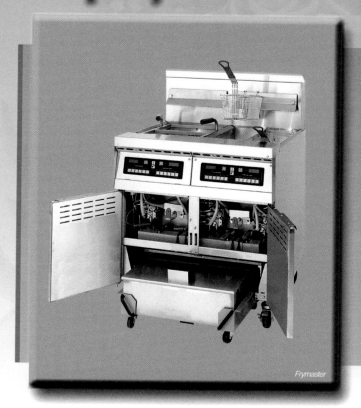

## Chapter 3

Frymaster

In the commercial kitchen, hand tools and equipment are used to prepare food items to be served. Hand tools are generally provided by the food service establishment. However, in the case of knives and other special tools, most chefs purchase their own. The accomplished chef takes pride in personal equipment used, keeping knives sharp and other tools in the best condition at all times.

Equipment is usually larger, heavier machinery placed in one specific location and is very seldom moved. Equipment used in the commercial kitchen is expensive and is always purchased by the establishment. New equipment requires that the food service staff have additional training to learn to use it. The manufacturers of the equipment are an excellent source of operation information. The safe and proper use of knives and other potentially dangerous tools and equipment is acquired after training and practice. With experience using tools and equipment, food preparation becomes increasingly more efficient.

HAND TOOLS

Hand tools are hand-held implements used in preparing food. Hand tools are generally supplied by the food service establishment and are stored on racks or in cabinets. However, knives and other special hand tools are commonly purchased by the chef. These knives and hand tools are for personal use. The experienced chef takes pride in personal equipment and in keeping knives sharp and other hand tools in the best condition at all times. Hand tools, especially cutting tools, are easier to keep in good condition if only one person uses and maintains the tools.

Knives and Other Cutting Tools

Knives are the chef's most important tool. A set of knives should consist of a French knife (sometimes known as a chef or sandwich knife), a carving knife (roast beef or ham slicer), a boning knife, a utility knife, and a paring knife. Other knives are available for specialized tasks.

Knives should be selected for quality and by personal preference. Like any tool, a high-quality knife may cost more initially, but will perform better and last longer. All knives must be kept sharp at all times. A sharp knife is safer than a dull knife because it requires less force to use and will not slip off the item being cut as easily. A steel or sharpening stone should be used periodically to keep the blade sharp.

Boning knife: This is a short, thin knife with a pointed blade. It is used to remove raw meat from bones with minimal waste. The blade may be either stiff or flexible. Popular lengths run 6″ to 8″.

Butcher knife: This knife has a slightly curved, pointed, heavy blade. It is used in cutting and sectioning raw meat.

Butcher's steel: This is a round steel rod approximately 18″ long with a handle. It is used to maintain an edge on a knife. The butcher's steel does not sharpen the edge of the blade, but hones it to maintain a sharp edge.

Cherry or *olive pitter*: This is a speciality tool used to remove the pits of cherries and olives.

Clam knife: This speciality tool has a short, flat, round-tipped blade. It is used to open clams.

Cleaver: A cleaver is a heavy, square blade knife made of carbon steel. It is used to chop bones.

Crown cutter: This tool is used to quickly cut grapefruit, oranges, lemons, tomatoes, and melons into decorative saw-toothed crown cut halves.

French knife: The French knife is the most popular knife and the most used cutting tool. Near the handle, the blade is wide and generally a bolster is present. The blade tapers to a point. It is used for slicing, chopping, mincing, and dicing. The most popular blade lengths are 8″, 10″, and 12″.

Slicer: The slicer has a narrow, flexible blade between 12″ to 14″ long. It is used to slice food such as ham or beef.

Hand meat saw: This saw has a thin, fine-toothed blade that is attached to a bow-shaped metal frame. The hand meat saw is used to saw through bones. Hand meat saws are available in different sizes.

Lettuce shredder/chopper: This speciality tool is used to cut large quantities of lettuce into uniform pieces for use in salad bar or catering operations.

Onion slicer: The onion slicer is used to slice and dice onions and other firm vegetables and fruits with minimum bruising and bleeding. It can be adjusted for slice sizes from 3/16″ to 1/2″ thick.

Oyster knife: This speciality tool has a short, slightly thin, dull-edged blade with a tapered point. It is used to open oysters.

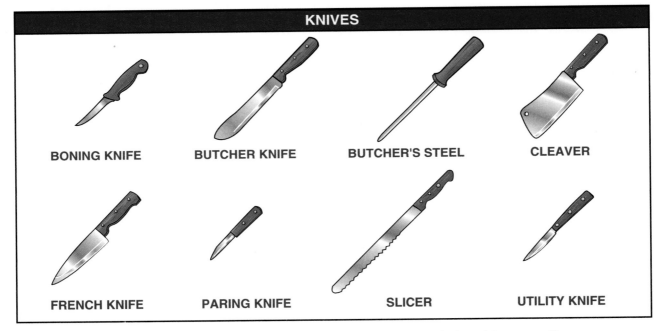

KNIVES

BONING KNIFE   BUTCHER KNIFE   BUTCHER'S STEEL   CLEAVER

FRENCH KNIFE   PARING KNIFE   SLICER   UTILITY KNIFE

Knives are the most important tool used by the chef or cook. Each knife is designed for a specific use.

Paring knife (vegetable knife): The paring knife is a short knife with a 2½″ to 3½″ blade. It is used for paring fruits and vegetables. The blade point can be used to remove eyes and blemishes in fruits and vegetables.

Pastry wheel: This round, stainless steel disk has a cutting edge that is rolled to cut all types of pastries.

Pie and *cake knives*: These offset knifes have a wide, flat blade that is tapered to a point and shaped like a wedge of cake or pie. The pie or cake knife is used to cut and serve pies and cakes without breaking the pieces.

Potato or *vegetable peeler*: This cutting tool has a metal blade attached to a metal handle. The blade is in the form of a loop, with sharpened edges, formed over a pin or axis attached to the handle. The blade can shift from side to side, allowing peeling in two directions.

Tomato slicer: This speciality tool is used to slice tomatoes quickly and uniformly. The blades can be adjusted for 3/16″, ¼″, and 3/8″ slices.

Utility knife: The utility knife is a pointed knife 6″ to 8″ long. It is used for a variety of cutting tasks, including cutting lettuce, fruits, and some meats.

Food Handling Tools

Food tongs: Spring-type metal consisting of two grippers with a saw-toothed grip on each end. The tongs are used to pick up and serve foods without touching the food with the hands.

Grill tender: The grill tender is used to clean grills quickly without burning the hands. A built-in splash guard catches splattered hot grease. Hardened carbon steel blades prevent gouging the grill surface when scraping.

Turner or metal spatula: This turner has a wide, flat, offset chisel-edged blade with a handle. It is used to turn hotcakes or hamburgers while grilling or broiling.

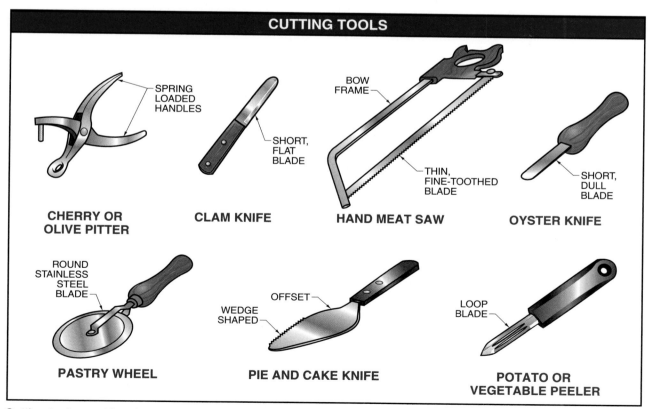

CUTTING TOOLS

SPRING LOADED HANDLES

CHERRY OR OLIVE PITTER

SHORT, FLAT BLADE

CLAM KNIFE

BOW FRAME

THIN, FINE-TOOTHED BLADE

HAND MEAT SAW

SHORT, DULL BLADE

OYSTER KNIFE

ROUND STAINLESS STEEL BLADE

PASTRY WHEEL

OFFSET

WEDGE SHAPED

PIE AND CAKE KNIFE

LOOP BLADE

POTATO OR VEGETABLE PEELER

Cutting tools must be cleaned properly after use to ensure good performance.

Kitchen fork: The kitchen fork is one of the most popular hand tools. It is a large two-pronged fork used for holding, slicing, turning, and broiling meats.

FOOD HANDLING TOOLS

WIDE, FLAT BLADE

GRIPPERS

PRONGS

FOOD TONGS

HOTCAKE OR MEAT TURNER

KITCHEN FORK

LONG HANDLE

DOES NOT HOLD LIQUID

FLAT PERFORATED DISK

HOLDS LIQUID

SOLID LADLE

PERFORATED LADLE

SKIMMER

RECTANGULAR METAL BLADE

PLASTIC SCRAPER

SCRAPER OR DOUGH CUTTER

American METALCRAFT, Inc.

Food handling tools are designed for specific uses.

Can opener: The can opener is mounted on a kitchen table for easy access and cleaning. It is used to open cans quickly.

Ladle: This is a stainless steel cup, solid or perforated, attached to a long handle used to stir, mix, and dip. It is also used to serve sauces, dressings, and other liquids when portion control is desired.

Ladles are available in many sizes. Table I lists ladle sizes and approximate portion weights.

TABLE I: LADLE SIZES AND APPROXIMATE WEIGHTS	
Ladle Size	**Approximate Weight of Portion**
¼ c	2 oz
½ c	4 oz
¾ c	6 oz
1 c	8 oz

Melon ball or *parisienne scoop*: The scoop has a stainless steel blade formed into a round half-ball cup attached to a handle. It is used for cutting various fruits and vegetables into small balls.

Potato scraper: The potato scraper is used to remove the skin from the potato quickly. The meat is cut into uniform wedges.

Plastic scraper: This flexible piece of plastic is approximately 4″ wide and 6″ long. It is used to scrape bowls when mixing batter.

Scraper or *dough cutter*: This wide, rectangular metal blade with a handle is used for scraping meat blocks and cutting dough.

Skimmer: This flat, stainless steel, perforated disk connected to a long handle is used to skim grease or food particles from soups, stocks, and sauces.

Spoons and *spatulas*: Spoons and spatulas are used for stirring, serving, scraping, and spreading.

Slotted kitchen spoon: A large stainless steel spoon with three to four slots cut into the bowl of the spoon. It is used to serve large cut vegetables or whole items without the liquid.

Solid kitchen spoon: A large, stainless steel spoon that holds about 3 oz. It is used for folding, stirring, and serving.

Spatula or *palette knife*: A broad, flexible, flat or offset blade knife with a round nose. It is used for mixing, spreading, and sometimes scraping. It comes in lengths from 3½″ to 12″ and is available from semi-flexible to highly flexible. It is most commonly used for spreading icing on cakes.

Spatula holder and *cleaner*: A device used for scraping and cleaning and as a holder for spatulas used on the griddle.

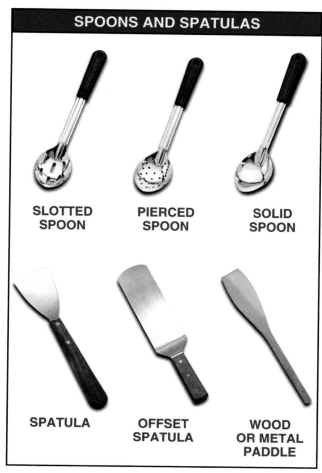

SPOONS AND SPATULAS

SLOTTED SPOON PIERCED SPOON SOLID SPOON

SPATULA OFFSET SPATULA WOOD OR METAL PADDLE

American METALCRAFT, Inc

Spoons and spatulas are among the most commonly used hand tools in the commercial kitchen.

Wood or *metal paddles*: These long-handled paddles are used to stir foods in deep pots or steam kettles. Wood utensils are not allowed by some local health authorities. Always check local codes.

Food Preparation Tools

Box grater: A metal box with various sized grids used to cut food into small particles.

China cap: A pointed strainer used to strain gravies, soups, sauces, and other liquids.

Colander: A bowl-shaped strainer usually made from stainless steel. Colanders are commonly used for washing cooked spaghetti and other pastas.

Hand meat tenderizer: This hammer-like tool is used to pound and break the muscle fibers of tough cuts of meat, making the meat more tender. The aluminum head is cast with a coarse pattern on one side and a fine pattern on the other.

Strainer: Perforated metal bowl used to strain and drain foods.

Wire whip: Wire whips are used for whipping eggs, cream, gravies, and sauces. The two types of wire whips commonly used in the commercial kitchen are the French whip and the piano wire whip. The piano wire whip is more flexible and is used for more delicate whipping procedures.

Cooking Tools

Bain-marie: A round, stainless steel food storage container with high walls. Bain-maries are available in many sizes from 1¼ qt to 11 qt. Also, a pan for holding hot water into which other pans, containing food, etc. are put for heating.

Bake pan: A rectangular aluminum pan with straight or sloped medium-high walls and loop handles. Bake pans are used for baking apples, macaroni, and certain meat and vegetable items.

Braiser: A shallow-walled, large, round pot used for braising, stewing, and searing meats. Braisers are available in sizes from 15 qt to 28 qt.

Double boiler: A double pot used to prepare items, such as cream pie filling and pudding, that scorch quickly if they come in contact with direct heat. The double boiler consists of two pots. The bottom pot resembles a stockpot and holds the boiling water. The top pot is suspended in the boiling water, which prevents contact with direct heat. Double boilers are available in sizes ranging from 8 qt to 40 qt.

Frying or *sauté pan*: A round, sloped, shallow-walled pan with a long handle ranging from 7″ to 16″ in the top diameter to sauté vegetables and meats.

Iron skillet: Iron skillets are made of thick, heavy iron to withstand high heat. They are used for pan broiling and frying such items as chicken, pork chops, veal cutlets, etc. Iron skillets are available in many sizes, with a top diameter of 6½″ to 15¼″.

Mixing bowls: Mixing bowls are used for mixing small or large amounts of ingredients. They are available in various sizes from ¾ qt to 45 qt and are made of stainless steel and aluminum. Stainless steel is not affected by foods that contain acid. Acidic foods discolor aluminum bowls, resulting in a metallic taste.

Roasting pan: A generally large, rectangular, medium- to high-walled metal pan. Roasting pans can be purchased with or without covers and come in various sizes.

Saucepan: A saucepan is similar to a sauce pot, but is smaller, shallower, and much lighter. It is used the same as a sauce pot but for smaller amounts.

FOOD PREPARATION TOOLS

BOX GRATER

American METALCRAFT, Inc.
CHINA CAP

COLANDER

American METALCRAFT, Inc.
HAND MEAT TENDERIZER

STRAINER

American METALCRAFT, Inc.
WIRE WHIPS

Food preparation tools mix and reshape food items in the preparation process.

Sauce pot: A fairly large, round, deep pot with loop handles for easy lifting. It is used for cooking on the top of the range when stirring and whipping is necessary.

Sheet pan: A very shallow, rectangular metal pan used for baking cookies, sweet cakes, and sheet pies. Sheet pans are available in various sizes.

Skewer: A pin of wood or metal used to hold foods together or in shape while broiling or sautéing.

Steel skillet: A lightweight steel skillet with sloping walls. It is used for frying eggs, omelets, potatoes, etc.

Steel skillets are available in various sizes, with a top diameter of 6½″ to 15⅞″. Steel or glass tops with handles are also available to fit the top diameters of the skillets.

Stockpot: A large, round, high-walled pot made of either heavy or light metal. It is used for boiling and simmering items such as turkeys, bones for stock, ham, and some vegetables. The stockpot has loop handles for easy lifting and is sometimes equipped with a faucet for drawing off contents. Sizes range from 2½ gal. to 40 gal.

COOKING TOOLS

1¼ QT – 11 QT

¾ QT – 45 QT

BAIN-MARIE

MIXING BOWL

SKEWER

IRON SKILLET

Cooking tools hold food during the cooking process.

Baking Tools

Bench brush: A brush with long bristles set in vulcanized rubber attached to a wood handle. It is used to brush excess flour from the bench when working with pastry dough.

Dough docker: An aluminum or stainless steel roller with stainless steel pins and a hardwood handle. It is used to perforate certain yeast/dough products to keep them from baking unevenly and blistering.

Flour sifter: A round metal container with a sieve or screen stretched across the bottom. The wire paddle wheel rotates to work the material being sifted through the sieve. Sifting removes lumps from powdered goods to make products light and fluffy.

Pastry bag: Cone-shaped bag made of duck cloth (water-repellent cloth) or other material. It is used for decorating cakes with icing, plank steaks with duchess potatoes, shortcakes with whipped topping, etc.

Pastry brush: A narrow-shaped brush with bristles fixed to a plastic, metal, or wood handle. It is used to brush on icing or egg wash (a mixture of egg and milk) when working with certain types of pastry.

Pastry tube: A metal canister with metal tips with various shaped openings. It is used to decorate cakes, canapés, and cookies.

BAKING TOOLS

BENCH BRUSH — BRISTLES — *Carlisle FoodService Products*

DOUGH DOCKER — STAINLESS STEEL PINS — *American METALCRAFT, Inc.*

PASTRY BAG — CONE-SHAPED BAGS

PASTRY BRUSH — BRISTLES

PASTRY TUBE TIPS — METAL TIPS

PEEL — WOOD PADDLE — *American METALCRAFT, Inc.*

PIE AND CAKE MARKER — GUIDE BARS — *American METALCRAFT, Inc.*

ROLLING PIN — HANDLE — ROLLER

Baking implements are used in preparation and baking of baked goods.

Peel: Long, flat, narrow piece of wood shaped like a paddle. It is used to place pizzas in and remove pizzas from the oven.

Pie and *cake marker*: A round, heavy wire disk with guide bars. It is used for marking of pies or cakes for cutting. Markers are available in various diameters and portion sizes.

Rolling pin: A roller made of wood, Teflon, or other materials, ranging in size from 10½″ to 25″. Handles are attached on each end of the roller. The rolling pin is used to roll dough to the required thickness.

Measuring Tools

Most recipe ingredients are given in weight. However, some may be given in measures. Liquids are usually given in liquid measure for ease in completing the recipes. Measures commonly used are the teaspoon, tablespoon, cup, pint, quart, and gallon. These quantities are usually abbreviated in the recipes. Table II lists the common abbreviations used. Table III lists the relationships of the various measures and weights.

TABLE II: ABBREVIATIONS FOR RECIPES	
tsp	teaspoon
tbsp	tablespoon
pt	pint
qt	quart
gal.	gallon
oz	ounce
lb	pound
bch	bunch

Table III can be used to convert from one measure to another. For example, if 2½ lb of water is called for, this converts to 1 fluid quart and 1 c (2 lb of water equals 1 fluid quart. A fluid pint equals 1 lb of water; but since 1 pt also equals 2 c, ½ lb of water would equal 1 c of water.) Occasionally a recipe will call for a "pinch" of some ingredient. This is roughly equivalent to ⅛ teaspoon.

Measures: Metal cups are used to measure liquids and some dry ingredients. Measures are graduated in quarters and are available in gallons, half gallons, quarts, and pints.

TABLE III: EQUIVALENTS OF MEASURES	
1 pinch	⅛ teaspoon (approx)
3 teaspoons	1 tablespoon
16 tablespoons	1 cup
1 cup	½ pint
2 cups	1 pint
2 pints	1 quart
4 quarts	1 gallon
16 ounces	1 pound
1 pound (water)	1 fluid pint
2 pounds (water)	1 fluid quart

Measuring cups: Measuring cups are used to measure liquids and some dry ingredients. A set consists of one-quarter, one-third, one-half, and one cup measures.

Portion Control Scoops: A scoop with a thumb-operated release lever. Scoops are available in specific sizes and are used to serve food in accurate amounts. Scoops are sized by numbers. Table IV lists scoop numbers and their approximate capacity in ounces. Table V relates the scoop numbers to the approximate content of each scoop size in cups or tablespoons. The numbers that identify scoops indicate the number of scoopfuls required to make 1 quart. Scoops can be used for portioning muffin batter, meat patties, potatoes, rice, bread dressing, croquette mixtures, and salads.

MEASURING TOOLS

MEASURING CUPS   **SCOOP**

American METALCRAFT, Inc.

Cans: Many recipes call for food ingredients by can size. Table VI lists the common can sizes and approximate weights. Table VII lists the substitutions that may be made for the basic can size (No. 10) used in commercial cooking.

TABLE IV: SCOOP SIZES AND APPROXIMATE WEIGHTS

Scoop No.	Approximate Weight
8	5 oz
10	4 oz
12	3 oz
16	2 to 2½ oz
20	1⅔ oz
24	1½ oz
30	1¼ oz
40	1 oz

TABLE V: SCOOP SIZES AND APPROXIMATE MEASURES

Scoop No.	Level Measure
8	½ c
10	⅖ c
12	⅓ c
16	¼ c
20	3⅕ tbsp
24	2⅔ tbsp
30	2⅕ tbsp
40	1⅗ tbsp

TABLE VI: COMMON CAN SIZES AND APPROXIMATE WEIGHTS AND MEASURES

Size	Approx. Contents	Approx. Measure	Products
No. 10	6½ to 7 lb	3 qt	Fruits and vegetables
No. 5	2 to 3 lb	1 qt	Fruit juices, chopped clams, and soups
No. 2½	1 lb 12 oz to 1 lb 14 oz	3½ c	Fruits and vegetables
No. 2	1 lb 4 oz	2½ c	Juices, soups, fruits, and a few vegetables
No. 303	1 lb	1 pt	Fruits, vegetables, and some soups
No. 300	14 oz to 16 oz	1¾ c	Cranberry sauce, pork and beans, and blueberries

TABLE VII: CAN SIZE SUBSTITUTIONS

One No. 10 can	Four No. 2½ cans
One No. 10 can	Seven No. 303 cans
One No. 10 can	Five No. 2 cans
One No. 10 can	Two No. 5 cans

EQUIPMENT

In addition to hand tools used in the commercial kitchen, many types of equipment are also required. *Equipment* is machinery or a larger tool that usually stays in one place. Unlike some hand tools, equipment is always purchased by the food service establishment. Equipment should only be used after proper instruction by the supervisor. Always follow the manufacturer's recommendations for operation of any equipment. Equipment commonly used in the commercial kitchen includes:

Baker's scale: The baker's scale ensures accuracy in measuring the proper amount of ingredients. The baker's scale has two platforms. The food to be weighed is placed on one platform, and a weight is placed on the other platform. Additional weights can be added if required by the food being weighed. The beam between the two platforms also has a weight that is used for fine adjustment of the scale.

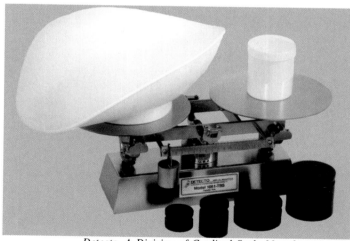

Detecto, A Division of Cardinal Scale Manufacturing Co.

The baker's scale is used to measure the weight of ingredients.

For example, to weigh 8 oz of egg whites, a container large enough to hold the egg whites is placed on the left platform of the baker's scale. Balance the scale, then move the weight on the beam an additional

8 oz. Add the egg whites until the scale balances again. With a balanced scale, the correct amount (8 oz) has been measured. The baker's scale can be used to weigh up to 10 lb.

Chafing dishes: Chafing dishes are used to keep food hot on the buffet or serving line. A chafing dish contains a pan of water over which the pan containing the food is placed to keep it hot. The water is kept hot using canned heat or an electric heating element.

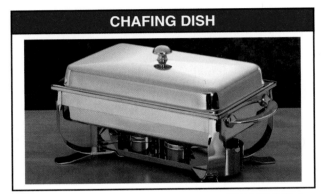

CHAFING DISH

American METALCRAFT, Inc.

Deep fryer: This is a large, automatic fry kettle used to deep-fry all foods. It holds from 25 lb to 50 lb of shortening, depending on the size of the kettle. Temperature controls adjust from 200°F to 400°F.

Frymaster

A deep fryer is used to prepare fried foods.

Food cutter or *chopper*: Attachments for the food cutter can be used for chopping, grinding, slicing, dicing, and shredding foods. It has a revolving stainless steel bowl and a revolving knife that chops food quickly and efficiently. *Extreme caution should be exercised when using the food cutter.*

Food storage boxes: Food storage boxes are used to safely store food while maximizing the use of shelf space in refrigerators, freezers, or on transport equipment. Some local boards of health require the use of these boxes or similar equipment when storing foods. The box lids are designed to snap on to create an airtight seal.

FOOD STORAGE BOXES

Carlisle FoodService Products

Fresh-O-Matic: This unit heats food with moist heat using distilled water and electricity. Quick reheating of many types of food is possible.

Stand mixer: The stand mixer is one of the most versatile pieces of equipment in the commercial kitchen. With different attachments, this machine whips, grinds, shreds, slices, and chops food. It is available in many different sizes.

Trade Tip

The life of frying oil can be prolonged by filtering the fat twice a day and adding 10% new oil, which will rejuvenate the original oil.

A stand mixer can be used to whip, grind, shred, slice, and chop food.

Portion scale: The scale is used for measuring food servings. It has a single steel platform where the food is placed. A large dial on the front of the scale is graduated from ¼ oz to 32 oz. The rotating needle on the dial indicates the weight of the item placed on the platform. This type of scale is used when exact serving portions are required. Portion scales are also available with a digital readout for more accuracy.

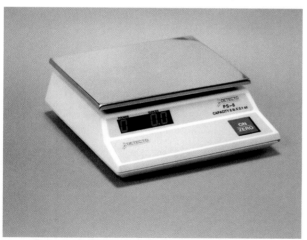

Detecto, A Division of Cardinal Scale Manufacturing Co.

A portion scale is used for measuring food servings.

Proofing/holding cabinet: The proofing/holding cabinet is used to keep food hot without drying it out, or for proofing dough. *Proofing* is the process of letting yeast dough rise in a warm (85°F), moist cabinet until it becomes double in bulk. This produces a soft, light-textured product after it is baked. All yeast dough products must be proofed before baking. Most proofing cabinets accommodate standard food service trays and pans. The proofing cabinet is mounted on wheels for easy cleaning and mobility.

Cres-Cor

A proofing cabinet is used to keep food hot without drying it out, or for proofing dough.

Slicing machine: The slicing machine can be manually or automatically operated. It has a regulator for providing a wide range of slice thicknesses up to ¾″ and has a feed grip that grips material firmly on top or serves as a pusher plate for slicing small end pieces. All slicing machines have safety features built in to help protect the user from the sharp revolving blade. The slicing machine is not as versatile as the mixer or food chopper; however, it can perform more than one operation. In addition to slicing, the slicing machine can be used for shredding lettuce and cabbage. *Exercise caution when using the slicing machine and check local health departments for age restrictions.*

Vertical cutter/mixer: This machine is used to cut and mix foods simultaneously for fast volume production. There are only two movable parts within the bowl: the knife blades, which move at a high speed, and the mixing baffle, which is operated manually to move the product into the cutting knives. Besides speed of production, another advantage of using a vertical cutter/mixer is that the product being processed is never bruised because the knife blades operate at high speeds and slice the product in mid-air.

Some of the products that can be quickly prepared in the vertical cutter/mixer include:

- 67 lb mayonnaise – 6 min
- 40 lb frozen beef chuck cut into hamburger – 40 sec
- 43 lb pie dough (mixed) – 20 sec
- 24 lb white cake (mixed) – 60 sec
- 12 heads lettuce (shredded) – 3 sec
- 224 portions cole slaw – 12 sec
- 151 portions meat loaf – 45 sec

Pressure steamer: Pressure steamers cook food with steam under pressure that is in direct contact with the item being cooked. Steam is used primarily for cooking vegetables, but can be used when cooking certain meats, such as corned beef. Cooking with steam provides many advantages. It is quick and preserves natural flavor, color, minerals, and vitamins. Foods do not stick to pans and cannot burn.

A solid stainless steel pan or perforated container (to allow more steam in) is used to hold the item to be cooked. The pans are placed in one of usually two stacked compartments that are approximately 8″ to 12″ high. The doors contain tight-fitting gaskets to seal in the steam. The pressure steamer can be operated automatically with preset timers or manually.

Always be careful when working with steam. *Steam can cause serious burns.* Before removing an item from a pressure steamer, make sure the steam is turned OFF. Open the door slowly to let all the steam escape before removing an item.

Convection steamer: A convection steamer functions like a convection oven, except a fan circulates steam around the food. Approximately six times as much heat as boiling water is applied. This results in more efficient heat transfer and shorter cooking time. The quicker the foods are cooked, the more nutrition, color, and flavor can be preserved.

Combination convection oven-steamer: A combination convection oven-steamer is very flexible by operating as a pressureless convection steamer and a general-purpose convection oven using a combination of hot air and steam.

Hobart Corporation

A pressure steamer cooks food with steam under pressure that is in direct contact with the item being cooked.

Henny Penny Corporation

The combination oven-steamer cooks with convection heat and steam, individually, in sequence, or in any combination.

Automatic steamer and *boiling unit*: These units are designed for larger cooking jobs, have a capacity of 158 gal., and are capable of cooking 1500 portions to 2000 portions of potatoes in 1 hr. They are ideal for food processors, large hospitals, commissaries, and other institutions. In addition to boiling potatoes, the automatic steamer and boiling unit gives excellent results when steaming vegetables, chicken, and seafood, and when boiling pasta products.

Convection ovens: These ovens, also known as *airflow ovens*, use air that is circulated throughout the interior. Heat is evenly distributed by this circulation method, allowing the oven to be loaded to capacity and still provide the required heat. Convection ovens are available in gas and electric models as floor ovens with roll-in dolly, table models, counter models, and stack ovens. A convection oven increases productivity, reduces shrinkage, and cooks more uniformly than a conventional oven. Cooking cycles are completely automatic. The same quantity of food can be cooked in less space and using less fuel than conventional ovens.

Combination oven-steamer: The combination oven-steamer offers versatility as it will steam, roast, or bake like a convection oven with hot air circulating around the food, or steam and bake with moist steam present during the baking period. With steam present in the oven, shrinkage is reduced, but a roast still acquires a brown surface. The steam in the oven produces excellent hard rolls or bread and a quality baked potato. The combination oven-steamer is available in several sizes designed to meet the needs of food service establishments.

Impinger conveyorized cooking system: In this system, foods travel through the baking chamber on a motorized conveyor at the required speed. Inside the baking chamber, high-velocity hot air strikes (impinges) the food surface for more efficient heat transfer at lower temperatures compared to other baking units. Cold foods do not reduce the operating temperature in the oven. This allows quicker baking, reheating, and finishing time.

Microwave oven: Microwave ovens utilize electromagnetic waves generated by a *magnetron*. The waves, when striking water molecules in a food, cause heat energy. Food is cooked from the inside out as the heat spreads throughout the food. Foods used in the microwave must have water present in them for heat to develop. Foods cooked in the microwave do not brown as in a conventional oven. In addition, larger foods must be turned as required to ensure even cooking. More cooking time must be allowed for additional food items cooked at the same time. Foods cooked in a microwave must have a minimum internal temperature of 165°F.

In the commercial kitchen, microwave ovens are used primarily for thawing, heating convenience foods, and reheating cooked foods. Any foods defrosted in a microwave must be cooked immediately. The microwave oven requires no preheating. Cooking times must be accurately controlled by a timer. Foods used in the microwave should be placed on china, plastic, or paper containers. Never use metal containers or objects in a microwave oven. Metal surfaces reflect microwaves and can cause damage to the oven.

Convection microwave: A convection microwave combines circulated hot air with microwave energy in the cooking chamber. This allows a great range of cooking times and temperatures. Food is browned, baked, broiled, and roasted using circulated conventional heat and heat from the inside from microwave heat energy.

Tilt fryer: The tilt fryer can be used for frying, braising, stewing, sautéing, simmering, boiling, grilling, and deep-fat frying. In addition to its versatility for different cooking jobs, its main features are a large cooking area, thermostatic heat control, a tilting feature, and the ease with which it can be cleaned and maintained.

Quartz-plate infrared oven: This equipment combines conventional heating with infrared rays for fast heating and reconstituting of meals that have been previously cooked and refrigerated or frozen. A specially fused silica plate that transmits high intensity infrared rays is combined with conduction heating to provide a uniform and controllable heating pattern, making it an ideal piece of equipment for preparing convenience foods.

Trade Tip

A tilt fryer is essential for rapid production in a large commercial kitchen. This versatile unit can be used for braising, sautéing, frying, steaming, holding foods, etc. Tilt fryers require little maintenance, are easy to clean, and can provide years of service. They are available in gas or electric models.

Automatic twin coffee urn: This fully automatic unit has two coffee liners built into a single body section. This allows an unlimited supply of hot water for making coffee and tea through its heat exchange system. It is equipped with a spray assembly for spreading the water evenly over the coffee grounds. It has both automatic and manual agitation, a built-in thermostat, and a timer to set for a brewing cycle. This equipment takes all guesswork out of coffee making. Each urn has a capacity of 3 gal., 6 gal., or 10 gal.

Automatic coffee brewer: Automatic coffee brewers using glass pots simplify coffee making. Steps required include pouring in cold water, plugging in, and turning ON. Coffee is finished in 4 min. Some automatic coffee brewers are connected to a water supply, which eliminates the need to add water in the coffee-making process.

Bunn-O-Matic Corporation

This automatic twin coffee urn can brew two different beverages at once.

Safety

Safety is a constant concern in the commercial kitchen. Accidents can easily occur because of the amount of activity and potentially dangerous tools and equipment in the kitchen. Commercial kitchens, like any plant in industry, require safety awareness by all workers. Accidents do not just happen; they are caused. When it comes to safety, prevention is much less costly than treatment. Unfortunately, safety programs are often enacted only after a serious accident.

Accidents can be prevented by identifying possible hazards and minimizing the potential for an accident occurring. Proper layout of the commercial kitchen, properly maintained tools and equipment, and safe work habits all contribute to a safe working environment. Respect for tools and equipment and the welfare of fellow food service workers is the standard that must be followed. Safety must be everyone's concern in the commercial kitchen.

KITCHEN SAFETY

Safety programs must be in place and safety must be practiced by all workers on a continuing basis. The Occupational Safety and Health Administration (OSHA) is responsible for setting and enforcing workplace safety standards in the U.S. Employers are responsible for safety training and ensuring that employees follow OSHA standards. It is important that all food service workers are informed of all safety procedures when they are hired. Employees must report all accidents and safety hazards to the employer or supervisor.

INJURIES

It is important to be aware of situations that may cause injury and take proper precautions to prevent injuries from

occurring. In the event that an accident or injury does occur, notify your supervisor and seek immediate medical attention. Following an injury, proper precautions must be taken to avoid further injury or wound contamination. Precautions must also be taken to protect food and kitchen equipment from contamination from cuts or personal injuries. The most common injuries resulting from accidents in the commercial kitchen are cuts, burns, falls, and strains and sprains.

Cuts

Cuts can occur easily in the commercial kitchen as knives and other cutting implements are constantly in use. The frequency and seriousness of cuts can be reduced by practicing proper cutting procedures and common sense. For instance, tucking fingers away from knife blades when cutting can reduce the risk of injuring fingers. When cuts do occur, they should be treated properly to prevent infection and more serious complications. For serious cuts, emergency medical help should be contacted immediately. The following procedure is for treatment of minor cuts:

1. Wear disposable gloves when treating any person with a bleeding injury.
2. Compress the wound by applying pressure to stop any bleeding. Use a clean cloth or bandage to apply light and steady pressure to a wound for 20 to 30 minutes, if necessary. Don't repetitively remove pressure to check on the wound status, as this will hinder clot formation and prolong bleeding time.

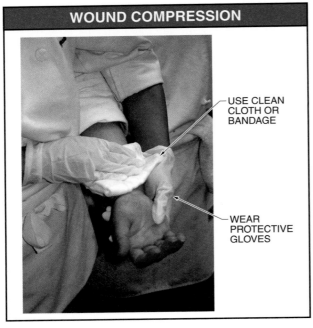

WOUND COMPRESSION

USE CLEAN CLOTH OR BANDAGE

WEAR PROTECTIVE GLOVES

Compressing a cut will help stop the bleeding.

3. If bleeding is severe or does not stop, have the injured person lie down. Elevate the wound to reduce blood flow to the area.
4. Remove any dirt or debris from the wound. Clean the bleeding area with fresh water. Soap can irritate a wound, so avoid contacting a cut directly with soap.
5. Apply an antibiotic to prevent infection and encourage the body's natural healing process.
6. Apply a bandage or sterile gauze to the wound area.
7. Carefully remove gloves and wash hands thoroughly after treating a wound.

All cuts must be properly covered and a waterproof covering such as a finger cot or disposable glove must be worn when working in the kitchen. A *finger cot* is a protective sleeve placed over the finger to prevent contamination of a cut.

Burns

Burns are generally more painful and take more time to heal than cuts. Burns that occur in the commercial kitchen are classified as either minor or serious. Minor burns are burns caused by popping grease or by handling hot pans with wet or damp towels. Pot holders should be used to protect hands when moving or handling hot cookware. Towels should not be used to handle hot cookware as they may not provide adequate insulation from heat and could also create cross contamination. The following procedure may be used for treating minor burns:

1. Cool the burn under cold running water. Do not put ice on the burn.
2. An aloe vera lotion or antibiotic ointment can be applied to the area to relieve pan after the wound has cooled.
3. Cover the burn with a bandage.

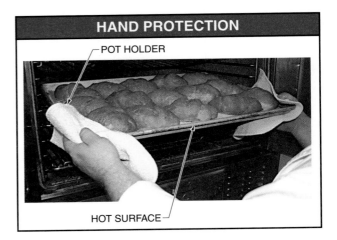

HAND PROTECTION

POT HOLDER

HOT SURFACE

Pot holders should be used to protect hands when moving or handling hot cookware.

Serious burns are burns caused by splashed grease, escaping steam, and gas ignited incorrectly. Serious burns are classified as first degree, second degree, or third degree burns. First degree burns affect only the top layer of skin, and appear red and swollen. First degree burns are treated as minor burns unless they occur on substantial areas of the hands, feet, face, groin, or buttocks, or on a major joint. Second degree burns occur when the burn area extends beyond the upper layer of skin into the second layer, known as the dermis. Blisters, intense pain, and swelling result from second degree burns. Second degree burns less than 3″ in diameter can be treated as minor burns. Third degree burns involve damage to body tissue well beyond the first and second layers of skin. Damage could affect fat, muscle, or even bones. If the burn is severe, it should be treated promptly by trained medical personnel. The following steps may be performed on severe burns while waiting for emergency medical personnel to respond to a burn victim:

1. Remove any smoldering materials and exposure to smoke and heat.
2. Cover the burn lightly with a cool, moist bandage or clean cloth.
3. Ensure that the victim is breathing. Cardiopulmonary Resuscitation (CPR) may be required if the victim stops breathing.

Falls

Falls can cause serious accidents in the commercial kitchen. Falls are caused by wet floors, spilled food or grease, and by torn mats or damaged floors. OSHA 29 CFR 1910.22, *Walking-Working Surfaces,* requires all places of employment to be kept clean, orderly, and in a sanitary condition. This includes keeping all floors clean and dry. If a spill occurs, it should be mopped up immediately. Wet floor areas should be designated with warning signs. Nonslip matting can help reduce the risk of falling in areas that tend to be wet. Step-up step-down areas can be indicated with hazard tape.

Falls can be minimized in the commercial kitchen by keeping floors clean and dry and by wearing proper footwear. Nonslip shoes will provide better traction in potentially slippery areas. Sandals, open-toe shoes, and high heeled shoes should not be worn in the commercial kitchen.

Strains and Sprains

Strains and sprains are not as serious as the other types of accidents, but are painful and can result in the loss of working hours. A strain is a bodily injury resulting from excessive tension, effort, or stretching of muscle and ligament tissue. A sprain is an injury resulting specifically from excessive stretching of ligaments. These injuries can be caused by lifting heavy or oversized objects, carrying large numbers of objects at once, or repetitive reaching across tables or counters. Strains and sprains can be prevented by not trying to carry loads that are too heavy, by following the proper lifting procedure, and by wearing slip-resistant shoes.

All cuts must be properly covered and a waterproof covering such as a rubber glove must be worn when working in the kitchen

When lifting objects from the ground, ensure the path is clear of obstacles and free of hazards. Bend the knees and grasp the object firmly. Next, lift the object, straightening the legs and keeping the back as straight as possible. Finally, move forward after the whole body is in the vertical position. Keep the load close to the body and steady. Using the proper lifting procedure will also prevent possible back injuries. Assistance should be sought when lifting heavy objects.

In the event that a sprain or strain occurs, seek proper first aid treatment to prevent further injury. Common first aid techniques for treating sprains and strains include the following:

- Isolation of the injured limb, preventing further use through restraining devices such as splints.
- Rest of the injured area. Avoid using sprained or strained limbs if possible.
- Use elastic wrap or a bandage to compress the injured area.
- Icing a sprained or strained area immediately after injury will reduce potential swelling. Avoid icing an injury for a prolonged period, as tissue damage can occur.
- Elevate the injury whenever possible to limit swelling.

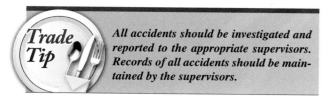

Trade Tip

All accidents should be investigated and reported to the appropriate supervisors. Records of all accidents should be maintained by the supervisors.

PROPER LIFTING PROCEDURE

KEEP BACK STRAIGHT

1. BEND KNEES AND GRASP OBJECT FIRMLY

2. LIFT OBJECT BY STRAIGHTENING LEGS

3. MOVE FORWARD AFTER WHOLE BODY IS IN VERTICAL POSITION

The proper lifting procedure will help avoid potential sprains, strains, or back injuries.

CHOKING

Choking is a medical emergency that may occur in a food service establishment. Choking often occurs as a result of solid food becoming lodged in a person's throat. A person who is choking may experience panic, which can complicate and elevate the situation. Indications that a person is choking include gasping for air, inability to breathe, or turning blue. The universal sign for choking is a hand to the throat with extended fingers and thumb. If a person can speak and has normal skin color, he or she is probably not choking.

If a person is choking, call for emergency medical assistance or perform the Heimlich maneuver. The Heimlich maneuver is an effective technique for stopping choking. It is recommended that a person learns the procedure for administering the Heimlich maneuver from a certified first aid course before performing it on a choking victim. The following is the basic procedure for performing the Heimlich maneuver:

1. Position yourself behind the choking victim and wrap your arms around the victim's waist.
2. Bend the victim slightly forward.

3. Make a fist with one hand and place it slightly above the victim's navel and below the rib cage.
4. With the other hand, grasp the fist and press it into the victim's abdomen with a hard, upward thrust. Repeat the thrusting motion until the blocked airway becomes clear.

This maneuver should not be performed on unconscious victims, pregnant women, or small infants. Special care should be taken in these cases to ensure the safety of the victims. In each case, first check the victim's mouth to see if the blockage can be cleared with a finger sweep of the throat. If the Heimlich is to be performed on an unconscious victim, lay the unconscious person on his or her back. Kneel over the victim and administer upward thrusts to the upper portion of the abdomen. For pregnant women, apply the normal Heimlich procedure but hold hands higher, at the base of the breastbone. Small infants should be laid face down and small thumps should be delivered to the center of the back. If choking does not stop, turn the infant over and apply five gentle chest compressions to the breastbone. Again, it is recommended that persons administering the Heimlich be trained in a certified first aid course.

HEIMLICH MANEUVER

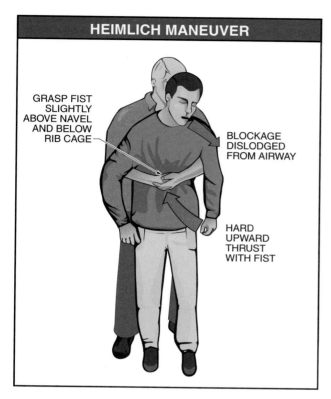

GRASP FIST SLIGHTLY ABOVE NAVEL AND BELOW RIB CAGE

BLOCKAGE DISLODGED FROM AIRWAY

HARD UPWARD THRUST WITH FIST

The Heimlich maneuver is an effective technique for stopping choking. It is recommended that a person learn the maneuver in a certified first aid course before administering it to a choking victim.

FIRE SAFETY

Work areas must be equipped with the correct number and type of fire extinguishers. The National Fire Protection Association (NFPA) classifies fires as Class A, B, C, D, and K, based upon the combustible material. The appropriate fire extinguisher must be used on a fire to safely and quickly extinguish the fire. The types of fires and appropriate fire extinguishers are as follows:

- Class A fires include combustible materials such as paper, wood, cloth, rubber, plastics, refuse, and upholstery. Class A fire extinguishers are commonly filled with water or water-based agents. Water should never be used to extinguish a flammable liquid fire. Water disperses the flammable liquid and flames over a larger area. Additionally, water is a conductor of electricity and should never be used to extinguish an electrical fire.
- Class B fires include combustible liquids such as gasoline, oil, grease, and paint. Class B fires are extinguished with smothering agents such as carbon dioxide and chemical foams.

FIRE EXTINGUISHER CLASSES

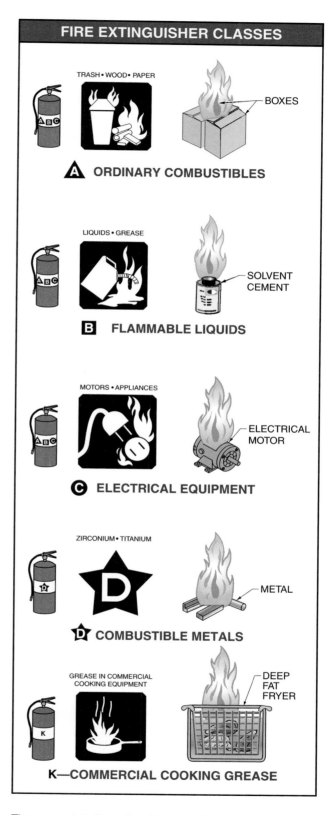

TRASH • WOOD • PAPER

BOXES

A ORDINARY COMBUSTIBLES

LIQUIDS • GREASE

SOLVENT CEMENT

B FLAMMABLE LIQUIDS

MOTORS • APPLIANCES

ELECTRICAL MOTOR

C ELECTRICAL EQUIPMENT

ZIRCONIUM • TITANIUM

METAL

D COMBUSTIBLE METALS

GREASE IN COMMERCIAL COOKING EQUIPMENT

DEEP FAT FRYER

K—COMMERCIAL COOKING GREASE

The appropriate fire extinguisher must be used on a fire to safely and quickly extinguish the fire.

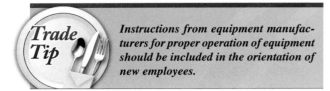

Trade Tip

Instructions from equipment manufacturers for proper operation of equipment should be included in the orientation of new employees.

- Class C fires include electrical equipment such as motors, appliances, wiring, fuseboxes, breaker panels, and transformers. Class C fires are extinguished with nonconductive dry chemical agents. Before extinguishing a Class C fire, electrical power should be shut off as quickly as possible.
- Class D fires occur with combustible metals such as magnesium, sodium, and potassium. Class D fires are not common in the food service industry.
- Class K fires occur with grease in commercial cooking equipment. Class K extinguishers coat the fuel with wet- or dry-base chemicals.

OSHA 29 CFR 1910.157, *Fire Protection,* details the fire protection requirements for the workplace. Fire extinguishers must be located and identified so they are readily accessible to employees without subjecting the employees to possible injury. Extinguishers are selected and distributed throughout the workplace based on the types of anticipated workplace fires and on the size and degree of hazard that would affect their use. The pressure gauge on top of the fire extinguisher must be checked periodically to verify the recommended pressure. In addition to OSHA requirements, state and local codes require a scheduled inspection of fire extinguishers.

A grease fire is a class K fire. Never use water to extinguish a grease fire.

Tuck in all apron strings.

The food service worker must know the location of every fire extinguisher in the kitchen and how to use them properly. OSHA 29 CFR 1910.157, *Fire Protection,* also requires employers to provide an educational program to familiarize employees with the general principles of fire extinguisher use and the potential hazards of fire fighting. Fire extinguishers are not to take the place of the local fire department. They are meant only to put out small fires or to contain larger fires until help arrives.

If extinguishers are provided but are not intended for employee use, OSHA requires the employer to have an emergency action plan (EAP) and a fire prevention plan per OSHA 29 CFR 1910.38, *Exit Routes, Emergency Action Plans, and Fire Prevention Plans.* An emergency action plan is a written document intended to facilitate and organize employer and employee actions in the event of an emergency. An employer with 10 or fewer employees may communicate the plan orally to employees. The plan must include the following:

- Evacuation procedures and emergency escape routes
- Procedures for employees who remain to operate critical facility operations prior to evacuation
- Procedures to account for all employees after an emergency evacuation takes place
- Any rescue or medical duties for employees to perform
- Procedures for reporting fires or other emergencies
- Names or titles of persons to contact for further explanation of procedures or duties described in the plan

Emergency exit diagrams are an important part of the emergency action plan. Emergency exit diagrams should be posted in locations accessible to employees in the event of an emergency. Primary and secondary exits are designated on the diagram. Escape routes marked on the diagram lead employees away from hazardous areas and toward the primary and secondary exits. Assembly areas where employees should gather following an emergency exit are also indicated on the diagram. Elevators should not be used to exit a building in an emergency.

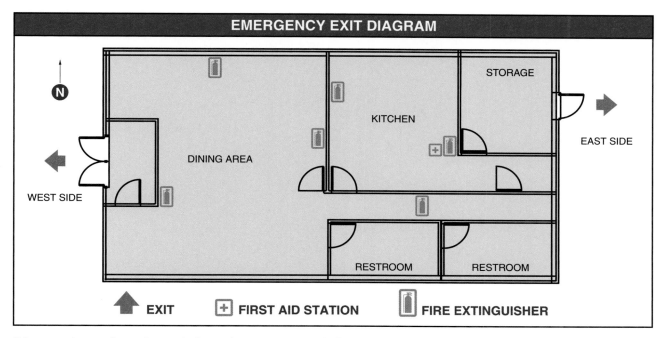

EMERGENCY EXIT DIAGRAM

STORAGE

KITCHEN

DINING AREA

WEST SIDE

EAST SIDE

RESTROOM RESTROOM

⬆ **EXIT**   ⊞ **FIRST AID STATION**   🔥 **FIRE EXTINGUISHER**

Primary and secondary exits are designated on emergency exit diagrams.

In addition to fire extinguishers, fire suppression systems such as an Ansul system located above cooking areas are used to extinguish fires rapidly. Such systems are designed to provide fire protection for kitchen hoods, plenums, ducts, filters, and cooking appliances such as deep-fat fryers, griddles, range tops, and broilers. When the system is actuated, the gas or electric supply to all protected appliances is cut off. A fire-suppressant liquid agent is discharged onto the fire through distribution piping, cooling the grease surface and forming a layer of soap-like foam on the surface of the fat. This layer serves as insulation between the hot grease and the atmosphere, helping to prevent the escape of combustible vapors.

Of the fires occurring in a commercial kitchen, grease fires are the most common. Grease fires can be prevented by avoiding splashing grease on top of the range and by cleaning ranges, ovens, hoods, and filters to eliminate grease buildup.

Fire Safety Rules

- Always use the proper fire extinguisher for the type of fire. *Never* use water on a grease fire or electrical fire.
- Know the locations of all fire extinguishers and fire exits. Keep areas around fire extinguishers and fire exits clear.
- Clearly post emergency telephone numbers.
- Clearly post evacuation routes.

- *Never* use flammable solvents or cleaners in the commercial kitchen.
- Periodically check the fire suppression system and fire extinguishers for proper pressures, use dates, and leakage or malfunction.
- Make sure fire suppression equipment instructions are clearly posted.
- Ventilate gas ovens for a few minutes before lighting pilot lights by letting the oven door hang open to remove residual gas. Place a lighted match to pilot light before turning on the gas.
- Keep the cooking area clean.
- Do not smoke in the kitchen.
- Remove all trash and other combustibles promptly.
- Always check to make sure the gas and electricity are shut off before leaving the kitchen.
- Keep towels away from the range. A dangling towel could catch on fire.

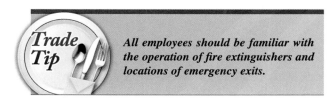

Trade Tip

All employees should be familiar with the operation of fire extinguishers and locations of emergency exits.

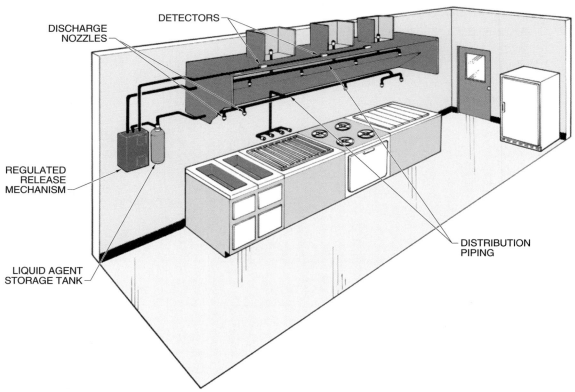

DETECTORS

DISCHARGE
NOZZLES

REGULATED
RELEASE
MECHANISM

LIQUID AGENT
STORAGE TANK

DISTRIBUTION
PIPING

Ansul Fire Protection

The Ansul fire suppression system provides fire protection for restaurant cooking equipment. Appliance energy sources are automatically shut off when a fire occurs.

Food Preparation Safety Rules

- Use clean, dry pot holders when handling hot skillets, pots, and roasting pans. (Wet cloth conducts heat more readily.)
- Remove the lids of pots slowly, lifting the side away from the hands and face to prevent steam burns.
- Always give notice of "HOT STUFF" to other employees when moving a hot container from one place to another.
- Avoid overfilling food containers to reduce the risk of spills. Overfilled food containers could also contribute to a fall.
- *Never* let the handles of saucepans and skillets extend into aisles. If hit or bumped, the pot may fall off the range.
- *Never* turn the handle of any pot toward the heat.
- When lifting is required, lift so the strain is absorbed in the legs and arms. Lift with the legs, not the back. *Never* lift when unbalanced.
- Get help in lifting or moving heavy pots or containers.
- When placing food in hot grease, always let the item slide *away* from the body so the grease will not splash and cause a serious burn.
- Keep workstation organized and free of excess food or other clutter.
- Pay attention to the job at hand.
- *Never* throw objects in the kitchen. Always pass them from hand to hand.
- Treat all injuries immediately. Seek medical assistance if necessary.
- *Never* leave a hot pot or pan on a counter top where someone may accidentally touch it and get burned.

Hand Tool Safety Rules

- Use the right knife for the job.
- Do not grab for falling knives. When a knife starts to fall, get out of the way.
- Always carry a knife with the tip pointing downward and with the blade parallel to the body.
- When cutting with a knife, always cut away from the body. The same applies to potato peelers or any implement with a cutting edge.

- *Never* leave knives in a sink covered with soapy water. If this occurs do not reach into soapy water in search of a knife, drain the sink instead.
- Use a cutting board at all times. Never cut on metal.
- Knives should always be placed in a knife rack for proper storage.
- When cleaning or wiping a knife, keep the sharp edge turned away from the body.
- Always use a sharp knife. A sharp knife is safer than a dull knife because less pressure is required.
- Pick up knives by the handle only.
- Always have a firm grip on the knife handle. Keep the handle free of grease or other slippery substances.
- When slicing round objects such as an onion or a carrot, cut a flat base so the object sets firmly and does not shift when being cut.
- *Never* force a meat saw. It may jump from the bone and cause injury.
- When using a cleaver, be sure the item to be chopped does not move easily.
- When grating foods, never work the foods too close to the cutting surface.

Use a NSF-approved cutting board at all times. Do not cut on metal or nonfoodsafe surfaces.

Stationary Equipment Safety Rules

- All utensils should be certified by NSF International and be nonporous. NSF International is a nonprofit organization responsible for standards development, product certification, education, and risk management for public health and safety.
- Use a food-safe plastic stomper (plunger) when feeding meat or other items into a grinder.

- Before cleaning or adjusting any machine, be sure all electrical switches are in the OFF position and pull the plug.
- Do not wear rings, wristwatches, or ties when operating electrical power equipment.
- All stationary electrical equipment must be properly installed and grounded.
- Keep hands to the front of the revolving bowl when operating the food cutter. The food cutter is one of the most dangerous pieces of equipment in the commercial kitchen.
- *Never* operate any machine unless trained to use it properly. Be familiar with the safety features and the operation of the emergency stop.
- When using electrical power equipment, always follow the manufacturer's instructions and recommendations.

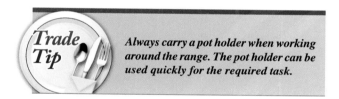

Trade Tip *Always carry a pot holder when working around the range. The pot holder can be used quickly for the required task.*

Clothing Safety Rules

- Wear shoes that prevent slipping and provide support, comfort, and safety. Tie shoelaces neatly.
- Wear long sleeves that cling tightly to the arms. This may prevent burns when frying food.
- Never wear loose-fitting clothing as it may get caught in a piece of equipment.
- Wear aprons at knee length. More protection is provided than with the half-length style.
- Tuck in all apron strings.
- Wear recommended headgear.

China and Glassware Safety Rules

- Discard chipped or cracked china and glassware.
- Never use glassware in forming or preparing food (such as for cutting biscuits or ladling liquids).
- Never place glassware in soapy water. Wash in a dishwasher using a compound recommended for glasses.
- When carrying china and glassware from one place to another, be alert and move cautiously. Keep complete control of the load at all times.

Floor Safety Rules

- Make sure the floor is dry before turning on any electrical equipment.

- If anything is spilled on the floor, clean it up immediately.

- Never leave any pots, pans, or utensils on the floor.

- Always walk in the kitchen; never run.

- Mop only a small area of the kitchen floor at a time.

- Use rubber mats around the range or other areas that may get wet. The mats must be kept in good condition and replaced when worn.

- Always use a wet floor sign when mopping or cleaning up spills to notify customers and employees to use caution.

Clean all spills immediately.

Nutrition

Food provides energy and nutrients to the body through consumption and digestion. Nutrition is the study of nutrients and how they nourish the body. In addition, nutrition is concerned with the type and amount of nutrients present in foods. Practicing healthy nutrition by choosing foods high in nutrients and low in fats, sugars, and sodium will promote physical health and longevity.

Obesity is a growing concern in America. Sixty-five percent of adult Americans are overweight, and 30% are considered obese. These numbers continue to rise both in adults and in children. The United States Department of Agriculture (USDA) recommends that Americans consume fewer calories, be more active, and make wiser choices among food groups. In 2005, the USDA revised their dietary guidelines to address the growing problem of obesity in America.

A healthy, balanced diet includes foods from all major food groups including grains and cereals, vegetables, fruits, milk and dairy products, and meats, fish, and beans. Saturated fats and trans fats, oils, salt, and refined sugars should be avoided as much as possible. Certain fats, such as polyunsaturated and monounsaturated fats, should be included as part of a healthy diet because they provide energy and help distribute fat-soluble vitamins in the body. The USDA MyPyramid Food Guidance System is one of many approaches to maintaining a healthy, balanced diet.

NUTRIENTS

Nutrients are the elements that nourish the body, providing energy and chemical resources to promote healthy living. Nutrients are found in all foods, although some foods are more nutrient-rich than others. Choosing foods that are high in nutrients and low in fats is essential to practicing good nutrition. Nutrients include vitamins, minerals, carbohydrates, proteins, fats, and water.

Vitamins

Vitamins are organic substances used by various metabolic processes in the body. They can be obtained from eating foods that contain vitamins, or they are sometimes produced within the body. Vitamins are either fat-soluble or water-soluble. Fat-soluble vitamins can be stored in the liver. Water-soluble vitamins must be replenished every day because they pass through the body and are not stored for later use. Vitamins are essential to cellular activities and body maintenance.

Vitamin A is a fat-soluble vitamin found in liver, eggs, and dark green vegetables. In addition, many foods such as breakfast cereals are fortified with vitamin A. Nutrients called carotenoids, which include beta carotene, are easily converted to vitamin A in the body. Vitamin A is important to vision, bone growth, cell division, and reproduction.

Vitamin B_6 is a water-soluble vitamin found in beans, meat, poultry, fish, and some fruits and vegetables. It is an essential nutrient for protein and red blood cell metabolism. The body also uses vitamin B_6 for hemoglobin production, maintenance of lymphoid organs that are part of the immune system, and blood sugar level maintenance.

Vitamin B_{12} is a water-soluble vitamin found in fish, meat, poultry, eggs, and dairy products. This vitamin helps maintain healthy nerve cells and red blood cells and is also needed in the construction of DNA.

Vitamin C is a water-soluble vitamin found in fruits and vegetables. Citrus fruits are particularly high in vitamin C. One function of vitamin C is to aid in the formation of collagen, a connective tissue found in skin, ligaments, joints, bones, and teeth. It also plays an important role in metabolism and is considered an antioxidant. Antioxidants protect against free radicals. Free radicals are damaging by-products of energy metabolism that can contribute to cardiovascular disease or cancer.

Vitamin D is a fat-soluble vitamin found in fortified milk, dairy products, egg yolks, and fortified cereals. It may also be obtained through exposure to ultraviolet (UV) rays from the sun. Sunshine contains UV rays that trigger synthesis of vitamin D in the skin. The function of vitamin D is to aid the body in calcium and phosphorus absorption. It also promotes bone mineralization.

Vitamin E is a fat-soluble vitamin found in vegetable oils, nuts, and leafy green vegetables. This vitamin is also considered an antioxidant. Vitamin E is used in the immune system, DNA repair, and metabolism.

Vitamin K is a fat-soluble vitamin found in leafy green vegetables. Its primary function is to facilitate blood clotting.

Minerals

Minerals are inorganic substances used in the body in vary small amounts for various processes. Some minerals are used in much larger quantities in the body than others. Minerals used in relatively large amounts are referred to as macro minerals. Minerals used in extremely minute amounts are referred to as trace minerals.

Calcium is the most abundant mineral in the body. Calcium is found in milk and dairy products. It is primarily stored in bones and teeth. Calcium aids in the formation of bones and teeth, muscle contraction, blood vessel contraction and expansion, secretion of hormones and enzymes, and sending messages in the nervous system.

Sodium is a mineral found most often in common table salt. The body uses sodium to help maintain blood pressure levels and fluid balance in the body. Too much sodium in the body causes high blood pressure (hypertension), which can lead to heart disease and stroke. The USDA recommends limiting sodium intake as much as possible. Choosing foods that are lower in sodium or eating more potassium-rich foods can lower sodium and blood pressure in the body.

Potassium is a mineral found in fruits, vegetables, dairy products, and seafood. Potassium works with sodium to maintain blood pressure and fluid levels in the body.

VITAMINS			
Name	**Type**	**Source**	**Function**
Vitamin A	fat-soluble	liver, eggs, dark green vegetables	vision, bone growth, cell division, reproduction
Vitamin B6	water-soluble	beans, meat, fish, poultry, some fruits and vegetables	hemoglobin production, immune system and blood sugar level maintenance
Vitamin B12	water-soluble	fish, meat, poultry, eggs, dairy products	nerve cell and red blood cell maintenance, DNA construction
Vitamin C	water-soluble	fruits and vegetables	collagen formation, metabolism
Vitamin D	fat-soluble	fortified milk and cereal, dairy products, UV rays	calcium and phosphorus absorption, bone mineralization
Vitamin E	fat-soluble	vegetable oils, nuts, leafy green vegetables	immune system, DNA repair, metabolism
Vitamin K	fat-soluble	leafy green vegetables	blood clotting

Vitamins are organic substances used by various metabolic processes in the body.

MINERALS		
Name	**Source**	**Function**
Calcium	milk, dairy products	muscle and blood vessel contraction and expansion, secretion of hormones sending messages in the nervous system
Sodium	table salt	maintain blood pressure and fluid levels
Potassium	fruits, vegetables, dairy products, seafood	maintain fluid levels
Magnesium	green vegetables, legumes, nuts, whole grains	muscle and nerve function, immune system maintenance, blood pressure and blood sugar regulation, metabolism, protein synthesis
Iron	red meat, fish, poultry, beans, whole grains, spinach	oxygen transport, cell growth, cell differentiation
Zinc	oysters, red meat, poultry	immune system, DNA synthesis
Selenium	fruits, vegetables, meat, seafood, nuts	antioxidant

Minerals are inorganic substances used in very small amounts for various processes in the body.

Magnesium is a mineral found in green vegetables, legumes, nuts, and whole grains. It is stored in bones and body tissues. Magnesium is used for muscle and nerve function, immune system maintenance, blood pressure and blood sugar regulation, metabolism, and protein synthesis.

Iron is a mineral found in red meats, fish, poultry, beans, whole grains, and spinach. Iron is essential for oxygen transport, and is therefore present in red blood cells. Iron also plays a role in cell growth and cell differentiation.

Zinc is a mineral found in oysters, red meat, and poultry. Almost every cell in the body contains zinc. It is used by the immune system to aid in healing wounds and in DNA synthesis. Zinc also helps maintain the senses of taste and smell. Zinc contributes to growth and development during pregnancy, childhood, and adolescence.

Selenium is a trace mineral that is needed in small amounts to make selenoproteins, which are important antioxidants. Selenium can be found in fruits and vegetables grown in areas with a high selenium content in the soil. It can also be found in some meats, seafood, and nuts originating from selenium-rich regions of the world.

Carbohydrates

Carbohydrates provide energy to the body, and are the body's main source of fuel. Carbohydrates include sugars, starches, and dietary fibers. Both sugars and starches

supply energy in the form of glucose. Glucose provides energy directly to red blood cells, the brain, and the central nervous system.

Sugars are considered simple carbohydrates. Natural sugars can be found in fruits and vegetables. Refined sugars are often added to foods as sweeteners during preparation. Foods high in added sugars tend to be higher in calories than foods with little or no added sugars. *Starches* are complex carbohydrates. Starches are contained in grains, pasta, beans, and some vegetables.

Dietary fibers are carbohydrates that do not provide energy to the body. Soluble fiber can be digested by the body, but insoluble fiber cannot. Insoluble fiber aids in digestive processes by helping remove harmful wastes from the digestive tract. The USDA *Dietary Guidelines for Americans* recommends consuming 14 grams of fiber per 1000 calories consumed.

Proteins

Proteins are complex substances formed from amino acids. They may contain carbon, nitrogen, hydrogen, oxygen, or other elements. Proteins are found in meats, poultry, eggs, fish, and dairy products. Some vegetables, grains, nuts, and beans contain incomplete proteins, which are proteins lacking certain amino acids. The body uses protein for muscle tissue, skin, bones, and hair. Protein may be used for energy if carbohydrate and fat sources are unavailable.

Fats

While fats may sound unhealthy, getting a moderate amount of the right kinds of fats is essential to a balanced diet. Different kinds of fats include unsaturated fats, saturated fats, trans fats, and cholesterol.

It is recommended that 20% to 35% of calories consumed each day come from fat. Good fats include unsaturated fats such as polyunsaturated fats and monounsaturated fats. Examples of these include liquid vegetable oils such as soybean oil, corn oil, safflower oil, canola oil, olive oil, and sunflower oil. Fish, shellfish, walnuts, and flaxseed are additional sources of good fats.

The USDA *Dietary Guidelines for Americans* recommends the following in regards to fats:

- Less than 10% of calories should come from fatty acids each day.
- Consume less than 300 mg of cholesterol each day.
- Avoid trans fats as much as possible.
- Total fat intake should be between 20% and 35% of calories each day.
- The majority of fats consumed should be from polyunsaturated and monounsaturated fatty acids found in fish, nuts, and vegetable oils.
- Choose low-fat or fat-free meat, poultry, dairy, and bean products.

Water

Sixty-six percent of the human body is water. Water is used by the body to transport materials and remove toxins. Replenishing the body's water supply is part of maintaining good nutrition. A normal, active adult should drink 2 quarts of water per day. Water is also consumed as part of many foods including fruits, vegetables, and fruit juices.

Whole grains provide nutrients such as dietary fiber, calcium, and potassium. These nutrients are lost in the refining process.

DIETARY GUIDELINES

The U.S. Department of Agriculture (USDA) publishes the *Dietary Guidelines for Americans* every five years to promote nutrition and healthy eating. Key recommendations of the publication are emphasized as important for lowering the risk of chronic diseases and promoting health. In addition to promoting healthy eating, the *Dietary Guidelines* emphasize active lifestyles and exercise to further reduce the risk of obesity and chronic diseases.

In recent revisions to the *Dietary Guidelines,* the USDA has significantly increased recommended daily consumption of fruits, vegetables, and milk and dairy products. Additionally, they have reduced recommended daily consumption of meats, grains, oils, solid fats, and sugars. They encourage whole grain foods over enriched grains.

Nutrients that are not being consumed by adults at recommended levels include calcium, potassium, fiber, magnesium, and vitamins A, C, and E. Choosing foods that contain these nutrients is necessary for promoting long-term health. Children and adolescents should be concerned with consuming daily recommended levels of calcium, potassium, fiber, magnesium, and vitamin E.

To help educate the public, the USDA has revised the food pyramid to more accurately reflect daily recommended consumption of the major food groups including grains, vegetables, fruits, milk and dairy products, and meat and beans. The new pyramid is called the MyPyramid Food Guidance System. The major groups are arranged from left to right along the bottom edge of the pyramid. A colored band represents the importance of each group in relation to the others by the width of the band, much like a pie chart. Orange represents grains, vegetables are green, fruits are red, oils and salts are yellow, milk and dairy products are blue, and meats and beans are purple. Oils and salts are the thinnest band and are not even titled like the rest of the food groups. This is to de-emphasize the consumption of this group. Oils and salts should be avoided as much as possible.

GRAINS

Whole grains are an important source of nutrients. One of the key nutrients provided by whole grains is dietary fiber. Whole grains use the whole grain seed, called the kernel. Refined grains lose much of the bran and germ from the kernel. What is left is called the endosperm. Most nutrients in grains are contained in the bran and germ. Because refined grains do not contain the nutrient-rich bran and germ, most refined grains are enriched with vitamins and minerals as part of the refining process.

The USDA revised food pyramid is called the MyPyramid Food Guidance System.

Whole grains contain more dietary fiber, calcium, magnesium, potassium, and niacin and fewer calories per serving than enriched grains. The average recommended daily portion of grains is between 5 oz and 7 oz. At least half of the grains consumed each day should be whole grains.

Whole Grains and Cereals

Whole grains contain essential vitamins and minerals. A variety of whole grains are available.

Brown rice: Brown rice is a hulled, unpolished rice.

Buckwheat: The seeds of the buckwheat plant are edible. Buckwheat is processed into flour for breads and pancakes.

Bulgur wheat: Bulgur wheat is a parched, cracked wheat.

Hominy: Hominy is hulled Indian corn. When coarsely ground or broken, it is known as *hominy grits*. Processed hominy is used as a cereal food.

Popcorn: Popcorn is an Indian corn. When exposed to heat, the kernels burst open to form a white, starchy food. Popcorn is a popular snack.

Quinoa: The starchy seeds of this annual herb can be eaten or ground into flour.

Sorghum: Similar to Indian corn, sorghum comes from a tropical grass. Syrup that is similar to corn syrup is made from the sorghum plant.

Triticale: Triticale is a protein-rich hybrid between wheat and rye.

Whole-grain barley: The grain of a cereal grass, barley is used in malt beverages and breakfast foods.

Whole-grain corn: The entire corn kernel can be ground into cornmeal.

Whole oats/oatmeal: Oatmeal is a meal made from rolled oats. Oatmeal is used as a baking ingredient, or can be cooked into a porridge-like cereal.

Whole rye: Seeds from a hearty grass.

Whole wheat: Flour made from the entire wheat kernel.

Wild rice: Rice derived from a tall North American grass. The grains are long and covered with a dark brown or black hull.

Make an effort to choose whole grains over enriched grains. For example, whole wheat bread is a healthier choice than white bread.

Grain Group
Make half your grains whole
MyPyramid.gov

The USDA recommends adults consume at least 3 oz. of whole-grain cereals, breads, crackers, rice, or pasta every day.

Flours and Thickening Agents

Wheat, rye, barley, and corn are commonly milled into flour. Only those flours that contain protein (gluten) can produce a raised bread. Flours that contain little or no gluten, such as rye flour, must have gluten flours mixed with them to produce a raised bread. Flours also contain a small amount of fat and must be stored properly to prevent them from becoming rancid.

Water is added to flour in order for the protein to form the gluten, which provides framework and elasticity. A small amount of salt is also required to slow down the action (fermentation) of the leavening agents and to enhance the flavor. Potato flour is also sometimes used and is mixed with high-protein flours to make some pastries, such as doughnuts. Flours, corn starches, and other agents are used in many recipes as thickening agents.

All-purpose flour: This blend of hard and soft wheat flours is designed primarily for home use. Preparations in the commercial kitchen generally use specific types of flours.

Arrowroot: Arrowroot is a small tropical plant from which starch is obtained. Arrowroot is used to thicken certain items when a high gloss is desired.

Bran: Bran is the outer coat or husks of wheat, rye, and other grains separated from the grain during the milling process. Bran is used in baked preparations such as bran muffins.

Bread flour: This flour is milled from hard wheat containing protein (gluten). Gluten is the elastic substance required in bread and roll making.

Breading: Breading is the procedure of passing an item through seasoned flour, egg wash (4 to 6 eggs to each quart of milk), and bread or cracker crumbs.

Cake flour: This flour is milled from soft wheat and contains all starch and no gluten.

Cornmeal: Cornmeal is coarsely ground kernels of corn. Yellow cornmeal is made from ground yellow corn. White cornmeal is made from ground white corn. Cornmeal is used in corn bread, corn sticks, mush, and corn muffins.

Cornstarch: Cornstarch is a starch made from Indian corn. It is used to thicken stews and liquids such as sauces and gravies.

Egg white stabilizer: This is a white powder mixture consisting of sugar, calcium sulfate, carrageen, and other ingredients. Egg white stabilizer is used when beating egg whites to create a stiff meringue.

Modified starch: Modified starch is a blend of starches such as arrowroot and cornstarch. It is used in fruit pie fillings and glazes because it holds a sheen longer than most starches, even if the item is refrigerated.

Pastry flour: This flour is milled from soft wheat and contains part starch and part gluten, which is important when preparing pie dough, cookies, and various pastries.

Pregelatinized starch: This starch is blended with sugar and added to a liquid for instant thickening. It reacts quickly without heat because the starch has been precooked and requires no additional heat to absorb liquid and gelatinize. Pregelatinized starch is an excellent product to use when speed is required.

Rye blend flour: This is a blend of rye flour and a high-gluten flour, usually consisting of 30% to 40% rye flour and 60% high-gluten flour. Rye blend flour eliminates the need to mix two separate flours when making rye bread and rye rolls.

Rye flour: Rye flour is milled from rye grain. Its composition is very much like wheat flour, but the protein is quite different. The protein of rye flour, when made into dough with the addition of water, does not produce gluten. Hard wheat flours must be added to rye flour to produce a porous, well-raised loaf of bread.

Trade Tip

The consumption of wheat may be a concern to people who have wheat allergies. It is important to know which menu items contain wheat or wheat products to safely serve patrons who are allergic to wheat.

Tapioca flour: This flour is made from the root of the tapioca plant. Tapioca flour is used as a thickening agent in pies and glazes.

Whole wheat flour: Whole wheat flour is made using the entire wheat kernel. It is used in making whole wheat bread and rolls.

VEGETABLES

The USDA *Dietary Guidelines for Americans* recommends that Americans increase their consumption of vegetables. A person eating a 2000 calorie daily diet should be eating 2 c to 3 c of vegetables every day. Adding vegetables to a diet can help reduce the risk of chronic diseases such as strokes, type 2 diabetes, and some forms of cancer. The USDA specifically recommends eating more dark green and orange vegetables and more dry beans and peas.

Asparagus: This thistle-like vegetable is a member of the lily family. It consists of tall, straight stalks with round, compact, bud-like tips.

Bamboo shoots: These young shoots of certain species of the bamboo palm are about 4″ thick at the base and about 1½″ long. Bamboo shoots are commonly used in oriental preparations. Bamboo shoots are a source of dietary fiber, iron, phosphorus, potassium, vitamin A, vitamin C, and vitamin E.

Bean sprouts: These are the sprouts of the mung bean, which is a small, round, green bean first grown in China and later brought to the United States. The sprouts are from 1½″ to 2″ long when picked. Bean sprouts have a delicate flavor and are commonly used in salads, chow mein, chop suey, and other oriental preparations.

Varying vegetables includes eating dark green and orange vegetables in addition to dry beans and peas.

Beets: Beets are large, red roots. They are typically fairly round in shape.

Broccoli: Broccoli is closely related to cauliflower, and is a member of the cabbage family. The stalk, leaves, and flower bud top are all edible parts of the broccoli plant.

Brussels sprouts: Brussels sprouts are a member of the cabbage family and resemble miniature cabbages. They have a tart, distinctive flavor.

Cabbage: This plant consists of a large head of tight leaves. Cabbage has a very mild flavor.

Carrots: Carrots are a root vegetable consisting of a long, tapered orange root.

Cauliflower: Cauliflower closely resembles broccoli, but is white or cream in color. It is a member of the cabbage family.

Corn: While corn is a cereal grain crop, it is commonly consumed as a vegetable. Corn must be served or preserved quickly after it has been picked. This will maintain the tenderness and sweetness of the kernels. Common varieties of corn include sweet corn, yellow corn, and shoepeg corn.

Eggplant: Eggplant is a purple-skinned, pear- or egg-shaped vegetable. Both the flesh and watery, grey-colored pulp are edible.

Green beans: Also known as string beans or snap beans, green beans consist of long green pods containing small, soft beans.

Kohlrabi: While related to cabbage, the edible stem portion of kohlrabi resembles a turnip.

Leeks: Leeks are members of the green onion family. They have long, wide, flat green stems and little or no bulb. Their very delicate flavor is used to enhance stews, soups, and sauces.

Lentils: These small, flat, round beans are light brown with a touch of green slightly visible. Lentils grow in a pod and are used only when ripe. They are used in soups and can be served as a vegetable.

Lima beans: Lima beans are large, flat, light green beans. They can be purchased frozen, canned, fresh, or dried.

Okra: Okra is a green, fuzzy, tapered pod vegetable. It contains seeds, has from 6 sides to 12 sides, and is generally 2″ to 3″ long. The shorter pods are preferred, as the longer pods are tough. Okra can be served as a vegetable (either boiled or fried) or used in soup preparations.

Pimientos: These large, sweet red peppers are peeled and canned with their stems, core, and seeds removed. Pimientos are used in many food preparations and are commonly used to decorate salads, deviled eggs, and canapés.

Snow peas: These small peas have nearly flat shells or pods. The pods are brittle and tender like snap beans. Snow peas are picked before the peas inside have developed.

The pods are usually from 3½″ to 4″ long. Snow peas are commonly used in oriental preparations.

Tomatoes: Tomatoes are classified as a fruit, but prepared as a vegetable.

Water chestnuts: This is an aquatic (grown in water) herb that produces a nut-like fruit. Water chestnuts are white in color and retain a crisp texture even when cooked. They are commonly used in hors d'oeuvres and in oriental preparations.

Wax beans: Closely related to the green bean, wax beans are yellow beans with a long, round pod containing small, soft beans.

FRUITS

The USDA *Dietary Guidelines for Americans* recommends that Americans also increase their consumption of fruits. A person eating a 2000 calorie daily diet should be eating 1½ c to 2 c of fruits every day. Eating a variety of fruits is best, and fresh fruit is preferred over fruit juices that often contain refined sugars.

Apples: Apples come in many varieties including Red Delicious, Golden Delicious, Gala, Granny Smith, Rome, and Fuji.

Apricots: Apricots are similar to peaches, but are much smaller and more delicate.

Avocados: Avocados are green pear-shaped tropical fruits containing a single seed, or pit.

Bananas: Bananas come in several varieties, and consist of a long, yellow fruit with a tough peel.

Blueberries: Blueberries are small blue berries that grow on shrubs native to the United States.

Cherries: Cherries are small, red fruits that come in several varieties of varying color and sweetness.

Coconut: Shredded coconut is long, thin particles of coconut used for cake decorating and other baking products.

Cranberries: Cranberries are round, tart fruits that vary in color from white to dark red.

Grapefruit: Grapefruit is a member of the citrus family, and consists of a large, round fruit with a tough yellow peel. The fruit can vary from white to deep red, and can be tart or sweet depending on the variety.

Grapes: Grapes vary in color from green to red and grow in bunches on vines. There are seeded and seedless varieties. Grapes are commonly classified as wine varieties and dessert varieties.

Kiwi: Kiwi are small, fuzzy brown fruit with bright green flesh.

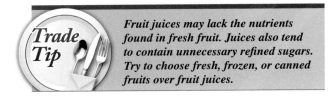

Trade Tip

Fruit juices may lack the nutrients found in fresh fruit. Juices also tend to contain unnecessary refined sugars. Try to choose fresh, frozen, or canned fruits over fruit juices.

Kumquats: These are a small citrus fruit.

Lemons: Lemons are medium-sized yellow citrus fruits that are very sour.

Limes: Limes are small, green citrus fruits that are rather sour.

Melons: Melons come in many varieties including watermelon, cantaloupe, and honeydew.

Oranges: Oranges are yellow to orange citrus fruits with a very sweet flavor. They come in many varieties, including navel oranges and Valencia oranges.

Peaches: Peaches are medium-sized, yellow-orange fruits with a fuzzy, edible skin and a large pit.

Pears: Pears are sweet, green, yellow, or reddish-brown fruit with a white, grainy flesh and core.

Pineapples: Pineapples are tropical fruits with a pinecone-like exterior and sweet, yellow flesh. The core and skin are not edible.

Plums: Plums are small fruits with delicate skins, juicy flesh, and pits. They come in many varieties and colors including green, red, and deep purple.

Raspberries: These small berries come in red or black varieties and are very sweet.

Strawberries: Strawberries are small, red berries with seeds embedded in the exterior skin.

Fruit Group
Focus on fruits
MyPyramid.gov

The USDA recommends eating a variety of fresh, frozen, and canned fruits while consuming fruit juice in moderation.

MILK AND DAIRY PRODUCTS

Milk and dairy products are an important source for calcium to help maintain bone strength. There are many fat-free and low-fat choices in this food group. In addition to various forms of milk and cream available, this group also includes butter, cheese, cottage cheese, sour cream, and yogurt. The USDA recommends choosing low-fat or fat-free dairy products. Lactose-free products and other sources of calcium should be consumed if a person does not or cannot consume milk products. The recommended daily serving of milk and dairy products is 3 c each day.

Butter: Butter is made from milk fats separated from other milk components by churning (agitation). The approximate fat content of butter is 80%, leaving 20% for water, salt, and curd. The two types of butter are sweet cream and sour cream butter. Most of the butter on the market is made from sour cream soured by natural or artificial processes. Sour cream butter is a little more flavorful than sweet cream butter. Sweet cream butter is sometimes marketed unsalted for people on special diets. Salted butter is most popular and has better keeping qualities.

Buttermilk: Buttermilk is the milk that remains after the butter has been separated from milk or cream, usually by the churning process. The milk may also be artificially curdled or separated using a culture. Some artificially treated buttermilk may contain small specks of butter to give the product the look of old-fashioned buttermilk. It is often used for baking.

If a recipe calls for buttermilk and none is available, fair results can be obtained by adding ¼ c of vinegar or lemon juice to each quart of milk. Cream can be converted into sour cream by using this same recipe.

Condensed milk: Condensed milk, like evaporated milk, is heated until part of the water evaporates. At this point, approximately 40% to 50% granulated sugar is added to act as a preservative. Condensed milk may be known as condensed or sweetened condensed, depending on the percentage of sugar added. Since the sugar acts as a preservative, the opened container does not need to be refrigerated if it is used within a safe period of time. It is purchased in cans. Condensed milk is used for preparing certain desserts.

Cottage Cheese: Cottage cheese is a combination of milk curds and whey. The fat content of cottage cheese can vary based on the size of the curds and the fat content of the whey.

Choose low-fat or fat-free dairy products, or include lactose-free calcium sources such as fortified foods and beverages.

Dry milk: This is whole milk with almost all the water and moisture removed. It should not contain more than 4½% moisture. The two types of dry milk are whole dry milk, which contains both milk solids and fat, and skim dry milk, also known as nonfat milk solids, which contains only the milk solids. Never boil soups or sauces that contain dry milk. In addition, mixtures containing dry milk solids are easily curdled. Some recipes will state when dry milk can be substituted for whole milk.

Evaporated milk: This is whole milk that has been heated until part of the water has evaporated. The resulting product is sterilized by heat and canned. Evaporated milk has a long shelf life.

Homogenized milk: This is whole milk that has been processed (homogenized) to break up fat globules into smaller ones and distribute them evenly in the milk. The fat globules remain suspended in the milk and never rise to the top of the milk, preventing separation of milk and fat. Homogenization of milk results in better appearance and richer taste. Homogenized milk may be purchased as skim milk, 1%, 2%, or whole milk, depending on the amount of butterfat required.

Liquid whole milk: This is milk as it comes from the cow.

It is important for people who are lactose intolerant or who do not consume milk or milk products to obtain the recommended daily allowances of calcium. Lactose-free, calcium enriched foods and beverages are available for people with specific dietary requirements.

Pasteurized milk: This is milk that has been heated to a high enough temperature to kill harmful bacteria, then chilled rapidly. Most milk today is pasteurized.

Half and half: This is cream that contains 18% butterfat. It is used in coffee, cereal, and in some food preparations. Coffee cream substitutes are being used in many food service establishments today for coffee service because they are convenient and have excellent keeping qualities. Some are made of nondairy ingredients such as soybean oil and derivatives. Others are prepared from dried dairy products.

Skim milk: Milk from which most of the butterfat content has been removed. To be labeled "milk," the butterfat content must be 3½%. Anything less than 3½% must be labeled "skim milk" or have the butterfat content emphasized, such as 1% or 2% milk.

Sour cream: This is cream that contains at least 18% butterfat and has been soured by special bacteria. Sour cream is commonly used to top baked potatoes and is also used in the preparation of entrees, salads, appetizers, and certain baked goods.

Whipping cream: Whipping cream contains from 30% to 36% butterfat. If the cream is too fresh, it will not whip well. Cream that is 24 hr to 48 hr old has improved whipping qualities. For best whipping results, the cream, bowl, and wire whip should be cold at the time of whipping. Overwhipping results in butter. Whipped cream is used in many desserts and entree items.

Yogurt: Yogurt is a semisolid food product made of milk and milk solids. Two forms of bacteria, *Lactobacillus bulgaricus* and *Streptococcus thermophilus*, are added. The result is a fermented and slightly acidic product which can be served plain or flavored with fruit.

Meat & Bean Group
Go lean with protein
MyPyramid.gov

Bake, broil, or grill low-fat or lean meats. Vary your protein routine with fish, nuts, and beans.

MEATS, FISH, AND NUTS

All foods made from meat, poultry, fish, dry beans or peas, eggs, nuts, and seeds are part of this food group. Fish, nuts, and seeds contain monounsaturated and polyunsaturated oils, and are a healthy choice. Lean or low-fat meats are preferred over more fatty cuts. The recommended daily portion size for adults is between 5 oz and 6 oz of foods from this group each day.

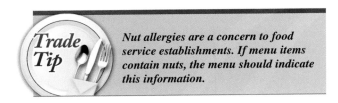

Trade Tip

Nut allergies are a concern to food service establishments. If menu items contain nuts, the menu should indicate this information.

Meats

Meats: Meats (beef, pork, veal, and lamb) are graded by government standards on the basis of quality and yield. Quality refers to the overall appearance of the flesh and judgment of eating qualities. Yield grades are numbers 1 to 5 and determine how much salable meat can be obtained from a carcass. The lower the yield number, the more usable the meat. The grades, in order of desirability and quality, are given in Table I.

Lower grades are also available, but are rarely used in the commercial kitchen. Poultry used in the commercial kitchen is also graded by the government as Grade A or Grade B. Grade A poultry is full-fleshed and meaty, is well-finished, and has an attractive appearance. Grade B is slightly lacking in fleshing, meatiness, and finish, or has some dressing defects.

Aspic: This is a clear meat, fish, or poultry jelly used for decoration on some preparations.

Canadian bacon: Canadian bacon is the trimmed, pressed, smoked loin of pork. It may be purchased cooked or uncooked.

Gelatin: Gelatin from animals is extracted by heat from bones, white connective tissues, and skins of food animals. Gelatin is odorless and tasteless. It can be purchased in three forms: sheet, powdered, or granulated. It is used in cold soups, aspics (meat jelly), and desserts.

Lard: Lard is the fat of pigs and hogs processed for cooking and baking. When cold, lard is semisolid. It is used for frying.

Poultry: Poultry includes chickens, turkeys, ducks, geese, Cornish hens, and other members of the bird family. Chicken is the most commonly consumed form of poultry.

Suet: Suet is the hard fat located around the kidneys and loin of beef and lamb. When rendered, it is usually used for frying.

Sweetbreads: These are the thymus glands found on each side of the throat of calves and lambs. They are used as a meat delicacy.

Tripe: This is the edible lining of a cow stomach. The most desirable tripe is honeycomb tripe, because it has a netted appearance. Tripe may be purchased fresh, pickled, or canned.

Fish and Seafood

Anchovies: These are small salted fish fillets of the herring family. Anchovies usually are purchased canned in olive oil. They have a very strong taste and are used in the preparation of hors d'oeuvres, canapés, and certain salads, such as Caesar salad.

Caviar: Caviar is the prepared and salted roe (eggs) of the sturgeon and certain other types of fish. Caviar is quite expensive and is considered a delicacy.

Finnan haddie: Finnan haddie is salted and smoked haddock. It is usually prepared by steaming.

Sardines: These are small fish of the herring family, including pilchards, sprats, bristlings, and young herrings. Sardines are usually canned in olive or cottonseed oil. Some larger sardines are packed in mustard or tomato sauce.

Fish and nuts are important sources of proteins and oils and should be included as part of a balanced diet.

Smoked salmon: Smoked salmon includes coho, chinook, or chum salmon that has been smoked, sliced thin, packed in olive or cottonseed oil, and canned. Smoked salmon is used in the preparation of hors d'oeuvres or canapés.

Surimi: This is a crabmeat product that looks, cooks, and tastes like crabmeat. Surimi is made from a mixture of pollock, snow crabmeat, turbot, wheat starch, egg whites, vegetable protein, and other ingredients. It is low in calories, sodium, fat, and cholesterol. It is high in protein. Surimi is marketed precooked and frozen to protect its flavor and can be purchased as legs, chunk meat, or flake meat.

Nuts

Nuts are an excellent source of proteins and oils. Common nuts include the peanut, pistachio, almond, walnut, filbert, and cashew. Nuts are typically roasted, and can be salted or unsalted.

Almond paste: This is a cooked mixture consisting of about 56 parts of ground, blanched almonds, 34 parts of sugar, and 10 parts of water and flavoring. Almond paste is used in baking and for making candies, macaroons, and marzipan (a type of candy).

Pecan pieces: These broken pieces of pecans are used for decorating cakes and for pecan pie filling.

Pistachio nuts: This is the kernel of the fruit of the pistachio tree. The nut is shaped like a bean and is covered with a grayish purple skin. It is used in the preparation of certain classical food preparations, such as gelatine of chicken.

Shaved almonds: These are sliced, toasted almonds used to decorate cakes and candies.

SUGARS AND SWEETENING AGENTS

Sugars and sweetening agents are considered discretionary calories in a balanced diet. Discretionary calories are the calories left over after all recommended daily allowances of the major food groups have been met. The average person is allotted between 100 and 350 discretionary calories per day.

Brown sugar: Brown sugars are refined at a lower temperature and contain more molasses and moisture than granulated sugar. Dark brown sugar contains more molasses and moisture than light brown or yellow sugar, both of which have been refined longer.

Chocolate: Preparation made by roasting and grinding cacao seeds. Chocolate is classified as bitter, semisweet, or sweet, depending on the amount of sugar added when it is made.

Chocolate is classified as bitter, semisweet, or sweet depending on the amount of sugar added when it is made.

Chocolate naps: This is bitter chocolate in small cakes (about 1 oz each) used for cooking and baking.

Chocolate shot: These are bits of sweet chocolate used for cake and cookie decorating.

Cocoa: Cocoa is pulverized chocolate with one-half of its butterfat extracted.

Dextrose: Dextrose is the sugar of vegetables (except beets). It is less sweet than cane sugar.

Glucose: Glucose is the heavy corn syrup used in preparing glazes and in candy making.

Granulated sugar: This is the sweet, white, crystalline substance obtained from sugar cane and beets. Granulated sugar is the most commonly used type of sugar.

Honey: Honey is the thick, sweet, slightly yellow syrup that bees make out of the nectar they collect from flowers. The color and flavor depend on the age of the honey and the source of the nectar. Sources of high-grade honey are white clover, orange blossoms, and alfalfa. Light-colored honey is usually graded highest. Honey is used in both cooking and baking.

Molasses: This is the dark, sticky, sweet syrup that cannot be crystallized when making sugar from sugar cane. Molasses is commonly used in certain cakes, cookies, and muffins.

Powdered or confectionery sugar: This is a very fine sugar made by grinding coarse granulated sugar and sifting through a fine silk cloth. Powdered sugar comes in three grades: 4X, 6X, and 10X. The higher the number, the finer the sugar.

Sanding sugar: This is a coarse sugar used by bakers to garnish sweet rolls, cream rolls, and other preparations, and that comes in a variety of colors.

Vanilla extract: This is made by extracting the flavor or oil from the vanilla bean and mixing it with diluted alcohol. Vanilla extract is used to flavor desserts, sweet rolls, coffee cakes, cookies, and other baked goods.

Verifine sugar: This is the same as granulated sugar but ground or rubbed finer. Verifine sugar is commonly used when making sugar molds for cake decorations.

FATS AND OILS

Solid fats and oils are also considered discretionary calories and should be avoided as much as possible. The maximum amount of saturated and trans fats consumed in a day should be under 24 g for people on a 2000 calorie diet. Choosing foods that are lean, low-fat, or fat-free will help eliminate trans fats and saturated fats from the daily diet.

CONVENIENCE FOODS

These foods, also known as frozen prepared foods, are often used in the food service industry. The commercial kitchen today uses many convenience foods, and that use is expected to increase as new products are developed.

The creation of these convenience foods started with seafood. The practice of freezing food began in 1912 when Clarence Birdseye journeyed to Labrador to investigate and study the methods used by the Eskimos to preserve seafood by freezing. When he returned to the United States, Birdseye refined these methods, and his products quickly became such a success that today his name is a household word.

The early success in freezing seafood led to overwhelming successes with frozen French fries and orange juice, which, in turn, started the search for other products and items that could be marketed in a frozen condition. World War II provided the opportunity to experiment with frozen beef and other products because the foods shipped overseas to American troops had to be in a condition that would prevent spoilage during the long journey. Today, this method of preserving food is almost limitless.

Trade Tip

When choosing convenience foods, it is important to read the nutrition label on the package to choose foods containing essential nutrients that are low in sodium and fats.

Presently available on the market are different soups, most vegetables, a large assortment of appetizers, including canapés and hors d'oeuvres, various potato preparations, meat, seafood, vegetable and fruit salads, pasta entrees, and almost any dessert that can be found on a menu, as well as complete plate combinations that only need to be heated.

Many of the companies in this business not only produce and market a complete line of frozen products, but also design complete food systems. This includes menu writing and design, work flow standards, production and equipment recommendations, cost comparison, labor standards, and cost.

Although there are many advantages of convenience foods and food systems, they are only another tool to help management control labor costs, increase production, and decrease storage space. They are not a replacement for the traditional food services offered. A disadvantage of convenience foods is that the cost is high. As with other products, the more that is done for the consumer, the higher the cost.

Many of the convenience foods have been well developed. Others still have shortcomings, such as a fibrous texture, a starchy taste, or a breakdown of sauces. Poor-quality preparations must be avoided. Well-prepared convenience foods can be of help to the chef in solving many production problems and providing a larger selection. These items include:

- Apple dumplings
- Beef Wellington
- Breaded seafood
- Chicken cordon bleu
- Chicken kiev
- Frozen cheesecakes

Stuffed ravioli is a popular convenience food. It may be purchased fresh or frozen.

- Frozen eggs
- Frozen toppings
- Prepared mousse
- Stuffed orange roughy
- Stuffed rainbow trout
- Stuffed shrimp
- Stuffed sole
- Imitation crab (surimi) sticks or chunk meat
- Veal cordon bleu

DIET CONSIDERATIONS

Recent trends in the United States show an increased awareness of nutrition and diet. This increased awareness has resulted in a change in the eating habits of the general public. The food service establishment must offer preparations that customers desire. In addition, food service establishments must consider their customers and any special diet needs. For example, hospitals, schools, and other food service operations may offer special menus. The chef or cook must be aware of what food items are necessary for proper nutrition. The U.S. Department of Agriculture, Department of Health and Human Services suggests a few dietary guidelines for proper nutrition:

- Eat a variety of foods.
- Avoid too much fat, saturated fat, and cholesterol.
- Eat foods with adequate starch and fiber.
- Avoid too much sugar.
- Avoid too much sodium (salt).

The recommended number of servings of the six major food groups per day is as follows:

- Bread, cereal, rice, and pasta: 5 oz to 7 oz
- Vegetables: 2 c to 3 c
- Fruit: 1½ c to 2 c
- Milk, yogurt, and cheese: 3 c
- Meat, poultry, fish, beans, eggs, and nuts: 5 oz to 6 oz
- Fats, oils, and sweets: avoid

One of the most common concerns with diet is the intake of sodium. Sodium is an ingredient used in a variety of foods, drinks, and medicines. Most sodium in the diet comes from sodium chloride (common table salt). Sodium is an essential nutrient and is required by the body in order to function properly. However, most Americans consume too much sodium. There are four major sources of sodium in the average diet:

- Table salt, which contains 40% sodium
- Foods processed with salt

- Sodium as naturally present in most foods; foods of animal origin are high in natural sodium while foods of vegetable origin are low
- Water supply, especially if softened

Prepared sauces often contain a high amount of sodium. Try to reduce the amount of sodium used in preparation of sauces.

Popular foods that contain a high percentage of sodium include the following:

- Fast foods
- Frozen entrees
- Prepared sauces (tomato, soy, Worcestershire, etc.)
- Canned soups and vegetables
- High-sodium meats (ham and bacon)
- Snacks (chips, pretzels, crackers, etc.)
- Cheese and cottage cheese

The sodium content of a food item can be determined by the label. Following are some examples of labels that are found and their explanation:

- Sodium free: Each serving has less than 5 milligrams (mg) of sodium.
- Very low sodium: Each serving has 35 mg of sodium or less.
- Low sodium: Each serving has 140 mg of sodium or less.
- Reduced sodium: The product has 75% less sodium than the item it is replacing.
- Unsalted: A product processed without salt that would normally be added.
- It is more important to note that no more than 2400 mg per day of sodium should be consumed. This is a little more than 1 tsp.

Cooking Methods and Techniques

A variety of different cooking methods and techniques are required for different food preparations. The type of preparation required for a food item before cooking depends on the nature of the food item, the size of the food item, and the method used to prepare the food item quickly and efficiently. However, preparation techniques may also require the use of hand tools by the chef. The hand tool used most often is the French knife.

Cooking techniques, as with preparation techniques, vary depending upon the type of food item to be cooked. Cooking techniques commonly used include roasting, baking, broiling, panbroiling, braising, steaming, boiling, sautéing, grilling, and deep fat frying; food items can also be cooked by microwave radiation.

Recipes specify the food items required for a given food preparation and list the steps or procedures required for preparation. Recipe quantities may have to be adjusted based on the number of servings required.

COOKING METHODS AND TECHNIQUES

Cooking is subjecting foods to heat in order to make them more digestible. Different methods are used to apply heat to foods. The method used depends on the nature of the food item. For example, if the food item is tough, a lengthy cooking method is required for the best results. If the food item is tender, a quick cooking method is used for the best results.

Equipment used in cooking has continually improved through the years. Most new equipment is designed to save time and speed production. Many pieces of new equipment have been accepted in commercial kitchens, such as convection ovens, microwave ovens, and convection microwave ovens. New equipment may require new techniques in the preparation and cooking methods used.

KNIVES

Knives are the chef's most important tool. Every tradesperson has tools of the trade to produce quality work. Likewise, the chef has a personal set of knives used to produce quality work efficiently. A set of knives should include a French knife (also known as a chef knife or sandwich knife), a carving knife (roast beef or ham slicer), boning knife, utility knife, and paring knife. These are knives required for general purposes. Other knives designed for special purposes may be required. All knives must be kept very sharp. Knives should be thoroughly cleaned and sanitized after each use to prevent the spread of bacteria.

The knife blade is moved across the sharpening stone with light, even strokes.

Sharpening a Knife

A sharp knife is always safer than a dull knife. Less force is required when cutting, and it will not easily slip off the item being cut. A sharpening stone and a butcher's steel are required to sharpen and hone a knife. The sharpening stone is used to restore the edge of the blade. The blade is positioned on the sharpening stone at a 15° to 20° angle. The knife blade is moved across the sharpening stone with light, even strokes. This operation is repeated on the other side of the blade. Use the sharpening stone only as required. Avoid oversharpening, which removes excess metal. If the blade is routinely honed on the butcher's steel, use of the sharpening stone will be reduced.

A butcher's steel is used as necessary to maintain a sharp edge on the blade of the knife.

The butcher's steel is used to maintain the sharp edge of the blade. The butcher's steel straightens irregularities of the edge of the blade and removes burrs that result from use. As with the sharpening stone, position the knife at about a 15° to 20° angle and make a full stroke over the entire length of the blade with even pressure applied. About four to five strokes on each side of the blade are sufficient. Too many strokes may dull the edge of the blade. The butcher's steel is used to maintain a sharp edge between sharpenings.

French Knife

The French knife is the most commonly used knife in the commercial kitchen. The French knife is very versatile and is used for slicing, dicing, mincing, chopping, julienning, and in some cases, shredding. The blade should always be very sharp.

The French knife should be held with a firm grip with the handle passing between the thumb and index finger. For chopping, mincing, dicing, and julienning, the point of the knife should be kept on the table while the blade is rotated rapidly with a forward up-and-down motion. The cutting is done with friction on the forward stroke. Never lift the point of the knife off the cutting board and never attempt to cut by pressing the knife straight down on the item to be cut. For slicing and shredding, the knife is held in the hand in the same manner, but the blade is passed across the item being cut with a smooth forward motion. Again, the cutting is done with friction on the forward stroke. For all cutting operations, the item being cut is firmly held with fingertips slightly tucked away from the blade of the knife.

Carving Knife

Carving or cutting with carving knives is with a slicing motion in either a horizontal or vertical position. Carving ham and roast beef is done by using the horizontal position. Turkey and sirloin of beef are carved using the vertical position. When carving in the horizontal position, the knife is held firmly between the thumb and index finger. The blade of the knife is held in a horizontal position. The cutting motion is backward and forward, similar to a sawing motion. The blade must be held straight to maintain a flat surface on the meat being carved and to avoid steps. Let the blade glide across the surface with the backward stroke doing most of the slicing. Always slice against the grain of the meat. Select the proper carving knife (ham or roast beef slicer, serrated or straight edge).

When carving in a vertical position, the meat on the cutting board is sliced in a downward motion. The knife is held firmly between the thumb and index finger. The motion of the knife is backward and forward, as well as downward. The cut is made on the forward downward stroke. Pressure is applied on the downward stroke, but the friction of the blade on the surface of the meat should do the cutting. When carving in a vertical position, the French knife is usually used. If preferred, a roast beef or ham slicer may also be used. Always cut against the grain of the meat.

Paring Knife

Paring or trimming with a paring knife is done with the food item held in the hand while being cut. The knife is cradled in the four fingers and the thumb is used to guide or direct the cut. Caution must be exercised to prevent cutting the thumb. Trimming celery and radishes and peeling an apple are examples of paring.

Boning Knife

Boning knives are used to bone, disjoint, separate, and skin meat. The knife is usually held in one of two positions, depending on the type of cut being made. Boning knives are held in the fist with the thumb on top, resting on the index finger with the blade pointing down, or cradled in the four fingers with the thumb resting on the opposite side of the handle to ensure a firm grip. With the exception of disjointing, cutting is done with short, quick, backward strokes. The forward portion of the blade does most of the cutting.

TABLE I: FOOD CUTTING		
Method	**Meaning**	**Techniques of Preparation**
Slice	A relatively thin, broad piece of food	Slice by using a slicing machine, French knife, or carving knife; Always slice against the grain, moving the blade of the knife in such a way as to cut by a sawing action
Chop	Cut into uneven bits, May be fine, medium or coarse	Cut on an NSF-approved cutting board; Use a French knife and cut with a rocking motion
Dice	Cut into cubes; May be small, medium, or large	Dice on an NSF-approved cutting board; Use a French knife and cut with a slicing motion; A small dice is ¼″, a medium dice is ½″, and a large dice is ¾″
Brunoise	Cut into very fine cubes	Dice on an NSF-approved cutting board; Use a French knife and cut with a slicing motion; Final dimensions are ⅛″
Mince	Chop into very fine pieces	Cut on an NSF-approved cutting board; Use a French knife or a power food cutter; Cut by applying short, sharp strokes; Meats may be minced by running through a meat grinder
Puree	Pound or mince fine and force through a sieve	Same as "mince" but finer, almost paste-like consistency
Julienne	Cut into long thin strips	Julienne on an NSF-approved cutting board; Use a French knife and cut with the grain with a slicing motion into very thin slices; Then cut a second time with the grain into very thin strips; Final dimensions are ⅛″ × ⅛″, × 2″
Grind	Crush into fine, medium, or coarse particles	Pass the item through a food grinder using the fine, medium, or coarse chopper plate; Do not force the item into the grinder, feed small amounts at a time
Grate	Pulverize by rubbing against a rough or indented surface	Grate by using a box grater; The mesh of the grate depends upon which surface is used; Grating is also done in a power food cutter
Shred	Cut into very fine strips	Shred by rubbing the item across the coarse grid of a box grater, by shaving with forward strokes of a French knife, or by passing the item across the revolving blade of a power slicing machine
Score	To mark the surface of certain foods with shallow slits	Cut with a French knife in parallel lines approximately ½″ apart; Cut about ⅛″ to ¼″ deep

Various food cutting methods are used for specific food preparation applications.

BASIC FOOD CUTS

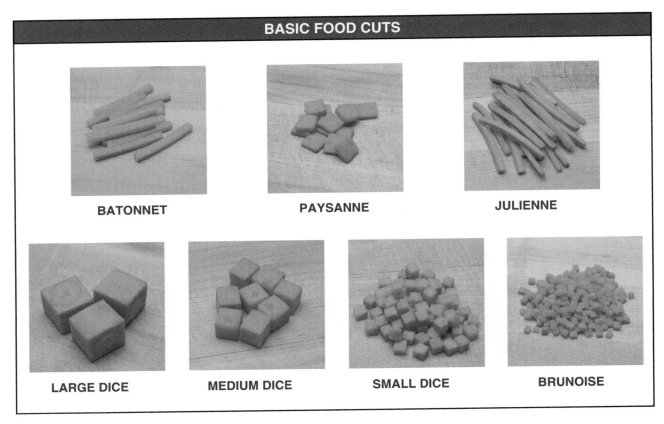

BATONNET   PAYSANNE   JULIENNE

LARGE DICE   MEDIUM DICE   SMALL DICE   BRUNOISE

Each cutting method results in a food item with specific size and shape.

Butcher Knife

Butcher knives and steak knives are primarily used to slice raw meats and fish into portion size. Cutting steaks and chops, slicing liver, and portioning fish are examples of this type of cutting. The knives are scimitar-shaped (sword-shaped) and come in various sizes. The knife is cradled in the four fingers with the thumb resting on the opposite side of the handle to provide support for a firm grip. The cutting motion is done with an arced, backward stroke to take full advantage of the scimitar shape. The cutting in most cases is against the grain.

PREPARATION OF FOODS FOR COOKING

Different food items require different preparation. Meats often must be cut into smaller pieces before they are cooked and served. The recipe states how the meat is to be cut. Vegetables usually must be cut before being used in a recipe. The cleaning and peeling is normally done before the chef uses the vegetables. Basic cuts include brunoise, dice, paysanne, julienne, and batonnet. Brunoise and

dice cuts result in evenly-shaped cube pieces of varying dimensions including large, medium, and small dice. A paysanne cut results in flat pieces ½″ square by ⅛″ thick. Julienne and batonnet cuts result in stick-shaped pieces. The various methods and techniques of cutting meats and vegetables are listed in Table I. Most recipes require one or more of these methods.

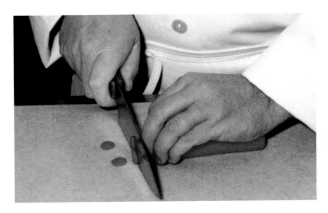

When cutting, fingers are tucked away from the blade to prevent injury.

MINCING ONIONS

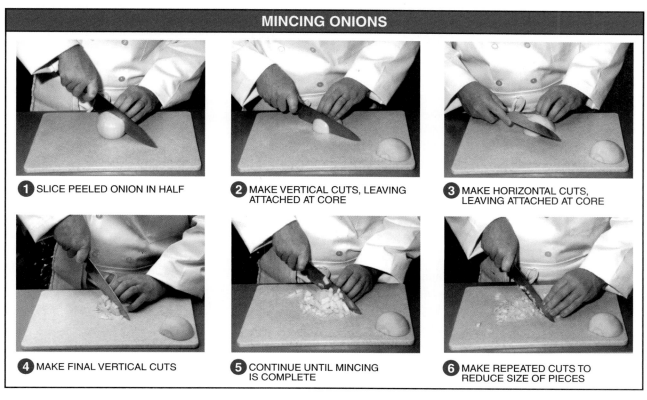

1. SLICE PEELED ONION IN HALF
2. MAKE VERTICAL CUTS, LEAVING ATTACHED AT CORE
3. MAKE HORIZONTAL CUTS, LEAVING ATTACHED AT CORE
4. MAKE FINAL VERTICAL CUTS
5. CONTINUE UNTIL MINCING IS COMPLETE
6. MAKE REPEATED CUTS TO REDUCE SIZE OF PIECES

A series of vertical and horizontal cuts efficiently produces minced onions.

CREAMING GARLIC

1. PRESS CLOVE WITH FLAT EDGE OF KNIFE AND PEEL OFF SKIN; REMOVE END AND SLICE CLOVE
2. CHOP GARLIC WITH ROCKING MOTION
3. HOLD TIP OF KNIFE WITH FLAT HAND AND CONTINUE ROCKING MOTION FOR FINE CHOP
4. CREAM BY DRAGGING FLAT EDGE OF KNIFE ALONG CUTTING SURFACE OVER CHOPPED GARLIC

Chopped garlic and creamed garlic are used in a variety of food preparations. Omitting step 4 results in chopped garlic.

MEAT COOKING

Meat cooking uses dry heat, moist heat, or a combination of cooking methods. Different cuts of meat require different cooking methods. *Dry heat cooking* is a cooking method that uses hot air, hot metal, a flame, or fat to conduct heat to the food without any moisture. A dry heat cooking method will brown foods, while a moist heat cooking method will not. Dry heat cooking methods for meat include deep frying, pan frying, roasting, broiling, panbroiling, griddling, sautéing, and grilling. Meats that are best cooked using dry cooking methods are those that are very tender, with little connective tissue, and that can be served medium-rare.

- *Deep frying*: Deep frying is a cooking method where food is completely covered with hot oil. Meat cooked in this manner should be battered or breaded. Battered items are deep fried using the swimming method, where items are slowly dropped into the hot oil without a basket. If a basket were used for frying battered items, they would stick to each other and the basket when cooking. Breaded items are deep fried using the basket method, where a fry basket is filled and submerged into the hot fat for the duration of the cooking time.

- *Pan frying*: Pan frying is similar to deep frying, but the quantity of fat used is much less. Foods are typically breaded and then fried in fat in a pan. The hot fat covers only half to two-thirds the thickness of the food, and the items are turned when the submerged side turns brown.

- *Roasting*: To roast is to surround food with dry, indirect heat in an oven. The roast is not covered. The term baking is used interchangeably with roasting, but for the most part, baking refers more to breads and pastries. In addition, roasting can also be used to describe *rotisserie* style cooking in which items are cooked on a rotating spit over or next to an open flame to avoid burning. It is common to *baste* when roasting by brushing or ladling juices or fat over the item during the cooking process.

- *Broiling*: The broiling of meat is similar to roasting, but in broiling, direct heat is used. The meat is exposed to the flame in gas cooking and to the heating element in electric cooking.

- *Panbroiling*: In panbroiling, cooking is accomplished by contact with a heated surface, such as a frying pan or sauté pan. No covering and no fat is used.

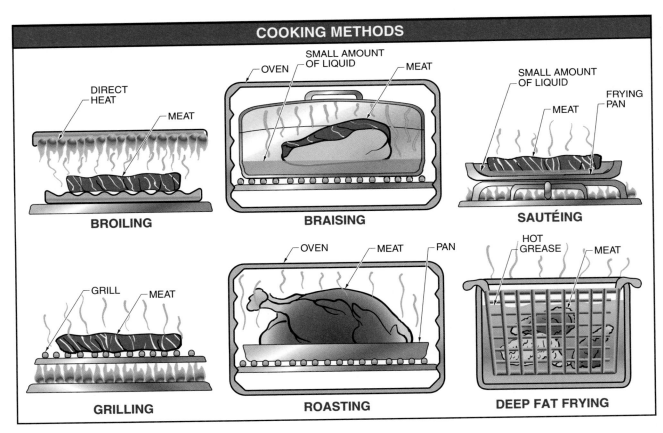

Different cuts of meat require different cooking methods.

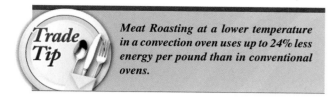

Trade Tip *Meat Roasting at a lower temperature in a convection oven uses up to 24% less energy per pound than in conventional ovens.*

- *Griddling*: Griddling is a method of cooking food on a solid metal surface called a griddle. A small amount of fat can be placed on the hot griddle to prevent foods from sticking to it. This method is similar to sautéing or panbroiling.

- *Sautéing*: Sautéing is a cooking method where heat is transferred to food through contact with a hot surface, such as a frying pan or sauté pan. To sauté lightly means "to brown." No covering and little fat is used. The difference between panbroiling and sautéing is that in panbroiling no fat is used, but in sautéing a small amount is used. Sautéing is also sometimes referred to as stir frying or pan frying; however, stir frying is an Oriental cooking style using a carbon steel wok, and pan frying uses significantly more fat in comparison.

- *Grilling*: Grilling is a method of cooking food over a heat source on open metal grates. Grilling foods will give them a smoky and charred flavor.

Moist heat methods are those in which heat is conducted by water (including stocks and sauces) or by steam. In moist heat cooking, foods will not brown through caramelization. Because of this, the natural flavor and smell of the food is heightened. Moist heat cooking methods commonly used for meats include steaming, simmering, and poaching. *Combination* methods use both dry heat and moist heat, and include stewing and braising. Meats high in connective tissue will be tough unless the tissue is broken down slowly by moist heat.

Roast chicken at 350°F for 1¾ hr to 2 hr or until a thermometer inserted in the thigh reads at least 165°F.

- *Steaming*: Steaming may be used in conjunction with either braising or cooking in liquid, or it may be used as a method by itself with or without pressure. The steam may be applied directly to foods as when using a steam pressure cooker.

- *Simmering*: To simmer, meat is put in a container, covered with a liquid (usually water), and then simmered (never boiled) until tender. When simmering, the bubbles of the liquid will break below the surface of the liquid as a temperature of 200°F is usually maintained. A *mirepoix* (onions, carrots, and celery, cut rough) may be added to improve the flavor of the meat and liquid. When blanching, the item is only partly cooked. The term *scald* is used when a liquid is heated to just below the boiling point.

- *Poaching*: To poach means to cook foods in a shallow amount of liquid held between 160°F and 180°F. Poaching is used for very delicate foods such as fish fillets.

- *Braising*: Meat is cooked at a low temperature in a small amount of liquid (water, stock, thin sauce, or a combination of these) in a covered container until done. Meat is usually browned before cooking.

- *Stewing*: When stewing, meat is first browned and then covered in liquid and cooked at a low temperature until tender. Vegetables are added to the meat and liquid near the end of the cooking time. Braising is used for larger cuts of meat while stewing is used for smaller cuts.

With any method, the length of the cooking time depends on the type of meat, the oven temperature, the degree of doneness desired, the quality of the meat, and the size and thickness of the meat.

Deep Frying

For best results use only hydrogenated vegetable shortening or oil designed for use in deep fat frying.

1. *Light the fry kettle and set the thermostat at 350°F*: The fat can be used much longer if not overheated and if it is strained after each day's use.

2. *Bread the item*: Breading prevents the surface of the meat from burning and drying out. Breading also adds to the flavor and appearance of the item. Various breadings may be used, such as flour, cornmeal, or bread crumbs.

3. *Shake off the breaded item before placing it in the fry basket*: Excess breading is shaken off so it does not settle to the bottom of the grease and burn, thus shortening the life of the fat. Do not overfill the fry basket.

4. *Fry until item is golden brown*: Overdone items are dry and tasteless. Some items float on top of the grease when done.

5. *Drain off excess grease*: This helps retain crispness and makes the item more digestible.

6. *Clarify fat as necessary*: Fat, if not overused, can be clarified by placing sliced raw potatoes in the cold fat and heating the fat gradually until the potatoes become brown. Some deep fat fryers are self-clarifying. However, for best results, oil should be changed frequently as required. Do not salt items (for example, French fries) over the deep fryer. This causes the fat to break down faster.

Roasting

1. *Season with salt and pepper*: It is best to season the day before roasting. This gives the seasoning a chance to penetrate the meat. If time does not permit, seasoning can be done immediately before roasting.

2. *Place meat in oven*: Opinions vary among chefs regarding the placement of fat side up or down. With the fat side up, juices penetrate the meat as it is cooked. With the fat side down, additional grease to prevent the roast from sticking to the pan is not required.

3. *Brown meat thoroughly*: This should be done in a hot oven at about 375°F. This helps develop better flavor for both meat and gravy and adds color to the meat. Do not sear the meat when browning.

4. *Add mirepoix*: Mirepoix is added when meat is browned to add flavor to both the roast and gravy. The mirepoix usually consists of onions, celery, and carrots.

5. *Add water only as necessary*: When roasting at a low or moderate temperature, the drippings should not evaporate. If there are not enough drippings, a small amount of water may be added.

6. *Do not cover the roast*: When a roast is covered, steam is created. This will result in a pot roast.

7. *Roast at a temperature of 325°F to 350°F*: Lower temperatures help reduce shrinkage.

8. *Turn the roast*: A rib roast should be turned only once and so that it rests on its rack (the arched rib bones), not on the bottom of the pan. A boneless roast should be turned frequently to prevent dryness. When turning a roast, the fork should be placed under the roast to support the meat. With a towel, carefully turn the meat over. Never stick a fork in the roast because the juices will cook out. Baste the roast often for best results.

9. *Roasting time*: The roasting time depends on four factors: type of meat, oven temperature, degree to which it is done, and the grade of the meat. The doneness of the meat can be determined by a meat thermometer.

Broiling

1. *Turn flame or heat to highest point*: The temperature desired for cooking is determined by the distance from the heat.

2. *Lightly brush meat with salad oil and season*: The meat should be brushed with salad oil and seasoned before placing it in the broiler to avoid sticking. This also helps the appearance of the meat. Season with salt and pepper except when broiling a steak. Salt is added just before the meat is removed from the broiler since salt has a tendency to draw out the juice.

3. *Place item on hot broiler*: Have broiler hot before placing an item on it. This creates the desired broiler markings. Place meat on the broiler at an angle with the fat side facing out. Before it is time to turn the meat, reposition the meat about 60°. This gives the desired crisscross markings on the meat.

4. *Broil until top of item is brown*: At this point the item should be half done.

5. *Turn item and continue to brown second side*: When broiling, the item should be turned only once. Reposition after turning to obtain the desired markings. Avoid sticking the fork in the meat to turn it. Use a pair of tongs to turn the meat.

6. *Broiling time*: This depends on the item, grade, size, degree to which it is done, and thickness.

7. *How to serve*: Broiled items should be served at once and always on hot plates or platters.

Panbroiling

1. *Season meat with salt and pepper*: For a better taste, season both sides before cooking.

2. *Place meat in hot sauté pan*: No fat is added when panbroiling. Do not cover. Covering causes steam to develop.

3. *Brown one side, then turn and brown the other side*: This helps develop flavor. Do not pierce meat when turning.

 **When carving pork roasts, allow the roast to cool for 10 to 15 min. Note the grain (direction in which muscle fibers run) and slice across the grain.**

BROILER POSITIONING

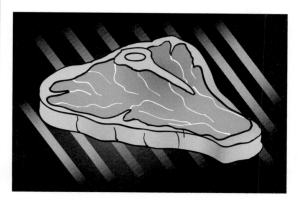

1 PLACE MEAT ON GRILL AT AN ANGLE

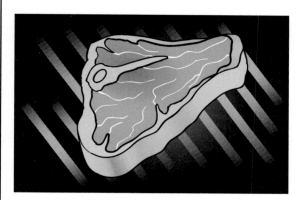

2 REPOSITION MEAT 60° ON SAME SIDE OF MEAT

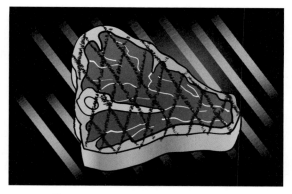

3 TURN WHEN MEAT IS HALF DONE; CRISSCROSS PATTERN IS MADE

Positioning of the meat on the broiler results in broiler markings.

4. *Cook at a moderate temperature*: This prevents excessive browning and makes the meat much juicier.
5. *Pour off any fat that appears in the sauté pan*: If fat is left in pan, this method would be sautéing, not panbroiling.
6. *Do not add liquid*: Adding liquid would classify it as braising, not panbroiling. Keep the pan as dry as possible.
7. *Cooking time*: Depends on the cut of meat, type of meat, thickness, degree to which it is done, and quality of meat.

Sautéing

1. *Heat sauté pan until hot*: The pan should be preheated before adding meat.
2. *Season with salt and pepper*: Seasoning should be done before cooking so the seasoning penetrates the meat.
3. *Pass through flour if desired*: Coat the meat thoroughly with flour so it browns evenly. Dust off any excess flour that may collect at the bottom of the pan and cause the fat to burn.
4. *Brown the meat in a small amount of fat on one side*: Brown the meat quickly on one side at a moderate temperature. To avoid sticking, the grease should be heated before the item is placed in sauté pan.
5. *Turn and brown second side*: Meat should be golden brown on both sides. The meat should be turned quickly to avoid shrinkage. Sautéing is a quick cooking method.
6. *Do not cover*: This causes steam to form.
7. *Cook at a moderate temperature*: This makes the meat crisper and gives it a more eye-appealing appearance.

Vegetables can be sautéed with the meat or separately for different preparations.

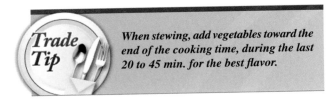

Trade Tip

When stewing, add vegetables toward the end of the cooking time, during the last 20 to 45 min. for the best flavor.

Poaching

1. *Bring liquid to boil*: Bring the appropriate amount of liquid (usually just enough to cover the item) to a boil and then lower to the desired temperature between 160°F and 180°F.
2. *Gently lower the item into poaching liquid*: Lowering the item gently ensures that it is not damaged or broken. Most foods cooked with the poaching method are delicate in nature.
3. *Poach the item until desired doneness*: Do not overcook, or the item will become tough and rubbery.
4. *Cool*: Usually the poached item can be cooled directly in the poaching liquid or can be served immediately.
5. *Prepare sauce if desired*: The poaching liquid is often reduced and incorporated into a sauce for the finished dish. To *reduce* refers to the process of gently simmering a liquid, allowing some of it to evaporate and therefore concentrate its flavors and possibly thicken.

Braising

1. *Heat braising pot until hot*: The pot should be preheated before adding meat.
2. *Season meat with salt and pepper*: For a better taste, season both sides before cooking.
3. *Place meat in braising pot*: A braising pot supplies quicker heat because more of the surface of the pot comes in contact with the heating unit. The meat should sizzle when it comes in contact with the bottom of the pot.
4. *Brown the meat thoroughly*: Browning the meat helps develop a richer color and better flavor.
5. *Add vegetable mirepoix*: Brown the mirepoix with the meat to increase the flavor of the dish.
6. *Add small amount of tomato product*: Adding diced tomatoes or tomato sauce will add flavor and increase the moisture of the meat. Stir the tomato product thoroughly with the meat and mirepoix to distribute flavor.
7. *Add liquid*: Add just enough liquid to cover the meat halfway. The meat and liquid will then be richer in flavor. The liquid used may be water or stock. Stock is preferred for better flavor.
8. *Cover the braising pot*: This keeps in the flavor and allows the meat to cook evenly throughout.

9. *Cook continuously on a range or in an oven*: Keep the brasier covered and the temperature consistent for the duration of the cooking time. This results in more tender meat, more pronounced flavor, and shorter cooking time.
10. *Simmer at a low temperature (between 200°F and 250°F)*: Lower temperatures result in less shrinkage and better flavor.
11. *Cook until the meat is tender*: The time required depends on the size, thickness, grade, and type of meat used.

Breaded chicken breasts are griddled to a golden brown. Griddling allows efficient preparation of large quantities of meat.

Stewing

1. *Cut meat into cubes (large, medium, or small)*: Cut the meat uniformly so it cooks evenly. A boneless stew is more desirable than one with a bone.
2. *Season with salt and pepper*: Season when starting to cook for best results. Herbs and spices may be used if desired.
3. *Heat pot until hot*: The pot should be preheated before meat is added.
4. *Brown the meat if desired*: A brown stew is more desirable and has more flavor than a white stew. However, if a white stew is desired, eliminate this step.
5. *Cover the meat with a liquid*: The meat should be entirely covered with the liquid so it cooks uniformly. The liquid may be either stock or water.
6. *Cover stewing pot*: The pot should be covered to reduce cooking time and to preserve the flavor.
7. *Cook at a low temperature (about 250°F)*: Cooking at a low temperature reduces shrinkage and preserves flavor.

8. *Add vegetables*: Vegetables should be added when the meat is about three-fourths done. Using this method of adding the vegetables, the stew will be more flavorful but will lack in appearance. The alternative method is to cook each vegetable separately. When the meat is tender, add the cooked, drained vegetables. Cooking each vegetable separately is the preferred method because each vegetable cooks at a different rate.

9. *Cooking time*: Cook until just tender. A stew that is overcooked lacks appearance and appetite appeal.

Other Cooking Methods

In recent years, two other cooking methods have become popular: Cajun style blackened cooking and Oriental stir-frying. Although these are a variation of sautéing, both methods have been widely accepted by the general public.

Cajun style blackened cooking is a method of preparing certain foods named after the Cajun people located in southern Louisiana. Blackened fish is the most popular blackened item. It uses a mixture of pepper and hot spices and is a form of sautéing, even though no fat is added to the skillet. The item being prepared is passed through melted butter, margarine, or fat before it is pressed into a mixture of spices. Dipping the item to be blackened into a melted shortening helps the Cajun spice adhere to the item and assists in the blackening process. Blackening is done in a very hot cast iron skillet.

Stir-frying is an Oriental cooking method of sautéing and stirring at the same time. Peanut oil is most often used. Stir-frying is a quick method of cooking that helps retain the natural crispness of meat and vegetables. Stir-frying is usually done in a carbon steel wok (a skillet shaped like a bowl) so the turning and stirring of the foods is simplified.

Blackening

1. *Dip the food in melted shortening*: Melted butter or margarine may also be used. This helps the Cajun spice adhere to the item being blackened and assists in blackening.

2. *Place food in the Cajun spice blend*: The blend may be extremely hot or slightly mild, depending on the supplier or recipe. This spice coating causes the surface of the item to blacken.

3. *Place the item in a very hot iron skillet*: Blacken one side, turn, and blacken the second side.

4. *Remove the item from the skillet*: This may be done using a meat turner or food tongs, depending on the tenderness of the item.

5. *Finish the cooking item in the oven*: This step depends on the type of meat, thickness, degree of doneness, and quality of the meat being blackened. Finish cooking is not required when blackening fish.

Note: The most popular blackened items are fish–redfish, orange roughy, and halibut; steaks–sirloin, rib, and tenderloin; pork-chops and cutlets cut from the loin; and chicken–boneless breast.

Stir-Frying

1. *Add oil to the wok or skillet*: A wok is preferred because of its bowl shape. More uniform heat is produced and the turning and stirring is made easier. Peanut oil is preferred, but salad oil may be used.

2. *Place the wok or skillet on the range and heat*: Heat for quick cooking. The oil should be hot before adding the food.

3. *Place the foods to be stir-fried in the hot oil*: The foods should be added at different times. The foods to be cooked the longest are added first.

4. *Sauté or fry and stir continuously*: Once the food or foods are added, stir continuously.

5. *Add the liquid, continue to stir*: The liquid added may be rice wine, soy sauce, soup stock, sherry, etc.

6. *Continue to simmer until desired doneness is obtained*: Never overcook vegetables; crispness should be maintained.

7. *Thicken liquid with starch*: If the item being prepared requires thickening in order to make a sauce, dilute the starch in liquid before adding it to the simmering preparation. Cornstarch or a modified or blended starch is used most often.

8. *When thickened, remove from the heat*: Starch begins to thicken at 205°F. Bring to a simmer and remove from the heat.

Note: Most seafood and meat can be stir-fried with excellent results if the item is tender in its natural state. Tough cuts of meat are not recommended for stir-frying.

VEGETABLE COOKING

Vegetables may be cooked using some of the same methods used for meats. In some cases, vegetables are cooked along with the meats. Dry heat cooking methods commonly used for vegetables include roasting, sautéing, and grilling. In addition, vegetables are commonly cooked using moist heat cooking methods such as simmering, blanching, or steaming.

Roasting Fresh Vegetables

Roasting is done using dry heat in an oven with little or no water. Fresh vegetables are preferred for this technique rather than frozen vegetables. Beans and potatoes are examples of vegetables that are roasted.

1. Prepare the vegetables according to the directions given in Table II.
2. Roast the vegetables until slightly soft using the suggested temperature listed in Table II.
3. Cooking time varies, depending on the size, variety and maturity, and cut size of the vegetable. Approximate cooking times are listed in Table II.

Sautéing Fresh Vegetables

Sautéing is a method in which food is cooked quickly in a small amount of fat or oil. Sautéing is done over high heat in a sauté pan, using caution to not allow the small amount of fat in the pan to burn. Sautéing and stir-frying are similar. The difference is that sautéing is done in a shallow pan with sloped sides and stir-frying is done in a Chinese wok.

1. *Heat sauté pan until hot*: The pan should be preheated before adding vegetables.
2. *Sauté the vegetables in a small amount of fat on one side*: Lightly sauté the vegetables quickly on one side at a moderate temperature. To avoid sticking, the grease should be heated before the items are placed in sauté pan.
3. *Turn and lightly sauté second side*: Vegetables should be evenly cooked on all sides. The vegetables should be turned quickly to avoid shrinkage. Sautéing is a quick cooking method.
4. *Do not cover*: This causes steam to form.
5. *Cook at a moderate temperature*: This makes the vegetables crisper and gives them a more eye-appealing appearance.

Grilling Fresh Vegetables

Grilling is a method of cooking vegetables over a heat source on open metal grates. Grilling vegetables will give them a smoky and charred flavor. For this method, vegetables are placed in a metal pan or arranged on skewers and placed directly on the grill. Turn the vegetables frequently to ensure all sides are evenly cooked.

TABLE II: ROASTING FRESH VEGETABLES			
Vegetable	**Preparation**	**Oven Temperature**	**Approximate Roasting Time**
White potatoes	Select potatoes that are uniform in size; Wash and scrub well; Potatoes may be wrapped in foil or brushed with oil so they will be soft when baked	400°F	1 hr 15 min to 1 hr 30 min
Sweet potatoes	Select potatoes that are uniform in size; Wash and scrub well; They may also be wrapped with foil or brushed with oil the same as white potatoes	400°F	45 min to 1 hr
Acorn squash	Wash and scrub well; Cut in half lengthwise, remove all seeds, brush the flesh with melted butter, and sprinkle with salt and brown sugar	375°F	1 hr 30 min (If the squash becomes too brown, cover the surface with foil or oiled brown paper)
Zucchini squash	Wash and scrub well; Cut lengthwise, season with salt and pepper, sprinkle with bread crumbs, and dot with butter	350°F	20 to 30 min
Tomatoes	Wash thoroughly; Remove stem, slice off bottom, rub with salad oil, and season	375°F	30 to 35 min

Vegetable roasting times vary based on size, variety, maturity, and cut size of the vegetable.

Boiling

Boiling is probably the most common of the moist heat cooking methods by name, but probably really one of the least actually used in a professional kitchen. When foods are boiled, a large amount of water is heated to 212°F, the boiling point of water at sea level. The water will circulate in the pot during boiling, which will keep the temperature consistent throughout the pot of water. At higher altitudes, for every 1000 feet above sea level, the boiling point of water drops 2°. This means that, because the boiling point is lower, the length of time something will need to be cooked will increase. The only items that are boiled in the professional kitchen are pasta, potatoes, and some other starches. Instead, simmering is used for most preparations.

Idaho Potato Commission

Potatoes may be brushed with oil so the skins will be crispy when roasted.

Simmering Fresh and Frozen Vegetables

To simmer means to cook in liquid that is bubbling very lightly. The liquid should be between 185°F and 205°F. The simmering method is performed by bringing the water to a simmer, adding the vegetables, and bringing the water back to a simmer. Little water is used and the vegetables are cooked for only a short time. Most vegetables may be simmered in water. Too much water or overcooking destroys the flavor and causes loss of nutrients in cooked vegetables. Many vegetables should be covered while cooking. Do not cover cauliflower, turnips, and green vegetables. This will allow gasses to escape completely and the vegetables will retain a bright color. It is important to maintain a constant and even temperature when cooking foods using the simmering method.

1. Most frozen vegetables should be cooked frozen, without thawing first. Cook in amounts no larger than 10 lb.

2. Use enough water to completely cover the vegetables. Bring the water to the desired temperature between 185°F and 205°F.

3. Add 1 tsp of salt for every quart of water used.

4. Add the vegetables to the simmering water and bring the water back to a simmer as quickly as possible. Cooking time starts when the water returns to a simmer. Check Table IV for approximate time. Cooking time will vary depending on the quantity of the vegetables prepared and desired tenderness.

5. After cooking, drain off part of the liquid and add 8 oz to 12 oz of butter or margarine to each 10 lb of vegetables. Ten pounds of frozen cooked vegetables will yield approximately 50 servings of 3 oz each.

Trade Tip *Be careful to keep simmering temperature below a full boil and keep cooking time to a minimum to prevent loss of nutrients.*

Blanching

Blanching is a quick method used to only partially cook an item. A chef might use a blanching method to prepare vegetables for use. Blanching can make vegetables easy to peel, partially soften hard vegetables, brighten or set color in produce, or eliminate bitter or undesirable flavors. Usually when produce items are blanched, they are immediately refreshed or shocked in ice water to stop the cooking process. *Refreshed* and *shocked* are words used interchangeably to describe the technique of quickly stopping foods from cooking by plunging them into ice water.

1. Clean and prepare item to be blanched.

2. Bring pot of water to boil.

3. Place the item in rapidly boiling water.

4. When desired effect is achieved (peel begins to loosen or color brightens on the vegetable), remove the item from the boiling water and immediately submerge in ice water to stop the cooking process.

A chef may also blanch an item to remove impurities. This is a method used when making soups or stocks from bones to rid the bones of blood proteins and impurities that would ultimately make a stock or soup cloudy if not removed.

1. Place the bones into a pot of cold water.

2. Turn heat on full and allow the water to come to a boil, reduce heat to a simmer, and simmer for a few minutes.

3. Remove the bones and plunge them into cold water.

4. Discard the blanching water.

LIBRARY **WITHDRAWN** WAUKESHA COUNTY TECHNICAL COLLEGE 800 MAIN STREET PEWAUKEE, WI 53072

TABLE III: SIMMERING FRESH VEGETABLES

Vegetable	Preparation	Boiling Water	Salt	Approximate Cooking Time
Beans, green (10 lb)	Trim ends, remove strings, and wash thoroughly; Cut into desired size pieces	2½ qt	1 tbsp	25 min to 30 min
Beets (10 lb)	Remove tops, wash thoroughly, and remove blemishes	To cover (approx 3 qt)	none	1 hr to 1 hr 30 min
Broccoli (10 lb)	Cut off tough woody stalk ends and wash; Peel stalks and cut in half lengthwise	3 qt	1 tbsp	20 min to 25 min
Cabbage, cut into wedges (10 lb)	Remove blemished outside leaves; Wash, cut into quarters, remove core, and cut into wedges	6 qt	2 tbsp	1 hr to 1 hr 15 min
Carrots (10 lb)	Scrape or pare, wash, slice or cut as desired	4 qt	1 tbsp	20 min to 30 min
Cauliflower (10 lb)	Remove outer leaves and stalks; Separate into flowerets and wash thoroughly	5 qt (plus 1 qt milk)	2 tbsp	20 min to 30 min
Celery (10 lb)	Trim; Cut into desired size pieces	4 qt	1 tbsp	20 min to 30 min
Corn on the cob (10 lb)	Husk, remove silk by brushing; Wash thoroughly and drain at once	4 qt (plus 1 qt milk)	1½ tbsp	10 min
Kale (10 lb)	Remove blemished leaves, strip leaves from stems; Wash at least four times, lifting out of the water each time so dirt will settle to the bottom and not cling to the leaves; When cooking, stir occasionally	4 qt	1 tbsp	30 min to 45 min
Kohlrabi (10 lb)	Pare, wash, and cut into 1" cubes as desired	4 qt	none	20 min to 30 min
Onions (10 lb)	Peel and wash, cut if desired	6 qt	1½ tbsp	30 min to 40 min
Potatoes, white (10 lb)	Peel, cut into uniform size, remove all eyes, and wash thoroughly	6 qt	2 tbsp	45 min to 1 hr
Rutabagas (10 lb)	Pare, wash, and cut into uniform pieces, preferably 1" cubes	4 qt	1 tbsp	15 min to 30 min
Spinach (10 lb)	Remove blemished leaves and coarse stems; Wash at least four times, lifting out of the water each time so dirt will settle to the bottom and not cling to the leaves; When cooking, stir occasionally	4 qt	1 tbsp	20 min to 30 min
Squash, winter (10 lb)	Wash and peel; If peel is too hard, steam or boil the whole squash for approximately 7 min; Cut in half, remove fibers and seeds; Cut into uniform pieces	5 qt	1½ tbsp	20 min to 30 min
Squash, summer (10 lb)	Wash and trim; Cut into uniform pieces	2 qt	1 tbsp	10 min to 20 min
Sweet potatoes (10 lb)	Select potatoes that are uniform in size; Wash and scrub well	5 qt	none	45 min to 1 hr
Turnips (10 lb)	Pare, wash, and cut into uniform pieces	3 qt	none	20 min to 30 min

Vegetables should be simmered in amounts not exceeding 10 lb.

TABLE IV: SIMMERING FROZEN VEGETABLES		
Vegetables	**Boiling Water**	**Approximate Cooking Time**
Asparagus, cut or tips (2½ lb box)	1½ qt	10 min to 12 min
Beans, lima, baby (2½ lb box)	2 qt	15 min to 20 min
Beans, lima, fordhook (2½ lb box)	2 qt	10 min to 15 min
Beans, green, cut (2½ lb box) or French cut (2½ lb box)	1 qt	10 min to 30 min
Broccoli, cut (2½ lb box)	1½ qt	10 min to 20 min
Broccoli, spears (2 lb box)	1½ qt	10 min to 15 min
Cauliflower (2 lb box)	1 qt (plus 1 qt milk)	10 min to 15 min
Corn, cut (2½ lb box)	1½ qt	5 min to 10 min
Kale (3 lb box)	2 qt	20 min to 25 min
Okra (2½ lb box)	1 qt	5 min to 8 min
Peas, green (2½ lb box)	1 qt	5 min to 10 min
Peas and carrots (2½ lb box)	1 qt	10 min to 12 min
Succotash (2½ lb box)	2 qt	10 min to 15 min
Turnip greens (3 lb box)	2 qt	25 min to 35 min
Vegetables, mixed (2½ lb box)	1 qt	25 min to 30 min

Cooking time for frozen vegetables begins when the water returns to a simmer.

This blanching method will remove all of the loose blood proteins and impurities from the bones that could cause a stock or soup to become cloudy. The bones would then be brought to a simmer again in the new, clean water, but this time the purpose would be to extract their flavor for a stock or soup.

Steaming Fresh and Frozen Vegetables

Steaming is another moist heat cooking method. Almost all vegetables may be steamed. Steaming is done by placing vegetables in a perforated kettle or on a rack inside a covered pot over boiling water, with steam forced into and through the container. The movement of the steam around the food will cook the food gently and evenly on all sides. Often, other ingredients are placed in the boiling water to add flavor to the food being steamed. Ingredients such as wine, herbs, or spices added to the boiling water will release flavors and smells into the hot, moist air and will be absorbed into the food as it cooks. These ingredients are referred to as aromatics. In a commercial kitchen, foods can also be cooked in a commercial convection steamer. A convection steamer will use steam in combination with pressure to cook foods even quicker. Unfortunately, because the process is much quicker and the steam is not generated from a mixture of ingredients, aromatics such as wine or herbs cannot be added to the steaming environment. Butter is often omitted for a more nutritious preparation.

Cook frozen vegetables in amounts not exceeding 5 lbs.

1. If a commercial steamer is not being used, in a shallow pan, prepare a steaming liquid and bring it to a boil.
2. If desired, add aromatics—herbs, spices, or wine—to the liquid to add to the finished aroma and flavor of the item being steamed.
3. Next, place another perforated pan or rack inside the first pan above the water.
4. Place the item to be steamed on the rack or perforated pan over the boiling liquid.
5. Cover the pan to keep the steam inside and cook to the desired degree of doneness.

TABLE V: STEAMING FRESH VEGETABLES IN A COMMERCIAL CONVECTION OVEN

Vegetables	Preparation	Type of Container	Approximate Cooking Time
Beans, green (10 lb)	Trim ends, remove strings, and wash thoroughly; Cut into desired size pieces	Solid (½ full) or perforated (¾) full	25 min to 35 min
Beets (10 lb)	Remove tops, wash thoroughly, and remove blemishes	Solid or perforated (full)	1 hr to 1 hr 30 min
Broccoli (10 lb)	Cut off though woody stalk ends and wash; Peel stalks and cut in half lengthwise	Bake Pan (single layer)	10 min to 12 min
Cabbage, cut in wedges (10 lb)	Remove blemished outside leaves; Wash, cut into quarters, remove core, and cut into wedges	Solid (½ full)	20 min to 25 min
Carrots (10 lb)	Trim, cut into desired size pieces, and wash thorougly	Solid (½ full) or perforated (¾ full)	25 min to 35 min 20 min to 25 min
Cauliflower (10 lb)	Remove outer leaves and stalks; Separate into flowerets and wash thoroughly	Solid (½ full) or perforated (½ full)	12 min to 15 min 8 min to 12 min
Celery (10 lb)	Trim, cut into desired size pieces, and wash thoroughly	Solid (⅓ full) or perforated (½ full)	15 min to 20 min 12 min to 15 min
Corn on the cob (10 lb)	Husk and remove silk by brushing; Wash thoroughly and drain at once	perforated (½ full)	5 min to 10 min
Kale (10 lb)	Remove blemished leaves and strip leaves from stems; Wash at least four times, lifting out of the water each time so dirt will settle to the bottom and not cling to the leaves	Solid (¼ full)	30 min to 35 min
Kohlrabi (10 lb)	Pare, wash, cut into 1″ cubes or as desired	perforated (½ full)	15 min to 20 min
Onions (10 lb)	Peel and wash and cut if desired	perforated (⅓ full)	25 min to 30 min
Potatoes, white (10 lb)	Peel, cut into uniform size, remove all eyes, and wash thoroughly	Solid (½ full) or perforated (½ full)	40 min to 60 min 30 min to 40 min
Rutabagas (10 lb)	Pare, wash, and cut into uniform pieces (preferably 1″ cubes)	Solid (½ full) or perforated (½ full)	25 min to 35 min 20 min to 30 min
Spinach (10 lb)	Remove blemished leaves and coarse stems; Wash at least four times, lifting out of the water each time so dirt will settle to the bottom and not cling to the leaves	Solid (½ full) or perforated (½ full)	8 min to 10 min 6 min to 8 min
Squash, winter (10 lb)	Wash and peel; If peel is too hard, steam or boil the whole squash for approximately 7 minutes; Cut in half, remove fibers and seeds; Cut into uniform pieces	Solid (½ full) or perforated (½ full)	20 min to 25 min 15 min to 20 min
Squash, summer (10 lb)	Wash and trim; Cut into uniform pieces	Solid (½ full) or perforated (½ full)	20 min to 25 min 15 min to 20 min
Sweet potatoes (10 lb)	Select potatoes that are uniform in size; Wash and scrub well	Solid (¾ full) or perforated (¾ full)	30 min to 45 min 25 min to 35 min
Turnips (10 lb)	Pare, wash, and cut into uniform pieces	Perforated (½ full)	15 min to 20 min

Commercial convection steamers decrease steaming time by adding pressure to the cooking process.

TABLE VI: STEAMING FROZEN VEGETABLES IN A COMMERCIAL CONVECTION OVEN

Vegetable	Approximate Cooking Time
Asparagus, cut or tips (5 lb)	6 min to 10 min
Beans, lima, baby (5 lb)	12 min to 15 min
Beans, lima, fordhook (5 lb)	15 min to 20 min
Beans, green, cut (5 lb) or French cut (5 lb)	10 min to 15 min
Broccoli, cut (5 lb)	12 min to 15 min
Broccoli, spears (5 lb)	5 min to 8 min
Cauliflower (5 lb)	5 min to 10 min
Corn, cut (5 lb)	5 min to 8 min
Kale (5 lb)	20 min to 30 min
Okra (5 lb)	4 min to 6 min
Peas, green (5 lb)	5 min to 8 min
Peas and carrots (5 lb)	5 min to 8 min
Succotash (5 lb)	12 min to 15 min
Turnip greens (5 lb)	15 min to 20 min
Vegetables, mixed (5 lb)	15 min to 20 min

Frozen vegetables can be steamed in a commercial convection steamer. Cooking times will vary depending on the size, variety, maturity, and cut size of the vegetable.

Steam cooking is very efficient when a large amount of food is being cooked at one time.

BAKING

Baking is the primary cooking method used in preparing breads, quickbreads, cookies, pies, cakes, and other pastries. Baking, like roasting meats, is cooking by surrounding the item with dry heat in an oven. Baking time varies depending upon the size of the item, the temperature of the oven, the type of item being baked, and the particular ingredients used. Before placing an item in the oven, always set the thermostat. Preheat the oven about 30 min before using to ensure correct oven temperature when ready to use. Baking instructions for most recipes must be followed closely. Table VII lists the common baking terms, their meaning, and techniques.

Baking Ingredients

The basic ingredients used in baked products are water, salt, yeast, and flour. In addition to the basic ingredients, sugar, shortening, and milk or milk solids are used. These ingredients give desirable qualities and enrichment to bread. Baking powder is used for cakes and quickbreads.

Ammonium carbonate: This is a powder leavening ingredient made by combining ammonia and carbonic acid. It is used in products to increase leavening and improve tenderness. Products such as cream puffs, éclairs, and certain types of cookies are improved when a small amount of ammonium carbonate is added. When ammonium carbonate is added to a product, it acts quickly, requiring baking as soon as possible.

Baking powder: This leavening agent is produced by mixing an acid-reacting material with common baking soda. Baking powder is used in the preparation of cakes, quickbreads, and other preparations that require a quick-acting leavening agent.

Butter-flavored vegetable shortening: This yellow shortening is made from the finest vegetable oils (soybean and sunflower). It is partially hydrogenated (hydrogen gas injected) for freshness and firmness. Yellow color and artificial butter flavor are added. This shortening is used as a roll-in shortening for butter flake and croissant rolls. It is also used in sweet doughs and streusel topping.

Cream of tartar: Cream of tartar is a white powdered chemical compound used in bakery products to retain whiteness. Examples of products that use cream of tartar are white cake batter and meringue mixtures.

Drifond®: Drifond® is a special blend of sugar and other dry ingredients that, when mixed with water, will produce an excellent fondant icing.

Emulsified vegetable shortening: This is purified oil that has been processed so it will absorb and retain moisture. This shortening does not cream well, but it will retain a higher percentage of sugar and liquid. It is excellent for use in high ratio cake mixes (high percentage of sugar) and icings.

TABLE VII: COMMON BAKING TERMS		
Term	**Meaning**	**Techniques for Performing**
Blend	To mix two or more ingredients thoroughly	Blending can be done by hand, on the mixing machine, or by using a kitchen spoon
Cut in	A part blended or rubbed into another	Cutting in is usually performed by rubbing the two ingredients together between the palms of the hands
Dissolve	To cause a dry substance to be absorbed in a liquid	Use a bowl or bain-marie when dissolving an item; Stir the liquid with a kitchen spoon until fluid
Dust	To sprinkle with flour or sugar	Dusting may be done with the hand or a flour sifter may be used when dusting with powdered sugar
Fold	A part doubled over another	Pass a spoon, skimmer, or the hands down through a mixture, run it across the bottom of the container, and bring up some of the mixture gently and place it on top
Ice	To cover a cake or some other item with frosting or icing	Using a spatula, apply the icing with smooth, even strokes; Dip the spatula in warm water at intervals for smoother, more even application
Knead	The manipulation of pressing, folding, or stretching the air out of dough	Knead the dough on a floured bench; Press the dough with the heel of the hand while at the same time stretching and folding it in an over and over motion
Make up	A method of mixing ingredients or handling when dividing an item into single units	Each item is usually made up in a different manner, depending on desired results, common practice, or creative skill
Mask	To cover an item completely in order to disguise or protect it	Masking is usually done with icing or a sauce; The item is placed on a wire rack with a sheet pan underneath; The icing or sauce is poured on in a smooth, even flow
Mix	To merge two or more ingredients into one mass	Mixing can be done in some cases by hand; however, the most practical way is by using the mixing machine
Proof	To let yeast dough rise by setting it in a warm, moist place	Proofing can be done by letting the item set in a warm room; however, for best results, use a proofing box; The proofing box is designed to maintain a warm, moist temperature of approximately 90°F
Punch	To knock air out of yeast dough after it reaches the right fermentation	Punching the dough is accomplished by applying sharp blows to the dough using the knuckles of both hands
Round	To shape dough pieces to seal ends and prevent bleeding	Rounding is done on a floured bench; The dough unit is rolled by running the palm of the hand in a forward motion across the base of the dough unit
Sift	To pass dry ingredients through a fine screen to make light	The dry ingredients can be tapped through a fine sieve or passed through a flour sifter
Tube	To press a substance through a pastry tube	The substance is placed in a pastry bag with a pastry tube in the small end; Pressure is applied by pressing the top of the bag until the substance flows in a steady stream
Wash	To apply a liquid to the surface of an unbaked product	Washing is usually done with a pastry brush; The liquid applied may be egg wash, water, milk, or a thin syrup

Common baking terms describe various methods and techniques used in baking.

Hydrogenated vegetable shortening: This vegetable shortening is processed by heating, and while the oil is still hot it is injected with hydrogen gas. The amount of gas injected controls the firmness of the finished product. This shortening is very versatile and can be used for frying, cooking, and baking.

Instant jelly: This is a dry blend of instant jelly powder that, when mixed with sugar and water, produces jelly excellent for use in coffee cakes, rolls, and other preparations.

Margarine: Margarine is made from vegetable or animal fat or a combination of both. It is churned with milk or cream to a spreading consistency. Margarine contains 80% fat and approximately 3% salt. It is commonly used as a substitute for butter.

Meringue powder: This is a blended powder mixture. It produces an excellent meringue when mixed with sugar and water.

Powdered lemon juice and corn syrup: This is a dry blend of corn syrup solids, lemon juice solids, and lemon oil. It is a very convenient product that is used to produce excellent lemon frosting, lemon cream, or lemon pie filling.

Puff paste shortening: This type of shortening was specially developed for use in preparing puff paste dough. Puff paste shortening has a plastic consistency when worked and a melting point of approximately 113°F.

Rye flavor: This is a manufactured liquid that contains the flavor of caraway seeds. It is used in the preparation of rye bread.

Yeast: Yeast is a microscopic plant grown in vats containing a warm mash made of ground corn, barley malt, and water. The standard-sized foil-wrapped cake of fresh yeast contains over 25,000,000 such plants compressed with a small amount of starch. When yeast plants are mixed with water, sugar, and flour and made into dough, they quickly begin to grow and multiply. In the process of growing, they produce small bubbles of leavening gas (carbon dioxide), which causes the dough to rise. Yeast is used in the preparation of dinner rolls, breads, and sweet doughs.

ADJUSTING RECIPES

Recipes may have to be adjusted to accommodate the number of expected guests on a given occasion. Recipes for commercial kitchens are commonly based on yields of 25, 50, or 100 servings. For example, in the given recipe, the approximate yield is 100 servings. However, there are 235 expected guests. To adjust the recipe from 100 servings to 235 servings, a *working factor* is required. A working factor is determined by dividing the required yield by the recipe yield.

$$working \ factor = \frac{required \ yield}{recipe \ yield}$$

$$working \ factor = \frac{235}{100}$$

$$working \ factor = \textbf{2.35}$$

To adjust the recipe for Beef a la Bourguignonne for 235 guests, the working factor is used.

Beef a la Bourguignonne

100 servings

Ingredients

36 lb beef tenderloin, cut into 1″ cubes
1¼ lb shortening
8 lb mushrooms, sliced
2 lb shallots or green onions, minced
6 oz flour
3 qt Burgundy wine
salt and pepper to taste

Ingredient	Amount		Working Factor		Adjusted Amount
Beef Tenderloin	36 lb	×	2.35	=	84.6 lb
Shortening	1¼ lb	×	2.35	=	2.94 lb
Mushrooms	8 lb	×	2.35	=	18.8 lb
Shallots	2 lb	×	2.35	=	4.7 lb
Flour	6 oz	×	2.35	=	14.1 oz
Wine	3 qt	×	2.35	=	7.05 qt

The fractional part of a pound represented by the decimal (as 0.6 lb of beef, 0.94 lb of shortening, 0.8 lb of mushrooms, 0.7 lb of shallots) is converted to ounces by multiplying 16 by the decimal number.

16 oz × 0.6 lb beef = 9.60 oz

16 oz × 0.94 lb shortening = 15.04 oz

16 oz × 0.8 lb mushrooms = 12.80 oz

16 oz × 0.7 lb shallots = 11.20 oz

TABLE VIII: ROUNDING OFF FRACTIONAL OUNCES	
0.00 – 0.09	0 oz
0.10 – 0.29	¼ oz
0.30 – 0.59	½ oz
0.60 – 0.79	¾ oz
0.80 – 0.99	1 oz

After converting fractions of a pound to ounces, use Table VIII to round off the fractional ounces. 9.60 oz of beef will be rounded to 9¾ oz; 15.04 oz shortening will be 15 oz; 12.80 oz of mushrooms will be 13 oz; and 11.20 oz of shallots will be 11¼ oz. Flour is already expressed as 14.1 oz, so it will only be necessary to round off to 14¼ oz.

The Burgundy wine, being a liquid, has been measured by volume, 7.05 qt. Since there are 32 fluid oz to a fluid quart, the fractional quart would be converted to ounces by multiplying 32 by the decimal.

$$32 \text{ oz} \times 0.05 \text{ qt} = \textbf{1.60 oz}$$

Using the table to round off the ounces, we would have 1¾ oz of wine. The adjusted recipe for Beef a la Bourguignonne yielding 235 servings is the following:

Beef a la Bourguignonne
235 servings

Ingredients

84 lb 9¾ oz beef tenderloin, cut into 1″ cubes
2 lb 15 oz shortening
18 lb 3 oz mushrooms, sliced
4 lb 11¾ oz shallots or green onions, minced
14½ oz flour
7 qt 1¾ oz Burgundy wine
salt and pepper to taste

If the yield of a recipe must be decreased, the recipe must be adjusted in a similar fashion. For example, the recipe yields 100 servings; however, only 60 servings are required. Determine the working factor.

$$working\ factor = \frac{required\ yield}{recipe\ yield}$$

$$working\ factor = \frac{60}{100}$$

$$working\ factor = \frac{60}{100} \div \frac{20}{20} = \frac{3}{5}$$

working factor = ⅗ (simplest form)

Using a working factor of ⅗, follow the same procedure used for increasing the recipe. For example:

Beef a la Bourguignonne
100 servings

Ingredients

36 lb beef tenderloin, cut into 1″ cubes
1¼ lb shortening
8 lb mushrooms, sliced
2 lb shallots or green onions, minced
6 oz flour
3 qt Burgundy wine
salt and pepper to taste

Ingredient	Amount	Working Factor		Adjusted Amount
Beef Tenderloin	36 lb	× ⅗	=	21.6 lb
Shortening	1¼ lb	× ⅗	=	¾ lb
Mushrooms	8 lb	× ⅗	=	4.8 lb
Shallots	2 lb	× ⅗	=	1.2 lb
Flour	6 oz	× ⅗	=	3.6 oz
Wine	3 qt	× ⅗	=	1.8 qt

Fractions of a pound in decimal form are converted to ounces by multiplying 16 by the decimal.

$$16 \times 0.6\ beef = 9.60\ oz$$

$$16 \times 0.8\ mushrooms = 12.8\ oz$$

$$16 \times 0.2\ shallots = 3.2\ oz$$

$$16 \times 0.6\ flour = 0.96\ oz$$

After fractions of a pound are converted to ounces, use Table VIII to round off the fractions of an ounce. For example, 9.6 oz of beef are rounded to 9¾ oz, 12.8 oz of mushrooms are 13 oz, 3.2 oz of shallots are 3¼ oz, and 0.96 oz of flour are 1 oz.

As in increasing the yield of a recipe, fractions of a quart are converted to ounces by multiplying 32 by the decimal (32 oz × 0.8 = 25.6 oz). Using the table to round off ounces, 25.6 oz = 25¾ oz or approximately ¾ of a quart (1 quart = 32 oz). The difference between ¾ quart (²⁴⁄₃₂) and 25.6 oz is not enough to make a difference. The adjusted recipe for Beef a la Bourguignonne for 60 servings is:

Beef a la Bourguignonne
60 servings

Ingredients

21 lb 9¾ oz beef tenderloin, cut into 1″ cubes
12 oz shortening
4 lb 3 oz mushrooms, sliced
1 lb 3¼ oz shallots or green onions, minced
4 oz flour
1¾ qt Burgundy wine
salt and pepper to taste

If a recipe uses volume rather than weights, the working factor would be determined and multiplied by cups, tablespoons, teaspoons, as required. For example, fractional cups would be multiplied by 16 since there are 16 tbsp in each cup. Fractional tablespoons are rounded off to the nearest fractional teaspoon.

When converting any recipe using a working factor, common sense must be used. Adjusting quantities, converting measurements, and rounding off amounts all require careful consideration to obtain the best results.

Kitchen Calculators

Using a kitchen calculator is a fast and convenient way to convert recipe yields. Calculators such as the KitchenCalc™ Pro are specially designed to quickly convert common kitchen measures and adjust recipe quantities as needed. To adjust a recipe, the original yield of a recipe and the new desired yield are stored in the calculator's memory. Ingredient measures for the original yield are then entered individually. The calculator will compute the new measures for each ingredient based upon the desired recipe yield.

For example, to convert the 84 lb 9¾ oz of beef tenderloin needed for the original Beef a la Bourguignonne recipe serving 235 to serve only 100, the following procedure is used:

1. Set the original recipe yield. Enter 235 (the original recipe yield) and press the RECIPE # SERVINGS button. This number is stored in the calculator's memory

2. Set the desired recipe yield. Enter 100 (the desired yield) and press the ACTUAL # SERVINGS button. This number is also stored in the calculator's memory

3. Convert any fractional parts of a pound from ounces to their fractional equivalents. Enter 9, 3, /, and 4 (9¾) and press the DRY OZ button. Press the blue CONV button followed by the LB button. This converts 9¾ oz to ⅝ lb. Pressing the LB button repeatedly will toggle between fraction and decimal equivalents.

4. Add total pounds to the fractional result. Enter +, 8, 4, and =. The result is 84⅝ lb.

5. Convert the original measure to the new measure needed for the adjusted recipe. Press the ADJUST RECIPE button. The resulting measure is 36 lb.

6. Press the ON/C button to clear the value.

Steps 3 through 6 are repeated for each ingredient in the recipe. If the calculator displays ^RND in the lower right corner, it is rounding up to the displayed value. If the calculator displays ∨RND in the lower right corner, it is rounding down to the displayed value. Pressing the unit button again—for example pressing LB if the units are in pounds—will display the rounded decimal equivalent.

The kitchen calculator can also convert common kitchen weights and measures. For example, to convert a dash into a teaspoon, the following procedure is used:

1. Enter the number. Press 1.

2. Enter the given measure. The dash measure is printed in blue on the . button. To access units printed in blue on a button, press the blue CONV button first. The operation is similar to a shift key on a keyboard. Press the CONV button and the . button. The display indicates that the dash unit is currently selected.

3. Convert to the desired measure. Press the CONV button and the TSP button. The display now reads 0⅛ tsp. One dash is equal to ⅛ of a teaspoon.

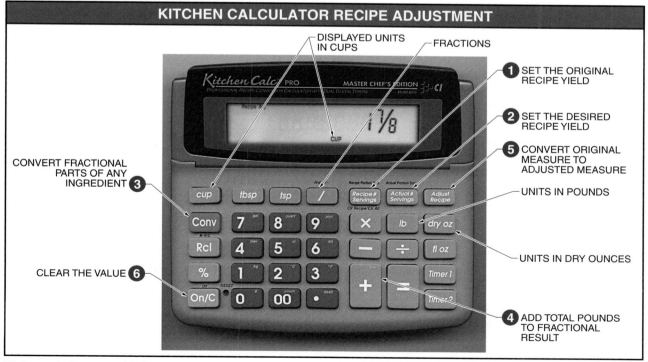

KITCHEN CALCULATOR RECIPE ADJUSTMENT

Kitchen calculators provide fast and accurate recipe conversions when changing the yield size of a recipe.

CONVERSION TABLE

Table IX lists information regarding approximate weights and measures of common foods. This information is useful because some recipes are given in weight and some in measures. The most accurate recipes are given in weight. However, for convenience, it may be necessary to convert weight to measure. For example, 1 lb of water may be converted to 1 pt, or 1 c of bread flour may be converted to 5 oz by weight.

TABLE IX: APPROXIMATE WEIGHTS AND MEASURES OF COMMON FOODS									
Food Product	Tbsp	Cup	Pt	Qt	Food Product	Tbsp	Cup	Pt	Qt
Allspice	$1/4$ oz	4 oz	8 oz	1 lb	Ginger	$3/16$ oz	$3^1/4$ oz	$6^1/2$ oz	13 oz
Apples, fresh, diced	$1/2$ oz	8 oz	1 lb	2 lb	Glucose	$3/4$ oz	12 oz	1 lb 8 oz	2 lb
Bacon, cooked, diced	$2/3$ oz	$10^1/2$ oz	1 lb 5 oz	2 lb 12 oz	Green peppers, diced	$1/4$ oz	4 oz	8 oz	1 lb
Bananas, sliced	$1/2$ oz	8 oz	1 lb	2 lb	Ham, cooked, diced	$5/16$ oz	$5^1/4$ oz	$10^1/2$ oz	1 lb 5 oz
Baking powder	$3/8$ oz	6 oz	12 oz	1 lb 8 oz	Horseradish, prepared	$1/2$ oz	8 oz	1 lb	2 lb
Baking soda	$3/8$ oz	6 oz	12 oz	1 lb 8 oz	Lemon juice	$1/2$ oz	8 oz	1 lb	2 lb
Beef, cooked, diced	$3/8$ oz	$5^1/2$ oz	11 oz	1 lb 6 oz	Lemon rind	$1/4$ oz	4 oz	8 oz	1 lb
Beef, raw, ground	$1/2$ oz	8 oz	1 lb	2 lb	Mace	$1/4$ oz	$3^1/4$ oz	$6^1/2$ oz	13 oz
Barley	—	8 oz	1 lb	2 lb	Mayonnaise	$1/2$ oz	8 oz	1 lb	2 lb
Bread crumbs, dry	$1/4$ oz	$4^1/2$ oz	9 oz	1 lb 2 oz	Milk, liquid	$1/2$ oz	8 oz	1 lb	2 lb
Bread crumbs, fresh	$1/8$ oz	2 oz	4 oz	8 oz	Milk, powdered	$5/16$ oz	$4^1/2$ oz	9 oz	1 lb 2 oz
Butter	$1/2$ oz	8 oz	1 lb	2 lb	Milk, powdered	4 oz + 1 qt water = 1 qt milk			
Cabbage, shredded	$1/4$ oz	4 oz	8 oz	1 lb	Molasses	$3/4$ oz	12 oz	1 lb 8 oz	3 lb
Carrots, raw, diced	$5/16$ oz	5 oz	10 oz	1 lb 4 oz	Mustard, ground	$1/4$ oz	$3^1/4$ oz	$6^1/2$ oz	13 oz
Celery, raw, diced	$1/4$ oz	4 oz	8 oz	1 lb	Mustard, prepared	$1/4$ oz	4 oz	8 oz	1 lb
Cheese, grated	$1/4$ oz	4 oz	8 oz	1 lb	Nutmeats	$1/4$ oz	4 oz	8 oz	1 lb
Chocolate, grated	$1/4$ oz	4 oz	8 oz	1 lb	Nutmeg, ground	$1/4$ oz	$4^1/4$ oz	$8^1/2$ oz	1 lb 1 oz
Chocolate, melted	$1/2$ oz	8 oz	1 lb	2 lb	Oats, rolled	$3/16$ oz	3 oz	6 oz	12 oz
Cinnamon, ground	$1/4$ oz	$3^1/2$ oz	7 oz	14 oz	Oil, salad	$1/2$ oz	8 oz	1 lb	2 lb
Cloves, ground	$1/4$ oz	4 oz	8 oz	1 lb	Onions	$1/3$ oz	$5^1/2$ oz	11 oz	1 lb 6 oz
Cloves, whole	$3/16$ oz	3 oz	6 oz	12 oz	Peaches, canned	$1/2$ oz	8 oz	1 lb	2 lb
Cocoa	$3/16$ oz	$3^1/2$ oz	7 oz	14 oz	Peas, dry, split	$7/16$ oz	7 oz	14 oz	1 lb 12 oz
Coconut, shredded,	$3/16$ oz	$3^1/2$ oz	7 oz	14 oz	Pickles, chopped	$1/4$ oz	$6^1/2$ oz	$10^1/2$ oz	1 lb 5 oz
Coffee, ground	$3/16$ oz	3 oz	6 oz	12 oz	Pickle relish	$5/16$ oz	$5^1/4$ oz	$10^1/2$ oz	1 lb 5 oz
Cornmeal	$5/16$ oz	$4^3/4$ oz	$9^1/2$ oz	1 lb 3 oz	Pineapple, diced	$1/2$ oz	8 oz	1 lb	2 lb
Cornstarch	$1/3$ oz	$5^1/3$ oz	$10^1/2$ oz	1 lb 5 oz	Pimientos, chopped	$1/2$ oz	7 oz	14 oz	1 lb 12 oz
Corn syrup	$3/4$ oz	12 oz	1 lb 8 oz	3 lb	Potatoes, cooked,	—	$6^1/2$ oz	13 oz	1 lb 10 oz
Cracker crumbs	$1/4$ oz	4 oz	8 oz	1 lb	Raisins, seedless	$1/3$ oz	$5^1/3$ oz	$10^3/4$ oz	1 lb 5 oz
Cranberries, raw	—	4 oz	8 oz	1 lb	Rice, raw	$1/2$ oz	8 oz	1 lb	2 lb
Currants, dried	$1/3$ oz	$5^1/3$ oz	11 oz	1 lb 6 oz	Sage, ground	$1/8$ oz	$2^1/4$ oz	—	—
Curry powder	$3/16$ oz	$3^1/2$ oz	—	—	Savory	$1/8$ oz	2 oz	—	—
Egg whites	$1/2$ oz	8 oz	1 lb	2 lb	Salt	$1/2$ oz	8 oz	1 lb	2 lb
Eggs, whole	$1/2$ oz	8 oz	1 lb	2 lb	Shortening	$1/2$ oz	8 oz	1 lb	2 lb
Egg yolks	$1/2$ oz	8 oz	1 lb	2 lb	Sugar, brown, packed	$1/2$ oz	8 oz	1 lb	2 lb
Eggs, dry (whole)	$1/4$ oz	4 oz	8 oz	1 lb	Sugar, granulated	$7/16$ oz	$4^3/4$ oz	15 oz	1 lb 14 oz
Eggs, dry (whole)	$1^1/2$ cup (6 oz) + 1 pt water = 1 doz eggs				Sugar, powdered	$5/16$ oz	$4^3/4$ oz	$9^1/2$ oz	1 lb 3 oz
Extracts	$1/2$ oz	8 oz	1 lb	2 lb	Tapioca, pearl	$1/4$ oz	4 oz	8 oz	1 lb
Flour, bread/pastry	$5/16$ oz	5 oz	10 oz	1 lb 4 oz	Tomatoes	$1/2$ oz	8 oz	1 lb	2 lb
Flour, cake	$1/4$ oz	$4^3/4$ oz	$9^1/2$ oz	1 lb 3 oz	Vanilla, imitation	$1/2$ oz	8 oz	1 lb	2 lb
Gelatin, flavored	$3/8$ oz	$6^1/2$ oz	13 oz	1 lb 10 oz	Vinegar	$1/2$ oz	8 oz	1 lb	2 lb
Gelatin, plain	$5/16$ oz	5 oz	10 oz	1 lb 4 oz	Water	$1/2$ oz	8 oz	1 lb	2 lb

Breakfast Preparation

Breakfast preparation offers the food service worker an opportunity for experience in handling many orders quickly. Eggs are the most popular breakfast item. Eggs offer variety for the customer because they can be prepared in many ways. Simple preparations, such as scrambled eggs, are quick to cook. More complex preparations, such as shirred eggs, require more time.

Pancakes are also popular on the breakfast menu. Pancake batter can be prepared before cooking, and cooking time is minimal. Pancakes are served with different toppings for variety.

Breakfast meats, usually bacon, sausage, or ham, can also be prepared quickly. Potatoes are usually fried for quick preparation. Cereals, hot or cold, can be prepared in individual servings. This assures freshness to the customer. Juices are served in small glasses and are an appetite stimulant. Fruits, such as grapefruit, oranges, and cantaloupe, are served chilled. Toast and pastry items are served on the side or eaten with coffee or juice.

BREAKFAST PREPARATION

Many nutrition experts have stated that breakfast is the most important meal of the day because it helps the body operate at maximum efficiency. Breakfast is usually the first food consumed for a period of twelve hours and must be composed of items that are easily digested.

Common breakfast preparations include eggs, pancakes, potatoes, waffles, hot cereals, toast (plain, cinnamon, and French), stewed fruit, and meats such as bacon, ham, and sausage. Of these, the most popular breakfast item is eggs. Eggs are very versatile and can be prepared in a variety of ways.

EGGS

Eggs are used in many preparations in the commercial kitchen. Knowledge about eggs and their various uses helps produce better results. Eggs are a complete protein food and are important in the daily diet. Eggs are very high in vitamin content and are easy to digest when cooked properly. Besides their importance in breakfast preparations, eggs can also be featured as luncheon and dinner entrees. In fact, eggs, like cheese, are a food that may be used in any part of the menu from appetizer to dessert. Eggs are relatively inexpensive and can help hold food costs down. The commercial kitchen should always have eggs available.

Eggs are used in a variety of ways in different preparations:

- As a thickening or binding agent. *Examples*: meat loaf, custard, pie filling, and croquettes.
- As an adhesive agent. *Example*: breading.
- As an emulsifying agent. *Examples*: mayonnaise and hollandaise.
- As a clarifying agent. *Examples*: consommé and aspic.
- As a leavening agent (incorporating air). *Examples*: soufflés, sponge cakes, and chiffon pies.
- As an entree. *Examples*: breakfast preparations, eggs Benedict, and quiche lorriane.

The quality of an egg is judged by many factors. The most important factor is the appearance and condition of the interior as revealed in the candling process (twirling the egg slowly before an ultraviolet light in picking plants). The cleanliness of the shell is also important. The size of an egg or the color of its shell has no bearing on its quality.

Eggs are classified according to quality. There are four grades. They are Grade AA, also known as U.S fresh fancy, a very fancy egg seldom found in retail food markets; Grade A, also known as U.S. Extra, a very fine egg and usually the top grade found in retail stores; Grade B, also known as U.S. Standard, a good quality egg suitable for most purposes; and Grade C, also known as U.S. Trade, suitable for cooking where flavor is not an important factor. Grades B and C are not usually available in retail markets, but may be used in bakeries.

Eggs are also classified according to size. Each of the four grades of eggs are sorted into various sizes. The size is determined by federal standards based on the number of ounces per dozen:

- Jumbo – 30 oz or more per dozen

Eggs are versatile and can be used in preparations for breakfast, luncheon, and dinner menus.

- Extra Large – 27 oz to 30 oz per dozen
- Large – 24 oz to 27 oz per dozen
- Medium – 21 oz to 24 oz per dozen
- Small – 18 oz to 21 oz per dozen

Recipes that use a number of eggs instead of a weight or measure are specifying the number of large eggs. In the commercial kitchen, fresh eggs are used more than any other form of egg. This is because eggs in fresh form are the most versatile. Eggs can be purchased frozen in 5 lb cartons and 30 lb cans as eggs, egg yolks, and egg whites. These forms and quantities are convenient only if the eggs are to be used in bulk food preparations and baked products. Dried eggs are also available, but are not often used in food service establishments. Dried eggs are occasionally used in some bakeshops.

Cooked eggs perish rapidly, mainly because of their delicate nature. It is difficult to cook eggs in large quantities even though it becomes necessary at times in the commercial kitchen. Whether frying, poaching, scrambling, shirring, or basting, best results can always be obtained by cooking eggs in small quantities and as close to serving time as possible. Cooking to order is the recommended procedure. This is not always possible. The next best technique is to undercook the eggs and finish in a warm oven or steam table. Always follow local health department regulations. Holding eggs over ten minutes destroys quality. Eggs that are scrambled or poached can be held with the best results if special holding techniques are followed. Uncooked eggs must be carefully stored to prevent spoilage. Tips for egg storage include:

- Store eggs in the refrigerator. If eggs are left in a warm place, they lose freshness rapidly. Take out only the number of eggs needed for an order or recipe at a time.
- When storing leftover egg yolks in the refrigerator, cover them with water and plastic wrap. If left uncovered, they form a surface crust and dry out rapidly.
- Leftover egg whites will keep for at least a week in the refrigerator if they are placed in a tightly covered container.

Eggs may be stored frozen, but this requires special handling before use. Before using frozen eggs, thaw them gradually, preferably in the refrigerator. To speed the thawing process, place the frozen eggs in cold water at room temperature. After they have thawed, stir thoroughly before using. One quart of frozen eggs equals approximately 24 fresh eggs.

Fried Eggs

Always fry eggs to order, using high-quality eggs. For best results, fry with butter, shortening, or margarine. Never fry with bacon grease as it produces strong flavors and is unhealthy.

Select the correct size skillet. For a single egg, the skillet should be 4″ in diameter at the bottom. For an order of eggs (two eggs), the skillet should be 6″ in diameter. The skillet should have sloped, shallow walls and a long handle. Place about ⅛″ of melted butter, shortening, or margarine in the bottom of the pan. Heat to a fairly moderate temperature and slide the eggs, which have been previously broken into a soup bowl, into the fat. The hot fat will solidify the eggs immediately so the whites will not spread. At this point, reduce the heat immediately to avoid a hard, brown surface under or around the edge of the eggs. Proceed to cook the eggs as requested by the customer.

Sunny side up eggs are lightly cooked with the yolks unbroken. *Eggs over* are eggs that are flipped over and cooked easy (lightly), medium, or hard (well). *Basted eggs* are cooked the same as sunny side up, but are finished under the broiler by cooking the top of the eggs until the whites are set and a cooked coating appears over the yolk.

Another method of frying eggs is known as *country style*. Country style fried eggs are served with ham, bacon, or sausage. The meat is precooked, placed in a greased egg skillet, and heated. The eggs are placed on top of the meat and cooked in the same manner as sunny side up; however, the skillet is covered during the frying period until the whites are set and the yolks are cooked slightly.

Eggs can also be fried on a well-greased grill. However, the results are not as attractive as pan-fried eggs. The grill must be very clean and well-conditioned for best results. An egg ring can be used to control spreading of the white. When cooking eggs on a grill, maintain a temperature of 300°F to 350°F.

Fried eggs are popular for the breakfast menu.

Fried Egg Faults

- Frying with too much fat. There is danger of burning oneself and eggs are greasy when served.
- Using a poorly conditioned pan or grill. Eggs stick to the pan or grill, causing them to burn and break.
- Frying at too low a temperature. Egg whites spread too rapidly.
- Frying at too high a temperature. Eggs burn and are usually overcooked.
- Frying with too little fat. Eggs stick and usually burn.

Scrambled Eggs

Scrambling is the easiest method to choose when preparing eggs in quantity. Scrambled eggs may be prepared in several ways. Eggs can be scrambled in a well-greased pan in the oven, in a steam jacket kettle, in a double boiler, in a steamer, or in a skillet on the range. Except for large quantities, unless one of the other methods proves more efficient, the best method is to scramble in the skillet.

Break the eggs into a stainless steel or china bowl. Never use an aluminum bowl because it will discolor the eggs. Beat the eggs slightly with a wire whip or kitchen fork. Add a small amount of milk or cream for tenderness if desired (about 4 oz to each pint of eggs). Too much milk will cause *weeping* (giving off water) after the eggs are cooked.

Trade Tip

Julian dates are commonly included on egg cartons to show the day the eggs were packed. Julian dates, beginning with January 1 as No. 1 and going through December 31 as No. 365, represent consecutive days of the year. Fresh eggs can be stored in their cartons in the refrigerator for 4 to 5 weeks past this date without significant loss in quality.

Pour the beaten eggs into a heated, greased, or buttered skillet so the eggs will start to coagulate immediately. Reduce heat and lift the eggs carefully from the bottom. At the same time, stir gently with a kitchen spoon so the uncooked portion will settle to the bottom and cook.

Scrambled eggs are properly cooked when they are soft and fluffy. Always undercook scrambled eggs slightly, as they will become firm when held for service. Never let the eggs brown or overcook because they will become dry, hard, and unpalatable. If scrambled eggs are to be held over 5 min before serving, add a medium cream sauce (combination of hot milk or cream to roux), using a ratio of 5:1. This will extend the holding time. The addition of the cream sauce prevents the eggs from drying and discoloring.

Scrambled Egg Faults

- Cooking at too high a temperature. Eggs usually burn and are overcooked.

- Excessive stirring when cooking. Egg particles become too fine, giving a poor appearance.

- Holding cooked eggs for too long in the steam table. The eggs develop off colors and lose flavors.

- Scrambling with too much fat in the skillet or pan. Eggs become greasy.

- Scrambling with too little fat in the skillet or pan. Eggs stick to pan, become tough, and burn.

Scrambled eggs can be served in pita bread with diced meat and vegetables for variety.

Boiled or Simmered Eggs

Although the common term is *boiled eggs*, the fact is that eggs should be simmered, never boiled, for best results. Boiling tends to toughen the texture and can create a green coating around the outside of the yolk. Simmering at a temperature of approximately 195°F is recommended.

Eggs should be room temperature before they are placed in the hot water, or they may crack. If the eggs are left in the refrigerator until it is time to cook them, run warm water over the eggs before placing them in the hot water. Two methods of "boiling" or simmering eggs are recommended:

1. Bring water to a boil (212°F). Place the eggs in the boiling water. Reduce the heat to a simmer by pulling the pot of water away from the heat. Cook to desired doneness:

   - Soft – 3 min to 5 min
   - Medium – 7 min to 8 min
   - Hard – 15 min to 17 min

2. Place the eggs in a pot and cover them with cold water. Place the pot on the range and bring the water to a boil, reducing the heat to a simmer by pulling the pot away from the heat. Cook to desired doneness:

   - Soft – 1 min to 2 min (local health departments may not allow soft cooked eggs as minimum internal temperatures cannot be achieved)
   - Medium – 3 min to 6 min
   - Hard – 8 min to 10 min

If eggs are boiled for breakfast service, they should be plunged immediately into slightly cold water and served in their shells. If the eggs are simmered for hard-boiled and held for use in kitchen preparations such as sandwiches, deviled eggs, or garnish, they should be cooled in ice cold water immediately after cooking for about five minutes.

A boiled egg is peeled by cracking the shell gently on a hard surface or by rolling it on a hard surface. Start to peel at the large end of the egg and peel downward, keeping the egg submerged in cold water. Holding the egg under cold running water will help loosen the shell.

Place the hard-boiled eggs in a *bain-marie* (container used for keeping foods hot) covered with water and store in the refrigerator. If the yolk of the egg is exposed, do not place it in the water; place it in a bowl and cover with plastic wrap or a damp cloth.

To coddle eggs for such preparations as Caesar salad, have the eggs at room temperature and place them in a pot. Add boiling water to the pot until the eggs are covered. Put a lid on the pot for 1 min to 2 min and let stand, without heating, until the eggs are cooked as desired.

Boiled Egg Faults

- Cooking at too high a temperature. Eggs become tough and rubbery. A green ring may appear around the yolk.

- Cooking at too low a temperature. Eggs are usually undercooked.

Note: Very fresh eggs are usually hard to peel. Eggs intended for boiling are preferably about one week old.

Poached Eggs

While poached eggs are fairly popular on the breakfast menu, they are also used extensively on the luncheon menu in such preparations as eggs a la Florentine, poached eggs on corned beef hash, and eggs Benedict.

Fill a fairly shallow pan with enough water to cover the eggs (about 2½″). Add 2 tbsp of distilled vinegar to each gallon of water. The salt and vinegar will cause the white to set firmly around the yolk when the egg is placed in the water, thus retarding the white from spreading. The acetic acid of the vinegar toughens the albumen contained in the egg white, and when the white is set firmly around the yolk, a more eye-appealing product is obtained. Acid will not affect the flavor of the egg when used in diluted quantities.

Bring the liquid to a boil, then reduce to a simmer (about 195°F to 200°F). Break the eggs into a bowl or saucer and slide them into the simmering liquid. The eggs should slide gently down the side of the pan so the yolk will remain in the center of the white.

Cook as desired; usually 3 min to 5 min is sufficient. Remove with a skimmer, perforated ladle, or slotted spoon. Drain well and serve on buttered toast.

About 12 eggs may be poached in each gallon of liquid. The water may be used for three different batches before it is discarded.

To prepare poached eggs in quantity, the eggs are slightly undercooked. The eggs are then placed immediately in cold water to stop further cooking and to hold until ready to serve. To serve, they are reheated in hot salt water. The quantity method is usually for luncheon preparations. For breakfast preparations, eggs are poached to order.

Poached Egg Faults

- Too much vinegar added to liquid. This toughens the eggs and affects their flavor.
- Cooking at too low a temperature. Eggs become too tender and difficult to handle when they are served.
- Cooking at too high a temperature. Eggs become tough and are usually overcooked.

American Egg Board

Frittatas are often served in the pan for visual effect.

Shirred Eggs

Shirred eggs are eggs that have been cooked in a shallow casserole with butter. When served on the breakfast or luncheon menu, shirred eggs present a very attractive dish. Shirred eggs must be prepared in a proper manner and not overcooked, as is often the case.

The eggs are broken into a bowl or saucer, then placed in a buttered shirred egg dish. Place the shirred egg dish on the range and cook at medium heat until the whites are set. Finish by transferring the dish to the oven or by basting the top of the eggs lightly under the broiler. In either case, never cook the eggs hard.

Poached eggs are prepared by simmering raw eggs in water and vinegar. The vinegar causes the white to set firmly around the egg.

Shirred eggs may be served with a variety of foods. Cooked bacon, ham, sausage, Canadian bacon, kidney, chicken liver, and cheese are the most popular. The meat or cheese can be placed around the edges of the eggs before or after cooking.

Shirred Egg Faults

- Cooking at too high a temperature. Eggs become hard and tough and usually are burned.
- Cooking at too low a temperature. Eggs spread and the yolk has a tendency to break.

Omelets

Another egg preparation that is popular on both the breakfast and luncheon menus is the omelet. Occasionally, omelets can also be found on dinner menus. The omelet is a versatile preparation because it blends well with other foods, producing almost limitless variations.

When an omelet is served with another item such as bacon, mushrooms, or Spanish sauce, it takes the name of that accompanying item; for example, bacon omelet, mushroom omelet, and Spanish omelet.

Omelets should be made to order for best results. If they are held for even a short period of time, they lose their tender fluffiness and become tough and rubbery. When preparing an omelet for the breakfast menu, two or three eggs are usually the rule. Buffet omelets are prepared using 12 to 15 eggs. When an omelet has a sweet accompaniment, such as jelly, jam, or marmalade, it is usually rolled or folded into the center of the omelet. When the omelet is completed, it is dusted with powdered sugar and scored or branded with a hot metal rod. A number of techniques can be used in preparing omelets. However, these variations still result in a rolled or folded covering of the accompanying food item inside.

PREPARING FOLDED OMELETS

1 MIX EGGS

2 ADD EGGS SLOWLY TO BUTTERED PAN

3 USE SPATULA TO ALLOW EGGS TO HEAT PROPERLY

4 ADD FILLING TO EGGS

5 HOLD PAN BY HANDLE AND SLIDE ONTO PLATE

6 FOLD OMELET OVER ON PLATE

Rolled Omelets

Break eggs into a bowl and whip slightly with a wire whip or kitchen fork. Place about ⅛″ of melted margarine or butter, fat, or oil in a conditioned egg skillet and heat to a fairly high temperature. Pour the beaten eggs into the hot grease. As the eggs bubble, tilt the skillet in all directions to spread the eggs to the sides of the skillet completely. Reduce heat by bringing the skillet to the side of the range, and cook the eggs at a moderate temperature.

As the eggs cook, lift up the cooked portion using a kitchen fork or palette knife, letting the uncooked portion run to the bottom and sides of the skillet.

When the omelet is slightly set but still in a very moist condition, tilt the skillet to about a 60° angle and, using a kitchen fork or palette knife, roll the eggs toward you. When the roll is completed, place it on a hot plate by grasping the handle of the skillet in the right hand, with fingers and thumb turned up. Bring the hot plate to the lip of the skillet with the left hand. Tilt the skillet over, letting the omelet slide onto the plate. Cover the omelet with a clean cloth to reshape it and also absorb excess grease. Serve immediately.

ROLLED OMELETS

Folded Omelets

For a folded omelet, the eggs are set but still in a very moist condition. Grasp the handle of the skillet in the right hand with fingers and thumb pointed up. Tilt the pan to slide the egg onto the plate while folding one half of the omelet over the other half. If an accompanying item is to be folded into the omelet, it is placed in the center of the eggs prior to folding.

Many variations of omelets can be created by the addition of one or more food items. Some of these variations are:

- Bacon omelet
- Cheese omelet
- Ham omelet
- Onion omelet
- Spanish or Creole omelet
- Western omelet (onions, green pepper, and ham)
- Lobster omelet
- Shrimp omelet
- Fine herb omelet

Extremely fluffy omelets have become popular in some areas, but as a rule the plain or French omelet is mainly featured on most menus. A fluffy omelet is prepared by separating the yolk from the white and beating each to a soft foam. The beaten white and yolk are then folded very gently together until well mixed. In this form, the eggs are poured into a hot, greased skillet and cooked in the same manner as a rolled or folded omelet. This type of omelet can also be finished in a 325°F oven. When finished, the fluffy omelet should be served at once.

Omelet Faults

- Too little fat in the pan. Eggs stick and burn.
- Poorly conditioned skillet. Eggs stick and the omelet will break when rolled or folded.
- Overcooking. Omelet becomes too brown and cracks when rolled or folded.
- Too much fat in skillet. Hot grease splatters when the eggs are added and excess grease will spill out when rolling or folding.
- Cooking omelet ahead of service. Omelet loses its fluffiness and becomes tough and rubbery.

PANCAKES

Another popular breakfast preparation is pancakes, also known as *hotcakes* or *griddle cakes*. Pancakes are popular because they are easy to digest, can be served in a variety of ways, and usually have a low menu price.

Pancakes must always be cooked to order and served piping hot on a hot plate or platter with butter and topping such as jam, jelly, or syrup. Maple syrup is the most popular topping; however, fruit or fruit-flavored syrups are also popular. Pancakes cost very little to make even when featured with a high-cost accompaniment such as strawberries, cherries, blueberries, or ice cream. Pancakes blend well with meat and, when served for breakfast, are usually accompanied by sausage, ham, or bacon.

Many quality pancake mixes are available on the market. Most food service establishments find that it is more profitable to use pancake mix than mixing their own. The amount of preparation depends on the type of mix used. Some mixes call for the addition of milk or water only. The better mixes call for milk, oil, and eggs. As the number of ingredients increases, the time saved decreases. Making a mix from scratch to save money becomes more attractive. The decision of whether to make the mix from scratch or use a prepared mix is usually made by the chef based on the number and variety of pancakes served. However, all cooks should know some of the basic pancake mixes and how to prepare them.

Buttermilk Pancakes

1 gal. or 15 orders of 3 cakes to 1 order

Ingredients

6 eggs
2 qt cultured buttermilk
2 lb flour
12 oz salad oil
3 tsp baking soda
1 oz baking powder
2½ oz sugar
3 tsp salt

Procedure

1. Place the eggs in a mixing bowl and mix in the mixing machine at slow speed using the paddle for 1 min.
2. Add the milk and oil, continuing to mix at slow speed for 1 min.
3. Combine the remaining dry ingredients, sift two times, and add gradually to the liquid mixture in the mixing bowl. Mix 1 min and scrape down bowl if necessary. Remove batter from mixing bowl and place in a bain-marie.
4. Let the batter rest at least 10 min.
5. Heat griddle to 375°F. Grease lightly.
6. Using a 3 oz ladle, spot the batter on the griddle. Cakes should spread to 5″ in diameter.
7. Brown one side until bubbles appear on top and the batter takes on a puffy quality.
8. Turn or flip and brown the second side.
9. Serve three cakes to each order with desired syrup.

Plain Pancakes or Hotcakes

1 gal. or 15 orders of 3 cakes to 1 order

Ingredients

8 eggs
2 qt milk
12 oz salad oil
2 lb flour
5 oz sugar
3 tsp salt
1½ oz baking powder

Procedure

1. Place eggs in a mixing bowl and mix in the mixing machine at slow speed using the paddle for 1 min.
2. Add the milk and oil, continuing to mix at slow speed for 1 min.
3. Combine the remaining dry ingredients, sift two times, and add gradually to the liquid mixture in the mixing bowl. Mix for approximately 1 min at slow speed. Scrape down bowl if necessary. Remove batter from mixing bowl and place in a bain-marie.
4. Let the batter rest at least 10 min.
5. Heat griddle to 375°F. Grease lightly.
6. Using a 3 oz ladle, spot the batter on the hot griddle. Cakes should spread to 5″ in diameter.
7. Brown one side golden brown, turn or flip, and brown second side.
8. Serve three cakes to each order with desired syrup.

French Pancakes or Crepes

1 gal. or 60 pancakes

Ingredients

1 qt milk
10 oz flour
6 eggs
½ tsp salt
3 oz melted butter
3 oz sugar

Procedure

1. Place the eggs in a mixing bowl and mix by hand using a wire whip, or mix in the mixing machine at medium speed using the paddle for approximately 1 min.
2. Add the milk and continue to mix until the milk is blended with the eggs.
3. Combine the dry ingredients, sift, and add gradually to the liquid mixture. Mix for approximately 1 min or until all the dry ingredients are blended with the liquid mixture.
4. Add the melted butter and mix until well blended.
5. Pour the batter into a bain-marie and place in the refrigerator until ready for use.
6. Using a well-conditioned egg skillet or omelet pan, coat with melted butter or shortening and heat slightly.
7. Using a 2 oz ladle, coat the bottom of the skillet with the crepe batter. While pouring the batter into the skillet, rotate the skillet in a clockwise direction so the batter will spread uniformly over the bottom of the skillet and the coating will remain very thin.
8. Place the skillet on the range and cook one side, then flip or turn by hand and cook second side. Brown the crepes very lightly.
9. Remove from the skillet and place on sheet pans covered with wax paper. If crepes are stacked, place a sheet of wax paper between the layers.
10. Spread each crepe with jelly, jam, preserves, marmalade, strawberries, or applesauce and roll up.
11. Dust with powdered sugar and serve three crepes to each order.

Waffles

1 gal.

Ingredients

10 eggs
2 qt milk
2 lb 10 oz cake flour
2 oz baking powder
8 oz sugar
1 lb melted butter

Procedure

1. Place the eggs in a mixing bowl and mix in the mixing machine at medium speed using the paddle for approximately 1 min.
2. Add the milk and continue to mix until the milk is blended with the eggs.
3. Combine the dry ingredients, sift three times, and add to the egg-milk mixture. Mix for 1 min.
4. Add the melted butter and mix until well blended.
5. Pour the batter into a bain-marie and place in the refrigerator until ready for use.
6. Brush the top and bottom of the waffle iron with salad oil. Heat to approximately 375°F.
7. Pour enough batter on the waffle grid to barely cover it. The amount used will depend on the size and shape of the waffle iron.
8. Let the waffle cook for about 1 min before lowering the top of the iron. When the top is lowered, cook about 1½ min to 2 min longer. Exercise caution while cooking because the top grid is usually hotter than the bottom grid.
9. Serve two waffles for each serving with jam, jelly, syrup, marmalade, or fruit.

BREAKFAST MEATS

Breakfast meats commonly include sausage, bacon, and ham. All of these breakfast meats come from pork. Bacon and ham are cured by smoking, whereas most sausage is fresh. Breakfast meats are available precut. These meats are usually precooked and reheated for service when the breakfast volume is large. When the volume is small, they are cooked to order. Precooking is done to speed service, since breakfast must be served quickly.

Breakfast sausage is available in the forms of patties and links. Patties may be purchased in bulk or prepared in the kitchen from ground fresh pork and spices. Sausage patties are usually portioned and formed into 1 oz or 2 oz servings by hand. Sausage patties can be precooked by baking on sheet pans in the oven at 350°F, cooking in a skillet on the range, broiling under the broiler, or grilling on a griddle. Whichever method is used, sausage patties should be cooked about three-quarters of the way and finished when ready to serve.

National Pork Producers Council

Bacon is a popular breakfast meat.

Sausage links, sometimes referred to as *little pigs*, average about 12 to the pound. The portion served for breakfast is generally three or four links. Sausage links are cooked by separating the links, placing them on sheet pans, and baking at 350°F until three-quarters done. Drain off excess grease and finish by browning under the broiler. After cooking, the links are moved to a hotel pan and held for service. Cook sausage links in small amounts. Sausage links held for the next day's service become dry.

Bacon may be purchased by the slab and sliced. The slab is cut on the slicing machine to the thickness desired after the rind has been removed. Sliced bacon (hotel slice or pack) contains about 20 to 22 slices per pound. Most commercial establishments prefer the hotel pack and are willing to pay a few more cents per pound for convenience.

Bacon is cooked by separating the slices and placing them on a sheet pan fat side down with each slice slightly overlapping the other. The bacon is baked at 350°F until three-quarters done. Remove the bacon from the oven, pour off the grease, and drape the bacon slices over a platter with a kitchen fork or offset spatula. This drains the bacon and keeps the slices from lying in grease until service. This method of cooking is recommended because it reduces shrinkage and curling, improves appearance, and makes the cooking more uniform. Bacon may also be cooked in a skillet on the range, under the broiler, or on a griddle.

Ham is usually purchased cooked in a form that is boneless or boned and rolled. This form gives su-

perior shape and allows the meat to portion into 3 oz or 4 oz pieces. Since the meat is already cooked, it is just a matter of heating it under the broiler, on a griddle, or in a skillet before serving.

Canadian bacon, the boneless, smoked, pressed loin of pork, is popular on breakfast menus. However, it is relatively expensive.

POTATOES

Potatoes are usually served fried at breakfast. Potatoes may be served a la carte or listed with featured breakfast combinations, such as "Two fried eggs with ham, hash brown potatoes, toast & beverage $3.45." Popular breakfast potato preparations are hash brown, home fried, Lyonnaise, and German fried. Recipes for potatoes are found in Chapter 15.

CEREALS

The two types of cereal commonly served in food service establishments are *cold cereal* and *hot cereal*. Cold cereal has been popular in the United States for many years. Most cereals come in a variety of textures and grains, including cornflakes, puffed rice, bran, and shredded wheat. Ready-to-eat cereals are purchased in small, individual-size boxes. The box is served with milk and opened by the customer. This assures a fresh, crisp cereal. Dry cereals can be served with milk, cream, or cereal cream (a blend of half milk and half cream).

Hot cereals are popular as a breakfast item during colder months. With the introduction of individual, instant hot cereals, only hot water has to be added. Hot cereals can be prepared by the guest or in the breakfast station. They are typically prepared by the kitchen staff.

JUICES

Both fruit and vegetable juices are common breakfast menu items. Although they may also be served on the luncheon and dinner menus, fruit and vegetable juices are very popular at breakfast as they stimulate the appetite.

Juices may be purchased fresh, frozen, or canned. Frozen juices should be allowed to stand for a while after mixing. Breakfast juices commonly include grapefruit juice, orange juice, pineapple juice, tomato juice, prune juice, and some mixed or blended juices. A standard serving of juice is a 4 oz glass.

Orange juice is the most popular breakfast juice. It is high in Vitamin C, which is an essential building block for good health. Oranges are picked only when ripe because they do not ripen further after they are picked. They are sized by the number of oranges per box.

Orange juice may be freshly squeezed, mixed from frozen concentrate, or in single-serve containers. Oranges may be stored for up to six weeks between 35°F and 50°F and squeezed as needed. Frozen concentrate, once thawed and mixed, keeps for 10 – 11 days at 34°F. Concentrate is available in cartons, cans, plastic, and bag-in-box packages. Single-serve containers have a shelf life of three weeks at 32°F – 35°F. They are available in glass, plastic cups, cartons, and cans.

FRUITS

Fresh, canned, frozen, and stewed fruits are usually good sellers on any breakfast menu. Grapefruit and oranges are the most popular citrus fruits and are available year-round. Melons, such as cantaloupe, honeydew, Persian, casaba, and crenshaw, are served in halves or wedges, depending on their size. Melons must be ripe and chilled when served.

Canned fruits are served chilled in cocktail glasses just as they come from the can. The most popular canned fruits are pears, peaches, Royal Ann cherries, kadota figs, apricots, and pineapple.

There are many fruit choices on a typical breakfast menu; including fresh, canned or stewed.

Stewed fruits prepared from dried fruits such as apples, apricots, and prunes are also served as standard breakfast items. Dried fruits must be cooked slightly in water to restore moisture.

TOAST

Toasted white bread is served with most egg preparations and as an a la carte item. The bread may be toasted in an automatic toaster, on a grill, or under the broiler. Bread is toasted on both sides, brushed with melted butter, and served with jelly or jam. Although toasted white bread is most popular, cinnamon toast and French toast are also popular. Additionally, toast may be made from wheat, rye, raisin, and sourdough breads.

Cinnamon Toast

25 servings of 3 slices per serving

Ingredients

75 slices white bread
melted butter to cover
1 c sugar
4 tbsp ground cinnamon

Procedure

1. Place the sugar in a shaker, add the ground cinnamon, and blend.
2. Toast the slices of white bread on both sides, three slices per a la carte order.
3. Brush each slice of toast with melted butter on top side.
4. Mix the cinnamon and sugar and sprinkle generously on buttered side of each slice.
5. Place toast on a sheet pan and place under the broiler until the sugar melts slightly.
6. Remove from the broiler, cut the slices diagonally in halves, and serve.

French Toast

25 servings of 2 slices per serving

Ingredients

50 slices white bread
20 eggs
1 qt milk or cream
3 oz sugar
1 tbsp vanilla

Procedure

1. Break the eggs into a stainless steel bowl and beat with a wire whip.
2. Add the milk or cream, sugar, and vanilla, and beat until well blended.
3. Pour the mixture into a hotel pan. Dip each slice of bread into the batter and coat both sides of the bread.
4. Remove the bread from the batter. Let drain slightly.
5. Brown the bread on both sides by placing it in the deep fat fryer at 350°F, in a greased skillet, or on a hot buttered griddle.
6. Serve two pieces to each order with desired syrup.

BREAKFAST BUFFET

The breakfast buffet is an excellent way to increase breakfast business. It is designed, like the regular buffet, to display foods in an appetizing manner to stimulate the appetite. In this case, the foods displayed are of the breakfast variety and fairly easy to digest. Many of the popular breakfast preparations, such as scrambled eggs, hash brown potatoes, ham, pancakes, and French toast contain colors that can be very attractive when arranged properly. The secret of a successful breakfast buffet is to offer a variety of foods that are well prepared, arranged to take full advantage of the natural colors, and attractively priced.

PASTRIES

Pastries are featured on the breakfast menu. The most popular types of pastries are sweet rolls, doughnuts, coffee cakes, and Danish pastry. If possible, always serve sweet rolls and Danish pastries warm. Some food service establishments that have their own bakeshops take great pride in producing pastries in their own baking facilities. However, in recent years, many food service establishments have purchased these items from commercial bakeries to reduce costs.

American METALCRAFT. Inc

Sweet rolls have different fruit topping.

CONTINENTAL BREAKFAST

The Continental breakfast, made popular in European countries, is growing in popularity in the United States. A Continental breakfast is a light breakfast consisting of fruit or juice, toast or pastry, and coffee. No heavy cooking is required. Continental breakfasts do not provide the balanced nutrition of more complete breakfasts, but they conserve labor costs.

Burgers' Smokehouse

American Egg Board

Breakfast can be as varied as fried eggs, country ham, and biscuits or breakfast burritos with tomato-basil topping.

Batter Cooking

Batter cooking involves preparing foods using a thick liquid mixture consisting primarily of eggs, milk, and flour. Cooking batters differ from baking batters. Cooking batters acquire their shape from the food item or the cooking surface. Baking batters acquire their shape from the form in which they are baked, such as muffins and corn sticks.

Pancakes and waffles use similar batters but different cooking methods. Crepes require a thinner batter than pancakes and waffles. Crepes are cooked in a shallow pan with sloping sides.

Batter-dipped foods include any food item that can be breaded or coated before cooking. The breading or coating protects the food item from overcooking around the edges.

Fritters are small cakes of batter with fruit, vegetables, or meat added and are deep-fried. Dumplings are small dough preparations that require a dough similar to the batter used in batter cooking. Batter-cooked foods can be found on breakfast, luncheon, and dinner menus.

BATTER

Batter is a thick liquid mixture consisting mainly of eggs, milk, and flour. It is thinner than dough because it contains a higher percentage of liquid (usually milk). Batter can be stirred, poured, or dropped from a spoon. The two basic types of batter are baking batters and cooking batters. Baking batters usually acquire their shape by the form of container in which they are baked, such as muffins, corn sticks, and popovers. These products are classified as quickbreads. Cooking batters are used in pancakes and waffles, batter dips, and fritter batters. Batter foods are versatile, present excellent eye appeal, and cost little to prepare. In addition, batter foods can be prepared quickly and return a high profit.

Pancakes and Waffles

Pancakes are the most popular of the batter preparations. Many countries in the world have created and popularized their own version of pancakes. For example:

- France – crepes
- Germany – potato pancakes
- Russia – blintzes and blinis
- Sweden – plattar (egg pancakes)

For best results, pancakes should always be cooked to order. However, if necessary, pancakes may be frozen and reheated in the microwave oven. Pancakes are usually served with butter and jam, jelly, or syrup. Maple syrup is most popular. Fruit or fruit-flavored syrups are served for variety.

Waffle batters are similar to pancake batters but contain more eggs, fat, and sugar and use melted butter instead of salad oil. Both pancakes and waffles are cooked on a hot iron. Pancakes are made on a seasoned griddle and waffles on a waffle iron that shapes the waffles. Waffles are usually served the way pancakes are served, with butter and jam, jelly, or syrup.

A number of quality mixes are available on the market that will produce both excellent pancake and waffle batters. These mixes are convenient because most only call for the addition of water, milk, or both. If a richer batter is desired, eggs may also be added. Pancake and waffle mixes are usually more expensive than preparing from scratch.

Crepes

Crepes are the most delicate and versatile of all pancakes. The crepe is thin and smooth-textured. The batter contains more egg than pancakes or waffles. It will not break and can be easily rolled or folded. This allows the crepe to be filled or stuffed with a variety of fillings and served as an appetizer, entree, or dessert.

Crepe batter is simple to prepare but requires careful mixing. If the batter is overbeaten, the crepes will be tough. Most cooks use a well-conditioned (seasoned) steel egg skillet or crepe pan for cooking crepes. These pans have a diameter of 6″ or 7″ and sloping sides, which make it easy to flip the crepes over and turn them out. Crepe batter should be made ahead and allowed to stand.

When crepes appear on the menu, mass-production techniques are required to keep up with the incoming orders. More than one skillet should be used. A good crepe cook can handle three or four skillets at one time. Crepes can also be cooked ahead of time, as much as a day ahead if necessary. As the crepes are turned out of the skillet, they are placed on a sheet pan or on wax paper, and each layer is separated by a sheet of wax paper. If they are not set up and served immediately, they should be covered with a cloth and refrigerated until needed.

Crepes may be rolled or folded in various ways. The type of roll or fold selected depends on the consistency of the filling. Always attempt to place the most attractive side of the crepe on the outside.

Fold and *roll crepes*: Spread the filling over the surface of the crepe to within ½″ of the edge. Fold sides of the crepe over the filling to create straight sides, then roll up from bottom to top. This method is preferred if the crepe is to be fried.

Rolled crepes: Spread the filling over the lower half of the crepe. Start at the bottom and roll away from the body toward the top of the crepe. This method is good for both thin and thick fillings.

Folded crepes: Place the filling in the center of the crepe, fold the crepe in half, then in half a second time, forming a four-layer triangle. This method is recommended for thin fillings and is used when preparing crepe suzettes.

Fold-over crepes: Place the filling in the center of the crepe, fold one side over to cover the filling, then fold the second side over to overlap the first fold. This method is excellent for thick fillings.

One-fold crepes: This method is used when two crepes are served in a casserole. One crepe is filled with the casserole filling, folded over, and placed in the casserole. The second crepe is placed on the opposite side of the first, and the same procedure is followed. When completed, the two filled half crepes in the casserole will present an eye-appealing entree.

Crepe preparations are often served aflame or flambéed. This is done to add flavor and create a unique effect. Flaming is a fairly simple procedure if the proper steps are taken. The crepes and liqueur must be warm or hot, but never boiling. If boiled, the alcohol will evaporate. The crepes and liqueur are usually heated before the guest in a blazer pan, then ignited. However, the liqueur may also be heated in a separate pan and poured over the warm crepes before igniting. Never add more liqueur when the crepes are aflame. Allow the flame to burn out before serving.

Potato Pancakes

Potato pancakes originated in Germany and are also popular in Jewish cuisine. Potato pancakes contain eggs and flour, much like regular pancakes, but milk is eliminated because of the large amount of moisture that is contained in the grated raw potatoes and onions. The secret of a successful potato pancake is crispness. This can be obtained by eliminating as much moisture as possible from the grated raw potatoes and onions. Fry potato pancakes as close to serving time as possible in shallow grease using a heavy frying pan. If left to set too long, they usually become greasy. Make the batter as close to frying time as possible. If made ahead of time, keep the batter covered and refrigerated until ready to use. Potato pancakes can be served as an accompaniment to other preparations, such as sauerbraten, pot roast, or braised beef rouladen. Potato pancakes may also be served with applesauce or lingonberry preserves.

Blintzes

Blintzes are the Russian form of pancakes. In Russia, blintzes are often filled with sour cream and served with caviar. In the United States, a cottage cheese filling is most popular. After preparing the small thin pancakes, the filling selected is placed in the center of each pancake. Each side is folded onto the center and then folded in half. After folding the blintzes, they are sautéed in butter until both sides are golden brown. Blintzes are commonly served hot with sour cream, jelly, or jam, or sprinkled with cinnamon and sugar. A blinnis are similiar to blintzes only they are smaller in size.

Swedish Pancakes

Swedish pancakes (plattar), or egg pancakes, are Sweden's favorite batter preparation. The batter differs from American pancake batter in the amount of eggs used in preparing the batter, thus the name "egg pancakes." Swedish pancakes have a smaller diameter than hotcakes and are served with lingonberries or other preserves.

Breaded and Batter-Dipped Foods

Most deep-fried foods are breaded or coated with batter before frying to protect the item from the hot fat. Breading also adds to the flavor and appearance of the item. Because the item being deep fried will absorb fat, the fat used should always be clean. Use high-quality shortening and strain the fat after each frying period to extend the life of the fat. Unstrained fat contains pieces of breading or batter that has fallen off the food. These pieces will eventually burn and shorten the life of the fat.

Breading techniques:

1. Foods to be deep fried should always be uniform in size and almost dry.
2. Pass the item to be fried through a mixture of flour, salt, and white pepper. Press the item into the flour so it will adhere tightly.
3. Remove the item from the flour and dip it into an egg wash (equal parts of beaten egg and milk).

Coat it thoroughly using only one hand; the other hand is kept dry.

4. Remove the item from the egg wash, letting the excess drip off. Place the item into fine bread crumbs, crackermeal, or any other dry mixture. The mixture used depends on the time required for the food to fry. For example, when the browning is to be slow, use crackermeal only. It does not brown as quickly as the other mixtures. Press the item into the crumbs with a fair amount of force using the dry hand so the crumbs adhere tightly to the item and do not drop off during the frying period. Never place the wet hand that removed the item from the batter into the crumb mixture. This would create lumps and also coat the fingers.
5. The item is now ready to be deep-fried, or it may be held in the refrigerator in its breaded state until it is needed. This is one advantage breaded deep-fried foods have over batter-fried foods. Batter-fried foods cannot be dipped in advance. Once dipped, they must be fried immediately. When using purchased battered or breaded foods, keep the food frozen before cooking to prevent batter from falling off.

Anchor Food Products, Inc.

Deep fried food items are breaded or coated with batter to prevent overcooking of the outer edges while cooking in the deep fat fryer.

 **Trade Tip** *Battered and breaded foods contain raw eggs so the amount of time the food is in the danger zone should be monitored.*

Batter techniques:

Batter dipping foods before they are deep fried can be a messy job if the proper techniques are not followed.

1. It is best to coat the item with flour before placing it in the batter. The coating of flour seals the item dipped into the batter. The batter then adheres more readily to the item and will not fall off as easily while frying.

2. Before removing the item from the batter, place the container holding the item and the batter as close to the deep fat fryer as possible. Remove the item from the batter by grasping it with the thumb and fingertips and inverting the hand to form a cup (palm up). The excess batter will drip into the cupped hand.

3. Once the item is removed from the batter, place it quickly into the hot fat. Avoid splashing the hot fat; serious burns could result.

4. Leftover batter should be discarded.

5. Use separate batter for each food to prevent cross-contamination.

Procter and Gamble Co.

Before onion rings are breaded, they are placed in ice water to prevent loss of flavor and juice.

Fritter Batters

Fritters are small cakes of batter that are fried to a golden brown in deep fat, very much like a doughnut.

They may contain fruit, vegetables, meat, or seafood. The most popular fritters prepared in the commercial kitchen are apple fritters and corn fritters. Fritters may be served as hors d'oeuvres, entrees, desserts, or as an accompaniment to other foods.

Fritters are usually prepared in one of two ways. The item is dipped in a basic fritter batter, fried to a golden brown, garnished, and served. The other method is using a batter mixture containing a leavening agent. The fruit, vegetable, meat, or seafood is mixed into the batter mixture. This mixture is then scooped out using a small scoop and deep fried to a golden brown.

Pan-Fried Fritters

Pan-fried fritters are prepared as a batter mixture, but instead of frying in deep fat they are pan fried. Using a kitchen spoon approximately half full, the batter is placed in a sauté pan containing a small amount of melted shortening. Deposit enough batter for each fritter to form a cake approximately 3″ in diameter. Each side is cooked to a golden brown. The fritters are then removed from the skillet, drained, and served. The fritters may also be cooked on a hot greased griddle using the same method used for pan frying.

DUMPLINGS

Dumplings are small starch preparations made from a soft dough or batter that is cooked by simmering or steaming. When cooked, they have a soft, light, spongy texture, making them an ideal preparation to serve with stews and fricassees. Dumplings contain ingredients used in batter preparations. Certain dumpling recipes require a dough consistency similar to a batter.

Dumplings are usually cooked by simmering in stock or steaming in the pressure steamer. The best results can be obtained by steaming. If the dumplings are steamed, they retain a light, fluffy texture longer.

Trade Tip

For onion rings, the batter must be thick, but not so thick that it will mask the flavor of the onion. The batter should be kept on ice during use. The oil for frying onion rings should be 375°F, frying a few at a time to maintain the temperature. Fry onion rings until light brown, about 1 min.

PANCAKES, WAFFLES, AND CREPES

Plain Pancakes or Hotcakes

1 gal. or 45 cakes

Ingredients

8 eggs
2 qt milk
12 oz salad oil
2 lb flour
5 oz sugar
1½ oz baking powder

Procedure

1. Place the eggs in the stainless steel mixing bowl. Mix in the mixing machine at slow speed using the paddle for approximately 1 min.

2. Add the milk and salad oil. Continue to mix at slow speed until incorporated.
3. Place the dry ingredients in a separate container, sift twice, and gradually add to the liquid mixture. Mix at slow speed for approximately 1 min until the batter is fairly smooth.
4. Remove the batter from the mixer and place in a stainless steel container. Let set for at least 15 min.
5. Heat the seasoned griddle to 375°F. Grease slightly.
6. Using a 3 oz ladle, spot the batter on the preheated griddle. The batter should spread to approximately 5″ in diameter.
7. Brown one side, turn using an offset spatula, and brown second side.
8. Serve three cakes with syrup to each order.

Buttermilk Pancakes

1 gal. or 45 cakes

Ingredients

6 eggs
2 qt cultured buttermilk
2 lb flour
12 oz salad oil
3 tsp salt
1 oz baking powder
3 oz sugar
3 tsp baking soda

Procedure

1. Place the eggs in the stainless steel mixing bowl. Mix in the mixing machine at slow speed using the paddle for approximately 1 min.

2. Add the buttermilk and salad oil. Continue to mix at slow speed until incorporated.
3. Place the dry ingredients in a separate container, sift twice, and gradually add to the liquid mixture while continuing to mix at slow speed. Mix until the batter is fairly smooth.
4. Remove the batter from the mixer and place in a stainless steel container. Let set for at least 15 min.
5. Heat the seasoned griddle to 375°F. Grease slightly.
6. Using a 3 oz ladle, spot the batter on the preheated griddle. The batter should spread to approximately 5″ in diameter.
7. Brown one side, turn using an offset spatula, and brown second side.
8. Serve three cakes with syrup to each order.

Griddle Cake Dry Mix

12 lb

Ingredients

9 lb cake flour
8 oz baking powder
2 oz salt
1 lb 8 oz sugar
1 lb 8 oz nonfat dry milk

Ingredients	30 Cakes	50 Cakes	100 Cakes	200 Cakes
Dry mix	2 lb	3 lb	6 lb	12 lb
Whole eggs	4	6	12	24
Water	1 qt	1½ qt	3 qt	1½ gal.
Fat, melted	4 oz	6 oz	12 oz	1 lb 8 oz

Preparation

- Blend the dry ingredients and store in a dry place until ready to prepare the batter. Follow the chart when preparing the batter.

Procedure

1. Using the stainless steel mixing bowl and paddle, combine all ingredients and mix until a smooth batter is formed. When cooking the cakes, follow the same procedure used for plain and buttermilk pancakes.

German Apple Pancakes

2 qt

Ingredients

1 lb sifted flour
½ oz salt
3 oz sugar
3 c milk
8 eggs
5 large apples
¼ c lemon juice
1 pinch grated nutmeg
1 tsp ground cinnamon

Procedure

1. Peel and julienne the apples. Place them in the lemon juice and toss gently.
2. Place the flour, salt, sugar, and spices in a stainless steel mixing bowl. Blend with a wire whip.
3. Break the eggs in a separate stainless steel mixing bowl. Beat slightly with a wire whip. Add the milk and continue to whip until thoroughly incorporated.
4. Pour the liquid mixture into the dry ingredients while working the whip vigorously. Whip until a smooth batter is formed.
5. Using a kitchen spoon, fold the apples and juice into the batter gently so the apples are not broken.
6. Place a small- or medium-sized sauté pan on the range. Add a small amount of butter or margarine and heat slightly.
7. Add enough batter to coat the surface of the pan. Cook until the bottom surface is light brown. Flip the cake and brown the other side slightly.
8. Sprinkle the surface of each cake with cinnamon-flavored sugar. Serve three pancakes with syrup to each order.

Note: Two ounces of soaked, drained raisins may be added to the batter if desired.

Crepes (French Pancakes)

½ gal. or 50 crepes

Ingredients

1 qt milk
10 oz flour
8 eggs
½ tsp salt
3 oz melted butter or margarine
4 oz sugar

Procedure

1. Place the eggs in a stainless steel mixing bowl. Using a wire whip, mix by hand until whites and yolks are blended.
2. Add the milk and continue to mix until thoroughly blended with the eggs.
3. Combine the dry ingredients, sift twice, and gradually add to the liquid mixture. Mix until the dry ingredients are incorporated into the liquid mixture and the batter is lump-free.
4. Add the melted butter and mix until blended into the batter.
5. Pour the batter into a plastic or stainless steel container. Refrigerate until ready to use.
6. Coat the surface of a seasoned egg skillet, omelet skillet, or non-stick egg skillet with melted butter or shortening and heat slightly.
7. Using a 2 oz ladle, coat the bottom of the skillet with crepe batter. While pouring the batter into the skillet, quickly tilt and rotate the skillet so the batter runs to the edges and coats the bottom thinly and evenly.
8. Place the skillet on the range and cook one side, then flip or turn by hand and cook the other side. Each side should be browned slightly.
9. Remove the crepes from the skillet and place on sheet pans covered with wax paper. Place a sheet of wax paper between each layer.
10. Fill with desired filling, roll or fold, and serve. If crepes are not served immediately, cover with a cloth and refrigerate.

Bintzes (Russian Pancakes)

½ gal.

Ingredients

12 eggs
1 qt milk
1 lb flour
3 oz sugar
5 oz melted butter
⅛ oz salt

Procedure

1. Place the eggs in a stainless steel mixing bowl. Using a wire whip, mix by hand until whites and yolks are blended.
2. Add the milk and continue to mix until thoroughly blended with the eggs.
3. Combine the dry ingredients, sift twice, and gradually add to the liquid mixture. Mix until the dry ingredients are incorporated into the liquid mixture and the batter is lump-free.

4. Add the melted butter and mix until blended into the batter.
5. Pour the batter into a plastic or stainless steel container. Refrigerate until ready to use.
6. Coat the surface of a 6″ seasoned egg skillet, omelet skillet, or non-stick egg skillet with melted butter or shortening and heat slightly.
7. Using a 2 oz ladle, pour enough batter into the skillet to cover the bottom of the pan, tipping the pan quickly in a circular motion to spread the batter evenly. The thinner the coating of batter is, the better the blintzes will be.

8. Hold the pan over a medium heat and let the cakes cook on one side only until they hold their shape without sticking to the pan. Avoid letting the cakes brown.
9. Remove the cakes from the pan and place them cooked side up on a sheet pan that has been covered with a damp cloth.
10. Continue to fry the pancakes until all the batter is used. Hold the cooked pancakes until the filling is prepared and ready for use.

Blinis: Prepare batter the same as for blintzes, but make in smaller 1″-2″ diameter form.

Cheese Filling

Ingredients

3 lb small curd cottage cheese
4 eggs
2 tbsp sugar
1 c raisins
2 grated lemon rinds
cinnamon to taste

Procedure

1. Put the cottage cheese in a china cap and force out as much liquid as possible.

25 servings

2. Place the eggs in a stainless steel bowl. Using a wire whip, whip slightly.
3. Add the remaining ingredients and mix using a kitchen spoon until thoroughly blended.
4. Place 1 tbsp of the filling in the center of each pancake, fold each side over onto the center, then fold in half.
5. Place butter or margarine in a heavy-bottom skillet and heat slightly.
6. Add the folded blintzes and fry quickly until golden brown on both sides.
7. Serve with sour cream, jam, or jelly, or dust with cinnamon and sugar.

Potato Pancakes (Latkes)

Ingredients

8 lb peeled red potatoes
10 oz onions
8 eggs
8 oz flour (variable)
1 oz salt
¼ c chopped parsley
pepper to taste

Procedure

1. Grate or grind the potatoes and onions. Strain off all liquid using a china cap.

25 servings (75 pancakes)

2. Beat the eggs slightly and blend into the potato/onion mixture.
3. Add the remaining ingredients and blend well.
4. Cover the bottom of an iron skillet with approximately ½″ of salad oil or shortening and heat.
5. Fill a kitchen spoon with the potato mixture and place the mixture in the shallow grease. Repeat this process until the skillet is filled.
6. Brown one side of each pancake, turn, and brown the other side.
7. Remove the pancakes from the skillet and allow to drain.
8. Serve three pancakes with applesauce, apple butter, or lingonberry preserves to each order.

Swedish Pancakes (Plattar) or Egg Pancakes

Ingredients

8 eggs
1 qt milk
1 lb flour
5 oz sugar
¼ oz salt
3 oz melted butter

½ gal.

Procedure

1. Place the eggs in the stainless steel mixing bowl. Using the wire whip, mix at medium speed for approximately 1 min.
2. Add the milk and continue to mix until incorporated with the eggs.
3. Place the dry ingredients in a separate container, sift twice, and gradually add to the liquid mixture. Mix in the mixing machine at slow speed for approximately 1 min or until the batter is fairly smooth.

4. Add the melted butter and mix until incorporated.
5. Remove the batter from the mixer and place in a stainless steel or plastic container.
6. Place in the refrigerator for 2 hr before using.
7. Heat the seasoned griddle to 375°F. Grease slightly.
8. Using a 1 oz or 2 oz ladle, spot the batter on the preheated griddle. Cakes should spread to approximately 3″ in diameter.

9. Brown one side, turn using an offset spatula, and brown the other side.
10. To serve, arrange six small pancakes overlapping each other on a plate. Place lingonberries or fruit preserves in the center. Dust with powdered sugar or place lingonberries or fruit preserves in the center of each cake and fold in half.

Corn Griddle Cakes (Johnny Cakes)

50 cakes

Ingredients

1 lb 12 oz flour
2¼ oz baking powder
½ oz salt
8 oz sugar
1 lb 10 oz yellow cornmeal
6 oz butter or shortening
12 oz eggs
2¼ qt milk

Procedure

1. Sift the flour, baking powder, salt, and sugar into the stainless steel mixing bowl.
2. Cut in the cornmeal and butter or shortening and mix in the mixing machine at slow speed using the paddle.
3. Combine the eggs and milk in a separate container. Add gradually to the flour mixture while continuing to mix at slow speed.
4. Mix until a fairly smooth batter is formed.
5. Remove the batter from the mixer and place in a stainless steel container. Let set for at least 30 min.
6. Using a 3 oz ladle, spot the batter on the preheated griddle. The cakes should spread slightly.
7. Brown one side until firm around the edges and full of bubbles. Turn and brown the other side.
8. Serve three cakes with syrup to each order.

Waffles

½ gal.

Ingredients

6 eggs
1 qt milk
1 lb 6 oz cake flour
1 oz baking powder
4 oz sugar
⅛ oz salt
8 oz melted butter

Procedure

1. Place the eggs in a mixing bowl. Mix in the mixing machine at medium speed for approximately 1 min using the paddle.
2. Add the milk and continue to mix until incorporated with the eggs.

3. Combine the dry ingredients, sift two times, and add to the egg/milk mixture. Mix for at least 1 min.
4. Add the melted butter and mix until well blended.
5. Pour the batter into a stainless steel or plastic container and place in the refrigerator until ready to use.
6. Brush the top and bottom of the waffle iron with salad oil. Heat to approximately 375°F.
7. Pour enough batter on the waffle grid to barely cover it. The amount used will depend on the size and shape of the waffle iron.
8. Let the waffle cook for about 1 min before lowering the top of the iron. When the top is lowered, cook about 1½ min to 2 min longer.
9. Serve two waffles with jam, jelly, syrup, or fruit to each order.

Crisp Golden Waffles

½ gal.

Ingredients

6 eggs
2 oz sugar
½ oz salt
1 lb cake flour
1 lb 4 oz milk
8 oz salad oil

1 oz baking powder

Procedure

1. Place the eggs, sugar, and salt in a mixing bowl. Mix in the mixing machine at slow speed using the paddle until blended.
2. Add the milk and flour alternately while continuing to mix at slow speed. Mix until smooth.

3. Add the salad oil and baking powder. Mix the batter until incorporated.
4. Pour the batter into a stainless steel or plastic container and refrigerate until ready to use.
5. Brush the top and bottom of the waffle iron with salad oil. Heat to approximately 375°F.
6. Pour enough batter on the waffle grid to barely cover the surface. The amount used depends on the size and shape of the waffle iron.
7. Let the waffle cook for about 1 min before lowering the top of the iron. When the top is lowered, cook about 1½ min to 2 min longer or until signal light indicates they are done, if iron is so equipped.
8. Serve two waffles with jam, jelly, syrup, or fruit to each order.

Variations

Bacon Waffles: Cut 1 lb of bacon into strips crosswise, place in a skillet, and cook until slightly crisp. Drain off the grease. Cover the treated waffle iron with batter. Immediately sprinkle a little of the cooked bacon on the uncooked batter, close the iron, and cook until waffle is done.

Cheese Waffles: Add 8 oz of grated cheddar cheese to the batter and mix thoroughly. Prepare the waffles using the same method used for bacon waffles.

Pecan Waffles: Sprinkle chopped pecans on the waffle batter before the iron is closed. Prepare the waffles using the same method used for bacon waffles.

Ham Waffles: Add 8 oz of finely diced ham to the batter and mix thoroughly. Prepare the waffles using the same method used for bacon waffles.

Sausage Waffles: Dice 8 oz of sausage (little pigs, hamilton mett, pork sausage, etc.). If the sausage is uncooked, cook in the same manner suggested for the bacon. Add the cooked sausage to the waffle batter and mix thoroughly. Prepare the waffles using the same method used for bacon waffles.

BATTER-DIPPED RECIPES

Onion Ring Batter

Ingredients

8 eggs
2 lb cake flour
2 tbsp baking powder
1½ tsp salt
1 qt milk

Procedure

1. Break the eggs into a stainless steel bowl and beat slightly with a wire whip.
2. Add the milk and continue to whip until eggs and milk are blended.
3. Sift the flour, baking powder, and salt in a separate bowl.
4. Add the dry ingredients slowly to the liquid mixture while whipping briskly with a wire whip.
5. When all the dry ingredients are added, continue to whip until a smooth batter is formed.
6. Peel onions and slice to desired thickness. Separate into rings and place in ice water to prevent them from losing flavor and juice.
7. Coat with batter and fry at 350°F to 375°F.

2 qt

Procter and Gamble Co.

Trade Tip

Batter foods such as pancakes, waffles, and crepes expand the breakfast menu for minimal cost and labor. By adding fruit and nuts to the batter foods, the menu can be expanded even further. Mixes are convenient and easy to use, although they are more costly than mixing from scratch.

Fish and Chicken Batter

Ingredients

8 oz (1 c) eggs
1 lb 2 oz milk
1 lb flour
¼ oz salt

Procedure

1. Break the eggs into a stainless steel bowl and beat slightly with a wire whip.
2. Add the milk and continue to beat.
3. Combine the flour and salt. Sift into the egg/milk mixture while whipping vigorously with a wire whip.
4. Whip until a smooth batter is formed. Let batter set for at least 1 hr before using.
5. Cut fish and chicken as desired. Coat and fry at 350°F.

Southern Pride Catfish

Beer Batter

Ingredients

1 lb 12 oz flour
1½ oz baking powder
½ oz sugar
½ oz salt
6 eggs
1 oz salad oil
12 oz beer

Procedure

1. Place the flour, baking powder, sugar, and salt in a bowl. Sift twice.
2. Place the eggs in the stainless steel mixing bowl. Beat slightly at second speed using the wire whip.
3. Reduce mixing speed to slow and add the salad oil and beer. Continue to mix until thoroughly incorporated.
4. Gradually add the sifted dry ingredients while continuing to mix at slow speed. Mix until a smooth batter is formed.
5. Cut or set up items to be fried. Coat and fry at 350°F.

Shrimp Batter

Ingredients

5 eggs
1 pt milk
1 tbsp prepared mustard
3 tbsp Worcestershire sauce
¼ c molasses
1 minced garlic clove
3 c flour
1 tbsp baking powder
1 tbsp celery salt
1 tbsp paprika

Procedure

1. Break the eggs into a stainless steel bowl and beat slightly with a wire whip.
2. Add the milk, mustard, Worcestershire sauce, and molasses. Blend with the wire whip.
3. Sift the flour, baking powder, celery salt, and paprika. Add gradually to the liquid mixture and continue to whip until a smooth batter is formed.
4. Lightly flour cleaned, raw shrimp and place into the batter. Fry at 350°F.

FRITTER BATTER RECIPES

Basic Fritter Batter No.1

2 qt

Ingredients

6 eggs
1 qt (2 lb) milk
1 c (8 oz) cooking oil
2 lb 4 oz flour
1 tsp salt
4 tsp sugar

Procedure

1. Place the eggs in a stainless steel bowl. Beat slightly with a wire whip.

2. Add the milk and continue to beat.
3. Sift in part of the flour and mix thoroughly.
4. Add the salt, sugar, and cooking oil. Blend thoroughly.
5. Sift in the remaining flour. Mix to a smooth batter, cover, and let set for at least 30 min before using.

Note: If a sweeter batter is desired, more sugar may be added. If a thicker batter is desired, more flour may be added. This basic batter is best when used for fruit or vegetable fritters. If fruit is used, pass the fruit through flour before placing it in the batter. Fry fritters at 350°F.

Basic Fritter Batter No. 2

1½ qt

Ingredients

8 eggs
2 tsp salt
2½ oz sugar
10 oz milk
1 oz baking powder
1 lb 6 oz sifted flour

Procedure

1. Place the eggs in a stainless steel bowl. Beat slightly with a wire whip.
2. Add the milk and continue to beat.
3. Add the salt and sugar. Blend thoroughly.

4. Add the sifted flour and baking powder and mix to a smooth batter.

Variations

Add 8 oz to the basic batter mixture:
Frozen whole kernel corn, thawed
Frozen blueberries, thawed
Cooked carrots, diced
Canned pineapple, diced
Cooked ham, diced
Cooked crisp bacon, crumbled
Cooked pork sausage, diced
Canned peaches, diced

Clam Fritters

36 fritters

Ingredients

3 c flour, sifted
¾ tsp salt
1½ tsp sugar
1 tbsp baking powder
3 eggs, beaten
¾ c milk
¾ c butter, melted
3 c canned clams, drained, chopped

Procedure

1. Place the eggs in the stainless steel mixing bowl. Mix slightly at slow speed using the paddle.
2. Add the milk and continue to mix until blended with the eggs.
3. Sift the flour, salt, sugar, pepper, and baking powder in a separate container. Add this dry mixture gradually to the liquid mixture in the mixing bowl. Mix until thoroughly blended and a fairly smooth batter forms.
4. Add the chopped clams. Mix until incorporated into the batter.
5. Drop the batter from a No. 24 scoop into deep grease and fry at 350°F until golden brown.

Chopped Apple Fritters

50 fritters

Ingredients

1 qt (2 lb) milk
6 eggs
4 oz melted butter or margarine
2 lb flour
1 tsp salt
2 oz baking powder
½ oz sugar
12 oz peeled, finely chopped apples
cinnamon and sugar, as needed

Procedure

1. Place the eggs in the stainless steel mixing bowl. Mix slightly at second speed using the paddle.
2. Add the milk and melted butter. Reduce the mixer speed to slow.
3. Sift the flour, salt, baking powder, and sugar in a separate container. Add the dry mixture to the egg/milk mixture gradually while continuing to mix at slow speed.
4. Add the chopped apples. Mix only until incorporated.
5. Drop the batter from a No. 24 scoop into deep grease and fry at 350°F until golden brown.
6. Remove from the deep grease and let drain.
7. Pass each fritter through a mixture of cinnamon and sugar. Serve three fritters with crisp bacon and brandy sauce to each order.

Corn Fritters

50 fritters

Ingredients

12 separated eggs
4 c drained whole kernel corn
3 c creamed corn
1 c milk
1 lb 14 oz cake flour
2 tbsp baking powder
½ tsp salt

Procedure

1. Place the egg yolks, corn, flour, milk, and baking powder in the stainless steel mixing bowl. Mix in the mixing machine at slow speed using the paddle until a batter is formed.
2. Remove the batter from the mixing bowl and place in a stainless steel bowl. Clean the mixing bowl thoroughly.
3. Place the egg whites in the mixing bowl. Whip in the mixing machine at high speed using the wire whip until the whites start to peak.
4. Add the salt and continue to whip until stiff peaks form. Remove from the mixer.
5. Fold the beaten egg whites gently into the batter with a rubber spatula or small skimmer.
6. Drop the batter from a No. 24 scoop into deep grease and fry at 350°F until golden brown.
7. Remove from the grease and let drain.
8. Serve three fritters with crisp bacon and maple syrup to each order or serve as an accompaniment with other foods.

Chicken Fritters

50 fritters

Ingredients

8 eggs
10 oz milk
1 lb 6 oz flour
¼ oz salt
¼ oz sugar
1 oz baking powder
1 pinch pepper
8 oz cooked, minced chicken or turkey
¼ oz grated onion

Procedure

1. Place the eggs in the stainless steel mixing bowl. Mix slightly at slow speed using the paddle.
2. Add the milk and continue to mix until blended with the eggs.
3. Sift the flour, salt, sugar, pepper, and baking powder in a separate container. Add this dry mixture gradually to the liquid mixture in the mixing bowl. Mix until thoroughly blended and a fairly smooth batter forms.
4. Add the minced chicken or turkey and onion. Mix until incorporated into the batter.
5. Drop the batter from a No. 24 scoop into deep grease and fry at 350°F until golden brown.
6. Serve as a hot hors d'oeuvre or as an accompaniment with other foods.

Note: For chicken/bacon fritters, use 5 oz of cooked, minced chicken or turkey and 3 oz of crisp, chopped bacon.

Ham Fritters

50 fritters

Ingredients

6 eggs
¼ oz salt
1 lb pastry flour
4 oz cornmeal
¼ oz baking powder
3 c milk
1 oz salad oil
1 lb ground ham
1 lb cooked rice

Procedure

1. Place the eggs in the stainless steel mixing bowl. Beat slightly using the paddle.
2. Sift the flour, cornmeal, baking powder, and salt in a separate container.
3. Add the milk and sifted dry ingredients alternately to the eggs in the stainless steel mixing bowl while mixing at slow speed. Mix until batter is smooth.
4. Add the salad oil and mix until incorporated.
5. Add the ham and rice. Continue to mix at slow speed until blended into the batter.
6. Place enough shortening in a large sauté pan to cover the bottom of the pan ¼". Place on the range and heat slightly.
7. Deposit thin patties, approximately 3" in diameter, in the sauté pan using a kitchen spoon. Fry until golden brown, turn, and brown the other side.
8. Remove from the sauté pan and drain. Serve three fritters with an appropriate sauce to each order or serve as an accompaniment with other foods.

Variations

Cheese Fritter: Substitute 8 oz of grated American cheese for the ground ham.
Salmon Fritter: Substitute 10 oz of flaked salmon for the ground ham.
Tuna Fritter: Substitute 10 oz of flaked tuna for the ground ham.

Yorkshire Pudding

50 servings

Ingredients

2 lb (1 qt) eggs
2 qt milk
2 lb bread flour
½ oz salt
1 lb melted butter or margarine
1 pt roast beef drippings

Procedure

1. Place the eggs in the stainless steel mixing bowl. Mix in the mixing machine at slow speed using the paddle.
2. Add the milk and mix until blended with the eggs.
3. Add the flour and salt gradually while continuing to mix at slow speed.
4. Add the melted butter or margarine and mix until blended.
5. Divide the roast beef drippings into two 15" pans. Place the pans on the bottom shelf of a hot oven (375°F). When the drippings start to smoke, add half of the batter to each pan.
6. When done, remove from oven and drain off excess grease. Cut the contents of each pan into 25 portions and serve with roast rib of beef.

Hush Puppies

100 hush puppies

Ingredients

2 lb yellow cornmeal
1 lb flour
12 oz milk
4 oz baking powder
¼ oz salt
2 oz bacon grease
1 lb diced onions
8 oz eggs
1¾ oz sugar
½ minced garlic clove
black pepper to taste

Procedure

1. Mix all the dry ingredients in a stainless steel mixing bowl.
2. Add the bacon grease.
3. Add the onions and garlic.
4. In a separate stainless steel bowl, whip the eggs slightly with a wire whip.
5. Add the milk to the eggs and blend.
6. Add the dry ingredients to the liquid mixture. Mix thoroughly to form a batter dough.
7. Form into fairly small round balls weighing approximately ¾ oz to 1 oz each. Fry in deep fat at 325°F until golden brown.

DUMPLINGS

Steamed Dumplings

80 dumplings

Ingredients

1 lb 14 oz bread flour
1 oz baking powder
¾ oz salt
9 oz shortening
14 oz eggs
14 oz milk
4 oz minced ham (optional)
2 oz minced chives

Procedure

1. Sift the flour, baking powder, and salt into a stainless steel bowl.
2. Add the shortening and cut into the dry ingredients by hand. Add the chives and minced ham.
3. Place the eggs in a second stainless steel bowl. Beat slightly with a wire whip.
4. Add the milk and blend with the eggs.
5. Pour the liquid mixture into the flour/shortening mixture. Mix by hand to form a soft dough or batter.
6. Turn the dough out onto a floured bench, cover with a cloth, and let rest 10 min.
7. Roll the dough out with a rolling pin to a thickness of approximately ½″.
8. Using a small diameter biscuit cutter, cut the dumplings into a half moon shape by filling only half the cutter with the dough or batter.
9. Place the dumpling on half sheet pans greased lightly with butter.
10. Place in the steamer and steam for 12 min to 15 min. Remove from the steamer and hold in a buttered steam table pan on the steam table until ready to serve.
11. Serve two dumplings with any type of stew or fricassee to each order.

Appetizers

## Chapter 9

Anchor Food Products, Inc.

Appetizers, usually a small amount of food, are the first course of a meal. They stimulate the appetite and prepare the customer for the courses of food to follow. Appetizers must be appealing in flavor and appearance. The flavor is the product of spicy or tangy ingredients. The appearance of an appetizer must be satisfying enough to create interest for the food to follow.

Appetizers include cocktails, hors d'oeuvres, canapes, relishes, dips, petite salads, soups, and garnishes. Cocktails are a chilled appetizer consisting of a juice or small bite-size seafood or fruit. Hors d'oeuvres are hot or cold preparations such as stuffed mushrooms or deviled eggs. Canapés are usually small servings of savory spreads on toasted or plain bread. Relishes are vegetables cut into small pieces and served chilled. Dips are usually made from a cheese base and are served with chips, crackers, or vegetables. Petite salads are small servings of a larger side salad. Soups offer a large variety of ingredients and may be served hot or cold. Garnishes are food items prepared in attractive forms to further enhance a food preparation.

APPETIZERS

An *appetizer* is a small serving of food prepared as the first course of a meal to stimulate the appetite for other foods to follow. An appetizer may be in liquid or solid form. The appearance of an appetizer must be eye-appealing and colorful. Appetizers are often prepared as finger foods, which can be eaten without utensils. Finger foods must be firm enough to handle easily. Appetizers are generally classified as cocktails, hors d'oeuvres, canapés, relishes, dips, petite salads, and soups. Terms commonly used in preparing appetizers include blend, chopper plate, salad tossing, grapefruit and orange sections, cheese and butter flowers, egg wash, bacon cracklings, and garnishes.

Blend: *Blend* or *blend thoroughly* is to mix two or more ingredients. Blending is done manually using a kitchen spoon unless a mixing machine is specified.

Chopper plate: A *chopper plate* is a perforated metal disk that is placed in front of a knife or blade on a food grinder. The food grinder comes equipped with a number of chopper plates with holes that vary in size. The food is first cut by the blade and then it passes through the chopper plate.

Salad tossing: *Salad tossing* is lifting or throwing salad items quickly. In the kitchen, tossing is usually done with the hands. Gloves should always be worn when handling ready to eat foods. In the dining room, before the guest, tossing is done with two large salad forks. Tossing should be done with a gentle motion so the ingredients, especially the greens, are not bruised.

Grapefruit sections: A *grapefruit section* is a portion of a grapefruit. Grapefruit can be purchased in cans or fresh in jars already sectioned. If the sections are to be removed from a fresh grapefruit, the fruit is peeled and each section is removed by cutting around the inside of the membrane of each grapefruit segment using a grapefruit knife (scallop edged knife) or paring knife. The sections are left whole. They are not broken.

Orange sections: An *orange section* is a portion of an orange. Oranges can be purchased in cans or fresh in jars already sectioned. If the fresh fruit is used, the orange is peeled and broken into natural sections. As much of the membrane as possible is scraped off with a paring knife. The sections are left whole for some cocktails.

Cheese flowers: *Cheese flowers* are small flowers formed by piping a cheese spread through a pastry bag and tube. To prepare a cheese flower, cream cheese is worked with a spoon to an elastic consistency. The cheese is colored as desired using liquid or paste food colors. To make a rosebud, red is used; for a sweet pea, blue is used; and for leaves, green is used. The colored cheese is placed in a pastry bag with the appropriate pastry tube. The flower and leaf are piped onto the item by squeezing the pastry bag with a steady pressure of the hand. Cheese flowers are used to garnish canapés and certain hors d'oeuvres.

Wisconsin Milk Marketing Board

Appetizers prepare the customer for foods to follow in the meal.

Butter flowers: *Butter flowers* are small flowers formed by piping a butter spread through a pastry bag and tube. Butter flowers are used to decorate or garnish canapés and certain hors d'oeuvres. They are formed in the same manner as cheese flowers; however, the consistency of the butter requires more attention than the consistency of the cream cheese. When forming flowers, work in a cool place because butter breaks down or melts fairly rapidly.

Egg wash: An *egg wash* is a mixture of eggs and milk (6 eggs to each quart of milk). The eggs are beaten slightly with a wire whip. The milk is poured into the beaten eggs while stirring with the whip. Egg wash is used for coating items that are to be cooked in hot grease. Another use of egg wash is to adhere sesame and poppy seeds to the surface of a baked good.

Bacon cracklings: *Bacon cracklings* are the crisp residue that remains after the grease has been cooked out of bacon. The bacon is ground on a food grinder using the desired chopper plate, placed in a saucepan, and cooked at a moderate temperature until the grease is cooked out of the bacon and the remaining residue becomes crisp. The cracklings are then drained to remove the grease.

Garnishes: There are many different garnishes that may go with appetizers. Usually the recipe suggests an appropriate garnish. Examples of common garnishes are sliced olives, chopped hard-boiled eggs, small pieces of pimiento, chopped chives, cheese flowers, paprika, onion rings, and anchovies. Garnishes should always be edible.

Cocktails

Cocktails are a chilled appetizer that may be fruit juice, vegetable juice, fruit, or seafood. Juice cocktails should be bright in appearance and tangy to the taste. Juice cocktails are served in a small chilled glass. Fruit or seafood cocktails are arranged for attractiveness and are cut bite size to avoid too much chewing.

Glasses used to serve cocktails generally have a round glass base attached to a glass stem that stands from $1\frac{1}{2}''$ high to $3\frac{1}{2}''$ high. The other end of the glass stem is attached to the glass cup or bowl, which holds the item or items being served. The cup or bowl varies in size, including sizes of 3 oz, $3\frac{1}{2}$ oz, 4 oz, $4\frac{1}{2}$ oz, and 6 oz glasses.

- Cut or slice all fruit in an attractive presentation. Appearance is important.
- Arrange all ingredients to maximize natural food colors to create eye appeal.

- Use crisp, clean, fresh vegetables called for in the recipe.
- Peel, clean, and devein shrimp thoroughly.
- Using the parisienne scoop, cut melon balls into complete properly formed balls for good appearance.
- Serve cocktails in cocktail glasses.
- Garnish all cocktails with an item that will enhance appearance and, if possible, improve the flavor.
- Serve all cocktails well chilled.

Hors d'oeuvres

Hors d'oeuvres are small servings of highly seasoned foods. Hors d'oeuvres are served either hot or cold and include such popular preparations as baked oysters on the half shell, stuffed mushrooms, stuffed celery, and deviled (stuffed) eggs. In addition, hors d'oeuvres may be the chef's own creation. Hors d'oeuvres may be served before dinner or as a snack.

Canapés

Canapés are slices of toasted or plain bread that are cut into various small shapes and decorated for eye appeal with a rich savory paste or butter spread. Crackers may be used as a base; however, toasted bread is more desirable as it will not absorb the moisture of the spread as quickly. In addition, canapés on toasted bread can be cut into interesting shapes.

Variety and imagination are the keys to success when preparing canapés. The canapé spread is an excellent way to use leftovers. This permits a great deal of flexibility by the chef.

When preparing canapés, start with eight to ten different spreads and four or five different types and colors of bread. For example, rye bread, white bread, whole wheat bread, sourdough bread, and pumpernickel bread could be used. After making spreads and slicing the bread (slice bread lengthwise for faster production), choose combinations that offer eye appeal and good taste. A garnish such as a cheese flower, chopped egg, pimiento, or olive may be added.

Hot canapés are prepared in the same way as cold canapés, but with less garnish. Hot canapés are heated in an oven or under the broiler and served hot.

Anchor Food Products, Inc.

Appetizers are made of small portions, which allow tasting of a variety of flavors.

- Adjust the consistency of all canapé spreads to the point that they can be applied or spread with ease.
- Select an extra sharp knife for trimming and cutting canapés.
- Keep all canapé spreads refrigerated until 15 min before using.
- When making canapés, purchase unsliced Pullman style bread (a square loaf referred to as *sandwich bread*) and slice it lengthwise, using a bandsaw or slicing machine. This will help speed up production.
- Toast the bread on both sides for canapés. This is done by placing it on sheet pans under the broiler. Toasted bread helps prevent a soggy canapé if the spread has a high moisture content.
- Decorate canapés with an item that improves appearance and enhances the taste.
- Keep canapés in the refrigerator covered with a damp cloth until ready to use.
- Arrange canapés on platters in such a way that they will excite the appetite and display a colorful assortment.
- Keep butter spreads refrigerated until 15 min before using. Let set at room temperature until of spreading consistency.

Relishes

Relishes are selected vegetables which are served chilled. They include celery hearts, radishes, stuffed olives, ripe olives, pickles, and vegetable sticks. Relishes are usually served in a deep, boat-shaped dish over crushed ice.

- Cut or slice vegetable relishes in an attractive manner. Appearance is important.
- Use a sharp French knife or fancy crinkle-edge cutter to cut vegetables.
- Keep vegetable relishes refrigerated until ready to serve. Keep items such as radishes, celery, and carrots refrigerated and covered with ice.
- Serve relishes covered with crushed ice for best results.

Dips

Dips are appetizers in which other food items are dipped. They are popular when served with crackers, chips, and vegetables. When preparing dips, the consistency required is based on the food item to be dipped. If the dip is too thick, the cracker or chip will crumble. If the dip is too thin, the dip will run. Most dips are prepared using a cheese base with various ingredients. This requires dips to be refrigerated until approximately 30 min before serving. The dip is then served at room temperature.

Petite Salads

Petite salads are small salads that are similar to large salads but are less filling. Petite salads consist of a base, body, and garnish. The purpose of a petite salad is to stimulate the appetite for the food that will follow.

- Use crisp, clean, fresh lettuce as the base for all petite salads.
- If the recipe calls for hard-boiled eggs, chop them on heavy paper, applying light blows with a French knife.
- Set up and garnish all petite salads for eye appeal.
- Serve petite salads well chilled.

Soups

Soups are considered appetizers because they are usually served before the main course of a dinner. Soups may be thick or thin and may be served hot or cold. Generally, bread or crackers are served with the soup.

Sandwiches

Generally, the five kinds of sandwiches made in food service establishments are the closed sandwich, the open-face sandwich, the combination sandwich, the double-decker sandwich, and the fancy sandwich. The *closed sandwich* consists of two slices of bread, one on top of the other, with a filling in the center. The *open-face sandwich* consists of two slices of bread laid side by side on a plate with the filling exposed on the surface of one or both slices.

The *combination sandwich* is similar to the closed sandwich, but instead of just one filling, two fillings are used. The two fillings used must combine well with each other. The *double-decker club sandwich* consists of three pieces of bread or toast and two fillings. The bread and fillings are stacked alternately, starting with a piece of bread and following with a filling. The two fillings used must combine well with one another.

Tea sandwiches are usually roll sandwiches or small, open-face sandwiches decorated for eye appeal. Roll sandwiches are made by slicing day-old Pullman bread (white, rye, or whole wheat) lengthwise on a bandsaw or slicing machine. If using a slicing machine, the bread must first be cut in half crosswise before slicing. The crust is trimmed and each slice is covered with a towel and rolled thin with a rolling pin to compress the bread. Spread the filling fairly thin for a more compact roll. At the end of the spread bread, where the roll will begin, place a wedge of sweet or dill pickle, a contrasting stick of bread, a row of stuffed olives, a stick of cheese, or another selected item to create an attractive center for the roll. Roll firmly, wrap each roll in waxed paper or plastic wrap, and refrigerate for a couple of hours before slicing into pinwheels.

- Prepare sandwich fillings just prior to use if possible and refrigerate until ready to use.
- Make sandwiches on the day they are to be served. If they must be made ahead, freeze them or keep them in the refrigerator covered with a damp cloth until ready to use.
- If using lettuce, wash, drain, refrigerate, and keep covered with a damp cloth to ensure crispness.
- Spreading sandwich bread with butter or margarine improves the eating qualities of sandwiches and prevents moist fillings from soaking through the bread. Soften the butter or margarine to make application easier and to keep from tearing the bread.

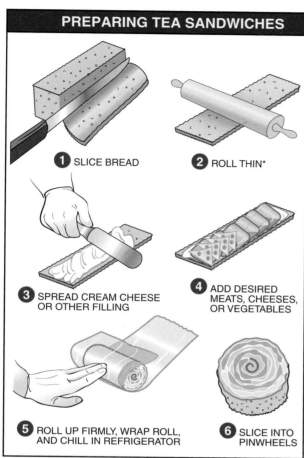

PREPARING TEA SANDWICHES

1 SLICE BREAD

2 ROLL THIN*

3 SPREAD CREAM CHEESE OR OTHER FILLING

4 ADD DESIRED MEATS, CHEESES, OR VEGETABLES

5 ROLL UP FIRMLY, WRAP ROLL, AND CHILL IN REFRIGERATOR

6 SLICE INTO PINWHEELS

*NOTE: Tortillas or pita bread can also be used in place of loaf bread.

Tea sandwiches require several steps but provide great eye appeal.

- Prepare sandwiches on a table or cutting board. This prevents the bread from slipping when being cut, trimmed, or spread.

- Have the necessary equipment readily available:
  - Spoon or scoop for portioning certain spreads
  - Sharp French knife for trimming and cutting
  - Spatula for spreading the butter, margarine, and filling mixtures
  - Pans, wax paper, and damp towels if necessary

- When preparing tea sandwiches to be rolled, place the trimmed bread on a damp towel before spreading. This keeps the bread moist to facilitate rolling.

- When using toasted bread for a sandwich, use only freshly toasted bread for best results.

Florida Department of Citrus

Roll up or wrap sandwiches are a popular alternative to traditional sandwich offerings and can be served hot or cold.

Trade Tip *When making cold sandwiches, remember to practice proper sanitation to avoid food borne illness. Gloves are worn when preparing food that will not be cooked.*

Preparing Large Quantities of Sandwiches

1. Arrange all ingredients within easy reach. Place bread supply to the left if the worker is right-handed, and to the right if left-handed. All spreads or filling ingredients should be directly in front of the worker. Tip containers forward for easy access.

2. Place bread or toast slices in rows directly in front of worker.

3. Spread the slices of bread or toast with soft butter, using a spatula.

4. Portion filling mixture on alternate rows of bread. If four rows of bread are used, place filling on two center rows.

5. Spread soft filling evenly. Bring filling to the edge of the bread. Arrange meat or cheese slices so the bread is well covered. Avoid extending beyond the edge of the bread.

6. Arrange crisp lettuce on the filling (if used).

7. Place the remaining slices of bread on the slices containing the filling.

8. If preparing tea sandwiches, trim the crust edges of the bread using a French knife. Cut the sandwich in half, thirds, or fourths. Sandwiches may be stacked so that several may be cut at the same time.

PREPARING LARGE QUANTITIES OF SANDWICHES

1 PLACE BREAD IN ROWS

2 SPREAD FILLING WITH SPATULA

3 ADD MEAT, CHEESE, OR VEGETABLES

4 COVER WITH BREAD OR TOP OF BUN

Preparation time is saved by using production techniques for large quantities of sandwiches.

APPETIZER RECIPES

The following recipes are some of the many popular appetizers. Their successful preparation enhances the menu of any food service establishment.

POPULAR COCKTAIL RECIPES

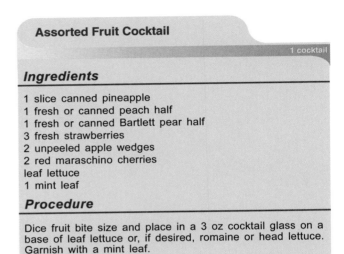

Assorted Fruit Cocktail

1 cocktail

Ingredients

1 slice canned pineapple
1 fresh or canned peach half
1 fresh or canned Bartlett pear half
3 fresh strawberries
2 unpeeled apple wedges
2 red maraschino cherries
leaf lettuce
1 mint leaf

Procedure

Dice fruit bite size and place in a 3 oz cocktail glass on a base of leaf lettuce or, if desired, romaine or head lettuce. Garnish with a mint leaf.

Grape-Melon Cocktail

1 cocktail

Ingredients

¼ c seedless grapes
¼ c diced honeydew melon
¼ c diced cantaloupe
juice of half a lemon
1 tsp powdered sugar
1 mint leaf

Procedure

Place fruit, juice, and powdered sugar in a stainless steel mixing container and toss. Place in a 3½ oz cocktail glass, garnish with the mint leaf, and serve chilled.

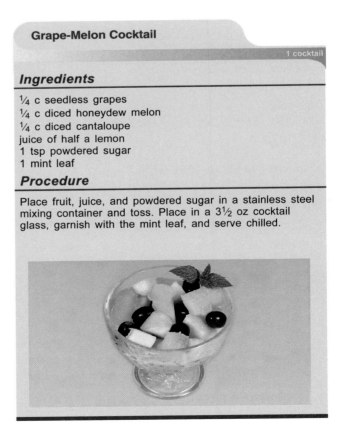

Shrimp Cocktail

1 cocktail

Ingredients

6 cooked medium-sized shrimp
¼ c diced celery
leaf lettuce

Procedure

Place the leaf lettuce and the diced celery in a 3 oz cocktail glass. Arrange the shrimp on top of the diced celery. Serve with a wedge of lemon and top with cocktail sauce.

Note: Lobster, seafood, and crabmeat cocktails are prepared in the same manner, only the shellfish is changed.

Mississippi Department of Marine Resources

POPULAR COLD HORS D'OEUVRE RECIPES

Deviled Eggs

40 stuffed eggs

Ingredients

20 hard-boiled eggs
¼ c mayonnaise
1 tsp prepared mustard
½ c soft cream cheese
¼ tsp white pepper
2 dashes Worcestershire sauce
2 dashes Tabasco sauce
1 tsp salt

American Egg Board

Procedure

1. Peel and cut the eggs into quarters lengthwise or cut in half crosswise. After cutting the ends off, stand the egg up.
2. Remove the yolks and pass through a china cap or sieve.
3. Combine all ingredients and blend thoroughly to a very smooth paste.
4. Place the yolk paste in a pastry bag with a star tube and refill the egg whites.
5. Decorate the top of each filled egg with a slice of stuffed olive, cheese flower, chopped parsley, paprika, pimiento, black olive, or slice of radish.

Variations

With Chives: Mash the yolks of eight hard-boiled eggs and combine them with 4 tsp chopped chives, 2 tbsp mayonnaise, ½ tsp mustard, and salt and pepper to taste. Fill the whites.

With Chicken: Mash the yolks of eight hard-boiled eggs and combine them with ¼ c cooked chicken pounded to a paste. Add 2 tsp mayonnaise, curry powder to taste, and a dash of cayenne pepper. Fill the whites.

With Chicken Liver: Mash the yolks of eight hard-boiled eggs and force them through a fine sieve. Sauté 6 oz of chicken livers in 2 tsp butter until just cooked. Add ½ tsp onion to the pan and cook for 2 min. Mash the onions and livers, force them through a sieve or grind very fine, and combine with the yolks. Add 3 tbsp butter and salt and pepper to taste. Fill the whites.

With Curry: Mash the yolks of eight hard-boiled eggs and force them through a fine sieve. Combine with 2 tsp minced onion, ¼ c finely chopped ham, curry powder to taste, salt and pepper to taste, and enough mayonnaise to bind the mixture. Fill the whites.

With Crabmeat: Mash the yolks of eight hard-boiled eggs and force them through a fine sieve. Combine with ½ c mashed crabmeat, 3 tbsp mayonnaise, 1 tsp prepared mustard, 1 tsp minced onion, salt and pepper to taste, and a dash of Worcestershire sauce. Fill the whites.

Smoked Salmon Rolls

30 rolls

Ingredients

1 lb cream cheese
¼ lb butter
1 tbsp minced onion
3 tsp lemon juice
2 lb thinly sliced smoked salmon
10 wedges dill pickle

Procedure

1. Blend the cream cheese, butter, onions, and lemon juice.
2. Place the thin slices of salmon on a towel and spread each slice with the cheese mixture.
3. Place the dill pickle wedge at one end of the salmon slice and roll.
4. Wrap in wax paper and refrigerate until firm.
5. Remove from the refrigerator and slice the roll about ½″ thick.
6. Insert a toothpick into each slice and serve.

Note: Thin slices of square luncheon meat may be substituted for the salmon.

Chopped Chicken Liver Mold

12 molds

Ingredients

2 lb chicken livers
2 c minced onions
6 hard-boiled eggs
3 tbsp butter
½ c chicken fat
pepper to taste
1½ tsp salt

Procedure

1. Season the chicken livers with salt and pepper. Sauté the livers and onions in the chicken fat until completely done. Let cool.
2. Add the eggs to the liver and onion mixture and put through the fine chopper plate.
3. Add the chopped parsley and butter and mix thoroughly.
4. Pack into greased molds and refrigerate.
5. Unmold and serve on a base of leaf and shredded lettuce. Garnish with onion rings and lemon.

Bleu Cheese Stuffed Celery

20 pieces

Ingredients

20 pieces celery, trimmed and cut about 4″ long
1 lb cream cheese
6 oz bleu cheese
2 dashes Tabasco sauce
1 tsp finely minced onions
juice of half a lemon

Procedure

1. Cut a thin strip off the back of each piece of celery so it will lay flat.
2. Blend the cream cheese, bleu cheese, Tabasco sauce, onions, and lemon juice to a smooth paste.
3. Place in a pastry bag with a star tube and fill the crevice in the celery with the cheese mixture.
4. Refrigerate until cheese is firm. Sprinkle with chopped parsley or paprika. The stuffed celery may also be decorated with cheese or butter flowers.

Puff Shells

60 shells

Ingredients

1½ c pastry flour
½ c shortening
¼ c butter
1½ c water
pinch of salt
6 eggs

Procedure

1. Place the water in a sauce pot and bring to a boil. Add the shortening and butter. Let melt.
2. Add the flour and salt, stir in thoroughly, and cook slowly until flour is cooked and mixture is slightly stiff. Remove from the range and let cool. Add unbeaten eggs one at a time, beating thoroughly after each addition. This can be done by hand or with the mixing machine at slow speed using the paddle.
3. Place dough in a pastry bag with a star tube and force out into very small spirals onto sheet pans covered with parchment paper or dusted with flour.
4. Bake in a preheated oven at 400°F for about 40 min or until golden brown. Let cool.
5. Cut the puff halfway through the center and fill with any kind of canapé spread.
6. Decorate the top of the puff with a cheese or butter flower.

Trade Tip

All employees must be aware of the danger of cross-contamination while preparing food. Do not cut different products on the same cutting surface without thoroughly cleaning the cutting surface.

Salami Horns

25 salami horns

Ingredients

25 thin slices of hard salami
25 toothpicks
1 lb cream cheese
2 dashes Tabasco sauce
2 dashes Worcestershire sauce
1 pinch salt

Procedure

1. Cut a slit in each slice of salami by cutting from the center to the outer edge.
2. Roll the cut slice of salami around the second finger of the hand to form a cornucopia. Secure with a toothpick.
3. Blend the remaining ingredients thoroughly until a smooth paste is acquired.
4. Fill each salami horn with the cheese mixture.
5. Place in refrigerator until cheese becomes firm.

Cheese Apples (or Pears)

40 apples (or pears)

Ingredients

2 lb longhorn or Cheddar cheese
paprika as needed
40 whole cloves

Procedure

1. Grind cheese in food grinder using medium chopper plate.
2. Form ground cheese into balls about the size of an English walnut.
3. Touch each side into the paprika to blush slightly.
4. Stick a clove upside down into the top.

Cheese and Bacon Balls

20 balls

Ingredients

1 lb sharp Cheddar or longhorn cheese
⅓ c mayonnaise
½ lb bacon

Procedure

1. Fry bacon until it becomes crisp. Drain well.
2. Crumble the bacon in a food grinder using the fine chopper plate. Grind cheese on food grinder.
3. Add the mayonnaise to the cheese and blend thoroughly.
4. Form into small balls approximately 1″ in diameter and roll in the crumbled bacon.
5. Serve on toothpicks.

Stuffed Mushrooms

20 mushrooms

Ingredients

20 medium mushroom caps
juice of one lemon
½ c shortening
¼ c butter

Procedure

1. Place the shortening and butter in a sauce pot. Heat slightly.
2. Add the mushroom caps and sauté until half done. Do not brown.
3. Add the juice of a lemon. Continue to cook until mushrooms are tender but still firm. Let cool.
4. Stuff with crabmeat spread, chicken liver spread, shrimp spread, etc.

Bouchees or Miniature Patty Shells

60 shells

Ingredients

1 lb 4 oz bread flour
¼ oz salt
2 oz eggs
5 oz cold water
2 oz butter
1 lb 4 oz Puff Paste Shortening*

*Puff Paste Shortening is a trade name given to this special shortening manufactured for making puff paste dough. This special shortening is rolled into a basic dough to create layers of fat in the dough so the dough expands when heated and baked, creating rich flaky layers of extremely tender crust. The shortening is rolled into the dough with a rolling pin. It is this rolling-folding process that creates the layers of fat that, in turn, produce the tender flaky layers of crust.

Procedure

1. Combine flour, salt, butter, eggs, and cold water. Mix into a dough.
2. Remove from mixer, place on a floured bench, round into a ball, and allow to stand 15 min. Keep covered with a towel.
3. Roll dough into a long rectangular shape about ½″ thick. Dot two-thirds of the dough with the Puff Paste Shortening. Fold three ways and roll Puff Paste Shortening into the dough. Use caution not to let shortening break through.

4. Keep dough covered and place in the refrigerator for 20 min.
5. Remove dough from refrigerator and repeat the rolling-folding process. In all, the dough should be rolled and folded (three folds each time) four times. Refrigerate for 20 min between rolls.
6. After the fourth roll and refrigeration are completed, roll again and proceed to make the bouchee. Roll dough about ⅛" thick. Cut one solid disk using a cutter about 1" in diameter. Cut a second solid disk using a cutter about 1" in diameter. Using a ½" cutter, cut the center of this disk and remove the center, leaving the washer-shaped dough. The center that was removed is discarded or baked separately and made into a special hors d'oeuvre. This dough can never be

reworked. Any leftover pieces are either discarded or baked and used wherever possible.
7. Wash the first disk with egg wash and place second disk on top.
8. Prepare the remaining dough in this fashion. Place the bouchees on sheet pans covered with parchment paper, let rest 10 min, and bake in a preheated oven at 360°F with parchment paper on top so the bouchees do not topple over.
9. Bake until golden brown. Let cool.
10. Fill the bouchee shells with shrimp spread, crabmeat spread, lobster spread, or chicken spread.
11. Decorate with chopped parsley, chopped hard-boiled eggs, or cheese flowers.

POPULAR HOT HORS D'OEUVRE RECIPES

Wisconsin Asiago Cheese Puffs

60 puffs

Ingredients

2 tbsp butter
2 tbsp olive oil
1 tsp salt
2 c water
2 c flour
8 eggs
1 c Wisconsin Asiago cheese, finely shredded
1 c Wisconsin Parmesan cheese, grated
Cayenne pepper to taste

Procedure

1. Combine butter, oil, salt, Cayenne pepper, and water. Bring to a boil.
2. Add flour and stir until mixture forms a smooth ball.
3. Cook over low heat until mixture is drier but still smooth.
4. Place in a mixing bowl. Beat in eggs, one at a time.
5. Stir in cheeses. Drop spoonfuls of batter onto greased cookie sheets.

Wisconsin Milk Marketing Board

6. Bake at 400°F for 20 min or until slightly browned and firm.

Sauerkraut Balls

12 doz balls

Ingredients

1 No. 10 can drained sauerkraut
1 lb peeled onions
6 oz butter
2 tbsp chopped parsley
3 oz flour
4 beaten eggs
salt and pepper to taste

Procedure

1. Grind the sauerkraut and onions on the food grinder using the fine chopper plate.
2. Place the butter in a sauce pot and melt. Add the ground onions and sauerkraut and cook about 8 min.
3. Add the flour and blend to make a thick paste. Continue to cook for 8 min more. Remove from the range.
4. Add the eggs, parsley, and seasoning. Let cool.
5. Form into small balls approximately 1" in diameter. Bread and fry in deep fat at 360°F until golden brown.

Simple Meatballs

3 doz meatballs

Ingredients

1 lb ground chuck
2 tbsp minced onions
2 slightly beaten eggs
salt and pepper to taste
1 c bread crumbs (moistened with milk)
1 pinch thyme
½ small minced garlic clove
1 tbsp shortening

Procedure

1. Sauté the onions and garlic in the shortening. Let cool.
2. Combine the ingredients and mix thoroughly.
3. Form into tiny meatballs and place on a greased sheet pan.
4. Bake in a preheated oven at 350°F until done.
5. Serve in chafing dish in barbecue sauce, curry sauce, or sour cream sauce.

Savory Meatballs

3 doz meatballs

Ingredients

1 lb ground chuck
1 egg
1 tsp salt
1 tsp monosodium glutamate
½ c bread crumbs
¼ c grated Parmesan cheese
1 tbsp minced onions
¼ tsp oregano
1 pinch nutmeg
1 pinch dry mustard
2 tbsp butter
½ c chili sauce
pepper to taste

Procedure

1. Combine all ingredients except the butter and chili sauce in a mixing container. Mix thoroughly.
2. Form into small meatballs.
3. Heat butter in skillet, add meatballs, and cook until slightly brown. Remove meatballs.
4. Add the chili sauce to the skillet and bring to a boil.
5. Serve meatballs from a chafing dish using the hot chili sauce as a dip.

Ham and Cheese Puffs

twenty puffs

Ingredients

Ham Mixture

1 c cooked ground ham
2 tsp prepared mustard
½ tsp Worcestershire sauce
¼ c mayonnaise
1 tsp minced onions
1 tsp baking powder

Procedure

Combine ham, mustard, Worcestershire sauce, mayonnaise, onions, and baking powder. Mix well.

Ingredients

Cheese Mixture

¾ c grated Cheddar cheese
1 beaten egg
1 tsp grated onion
1 tsp baking powder

Procedure

Combine all ingredients and blend thoroughly. Prepare twenty 2″ toasted bread rounds, but toast on one side only. Spread untoasted side with the ham mixture. Top with the cheese mixture. Place on sheet pans and broil slowly until topping puffs and becomes golden brown.

Barbecued Wiener Tidbits

72 tidbits

Ingredients

12 wieners
1 qt barbecue sauce

Procedure

1. Cut each wiener into six pieces.
2. Place the barbecue sauce in a sauce pot and bring to a boil. Move pot to side of the range.
3. Add the wiener tidbits to the sauce and simmer for 2 min.
4. Serve in a chafing dish with toothpicks.

Trade Tip

Appetizers provide small samples of a variety of food items and promote the purchase of items customers may not have considered. Appetizers can easily average 10% of gross sales, but as much as 30% in profit.

Oysters Rockefeller

24 oysters

Ingredients

24 blue point oysters
⅓ c butter
1½ c raw spinach
3 tbsp minced onion
¾ c bread crumbs
½ tsp salt
1 small pinch nutmeg

Procedure

1. Open oysters and drain. Leave the oysters in the deepest half shell.
2. Place the oysters in a pan covered with rock salt.
3. Melt the butter in a sauce pot and add all the remaining ingredients. Cook until ingredients are soft, stirring constantly.
4. Spread the spinach mixture over the oysters.
5. Bake in a preheated oven on rock salt at 400°F for about 10 min. Do not overbake.
6. Serve at once.

Note: A small piece of bacon is often placed on the oyster before baking.

Baked Oysters Casino

24 oysters

Ingredients

24 blue point oysters
12 slices of bacon, cut in half
½ lb butter
1 tbsp minced onions
3 tbsp minced green pepper
2 tbsp minced pimientos
1 tbsp minced chives
1 pinch pepper
1 tsp lemon juice

Procedure

1. Open oysters and drain. Leave the oysters in the deepest half shell.
2. Place the oysters in a pan covered with rock salt.
3. Combine all the remaining ingredients except the bacon. Mix thoroughly.
4. Dot each oyster with the butter mixture and top with a piece of bacon.
5. Bake in a preheated oven at 400°F for about 10 min. Serve at once.

Crabmeat Balls

1 doz balls

Ingredients

2 c king or blue crabmeat
3 tbsp minced onions
1 beaten egg
3 tbsp shortening or butter
1 c milk or cream
½ c flour
1 tsp sherry
½ tsp prepared mustard
1 tsp salt
1 dash Tabasco sauce
2 dashes Worcestershire sauce
½ c bread crumbs

Procedure

1. Place shortening or butter in sauce pot and heat.
2. Add the minced onions and cook without color.
3. Add the flour, making a roux. Continue to cook slightly.
4. Add the milk or cream, making a thick paste.
5. Add the crabmeat, mustard, salt, Tabasco, sherry, and Worcestershire. Blend thoroughly. Remove from the range and let cool.
6. Add the eggs and bread crumbs. Mix thoroughly.
7. Form into small balls approximately 1″ in diameter. Bread and fry in deep fat at 350°F until golden brown.
8. Serve in a chafing dish with toothpicks. Use cocktail sauce as a dip.

Salmon Nuggets

48 nuggets

Ingredients

1 16 oz can red salmon
½ c mashed potatoes
1 tbsp finely chopped celery
1 tbsp grated onions
1 tbsp melted butter
¼ tsp salt
1½ tsp Worcestershire sauce
1 beaten egg
¼ lb longhorn cheese
½ c bread crumbs
pepper to taste

Procedure

1. Drain salmon. Flake and remove bones.
2. Add all other ingredients except the bread crumbs and cheese. Mix well.
3. Cut the cheese into 48 cubes (approximately ⅜″ each).
4. Shape the salmon mixture around the cheese cubes to form small balls.
5. Roll in bread crumbs and fry in deep fat at 375°F until golden brown.

Shrimp Stuffed Mushroon Caps

25 caps

Ingredients

25 fresh mushroom caps, medium to large
2 c cooked chopped shrimp
2 c cooked rice
1½ tbsp chopped parsley
1½ tbsp chopped chutney
1 tsp salt
¼ tsp thyme
½ c grated Cheddar cheese

Procedure

1. Wash mushroom caps and dry.
2. Combine remaining ingredients, except the Cheddar cheese, by mixing in a stainless steel bowl using a kitchen spoon.
3. Press the shrimp mixture firmly and generously into the mushroom caps.
4. Sprinkle the stuffed mushrooms with the grated cheese.
5. Bake in a preheated oven at 375°F for about 10 min.
6. Serve hot.

Pizza Puffs

50 puffs

Ingredients

2 c shredded mozzarella cheese
1 c cracker meal
½ c cornflake crumbs
½ tsp oregano
1 minced garlic clove
¾ tsp salt
1 pinch basil
2 eggs

Procedure

1. Separate the egg yolks from the whites. Whip the whites until stiff. Set aside.
2. Mix the remaining ingredients and fold into the stiff egg whites.
3. Shape into balls, roll in additional cracker meal, and refrigerate until firm.
4. Fry in deep fat at 350°F until golden brown.

Chinese Egg Rolls

24 egg rolls

Ingredients

Shrimp Filling For Egg Rolls

½ lb cooked, finely chopped shrimp or crabmeat
2 tbsp minced onions
1 tbsp chopped scallions
1 tbsp chopped bamboo shoots
3 tbsp cornstarch
2 well beaten eggs
¼ tsp soy sauce
salt and pepper to taste

Carlisle FoodService Products

Procedure

1. Place all ingredients in a round bottom mixing bowl. Using a kitchen spoon, mix until thoroughly blended.
2. Refrigerate until ready to use.

Note: If variety is desired, the shrimp or crabmeat may be replaced with cooked chicken, lobster, or tuna.

Ingredients

Skins

2 lb bread flour
4 beaten eggs
½ tsp salt
1 lb cold water

Procedure

1. Sift the flour and salt. Place in the bowl of the mixing machine.
2. Add the eggs and water. Mix at slow speed using the paddle until the dough is firm and smooth.
3. Turn the dough onto a floured bench, let rest 10 min, and keep covered with a damp cloth.
4. Using a rolling pin, roll the dough to a thickness of approximately ⅛". Cut into 6" squares.
5. Place 1 oz to 1 ½ oz of filling on each 6" square of dough. Fold the two sides so the filling cannot flow out. Roll the filled dough tightly. Dampen the end with water and secure.
6. Fry the rolls in deep fat at 350°F until golden brown. Let drain.
7. Cut each roll into four pieces and serve in a chafing dish with toothpicks.

POPULAR CANAPÉ SPREAD RECIPES

Egg Spread

2 c

Ingredients

12 hard-boiled and strained egg yolks
2 tsp horseradish
2 tsp minced onions
2 tsp Worcestershire sauce
1 dash Tabasco sauce
½ c mayonnaise
½ tsp salt
⅓ c cream cheese

Procedure

Combine all ingredients and blend to a smooth paste of spreading consistency. Refrigerate until ready to use.

Cheddar Cheese Spread

3 c

Ingredients

1 lb ground sharp Cheddar cheese
2 tsp Worcestershire sauce
2 tsp minced onion
¼ tsp Tabasco sauce
1 tsp tarragon vinegar
½ tsp prepared mustard
4 oz cream cheese

Procedure

Place all ingredients in mixing bowl. Mix in the mixing machine and blend to a smooth paste of spreading consistency using the paddle.

Bleu Cheese Spread

2½ c

Ingredients

½ lb bleu cheese
1 lb cream cheese
¼ lb butter
1 tbsp fresh lemon juice
2 tbsp fresh chopped dill

Procedure

Place all ingredients in mixing bowl and blend on the mixing machine until of spreading consistency. Refrigerate until ready to use.

Pineapple-Cheese Spread

2½ c

Ingredients

2 c cream cheese
½ c drained crushed pineapple
pinch of salt
yellow color as desired

Procedure

Place ingredients in mixing container and blend until of spreading consistency.

Pimiento-Cheese Spread

2 c

Ingredients

1 lb cream cheese
2 tbsp drained, chopped pimientos
2 drops Tabasco sauce

Procedure

Place all ingredients in mixing container and blend until of spreading consistency. Refrigerate until ready to use.

Olive-Cheese Spread

2 c

Ingredients

1 lb cream cheese
2 tbsp finely chopped stuffed olives
2 drops Tabasco sauce

Procedure

Place all ingredients in a mixing container and blend to spreading consistency. Refrigerate until ready to use.

Trade Tip

Back and spine injuries are very costly. Employees should transport heavy loads with carts designed for the task.

Avocado Spread

2 c

Ingredients

2 large ripe avocados, peeled, pitted, and mashed
2 tbsp minced onions
1 minced small garlic clove
1 tsp salt
2 tsp lemon juice
1 tsp catsup
pepper to taste

Procedure

Place all ingredients in mixing container and blend to spreading consistency. Refrigerate until ready to use.

Bacon-Cheese Spread

2 c

Ingredients

1 lb cream cheese
¼ c fine bacon cracklings

Procedure

Place ingredients in mixing container and blend until of spreading consistency. Refrigerate until ready to use.

Chipped Beef and Cheese Spread

2 c

Ingredients

1 lb cream cheese
½ c finely chopped, dried chipped beef
2 drops Tabasco sauce

Procedure

Place all ingredients in a mixing container and blend to spreading consistency. Refrigerate until ready to use.

Peanut Butter and Bacon Spread

4½ c

Ingredients

3 c peanut butter
½ c bacon cracklings
1 c minced celery
sour cream as needed

Procedure

Thin the peanut butter to spreading consistency with sour cream. Add the bacon cracklings and minced celery. Mix well. Refrigerate until ready to use.

Deviled Ham Spread

1½ c

Ingredients

1 c packed ham trimming
2 tbsp dill pickles
1½ tbsp peeled quartered onions
1 tsp chopped parsley
1 tbsp prepared French mustard
1½ tbsp mayonnaise
½ tsp Worcestershire sauce
2 dashes Tabasco sauce
red food color as desired

Procedure

Combine the ham, onions, and pickles. Put through food grinder. Add remaining ingredients and blend thoroughly. Refrigerate until ready to use.

Chicken Liver Spread

1½ c

Ingredients

12 fresh chicken livers
3 tbsp salad oil
3 tbsp butter
½ c minced onions
1 bay leaf
2 tsp leaf sage
⅔ c white wine
1 tsp chopped parsley
½ tsp salt

Procedure

1. Sauté the onions and livers in the butter and salad oil until slightly brown.
2. Add the bay leaf, sage, parsley, salt, and wine. Continue to cook until the wine evaporates slightly.
3. Remove from heat, let cool, and remove bay leaf.
4. Put mixture through fine food grinder and blend well.
5. Refrigerate until ready to use.

Chicken-Bacon Spread

2 c

Ingredients

2 c minced boiled chicken
6 slices of bacon, cooked crisp, minced
½ tsp salt
¾ c mayonnaise
½ minced peeled apple

Procedure

Place all ingredients in a mixing container and blend to spreading consistency. Refrigerate until ready to use.

Shrimp Spread

2 c

Ingredients

2 c peeled, cooked, deveined shrimp
1 tbsp finely cut onions
1 tbsp finely cut celery
1 tsp salt
1 tbsp lemon juice
½ tsp Spanish paprika
1 tsp Worcestershire sauce
1 tsp prepared mustard
1 tsp chopped parsley
3 tbsp mayonnaise

Procedure

1. Mix the shrimp, onions, celery, and lemon juice.
2. Put through the fine grinder.
3. Add the mayonnaise, mustard, parsley, salt, Worcestershire sauce, and paprika. Blend until of spreading consistency.
4. Refrigerate until ready to use.

Sardine Spread

2 c

Ingredients

2 c mashed canned sardines
1 tbsp minced onions
1 tbsp lemon juice
1 tsp chopped parsley
1 tsp horseradish
¼ c mayonnaise

Procedure

1. Place all ingredients in a mixing container and blend to a paste of spreading consistency.
2. Refrigerate until ready to use.

Anchovy Spread

1 c

Ingredients

1 c soft butter
4 tbsp chopped anchovies
1 tbsp minced onions
1 tsp chopped parsley

Procedure

Blend all ingredients into a smooth paste of spreading consistency. Refrigerate until ready to use.

Crabmeat Spread

1½ c

Ingredients

1 lb canned or frozen back fin lump crabmeat
2 tbsp finely cut onion
4 tbsp mayonnaise
1 tbsp celery
1 tsp finely chopped pimientos
½ tsp lemon juice
2 tbsp chopped parsley
½ tsp salt

Procedure

1. Mix the crabmeat, onions, and celery. Put through the fine food grinder.
2. Add remaining ingredients and blend until of spreading consistency.
3. Refrigerate until ready to use because crabmeat spoils quickly.

King Crabmeat Spread

1½ c

Ingredients

1 lb frozen king crabmeat
1 tbsp finely chopped parsley
1 tbsp finely cut celery
1 tbsp finely cut onions
3 tbsp mayonnaise
½ tsp lemon juice
1 pinch curry powder

Procedure

1. Mix the crabmeat, onions, celery, and lemon juice. Put through the fine food grinder.
2. Add the mayonnaise, chopped parsley, and curry powder. Blend until of spreading consistency.
3. Refrigerate until ready to use because crabmeat spoils quickly.

Lobster Spread

Ingredients

2 c cooked lobster meat
1 tbsp finely cut onions
1 tbsp finely cut celery
1 tsp finely cut green pepper
2 tbsp mayonnaise
1 tsp lemon juice
1 tsp chopped parsley

Procedure

1. Mix the lobster, onions, celery, and green pepper. Put through the fine food grinder.
2. Add remaining ingredients and blend until of spreading consistency.
3. Refrigerate until ready to use.

Salmon Spread

Ingredients

2 c drained canned salmon
1 tbsp lemon juice
1 tsp chopped parsley
1 tbsp finely cut onions
2 tbsp mayonnaise
1 tbsp finely cut dill pickles
1 hard-boiled egg
½ tsp salt

Procedure

1. Mix the salmon, dill pickles, onions, and hard-boiled egg. Put through fine food grinder.
2. Add remaining ingredients and blend until of spreading consistency.
3. Refrigerate until ready to use.

Tuna Spread

Ingredients

2 c drained white meat tuna
1 tbsp finely cut onions
1 tbsp finely cut celery
2 tbsp mayonnaise
1 tsp lemon juice
1 tsp salt
1 hard-boiled egg
1 tbsp finely chopped pimientos

Procedure

1. Mix the tuna, onions, celery, and hard-boiled egg. Put through the fine food grinder.
2. Add the remaining ingredients and blend until of spreading consistency.
3. Refrigerate until ready to use.

POPULAR BUTTER SPREAD RECIPES

Butter spreads are used in the preparation of canapés not only as a spread, but in many cases as filling or decoration. Specialty butter spreads can be a signature item for restaurants. These butter spreads can also be used for sautéing, or added to a cooked dish. It is recommended that a high-quality butter be used. Before other ingredients are added, the butter should be brought to room temperature. This will make it easier to work with when mixing in the other ingredients. All ingredients mixed into the butter should be pureed (pounded or finely minced and forced through a sieve) to make a smooth product that will not clog pastry tubes when using to decorate. The ingredients may be mixed into the butter by hand using a kitchen spoon, or a mixing machine may be used. In such case, use the paddle and mix at slow speed.

Tuna butter: Add 8 oz of pureed canned tuna to 1 lb of butter. Blend until smooth.

Chive butter: Add 1 small bunch of finely chopped chives to 1 lb of butter. Blend until smooth.

Pimiento butter: Add 8 oz of pureed canned pimientos to 1 lb of butter. Blend until smooth.

Garlic butter: Add 1 clove of pureed garlic to 1 lb of butter. Blend until smooth.

Horseradish butter: Add 4 oz of horseradish to 1 lb of butter. Blend until smooth.

Mint butter: Add 6 tbsp of finely chopped mint to 1 lb of butter. Blend until smooth.

Roquefort butter: Add 4 oz of pureed Roquefort cheese to 1 lb of butter. Blend until smooth.

Onion butter: Add 1 small, finely minced Bermuda or Spanish onion to 1 lb of butter. Blend until smooth.

Shrimp butter: Add 12 oz of pureed, cooked, and cleaned shrimp to 1 lb of butter. Blend until smooth.

Lobster butter: Add 12 oz of pureed cooked lobster to 1 lb of butter. Blend until smooth.

Anchovy butter: Add 2 oz of canned pureed anchovy filets to 1 lb of butter; oil should be drained from the filets. Blend until smooth.

Lemon butter: Add 4 tbsp of fresh lemon gratings to 1 lb of butter. Blend until smooth. Fresh lemon gratings are made by rubbing the peel of the fresh lemon across the medium grid of a box grater. Do not cut too deep into the peel or the gratings may be bitter.

POPULAR RELISH RECIPES

Radish roses: Plump, solid radishes are selected. The green stem and leaves are cut off approximately 1″ from the base of the radish, leaving a 1″ green stem attached to the radish. The opposite end of the radish is sliced off, leaving a slight part of the white meat visible. Using a paring knife, five very thin vertical slices approximately ½″ wide are cut into the sides of the radish. The radish is placed in ice water until the slices curl outward.

Celery sticks: Pascal or white celery may be used. The sticks are separated from the stalk and cut into fairly uniform pieces approximately 4″ long and ¼″ wide to ½″ wide. They are placed in ice water to keep them crisp until they are served.

Carrot sticks: The fresh carrots are peeled and cut into fairly uniform pieces approximately 4″ long and ¼″ wide to ½″ wide. They are placed in ice water to keep them crisp until they are served.

Carrot curls: The fresh carrots are peeled and sliced lengthwise paper-thin on the automatic slicing machine. These thin slices are then rolled and secured with a toothpick, placed in ice water, and left to set for approximately 2 hr or longer. When ready to serve, the toothpicks are removed. The roll will loosen slightly, taking on the appearance of a curl.

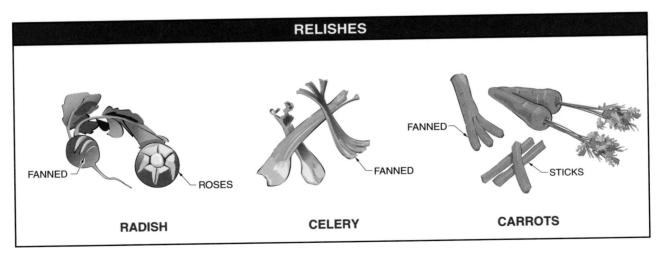

RELISHES

FANNED — ROSES — **RADISH**

FANNED — **CELERY**

FANNED — STICKS — **CARROTS**

Trade Tip

With today's desktop publishing systems, food service personnel can easily develop step-by-step procedures for cutting fancy relishes and garnishes. A digital camera is used to photograph sequential photographs. Captions are written to explain each step and a final photograph shows the completed relish or garnish.

POPULAR DIP RECIPES

Cheddar Cheese Dip

2½ c

Ingredients

8 oz cream cheese
8 oz grated sharp Cheddar cheese
2 tbsp cider vinegar
½ minced garlic clove
½ tsp salt
1 tsp Worcestershire sauce
¼ tsp prepared mustard
½ c sour cream

Procedure

1. Place all the ingredients in the bowl of the rotary mixer and mix at second speed using the paddle. Blend thoroughly.
2. Use extra coffee cream if necessary to obtain proper consistency.

Pineapple-Cheese Dip

3 c

Ingredients

1 lb cream cheese
1 c sour cream
½ c crushed, drained, canned pineapple (save juice)
3 tbsp pineapple juice
½ tsp salt
yellow color as desired
1 pinch nutmeg

Procedure

1. Place all the ingredients in the bowl of the rotary mixer and mix at second speed using the paddle. Blend thoroughly.
2. Use extra coffee cream if necessary to obtain proper consistency.

Onion-Cheese Dip

3 c

Ingredients

1 lb cream cheese
1 c sour cream
2 tbsp onion juice
½ tsp salt
2 dashes Tabasco sauce
1 tbsp chopped chives

Procedure

1. Place all ingredients in the bowl of the rotary mixer and mix at second speed using the paddle. Blend thoroughly.
2. To make onion juice, grind onions in food grinder, place in a towel, and squeeze out juice.

Bacon-Cheese Dip

3 c

Ingredients

1 lb cream cheese
1 c sour cream
½ tsp salt
3 tbsp slightly warm bacon grease
⅓ c finely crumbled bacon cracklings

Procedure

1. Place all the ingredients in the bowl of the rotary mixer and mix at second speed using the paddle. Blend thoroughly.
2. Use extra coffee cream if necessary to obtain proper consistency.

Mint-Cheese Dip

3 c

Ingredients

1 lb cream cheese
1 c sour cream
¼ c finely chopped mint leaves
1 tsp sugar
1 tsp lemon juice
1 tsp salt

Procedure

1. Place all ingredients in the bowl of the rotary mixer and mix at second speed using the paddle. Blend thoroughly.
2. Use extra coffee cream if necessary to obtain proper consistency.

Garlic-Cheese Dip

3 c

Ingredients

1½ c sour cream
1 lb cream cheese
6 pureed garlic cloves
½ tsp salt

Procedure

1. Place all ingredients in bowl of rotary mixer and mix at second speed using the paddle. Blend thoroughly.
2. Use extra coffee cream if necessary to obtain proper consistency.

Bleu Cheese Dip

3 c

Ingredients

1 lb cream cheese
5 oz bleu cheese
½ c sour cream
1 tbsp onion juice
1 dash Tabasco sauce
½ tsp salt

Procedure

1. Place all the ingredients in the bowl of the rotary mixer and mix at second speed using the paddle. Blend thoroughly at slow speed.
2. Use extra coffee cream if necessary to obtain proper consistency. To make onion juice, grind onions in the food grinder, place in a towel or cloth, and squeeze out the juice.

Shrimp-Cheese Dip

3 c

Ingredients

1 lb cream cheese
½ c cooked, chopped shrimp
3 tbsp chili sauce
1 tsp onion juice
½ tsp lemon juice
½ tsp salt
½ tsp Worcestershire sauce
1 tsp horseradish

Procedure

1. Place all the ingredients in the bowl of the rotary mixer and mix at second speed using the paddle. Blend thoroughly.
2. Use extra coffee cream if necessary to obtain proper consistency.

Avocado Dip

3 c

Ingredients

8 oz cream cheese
1 medium-sized ripe avocado, peeled, pitted, mashed
1 tbsp lemon juice
½ tsp minced onion
3 tbsp coffee cream

Procedure

1. Place all ingredients in the bowl of the rotary mixer and mix at second speed using the paddle. Blend thoroughly.
2. Use extra coffee cream if necessary to obtain proper consistency.

Clam Dip

3 c

Ingredients

1 lb cream cheese
1 c canned, drained, minced clams (save juice)
3 tsp lemon juice
2 tsp Worcestershire sauce
1 small minced garlic clove
¼ c clam juice (from canned clams)

Procedure

1. Place all ingredients in the bowl of the rotary mixer and mix at second speed using the paddle. Blend thoroughly.
2. Use extra clam juice if necessary to obtain proper consistency.

POPULAR PETITE SALAD RECIPES

Seafood Green Goddess

20 servings

Ingredients

1 c cooked, deveined, diced shrimp
1 c cooked diced lobster
1 c king or blue crabmeat
2 c minced celery
2 tsp salt (variable)
2 tbsp minced onions
1 tsp lemon juice
2 dashes Tabasco sauce
1 pt green goddess dressing
20 thick tomato slices

Procedure

1. Combine the seafood, celery, salt, onions, lemon juice, and Tabasco. Toss gently.
2. Place a slice of tomato on a 4″ plate with a base of leaf lettuce and shredded head lettuce.
3. Place a mound of the seafood mixture on the tomato.
4. Top with the green goddess dressing.
5. Serve immediately after dressing is applied.

Note: Seafood can be omitted to make a shrimp green-goddess salad.

Tuna Fish Ravigote

15 servings

Ingredients

2 13 oz cans white tuna
2 tbsp minced onions
1 tsp lemon juice
1 c minced celery
1 pt ravigote sauce

Procedure

1. Combine the tuna, onions, lemon juice, and celery. Toss gently so as not to break up the tuna too much.
2. Place a small mound of the mixture on a 4″ plate with a base of romaine lettuce and shredded head lettuce. Top with ravigote sauce and a piece of pimiento.
3. Serve with a wedge of lemon.

Note: Salmon may be substituted for the tuna.

Marinated Herring

12 servings

Ingredients

1 qt jar pickled herring, drained
2 c sour cream
1 tsp salt
2 tbsp lemon juice
½ tsp ground white pepper
1 sliced, medium-sized onion

Procedure

1. Combine all ingredients in a stainless steel mixing container.
2. Refrigerate overnight.
3. Serve four pieces of herring with onions and some liquid on a 4″ plate with a base of leaf lettuce and shredded head lettuce.
4. Garnish with a twist of lemon and chopped parsley.

Eggs A La Russe

24 servings

Ingredients

12 hard-boiled eggs, cut in half
1 pt Russian dressing
chopped parsley as needed

Procedure

1. Place half of a hard-boiled egg on a 4″ plate with a base of leaf lettuce and shredded head lettuce.
2. Cover half of the egg with Russian dressing.
3. Top with chopped parsley and serve.

Deviled Lobster or Crabmeat

12 servings

Ingredients

1 lb cooked lobster or crabmeat
1 tsp salt
1 tsp dry mustard
1 tsp chopped parsley
3 chopped hard-boiled eggs
3 tbsp French dressing
1 tbsp minced scallions
1 large cucumber

Procedure

1. Combine the lobster or crabmeat, salt, dry mustard, chopped eggs, parsley, scallions, and French dressing. Toss lightly.
2. Score cucumber with the tines of a fork and slice about ¼″ thick.
3. Place the cucumber on a 4″ plate with a base of leaf lettuce and shredded head lettuce.
4. Top the cucumber with the lobster or crabmeat mixture. Garnish with chopped parsley or a piece of pimiento.

Shrimp Delight

20 servings

Ingredients

2 lb chopped, cooked, cleaned, deveined shrimp
2 ½ tbsp anchovy paste
2 tbsp lemon juice
1 ½ c mayonnaise
4 chopped hard-boiled eggs
2 tbsp chopped parsley
20 cooked, deveined whole shrimp
4 medium-sized avocados

Procedure

1. Combine the chopped shrimp, anchovy paste, lemon juice, and mayonnaise. Blend thoroughly.
2. Place a small mound of the mixture on a 4″ plate with a base of leaf lettuce, shredded head lettuce, and two small wedges of avocado.
3. Top with mayonnaise, a shrimp, and the chopped eggs and parsley.

Trade Tip *For maximum safety, all commercial kitchen employees should wear low-heeled, properly fitting shoes with non-skid soles and completely enclosed toes and heels.*

Chicken Liver Pâte

12 servings

Ingredients

1 lb chicken livers
½ c sliced onions
½ c butter
4 hard-boiled eggs
2 tsp salt
¼ tsp thyme
¼ tsp black pepper
1 bay leaf
3 tbsp white wine
12 small uncooked onion rings

Procedure

1. Place the butter in a skillet and heat slightly. Add the chicken livers, onions, thyme, and bay leaf. Sauté until brown, stirring frequently. Remove from range and let cool. Remove bay leaf.
2. Add salt, pepper, and two of the hard-boiled eggs. Grind twice on the food grinder using the fine chopper plate.
3. Add the white wine and blend thoroughly.
4. Cover with wax paper and refrigerate.
5. Form mixture into balls the size of a golf ball and serve on a 4″ plate with a base of leaf lettuce and shredded head lettuce.
6. Garnish with remaining chopped eggs, chopped parsley, and a ring of onion.

Italian Antipasto

20 servings

Ingredients

2 c peeled eggplant, cut into cubes
½ c medium diced onions
½ c medium diced fresh mushrooms
2 cloves minced garlic
½ c olive oil or salad oil
1 c tomato paste
½ c water
¼ c wine vinegar
⅓ c medium diced green peppers
¼ c sliced stuffed green olives
¼ c pitted, sliced, ripe olives
1 ½ tsp sugar
1 tsp oregano
1 c cauliflower
1 tsp salt

pepper to taste
¼ c precooked, medium diced celery

Procedure

1. Place the eggplant, onions, mushrooms, garlic, oil, green peppers, cauliflower, and the precooked celery in a braising pot. Cover and cook gently for about 10 min, stirring gently.
2. Add all remaining ingredients, blend well, and continue to cook covered until all ingredients are tender. Remove from heat and let cool.
3. Place in refrigerator overnight for all flavors to blend.
4. Serve on a 4″ plate with a base of leaf lettuce and shredded head lettuce accompanied by two Brisling sardines and a cornucopia of hard salami. Garnish with parsley or watercress.

GARNISHES

Beet and Turnip Asters

Beet and turnip asters are a vegetable flower that can be used to garnish an assortment of foods and platters on the buffet or smorgasbord. Like many of the vegetable flowers, they are quite simple to make. Select a firm, well-rounded beet or turnip and use a very sharp paring or utility knife.

1. Cut the top off the vegetable selected.

2. Using a vegetable peeler, peel the vegetable completely.

3. Holding the vegetable firmly with one hand, cut a series of parallel cuts using the sharp utility or paring knife with the other hand. Start these parallel cuts at one side of the cut surface created when the top of the vegetable was removed and cut completely across the cut surface. Each cut should extend to the root of the vegetable, but not through it. Stop the cut about ¼″ above the root end. Make each cut as close to the previous cut as possible.

4. When this series of cuts is completed, give the vegetable a quarter turn and start a new series of cuts across the first series of cuts, forming very small squares. Cut to the same depth as the first series of cuts.

Place the cut vegetable aster in ice water and let it soak until ready to use. When ready to use, spread the tiny square cuts outward to form the aster.

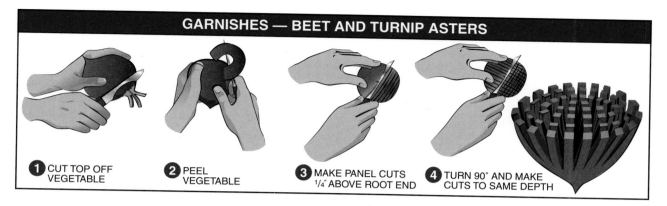

GARNISHES — BEET AND TURNIP ASTERS

1 CUT TOP OFF VEGETABLE

2 PEEL VEGETABLE

3 MAKE PANEL CUTS ¼" ABOVE ROOT END

4 TURN 90° AND MAKE CUTS TO SAME DEPTH

Beet, Carrot, or Turnip Rose

The beet, carrot, or turnip rose is one of the more difficult vegetable garnishes to prepare because it is hand-carved from a fairly hard, raw vegetable. It is, however, a very satisfying experience and achievement for a chef to be able to create one of these elegant vegetable flowers. Vegetable roses can be used to garnish a number of different foods, trays, and platters. They can be kept for days by storing them in the refrigerator in plain or colored water. They can be coated with a plain gelatin solution to keep them from discoloring and to enhance their appearance. The turnip rose may be left white, its natural color, or changed to a red or yellow rose by placing it in colored water. With a little artistic ability, a firm fresh vegetable, and a very sharp firm blade paring knife, the rose can easily be carved.

1. Select a firm, well-shaped beet, carrot, or turnip. Peel the vegetable selected. If a carrot is used, select one with a large diameter at the top and cut off a 1½″ piece crosswise.
2. Turn the peeled vegetable so the bottom is facing up. Using the paring knife, mark the bottom with a cross. This is done so the four-petal cut at the base of the vegetable will be uniform.
3. Holding the vegetable in one hand and the paring knife in the other, proceed to cut the four petals around the circumference of the base. Cut these petals close to but not all the way through the base. Attempt to keep the petals as thin as possible.
4. Place the point of the knife between the petals and the body of the vegetable and cut out a thin ring around the circumference of the vegetable. Cut this ring in a continuous strip. This is the first row of petals.

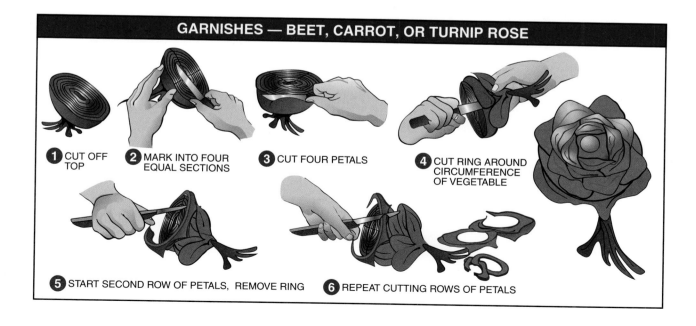

GARNISHES — BEET, CARROT, OR TURNIP ROSE

1 CUT OFF TOP

2 MARK INTO FOUR EQUAL SECTIONS

3 CUT FOUR PETALS

4 CUT RING AROUND CIRCUMFERENCE OF VEGETABLE

5 START SECOND ROW OF PETALS, REMOVE RING

6 REPEAT CUTTING ROWS OF PETALS

5. Start the second row of petals by cutting between the two outside petals of the previous row. Repeat cutting a thin ring around the inside circumference of this row of petals. This is the second row of petals.

6. Repeat cutting the rows of petals and thin rings around the circumference of the vegetable until the center is reached and only a small piece remains. Shape this small piece to resemble the center of a rose. Note that as the cutting of the petals and rings proceeds, the circumference gets smaller and smaller. The more rows of petals carved, the more natural-looking the flower is.

Place the completed vegetable rose in plain or colored water in the refrigerator until ready to use.

Beet Strip Rose

This rose is very similar to the tomato rose but has a deeper red color. The ease in forming this rose depends, to a degree, on how thin the beet strips are cut. Select a medium-sized, blemish-free, firm beet.

1. Peel the outside skin from the beet using a paring knife or vegetable peeler. Using the selected cutting implement, cut around the circumference of the beet until the strip is approximately 4″ long.

This is the same cut used when making a tomato strip rose. Keep the strip as thin as possible.

2. Cut a second strip of beet approximately 3″ long. Place both strips in a solution of salt water (1 tbsp salt to 1 qt water) for approximately 5 min. Soaking will make the strip more flexible.

3. Using the first beet strip cut, form a loose coil. Form a tight coil using the second or shorter beet strip cut.

4. Place the tight coil in the center of the loose coil. Shape the two coils to resemble a rose.

This step is optional, but the rose will last longer and be more attractive. Place the formed rose in the refrigerator and chill thoroughly. Set it on a screen with a tray underneath. Drip a plain gelatin solution over the rose until it clings and adheres to the coiled strips. The plain gelatin solution is prepared by dissolving 1 tbsp of plain gelatin in ¾ c of boiling water.

Cucumber Boat

This is an interesting garnish that can add eye appeal to a buffet or smorgasbord and at the same time prove practical when used as a container to hold sour cream, tartar sauce, or other condiments. To make the cucumber boat, select a fairly large, fresh, firm cucumber.

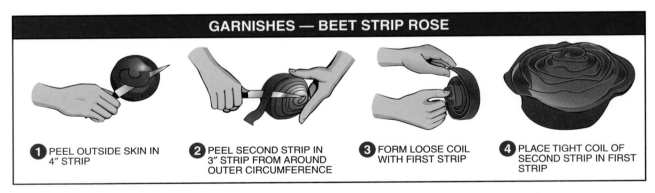

GARNISHES — BEET STRIP ROSE

1 PEEL OUTSIDE SKIN IN 4″ STRIP

2 PEEL SECOND STRIP IN 3″ STRIP FROM AROUND OUTER CIRCUMFERENCE

3 FORM LOOSE COIL WITH FIRST STRIP

4 PLACE TIGHT COIL OF SECOND STRIP IN FIRST STRIP

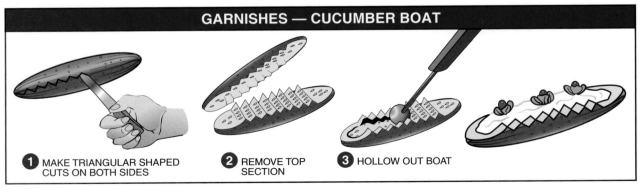

GARNISHES — CUCUMBER BOAT

1 MAKE TRIANGULAR SHAPED CUTS ON BOTH SIDES

2 REMOVE TOP SECTION

3 HOLLOW OUT BOAT

1. Using a sharp paring or utility knife, cut a thin strip down the length of the cucumber to form a flat surface so it will sit firm. Make a cut approximately 1″ deep in the center of each end of the cucumber.

Starting at one end of one of the 1″ cuts, insert the point of the knife into the center of the cucumber to cut a slanting slit. Cut the second slanting slit in the opposite direction to connect the two, forming a triangular-shaped cut. Continue these cuts along the length of the cucumber until reaching the 1″ cut at the opposite end. Repeat this procedure on the opposite side of the cucumber.

2. Separate the two halves.

3. With a parisienne or melon ball scoop, remove seeds and pulp in the center to hollow out the boat.

When ready to serve, fill the center of the boat with a meat or cheese spread, choice of condiment, or sour cream.

Cucumber Tulip

The cucumber tulip is more versatile than most vegetable flowers because it is not only eye-appealing, but it can also be used to hold various condiments or an assortment of spreads when self service is featured. Select a solid, well-formed cucumber.

1. Using a sharp utility or paring knife, cut a thin slice off the end of a cucumber. From the end of the cucumber, measure approximately 2″ along the side of the cucumber and cut the cucumber crosswise, removing the 2″ piece. Test to see that this piece will stand upright.

Cut the outside skin of the cucumber around the circumference into scallop shape. This forms the petals.

2. Scoop out the center of the cucumber using a melon ball or parisienne scoop. Place in ice cold water for approximately 20 min or until the cut petals open slightly. Remove from the water and dry.

3. When ready to serve, fill the tulip center with choice of condiment, cream cheese, or meat spread. If it is to be used as just an attractive garnish, place a little cream cheese in the bottom of the scooped out center and stick very thin strips of carrots into the cheese so they stand up to resemble stigmas.

Leek Flower

A leek is a vegetable of the green onion family. It has long, wide, flat green stems and a white base that has little or no bulb. The two-tone color of the leek is ideal for making an impressive flower that can be used to garnish an assortment of food trays and platters. To make the leek flower, select a fresh, bright-colored leek that has a fairly large base (about 1″ in diameter).

1. Cut the roots off the leek and wash the leek thoroughly. From the base of the leek, measure approximately 5″ and cut crosswise to remove the remaining part of the leek.

2. Using a sharp utility knife, make cuts the length of the leek, starting approximately 1″ above the root end. Make these cuts close together around the circumference of the leek with each cut going through all the layers of the leek.

3. Insert a toothpick or skewer in the base of the leek and spin the flowered end to open the petals. Place the petal end in ice water for approximately 10 min to open the petals further. Placing a few drops of food color in the water creates an assortment of colorful flowers.

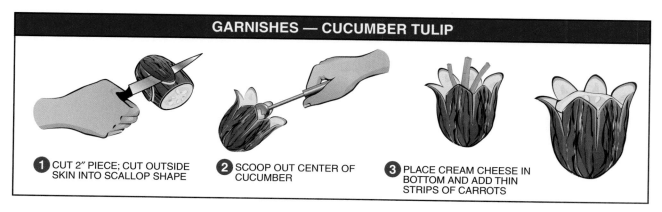

GARNISHES — CUCUMBER TULIP

1 CUT 2″ PIECE; CUT OUTSIDE SKIN INTO SCALLOP SHAPE

2 SCOOP OUT CENTER OF CUCUMBER

3 PLACE CREAM CHEESE IN BOTTOM AND ADD THIN STRIPS OF CARROTS

GARNISHES — LEEK FLOWER

1. CUT OFF ROOTS
2. MAKE CUTS THROUGH THE LENGTH OF LEEK
3. INSERT SKEWER IN BASE; SPIN TO OPEN PETALS

Tomato, Orange, or Lemon Crown

The tomato, orange, or lemon crown is a simple standard garnish used to add eye appeal to buffet or smorgasbord presentations. The garnish selected should have some association with the food or foods it is used to garnish. For example, the tomato crown would be a good selection when garnishing a tossed green salad bowl or chicken salad bowl; the orange crown to garnish roast duck; and the lemon crown to garnish a fish platter or seafood salad. To prepare a fruit crown, select a blemish-free, well-shaped, firm fruit.

1. Using a sharp utility or paring knife, start in the center of the fruit and cut uniform V-shaped cuts around the middle of the circumference of the fruit.

2. Make sure the cuts are uniform and straight. They should penetrate the center of the fruit. The last V cut should connect evenly with the first.

3. When the cuts around the fruit are completed, separate the two halves. Place a cherry, a spot of cream cheese, or some other attractive item in the center of each crown.

Note: A cantaloupe or grapefruit can be cut using this same method.

Tomato Rose

The tomato rose is simple to make and can be completed in a short period of time. It is an ideal garnish for buffet salads, trays, or platters when time is limited. To prepare a tomato rose, select a tomato that is ripe, yet fairly firm.

1. Using a paring or utility knife and starting at the stem end of the tomato, cut a flat surface. Continue this cut around the circumference of the tomato until the strip is approximately 4″ long. The cutting action is similar to peeling an apple. Keep the 4″ strips as thin as possible.

2. Start cutting a second strip where the first strip ended. Cut this second strip approximately 3″ long.

3. Using the first strip, form a loose coil around the stem base. Form a tight coil using the second strip and place this coil into the center of the loose coil. Shape the two coils to resemble a rose.

This step is optional, but the rose will last longer and be more attractive. Place the formed rose in the refrigerator until it is thoroughly chilled. Set it on a screen with a tray underneath. Drip a plain gelatin solution over the rose until it clings and adheres to the coiled strips. The plain gelatin solution is prepared by dissolving 1 tbsp of plain gelatin in ¾ c of boiling water.

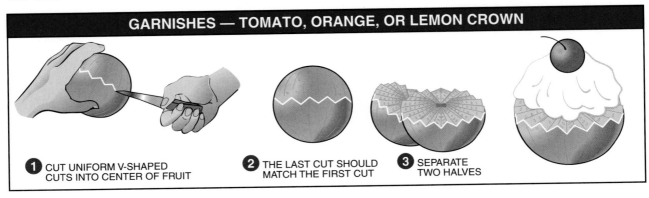

GARNISHES — TOMATO, ORANGE, OR LEMON CROWN

1. CUT UNIFORM V-SHAPED CUTS INTO CENTER OF FRUIT
2. THE LAST CUT SHOULD MATCH THE FIRST CUT
3. SEPARATE TWO HALVES

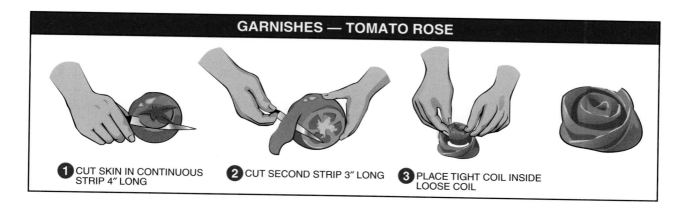

GARNISHES — TOMATO ROSE

1 CUT SKIN IN CONTINUOUS STRIP 4″ LONG

2 CUT SECOND STRIP 3″ LONG

3 PLACE TIGHT COIL INSIDE LOOSE COIL

Turnip Lily

The turnip lily is another attractive vegetable flower used to garnish food platters and trays on the buffet or smorgasbord. It is appropriate when used to garnish during the Easter season.

To make this elegant flower, select a firm, well-rounded, raw turnip with a diameter of approximately 3″.

1. Peel the turnip using a vegetable peeler. Slice it in half crosswise using a French knife. On the electric slicing machine, with the cut part of the turnip facing the blade, slice two very thin slices for each lily.
2. Using one of the thin slices of turnip, shape it to form a cone. Holding the cone-shaped slice in one hand, curve a second slice around the first with the other hand.
3. Secure these two curved slices with a toothpick. Place in ice cold water for approximately 10 min.
4. Peel a small carrot. Cut a strip to place in the center of the turnip lily to act as a stamen. Keep the lily in cold water until ready to use.

Vegetable Palm Tree

The vegetable palm tree is an attractive centerpiece for meat platters, cheese trays, or relish trays. The trunk can be made using carrots, a cucumber, or zucchini. A green pepper is always used for the leaves. To make the vegetable palm tree, select fresh, firm vegetables.

1. Using a sharp paring or utility knife, cut small gashes into the skin of the vegetable selected. Cut these gashes around and up the complete surface of the vegetable.
2. At one end of the vegetable, where the circumference is the largest, cut a flat surface so the trunk will stand up. Some support may have to be used so it will stand properly. Place the trunk in ice water for approximately 2 hr for the gashes to curl open.
3. The green pepper selected for the leaves should have three round surfaces on the bottom to give a more natural effect to the leaves. Cut these three rounds into an oval shape, but do not cut into the top section

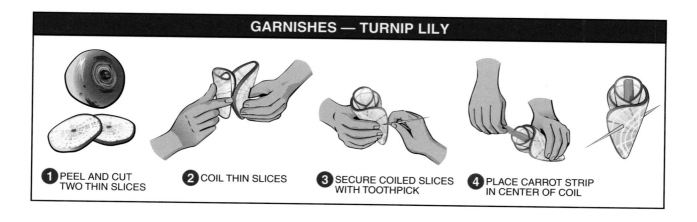

GARNISHES — TURNIP LILY

1 PEEL AND CUT TWO THIN SLICES

2 COIL THIN SLICES

3 SECURE COILED SLICES WITH TOOTHPICK

4 PLACE CARROT STRIP IN CENTER OF COIL

that holds the rounds together. Using the knife or a pair of scissors, cut slanting slits close together into each side of the three oval-shaped pepper leaves. Cut toward the center but not all the way through.

To assemble the tree, stand the trunk on its flat end. Some support may be needed. Place the green pepper leaves on the top of the trunk and secure with a toothpick.

GARNISHES — VEGETABLE PALM TREE

1 CUT SMALL GASHES

2 CUT BOTTOM FLAT

3 CUT THIN SLITS IN PEPPER FOR LEAVES

4 MOUNT TO TRUNK WITH TOOTHPICK

Herbs, Spices, and Seasonings

Chapter 10

Generally speaking, herbs are from the leaves of certain plants, and spices are from the flowers, fruits, seeds, and roots of trees and plants. Herbs and spices are used in virtually every recipe to add to the flavor in a food preparation. Food preparations would be very bland without these important ingredients.

Although herbs and spices are an important part of a recipe, their purpose in the preparation is only to enhance the natural flavors of the food items, never to overpower the flavor of the preparation. Herbs and spices are added before, during, and after the cooking process and produce different results when used on different types of food items.

Herbs and spices should be ordered only as required for use for a limited time period. Whole spices store better and lose less flavor than ground spices. Imitation spices are less expensive, but do not provide the same quality of taste as natural herbs and spices. Herbs and spices are added in small amounts to a preparation; more can be added if required.

HERBS, SPICES, AND SEASONINGS

Herbs and spices are an important ingredient in all food preparations. Without them, most food preparations would be bland. *Seasoning* is the addition of ingredients to enhance the taste and smell of food. The proper use of herbs and spices is learned from experience and offers ways of creating different and more exciting preparations. Many traditional food preparations have been given new life with the addition of a special aroma or flavor.

Herbs and spices, like cheese, are as old as civilization. They are discussed in the Bible, in history books, and in medical books. Many historical events would not have occurred if there was not a desire for tastier foods. Columbus would not have discovered America if he had not been searching for a shorter route to the East for herbs and spices. The people of Europe during this period had become tired of bland, tasteless food. Once aware of the exotic flavors and aromas that could be produced by using exotic spices and herbs, they had to find ways of acquiring them. This desire was a key factor in many voyages during this time.

The difference between herbs and spices is in the aroma and flavor. Spices are pungent (strong) in aroma and often in flavor. Herbs are more delicate in both aroma and flavor. The source of herbs and spices also differs. An *herb* is a seasoning that comes from the leaf, stem, or flower of small annual, perennial, or biennial plants in temperate climates. A *spice* is a seasoning that comes from the fruit, berry, root, or leaf of a tree or plant. A spice, by itself, is pungent, zesty, and aromatic. Today, spices may include other aromatic seeds that are found on the market. Spices are commonly grown in tropical climates.

Herb and Spice List

The following is a list of the popular herbs and spices used in commercial kitchens throughout the world. The chef should be familiar with all of these. Herbs and spices are listed including their origin, characteristics, and some of the common uses of each.

Allspice: This spice is the dried, unripened fruit of the small pimiento tree of the clove tree family, which is grown in the west Indies. Many people think, because of its name, that allspice is a blended spice. This is not true, although this pea-shaped spice does possess a flavor that suggests the combined flavors of cinnamon, nutmeg, and cloves. It is because of its flavor, in fact, that it is known as "allspice." Allspice is used in both whole and ground form and is used in such preparations as mincemeat pie, pumpkin pie, puddings, stews, soups, preserved fruit, boiled fish, relishes, and gravies.

Anise: This spice is a small annual plant that stems from the parsley family and produces a comma-shaped seed known as anise seed. This spice has a licorice flavor. The use of anise dates back to ancient Egypt. Historically, anise has had many uses. It has been used to prevent indigestion, as an antidote for scorpion bite, as a safeguard against evil, and even as a perfume. Today, this spice is used in coffee cake, sweet rolls, cookies, sweet pickles, licorice products, candies, cough syrups, and certain fruits. Anise is grown in India, southern and northern Europe, Chile, Mexico, and the United States.

Basil or sweet basil: This herb is one of the most savory and popular herbs because it blends well with many different foods. It is a native of India, where it is considered to be holy. In many countries throughout southern and eastern Europe, this fragrant herb is considered a token of love. Basil consists of the dried leaves and stems of an annual plant of the same name. It is grown in Europe and the United States and is used to flavor tomato paste products, spaghetti sauces, vegetables, and egg dishes.

Bay or laurel leaf: This spice has always been held in high esteem. Roman emperors, such as Julius Caesar, Tiberius, and Claudius, wore a wreath of laurel. Celebrated scholars and athletes were crowned with a wreath of laurel because it was considered a symbol of the triumphant leader or champion. From this custom has arisen our modern expressions "to rest on one's laurels" or "to win one's laurels." The bay or laurel leaves are the thick, aromatic leaves of the sweet-bay tree grown in Italy, Greece, and other Mediterranean countries. They are used for flavoring soups, roasts, stews, gravies, and meats, and for pickling.

Cajun spice: This is a blended spice that consists of red and white pepper, salt, and natural herbs and spices. Each manufacturer has its own blend for flavor. Cajun spice was introduced on the market when Cajun-style cooking became popular nationally. Cajun spice is used in blackened preparations including redfish, orange roughy, steaks, pork chops, and chicken.

Capers: These are the green unopened flower buds of a European plant very similar to the nasturtium plant. The small buds are dried and added to a vinegar solution and are also sold salt cured. Capers are used in sauces and as a garnish on certain salads.

Caraway seed: This spice is the dark brown dried seeds of the caraway plant, a biennial plant that grows in Holland, Germany, England, and Poland. More caraway seed is used in food preparations in Germany than in any other nation. The Germans like its flavor not only in rye bread, but also in sauerkraut, pork, and cabbage. In the United States, caraway seed is thought of as "rye seed" because it is always used in the preparation of rye bread. There are, of course, many other uses for this flavorful seed. It can be used in cheese, potatoes, stews, and soups with excellent results. Caraway seed is also one of the major flavoring ingredients in kümmel, a popular liqueur.

Cardamom: This spice is a member of the ginger family. It is the dried, immature fruit of a tropical bush that grows to a height of about 10′. The fruit consists of a yellowish colored pod about the size of a small grape, which holds the dark, aromatic cardamom seeds. Cardamom is considered the world's second most valued spice. Only saffron is considered more valuable. Cardamom is available in whole or ground form and is used in pickling, coffee cake, curries, and Danish pastry. It is grown in the Far East.

Cassia: This spice is very similar to cinnamon. It takes an expert to distinguish between the two when in ground form. Cassia, like cinnamon, is the bark of a tree; however, cassia bark is thicker and the color is darker. The flavor of cassia is much stronger than the flavor of cinnamon. Cassia is used in pickling, preserving, and in many of the same preparations as cinnamon.

In fact, in many cases, it is used as a substitute for cinnamon. Cassia is grown in Malaysia and China.

Cayenne pepper: This spice is the hottest of all peppers. It is ground from the small pods of certain varieties of hot peppers. It is not as bright in color as red pepper, but it is much hotter. Cayenne peppers are grown in South America and Africa. Cayenne pepper is used in cream soups, cream dishes, meat, fish, cheese, and egg dishes. Be sure to use cayenne pepper in moderation; it is very hot.

Celery seed: This spice has little connection with the celery stalk that is classified as a vegetable. It does have a similar flavor, but celery seed is actually a wild variety of celery. The seeds are very small and brown in color. Celery seed can be purchased whole or in ground form. This spice is grown in France, India, and the United States. It is a delightful seasoning for cole slaw, potato salad, sauces, soups, dressings, fish, and certain meats.

Chervil: This herb has always been admired for its lovely fern-like leaves and its delicate flavor, which is similar to parsley. Chervil is used freely in French kitchens as a substitute for parsley because it is considered more delicate in flavor. It is grown in England, northern Europe, and the United States. It can be used in salads, soups, and egg and cheese dishes.

Chili powder: This is a blended spice consisting of cumin, Mexican peppers, oregano, and other spices. It most likely originated in Mexico. Today, it is used throughout the world, especially in the United States. It is used in preparing chili con carne, tamales, stews, Spanish rice, gravies, and appetizers.

Chives: This herb is small and delicate with onion-like sprouts that are long and green. Chives, which are very delicate, can be grown indoors or outdoors. The chefs of many establishments grow their own chives in small flower pots in their kitchens. Chives add color and flavor to cottage cheese, cream cheese, egg dishes, soups, salads, and potato dishes.

Cinnamon: This spice is the dried, thin, inner bark of a medium-sized evergreen tree grown mainly in the Far East. The bark is harvested during the rainy season because the damp atmosphere makes the bark easier to manage. After the bark has been removed from the tree, it is rolled into moderately long quills known as cinnamon sticks. Cinnamon can be purchased whole or in ground form. It is used in baking apples, pickling, preserving, pies, cakes, puddings, stewed fruits, custard, and sweet doughs.

Cloves: This spice consists of the dried, unopened buds of the clove tree. The tree is very thick and has leaves similar to the laurel leaf. The cloves grow at the end of the twigs in clusters of about twenty. When the buds start to sprout, they are white in color. When they are ready to be picked, they are red. When they are dried, they become a dark brown. The clove tree only grows in mountainous regions and requires a tropical climate to survive. The clove is referred to as the "nail-shaped spice" and is thought to possess the most pungent flavor of all the spices. In fact, the clove is so rich and pungent in flavor that it is often used in toothache medicines to deaden pain. The clove tree is grown in the East Indies and the islands off the coast of Africa. Cloves are used in pickling and in the preparation of roast pork, corned beef, baked ham, soups, applesauce, fruitcakes, pumpkin pie, and cakes.

Idaho Potato Commission

Chives are often used as a topping for baked potatoes.

Coriander: This spice dates back to the beginning of civilization. Coriander grew in the hanging gardens of Babylon and was placed in ancient Egyptian tombs. Coriander is a small seed and is light brown in color. To many, the whole seed is similar in appearance to whole white pepper. It has a very pleasant taste that suggests the combined flavors of sage and lemon peel. Coriander is grown in Morocco and in many of the Mediterranean countries. It is used in candies, pickles, frankfurters, baked goods, meat products, and curry dishes.

Cumin or comino: This spice is a member of the carrot family. It was said of cumin that when it was in the possession of a wife, it would keep the husband from wandering. As a result, most ancient wives kept it handy at all times. Cumin, sometimes known as comino, originated in Egypt. It is the dried, aromatic seeds of a plant similar to the caraway plant that grows to a height of about 1′. The spice has a slightly bitter, warm flavor and is used quite freely in Italian

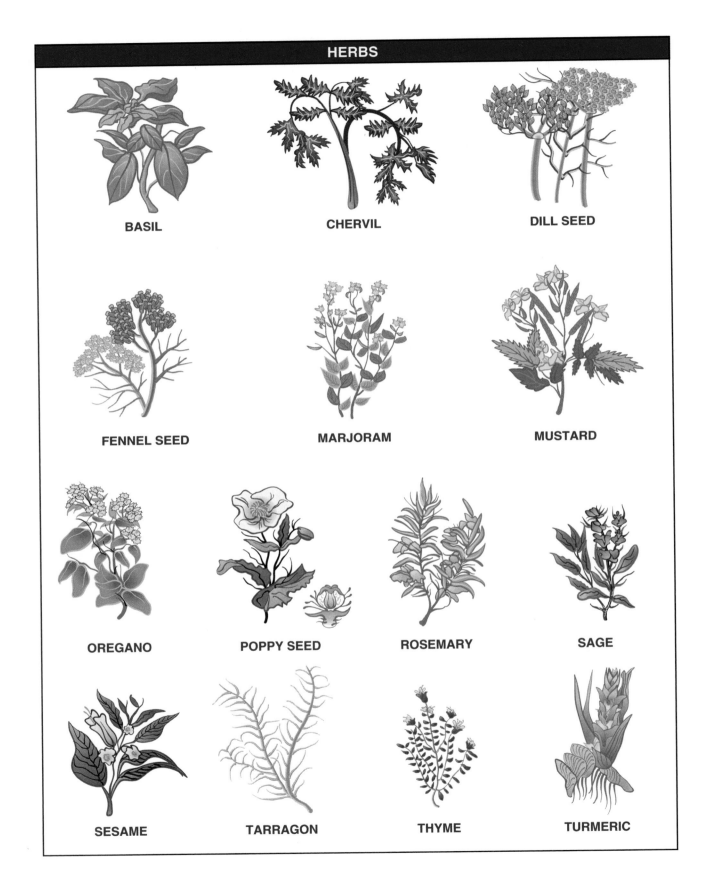

HERBS

BASIL

CHERVIL

DILL SEED

FENNEL SEED

MARJORAM

MUSTARD

OREGANO

POPPY SEED

ROSEMARY

SAGE

SESAME

TARRAGON

THYME

TURMERIC

SPICES

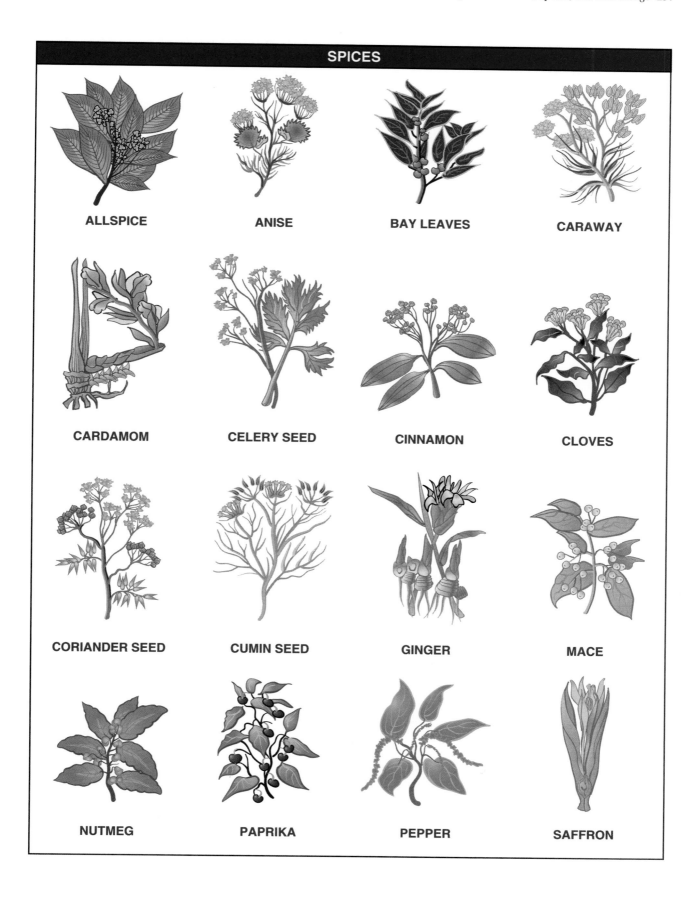

| ALLSPICE | ANISE | BAY LEAVES | CARAWAY |

| CARDAMOM | CELERY SEED | CINNAMON | CLOVES |

| CORIANDER SEED | CUMIN SEED | GINGER | MACE |

| NUTMEG | PAPRIKA | PEPPER | SAFFRON |

and Mexican cuisine. Cumin is grown in Morocco, India, Egypt, and South America. It is an essential ingredient in the blending of curry powder and chili powder. This spice is used to flavor chili, soup, tamales, and rice and cheese dishes.

Curry powder: This blended spice originated in India many years ago, but it has only become popular in America during the twentieth century. It consists of 12 or more spices blended in the proper amounts to create the flavor often associated with India. The color and, to some degree, the flavor of curry vary with different manufacturers. Since India is the home of curry, it is only natural that they would excel in curry preparation. There are, however, many different forms of curry in use. For example, the curry dishes served in some regions of India are hot because the people like to use generous amounts of red pepper. In other regions, however, they desire a milder curry. Curry powder is used to make curries of meat, fish, chicken, and eggs. It is also used to season rice, soups, and some shellfish preparations.

Dill: This herb is a member of the parsley family. The small, flat aromatic seeds, as well as the dried leaves of the dill plant, are used. The seeds, however, are preferred over the dried leaves. Dill is grown throughout the countries of Europe and also in the United States. It has become famous because it is used to flavor dill pickles. Dill is also a welcome addition to such preparations as green beans, potato salad, poached fish, marinated cucumbers, cauliflower, and some lamb dishes.

Fennel seed: This herb is the small seedlike fruit of the fennel plant. Fennel seed is light brown and resembles anise seed in flavor and aroma. Fennel is grown chiefly in India and eastern Europe and is used freely in Scandinavian cooking and Italian baking. Fennel seed also enhances the flavor of roast duck and some chicken preparations.

Garlic: This herb is a member of the onion family. It has been used for centuries by French, Spanish, Italian, and Mexican cooks. However, the use of garlic did not become popular in the United States until after World War II. This popularity was created when the soldiers returned home and began to prepare some of the dishes they had eaten in Europe – dishes that required a hint of garlic.

Garlic is a potently flavored bulb of the onion family. The bulb contains about a dozen compactly arranged cloves that are covered with a thin skin. If the skin is white, it is known as white garlic. If it is red, it as known as red garlic. Garlic grown in a warm climate differs from garlic grown in a cold climate. Garlic grown in a cold climate is stronger in flavor. Garlic is grown throughout the world and is available in many forms: whole, dehydrated, garlic salt, garlic powder, and instant minced garlic. Be discreet when using garlic. A little is very helpful; too much is overpowering. Garlic helps accent the flavor of sauces, soups, salads, pickles, meat preparations, and salad dressing.

Ginger: This spice is one of the oldest, if not the oldest, known spice. It is an essential ingredient in the preparation of gingerbread. The Romans, for example, loved the honey-sweet flavor of gingerbread and carried the recipe throughout their empire. Ginger is the root of a subtropical plant grown in China, India, and Jamaica. When the plant is about a year old, the roots are dug up, dried in the sun, and ground. Ginger has a very warm, pungent, spicy flavor that adds zest to cakes, cookies, pies, fruits, puddings, and some meat preparations. Ginger is available in whole or ground form.

Mace: This spice is the lacy covering of the nutmeg shell. It lies between the nutmeg shell and the outer covering. Mace is reddish orange in color when removed from the shell, but turns to a dark yellow when dried. Its flavor resembles to some degree that of the nutmeg, although it is not as pungent as the nutmeg. Mace is the traditional spice for pound cake and sweet doughs. It is also used in doughnuts, cherry pie, chocolate dishes, oyster stew, spinach, and pickling.

Marjoram: This herb is a member of the mint family. The perennial grows about 2′ high and the dried leaves and flowering tops are used to produce the sweet, minty flavor that is similar to oregano. The marjoram plant originated in regions of western Asia and has always been a symbol of honor and happiness in many countries of the world. Marjoram is grown in France, Chile, Peru, and England. It is used to flavor soups, stews, sausage, cheese dishes, and lamb preparations.

Mint: This herb is a member of the mint family. It originally came from the Mediterranean area. It was held in high esteem by the people of the ancient world. Today it is widely known and has become one of the most popular herbs because of its cool, refreshing flavor. It is grown throughout the world and can be found growing in the backyards of city and suburban homes. There are many varieties of mint, but only spearmint and peppermint are found on the spice shelf. Mint is used in the preparation of lamb dishes, vegetables, fruit salads, iced tea, fruit drinks, and poached fish.

Monosodium glutamate (MSG): This seasoning is a white crystal chemical compound used to enhance and intensify the flavor of certain foods. MSG was originally extracted from seaweed and certain dried fish by the Chinese and Japanese to enhance the flavor of rice, seafood, and vegetables. MSG is sometimes used as a flavor enhancer in a variety of food preparations. However, MSG can cause allergic reactions in some people and as a result it is used increasingly less in commercial kitchens. Items containing MSG should be identified on a menu.

Mustard seed: This herb comes in two forms: the white or yellow seeds, which are mild in flavor, and the dark brown seeds, which are more pungent in flavor. The dark brown seed is most often used in Chinese cuisine. Mustard seeds are very small. Mustard, when in ground form, is unique in that its flavor is not released until it is blended with water. After the powdered mustard stands in the liquid for about 10 min it is at its best. Mustard can be purchased in three forms: whole, ground (powdered), and the ever-popular prepared mustard, which is a blend of ground mustard seed, other spices, and vinegar. Mustard seed is grown in Canada, Denmark, the United Kingdom, the Netherlands, and the United States. It enhances the flavor of pickles, cabbage, beets, sauerkraut, sauces, salad dressings, ham, frankfurters, and cheese.

Nutmeg: This spice is the kernel of the nutmeg fruit. The tree that produces this fruit is bushy and reaches a height of about 40′. It is an evergreen and grows best in tropical climates near the sea. The tree produces its first fruit when it is 6 years or 7 years old and continues to produce for about 60 years. Good production for a nutmeg tree is around 1000 nutmegs per year. When the nutmeg fruit is ripe, the outer hull splits open, exposing the sister spice *mace*, which partially covers the nutmeg kernel. Nutmeg is grown in the Dutch East Indies and British West Indies. Its sweet, warm, spicy flavor can be used to enhance such preparations as cream soups, doughnuts, puddings, baked goods, potato preparations, custards, cauliflower, sauces, hash, and stews.

Oregano: This herb is very popular in the United States today because of its use with pizza. The American soldiers of World War II brought the taste for pizza home with them. Oregano is obtained from a small plant similar to the marjoram plant. The leaves are slightly curly and small. The flavor is pleasingly pungent and, to some degree, resembles the flavor of marjoram. Oregano is native to the Mediterranean region and is grown extensively in Greece and Italy. In Mexico it grows wild and, for this reason, is sometimes referred to as *Mexican sage*. It can be used in tomato products, Mexican and Italian dishes, cheese preparations, and pizza.

Paprika: This spice is a blend obtained by grinding and mixing together various sweet red peppers after the seeds and stems have been removed. There are two kinds of paprika used in the commercial kitchen: Spanish and Hungarian. Spanish paprika is slightly mild in flavor and has a bright red color. Hungarian paprika is darker in color and more pungent in flavor. Paprika is grown in Spain, central Europe, and the United States. It is used as a colorful garnish for many foods. In some cases it is used to help brown food. Paprika is a necessary ingredient in such preparations as Hungarian goulash, chicken paprika, Newburg sauce, French dressing, and veal paprika.

Parsley: This herb is a very popular garnish. It is one of the delicate herbs. It has been cultivated for thousands of years, but its origin seems to have been in ancient Greece. Parsley is mentioned in the stories of Hercules. The Romans also used parsley to make crowns for their guests, which they thought prevented excessive drinking at banquets. Parsley is a garden plant. The leaves are used as a garnish or to flavor other foods. Parsley is grown throughout the world and is used in soups, salads, stuffing, stews, sauces, potatoes, and vegetable dishes.

Pepper (black and white): This spice is the most common of all spices. It is a native of the tropics and never grows further than 20° from the equator. The two kinds of pepper are black and white, and both are produced by the climbing vine known as the pepper plant. Black pepper is the dried, immature berries. They are picked when still slightly green, left to dry in the sun, and either sold whole as peppercorns or ground. Black pepper is quite pungent in flavor.

White pepper is the mature berry of the pepper plant after the outer covering, which contains most of the hotness, has been removed. It is much milder in flavor than black pepper. It, too, is sold in either whole or ground form. White pepper is generally more costly than black pepper because the peppercorns used for white pepper are usually more carefully cultivated.

Pepper is grown in several far eastern countries. The rule to follow when seasoning with pepper is that if the item is light in color use white pepper; if dark, use black pepper. Fresh ground pepper always supplies more flavor and aroma than pepper purchased in ground form.

Pickling spice: This spice is sometimes known as a mixed spice. It is a blend of ten or more whole spices used mainly for pickling purposes. It is also an excellent addition to stocks, soups, relishes, sauces, and some meat preparations such as pot roast and sauerbraten. In most cases, when pickling spices are used in cooking, they are added in the form of a spice bag. The spice is tied in a piece of cheesecloth, added to the preparation, and removed when the desired amount of flavor is acquired.

Poppy seed: This herb is very small and light. It would take about 900,000 of the seeds to make a pound. These blue-colored flavoring seeds are the seeds of a specially cultivated poppy plant that grows chiefly in Holland. However, it can also be found growing in other parts of the world. Poppy seeds are best known for garnishing rolls and bread. They can also be used as a topping for cookies and in butter sauces for fish, vegetables, and noodles.

Poultry seasoning: This seasoning is a ground mixture of sage, thyme, marjoram, savory, pepper, onion powder, and celery salt. It is most often used in seasoning bread dressing, but has other uses as well. It will also help the flavor of meat loaf and dumplings.

Rosemary: This herb is the dried, somewhat curved, needle-like leaves of an evergreen plant of the mint family. It is considered to be one of the two very strong herbs. The other is sage. The flavor is fragrant and sweet-tasting. Rosemary has been used for centuries as a symbol of fidelity for lovers. In many European countries today, the herb is used to stuff pillows and to supply the fragrance for soap, toilet water, and cosmetics. Rosemary is grown in France, Spain, and Portugal, and is a great asset when used sparingly with lamb, chicken, pork, and duck. It also enhances the flavor of tomato and cheese dishes, stuffing, and soups.

Saffron: This spice is very costly, but a little will go a long way. Saffron is the dried, bright red stigmas of the purple, crocus-like flower of the saffron plant. It takes about 225,000 stigmas from over 75,000 flowers to produce 1 lb of this very desirable spice. There are only three stigmas to each flower and these stigmas are removed by hand.

Saffron is grown in Spain and the Mediterranean region. It imparts a very agreeable flavor as well as a rich, deep yellow color that is desired in such preparations as rice and fine bakery goods. Saffron is used quite freely in Scandinavian and Spanish cuisine.

Sage: This herb is the greenish-white leaves of the sage plant. It is a low-growing, perennial shrub that possesses a minty spiciness. It is considered to be one of the two very strong herbs. The other is rosemary. Sage is grown throughout the world, but the most choice sage comes from eastern Europe. Sage can be purchased either in leaf form (whole), rubbed (crushed), or in ground form. It is often used for poultry stuffing. It is also used in the making of sausage and in bean and tomato preparations.

Salt: This seasoning is the most popular seasoning in the world. It is a white crystalline substance found in seawater, mineral springs, and subterranean beds. Salt is primarily used to season foods and preserve meats. Salt is chemically classified as sodium chloride. Sodium is an essential nutrient required by the body. However, most Americans consume too much salt, which can be harmful. Follow the recipe for the correct amount of salt. Additional salt can always be added by the consumer. Flavored salts containing predominant flavors such as garlic, onion, and celery have become popular.

Savory: This herb is the dried, smooth, slightly narrow leaves of the savory plant. It is a member of the mint family. The two kinds of savory found on the market are *summer savory* and *winter savory*. The summer savory is the best because during the summer the flavor of the leaves is at the peak of quality. Savory is grown in France and Spain. It possesses a delicately sweet flavor similar in some respects to thyme. Savory is an important ingredient in flavoring green or dried beans. It is also used in meat sauces, fish sauces, egg dishes, and meat stuffings.

Scallions: These members of the green onion family have a very small bulb or, in some cases, no bulb at all. They resemble leeks, but the stems are much smaller and the flavor is much stronger. They are used in salads, as a relish, in soups, and in sauces.

Sesame seed: This herb is a small, honey-colored seed with a toasted almond flavor and a high oil content. Sesame seeds come from pods that grow on the sesame plant, an herb growing about 2′ high. It is native to tropical and semitropical countries. The sesame plant is grown in Central America, Egypt, and the United States. Sesame seeds are baked on rolls, bread, and buns to supply a nut-like flavor to the product. The seeds are also toasted and stirred into butter and served over fish, noodles, and vegetables.

Shallots: These members of the green onion family have a bulb that consists of several cloves, similar to the garlic bulb. Shallots are very pronounced in flavor and are a favorite of many chefs in such preparations as stews and sauces.

Tarragon: This herb is native to the vast wastelands of Siberia. It is a small perennial plant, the leaves of which are slightly long and smooth. It is best known as a flavoring for vinegar, and the fresh leaves are often used to decorate aspic and chaud-froid pieces. Tarragon has a flavor that suggests a touch of licorice, similar to anise. Tarragon blends well with seafood and is a delightful addition to salads and salad dressings. This herb seems to be a favorite of French chefs and is used more often in French cuisine than any other herb.

Thyme: This herb is a member of the mint family. It has been popular in the United States for many years, mainly because it is the finest herb to use with fish and shellfish. Thyme is the leaves and tender stems of a low-growing shrub. The leaves and tender stems are picked just before the blossoms start to bloom, then cleaned and dried. Thyme is available in whole or ground form and is a welcome addition to beef stew, clam chowder, oyster stew, meat loaf, poultry seasoning, and vegetable preparations. Thyme is grown principally in southern France and Spain. There are nearly 100 varieties of thyme.

Turmeric: This herb is the root of a lily-like plant of the ginger family. It is native to Asia and is used not only in food preparation, but also in medicine as a dye. When turmeric roots are ground, a bright yellow powder is produced that has a taste similar to mustard. Turmeric is an important ingredient in blending curry powder and is also used in making some prepared mustards. To some degree, turmeric is associated with the spice saffron because both possess a deep yellow color and are used in much the same way in food preparation. Turmeric is grown in India, Peru, Haiti, Jamaica, and wherever ginger thrives.

Vinegars, Dressings, and Seasonings

Beef base: This concentrated beef mixture is used to provide a rich beef flavor for various items, such as beef stock and gravy.

Chicken base: This concentrated chicken mixture is used to provide a rich chicken flavor for various items, such as chicken soup and sauces.

Chutney: This East Indian pickle relish is prepared from currants, cucumbers, apples, ginger, mustard seed, etc. It is usually served with curry dishes.

Cider vinegar: Cider vinegar is made by fermenting apple juice. It has a light or slightly dark brown color and is used more often than any other type of vinegar in the commercial kitchen. Cider vinegar is used as a flavoring on salads and in salad dressing and as a pickling agent.

Distilled or white vinegar: White vinegar is made by fermenting diluted distilled alcohol. It is used most often in pickling and when a weaker vinegar is desired.

Mayonnaise: This is the thick, uncooked emulsion formed by combining salad oil with egg yolks, vinegar, and seasoning. Mayonnaise may be made in the commercial kitchen or purchased in convenient size jars. It is used in salad dressings, salads, and as a sandwich spread.

Salad dressing: This is a cooked product with a mayonnaise base that contains a filler or stretcher consisting of water and starch. This filler or stretcher is whipped into the mayonnaise base until it is smooth and creamy. Salad dressing is much sweeter than mayonnaise and is used in salads, sandwich spreads, and other preparations where mayonnaise is used.

Salad oil: Salad oil is obtained from the kernel of the corn or seed of the cotton plant. Both oils are golden in color, bland, and can withstand a very high degree of heat without smoking. Salad oil is used in the preparation of mayonnaise and other salad dressings, as well as for frying and sautéing certain items.

Soup base: This concentrated mixture made from evaporated stock, meat juices, and vegetables is used to help provide a rich ham flavor for certain soups and vegetable preparations.

A convenient and easy way to store fresh stock is to pour it into ice cube trays and freeze. These stock cubes can then be stored in a plastic bag for later use.

Soy sauce: This dark brown sauce is made by mixing mashed soybeans, roasted barley, salt, and water. A culture is added and the mixture is left to ferment for 6 months to 18 months in vats. At the end of the fermentation period, this mixture is pressed and strained to produce soy sauce. Soy sauce is commonly used in oriental preparations.

Tarragon vinegar: This is cider vinegar flavored with tarragon (an herb). Tarragon vinegar has a distinctive flavor and is generally used in salads or salad dressings. It can be purchased premade or it can be made by letting the herb soak in cider vinegar for a few days.

Tabasco sauce: This is a very hot, red-colored sauce made from red peppers, vinegar, and salt. Tabasco sauce is usually packaged in small shaker-top bottles and used in flavoring meat sauces, salads, and soups.

Worcestershire sauce: This is a pungent, dark-colored sauce. It is used in cooking and for seasoning prepared meats, such as steak.

COOKING TECHNIQUES

Successful cooking with herbs and spices requires patience, practice, and following some basic rules.

- Herbs and spices should be used to enhance natural flavor.
- Herbs and spices should not disguise the flavor of any food.
- A specific herb or spice should not appear dominant in a food preparation. Exceptions to this rule include curry or chili dishes where the character of the preparation depends on the specific spice or herb.
- Always season in moderation. More seasoning can always be added, but removal is impossible.
- If the cooking time is long, add spices and herbs near completion if in ground form. If the herb or spice is whole, add at the start of the preparation.
- If the cooking time is short, add the spices and herbs at the start.
- Rub herbs between the palms before adding to a preparation to release their flavor.
- When using fresh herbs or spices in soups or sauces, tie them in a bunch for easy removal after use or mince fine to leave in soup or sauce.
- In uncooked dishes, the herbs or spices should be added hours before serving for the flavor to develop. This is especially important in salad dressings, fruit juices, and marinades.
- Pepper, the most common of all spices, should be used in moderation. Pepper may be added at the dinner table to suit individual taste.

- Bay leaves used in food preparation must be removed when the desired flavor is achieved.
- In some preparations, an herb or spice can ruin the appearance of the preparation. In addition, sometimes it is impossible to remove the herb or spice when the desired flavor is achieved. To avoid these problems, simmer the herb or spice in a small amount of water for a short period of time. Strain the flavored liquid through a cheesecloth. Discard the herb or spice. Use the strained flavored liquid to season the item being prepared.

PURCHASING

When purchasing herbs and spices, buy in small amounts as flavor is lost even when stored under proper conditions. The best storage condition is a dry and cool place where heat will not remove any flavor and dampness will not cause caking. As soon as spice is ground it will start to lose its flavor. However, if kept in a tightly closed container, deterioration will be retarded.

Herbs lose their flavor faster than spices if in rubbed or ground form. As a result, most users purchase herbs in whole form because they keep better. Good color, strong flavor, and aroma are the important points to consider when buying herbs.

Herbs and spices should be tested for freshness when delivered. A good test is to examine the color for a bright, fresh, and rich appearance. In the case of herbs, rub a small amount in the palm of the hand and smell. The scent should be fairly strong and fresh. Test spices by placing some in the palm and bringing it up slowly to the nostrils. The aroma should be fresh and come up to meet you.

Imitation spices are less expensive than natural spices and are made by spraying the oil of the pure spice on a carrier such as ground soya, buckwheat, or cottonseed hulls. However, imitation spices do not possess the same strength and quality as natural spices. The savings in cost does not make up for the loss in quality. The most common imitation spices are pepper, nutmeg, cinnamon, and mace.

All spices are generally strong in flavor. Herbs vary in flavor strength. For simplicity, herbs may be divided into three groups based on flavor strength: very strong, fairly strong, and delicate.

- *Very strong herbs:* sage and rosemary leaves.
- *Fairly strong herbs:* basil, mint, marjoram, tarragon, dill, thyme, etc.
- *Delicate herbs:* chives, parsley, and chervil.

Salads and Salad Dressings

## Chapter 11

The National Chicken Council

Salads are a popular menu item and can be served as an appetizer, an accompanying dish, or a main dish, depending on the ingredients used. The four basic parts of a salad are the base, body, dressing, and garnish. The base usually consists of a salad green and provides contrast in color. The body is the main or predominant ingredients used in the salad. The dressing adds flavor to the salad. Garnish adds eye appeal and enhances the flavor of the salad.

Salads are nutritious, providing many essential vitamins and minerals. Salads are categorized as fruit, vegetable, leafy green, meat, seafood, gelatin, and pasta salads. Fruit salads use canned, frozen, or fresh fruits with other ingredients. Vegetable salads include cooked or raw vegetables. Leafy green salads use salad greens as the body or main ingredient. Meat salads use prepared cooked meat. Seafood salads use meats that have a delicate flavor or meats that can be blended with other ingredients. Gelatin salads can be prepared in many molded forms and offer a variety of colors. Pasta salads use various pasta shapes as a body for the salad.

SALADS AND SALAD DRESSINGS

Salads are a preparation consisting of a combination of cold ingredients served with a dressing. The cold ingredients usually include salad greens, meat, fruit, vegetables, poultry, dairy products, seafood, and/or pasta. Salads may be served as a main dish (entree), an accompanying (side) salad, garnish salad, or as a dessert salad. Salads as a main dish have become increasingly popular as people are more conscious of watching their diet. Such salads are usually a large portion garnished with other foods, such as meat and cheese. The side salad may be served as an appetizer or as an accompaniment to the main course of the meal. The size of the side salad is reduced when used as an appetizer.

Salad bars, which allow the customer to make salads using a variety of ingredients, can be used for main dish salads or side salads. The arrangement and placement of ingredients should allow the customer to easily gather ingredients and should conform to local health standards.

The most common method of classifying salads is by the ingredients used. Most salads are classified under one of the following categories:

- Fruit
- Vegetable
- Leafy green
- Meat
- Seafood

- Gelatin
- Pasta

For the best results in salad preparations:

- Ingredients should be fresh.
- Arrange ingredients to take full advantage of their natural colors.
- Place or mold foods to create various levels. Flat surfaces have little or no beauty.
- Use the best ingredients.
- Salads should be well-chilled.
- Keep the preparation and garnish simple.
- Choose food combinations that blend well together.
- All salad preparations must be neat, attractive, and appetizing.
- A good salad has a balance of flavor, color, and texture.
- Cut ingredients to retain crispness.
- Prepare salads as close to serving time as possible.
- When tossing a salad, use a very light, gentle motion. Use large forks or a large spoon and fork. Toss close to serving time and never overwork the ingredients.

Parts of a Salad

Salads commonly contain four basic parts: base, body, dressing, and garnish. Each part contributes to the total preparation.

Base: The base usually consists of salad greens such as leaf lettuce, romaine, head lettuce, or Bibb. It can be eaten, but in most cases the guest will choose to let it remain on the plate or in the bowl. The main purpose of the base is to keep the plate or bowl from looking bare and to provide contrast in color when the body is added.

Body: The body is the main part of the salad. The type of salad determines the ingredients used.

Dressing: The dressing is poured over or mixed in the salad and is usually served with all salads. It adds flavor, acts as a binder, provides food value, helps digestion, improves palatability, and in some cases acts as a garnish.

Garnish: The main purpose of the garnish is to add eye appeal to the finished product. In some cases a garnish may help improve the taste of the salad. The garnish should be kept simple at all times. It should attract the attention of the diner and help stimulate the customer's appetite. Table I lists some common garnishes.

TABLE I: GARNISH SUGGESTIONS FOR SALADS	
Beets, slices or julienne	Lemon slices or wedges
Bonbons – Marshmallows rolled in coconut	Melon balls or wedges
Carrot curls or sticks	Nuts, whole or chopped
Celery curls, celery hearts	Olives, green and ripe
Cheese – American, Swiss-julienne, bar, sliced, shredded	Onion rings
Cheese balls or bars rolled in chopped parsley or paprika	Orange twists or slices
Cherries, canned or maraschino	Paprika, dash of
Coconut, plain or colored	Parsley, sprig or chopped
Cranberry relish	Pickles, all kinds
Croutons	Pimiento, strips or chopped
Cucumber slices or curls	Pineapple fans or fingers
Dates or prunes, stuffed	Poppy seed
Eggs, hard cooked, sliced, quartered, or stuffed	Radishes, plain, roses, accordion, or sliced
Fresh berries or fresh fruit	Strawberries, whole or sliced
Grapes, fresh, sugared	Tomato slices and wedges
Green pepper rings or strips	

Salad Greens

All salad greens must be washed, drained, trimmed, and usually stored before they are used in salad preparations. These tasks must be done properly or the greens will bruise, start rusting rapidly, and lose their desired crispness.

Wash all greens thoroughly two or three times in cold water. The loosely-packed heads should be separated and each individual leaf washed separately. The task must be done gently because certain tender greens bruise easily. It may even be necessary to cut some of the elongated heads in half lengthwise in order to remove all dirt, grit, and sand.

The solid-packed heads, such as head lettuce, may be left in their natural form and only the core removed before washing and draining. To remove the core and avoid bruising, hold the head of lettuce in the palm of one hand with the core facing down. Apply pressure against the core on a counter or flat surface. The core should pop free so it can be lifted out. It can also be removed by cutting around it with the tip of a very sharp utility or paring knife.

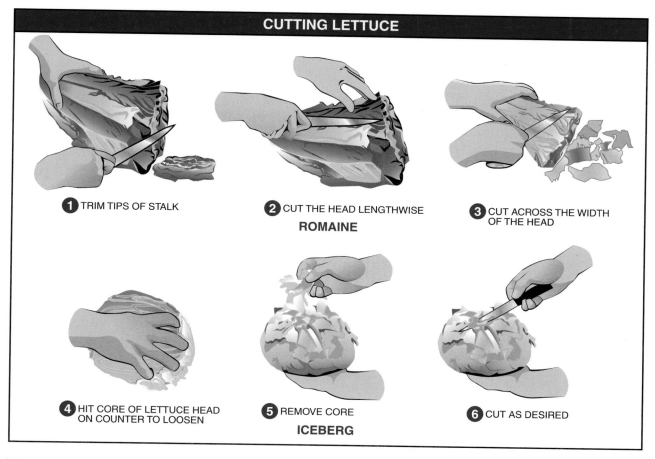

CUTTING LETTUCE

1 TRIM TIPS OF STALK

2 CUT THE HEAD LENGTHWISE
ROMAINE

3 CUT ACROSS THE WIDTH OF THE HEAD

4 HIT CORE OF LETTUCE HEAD ON COUNTER TO LOOSEN

5 REMOVE CORE
ICEBERG

6 CUT AS DESIRED

Using proper techniques to remove cores and cut lettuce will avoid bruising and extend keeping times.

All washed greens should be drained thoroughly and stored in a perforated stainless steel pan with a second solid pan as an underliner to hold the drippings. Cover the washed greens with a damp cloth or plastic wrap to retain crispness. Crispness may be improved by adding slices of lemon to cold water and letting the greens soak for a short period of time. When removed from this solution, again drain thoroughly. Always remember to be very gentle when handling extremely tender and fragile greens. Check the temperature of the refrigerator before storing the greens. Make sure the temperature is well above freezing (38°F to 40°F) and place the greens on an upper shelf in the refrigerator because the lower area is the coldest. If any part of the lettuce is frozen, it must be discarded because its appearance and crispness will have been damaged.

Salad greens comprise the body or main ingredient of leafy green salads. Leafy green salads are the most popular salads served. Many different kinds of salad greens are used in salads.

Belgian endive has a slender, tightly-packed elongated head that forms a point. The size is usually about 4″ to 6″ in length. The leaves are creamy white and possess a slightly bitter taste. The head is split in half lengthwise before it is cleaned by washing thoroughly. A half head portion set on another contrasting green as a base makes one of the most attractive salads. The slightly bitter taste is desired by many gourmets when it is served with a dressing that enhances its flavor. Belgian endive is generally expensive.

Bibb or *limestone lettuce* is similar in many ways to Boston lettuce. It has the same size loosely-packed round head, but the leaves are darker green and crisper. It grows best in soil containing a high percentage of limestone, hence its other name. In the eastern states, it is known as Bibb lettuce, taking its name from John Bibb, a retired Army major who supposedly perfected the plant. The name limestone prevails in the western states. This lettuce blends well with other greens, or it may be served alone with an appropriate dressing. In almost all cases, this lettuce is raised close to market because it does not hold up well when shipped. When washing and cutting Bibb lettuce, be very gentle because it bruises and deteriorates rapidly. Remove bruised or blemished spots by clipping with salad shears or scissors. When refrigerated, keep covered.

Boston lettuce grows into a very loosely-packed round head. The leaves are light green in color and very easy to remove from the head. When removed, the leaves seem to form a cup. For this reason, the lettuce is also known as buttercup lettuce. It is a very tender lettuce with a mild, sweet flavor. Its general appearance resembles Bibb lettuce, but its outer leaves are a lighter green and the inner leaves are light yellow. When cleaning and cutting, one must be very gentle because the leaves are very fragile and bruise easily. This lettuce does not ship well; therefore, it must be grown close to its market. Boston lettuce is often served alone, or it can be blended with other greens. Because the leaves have a cup shape, it also makes an excellent salad base. Before using, wash thoroughly and remove blemished leaves or spots on leaves by clipping with salad shears or scissors. When refrigerated, keep covered.

Chinese or *celery cabbage* grows into an elongated head approximately 12″ long, quite similar to romaine lettuce in appearance. The tightly-packed leaves have a coarse texture with light green outer leaves, but the inner leaves are almost entirely white. The outer leaves are more pronounced in flavor than the inner leaves. The taste resembles a milder form of cabbage. This association with cabbage gave it its most popular name. The reference to Chinese in its other name came about because it is sometimes used in Chinese cooking. The slight cabbage flavor requires this salad green to be used in moderation when mixed or blended with other greens. Its crisp texture enhances any mixed green preparation.

Chicory, also known as endive and curly endive, has curly, twisted, thin leaves that grow into a loose, spread-out bunch. The leaves vary in color from dark green on the outer leaves to pale green or white in the center and base. It has a bitter taste, and for this reason is blended with milder greens when used in a salad. Chicory can also be used effectively as a garnish when setting up salad bowls for buffets. Before using, wash thoroughly and remove blemished leaves or spots on leaves by clipping with salad shears or scissors. Cut off about 2″ of the base because the white leaves that usually appear in this area are extremely bitter and, to a degree, unfit for consumption.

Dandelion greens grow wild or they may be cultivated. The cultivated greens are more tender and mild in taste. The greens are fairly smooth but have a slightly rough irregular edge. The wild dandelions are tasty until the yellow flower blooms, then they become bitter and tough. This green is an early spring favorite. Dandelion greens can be blended with other salad greens or they may be served alone with a hot bacon dressing and chopped hard-boiled eggs.

Escarole lettuce, also known as broad leaf endive, is similar in taste to chicory, although not quite as bitter. It has a dark green color and broad thick leaves

WASHING SALAD GREENS

1 PLACE CUT GREENS IN SINK OR BOWL FILLED WITH COLD WATER

2 GENTLY STIR AND REMOVE THE GREENS WITHOUT ALLOWING THEM TO SOAK

3 DRY THE GREENS IN A SALAD SPINNER FOR 30 SECONDS

All greens must be washed, drained, trimmed, and usually stored before they are used in salad preparations.

with an irregular shape that grow into a loose fan-shaped head or bunch. Like endive, the base of the head where the leaves are lighter in color is sometimes very bitter and tough. It is advisable to discard this part. Because of its bitter taste, escarole is seldom eaten alone. It is almost always blended with milder and sweeter greens. Escarole is grown in places that have slightly mild winters, such as the northern part of Florida and the Texas panhandle. Before using, wash thoroughly and remove blemished leaves or spots on leaves by clipping with salad shears or scissors.

Head lettuce, also known as New York head lettuce or iceberg lettuce, is the most popular salad green. It has a very round, compact head. The leaves are pale green, and it has a very crisp texture, a very mild flavor, and excellent keeping qualities. It retains crispness even when roughly tossed or bruised by improper cutting. It is a clean lettuce and usually only requires a light

washing because the head is so compact that dirt is unable to penetrate. After washing, only the core and a few of the outside leaves are removed before using.

Leaf lettuce is a salad green, but one that is not recommended for use in the preparation of the body of a salad. It is used mostly as a salad base, and in most cases it is left on the plate after the body is consumed. It is popular when used on a sandwich to add color and separate the meat or cheese from the bread. Leaf lettuce has a rich green color, soft-textured leaves, and a very mild flavor. It grows into a loose bunch, which makes it very easy to wash and clean. It should be kept covered in the refrigerator until ready to use.

Romaine lettuce has long, fairly dark green leaves that grow into a loose elongated head. It has a very mild, sweet flavor and blends extremely well with other greens. Because of its loose head, dirt collects in the ridges of the leaves during

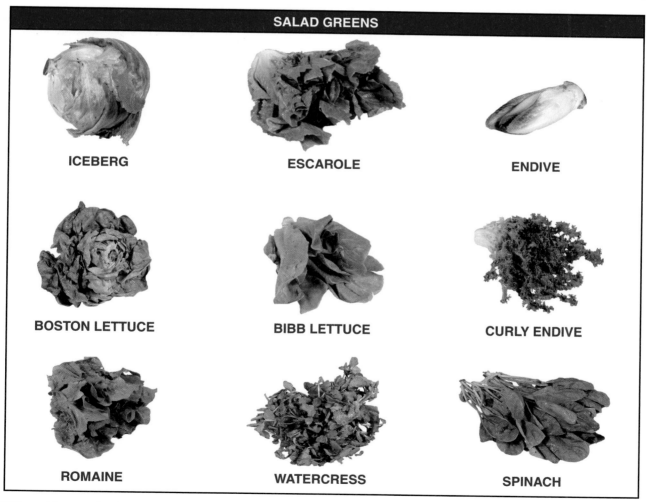

SALAD GREENS

ICEBERG · ESCAROLE · ENDIVE

BOSTON LETTUCE · BIBB LETTUCE · CURLY ENDIVE

ROMAINE · WATERCRESS · SPINACH

Different types of salad greens may be used for the base and/or the body of the salad.

growth, so it must be washed thoroughly before it is used or stored. Romaine has excellent keeping qualities and does not bruise easily when cut.

Radicchio is a small, firm, compact head of ruby red leaves. The leaves are slightly bitter in taste and are used for flavor and color with other salad greens. It is very expensive, but a little goes a long way.

Spinach is classed as a vegetable and is a popular one. The dark green, tender, pleasant-tasting leaves are also desirable as a salad green. They can be served alone or blended with other greens to supply color and flavor. Before using, the long tough stem at the base of each leaf must be removed and each leaf washed two or three times to remove dirt and grit that collects in the ridges of the leaves during growth. Spinach leaves bruise easily, so they must be washed gently. Store covered in the refrigerator until ready to use.

Watercress has small, round, dark green leaves that grow on thin stalks. They are bound in small, loose bunches. The leaves are very tender and fragile. It has a slight peppery taste, similar to the turnip. Its taste has a tendency to stimulate the appetite. It is an excellent addition to any mixed green preparation. However, it is most popular as a garnish for salads, broiled steaks, fruit, or soup preparations. Because it is grown in sandy streams, it must be washed thoroughly but gently under cold running water. Store in the refrigerator in a covered container. When ready to use, set in a pan of ice water with a little lemon juice.

Vegetable Salads

Vegetable salads are either cooked or raw. They contain foods that have a variety of shapes and natural colors. Cooked vegetable salads include vegetables that require cooking before using in a salad. Salads such as potato, bean, and beet are examples of cooked salads. Raw vegetable salads are more popular with the dining public. This popularity came about because most raw vegetables are very crisp, a key element in a popular and successful salad. Cole slaw, cucumber and onion, and carrot and raisin salads are examples of raw vegetable salads.

Fruit Salads

Fruit salads are colorful and contain many essential vitamins and minerals required for proper nutrition. However, fruit salads are fragile and discolor rapidly when cut and exposed to air. Many fruits break down if tossed or handled improperly.

The vegetables in bean salad are cooked before being placed into the salad.

Fresh fruits are preferred when making fruit salads. Canned or frozen fruit may also be used.

With few exceptions, fresh fruit requires refrigeration. If fruits arrive unripened, they may be left at room temperature until ripe and then refrigerated. Bananas are an exception to the refrigeration requirement. They turn dark when refrigerated and ripening is retarded.

Fresh fruits, like fresh vegetables, are superior to the canned or frozen product when taste and texture are considered. However, the food service industry demands the production and convenience of canned or frozen products. Store canned fruit in a cool dry place. Refrigerate overnight for best results when using canned fruit in salad preparation. Once the can is open, place the contents in a glass or plastic container and refrigerate the unused portion.

Some fresh fruits, such as apples, avocados, bananas, and pears, discolor when cut and exposed to air. To prevent a rapid discoloration and to improve flavor, dip or marinate the fruit in liquids that contain citric acid. Lemon, orange, pineapple, and lime juices are ideal for this treatment. Also, use a stainless steel knife when cutting fresh fruit. Carbon steel has a tendency to discolor the fruit and acids stain the knife blade.

Some fruit salads can be mixed and tossed with success, but most depend on arranging the fruit in an attractive manner. Set up or prepare fruit salads as close to serving time as possible and serve while still slightly chilled.

Meat Salads

Meat salads are limited in variety because all meats do not produce quality salads. Ham, turkey, and chicken are the usual choices for a meat salad. These meats have the moistness, tenderness, color, and delicate flavor required for desirable salads.

The meat for these salads can be purchased raw or cooked. The type and preparation of meat used in a salad depends upon production time, price, and quality of the meat.

When preparing a meat salad there are two suggested requirements. Be sure the meat is tender, and dice the meat fairly small and uniform. Appearance of ingredients is very important in salad preparation.

National Pork Producers Council

Crunchy Chinese pork salad is a meat salad that adds variety to the menu.

Seafood Salads

Seafood salads require meat with a delicate flavor or meat that can be blended with other ingredients for variety in taste. Fish such as halibut, sole, codfish, and orange roughy have a mild flavor and are commonly used. Fish may be purchased to be placed on the menu as an entree. Leftovers can be utilized in a salad preparation. Fish that make extremely popular salads are tuna and salmon. The meat of these two saltwater fish contains a flaky texture, enough oil for proper moisture, and a very agreeable taste. Blending tuna or salmon with celery, onions, mayonnaise, and lemon juice is a popular way of preparing tuna and salmon in a seafood salad.

Canned tuna and salmon are used most often in salad preparation. Fresh fish can be used if cooked by poaching or steaming. In the case of salmon, the deeper the color of the flesh, the better the quality. Red salmon is of better quality than pink salmon. Canned salmon is canned in its raw state and cooked by steaming. This is why the skin and bones of the fish are still in the can when opened. The bones are tender and can be eaten. Most salad preparations do not use the skin and bones. These are removed from the flesh before use in most preparations.

Canned tuna can be purchased as both light and dark meat, packed in oil or water. The most desirable product to use in salad preparation is light meat packed in water.

Shellfish commonly used in seafood salads include shrimp, crabmeat, and lobster. Shrimp is the most popular and must be cooked before it can be used in salad preparations. The frozen product, cooked or raw, is usually selected. Frozen shrimp, peeled, deveined, and cooked, is the most convenient shrimp product to select. However, if cost is a factor, frozen shrimp that are steamed or boiled then peeled and deveined would be best to use. Avoid overcooking shrimp. Overcooked shrimp are soft and tasteless.

The meat of the blue crab from the Atlantic coast or the king or Dungeness crab from the Pacific coast produces the best crabmeat salad. The meat of these crabs can be purchased both frozen and canned Canned meats should be chilled prior to preparation. Synthetic crabmeat can be used for a crabmeat salad if cost is a factor. Fresh cold water lobster or frozen warm water lobster tails are the best choice for a lobster salad, but are very expensive. Lobster is cooked by steaming or boiling in water. It is then cooled and the meat is removed from the shell.

Gelatin Salads

Gelatin salads can be presented in many different forms, offer a variety of colors, are easy to prepare, and are inexpensive. In any gelatin preparation, the correct ratio of gelatin to the amount of liquid must be used. The gelatin is dissolved thoroughly in hot liquid before cold liquid is added. The most commonly used recipe for gelatin is 1 c fruit-flavored gelatin to each quart of liquid. Fruit-flavored gelatin packaged for commercial use contains 1 lb 8 oz of gelatin powder. It is recommended that 1 gal. of water or fruit juice be added: ½ gal. hot to dissolve the gelatin, and the other ½ gal. cold.

Mix the gelatin in a stainless steel container. Be sure all the gelatin is thoroughly dissolved in the hot liquid before the cold liquid is added. If speed is required in forming the gel, place ice in the cold liquid to facilitate this action. Gelatin sets more quickly at a cold temperature, so placing it in the coldest area of the refrigerator (lower area) will produce a quicker gel. Never attempt to speed the gel by placing it in the freezer. Freezing causes crystallizing and upon defrosting the gelatin will become liquid.

The recipe for using plain, pure gelatin is 3¾ oz to each gallon of liquid. Plain gelatin is required to prepare aspics and chaud-froid (jellied white sauce) used in decorating certain foods that are to be displayed on buffets. Plain gelatin should be soaked in a small amount of cold water prior to its use. This causes the gelatin to dissolve more rapidly when added to the hot liquid and keeps it from settling to the bottom of the container and perhaps from scorching if the preparation must be reheated or cooked.

When preparing a gelatin mold for display on a buffet, the volume of the mold size selected must hold the amount prepared. To do this, select a mold that will hold the amount the recipe yields. Fill the mold with water and measure the amount of water used. If the amount of water varies more or less than one cup from the recipe yield, the mold will hold the required amount.

When adding mayonnaise, sour cream, whipped cream, beaten egg whites, fruits, or vegetables to the gelatin, the gelatin mixture must be chilled until slightly thickened before additions are made. This ensures even distribution of the added ingredients. To obtain the syrup-like consistency when chilling gelatin, stir the mixture occasionally so it does not lump or become firm. If the gelatin becomes firm, it must be converted back to liquid form. This is done by pouring the gelatin mixture into a stainless steel bowl over hot water. The mixture is stirred until it liquefies.

Many gelatin recipes call for folding in whipped cream or beaten egg whites. This procedure is used to give a fluffy, spongy texture to the gelatin preparation. Both products should be whipped at high speed using a wire whip or an electric mixer until soft peaks are formed and stiff enough to hold their shape before they are folded into the slightly thickened gelatin. If they are underwhipped or overwhipped, the texture of the finished product is impaired. Both products should be whipped just before they are added to the gelatin, never in advance. Cream whips best when utensils and cream are cold. Egg whites whip best when they are at room temperature. For both, all utensils must be clean and free of oil or grease.

When tiering two or more gelatin mixtures in the same pan or mold, each layer poured into the pan or mold must be chilled until slightly firm before adding the next layer. This is done to prevent one layer from running into another. If one layer is set too firm, the layer placed on top will probably slip off or separate from the completed mold when it is unmolded. If a layer should become too firm, remove it from the refrigerator and let it set at room temperature for approximately 15 min or until it loses some of its firmness before adding the next layer. Before adding more gelatin to a layer that has already set, be sure the gelatin is completely cool. If the gelatin being added is too warm, it will melt the set layer and the two mixtures will run together.

For certain occasions, it may be necessary or desirable to place a design, logo, or pattern in the bottom of the mold. When unmolded, this design, logo, or pattern will appear on the top of the molded preparation. To accomplish this, the procedure for tiering or setting gelatin in layers must be followed, although smaller quantities of gelatin would be used. Place the mold in a bowl that is half full of ice and water. Add the first layer of cold gelatin, which is still in liquid form. Let the layer set. Lay out the pattern, logo, or design and add a very small amount of the cold gelatin to set the design. If too much is added, the design pieces will float. Keep adding small amounts of the gelatin mixture and let set after each addition. When the pattern is firmly set in solid gelatin, complete the mold.

Gelatin molds placed in the refrigerator take approximately 1½ hr for each pint of liquid gelatin to set. Once it has set and becomes firm, cover the mold with plastic wrap. Covering the preparation keeps the mold from shrinking and those made with mayonnaise, cream, whipped cream, and beaten egg whites from drying out and discoloring.

To unmold large molds, usually used on buffet service, place about 6″ of warm water in a large sink. Dip the mold in the warm water up to about ½″ from the top of the mold. Slightly tilt the mold from side to side until the gelatin mixture separates from the sides of the mold. Remove at once from the warm water. Place the serving dish or tray over the top of the mold with one hand holding the mold. Hold the serving dish or tray with the other hand and invert the mold onto the tray. Shake the mold slightly from side to side until the gelatin releases itself from the mold.

To unmold small individual molds, place the mold in the palm of one hand with the open end resting in the palm. Place under warm running water for just a second or two. As soon as the mold is removed from the water, tap the top of the mold with the palm of the free hand. The gelatin preparation should drop out of the mold and into the palm of the hand. If the gelatin preparation does not drop out easily, place it under warm running water a second time. Gelatin molds can be oiled or sprayed to make unmolding easier.

Pasta Salads

Pasta salads can be prepared in advance. They are tasty, inexpensive, colorful, and attractive. These characteristics qualify them for salad bar and buffet service. Presenting them in this way provides the opportunity to offer the guest a food item that is a little different or unusual.

The secret of a good pasta salad is cooking the pasta *al dente*, which means a little on the firm or chewy side. Never cook pasta to the point where it becomes soft or mushy. Other ingredients added to the pasta should add flavor and crispness to the salad preparation. Common ingredients for pasta salads include cherry tomatoes, sun-dried tomatoes, onions, green olives, black olives, pine nuts, various cheeses, basil, and olive oil.

Roasted red peppers and red pepper flakes are combined to make a colorful and spicy bow tie pasta salad.

Serving Salads

A variety of salads should be offered on the menu. Salads are usually served in small salad bowls. Small salad bowls are small plastic or wood bowls that usually have a diameter of 5½″ and a depth of 1¾″. They usually hold about 12 oz and are used for side salads. Larger bowls are used if the salad is served as an entree or main course.

SALAD DRESSINGS

A salad may possess all the characteristics of a successful preparation, but if the dressing is not of high quality, the salad will be a failure. In most cases, the flavor of the dressing is the first flavor that is tasted. Therefore, the dressing must be prepared with the finest ingredients available and with the utmost care. Purchased dressings can save preparation time. Table II gives some of the popular salad dressings. Three basic salad dressings are oil and vinegar, mayonnaise, and boiled or cooked dressing.

Oil and vinegar is a mixture of oil, vinegar, and seasonings. It is the basis for preparing French dressing. French dressing may be prepared by forming either a permanent emulsion or a temporary emulsion. When the French dressing is to form a permanent emulsion, the oil is held in suspension using egg yolks or a combination of yolks and eggs. This dressing is thicker and clings to and coats the salad ingredients better than the temporary dressing. When the dressing is to form a temporary emulsion, no eggs are used. The oil is dripped slowly into the vinegar and flavoring ingredients. After it sets, the oil and acid (vinegar or lemon juice) usually separate. This French dressing must be stirred thoroughly or shaken before serving.

Mayonnaise is a semisolid dressing prepared by forming an emulsion by dripping salad oil into egg yolks or yolks and eggs, depending on how quickly the mayonnaise is needed. The quality of the mayonnaise depends on the quality of the oil used. A number of oils, such as olive, cottonseed, soybean, peanut, and corn oil, are available on the market. Olive oil is the most expensive but does not make the best mayonnaise because it has a strong flavor. A good-tasting mayonnaise requires an oil that is more bland, such as corn or cottonseed oil. If olive oil is used, it is best to blend it with other oils. Mayonnaise is an extremely important dressing because it is the basis for many of the other very popular dressings, such as thousand island, green goddess, and bleu cheese. In some cases when mayonnaise is prepared in the commercial kitchen, a stretcher is added. The stretcher is usually prepared by thickening boiling water with diluted cornstarch and tinting it with yellow coloring. A stretcher lessens the cost of the mayonnaise and increases the quantity, but once the stretcher is added the flavor and eating qualities suffer.

Boiled or *cooked dressing* is not used to any great extent in the commercial kitchen. It is used in the preparation of some cole slaws, fruit dressing, and potato salad. Boiled dressing is simple to prepare, but care must be taken not to scorch or curdle the mixture.

TABLE II: SALAD DRESSING COMBINATION SUGGESTIONS...	
Salads	**Dressings**
Fruit Salads	
Waldorf	Mayonnaise, Boiled Dressing
Diplomat	Mayonnaise, Boiled Dressing
Tossed Fruit	Mayonnaise, Boiled Dressing, French, Sour Cream, Honey Cream, Honey Dressing
Pear and American Cheese	Mayonnaise, Boiled Dressing, French, Sour Cream, Honey Cream, Honey Dressing
Banana-Pecan	Mayonnaise, Boiled Dressing, French, Sour Cream
Fruit	Mayonnaise, Boiled Dressing, French, Sour Cream, Honey Cream, Honey Dressing
Belgian Endive and Orange	Mayonnaise, French, Boiled Dressing, Sour Cream, Honey Cream, Honey Dressing
Orange and Grapefruit	Mayonnaise, French, Boiled Dressing, Sour Cream, Honey Cream, Honey Dressing
Vegetable Salads	
Asparagus and Tomato	Mayonnaise, French, Thousand Island, Louis, Italian, Sour Cream, Vinegar and Oil, Bleu or Roquefort Cheese, Russian, Vinaigrette, Bleu Cheese-Sour Cream, Chiffonade
Tomato and Cucumber	Mayonnaise, French, Thousand Island, Louis, Italian, Sour Cream, Vinegar and Oil, Bleu or Roquefort Cheese, Russian, Vinaigrette, Bleu Cheese-Sour Cream, Chiffonade, Green Goddess, Zippy Italian
Combination	Mayonnaise, French, Thousand Island, Louis, Italian, Sour Cream, Vinegar and Oil, Bleu or Roquefort Cheese, Russian, Vinaigrette, Bleu Cheese-Sour Cream, Chiffonade
Leafy Green Salads	
Mixed Green	French, Thousand Island, Louis, Italian, Sour Cream, Vinegar and Oil, Bleu or Roquefort Cheese, Russian, Vinaigrette, Bleu Cheese-Sour Cream, Chiffonade, Green Goddess
Julienne	French, Thousand Island, Louis, Italian, Sour Cream, Vinegar and Oil, Bleu or Roquefort Cheese, Russian, Vinaigrette, Bleu Cheese-Sour Cream, Chiffonade, Green Goddess
Spring	French, Thousand Island, Louis, Italian, Sour Cream, Vinegar and Oil, Bleu or Roquefort Cheese, Russian, Vinaigrette, Bleu Cheese-Sour Cream, Chiffonade
Garden	French, Thousand Island, Louis, Italian, Sour Cream, Vinegar and Oil, Bleu or Roquefort Cheese, Russian, Vinaigrette, Bleu Cheese-Sour Cream, Chiffonade, Green Goddess, Bacon Bit, Zippy Italian
Western	French, Thousand Island, Louis, Italian, Sour Cream, Vinegar and Oil, Bleu or Roquefort Cheese, Russian, Vinaigrette, Bleu Cheese-Sour Cream, Chiffonade
Italian	Italian, French, Vinegar and Oil, Chiffonade, Bleu Cheese-Sour Cream, Bleu or Roquefort Cheese, Sour Cream, Zippy Italian
Chef's	French, Thousand Island, Louis, Italian, Sour Cream, Vinegar and Oil, Bleu or Roquefort Cheese, Russian, Vinaigrette, Bleu Cheese-Sour Cream, Chiffonade
Gelatin Salads	
Perfection	Mayonnaise, Sour Cream, French, Boiled Dressing
Spicy Peach Mold	Mayonnaise, Sour Cream, French, Boiled Dressing, Sour Cream, Honey Cream
Cranberry-Orange	Mayonnaise, Boiled Dressing, Sour Cream, Honey Cream
Lime Glow	Mayonnaise, Boiled Dressing, Sour Cream, Honey Cream
Molded Tuna Fish	Mayonnaise, Green Goddess
Cranberry Snow	Mayonnaise, Boiled Dressing, Honey Cream, Sour Cream
Jellied Cole Slaw	Mayonnaise
Molded Spring	Mayonnaise, Thick French, Sour Cream, Boiled Dressing
Jellied Tangy Fruit	Mayonnaise, Boiled Dressing, Honey Cream, Sour Cream
Pineapple-Strawberry Souffle	Mayonnaise, Boiled Dressing, Honey Cream, Sour Cream

Salads	Dressings
Cranberry Souffle	Mayonnaise, Boiled Dressing, Honey Cream
Cinnamon-Apple	Mayonnaise, Boiled Dressing, Honey Cream
Lime-Pear Aspic	Mayonnaise, Boiled Dressing, Honey Cream
Peach and Raspberry Mold	Mayonnaise, Boiled Dressing
Cinnamon Applesauce Mold	Mayonnaise, Boiled Dressing, Honey Cream
Fruited Cheese Mold	Mayonnaise, Boiled Dressing, Honey Cream
Jellied Diplomat	Mayonnaise, Boiled Dressing, Honey Cream
Tomato Aspic	Mayonnaise, Thick French, Louis, Sour Cream, Chiffonade
Fruited Cranberry	Mayonnaise, Boiled Dressing, Sour Cream, Honey Cream
Green Island	Mayonnaise, Sour Cream, Honey Cream, Boiled Dressing
Cider	Mayonnaise, Boiled Dressing, Honey Cream

. . . TABLE II: SALAD DRESSING COMBINATION SUGGESTIONS

FRUIT SIDE SALAD RECIPES

Waldorf Salad

25 servings (No. 16 scoop)

Ingredients

4 lb eating apples
1 lb celery, diced
4 oz raisins
¼ c lemon juice
1 pt salad dressing
3 oz walnuts, chopped
25 lettuce leaves
25 parsley sprigs
salt and sugar to taste

Procedure

1. Wash, core, and cut apples in half. Do not peel.
2. Dice into ½″ cubes and place in a mixing container.
3. Add diced celery, raisins, lemon juice, and salad dressing. Toss gently until thoroughly blended.
4. Season with salt and sugar.
5. Place on a leaf of lettuce. Garnish the top with additional salad dressing. Sprinkle with chopped nuts and top with a maraschino cherry.
6. Garnish with a sprig of parsley.

Diplomat Salad

25 servings (No. 12 scoop)

Ingredients

4 lb eating apples
1 lb celery
1½ lb canned pineapple tidbits
¼ c lemon juice
4 oz sugar
3 c mayonnaise
1 tsp salt
3 oz pecans, chopped
25 lettuce leaves
25 parsley sprigs

Procedure

1. Wash, core, and cut the apples in half. Do not peel.
2. Dice into ½″ cubes and place in a mixing container. Add the lemon juice.
3. Dice the celery and pineapple slightly finer than the apples. Place in the mixing container.
4. Add the sugar, mayonnaise, and salt. Toss gently until all ingredients are thoroughly blended.
5. Place on a leaf of lettuce. Garnish the top with additional mayonnaise. Sprinkle with chopped nuts and top with a maraschino cherry.

Trade Tip

Salads are a vital part of healthy eating. The vegetables, fruit, cereals, etc. in salads provide good nutrition with reduced calories. Salads may be placed on the menu as "health conscious" foods. Dressings should be used sparingly if monitoring fat and calories.

Tossed Fruit Salad

25 servings (No. 6 scoop)

Ingredients

1 qt fresh strawberries, stem removed
1 qt orange sections
1 qt grapefruit sections, cut in half
1 qt seeded grapes
1 qt canned pineapple, diced large
¼ c chopped mint
1 pt mayonnaise
1 c heavy cream, whipped
25 lettuce leaves
3 heads iceberg lettuce, shredded

Procedure

1. Drain all fruits and place in a mixing container. Toss gently.
2. Blend the mayonnaise and whipped cream.
3. Pour the dressing over the fruit mixture and fold in gently.
4. Serve on a base of leaf and shredded iceberg lettuce. Garnish with chopped mint.

Deluxe Waldorf Salad

25 servings (No. 12 scoop)

Ingredients

1½ qt eating apples, diced
1 qt lean ham, diced
1½ qt celery, diced
3 c mayonnaise
1 tsp lemon juice
salt and white pepper to taste
25 leaves leaf lettuce
3 heads iceberg lettuce, shredded
25 parsley sprigs

Procedure

1. Place the apples, ham, celery, and lemon juice in a mixing container.
2. Add the mayonnaise and toss gently until thoroughly mixed.
3. Season with salt and white pepper.
4. Line each cold salad plate with a leaf of lettuce. Sprinkle on the shredded iceberg lettuce.
5. Place a mound of salad in the center of each plate. Top with additional mayonnaise and garnish with a sprig of parsley.

Fruit Salad

25 servings (No. 12 scoop)

Ingredients

8 oranges, peeled and diced
8 slices canned pineapple, diced
4 bananas, diced
1 cantaloupe, peeled, seeded, and diced
1 lb grapes, cut in half and seeded
12 canned pear halves, diced
12 canned peach halves, diced
25 medium-sized fresh strawberries
¼ c lemon juice
1 qt fruit juice, drained from canned fruit
25 crisp iceberg lettuce leaves
25 mint leaves

Procedure

1. Combine all the ingredients except the lettuce in a mixing container, toss gently, and chill thoroughly.
2. Place a leaf of lettuce on cold salad plates.
3. Portion out the fruit salad and place a mound on each salad plate.
4. Top with honey cream dressing and garnish with mint leaves.

Banana-Pecan Salad

25 servings

Ingredients

25 bananas
⅓ c lemon or pineapple juice
10 oz pecans, finely chopped
25 maraschino cherries, coarsely chopped
3 c mayonnaise
1 pt heavy cream, whipped
3 heads iceberg lettuce, shredded
25 leaves leaf lettuce

Procedure

1. Peel the bananas, cut in half crosswise, and dip in the fruit juice.
2. Blend the mayonnaise and whipped cream.
3. Dip each banana half into the mayonnaise-whipped cream mixture.
4. Roll into the finely chopped pecans.
5. Place a leaf of lettuce on cold salad plates. Sprinkle shredded iceberg lettuce on the leaf lettuce.
6. Place the banana halves in the center of each plate and sprinkle with the chopped maraschino cherries.
7. Serve with a small mound of dressing and a sprig of parsley.

Pear Saint Charles

25 servings (No. 20 scoop)

Ingredients

50 canned, drained small pear halves
3 lb cottage cheese
25 red maraschino cherries (with stem)
25 leaves leaf lettuce
3 heads iceberg lettuce, shredded
25 sprigs parsley
red food coloring

Procedure

1. Place a leaf of lettuce on cold salad plates.
2. Cut the iceberg lettuce into quarters and shred. Sprinkle the shredded iceberg lettuce on the leaf lettuce.
3. Place a scoop of cottage cheese in the center of the shredded lettuce.
4. Blush the outside of each pear half with red color and lean two of the halves against the cottage cheese.
5. Place a maraschino cherry on top of the cottage cheese and garnish with a sprig of parsley.

Pear halves are placed on lettuce leaves, sprinkled with bleu cheese and walnuts, drizzled with low-fat Italian, French, or Ranch dressing and garnished with parsley in this low-calorie salad.

Pear and American Cheese Salad

25 servings

Ingredients

50 small or 25 large canned drained pear halves
1½ lb coarsely grated longhorn cheese
3 heads iceberg lettuce, shredded
25 leaves leaf lettuce
1 pt mayonnaise
25 parsley or watercress sprigs

Procedure

1. Line each cold salad plate with a leaf of lettuce. Sprinkle on the shredded iceberg lettuce.
2. Place one or two pear halves, depending on the size, on the shredded iceberg lettuce.
3. Spot a small amount of mayonnaise in the cavity of each pear and sprinkle the grated cheese over each pear.
4. Garnish with a sprig of parsley or watercress.

Avocado, Grapefruit, and Orange Salad

25 servings

Ingredients

8 fresh grapefruits, sectioned
8 fresh oranges, sectioned
4 ripe avocados
25 leaves leaf lettuce or romaine
3 heads iceberg lettuce, shredded
¼ c lemon juice
25 parsley or watercress sprigs

Procedure

1. Cut avocados in half lengthwise. Remove seed and peel. Cut slices crosswise and dip each slice in the lemon juice.
2. Place a leaf of lettuce on cold salad plates. Sprinkle shredded iceberg lettuce on the leaf lettuce.
3. Alternate two avocado slices, two orange sections, and two grapefruit sections on the shredded lettuce.
4. Serve with French dressing and garnish with a sprig of watercress or parsley.

Trade Tip

Add fruits to salads. Fruit is a natural thirst quencher with most fruits being 80% to 90% water. On an ounce-for-ounce basis, fruits are lower in calories than most other foods. Additionally, fruits contain a large amount of fiber.

Belgian Endive and Orange Salad

26 servings

Ingredients

13 heads Belgian endive lettuce, cut in half lengthwise
13 peeled oranges, sliced into cartwheels (6 per orange)
26 leaves leaf lettuce
26 parsley or watercress sprigs

Procedure

1. Line each cold salad plate with a leaf of lettuce. Place a half head of Belgian endive on the leaf lettuce base.
2. Line 3 cartwheels of orange on top of the endive, overlapping the cartwheels slightly.
3. Serve with bleu cheese dressing and garnish with a sprig of watercress or parsley.

Cranberry Relish Salad

25 servings (No. 16 scoop)

Ingredients

3 unpeeled oranges, cut into wedges
3 lb unpeeled apples, cored and cut into wedges
2 lb raw, fresh cranberries
1½ lb sugar
25 leaves leaf lettuce
3 heads iceberg lettuce, shredded
25 parsley sprigs

Procedure

1. Grind the apples, oranges, and cranberries through the food grinder using the coarse chopper plate. Mix thoroughly.
2. Add the sugar and mix.
3. Line each cold salad plate with a leaf of lettuce. Sprinkle on the shredded iceberg lettuce.
4. Drain the juice off the salad and place a mound of salad in the center of each salad plate.
5. Top with a small amount of salad dressing and garnish with a sprig of parsley.

Orange-Grapefruit Salad

25 servings

Ingredients

10 fresh oranges, sectioned
10 fresh grapefruits, sectioned
3 heads iceberg lettuce, shredded
25 leaves leaf lettuce or romaine
25 mint leaves, watercress, or parsley sprigs

Procedure

1. Place a leaf of lettuce on cold salad plates. Sprinkle shredded iceberg lettuce on the leaf lettuce.
2. Alternate three orange sections and three grapefruit sections on the shredded iceberg lettuce.
3. Serve with fruit or French dressing. Garnish with mint leaves, watercress, or a sprig of parsley.
Note: Shredded coconut may be sprinkled on top of the fruit sections for additional eye appeal.

Peach and Cottage Cheese Salad

25 servings (No. 20 scoop)

Ingredients

25 large, canned peach halves, cut in half
3 lb cottage cheese
25 red maraschino cherries (with stem)
25 leaves leaf lettuce
3 heads iceberg lettuce, shredded
25 parsley sprigs

Procedure

1. Line each cold salad plate with a leaf of lettuce. Sprinkle on the shredded iceberg lettuce.
2. Place a scoop of cottage cheese in the center of each salad plate.
3. Place a peach wedge on each side of the cottage cheese.
4. Place a maraschino cherry on top of the cottage cheese and garnish with a sprig of parsley.

VEGETABLE SIDE SALAD RECIPES

Asparagus-Tomato Salad

25 servings

Ingredients

7 fresh tomatoes (4 slices each)
50 canned or fresh cooked asparagus spears
3 heads iceberg lettuce, shredded
25 leaves leaf or romaine lettuce
½ 7 oz can pimientos
25 parsley or watercress sprigs

Procedure

1. Line each cold salad plate with a leaf of lettuce. Sprinkle on the shredded iceberg lettuce.
2. Place a slice of tomato in the center of each salad plate and two asparagus spears on top of the tomato.
3. Cut the pimientos into 25 strips and lay one strip across the asparagus. Garnish with a sprig of watercress or parsley.

Assorted Vegetable Salad

25 servings

Ingredients

2 lb 4 oz beets, cooked and diced
1 lb 12 oz green beans, cooked and diced
1 lb 4 oz peas, cooked
1 lb 8 oz celery, diced
4 oz minced onions
8 oz head lettuce, diced
1 pt mayonnaise
salt and white pepper to taste

Procedure

1. Place the beets, green beans, peas, celery, and onions in a stainless steel mixing container.
2. Add the mayonnaise and toss lightly until well mixed.
3. Season with salt and white pepper. Toss lightly a second time.
4. Blend in the diced lettuce and serve immediately by placing a mound of salad on plates covered with a leaf of crisp lettuce. Garnish with a sprig of parsley.

Western Garden Slaw

25 servings (No. 12 scoop)

Ingredients

4 qt cabbage, finely shredded
1 qt celery, finely diced
½ c green peppers, finely diced
½ c pimientos, finely diced
½ c onions, minced
2 tsp celery seed
1 pt mayonnaise
1 pt sour cream
½ c sugar
½ c lemon juice
2 tsp dry mustard
2 tbsp salt
½ tsp pepper
25 leaves leaf or romaine lettuce
⅓ c parsley, chopped

Procedure

1. Blend the cabbage, celery, green peppers, pimientos, and onions in a mixing container.
2. In a separate container, combine the sour cream, celery seed, mayonnaise, sugar, lemon juice, dry mustard, salt, and pepper. Blend thoroughly.
3. Pour the dressing over the vegetable mixture and toss gently until thoroughly blended.
4. Line each cold salad plate with a leaf of lettuce.
5. Place a mound of the slaw in the center of each plate.
6. Garnish with chopped parsley.

Pickled Beet Salad

25 servings

Ingredients

1 No. 10 can sliced beets
12 oz small onions, cut into rings
2 bay leaves
1½ pt cider vinegar
1 tbsp salt
4 cloves
1 c sugar
25 iceberg lettuce cups
⅓ c parsley, chopped

Procedure

1. Place the beets (with juice) and onions in a bain-marie.
2. Blend the vinegar, salt, and sugar in a separate container. Stir until the sugar has dissolved.
3. Pour over the beets and add spices.
4. Cover and place in the refrigerator to marinate overnight.
5. Place a crisp lettuce cup on each cold salad plate.
6. Drain juice from beets, remove cloves and bay leaves, and place a serving portion in the center of each lettuce cup.
7. Top with onion rings and garnish with chopped parsley.

Garden Cole Slaw

25 servings (No. 12 scoop)

Ingredients

4 lb cabbage, shredded
1 c green peppers, finely chopped
½ c green onions, finely chopped
1 pt shredded carrots
1 tbsp celery seed
1 pt sour cream
1 pt mayonnaise
2 tbsp lemon juice
1 tbsp cider vinegar
1 tsp horseradish
salt and white pepper to taste
25 iceberg lettuce cups
⅓ c parsley, chopped

Procedure

1. Combine the cabbage, green peppers, green onions, carrots, and celery seed in a mixing container.
2. Whip the cream slightly and blend in the mayonnaise, lemon juice, vinegar, and horseradish. Season with salt and pepper.
3. Pour the mixture over the vegetables and blend thoroughly. Cover and place in the refrigerator for at least 2 hr.
4. Place a crisp lettuce cup on each cold salad plate.
5. Place a mound of the slaw in the center of each plate.
6. Garnish with chopped parsley.

Potato Salad

Ingredients

7 lb red potatoes, peeled, boiled, and cooled
12 oz celery, diced
3 oz pimientos, diced
3 oz onions, minced
4 chopped hard-boiled eggs
1 oz parsley, chopped
½ c drained bacon or ham cracklings
½ c sweet relish
3 c mayonnaise
½ c cider vinegar
1 tbsp sugar (variable)
salt and white pepper to taste
25 iceberg lettuce cups
25 parsley sprigs

Procedure

1. Slice or dice the cold boiled potatoes and place in a mixing container.
2. Add the remaining ingredients and toss gently so the potatoes do not break.
3. Season with salt and pepper.
4. Line each cold salad plate with a crisp iceberg lettuce cup.
5. Place a mound of the salad in the center of each lettuce cup.
6. Serve with a sprig of parsley.

German Potato Salad

Ingredients

7 lb raw red potatoes
1 lb bacon, diced
1 c cider vinegar
8 oz onions, diced
3 oz pimientos, diced
1 pt hot ham stock
1 tbsp sugar (variable)
2 tbsp parsley, chopped
salt and pepper to taste

Procedure

1. Boil the potatoes in their jackets. Peel and dice or slice thick while still warm.
2. Fry the diced bacon in a saucepan until crisp. Add the diced onions and continue to cook until the onions are slightly tender.
3. Add the hot ham stock, vinegar, and sugar. Bring to a boil and pour over the potatoes.
4. Add pimientos and chopped parsley, season with salt and pepper, and toss gently until all ingredients are blended thoroughly.
5. Check seasoning for desired taste.
6. Serve warm.

Cucumber and Onion Salad

Ingredients

5 lb cucumbers, peeled, scored, and thinly sliced
1½ lb onions, peeled and thinly sliced
1 c water
1 pt cider vinegar
1 c salad oil
2 oz sugar
1 tbsp salt
1 tsp pepper
¼ c parsley, chopped
25 iceberg lettuce cups
25 parsley sprigs

Procedure

1. Place the onions and cucumbers in a mixing container.
2. Add the water, salad oil, vinegar, sugar, salt, and pepper. Blend thoroughly.
3. Cover and let marinate in the refrigerator for at least 2 hr before serving.
4. Line each cold salad plate with a crisp iceberg lettuce cup.
5. Place a mound of the salad in the center of each lettuce cup. Garnish with chopped parsley.
6. Serve with a sprig of parsley.

Carrot and Raisin Salad

25 servings (No. 16 scoop)

Ingredients

5 lb carrots, peeled and coarsely grated
1 lb raisins
3 c water
1 tbsp sugar
1 tsp cider vinegar
1 pt mayonnaise
1 pt French dressing
25 iceberg lettuce cups
25 parsley sprigs

Procedure

1. Place the raisins, water, sugar, and vinegar in a saucepan. Bring to a boil. Remove from the range and let set for 5 min, then drain thoroughly and let cool.
2. Place the carrots and raisins in a mixing container.
3. Blend the mayonnaise and French dressing. Pour over the carrot-raisin mixture. Toss until all ingredients are thoroughly blended.
4. Line each cold salad plate with a crisp iceberg lettuce cup.
5. Place a mound of the salad in the center of each lettuce cup. Garnish with a sprig of parsley.

Stuffed Tomato with Cottage Cheese

25 servings

Ingredients

25 fresh, medium-sized tomatoes, peeled
¼ c chives, minced
¼ c radishes, minced
3 lb cottage cheese
4 heads iceberg lettuce, shredded
25 leaves leaf lettuce
25 parsley or watercress sprigs

Procedure

1. Place the tomatoes in hot water. Allow to set until the skin becomes slightly loose. Remove, peel, and cut cores out of the tomatoes.
2. Cut a slice off the top of each tomato and hollow out the center. Save pulp for use in some other preparation such as soup or stew.
3. Place the cottage cheese in a mixing container. Add the minced chives and radishes. Mix thoroughly.
4. Line each cold salad plate with a leaf of lettuce. Sprinkle with shredded iceberg lettuce.
5. Place a hollowed-out tomato in the center of each plate. Fill the cavity with the cottage cheese mixture.
6. Garnish with a sprig of watercress or parsley.

Sour Cream Cucumber Salad

25 servings

Ingredients

4 lb cucumbers, peeled and thinly sliced
1½ lb onions, peeled and thinly sliced
1 lb julienne cut tomatoes
1 pt sour cream
1 pt cider vinegar
1 pt water
1 tbsp salt
1 c mayonnaise
25 iceberg lettuce cups
⅓ c parsley, chopped
25 parsley sprigs

Procedure

1. Place the cucumbers, onions, and tomatoes in a mixing container. Toss gently.
2. Add the vinegar, sour cream, water, salt, and mayonnaise. Blend thoroughly.
3. Cover and let marinate in the refrigerator for at least 2 hr before serving.
4. Line each cold salad plate with a crisp iceberg lettuce cup.
5. Place a mound of the salad in the center of each lettuce cup. Garnish with chopped parsley.
6. Serve with a sprig of parsley.

Trade Tip

To make salad taste extra fresh, add lemon juice and toss gently. This will also preserve and enhance the natural color of the vegetables.

Florida Tomato Commission

Tomatoes may be sliced, chopped, or diced and used to add flavor and color to salads.

LEAFY GREEN SIDE SALAD RECIPES

Caesar Salad

25 servings

Ingredients

Salad

3 lb romaine lettuce
2 lb Bibb or Boston head lettuce
1 lb bacon, cut into 1" squares
50 anchovy filets
1 qt croutons (toasted or fried bread cubes)
½ c Parmesan cheese, grated
8 coddled eggs (set in very hot water for 1 min)
6 cloves garlic, finely chopped

Dressing

3 c salad oil
¼ c oil from anchovies
1 tsp salt
1 pinch black pepper
¼ c lemon juice

Procedure

1. Wash all greens thoroughly. Chill in the refrigerator.
2. Cut the greens into bite-size pieces, place in a mixing container, and toss gently.
3. Cook the bacon squares until slightly crisp and drain.
4. Break the coddled eggs in a separate container. Beat slightly. Pour over the greens.
5. Add the Parmesan cheese and garlic. Toss gently until thoroughly blended.
6. Blend the salad oil, anchovy oil, salt, pepper, and lemon juice in the mixing machine at medium speed.
7. Fill the small salad bowls with the tossed greens. Refrigerate.
8. To serve, ladle the dressing over the salad. Garnish with two or three crisp squares of bacon, two curled anchovies, and croutons.

Note: This salad can also be served as an entree or main course salad by increasing the serving portion and serving it in a larger salad bowl. Caesar salad, when served a la carte, is prepared and tossed at the diner's table.

Seven Layer Salad

25 servings

Ingredients

2 heads iceberg lettuce, coarsely shredded
1 pt celery, finely diced
½ c onions, finely diced
½ c green peppers, finely diced
1 c carrots, shredded
3 c salad dressing (variable)
3 c Cheddar or longhorn cheese, shredded
1 lb bacon, diced and cooked to a crackling

Procedure

1. In a 10" × 18" × 2½" pan, layer the ingredients in the following order: shredded lettuce, celery, onions, green peppers, and carrots.
2. Spread the salad dressing over the vegetables completely. Spread to the edges of the pan to seal.
3. Sprinkle the shredded cheese over the salad dressing to cover completely.
4. Sprinkle the bacon crumbs over the surface of the cheese.
5. Refrigerate until ready to serve.

Note: This salad is an excellent choice for buffet service.

An extensive listing of a variety of salad recipes can be accessed at www.recipesource.com/fgv/salads.

Spring Salad

25 servings

Ingredients

3 heads iceberg lettuce
2 heads romaine lettuce
3 heads Bibb lettuce
1 bunch watercress
1 head chicory lettuce
6 oz dandelion greens
½ stalk diced celery
1 bunch radishes, sliced
½ bunch carrots, sliced
1 bunch green onions, diced
1 lb white turkey meat, cooked and diced
1 lb ham, cooked and diced
50 tomato wedges
50 hard-boiled egg quarters

Procedure

1. Wash all greens thoroughly. Chill in the refrigerator.
2. Cut the greens into bite-size pieces. Place in a mixing container and toss gently.
3. Add the celery, radishes, carrots, and green onions. Toss gently a second time.
4. Fill the small salad bowls. Arrange the diced ham and turkey over the greens.
5. Garnish with two tomato wedges and two hard-boiled egg quarters.
6. Serve with an appropriate salad dressing.

Note: This salad can also be served as an entree or main course salad by increasing the serving portion and serving it in a larger salad bowl.

Julienne Salad Bowl

25 servings

Ingredients

3 heads iceberg lettuce
2 heads romaine lettuce
1 head escarole lettuce
½ lb spinach
½ stalk celery, julienne cut
½ bunch carrots, peeled and julienne cut
1 bunch radishes, julienne cut
1 cucumber, 4″ julienne cut
1 bunch green onions, julienne cut
1 lb white turkey meat, cooked, julienne cut
1 lb ham, cooked, julienne cut
1 lb bacon, cooked, julienne cut
3 tomatoes, julienne cut
6 hard-boiled eggs, coarsely chopped

Procedure

1. Wash all greens thoroughly. Chill in the refrigerator.
2. Shred the greens into fairly fine strips. Do not bruise. Place in a mixing container and toss gently.
3. Add the celery, carrots, radishes, cucumber, and green onions. Toss gently a second time.
4. Fill the small salad bowls. Arrange the julienned meat and tomatoes over the greens.
5. Garnish by sprinkling the chopped eggs over the salad.
6. Serve with appropriate salad dressing.

Note: This salad can also be served as an entree or main course by increasing the serving portion and serving it in a larger salad bowl.

Garden Salad

25 servings

Ingredients

3 heads iceberg lettuce
2 heads romaine lettuce
4 heads Bibb lettuce
2 bunches watercress
½ bunch carrots, peeled and sliced
1 bunch radishes, sliced
1 cucumber, scored, cut in half lengthwise, and sliced
1 bunch green onions, sliced
½ stalk celery, diced
1½ lb ham, cooked and diced
1 lb Swiss cheese, diced
50 tomato wedges
50 hard-boiled egg quarters

Procedure

1. Wash all greens thoroughly. Chill in the refrigerator.
2. Cut the greens into bite-size pieces. Place in a mixing container and toss gently.
3. Add the carrots, radishes, cucumber, green onions, and celery. Toss gently a second time.
4. Fill the small salad bowls. Arrange the diced ham and cheese over the greens.
5. Garnish with two tomato wedges and two hard-boiled egg quarters.
6. Serve with an appropriate salad dressing.

Note: This salad can also be served as an entree or main course salad by increasing the serving portion and serving it in a larger salad bowl.

Chef's Salad

25 servings

Ingredients

3 heads iceberg lettuce
1 head chicory lettuce
1 head escarole lettuce
2 heads Bibb lettuce
1 head romaine lettuce
1½ lb white turkey meat, julienne cut
½ lb ham, julienne cut
½ lb Swiss cheese, julienne cut
½ bunch carrots, peeled and sliced
2 bunches radishes, sliced
1 bunch green onions, diced
½ stalk celery, diced
50 tomato wedges
50 hard-boiled egg quarters

Procedure

1. Wash all greens thoroughly. Chill in the refrigerator.
2. Cut the greens into bite-size pieces, place in a mixing container, and toss gently.
3. Add the carrots, radishes, onions, and celery. Toss gently again.
4. Fill the small salad bowls. Arrange the julienned cheese and meat over the greens.
5. Garnish each salad with two tomato wedges and two hard-boiled egg quarters.
6. Serve with any appropriate salad dressing.

Western Salad

25 servings

Ingredients

3 heads iceberg lettuce
2 heads romaine lettuce
1 head escarole lettuce
½ lb spinach
2 heads Bibb lettuce
2 bunches radishes, sliced
1 stalk celery, diced
1½ lb beef, cooked and cut into slightly thick strips
1 lb ham, cooked and cut into slightly thick strips
1 lb longhorn cheese, cut into slightly thick strips
50 tomato wedges
50 hard-boiled egg quarters

Procedure

1. Wash all greens thoroughly. Chill in the refrigerator.
2. Cut the greens into bite-size pieces. Place in a mixing container and toss gently.
3. Add the celery and radishes. Toss gently a second time.
4. Fill the small salad bowls. Arrange the strips of meat and cheese over the greens.
5. Garnish with two tomato wedges and two hard-boiled egg quarters.
6. Serve with an appropriate salad dressing.

Country Salad

25 servings

Ingredients

3 heads iceberg lettuce
2 heads romaine lettuce
1 head chicory lettuce
8 oz spinach
6 stalks celery, diced
1 lb 8 oz bacon, cut into ½" cubes, cooked to a crackling
1 lb 8 oz little pig sausages, cooked and cut into ½" pieces
1 lb 8 oz potatoes, cooked and cut into ½" cubes
1 bunch radishes, sliced
3 leek stems, finely diced
25 tomato wedges
25 hard-boiled egg quarters

Procedure

1. Cut the washed and drained greens into bite-size pieces using a sharp knife or salad scissors.
2. Place in a stainless steel mixing bowl. Add the celery, potatoes, radishes, and leeks. Toss gently.
3. Fill the small salad bowls. Arrange the bacon cracklings and diced sausage over each salad.
4. Garnish each with a wedge of tomato and hard-boiled egg.
5. Top with an appropriate salad dressing just before serving.

Italian Salad

25 servings

Ingredients

3 heads iceberg lettuce
1 head chicory lettuce
2 heads romaine lettuce
1 lb spinach
½ stalk celery, diced
1 green pepper, diced
2 bunches radishes, sliced
1 bunch green onions, diced
1 pt croutons
1 c Parmesan cheese
25 red onion slices
50 tomato wedges
50 hard-boiled egg quarters
1 lb salami, julienne cut
1 lb pepperoni, julienne cut

Procedure

1. Wash all greens thoroughly. Chill in the refrigerator.
2. Cut the greens into bite-size pieces. Place in a mixing container and toss gently.
3. Add the radishes, celery, green pepper, and green onions and toss gently again.
4. Fill the small salad bowls. Arrange the julienned meat and onion rings over the greens and sprinkle with Parmesan cheese.
5. Garnish with croutons, two tomato wedges, and two hard-boiled egg quarters.
6. Serve with an Italian salad dressing.
Note: This salad can also be served as an entree or main course salad by increasing the serving portion and serving it in a larger salad bowl.

Trade Tip

Toasted pine nuts add a nice crunch to salads. To toast the pine nuts, heat a dry, nonstick skillet over medium heat. Add the pine nuts and sauté until toasted (about 2 min). Make sure the pine nuts do not burn.

MEAT SIDE SALAD RECIPES

Chicken Salad

Ingredients

3 lb chicken, cooked and diced into ½″ cubes
1½ lb celery, diced
1 pt mayonnaise (variable)
juice of one lemon
salt and white pepper to taste
25 leaves leaf lettuce
3 heads iceberg lettuce, shredded
2 pimientos, cut into 25 small strips
25 parsley sprigs

Procedure

1. Place the chicken and celery in a mixing container.
2. Add the lemon juice and mayonnaise. Toss gently to blend all ingredients.
3. Season with the salt and white pepper. Toss gently a second time.
4. Line the cold salad plates with a leaf of lettuce. Sprinkle on the shredded iceberg lettuce.
5. Place a mound of the salad in the center of the salad plate. Top with mayonnaise and a strip of pimiento.
6. Garnish with a sprig of parsley.

Note: Chicken salad is an extremely popular salad in the commercial kitchen. It can be prepared with all white meat or a combination of white and dark meat. Chicken salad is most popular when served as an entree or main course salad. For a main course salad, the serving portion is increased, a salad bowl is used, and in most cases it is garnished with tomato wedges and hard-boiled egg quarters.

Ham Salad

Ingredients

3 lb ham, cooked, julienne cut
1½ lb celery, sliced fine diagonal
1 head iceberg lettuce, shredded
1 pt mayonnaise (variable)
½ c sweet relish
salt to taste
50 tomato wedges
5 hard-boiled eggs, medium chopped
¼ c parsley, chopped
25 leaves romaine lettuce

Procedure

1. Place the julienned ham and celery and the shredded lettuce in a mixing container. Toss gently.
2. Add the sweet relish and mayonnaise. Toss gently a second time and season with salt.
3. Line the cold salad plates with a leaf of romaine lettuce.
4. Place a mound of ham salad in the center of each plate. Top with additional mayonnaise, chopped parsley, and chopped eggs.
5. Garnish with two wedges of tomato and serve.

Note: Ham salad can be served as an entree or a main course. It can be presented in a large salad bowl or on a 9″ plate. The serving portion is larger and the garnish should be increased enough to enhance the appearance.

Tex-Mex Salad

Ingredients

3 lb lettuce leaves
3 lb torn lettuce
choice of Italian-style dressing
3 lb kidney beans, rinsed and drained
1½ lb ripe olives, sliced
3 lb chicken, cooked and shredded
6 avocados, peeled and seeded
2 or 3 tsp lemon juice
8 oz sour cream
1 lb Cheddar cheese, shredded
3 lb tortilla chips

Procedure

1. Line individual salad plates with 2 oz of lettuce leaves. Cover with 2 oz of torn lettuce.
2. For each salad, spoon about 1 tbsp of dressing over the lettuce.
3. Toss kidney beans and olives with ¾ c dressing. Place 3 oz of bean-olive mixture on top of the lettuce.
4. Toss shredded chicken with ¾ c dressing. Place 2 oz chicken on each salad.
5. Mash avocados with lemon juice. Place a spoonful of mashed avocado on top of the shredded chicken.
6. Place ½ tbsp of sour cream on the avocado.
7. Sprinkle about ¾ oz of shredded cheese over each salad.
8. Arrange tortilla chips (about 2 oz per salad) around salad.

Artichoke hearts, snow peas, red peppers, rice, and zucchini are combined with lettuce leaves and grilled chicken. The salad is drizzled with Italian dressing and sprinkled with chopped basil.

Florida Department of Citrus

Grilled Gulf Coast chicken salad combines grapefruit, red onions, and cherry tomatoes with chicken breasts to make an excellent meat salad.

Fruited Turkey Salad

25 servings (No. 12 scoop)

Ingredients

3 lb turkey, cooked and diced into ½″ cubes
1½ lb celery, diced
10 oz canned pineapple, drained and diced
8 oz red grapes, cut in half and seeded
juice of 1 lemon
1 pt mayonnaise (variable)
salt to taste
25 red maraschino cherries with stem
25 leaves leaf or romaine lettuce
3 heads iceberg lettuce, shredded
25 parsley sprigs

Procedure

1. Place the turkey, celery, pineapple, and grape halves in a mixing container. Toss gently.
2. Squeeze the lemon juice over the mixture. Add the mayonnaise and toss gently a second time. Season with salt.
3. Line each cold salad plate with a leaf of lettuce. Sprinkle the shredded iceberg lettuce over the leaf lettuce.
4. Place a mound of salad in the center of each plate. Top with additional mayonnaise.
5. Garnish each salad with a maraschino cherry and a sprig of parsley.

Note: This salad may be served as an entree or main course salad by serving it in a large salad bowl and increasing the serving portion and garnish.

Chicken and Bacon Salad

25 servings

Ingredients

4 lb chicken, cooked and diced into ½″ cubes
1 lb bacon, cut crosswise into ½″ pieces, cooked to a crackling, drained
1 lb 8 oz celery, finely diced
1 tbsp onion, finely minced
1 pt mayonnaise (variable)
1 tbsp parsley, finely chopped
salt and white pepper to taste
25 c head lettuce
25 tomato wedges
25 hard-boiled eggs, quartered

Procedure

1. Place the chicken, bacon cracklings, celery, onion, and parsley in a stainless steel mixing bowl. Toss gently.
2. Add the mayonnaise. Toss a second time until thoroughly incorporated.
3. Season with salt and white pepper.
4. Line each cold plate with a cup of crisp head lettuce. Place a mound of salad in the center of each plate.
5. Garnish each salad with a wedge of tomato and hard-boiled egg.

Note: Turkey meat may be used in place of the chicken with excellent results.

Ham and Turkey Salad

25 servings (No. 16 scoop)

Ingredients

1½ lb white meat turkey, cooked and diced into ½″ cubes
1½ lb ham, cooked and diced into ½″ cubes
1½ lb celery, diced
juice of lemon
1 pt mayonnaise (variable)
salt and white pepper to taste
50 tomato wedges
50 hard-boiled egg quarters
25 leaves leaf lettuce
3 heads iceberg lettuce, shredded
25 parsley sprigs

Procedure

1. Place the ham, turkey, and celery in a mixing container. Toss gently.
2. Squeeze lemon juice over the ham and turkey mixture.
3. Add the mayonnaise and toss gently a second time. Season with salt and pepper.
4. Line each cold salad plate with a leaf of lettuce. Sprinkle the shredded iceberg lettuce over the leaf lettuce.
5. Place a mound of salad in the center of each plate. Top with additional mayonnaise.
6. Garnish each plate with two tomato wedges and two hard-boiled egg quarters.
7. Serve with a sprig of parsley.

Note: This salad can be served as an entree or main course salad by serving it in a large salad bowl and increasing the serving portion.

SEAFOOD SIDE SALAD RECIPES

Note: Seafood salads can also be served as an entree or main course salad by increasing the serving portion and serving in a large salad bowl.

Tuna Salad

25 servings (No. 12 scoop)

Ingredients

5 lb canned tuna, drained and flaked
2½ lb celery, diced
juice of 2 lemons
⅓ c onions, minced
1 qt mayonnaise (variable)
salt and white pepper to taste
3 heads iceberg lettuce, shredded
25 leaves leaf or romaine lettuce
25 slices or wedges of lemon

Procedure

1. Place the tuna and celery in a mixing container. Toss gently.
2. Add the onions, juice of two lemons, and mayonnaise. Toss gently a second time.
3. Season with salt and white pepper.
4. Line each cold salad plate with a leaf of lettuce. Sprinkle on the shredded iceberg lettuce.
5. Place a mound of the salad in the center of each plate. Top each salad with a small amount of additional mayonnaise.
6. Garnish with a slice or wedge of lemon.

Trade Tip

For convenience and variety, salad kits that contain greens, dressing, croutons, etc. may be used. These kits, though costly, reduce the preparation time of the kitchen staff and allow a larger variety of salads to be offered by the establishment.

Shrimp Salad

25 servings (No. 12 scoop)

Ingredients

5 lb shrimp, cooked, peeled, deveined, and cut into ½″ pieces
2½ lb celery, diced
juice of 2 lemons
1 qt mayonnaise (variable)
salt and white pepper to taste
3 heads iceberg lettuce, shredded
25 leaves leaf or romaine lettuce
25 slices or wedges of lemon

Procedure

1. Place the shrimp and celery in a mixing container. Toss gently.
2. Squeeze the lemon juice over the shrimp-celery mixture. Add the mayonnaise and toss gently a second time.
3. Season with salt and white pepper.
4. Line each cold salad plate with a leaf of lettuce. Sprinkle on the shredded iceberg lettuce.
5. Place a mound of the salad in the center of each plate. Top each salad with a small amount of additional mayonnaise.
6. Garnish with a slice or wedge of lemon.

Lobster Salad

25 servings (No. 12 scoop)

Ingredients

5 lb lobster meat, cooked and cut into ½″ pieces
2½ lb celery, diced
juice of 2 lemons
1 qt mayonnaise (variable)
salt and pepper to taste
3 heads iceberg lettuce, shredded
25 leaves leaf or romaine lettuce
25 slices or wedges of lemon

Procedure

1. Place the lobster and celery in a mixing container. Toss gently.
2. Squeeze the lemon juice over the lobster-celery mixture. Add the mayonnaise and toss gently a second time.
3. Season with salt and white pepper.
4. Line each cold salad plate with a leaf of lettuce. Sprinkle on the shredded iceberg lettuce.
5. Place a mound of the salad in the center of each plate. Top each salad with a small amount of additional mayonnaise.
6. Garnish with a slice or wedge of lemon.

Trade Tip *To prepare shrimp salads quickly, use value-added shrimp. This refers to any further processing of the shrimp beyond deveining and includes peeled, peeled and deveined, peeled undeveined, individually quick frozen (IQF), peeled and deveined IQF, cooked and peeled IQF, easy to peel, etc.*

Salmon Salad

Ingredients

5 lb canned red salmon, drained, flaked, boned, and skinned
2½ lb celery, diced
juice of 2 lemons
1 qt mayonnaise (variable)
⅓ c onions, minced
salt and white pepper to taste
3 heads iceberg lettuce, shredded
25 leaves leaf or romaine lettuce

Procedure

1. Place the salmon, celery, and onions in a mixing container. Toss gently.
2. Add the juice of two lemons and the mayonnaise. Toss gently a second time.
3. Season with salt and white pepper.
4. Line each cold salad plate with a leaf of lettuce. Sprinkle on the shredded iceberg lettuce.

5. Place a mound of salad in the center of each plate. Top with a small amount of additional mayonnaise.
6. Garnish with a slice or wedge of lemon.

King Crabmeat Salad

25 servings (No. 12 scoop)

Ingredients

5 lb king crabmeat, cooked and cut into ½" pieces
2½ lb celery, diced
juice of 2 lemons
1 qt mayonnaise (variable)
salt and white pepper to taste
3 heads iceberg lettuce, shredded
25 leaves leaf or romaine lettuce
25 slices or wedges of lemon

Procedure

1. Place the crabmeat and celery in a mixing container. Toss gently.
2. Squeeze the lemon juice over the crabmeat-celery mixture, add the mayonnaise, and toss gently a second time.
3. Season with salt and white pepper.
4. Line each cold plate with a leaf of lettuce. Sprinkle on the shredded iceberg lettuce.
5. Place a mound of the salad in the center of each plate. Top each salad with a small amount of additional mayonnaise.
6. Garnish with a slice or wedge of lemon.

Assorted Seafood Salad

25 servings (No. 12 scoop)

Ingredients

1 lb lobster meat, cooked and cut into ½" pieces
2 lb shrimp, cooked, deveined, and cut into ½" pieces
1 lb canned tuna, drained and flaked
1 lb king crabmeat, cooked and cut into ½" pieces
2 lb celery, diced
⅓ c onions, minced
juice of two lemons
1 qt mayonnaise (variable)
salt and white pepper to taste
3 heads iceberg lettuce, shredded
25 leaves leaf or romaine lettuce
25 slices or wedges of lemon

Procedure

1. Place the lobster, shrimp, tuna, crabmeat, celery, and onion in a mixing container. Toss gently.
2. Add the juice of two lemons and the mayonnaise. Toss gently a second time.
3. Season with salt and white pepper.
4. Line each cold salad plate with a leaf of lettuce. Sprinkle on the shredded iceberg lettuce.
5. Place a mound of salad in the center of each plate. Top with a small amount of additional mayonnaise.
6. Garnish with a slice or wedge of lemon.

Orange Caesar Salad

12 servings

Ingredients

2 tsp anchovy paste
¼ c Dijon mustard
¼ c garlic, finely chopped, divided
1½ qt orange juice, divided
1½ c olive oil
3 c grapefruit juice
½ c lime juice
2 tsp Old Bay® seasoning
96 shrimp, peeled and deveined
3 lb Romaine lettuce, 1" × 1"
6 c croutons
1 qt orange sections
¾ c Parmesan cheese, grated

Florida Department of Citrus

Procedure

Dressing

1. Combine anchovy paste, mustard, and 1 tbsp garlic in a food processor and puree until smooth.
2. Add 2 c orange juice and the olive oil. Mix until incorporated, cover, and refrigerate.

Marinade

1. Combine grapefruit juice, lime juice, Old Bay® seasoning, remaining garlic, and remaining orange juice.
2. Place 4 shrimp on each skewer and marinade in refrigerator for 4 hr.

3. Discard the marinade and cover. Refrigerate the shrimp until needed.

Salad

1. Grill skewered shrimp 2 min – 3 min per side.
2. Toss lettuce with dressing and place 4 c of greens on each plate.
3. Top each plate with ½ c croutons, ⅓ c orange sections, 1 tbsp Parmesan cheese, and the shrimp from two skewers.

Shrimp and Tuna Salad

25 servings (No. 12 scoop)

Ingredients

3 lb canned tuna, drained and flaked
2 lb shrimp, cooked, deveined, and cut into ½″ pieces
2½ lb celery, diced
juice of 2 lemons
⅓ c onions, minced
1 qt mayonnaise (variable)
salt and white pepper to taste
3 heads iceberg lettuce, shredded
25 leaves leaf or romaine lettuce
25 slices or wedges of lemon

Procedure

1. Place the tuna, shrimp, celery, and onions in a mixing container. Toss gently.
2. Add the juice of two lemons and the mayonnaise. Toss gently a second time.
3. Season with salt and white pepper.
4. Line each cold salad plate with a leaf of lettuce. Sprinkle on the shredded iceberg lettuce.
5. Place a mound of salad in the center of each plate. Top with a small amount of additional mayonnaise.
6. Garnish with a slice or wedge of lemon.

GELATIN SIDE SALAD RECIPES

Spicy Peach Mold

25 servings

Ingredients

1 qt canned peaches, sliced and drained
1 pt peach syrup
1 pt hot water
1 qt cold water
1 c vinegar
1½ c sugar
1 oz cinnamon stick
1 tbsp cloves
14 oz orange gelatin

Procedure

1. Combine the peach syrup, pint of hot water, vinegar, sugar, cinnamon stick, and cloves in a saucepan. Simmer for 15 min. Remove from the heat and strain.
2. Add the gelatin to the hot liquid. Dissolve thoroughly.
3. Add the cold water and stir.
4. Place a small amount of gelatin in each mold. Chill until firm and then remove from the refrigerator.
5. Place the sliced peaches in each mold and cover with the remaining gelatin. Return to the refrigerator and chill until firm.
6. Unmold and serve on crisp salad greens.

Cranberry-Orange Salad

25 servings

Ingredients

12 oz orange gelatin
1 pt boiling water
3 c cold water
1 c celery, minced
2 tbsp sugar
2 oranges
3 cans whole cranberry sauce
3 oz pecans, chopped

Procedure

1. Dissolve the orange gelatin in the boiling water.
2. Add the cold water, place in a pan, and chill until the gelatin is partly set. Remove from the refrigerator.
3. Grind the oranges on the food grinder. Use the medium chopper plate.
4. Blend the ground oranges, celery, whole cranberry sauce, sugar, and pecans.
5. Fold this mixture into the gelatin mixture.
6. Pour into individual molds. Refrigerate until firm.
7. Unmold and serve on crisp salad greens.

Trade Tip

Dry gelatin has an indefinite shelf life. It should be stored in a cool, dry environment, protected from odors and extremes of temperature and humidity. Once prepared, it should be used within 4 hours. Always prepare only the amount of gelatin that will be served promptly.

Green Island Salad

25 servings

Ingredients

12 oz lime gelatin
3 c hot water
3 c pear juice
1 tsp cider vinegar
½ tsp salt
1 lb cream cheese
½ tsp ginger
1½ lb canned pears, drained and diced

Procedure

1. Dissolve the gelatin in hot water. Add the pear juice, vinegar, and salt. Stir.
2. Fill individual molds one-third full and place in the refrigerator until firm.
3. Chill the remaining gelatin until slightly thickened. Remove from the refrigerator and place in the mixing bowl.
4. Whip at medium speed until light and fluffy.
5. Add the cream cheese and ginger. Continue to whip until the cheese is blended with the gelatin.
6. Remove from the mixer and fold in the diced pears.
7. Spread this mixture over the firm layer of gelatin in the molds. Refrigerate until firm.
8. Unmold and serve on crisp salad greens.

Tomato Aspic Salad

Aspic salads can be used in a cold food preparation or as a garnish. An herb-flavored aspic is a possible alternative to the traditional tomato aspic.

25 servings

Ingredients

2 oz plain gelatin
1 pt cold water
2 qt tomato juice
½ c onions, diced
1 bay leaf
½ c celery, diced
4 cloves
1 tsp dry mustard
6 oz sugar
¼ oz salt
1 c lemon juice

Procedure

1. Add the plain gelatin to the cold water and allow to set for 5 min.
2. Place the tomato juice, onions, bay leaf, celery, cloves, dry mustard, sugar, and salt in a saucepan. Bring to a boil and let simmer for 10 more minutes. Remove and strain.
3. Add the soaked gelatin to the hot tomato liquid and stir until gelatin is thoroughly dissolved.
4. Add the lemon juice and stir.
5. Pour into individual molds. Refrigerate until firm.
6. Unmold and serve on crisp salad greens.

Jellied Diplomat Salad

25 servings

Ingredients

1 c lemon gelatin
1 pt hot water
1 c cold water
12 oz mayonnaise
1 lb 4 oz celery, diced
6 oz pineapple, diced
1 lb 6 oz unpeeled apples, diced
½ tsp salt
2 tbsp cider vinegar

Procedure

1. Dissolve the gelatin in the hot water. Add the cold water and stir.
2. Chill the gelatin until it begins to thicken.
3. Place the apples, celery, pineapple, mayonnaise, salt, and vinegar in a mixing container. Toss until all ingredients are thoroughly blended.
4. Place the slightly thickened gelatin in the mixing bowl. Whip until light and fluffy.
5. Fold the gelatin mixture into the diplomat salad until well blended.
6. Place into individual molds. Refrigerate until firm.
7. Unmold and serve on crisp greens.

Cider Salad

25 servings

Ingredients

12 oz apple-flavored gelatin
3 pt apple juice or cider
⅓ c lemon juice
½ tsp salt
1 c celery, minced
1 c carrots, grated
1 c unpeeled apples, minced
2 oz walnuts, finely chopped

Procedure

1. Heat 2 c of the cider or apple juice. Add the gelatin and dissolve thoroughly.
2. Add the remaining cup of cider or apple juice, lemon juice, and salt. Allow to cool.
3. Place in the refrigerator until the gelatin starts to thicken, then remove.
4. Fold in all the remaining ingredients. Pour into individual molds and refrigerate until firm.
5. Unmold and serve on crisp salad greens.

SALAD DRESSING RECIPES

Mayonnaise

Mayonnaise is a semisolid dressing prepared by forming an emulsion with eggs and salad oil. This dressing is important in the preparation of other dressings popular in the commercial kitchen.

1 gal.

Equipment

- Mixing machine and wire whip attachment
- Baker's scale
- Quart measure
- Cup measure
- Bain-marie

Ingredients

4 egg yolks
4 eggs
4 qt salad oil
¼ oz dry mustard
⅓ c cider vinegar
¼ oz salt (variable)
4 dashes hot sauce
½ c water
white pepper to taste

Preparation

1. Break the eggs. Separate four of the yolks from the whites.

Method No. 1

Tap the shell on the edge of a bowl until it cracks. Break the shell in half. Holding half the shell in the right hand and half in the left hand, pass the yolk back and forth from one half shell to the other until the white runs off. The action of passing the yolk from one shell to the other is done over a bowl so the white can be caught as it runs off.

Method No. 2

Tap the shell on the edge of a bowl until it cracks. Break the shell in half and let the egg run into the hand. Spread the fingers back and forth slowly letting the white run off the yolk and into a bowl that is placed below the hand.

Procedure

1. Place eggs and yolks in a mixing bowl.
2. Add salt, dry mustard, and pepper. Whip slightly.
3. Add half of the oil, pouring in a very slow stream with the mixer running at high speed. This forms the emulsion.
4. Add water, vinegar, and remaining oil alternately, one-third at a time.
5. Check seasoning, pour into a bain-marie, and refrigerate.

Precautions

- When adding the oil, pour in a very slow stream or the emulsion will not form. After the emulsion has formed, continue to pour slowly because there is still a chance it may break.
- Have the mixing machine running at high speed throughout the entire operation.
- When adding the salad oil to the eggs forming an emulsion, the oil must be added in a very slow stream with the mixer running at high speed or the emulsion may break (return to a liquid). If this should happen, the process must be repeated.

Note: Most commercial kitchens today have a rotary mixer; however, mayonnaise can be also be prepared by hand with a French or piano whip.

Thick French Dressing

Thick French dressing is prepared by forming an emulsion with eggs and salad oil similar to the mayonnaise preparation, but not as thick.

1 gal.

Equipment

- Mixing machine and wire whip attachment
- Quart measure
- Bain-marie
- Spoon measure

Ingredients

4 eggs
1 pt cider vinegar
⅓ c paprika
1 tsp salt
1 c sugar
1 tbsp Worcestershire sauce
½ tsp dry mustard
4 qt salad oil
½ c catsup
1 c lemon juice
3 dashes hot sauce

Procedure

1. Blend the lemon juice, vinegar, paprika, salt, sugar, Worcestershire sauce, hot sauce, catsup, and dry mustard.
2. Place the eggs in the mixing bowl. Beat at high speed with the rotary mixer.
3. Pour the oil in a very slow stream while continuing to beat at high speed.
4. As the emulsion forms and the mixture thickens, add the mixture from Step 1 to thin it down. Continue this process until all the ingredients are incorporated.
5. Check seasoning. Pour into a bain-marie and refrigerate.

Precautions

- When adding the oil, pour in a very slow stream or the emulsion will not form. After the emulsion has formed, continue to pour slowly because there is still a chance it may break.
- Operate the mixing machine at high speed until at least half of the oil is added.

Preparation

1. Squeeze juice from lemons.

Ranch Dressing

This dressing has gained popularity because of its creamy consistency and tangy sour taste. It is a blend of buttermilk, mayonnaise, garlic, and onion flavors. It is recommended for a leafy green salad.

1 gal.

Equipment

- Quart measure
- Spoon measure
- Wire whip
- Stainless steel bowl
- Bain-marie
- French knife

Ingredients

2 qt buttermilk
2 qt mayonnaise
1 tsp garlic powder
1 tbsp onion powder
1 tbsp parsley
salt and white pepper to taste

Procedure

1. Place the garlic powder and onion powder in a stainless steel mixing bowl. Add the buttermilk and whip with a wire whip until all ingredients are blended.
2. Add the mayonnaise and whip until dressing is smooth.
3. Add the chopped parsley and season with salt and white pepper. Continue to whip until all ingredients are incorporated.

Preparation

1. Chop the parsley using a French knife.

Precaution

- Exercise caution when chopping the parsley.

Thousand Island Dressing

Thousand Island dressing is prepared by using mayonnaise as a base. It is a sweet dressing because sweet pickle relish is one of the main ingredients.

1 gal.

Equipment

- Bain-marie
- Kitchen spoon
- Quart measure
- French knife
- Towel

Ingredients

3 qt mayonnaise
3 c chili sauce
3 hard-boiled eggs
1 pt sweet pickle relish, drained
¼ c parsley
2 tsp paprika

ter of a kitchen towel. Draw the four corners of the towel to completely envelope the chopped parsley. Holding the towel securely at the top of the enveloped parsley, place under running cold water and wash thoroughly. Twist the towel until all water is wrung from the parsley. Place the chopped parsley in a bowl.

Procedure

1. Place the mayonnaise in a bain-marie.
2. Add the remaining ingredients. Blend thoroughly with a kitchen spoon.

Preparation

1. Chop the eggs fine.
2. Place the parsley on a cutting board and chop very fine. Place the very finely chopped parsley in the cen-

Precautions

- Chop the eggs on heavy paper for best results.
- Drain the sweet relish thoroughly or the dressing will be too runny.

Louis Dressing

Louis salad dressing is similar in appearance to Thousand Island but has a more tart, sharp taste.

1 gal.

Equipment

- Bain-marie
- Quart measure
- Kitchen spoon
- Spoon measure
- French knife

Ingredients

2 qt mayonnaise
¼ c horseradish
1 c dill pickles
3 pt chili sauce
1 c celery
2 tbsp lemon juice

2. Chop the dill pickles very fine.

Procedure

1. Place the mayonnaise in a bain-marie.
2. Add the remaining ingredients. Blend thoroughly with a kitchen spoon.

Preparation

1. Chop the celery very fine.

Precaution

- Exercise caution when chopping the celery and pickles.

Bacon Bit Dressing

This is an emulsified dressing with a flavor of hickory-smoked bacon. This dressing is often served with leafy green salads.

1 gal.

Equipment

- Mixing machine and wire whip attachment
- Quart measure
- Bain-marie
- French knife
- Spoon measure
- Baker's scale
- Food grinder
- 1 qt saucepan
- Kitchen spoon
- China cap

Ingredients

- 3 eggs
- 3 egg yolks
- 3 qt salad oil
- 6 oz cider vinegar
- 1 tsp salt
- ½ tsp white pepper
- 1 tbsp Worcestershire sauce
- 1 tbsp sugar
- 6 oz bacon
- 1 tbsp chives, fresh, frozen, or freeze-dried

Preparation

1. Mince the chives with a French knife.
2. Remove rind from bacon with a French knife. Cut into strips and grind in the food grinder using the medium chopper plate. Place ground bacon in a saucepan.

Cook at medium heat until a crackling is formed. Drain in china cap. Save drained grease for use in another preparation.

Procedure

1. Place the eggs in the mixing bowl and beat on the rotary mixer at high speed using the wire whip.
2. Pour the oil in a very slow stream while continuing to beat at high speed.
3. As the emulsion forms and the mixture thickens, add the vinegar to thin it down. Continue this process until all the vinegar and oil have gone into the mixture.
4. Add all the remaining ingredients and blend into the mixture at slow speed until thoroughly blended. Remove from the mixer, place in bain-marie, and refrigerate.

Precautions

- Pour the oil in very slowly.
- Keep the mixing machine running at high speed until all the oil and vinegar have been added.
- Exercise caution when mincing the chives and grinding the bacon.
- If dressing is too thick, it can be thinned by adding water.

Italian Dressing

Italian dressing is a thin, tart, spicy dressing. It can be served with any leafy green salad. Shake or stir well before using.

1 gal.

Equipment

- Bain-marie
- Quart measure
- Cup measure
- Spoon measure
- Mixing machine and whip attachment
- Saucepan
- China cap
- Cheesecloth
- French knife

Ingredients

- 2 qt salad oil
- 3½ c catsup
- 3½ c cider vinegar
- 4 cloves garlic
- ¼ c chives
- ⅓ c Parmesan cheese, grated
- 3 tsp salt
- 4 tsp paprika
- ¼ c pickling spices
- 3 tsp dry mustard

Procedure

1. Place the vinegar, pickling spices, and garlic in a saucepan. Bring to a boil, then reduce to a simmer and cook for 3 more minutes.
2. Remove from the range and allow to cool. Strain through a china cap covered with a cheesecloth.
3. Combine the paprika, salt, dry mustard, sugar, and Parmesan cheese in the mixing bowl. Blend in the strained vinegar mixture and mix thoroughly in the rotary mixer until smooth.
4. Pour in the salad oil in a very slow stream while beating briskly.
5. Add the catsup and chives and blend thoroughly.
6. Remove from the mixer, place in a bain-marie, and refrigerate.

Precautions

- Pour the oil in very slowly.
- If the mixture starts to splash when the oil is only half-added, reduce the speed of the mixer to medium.

Preparation

1. Mince the garlic and chives.

Trade Tip

In most salads, the dressing contains the largest number of calories. To reduce the calories, use a light dressing and serve moderate-sized salads. Also, consider the use of lemon juice, balsamic vinegar, etc. to reduce calories in the salad. Using a lower fat mayonnaise when making a dressing will have a significant impact on calories and fat.

Vinaigrette Dressing

Vinaigrette dressing is prepared by adding finely-chopped herbs, pickles, hard-boiled eggs, etc., to a vinegar and oil dressing.

1 qt

Equipment

- Quart measure
- Spoon measure
- Kitchen spoon or ladle
- Bain-marie
- French knife

Ingredients

1 qt vinegar and oil dressing
1 tbsp parsley
1 tbsp chives
1 tbsp olives
1 tbsp capers
1 hard-boiled egg
1 tbsp sweet pickles
1 tbsp pimientos

Preparation

1. Chop the parsley, chives, olives, capers, hard-boiled egg, sweet pickles, and pimientos very fine.
2. Prepare the vinegar and oil dressing.

Procedure

1. Place all the ingredients in a bain-marie and blend thoroughly.
2. Place in the refrigerator until ready to use.

Precaution

- Chop the hard-boiled egg gently on heavy paper for best results.

Russian Dressing

Russian dressing is similar to Thousand Island dressing, but caviar is generally added. This dressing blends well with most leafy green salads.

1 gal.

Equipment

- Quart measure
- Kitchen spoon
- Bain-marie
- Spoon measure

Ingredients

3 qt mayonnaise
3 c chili sauce
¼ c paprika
1 c pimientos
1 c caviar, if desired

Procedure

1. Place all the ingredients in a bain-marie and blend thoroughly using a kitchen spoon.
2. Place in the refrigerator until ready to use.

Preparation

1. Chop the pimientos very fine.

Precaution

- If imitation caviar is used, drain before blending.

Bleu or Roquefort Cheese Dressing

Bleu or Roquefort cheese dressing is one of the most popular salad dressings. Roquefort is a product of France and is superior in taste and quality to the American bleu cheese.

1 gal.

Equipment

- Quart measure
- Cup measure
- Spoon measure
- Bain-marie
- Mixing machine and whip attachment
- French knife
- Kitchen spoon

Ingredients

6 eggs
3 qt salad oil
½ c lemon juice
1 c vinegar
½ c sugar
1 tsp dry mustard
2 tsp salt
1 tbsp Worcestershire sauce
3 drops hot sauce
12 oz bleu or Roquefort cheese

2. Break the eggs.

Procedure

1. Place the dry mustard, salt, sugar, and lemon juice in the mixing bowl. Whip until thoroughly blended.
2. Add the eggs and continue to whip at slow speed.
3. Increase the speed of the mixer to high and pour in the oil in a very slow stream to form a permanent emulsion. Add the vinegar at intervals to thin slightly.
4. Reduce the speed of the mixer to slow. Add the Worcestershire sauce and hot sauce and blend.
5. Remove from the mixer and fold in the bleu or Roquefort cheese.
6. Check the seasoning. Pour into a bain-marie and refrigerate until ready to use.

Preparation

1. Chop the bleu or Roquefort cheese into fairly fine chunks with a French knife.

Precautions

- Before attempting to chop the cheese, freeze it to make the job less difficult.

Hot Bacon Dressing

Hot bacon dressing is a sweet-sour preparation containing minced, cooked onions and crisp bacon bits. It is eaten over shredded cabbage for a hot slaw and over lettuce for a wilted lettuce salad.

2 qt

Equipment

- Quart measure
- Baker's scale
- 1 gal. sauce pot
- French knife
- Kitchen spoon

Ingredients

1 qt vinegar
1 qt water
10 oz sugar (variable)
½ oz salt
2 lb bacon
4 oz onions

Procedure

1. Place the diced bacon in a sauce pot, place on the range, and cook to a crisp crackling.
2. Add the minced onions and cook just slightly.
3. Add the vinegar, water, salt, and sugar. Bring to a simmer and simmer for 2 min.

Precautions

- Exercise caution when dicing the bacon and mincing the onions.
- Be alert when cooking the bacon. Do not let it become too dark or burn.

Preparation

1. Dice bacon into small cubes with a French knife.
2. Mince onions with a French knife.

Green Goddess Dressing

Green goddess dressing is prepared with a mayonnaise base. It has a flavor of anchovy. The word green is used because finely chopped green onion stems flow through the dressing. This dressing is an excellent choice for seafood or leafy green salads.

1 gal.

Equipment

- Bain-marie
- French knife
- Spoon measure
- Quart measure
- Kitchen spoon

Ingredients

2 qt mayonnaise
1 qt heavy sour cream
1 c tarragon vinegar
1 c scallions
1 c anchovies, drained
3 cloves garlic
1 tsp sugar
salt and white pepper to
    taste

Preparation

1. Mince the scallions and garlic
2. Chop the anchovies very fine with a French knife.

Procedure

1. Place all the ingredients in a bain-marie and blend thoroughly.
2. Season with salt and white pepper.
3. Place in the refrigerator.

Precautions

- Drain the anchovies before chopping.
- Exercise caution when chopping with the French knife.

Vinegar and Oil Dressing

Vinegar and oil dressing is a blend of vinegar and salad oil. Cider vinegar is used most often; however, wine vinegar also produces excellent results. This dressing should be shaken or stirred well before using.

1 qt

Equipment

- Cup measure
- Bain-marie
- Spoon measure
- Wire whip (hand)

Ingredients

3 c salad oil
1 c cider vinegar
1 tbsp salt
½ tsp white pepper
1 tsp sugar

Procedure

1. Place all the ingredients in a bain-marie and whip briskly until thoroughly blended.

Precaution

- Shake or stir well before serving.

Cheshire cheese is similar to Cheddar cheese, but has a more crumbly texture and is not as compact as Cheddar. It is sold in its natural white color or in deep yellow. The yellow is the result of adding annatto (a dye) to the curd. Cheshire cheese cannot be imitated because of the soil and the kind of grass that grows in England. The grazing lands have rich deposits of salt, which is passed along to the cows. The cows in turn, produce a salty milk. Cheshire cheese is salty, but this salty taste is not very noticeable when the cheese is consumed. Cheshire cheese is used in the same way as Cheddar and makes an excellent Welsh rarebit or fondue.

Cold pack cheese: This blended cheese is made without the aid of heat and pasteurization as with other cheeses. Cold pack cheese is creamy and ranges from white to orange in color. It is available in many different flavors and is often mixed with a variety of spices and seasonings.

Cottage cheese: Cottage cheese is a soft cheese and perhaps the simplest of all cheeses. It is known by many names: Dutch cheese, pot cheese, smearcase, and, in some localities, popcorn cheese (because of its large curds). It is marketed in about five different varieties: small curd, large curd, flake curd, homestyle, and whipped. It is either plain or creamed. Cottage cheese can be made in the home as well as in the factory with fine results.

Large quantities of cottage cheese are consumed in the United States today because of its very fine, mildly sour taste. In the commercial kitchen, cottage cheese is put to many uses. It is used for appetizers, salads, cheesecakes, pies, and also in some cooked dishes. It is very perishable and should always be stored at a low temperature.

Cream cheese: This is a soft, mild cheese that is rich in flavor. It is an uncured cheese made from cream or a mixture of cream and milk. It is similar to the unripened French Neufchâtel, but higher in fat content. Cream cheese is one of the most popular cheeses in the United States. There are many brands of cream cheese on the market today and all have good eating qualities; however, they are not necessarily the same. The difference lies in the use of gum arabic, a stabilizer used to extend the keeping qualities of the cheese. Cream cheese that does not contain gum arabic has a lighter, more natural texture, but it does not keep as well. Cream cheese is used extensively in the commercial kitchen in the preparation of such items as canapé

spreads, sandwiches, salads, salad dressings, and numerous desserts.

Edam cheese: Edam is another cheese that is named after its birthplace, Edam, which is in the province of North Holland, Netherlands. Edam is a hard, mild cheese and is made from cow's milk. It possesses a rather firm and crumbly texture and is usually shaped into what might be described as a cannon ball shape. In the Netherlands, the cheese for export is colored red on the outside, rubbed with oil, wrapped, and shipped. The red coating is one of the chief characteristics of the cheese. However, the cheese made for consumption within the country is rubbed with oil but is not colored. Edam cheese made in the United States is covered with a thin coating of red paraffin to give it its characteristic color. It is used most often as a dessert cheese on platters and on buffet tables where its color helps stimulate the appetite.

Feta cheese: Feta is a cheese of Greek origin made from sheep's or goat's milk. The cheese is slightly cured from a few days to four weeks. It has a salty taste and, when aged for a long period of time, becomes very salty and dry. Because of this condition, it is always wise to taste, if possible, before making a purchase. When aged for the average consumer, Feta can have a creamy texture, be pleasantly salty, and have a soft to semisoft consistency. The smell is similar to cider vinegar and its taste resembles a faint taste of olives. It has a creamy white interior and can be used for snacks, salads, and certain cooked dishes. It can be used in a salad bar, lasagna, or omelet. Feta is available on the market in jars, cans, plastic wrap, or fresh.

Gorgonzola cheese: Gorgonzola, which is from Italy, is in the blue-green veined cheese family. It is the Italian cousin of the French Roquefort, the English Stilton, and the American Bleu cheese. It originated in the village of Gorgonzola, near Milan; however, very little of the cheese is made there now. Today, it is made chiefly in the regions of Lombardy and Piedmont. The cheese is mottled with characteristic blue-green veins produced by a mold known as *penicillium glaucum*. The surface of the cheese was originally protected with a reddish coat made by mixing brick dust, lard, and coloring together, and smearing it on the cheese. Today, this method has been eliminated and the cheese is protected with aluminium foil. Gorgonzola cheese is generally cured for a period of six months to a year. Gorgonzola is used in salads, salad dressings, and as a dessert and buffet cheese.

POPULAR CHEESES

Variety	Characteristics	Usage
Brie	Soft; thin, white, edible crust, creamy interior; slightly firm and mild when young and creamy and pungent when aged	Appetizers, desserts, toppings, fillings; feature on buffet tables, serve in warm puff pastry of baked with almonds
Feta	Soft; flaky white interior; salty, "pickled flavor"	Salads, cooked dishes, snacks; add as a Greek accent to the salad bar or lasagna, use as an omelet filling
Cream Cheese	Soft; smooth texture; white color throughout; creamy mild flavor	Spreads, sauces, desserts, sandwiches; serve with specialty breads, use in cheesecakes, pastries, and frostings
Ricotta	Soft; moist, grainy; white; mild, slightly sweet flavor	Filings, cooked dishes, desserts, dips; use as filling for stuffed pasta, cannoli
Port du Salut	Semisoft; smooth, buttery; creamy yellow; mild to robust	Appetizers, desserts, snacks; serve with fresh fruit, crackers, or wine
Muenster	Semisoft; waxy open texture; creamy white with orange exterior; mild to mellow	Cheese plates, snacks, sandwiches, toppings, appetizers; melt over open-faced sandwiches or entrees, serve cubed with cocktail sauce
Brick	Semisoft; waxy open texture; creamy white; mild to mellow, pungent when aged	Sandwiches, snacks, salads; use for a hearty grilled cheese sandwich
Baby Swiss	Semisoft; smooth texture with well-distributed eyes; creamy white; mild, sweet, nutty flavor	Appetizers, snacks, sandwiches, cooked dishes; use on cheese trays, deli sandwiches, and quiches
Monterey Jack	Semisoft; smooth open texture; creamy white; mild to mellow	Entrees, sandwiches, toppings, salads; use in Mexican dishes, excellent with avocado salads, great for kids
Bleu	Semisoft; crumbly; blue-green mold marbled or streaked-white interior; sharp, piquant, spicy flavor	Dips, salads, dressings, desserts; crumble over vegetables, use to create gourmet burgers and salads
Mozzarella; String	Semisoft; smooth plastic body; creamy white; mild, delicate flavor; mozzarella in the shape of a string or rope	Pizza, entrees, appetizers, sandwiches, snacks; use on antipasto plates and subs, try string cheese with cocktail sauce or plain, great for kids
Gouda and Edam	Semisoft to firm; creamy with small holes; light yellow; mild, nutlike	Appetizers, sandwiches, desserts, cooked dishes; ideal for cheese plates and take-out box lunches
Cheddar	Firm; smooth body; color ranges from nearly white to orange; varied shapes and styles; mild to sharp	Entrees, side dishes, soups, salads, sandwiches, snacks; feature Wisconsin Cheddar burgers, serve warm on pie
Colby	Firm; open texture; light yellow to orange color; mild to mellow	Cooked dishes, snacks, sandwiches, salads; delicious in omelets and quiches, serve with fresh fruit
Colby-Jack/Co-Jack	Firm; smooth body; marbled white and orange; mild to mellow	Appetizers, sandwiches, cooked dishes; adds extra eye appeal on cheese trays and open-faced sandwiches
Swiss	Firm; smooth with large, shiny eyes; pale yellow; mellow, nutlike flavor	Sandwiches, toppings, entrees, sauces, salads; use in quiches and fondue, great in melted sandwiches
Provolone	Firm; smooth, plastic body; creamy white; mild to piquant or smoky in flavor	Entrees, pizza, salads, snacks; melt over Italian bread, cube for pasta salads, add julienned strips to green salad
Parmesan/Romano	Hard; granular; light yellow; sharp, piquant flavor	Entrees, side dishes, soups, salads, seasoning; add to pizza, sprinkle over omelets, breads, and pasta dishes
Pasteurized Process	Blended with the aid of heat; semisoft; smooth, uniform body; white to orange; mild to mellow; available in many flavors	Sandwiches, sauces, entrees, snacks; melt over hash browns, try cheese on fries, take advantage of flavors for sandwiches
Cold Pack	Blended without the aid of heat; soft; creamy; white to orange; mild to sharp; available in many flavors	Sauces, dips, sandwiches; offer at the salad bar and with breads and crackers; serve on sandwiches and vegetables

Wisconsin Milk Marketing Board

Gouda cheese: Gouda is very similar to Edam cheese. Like Edam, it is a hard, sweet curd cheese. It also originated in Holland, but in a different province. Gouda was first produced in the Dutch province of Gouda and, as is the custom, was named after its place of birth. The main difference between Edam and Gouda is that Gouda contains more fat. Gouda is usually shaped like a flattened ball or formed into a loaf. Neither Edam nor Gouda are recommended for cooking. They are intended to be dessert or buffet cheeses.

Gruyère cheese: This cheese originated in western Switzerland in the village of Gruyère. Gruyère is a hard cheese similar in many ways to Swiss cheese. Gruyère has smaller holes and a sharper taste than Swiss cheese. It is also manufactured in smaller wheels.

Gruyère is an excellent cheese to use for cooking. In commercial establishments, it is used in fondue, veal cordon bleu, and sautéed veal chops Gruyère. Gruyère is also one of the many cheeses used in process cheese, but the result is an entirely different product from true Gruyère.

Liederkranz® cheese: Liederkranz® is a soft, surface-ripened cheese made from cow's milk. It is very similar to Limburger cheese in body, flavor, aroma, and method of ripening. Liederkranz, like brick cheese, is strictly American. Liederkranz was discovered by a New York cheese maker in Monroe, New York in 1882. It was named after Liederkranz Hall in New York, where it was first enjoyed by a singing group to which the maker belonged. Liederkranz was made for years in the state of New York, but in recent years the plant was moved to Van Wert, Ohio. Ohio, thus, is the largest producer of this cheese.

Liederkranz is packaged in small, oblong loaves weighing 4 oz each and, like Camembert, is never marketed in bulk. Liederkranz spoils rapidly, so it must be watched closely once it is placed in a store's dairy case. To be enjoyed to its fullest extent, the cheese should be brought to room temperature before serving. Liederkranz is always served as a dessert or buffet cheese accompanied with crackers and onions.

Limburger cheese: Limburger is a soft, ripened cheese with a characteristic strong aroma and flavor. Limburger cheese was first made in Liege and marketed in Limburg, Belgium. Much of this cheese, however, is made in Germany as well as in the United States. Limburger cheese is either made from whole milk or skim milk. It has a very creamy texture that is brought about by ripening in a damp atmosphere for a period of two months. Limburger cheese is served most often as a dessert or buffet cheese, and always with crackers and onions.

Monterey Jack: This is a cheese that displays a smooth open texture, a creamy white color, and a mild taste. It is sometimes described as having a taste similar to American Muenster. Monterey Jack that is aged for a longer-than-average period of time becomes harder and more zesty in flavor, but is not as popular as the surface-aged cheese. It is used in the commercial kitchen for sandwiches, salads, and certain entree dishes, especially Mexican dishes.

Mozzarella cheese: Mozzarella is a very tender cheese with a soft, plastic curd. It was made originally in southern Italy from buffalo milk. Today, Mozzarella is primarily made from cow's milk. Mozzarella is an unripened cheese and, when eaten fresh, is still dripping with whey. In making Mozzarella cheese, the whey is ordinarily drained from the curd and used in making Ricotta cheese. Mozzarella, when melted, has a very elastic or rubbery consistency and is commonly used in pizza and lasagna.

Muenster cheese: Muenster cheese is a semisoft cheese made of cow's milk. It has a flavor between that of brick cheese and Limburger. It was first produced in the vicinity of Muenster, near the western border of Germany. The French also produce a Muenster known as Gerome. The European Muenster cheese is unlike the product produced extensively in the United States today because it is much sharper in taste and has a strong aroma. This is mainly due to a longer aging period. Muenster is marketed in cylindrical form and is used as a buffet or sandwich cheese.

Wisconsin Milk Marketing Board

Cheese is available in a variety of sizes, shapes, and textures.

Neufchâtel cheese: Neufchâtel is very similar to cream cheese but possesses a higher moisture content and a lower fat content. It is a cheese of French origin and is made extensively in the Department of Seine Inferieure, France. Neufchâtel has a very soft texture and a mild flavor. It is made from whole or skim milk or a mixture of milk and cream. Neufchâtel is generally marketed as a fresh cheese, although it can be cured. Because of the smooth texture of this cheese, it spreads and blends well and is used in canapé spreads, salads, salad dressing, and many dessert items.

Parmesan cheese: Parmesan is a hard cheese produced in Italy and is thought of as one of the six best cheeses of the world. It was first made in the vicinity of Parma, in Emilia, hence the name. There are many cheeses of this type made in Italy, but Parmesan is the most famous. Parmesan has a granular texture when properly cured and because of this, it is classified with a group of Italian cheeses known as *grana* (meaning grain). Parmesan cheese is made in great quantities in the United States and Argentina. This product is not as good as the true Italian product. Parmesan cheese is rubbed with oil and dark coloring from time to time through the aging period. When properly cured, it is very hard and will keep indefinitely. It is sold mostly in grated form but can be used as a table cheese when still slightly moist. Parmesan could be considered a seasoning cheese because it is used to season such preparations as onion soup, spaghetti and meatballs, macaroni, and lasagna. It, along with olives and wine, is the foundation of true Italian cuisine and is popular with cooks all over the world. Two other Italian cheeses of the grana family, Parmigiano and Reggiano, are very similar to Parmesan and are fairly popular in the United States.

Port du Salut: Port du Salut was first made around 1865 by Trappist monks at an abbey in Port du Salut, France and was named after its birthplace. It is now manufactured in abbeys in various parts of Europe and one monastery in Kentucky. The Trappists have kept the exact process a secret, but a similar cheese is made outside the monasteries in Europe and the United States.

Port du Salut has a soft, smooth, orange-colored rind and a glossy ivory, cream-colored interior. Its flavor may range from mellow to robust, depending upon the age of the cheese. However, the flavor has also been compared to that of Gouda cheese. In some instances, the aroma of the cheese is like a very mild Limburger. The cheese is used as a dessert, for appetizers, and served with apple pie.

Process cheese: Process cheese, or pasteurized cheese, is made by combining one or more cheeses of the same variety, or by combining two or more varieties and adding an emulsifying agent. Vinegar or lactic acid, cream, salt, coloring, and/or spices can be added as flavoring. The end result is a cheese product that is uniform in body, flavor, and texture. It can be packaged in just about any shape or size. For the best process cheese, care must be taken to select cheeses that are fully cured and sharp in flavor.

Process cheese was made in Germany and Switzerland as early as 1895, but the first patent for process cheese in the United States was not issued until 1916. Approximately one-third of all cheeses made in the United States today are marketed as process cheese.

Process cheese has certain advantages over other cheeses. It is more economical, it does not require refrigeration until it is opened, it melts fairly easily and evenly, and it has unusual keeping qualities. Numerous varieties of process cheese are found in the supermarket. Its use in the commercial kitchen is extensive.

Provolone cheese: Provolone is a hard Italian cheese with a mild to sharp taste and a stringy texture. It was first made in southern Italy, but is now made in other parts of Italy as well as in Wisconsin and Michigan. Provolone is light in color and cuts without crumbling. Its most distinguishing characteristic is that it is formed into the shape of a sausage and corded. The cording is done for easy hanging when the cheese is smoked. One of the unusual steps in the making of Provolone is the kneading and stretching of the curd until it is smooth and free of lumps. Provolone is used in the preparation of many Italian dishes, but its most popular use is in pizza.

Ricotta cheese: Ricotta is the Italian version of our cottage cheese; however, it is not lumpy and is made from the whey of other cheeses instead of milk. It is white and creamy with a bland yet sweet flavor. Today, ricotta is made in the countries of central Europe and in some parts of southern Europe where the whey of other cheeses is considered too nutritious to be discarded. This cheese is also made in the United States, but a mixture of whey and whole milk is used in the preparation. Ricotta blends well with the flavor and textures of other foods. It is an important ingredient in lasagna and manicotti.

Romano cheese: Romano is another of the famous hard Italian cheeses. It is similar to Parmesan in many respects, but is softer in texture. In Italy, it is used both as a grated cheese and a table cheese. It was first made in the vicinity of Rome from ewe's milk. Today it is made in other parts of Italy, as well as in the United States, from cow's milk and goat's milk. Romano has a granular texture, a sharp flavor, and a hard, brittle black rind. The cheese is aged for a period of five to eight months if it is to be used as a table cheese. If it is to be grated, it is aged about one year. A longer aging period sharpens the flavor. In the commercial kitchen, Romano is used in the same fashion as Parmesan, sometimes as a topping for au gratin dishes (browned covering of cheese and/or bread crumbs) and sometimes as a seasoning.

Roquefort cheese: Roquefort is the most famous of the blue-green vein cheeses and the finest of the many French cheeses. It was first made in the Department of Aveyron, France in the village of Roquefort. It was discovered by accident, which has been the case of many popular foods down through the centuries.

Roquefort is made from ewe's milk. Although it is made in other countries from cow's milk, the word Roquefort cannot be used. A French regulation limits the use of this word in connection with any other cheese product. Roquefort is characterized by its sharp, tangy flavor and by the blue-green veins that flow through the white curd. The blue-green veins are created by spreading a powdered bread mold over the curd as it is being packed into the hoops. The cheese is cured for a period of two to five months, depending on the sharpness desired. Roquefort is used principally as a dessert cheese; however, it is also used in salads. Roquefort dressing is famous around the world.

Stilton cheese: Stilton is the English relative of Roquefort. It was first made in the village of Stilton around the mid-eighteenth century. It is rich and mellow and has a piquant flavor, although it is milder than Roquefort or Gorgonzola. It is made from cow's milk, and the curing period is from four to six months. The distinguishing characteristics are the blue-green veins of mold running through the curd and the wrinkled rind. Stilton, like the other blue-green veined varieties, has a very crumbly texture. Some Stilton is imported into the United States, but not in great quantities. It is used chiefly as a dessert and buffet cheese.

Swiss cheese: Swiss cheese originated in Switzerland and is known there by the name Emmentaler rather than Swiss. This cheese was first made during the fifteenth century, and the traditional methods of making it have been handed down from generation to generation since that time.

Swiss cheese is a large, hard, pressed-curd cheese with an elastic body and a mild, sweet flavor. Its chief characteristic is the large eyes or holes found throughout the body of the cheese. These holes are developed by special gas-producing bacteria released during the ripening period.

A large part of the milk produced in Switzerland is used in the production of this cheese. Swiss cheese was brought to the United States by Swiss immigrants in 1850, and since that time has become a very popular cheese in this country.

The curing period for Swiss cheese varies depending on where it is made. In the United States, it is placed on the market after a curing of three to four months. In Switzerland, the cheese made for export is cured for six to ten months and has a more pronounced flavor. Swiss cheese is used in many preparations in the commercial kitchen, from sandwiches to stuffing veal chops, but the most desirable preparation is fondue, which can be created using this kind of cheese.

Anchor Food Products, Inc.

Cheese sticks and dip can be served as an attractive appetizer.

Storing Cheese

- Cheese is perishable and should be kept refrigerated. There are a few exceptions to this rule, such as cheese in aerosol cans and squeeze packs.
- Store cheese in plastic bags or wrap tightly before placing in the refrigerator to keep the air out and moisture in.
- When refrigerated properly, natural and processed cheeses retain their freshness for approximately four to eight weeks. Fresh cheeses, such as cottage cheese and cream cheese, are more perishable and should be used within a week to 10 days.
- When wrapped and sealed properly in moisture vapor-proof plastic or aluminum wrap, cheese can be frozen for six weeks to eight weeks and still maintain excellent eating qualities.
- Baker's cheese, the cheese used in quality cheesecakes, can be kept moist and also delay molding if placed in an earthenware or tin container. After the cheese is placed in the container, smooth the surface until even, cover the surface with a thin covering of fine sugar, and store in the refrigerator.

Cooking and Molding with Cheese

- When baking a dish containing cheese, use a moderate oven heat of 325°F to 350°F.
- Pasteurized process cheese products can withstand heat better than other cheese products; however, use a low temperature and avoid overcooking.
- When broiling a cheese preparation, keep the cheese several inches below the fire and broil only until the cheese melts. Excessive heat toughens the cheese.
- When preparing au gratin dishes, a cream sauce (hot milk added to roux) is prepared. The cream sauce is then added to the cooked main ingredient (macaroni, potatoes, broccoli, cauliflower, etc.), placed in a pan, covered with cheese, and baked brown. To create a creamier preparation, add approximately 1 lb of cream cheese to each gallon of cream sauce before adding it to the main ingredient. Adding 6 oz of grated longhorn or Cheddar cheese to each gallon of cream sauce also produces a superior finished product.
- For a rich cheese flavor in breads and quickbreads, use a sharp-flavored cheese such as Parmesan, Cheddar, or process cheese.

- When cooking with cheese, use a low or moderate temperature and minimum cooking time. Excessive heat and prolonged cooking toughens the product. A double boiler is recommended when cooking on the range.
- When adding cheese to a starch-thickened mixture that also requires the addition of eggs (cheese soufflé), add the eggs first, then the cheese. The eggs aid in bringing the cheese into solution in the sauce.
- When adding Parmesan to Italian sauce or minestrone, sift the cheese and stir while it is being added to avoid lumps. Add the cheese just before it is removed from the heat.
- Forming or molding cheese in such forms as apples, pears, and pumpkins to be served as an hors d'oeuvre or garnish has become a common practice. Select a hard cheese such as American or Cheddar, Swiss or Provolone, and grind it on the food grinder using a fine chopper plate. After grinding the cheese, it is usually moist enough to form by hand into various objects. If the cheese is too dry for forming, place it in the stainless steel mixing bowl, add some cream cheese, and mix using the paddle until the desired consistency is achieved. If color is added to the cheese mixture, as is the case in forming pumpkins, use paste colors for best results.
- Shredding 1 lb of hard or semihard cheese produces 1 qt of shredded cheese.

Wisconsin Milk Marketing Boar

Cheese casseroles combine cheese and a variety of ingredient

CHEESE RECIPES

Wisconsin Cheese and Walnut Spread

This spread has a slight crunch from the walnuts. It may be served in individual portions or placed on the table in a single bowl.

Equipment

- Mixing bowl
- French knife
- Kitchen spoon
- Food grater
- Serving bowl

Ingredients

2 c cream cheese
3 c Wisconsin Swiss cheese
1 c sour cream
⅔ c walnuts
⅔ c fresh parsley
½ c green onions
4 tbsp Dijon mustard

Wisconsin Milk Marketing Board

Preparation

1. Chop the walnuts, parsley, and green onions.
2. Grate the Wisconsin Swiss cheese.

Procedure

1. Combine all ingredients in the mixing bowl.
2. Cover and refrigerate 1 hr.
3. Transfer to serving bowl.
4. Serve with toasted pita triangles, crackers, or bagel chips.

Cheese Soufflé

This soufflé, like any other soufflé, is very light and puffed up. Success or failure generally lies in the care that is taken when folding the beaten egg whites into the cheese mixture. Cheese soufflé is a specialty item usually served a la carte.

12 servings

Equipment

- Thin wire whip
- 2 qt casseroles (2)
- Food grater
- Cup measure
- Spoon measure
- Sauce pot
- Kitchen spoon
- Mixing machine and whip

Ingredients

½ c butter
½ c flour
3 c milk
½ tsp Worcestershire sauce
1 lb sharp Cheddar
10 eggs
salt and white pepper to taste

Preparation

1. Grate the Cheddar fine.
2. Grease the casseroles lightly with butter.
3. Separate the eggs.
4. Heat the milk.
5. Preheat the oven to 375°F.

Procedure

1. Place the butter in the sauce pot, melt over low heat, and add the flour, making a roux. Cook slightly.
2. Add the hot milk, stirring vigorously until smooth.
3. Add the Worcestershire sauce and season with salt and pepper. Blend thoroughly. Remove the mixture from the heat and let cool slightly.
4. Add the grated cheese and blend thoroughly until cheese is melted and mixture is smooth.
5. Beat in the egg yolks one at a time with the wire whip and let cool.
6. Beat the egg whites on the mixing machine until they form soft peaks. Fold the beaten egg whites into the cheese mixture using a very gentle motion.
7. Place the mixture into the greased casseroles and bake in a preheated oven at 375°F for about 35 min to 45 min until light, puffy, and golden brown. Serve at once.

Precautions

- Do not open oven until the soufflé has been in at least 20 min to 25 min.
- Do not overbeat the egg whites. Keep the peaks soft and moist.

Macaroni and Cheese

This is a very popular item in many restaurants. The macaroni is cooked using the boiling method, placed in a thin cream sauce, covered with a good-quality Cheddar, and baked brown.

50 servings

Equipment

- Bake pans
- Sauce pot
- Stockpot
- Food grater
- Wire whip
- Kitchen spoon
- Colander
- Baker's scale

Ingredients

3 lb macaroni
2 lb Cheddar
6 oz butter or shortening
6 oz flour
3 qt milk
salt and white pepper to taste

Preparation

1. Grate the Cheddar on the coarse grid of the food grater.
2. Heat the milk.
3. Preheat the oven to 375°F.

Procedure

1. Place water in a stockpot, add salt, and bring to a boil.
2. Add the macaroni, stirring occasionally with a kitchen spoon until the water comes back to a boil. Boil about 7 min until the macaroni is tender.
3. Drain in a colander. Wash the cooked macaroni with cold water. Reheat with running hot water, let drain, and set aside.
4. Place the butter or shortening in a sauce pot, heat slightly, add the flour, and blend it with the wire whip into the shortening. Cook slightly.
5. Add the hot milk, stirring constantly to make a smooth cream sauce.
6. Blend the cream sauce into the cooked macaroni with a kitchen spoon. Season with salt and white pepper.
7. Stir in half the grated cheese and blend thoroughly.
8. Place in bake pans and top with the remaining cheese and paprika. *Note*: A crumbled cracker topping is also frequently used.
9. Bake in a preheated oven at 375°F until cheese is melted and top is golden brown. Serve while hot.

Precautions

- When boiling the macaroni, stir occasionally to avoid sticking.
- When mixing the cream sauce with the cooked macaroni, if mixture is too thick, add more warm milk. Mixture should not be too stiff.

Swiss Fondue

This is a melted cheese preparation served in a chafing dish as an appetizer. The cheese mixture must be kept warm at all times. Special fondue forks are used when served. For proper eating, a cube of hard crust bread is speared with a fork and dipped into the fondue.

4 servings

Equipment

- Spoon measure
- Kitchen spoon
- French knife
- Heavy-bottom sauce pot
- Fondue forks
- Chafing dish
- Cup measure
- Baker's scale
- Food grater

Ingredients

1 lb Swiss cheese
1 tsp cornstarch
2 c dry white wine
salt, pepper, and paprika to taste
2 loaves French bread
3 tbsp kirsch (Kirschwasser)
1 garlic clove

Procedure

1. Cut the garlic clove in half and rub the bottom and sides of the heavy bottom sauce pot with the garlic
2. Add the wine and heat, but do not boil.
3. Add the shredded cheese slowly to the wine, stirring constantly with the kitchen spoon until the cheese has melted and blended with the wine.
4. Dissolve the cornstarch in the kirsch. Add to the cheese mixture, stirring vigorously until mixture starts to bubble.
5. Season with salt and pepper. Add a dash of paprika, remove from the range, and place in the chafing dish at once.
6. Serve with the cubes of French bread and the fondue forks.

Preparation

1. Shred the Swiss cheese on the coarse grid of the food grater.
2. Cut the French bread into bite-size cubes.

Precautions

- Do not let the wine boil.
- If the fondue is too thick, add a little more warm wine.
- Use caution at all times so as not to scorch the fondue.

Welsh Rarebit

This is a main course cheese preparation. Cheddar is blended with beer and seasoning to create a preparation which is generally served over toast on either the luncheon or a la carte menu.

25 servings

Equipment

- Heavy-bottom sauce pot
- Baker's scale
- Spoon measures
- Food grater
- French knife
- Kitchen spoon

Ingredients

8 lb sharp Cheddar
5 bottles (60 oz) dark beer
3 tbsp Worcestershire sauce
2 tbsp dry mustard
2 tsp paprika
½ tsp Tabasco sauce
1 tbsp salt
25 slices sandwich bread

Preparation

1. Grate the Cheddar on the coarse grid of the food grater.
2. Toast and trim the bread.

Procedure

1. Place the beer in the heavy-bottom sauce pot and bring to a boil.
2. Using a kitchen spoon, blend the Worcestershire sauce, dry mustard, paprika, Tabasco sauce, and salt to a smooth paste. Add to the beer, blending thoroughly.
3. Add the grated cheese, a little at a time, until thoroughly blended. Stir constantly with the kitchen spoon.
4. Ladle the hot cheese mixture over the toast and serve.

Variations

Welsh rarebit may be served with a slice of tomato on top of the toast or with tomato and asparagus, tomato and bacon, or just bacon.

Precautions

- When the cheese is added to the beer mixture, keep on low heat. If heat is too high, the mixture will become rubbery.
- If the Welsh rarebit is to be held for any length of time, keep warm in a double boiler.
- Welsh rarebit is at its best when served at once.

Baked Wisconsin Brie

Baked Wisconsin Brie is an elegant and light appetizer with nuts and seasonal fruits added for color.

8 servings

Equipment

- 9″ pie plate
- Kitchen spoon
- Glass dish

Ingredients

2 tbsp butter
½ c almonds
8 oz Wisconsin Brie

Preparation

1. Slice the almonds.

Procedure

1. Place the butter in a 9″ pie plate and microwave on high for 30 sec until melted.
2. Stir in the almonds and microwave on high for 3 min to 4 min or until golden brown. Stir every two min.
3. Place cheese on a decorative microwave-safe glass dish and microwave on medium an additional 30 sec to 40 sec.

Wisconsin Milk Marketing Board

4. Serve immediately with thin slices of toasted French bread, crackers, or apple and pear wedges.

Portion control and the financial savings that can result are important in today's food service operations. Use color-coded handles on dishes and scoops to quickly identify the proper portion. For example, a dark blue handle indicates 2 oz portions.

Quiche is a custard cheese pie that may be flavored by adding other ingredients such as spinach, crisp bacon, onions, and broccoli. Besides its two main ingredients, cream and eggs, Quiche Lorraine contains diced crisp bacon, onions, and Swiss cheese. It is most popular as a luncheon entree or served as a warm appetizer when cut into small serving portions.

Quiche Lorraine

one 9″ pie (6 servings)

Equipment

- 9″ pie pans (2)
- Rolling pin
- Pint measure
- Small saucepans
- Spoon measure
- French knife
- Stainless steel bowl
- Wire whip
- Baker's scale

Ingredients

8 oz pie dough
1 pt light cream
5 eggs
2 oz finely minced onions
1 tbsp flour
6 oz Swiss cheese
6 oz bacon
salt and white pepper to taste
extra flour for rolling pie dough

Preparation

1. Dice the bacon and Swiss cheese medium-sized using a French knife.
2. Cook the diced bacon to a crackling in a small saucepan and drain.

Procedure

1. Using a rolling pin, roll the pie dough on a floured bench until it is approximately ⅛″ thick.
2. Cover the pie tin with the dough. Shape to the pan and flute the edges.
3. Insert the second pie pan in the shell and place in a 350°F oven. Weight the top of the pie pan so it will not rise. Bake the shell slightly. Remove from the oven.

4. Break the eggs into a stainless steel bowl. Beat slightly using a wire whip.
5. Add the cream slowly while continuing to beat.
6. Add the flour and beat until it is blended into the egg/cream mixture.
7. Place the Swiss cheese, bacon crackling, and minced onions in the slightly-baked pie shell.
8. Pour in the egg/cream mixture. Bake in a preheated oven at 350°F until the custard is set and the pie crust is golden brown.
9. Remove from the oven and let set approximately 5 min. Cut into six serving portions and serve warm.

Variations

Seafood quiche: Omit the bacon and add 12 oz cooked seafood (shrimp, crabmeat, etc.).
Broccoli quiche: Replace the Swiss cheese with Cheddar or longhorn cheese and add 10 oz cooked broccoli.
Spinach quiche: Swiss, Cheddar, or Longhorn cheese may be used. Add 8 oz cooked spinach.

Precautions

- Be alert when cooking the bacon to a crackling. Do not burn.
- Exercise care when placing the pie in the oven. Do not let the filling spill or coat the fluted edges of the crust.
- Bake only until the filling is set. Overbaked custard is watery and undesirable.

This omelet is a combination of Cheddar or some other high-quality cheese and eggs. The eggs are whipped, combined with the cheese, and formed into a roll or fold while cooking in a skillet.

Cheese Omelet

1 serving

Equipment

- Steel skillet
- Kitchen fork
- Mixing bowl
- Food grater

Ingredients

3 eggs
1 oz Cheddar
salt and pepper to taste

Preparation

1. Break the eggs into a small mixing bowl and whip with a kitchen fork.
2. Grate the cheese on the coarse grid of the food grater.
3. Clean the skillet by rubbing with a cloth.

Procedure

1. Place the skillet on the range, add a small amount of shortening, and heat slightly at a temperature of about 275°F.

2. Pour in the beaten eggs, shaking the skillet back and forth with a quick motion to keep the egg mixture turning over in the pan.
3. When the egg mixture starts to set, but is not firm, season with salt and pepper and add the cheese.
4. Using the kitchen fork, roll or fold the egg mixture in the skillet and, at the same time, give quick backward snaps until the mixture is completely rolled or folded.
5. Let brown slightly and invert on a warm plate. Serve while hot.

Precautions

- Do not overcook the cheese. It will become tough and rubbery.
- Serve the omelet at once. An omelet that is left standing has very poor eating qualities.

Cheesecake

This is a tender, tasty cake preparation. It is generally prepared with a graham cracker or cookie crust. It is a very popular item on the dessert menu.

Equipment

- Baker's scale
- Mixing bowl
- 8″ cake pans (8)
- Plastic scraper
- Bun or sheet pans
- Mixing machine with paddle and whip
- Skimmer

Ingredients

Filling

3 lb baker's cheese
3 oz cornstarch
3 oz bread flour
12 oz emulsified vegetable shortening
1 lb egg yolk
2 lb milk
½ oz vanilla
1 lb egg whites
1 lb 8 oz sugar

Graham Cracker Crust

3 lb graham cracker crumbs
10 oz shortening
4 oz eggs

Preparation

1. Grease bottoms and sides of cake pans heavily with additional shortening.
2. Prepare graham cracker crumbs.

Procedure

Filling

1. Place the cheese in the mixing bowl. Mix smooth using the paddle.
2. Add cornstarch and flour. Mix smooth with mixer at first speed.
3. Add the emulsified vegetable shortening, blending to a smooth paste.
4. Add egg yolks gradually while creaming with mixer at second speed.
5. Add milk slowly and mix smooth.
6. Add vanilla. Remove this mixture from the mixer with plastic scraper.
7. Place egg whites in mixer, whip to soft peaks, then add sugar gradually.
8. Fold meringue mixture into cheese mixture using a skimmer. Hold until pans are set up with the graham cracker crust.

Graham Cracker Crust

1. Thoroughly mix crumbs by hand with the shortening and eggs.
2. Line the heavily-greased cake pans with the graham cracker crust mixture.
3. Fill the pans with the cheesecake mixture to about ¼″ from the top. Set pans in a bun pan with about ½″ of water.
4. Bake in a preheated oven at 350°F until the filling is set. Remove and let cool.
5. Top cheesecakes with strawberries, blueberries, or sour cream if desired.

Precautions

- Use caution when whipping the egg whites.
- Do not overbake or the top of the cheesecakes will crack and become too brown.

Cheese Danish Pocketbooks

These are a delicious pastry consisting of Danish pastry with a rich filling of cheese. It is an excellent selection for the dessert menu.

Equipment

- Rolling pin
- Baker's scale
- Sheet pans
- Pastry wheel
- Parchment paper
- Wire whip
- Mixing machine and paddle
- 1 qt stainless steel bowl

Ingredients

1 qt Danish pastry dough
2 lb cream cheese
5 lb baker's cheese
4 egg yolks
6 eggs
8 oz sugar
juice of 4 lemons
1 tsp vanilla
1 pinch nutmeg

Preparation

1. Prepare 1 qt of Danish pastry dough.
2. Separate egg yolks from the whites.

3. Preheat oven to 425°F.
4. Break the eggs into a stainless steel bowl and beat slightly with a wire whip.

Procedure

1. Place all the ingredients except the Danish pastry dough and the beaten eggs in the stainless steel mixing bowl. Blend thoroughly on the mixing machine at low speed using the paddle until the cheese mixture is smooth. Place the mixture in the refrigerator overnight for best results.
2. Roll out the Danish pastry dough to a thickness of about ¼″.
3. Using the pastry wheel, divide the dough into 4″ squares and place approximately 2 oz to 3 oz of the cheese mixture in the center of each square of dough.

4. Brush the edges of the dough with the beaten eggs and fold the four corners of the dough over the cheese filling.
5. Brush again with the beaten eggs and sprinkle with sliced almonds.
6. Place on a sheet pan covered with parchment paper, proof, and bake in the preheated oven at 425°F for approximately 30 min.
7. Garnish slightly with powdered sugar and serve one pocketbook per order.

Precautions

- Do not overproof the dough.
- Use caution when baking. Do not overbrown.

Cheese Pizza

Pizza is a very popular Italian preparation. It is rich, highly seasoned with tomato sauce and a variety of cheeses, and served on a crust.

12 pizzas (approx 12″ diameter)

Equipment

- Rolling pin
- Food grater
- Peel
- Pizza cutter
- Mixing machine and dough hook

Ingredients

Dough

5 lb bread flour
3 lb water
¾ oz salt
¾ oz compressed yeast
3 oz salad oil
½ oz sugar

Topping (for one pizza)

5 oz canned pizza sauce
2 oz Mozzarella or Provolone
Parmesan to taste
4 drops olive oil
black pepper, oregano, and basil to taste

Preparation

1. Preheat oven to 550°F to 600°F.
2. Grate the Mozzarella or Provolone on the coarse grid of the food grater.

Procedure

1. Dissolve the yeast in the water.
2. Place all the dough ingredients, including the dissolved yeast, in a mixing bowl. Using the dough hook, mix at low speed until the dough leaves the side of the bowl and becomes smooth.
3. Turn out the dough on a floured bench, knead, and place in a greased container. Place in the refrigerator overnight. Cover the dough with a damp cloth.
4. Remove the dough from the refrigerator and knead on a floured bench. Divide into 10 oz units.
5. Round the units into balls and let rest 5 min.
6. Roll out a unit of dough into a circle, stretching the dough as much as possible without creating tears or holes in the dough.

Cres-Cor

7. Place the circle of dough on the peel which has been sprinkled with cornmeal to act as a roller.
8. Cover the surface of the dough with the pizza sauce. Season with oregano, basil, and black pepper. Sprinkle on the Mozzarella or Provolone and the Parmesan. Dot with olive oil.
9. Slide the pizza off of the peel onto the hearth of the oven. Let bake until the dough is slightly brown and crisp. Remove, cut into pie-shaped wedges with a pizza cutter, and serve at once.

Variations

The following items may be added to the pizza: anchovies, pepperoni, green peppers, mushrooms, sausage, salami, etc. Add only vegetables to make a vegetarian pizza. The garnishes are limitless.

Precautions

- The oven must be cleaned out often or the cornmeal will burn.
- Be alert when pizza is in the oven. It will brown quickly.

Cheese Blintzes

Blintzes are thin pancakes with a cheese filling served with sour cream, applesauce, or jam. They are of Jewish and Russian origin and are served most often in restaurants catering to a Jewish clientele.

10 servings

Equipment

- 6″ pancake skillets
- Stainless steel mixing bowls (2)
- Sheet pan
- Spoon measure
- Wire whip
- Flour sifter

Ingredients

Pancakes

4 eggs
4 egg yolks
1 c cake flour
1 tbsp sugar
4 c milk
½ c butter
2 tsp salt

Filling

3 lb dry cottage cheese
1 beaten egg
1 pinch of nutmeg
salt to taste

Preparation

1. Melt the butter.
2. Prepare skillets for cooking the pancakes.

Procedure

Pancakes

1. Beat eggs and egg yolks slightly with a wire whip.
2. Sift the flour, sugar, and salt with a flour sifter and blend thoroughly.
3. Add the melted butter and milk. Beat well.
4. Heat the pancake skillets. Add enough shortening to coat the bottom and sides of the skillet. Hold the handle of the skillet with the left hand when pouring enough batter into the skillet with the right hand to make a thin layer that just covers the pan. Turn the left hand back and forth while pouring so the pan will be covered quickly and evenly. Place on the heat just enough to let the pancake set. Turn out onto wax paper. Repeat this process until all the pancakes are prepared.

Filling

1. Combine all ingredients until thoroughly blended.
2. Place about 2 tbsp of the cheese mixture on each pancake, cooked side up. Fold each side to form a square. Turn over and place folded side down on a sheet pan. Repeat this process until all the pancakes are filled and placed on the sheet pan.
3. Sprinkle tops of pancakes with sifted powdered sugar. Glaze lightly under the broiler.
4. Serve two to each order with sour cream, cinnamon, applesauce, or apricot jam.

Precautions

- Do not have the skillet too hot when adding the batter.
- Do not attempt to brown the pancakes.
- Do not overcook the pancakes.

Cheese Biscuits

These are a favorite American quickbread with a Cheddar cheese flavor. They are an excellent choice for either the luncheon or dinner menu when hot breads are desired.

7 doz

Equipment

- Mixing container
- Sheet pans
- Biscuit cutter
- Baker's scale
- Pint measure
- Rolling pin
- Pastry brush
- Parchment paper
- Food grater
- Kitchen spoon
- Flour sifter

Ingredients

1 lb 8 oz cake flour
1 lb 8 oz bread flour
2½ oz baking powder
½ oz salt
1 lb butter or shortening
6 oz sugar
6 egg yolks
½ pt cold milk
8 oz Cheddar

Preparation

1. Grate the cheese on the coarse grid of the food grater.
2. Preheat oven to 450°F.
3. Place parchment paper on sheet pan.

Procedure

1. Place the butter or shortening and sugar in the mixing container and cream with a kitchen spoon.
2. Add the egg yolks and blend well by stirring with a kitchen spoon.
3. Continue to stir while adding the milk.
4. Combine and sift the flours, baking powder, and salt. Add the grated Cheddar and blend all ingredients using a gentle motion.

5. Place the dough in the refrigerator to chill for about 45 min.
6. Place the dough on a floured bench, roll out with a rolling pin to a thickness of about ¾″, and cut into units with the biscuit cutter.
7. Place the units on the parchment-covered sheet pan fairly close together.
8. Using a pastry brush, brush the tops of the biscuits with melted butter or egg wash. Let rest 5 min.

9. Bake in the preheated oven at 450°F for approximately 15 min or until done. Serve while hot.

Precautions

- Do not overwork the dough.
- When placing the biscuits on the sheet pans, leave just enough space for the heat to penetrate properly.

Peppery Wisconsin Cheese Squares

This recipe is particularly spicy. For a milder version, reduce the amount of jalapeno peppers. For a tangier flavor, substitute Wisconsin Asiago cheese for Parmesan cheese.

48 servings

Equipment

- 9″ square pan
- Food grater
- French knife

Ingredients

½ c cornmeal
2 c Wisconsin Sharp Cheddar cheese
1 c Wisconsin Onion Cold Pack cheese, softened
1 c Wisconsin Parmesan cheese
¼ tsp garlic powder
¼ tsp dried oregano
¼ tsp cumin
4 eggs, beaten
2 oz pimentos
4 oz jalapeno peppers, drained
Cayenne pepper, to taste

Preparation

1. Dice the pimentos with the French knife.
2. Chop the jalapeno peppers with the French knife.
3. Shred the Cheddar and Parmesan cheese with the food grater.

Procedure

1. Sprinkle cornmeal in lightly greased 9″ square pan, coating pan well.
2. Cream together cheeses until well-blended.
3. Add garlic powder, oregano, cumin, and eggs.

Wisconsin Milk Marketing Board

4. Stir in pimentos and jalapenos and add Cayenne pepper. Mix ingredients well.
5. Spread mixture into pan, smoothing top.
6. Bake at 350°F for 30 min.

Fruit Preparation

Chapter 13

Fruits are important in the menu planning of food service establishments. Fruits add variety, color, and flavor to any preparation. Although fruit preparations are most commonly featured as a dessert, fruit is also served as an appetizer, salad, and garnish. Fruits are available fresh, frozen, canned, and dried.

Fresh fruit, like vegetables, must be carefully purchased and properly stored. Fresh fruit is examined for size, color, firmness, and blemishes and bruises. Most fresh fruits require refrigerated storage. Soft fruits that do not have a protective skin do not store as well as fruits with a skin. Fruits that are ripened after they are picked permit longer storage times than those purchased already ripened on the tree or plant.

Canned and frozen fruits are commonly used in the commercial kitchen because of availability. However, whenever possible, fresh fruit should be used for best results. Dried fruits are rarely used in the commercial kitchen.

FRUIT

Fruits are purchased fresh, frozen, canned, or dried. Fresh fruit produces the best results in fruit preparations. However, fresh fruit is not always in season. In addition, convenience, spoilage, and cost must always be considered when making a purchase. Canned and frozen fruits are commonly used in pies, fritters, cobblers, sauces, and fillings. Dried fruit is seldom used in the commercial kitchen. Choose medium-sized fruit for the best flavor and texture.

Fresh fruit is perishable and must be stored at a temperature of 36°F to 40°F. The length of time and the storage temperature at which fruit can be stored vary greatly. For example, bananas are stored best at a room temperature of 68°F to 70°F. Canned fruits are stored best in a cool, dry area. Frozen fruits are stored frozen. They should be thawed slowly by placing them in the refrigerator at 34°F to 38°F. Fresh fruit should be washed thoroughly before using.

Fruits commonly used in the commercial kitchen include the following:

Apples are a very popular fruit that can be prepared many ways using different cooking methods. The characteristics of the apple are subject to many variations. The skin color may range from green to a very deep, dark red. The shape varies from oblate to oblong with varying diameters. The best eating apples are Red Delicious, Golden Delicious, Jonathan, and McIntosh. The best apples to select for cooking are Granny Smith, Winesap, and Rome Beauty. Avoid purchasing apples that are bruised, soft, or shriveled from overripeness.

To peel apples quickly, dip them in and out of boiling water. The skin can be removed more easily. If speed is required when removing the core, cut the apple in half with a stainless steel knife to prevent discoloration, and scoop out the core from each half using a melon ball scoop. A tubular apple corer can be used if speed is not required. To prevent discoloration of apples while peeling, dicing, or slicing, place them in a solution consisting of a quart of water, a pinch of salt, and a cup of bottled lemon juice.

Rome Beauty apples are best for baking because they hold and retain a little firmness. Other apples often turn to mush when baked. To prevent apple skins from wrinkling too much when baked, slit the skin in several places or peel approximately ½" of the skin from the top of the apple. To acquire exceptional flavor and color, use the following procedure when baking:

1. Remove core and slit skin or peel apple.
2. Place apples in bake pan. Using sugar, fill holes created when core was removed.
3. Add red colored water or a mixture of colored water and red maraschino cherry juice to a depth of approximately 1".
4. Add three or four cinnamon sticks and place in the oven at a temperature of 350°F.
5. Bake until slightly tender. Baked apples may be served hot or cold and plain or with an appropriate sauce.

When sautéing or grilling fresh apple rings or slices, select any apple recommended for cooking. The preparation is similar to that for baking, but cooking methods differ. Apples are sautéed following these steps:

1. Process the apples into rings or slices. The skin may be left on or removed.
2. Place a small amount of butter, margarine, shortening, or oil in a sauté pan and heat.
3. Add the rings or slices of apple and sprinkle with a mixture of cinnamon and sugar and a small amount of paprika for color.
4. Sauté until the apples are just slightly tender.
5. Place in a 2" steam table pan and hold for service. Sautéed apples may be served on the breakfast, luncheon, or dinner menu and are most popular when served with pork preparations.

APPLES

RED DELICIOUS
bright to dark red, sometimes striped, favorite eating apple, mildly sweet, juicy
— available year-round

GOLDEN DELICIOUS
mellow, sweet all purpose apple, for baking, salads and fresh eating, flesh stays white longer than other apples
— available year-round

FUJI
spicy sweetness, firm flesh and tender skin. The "eating apple of the 21st century"
— available year-round

GRANNY SMITH
green, tart, crisp, juicy, excellent for cooking, salads, fresh eating
— available year-round

GALA
yellow to red, rich full flavor, delightful out of hand, salads
— available Sept.-Dec.

WINESAP
dark red, spicy, slightly tart, appropriate for cider, snacking and cooking
— available Oct.-Aug.

BRAEBURN
high-impact flavor! Sweet, tart, crisp, aromatic and as juicy as they come
— available Oct.-July

ROME BEAUTY
brilliant red, round, holds shape well when cooked, great for baked apples
— available Oct.-June

JONAGOLD
tangy-sweet flavor excellent for eating fresh and for cooking
— available Sept.-March

Washington Apple Commission

Apples may be eaten or prepared by different cooking methods such as baking, grilling, sautéing, or broiling.

Grill apples according to the following steps:

1. Process the apples into rings or slices. The skin may be left on or removed.
2. Place a small amount of butter, margarine, shortening, or oil on a hot griddle at 350°F.
3. Add the rings or slices of apples and sprinkle with a mixture of cinnamon and sugar and a small amount of paprika for color.
4. Cook, turning the apples gently from time to time using a meat or pancake turner. Cook until just tender.
5. Place in a 2″ steam table pan and hold for service.

Grilled apples may be served the same way as sautéed apples.

Apricots have characteristics similar to plums and peaches. Apricots are available in many varieties differing in hardness, texture, and size. Apricot colors range from pale yellow to deep reddish orange. Some apricots are sun-freckled with a brick or crimson color. The flesh is usually a shade of yellow or orange. They have a thin, tender skin that makes them difficult to peel, but because of their tenderness, peeling is usually not necessary when they are served fresh, canned, or dried. In the United States, most apricots are grown on the Pacific coast. Apricots are best when tree-ripened. Avoid purchasing apricots that are bruised or too soft.

Avocado or *alligator pear*, as it is sometimes known, is a greenish, thick-skinned, pear-shaped tropical fruit that contains a large, hard seed. When it is ripe, its flesh has the consistency of firm butter. It has a very delicate nut-like flavor that blends well with other foods. Its most popular use is in salads, sandwiches, and dips. Avocados have a relatively high fat content, containing 10% to 20% oil; however, this fat is monounsaturated. Among fruits, only ripe olives contain a higher percentage of fat.

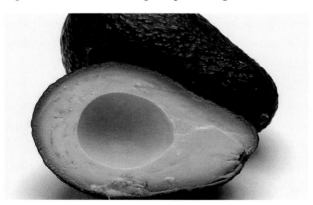

To test the ripeness of an avocado, hold it gently in both hands and squeeze slightly. If it yields to the slight pressure, it is ripe and ready to serve. Ripening time may be reduced by keeping the fruit in a warm room for two to five days or placing it in a paper bag with an apple. The length of time depends on its firmness when placed in the room. To retard ripening, the fruit should be kept in the refrigerator or a cool, dry place but not below 40°F. Flavor is harmed at a temperature below 40°F and enhanced at room temperature (70°F). Avoid purchasing avocados with bruises or soft spots.

When peeling an avocado, start at the narrow end and work toward the larger end. This method simplifies the task. Lemon juice prevents the flesh from discoloring.

When only half an avocado is used, the seed should be left in the remaining half, wrapped in wax paper or plastic wrap, and stored in the refrigerator until ready to use. This procedure extends its keeping qualities. Avocado halves can be stuffed with chicken, shrimp, tuna, and other salads. Guacamole, a popular avocado preparation, is served as a dip or as a salad topping.

Bananas are grown in tropical countries. Not all banana varieties can be eaten raw; some require cooking. Small bananas are best when eaten raw. Cooking varieties are larger, not as sweet, and very firm. Cooking bananas are ready to be cooked when the peel is light yellow and the tip is still green.

International Banana Association

Bananas to be transported are cut when they are full size but still green. They are packed carefully in cartons to avoid bruising and stored at a temperature of 54°F to 56°F during the shipping and holding period. Colder temperatures cause the banana skins to turn black. Before using, bunches are hung to ripen in a warm area. A banana is fully ripe and ready to be eaten as a fruit when the peel is a deep yellow, flecked with brown spots, and there is no trace of green at the tips. Bananas are very seldom ripened on the vine even in the tropical areas where they are grown. They are picked green and ripened in the shade. Bananas ripened on the plant are dull, colorless, and weak in flavor.

To prevent bananas from discoloring after they are peeled, coat them or dip them in lemon or pineapple juice. For best results, if it is possible to do so, peel and cut bananas when ready to serve. A stainless steel knife should be used when slicing bananas to prevent discoloration.

Bananas can be prepared using many different cooking methods. They can be used to flavor preparations such as banana cream pie, banana cake, banana bread, banana muffins, and banana pudding.

International Banana Association

Banana pancakes may be served on the breakfast or luncheon menus.

Bananas are sautéed using the following steps:

1. Select a firm, slightly green, or unripened banana. Peel and cut in half crosswise, then cut in half lengthwise.

2. Place a small amount of butter, margarine, or oil in a sauté pan and heat.

3. Press each side of the sliced bananas into sugar and place in the saucepan.

4. Brown on one side until golden. Turn, using a kitchen fork or spatula, and brown the second side.

5. Remove from the pan and place in a 2″ steam table pan and hold for service. Sautéed bananas may be served with an entree as a garnish or as a dessert topped with whipped cream or topping.

Bananas are grilled using the following steps:

1. Select a firm, slightly green, or unripened banana. Peel and cut in half crosswise, then cut in half lengthwise.

2. Place a small amount of butter, margarine, or oil on a 350°F griddle.

3. Press each side of the sliced bananas into sugar and place on the griddle.

4. Brown on one side until golden. Turn, using a kitchen fork or spatula, and brown the second side.

5. Remove from the griddle and place in a 2″ steam table pan and hold for service. Grilled bananas may be served with an entree as a garnish or as a dessert topped with whipped cream or topping.

Bananas are baked using the following steps:

1. Select a firm, slightly green, or unripened banana. Peel and cut in half crosswise.

2. Grease a bake pan with butter or margarine. Place the banana halves in the pan and brush them with melted butter or margarine.

3. Sprinkle granulated or brown sugar lightly on the surface of the bananas.

4. Bake in a preheated oven at 375°F until tender. Test for doneness by gently pressing with the fingers or by piercing with a fork.

5. Remove from the oven and finish by browning lightly under the broiler.

6. Remove from the pan and place in a 2″ steam table pan and hold for service. Baked bananas may be served as a garnish with beef, ham, chicken, or turkey or as a dessert with vanilla, custard, or lemon sauce.

A variation of baked bananas is *maple baked bananas*. Place peeled and cut bananas in a greased bake pan. Brush bananas with lemon juice. Pour maple syrup over the bananas, allowing 1 c for every 8 bananas. Bake the same way as for baked bananas.

Blueberries grow on shrubs that are native to the United States. In the commercial kitchen, blueberries are used in many preparations such as muffins, pies, breads, fruit salads, and desserts. Blueberries are extremely perishable and must be carefully inspected before purchasing. Select plump, firm blueberries for best results.

Cherries grow on trees and are classified as sweet and sour. Sweet cherries are available in black or white varieties. Cherries are susceptible to damage from insects on the tree. Avoid blemishes and bruises when purchasing. The most popular black cherry is the bing cherry, which is used in salad and dessert preparations. The most popular white cherry is the Royal Anne, used in flaming and other dessert preparations. Sour cherries are most commonly used in pie production.

National Cherry Growers and Industries Foundation

Cherries are used in popular dessert preparations such as cherry pie and cherries jubilee. Cherries jubilee is a unique preparation because it is served aflame and attracts attention when ignited and served over ice cream.

Cranberries grow on vines in bogs with rich soil. The bogs are flooded and drained to protect the vines from freezing temperatures and to destroy insects.

Cranberries are a round, oblong, or pear-shaped berry, varying in color from white to dark red. Cranberries ripen during August, September, and October. They are marketed fresh, dried, canned, or frozen. In the United States, at least 50% of the crop each year is canned as sauce or jelly. The fruit is most popular during fall and winter.

The most popular cranberry preparation is whole cranberries or jellied sauce with roast turkey. Cranberries are also used in salads, relishes, and sauces. The quality of a fresh cranberry can be tested by its bounce. The higher they bounce, the better the quality. When refrigerated, cranberries will keep for as long as eight months.

Grapefruit grows in clusters on grapefruit trees. Grapefruit has a yellow skin and grows to approximately 4″ to 6″ in diameter. It has a juicy, acidic pulp surrounded by a leathery rind. The color of the pulp may be light yellow or pink, depending on the variety. Heavy grapefruit is most desirable because a heavy weight indicates high juice content. Grapefruit is low in calories but high in vitamin C. It is used most often in commercial food service operations as a breakfast fruit appetizer. To peel a grapefruit, place it in boiling water. Remove the pot from the range and let set for approximately 5 min. Peel the grapefruit using a paring knife.

To section a half grapefruit, cut a thin sliver off each end of the fruit using a French knife so the fruit will stand without rolling. Cut in half crosswise, making two equal halves. Using a serrated, curved grapefruit knife or paring knife, cut around each section, freeing it from the pulp. Cut fibers out of the center and place a red maraschino cherry in this depression.

To prepare a broiled grapefruit half, which is usually served as an appetizer, the grapefruit is cut and sectioned the same way as for peeling. The exposed flesh is sprinkled with sherry, coated lightly with brown sugar, placed under the broiler, and browned. Serve while still warm.

Grapes are grown on a vine and are classified according to their use: wine grapes and dessert grapes. In general, wine grapes are small, often tough-skinned, very sweet, and fairly acidic. Wine grapes may be black, red, or white and strongly influence the character of the wine they produce. Dessert grapes come from many varieties, but, in general, they are low in acidity and sugar content and must conform to certain standards of size, shape, and color. Most grapes have seeds, but seedless grapes are very popular and easier to prepare. Raisin grapes are seedless with a very high sugar content and low acidity. The most popular raisin grape is the Thompson seedless grape, grown in California. Grapes are best when ripened on the vine. Both varieties of grapes (with seeds and seedless) are commonly used in salad preparations. Purchase grapes that are firm and do not fall off the stems when shaken.

Kiwi is similar in size to a lemon. It has fuzzy brown skin, and the interior consists of a green- and cream-colored flesh. Kiwi contains a high percentage of vitamin C and fiber, making it an excellent breakfast fruit. It is available year-round from U.S. crops and crops around the world.

Kiwi has excellent storage life if stored and handled properly. It is ripe when it is soft but still slightly firm. Kiwi is easy to peel using a sharp knife. It holds its attractive color for hours, making it an excellent garnishing fruit. Garnish such items as fruit plates, breakfast plates, cocktails, sandwiches, certain salads, and desserts with kiwi. Kiwi can be used in chicken salad, ham salad, seafood salad, and to fill meringue shells and tart shells.

Kumquats are a citrus fruit that are small in size. They are used primarily for the preparation of preserves.

Lemons grow on small, thorny trees that are sparsely covered with foliage. Fruit is picked from each tree six to ten times a year. The lemon reaches full size while still green. The fruit is then ripened under controlled conditions. A fresh lemon keeps for as long as three months. In the U.S., 50% of the crop is marketed as fresh lemons. Much of this fresh fruit is used in commercial food service establishments in the preparation of beverages, sauces, desserts, flavoring, and garnishes.

To acquire more juice from old lemons, soak them in hot water for 20 min before cutting and squeezing the juice. This method helps restore freshness and increases the amount of juice. To keep a cut lemon from drying out, place a little vinegar on a plastic lid and place the cut side of the lemon on top of the vinegar. When storing slices or wedges of lemons, place them in a shallow container and cover them with a mixture of cider vinegar and sugar. Keep them in the refrigerator. This storing method extends their keeping qualities and helps retain their true color.

Citrons are a variety of the lemon with a thicker skin. They are larger in size than traditional lemons and are less acidic. Their principal use is in candied peel.

Limes are grown on thorny trees that have pale green leaves. The fruit is oval to nearly round in shape, has a thin green-yellow skin, and has a tender pulp that contains one-third more citric acid and more sugar than lemons. Limes are rich in vitamin C. They are used primarily to flavor drinks, food, and confections. Lime juice may be concentrated, dried, frozen, or canned.

Melons are members of the gourd family, including cucumbers, squash, and pumpkins. Generally, melons are classified as either muskmelon or watermelon. Muskmelons include cantaloupe, Persian, casaba, and honeydew melons.

The muskmelon plant is a vine with coarse leaves. Cantaloupe is the most popular and well-known of the muskmelon species. It is fairly round, has a thin-netted skin, and a very sweet flesh that is peach-orange in color. Cantaloupes require a warm climate and sunlight and grow best in subtropical regions. The Persian melon has a much larger diameter than the cantaloupe, but the netted skin, sweet flesh, and peach-orange color are similar. The casaba melon has a shape that is almost round, and a furrowed, tough skin that varies in color from lemon yellow to dark green on the same melon. It is very sweet and has fine-grained, juicy, creamy white flesh.

The honeydew melon has a slightly oval shape with thin, smooth, creamy yellow skin. It is juicy and sweet when ripened properly. Like the casaba, it is usually ripened off the vine. The honeydew melon keeps for a fairly long period of time. It averages from 4 lb to 6 lb. All the muskmelons are used in the commercial kitchen on the breakfast or luncheon menus. Occasionally, they appear on the dinner menu. Muskmelons are commonly used in appetizers, salads, and desserts.

The watermelon is related to the muskmelon only in that it is a member of the gourd family. This smooth, green-skinned melon has a pink to red flesh that contains many seeds and a very high percentage of juice, but has less flavor than the muskmelon species. The name "watermelon" came about because of this high percentage of juice or water. It is widely cultivated in all temperate zones and grows on vines very similar to those of the muskmelons. The watermelon is a very large melon. Some varieties have been known to weigh as much as 50 lb. In the commercial kitchen, it is used on buffet tables, on salad plates, or sliced and served as a dessert.

Oranges are a round, reddish-yellow, juicy citrus fruit that grow in warm climates. Two types of oranges commonly used in food service establishments are navel and Valencia. The navel orange is used mostly in desserts and on fruit plates when a seedless orange with easy-to-free segments is required. The Valencia orange is used in an item that requires an orange with a high juice content.

Oranges are picked when ripe. They do not ripen after picking. The pulp or flesh of the orange is arranged in segments. Each segment provides a high percentage of juice, which is the most desirable part of the orange. In the U.S., 40% of the orange crop is processed into frozen orange juice. When serving fresh oranges, the orange is usually sectioned. To section an orange, it is first peeled using a paring knife, then each individual segment is cut free and the membrane discarded.

Peaches are classified as Freestone and Clingstone. In the Freestone peach, the pit (stone) pulls away or can be freed easily from the flesh. In the Clingstone peach, the pit clings fairly tightly to the flesh. Both types include yellow flesh and white flesh varieties. Freestone peaches are generally preferred when the peaches are being served fresh. Clingstone peaches are used for commercial canning and freezing.

Peeling fresh peaches is made easier by placing them in boiling water, to which a small amount of potassium (if available) is added, for a short period of time. The skin is then removed using a skinner. When preparing or serving sliced fresh peaches, the flavor can be enhanced by adding a touch of lemon juice and a sprinkle of ground cinnamon. Fresh sliced peaches served in a syrup over cake, ice cream, bread or rice pudding, and hotcakes or pancakes is a popular preparation. Fresh peaches in syrup are prepared as follows:

1. Peel and slice the ripe peaches and place in a plastic or stainless steel container. Set aside.

2. Place 3 lb sugar, 1 lb 8 oz corn syrup, and 1½ qt water in a sauce pot. Bring to a boil.

3. Let the sauce cool.

4. Pour enough sauce over the sliced peaches to cover. Let stand approximately 1 hr before using. Stir occasionally.

Pears are available in many varieties. These varieties are divided into two major types: thin, smooth-skin pears and coarse-skin pears. Of the smooth pears, the Bartlett pear is the most popular. The Bartlett pear is large and has a handsome appearance. The flesh is white, has a fine grain, and is very juicy. The skin is thin and golden yellow in color with a blush of red on the side that faced the sun. Of the coarse-skin pears, the Kieffer is the most popular.

The Kieffer pear is smaller than the Bartlett pear and is slightly rounder. The color varies from yellow to russet to brown. The flavor, texture, and juiciness depend on proper ripening. Pears ripened while in storage usually have a better flavor than those ripened on the tree.

Canned Bartlett pear halves are the most popular form used in commercial food service operations. They are used mainly in salads and dessert preparations. When fresh pears are used, they are usually stewed and served as a breakfast or dessert item.

Pineapples resemble a pine cone in their outer appearance. The pineapple has a number of smooth- or serrated-edged, pointed, rigid leaves growing from the root of the plant. In the center, a short flower stem sprouts up, bearing a single spike of flowers that produces a single fruit. Pineapples ripened on the vine produce the best quality. Most pineapples are canned in various styles, made into juice, or frozen. Pineapples are grown in tropical climates where frost never occurs.

To test the ripeness of fresh pineapple, pull out a leaf or two growing out of the crown on top of the fruit. If they are easy to pull out, the pineapple is ripe; if difficult, it is still unripe. To peel a fresh pineapple, cut into crosswise sections approximately 1″ to 2″ thick. Using a paring or utility knife, peel each section separately and remove the core.

Grilled pineapple is often used as a garnish in food service operations. It improves the appearance and, in some cases, the taste of pork and poultry preparations. When pineapple is used in other preparations, the descriptive term often used is "Hawaiian," suggesting that the food is prepared and served in a Hawaiian style or manner.

To grill a pineapple, use canned, sliced pineapple and drain thoroughly. Place the slice directly on a hot griddle coated with a thin layer of butter or oil. Brown one side, turn using an offset spatula, and brown the second side. If the pineapple needs more color, sprinkle a small amount of paprika on the surface of each side. Place a cherry in the center of the ring and serve as a garnish.

Pineapple cup is another unique and attractive way to feature the pineapple. Cut the pineapple in half lengthwise. Scoop out the center, leaving a shell approximately 1″ thick. Remove the core from the scooped-out meat and cut the rest into cubes. Chill the pineapple shell. Place the pineapple cubes, 1 c whole fresh strawberries, 1 c honeydew melon balls, 1 c cantaloupe balls, ½ c seedless green grapes, and 1 c cherry liqueur in a stainless steel container. Toss until thoroughly incorporated and chill. To serve, spoon the fruit mixture and marinade into the chilled pineapple shell. The pineapple cup serves approximately six people and can be featured on a buffet table or as a special dessert for a small group of people.

PINEAPPLE

PINE CONE APPEARANCE

Plums are a cousin of the cherry and peach. They have smooth cherry-like skin and a size similar to a peach. Plums are further distinguished by the slight powdery coating or frosting that appears on the skin of some varieties. Plums vary in color, size, taste, and fruiting season. There are Clingstone and Freestone varieties. Some plums are classified as cooking plums, others as dessert plums. Some varieties always remain sour even when fully ripe, while others become very sweet. Plums provide the best flavor when picked just before they reach peak ripeness. When purchasing plums, avoid those that are overripe or bruised. Select plums that are slightly underripe because they ripen fairly quickly when placed in a warm room. Store ripe plums in the refrigerator.

Strawberries are a unique fruit because the seeds are dotted around the outside rather than enclosed within. Strawberries are grown in temperate regions throughout the world and are planted in early spring, but do not bear fruit until the following year. A class of strawberries known as "everbearing" usually produces twice a season. The best way to check the quality of strawberries is by tasting. Appearance is not a very good method to use to determine the quality of strawberries. Strawberries are canned, frozen, and are made into popular jellies and jams. However, strawberries are best when used fresh in preparations.

Strawberry shortcake and strawberries in cream are two of the great summer treats that people look forward to year-round.

Strawberries are very perishable. To avoid damage when picking and improve the appearance of the fruit when it gets to market, farmers gather strawberries before they are completely ripe. This affects the flavor because strawberries provide the best flavor when eaten right after picking. Stems are removed from the strawberries before serving.

Tomatoes by definition are a fruit, meaning a food item associated with the edible part of a seed of a tree or plant. However, tomatoes are commonly considered a vegetable.

Florida Tomato Commission

Coloring Fruit

Certain fruits and preparations using fruit can be enhanced by coloring the fruits. Fruits commonly colored include the following:

Minted pears are colored green when used as a garnish for lamb dishes or on a fruit plate. Add liquid green color to the pear juice, add mint flavor, and let the canned pear halves soak in the juice until the desired shade of green is obtained. Remove the pears from the juice and let drain before using.

Cinnamon pears are colored red and may be used on fruit plates or served as a garnish with those items

that blend well with the flavor of cinnamon such as ham or pork sausage. Drain the juice from canned pear halves and place the juice in a saucepan. Add cinnamon stick and simmer until the desired flavor is extracted. Remove from the heat and add liquid red color. Return the pear halves to the juice. Let soak until the desired flavor and color are absorbed by the pear halves. Remove from the juice and let drain before using.

Blushing pear and *peach halves* are commonly used in fruit plates or tarts. To blush a pear or peach half, the fruit is dried thoroughly after it is removed from the can and placed on a sheet pan that has been covered with a clean cloth. Some red color is blended with a small amount of fruit juice. This solution is brushed on the pear and peach half very lightly with a small pastry brush until a blushed effect is achieved. The effect should suggest a perfectly ripened piece of fruit.

Canned fruit, such as canned pineapple, often has an unappealing color. The color can be improved by adding a little yellow color to the canned juice and letting the pineapple slices soak until the desired color is obtained.

Color may also be added to fruit pie fillings such as peach, pineapple, cherry, and blueberry to enhance appearance. Always keep in mind that the first bite is usually taken because the item appeals to the eye.

Fruit Plates or Platters

Fruit plates or platters are popular cold entrees on food service menus. Fruit plates possess natural colors that are pleasing to the eye and stimulate the appetite. During the summer, many fruits that provide a good selection for taste and color are in season. In the winter, less fruit is available, creating a challenge for the chef or cook. Preparing fruit dishes during the winter requires using more canned fruit and substitutes for fresh fruit.

The following are ideas for creating eye appeal:
- Fill the cavity of a canned peach or pear half with a cube of red gelatin or a cream cheese ball coated with colored coconut or finely chopped nuts.
- Use cinnamon or minted pears.
- Create a center of attraction by using molded fruit gelatin.
- Add petite sandwiches with tasty, colorful fillings.
- Colored marshmallows can be used in various ways.

- Raspberry, orange, and lemon sherbet not only add color but, because of their moist composition, can take the place of dressing.

Some fruit plates or platters use fried fruit. Frying fruit is accomplished by coating the fruit with a batter and frying it in deep fat until golden brown. The batter is known as a *fritter batter* and can be made ahead of time to speed preparation. Both canned and fresh fruits can be made into fritters. The popular canned fruits used are pears, apricots, peach halves, and sliced pineapple. The popular fresh fruits are bananas, apples, and peaches. If using canned fruit, drain and dry the fruit thoroughly. How fresh fruit is processed depends on the fruit being used. Apples must be cored, peeled, and sliced; peaches peeled, cut in half, and stone removed; and bananas peeled and cut into desired portions.

When fruit has been prepared for frying, it is first coated with flour to seal the surface and prevent excess moisture from seeping through. Batter adheres better to a dry surface and does not have a tendency to loosen or fall off during the frying period. Place the fruit in the batter and bring it as close as possible to the deep fry kettle. Coat the fruit thoroughly with batter. To remove the fruit from the batter, hold it with the thumb and forefinger, and turn hand right side up to form a cup, catching the dripping batter. Place it in the preheated deep fat at 350°F and fry until golden brown. Remove and drain thoroughly.

After frying, apple fritters are usually passed through a mixture of cinnamon and sugar. Other fruits are dusted with powdered sugar and served, or dusted with powdered sugar and glazed lightly under the broiler.

FRUIT RECIPES

Apple Crisp

25 servings

Ingredients

1 No. 10 can sliced apples
4 oz lemon juice
12 oz sifted cake flour
1 lb 8 oz light brown sugar
¼ oz salt
¼ oz ground cinnamon
12 oz butter or margarine

Procedure

1. Place the canned apples in a bake pan that has been generously greased with extra butter or margarine. Sprinkle with lemon juice.
2. Combine the flour, brown sugar, salt, and cinnamon in a separate mixing container.
3. Add the butter or margarine and cut into the mixture by rubbing between the palms of the hand to create a crumb mixture.
4. Sprinkle the crumb mixture over the apples, covering well.
5. Bake in a preheated oven at 375°F for approximately 35 min to 40 min or until the surface is slightly brown.
6. Serve hot or cold, with or without a sauce.

Apple Scallop

25 servings

Ingredients

1 No. 10 can sliced apples
4 lb dry cake crumbs
8 oz brown sugar
⅛ oz cinnamon
¼ oz lemon juice
8 oz butter or margarine

Procedure

1. Spread some of the butter or margarine over the bottom and sides of a bake pan. Spread two-thirds of the cake crumbs over the bottom of the pan.
2. Cover the cake crumb surface with the apples.
3. Combine the brown sugar, cinnamon, and remaining cake crumbs. Sprinkle the mixture over the layer of apples.
4. Dot the remaining butter or margarine on top of the cinnamon and sugar.
5. Bake in a preheated oven at 350°F until the surface is brown and the apples are tender.
6. Serve in an appropriate dessert dish topped with whipped cream or a custard sauce.

Trade Tip

Apples are picked by hand during late summer or early fall. They are ripe and ready to be eaten when picked. Apples may be stored in cool, humid rooms where they can remain fresh and ready to be eaten for up to 1 year.

Applesauce

25 servings

Ingredients

6 lb slightly tart apples
1 lb 4 oz sugar
1 qt (2 lb) water
2 oz lemon juice
¼ tsp cinnamon
¼ tsp nutmeg or mace

Procedure

1. Wash, core, peel, and cut apples into quarters.
2. Place the sugar, water, and apples in a sauce pot. Bring to a simmer.
3. Simmer, stirring frequently, until apples are soft. Remove from the range.
4. Pass the mixture through a food mill or press through a medium-hole china cap.
5. Add the lemon juice, cinnamon, and nutmeg or mace. Stir thoroughly, blending with the apple mixture. Place in the refrigerator until chilled.

Stewed Apricots

25 servings

Ingredients

3 lb dried apricots
7 lb hot water
1 lb 4 oz sugar
3 oz lemon juice

Procedure

1. Wash the apricots in hot water and drain. Place in a braising pot, cover with the hot water, and let set for approximately 2 hr.
2. Add the sugar and lemon juice. Place on the range and bring to a boil.
3. Reduce heat to a simmer and cook for 20 min or until just tender.
4. Remove from the range and cool. Place in the refrigerator until ready to use.

Apricot Sauce

1 gal.

Ingredients

1 No. 10 can drained apricots
4 lb juice and water
4 lb sugar
3 oz waxy maize starch (clearjel)
3 oz lemon juice
¼ oz salt
1 lb cold water

1. Grind apricots into a pulp on the food grinder using the medium-sized chopper plate or chop by hand using a French knife. Place in a sauce pot.
2. Add the juice and water, sugar, salt, and lemon juice. Place on the range and bring to a boil.
3. Place the starch in a stainless steel or plastic bowl. Add the second amount of water (1 lb) and, using a wire whip, dissolve the starch in the water.
4. Pour the dissolved starch into the boiling apricot mixture, while at the same time whipping rapidly with a wire whip.
5. Bring back to a boil, reduce to simmer, and cook for 2 min or 3 min until mixture is slightly thickened and clear. Remove from the range.
6. This sauce may be served hot or cold with an appropriate dessert or entree.

Procedure

Guacamole

1 qt

Ingredients

8 ripe avocados
1½ oz grated onion
1 pressed small garlic clove
1 oz lemon juice
4 medium-sized tomatoes
⅛ oz chili powder
2 or 3 drops hot sauce
¼ oz sugar
salt to taste

Procedure

1. Peel, seed, and dice the avocados.
2. Place the avocados in a stainless steel or plastic bowl and mash with a fork into a coarse puree.
3. Peel and chop the tomatoes.
4. Add the remaining ingredients and stir using a kitchen spoon until thoroughly blended.
5. Place in the refrigerator. Chill for approximately 2 hr before serving as an appetizer, salad, or salad topping.

Cream Chicken and Avocado

25 servings

Ingredients

8 lb boiled chicken
1 lb shortening, butter, or both
14 oz flour
4 lb hot, rich chicken stock
2 lb hot milk
1 lb hot light cream
8 oz sherry
3 peeled avocados
salt to taste

Procedure

1. Place the shortening, butter, or a mixture of both in a sauce pot and heat.
2. Add the flour, making a roux. Cook for approximately 5 min and stir with a kitchen spoon.
3. Add the hot chicken stock, whipping rapidly with a wire whip until thickened and smooth.
4. Add the hot milk and cream, continuing to whip until the sauce is smooth.
5. Dice chicken into 1″ cubes. Peel and dice the avocados in medium-size pieces.
6. Add the chicken, diced avocado, and sherry. Stir carefully with a kitchen spoon until thoroughly blended.
7. Serve with rice or in a patty shell.

Note: Avocado can also be diced and placed in curried chicken.

Bananas Foster

4 servings

Ingredients

3 oz butter
5 oz brown sugar
4 oz rum (high proof to flame)
1 oz crème de banane or other banana-flavored liqueur
4 large scoops vanilla ice cream
4 slightly ripe bananas
dash of cinnamon

Procedure

1. Place the butter in the blazer of a chafing dish and melt.
2. Add the sugar and cook while stirring frequently with a stainless steel spoon. Cook until sugar caramelizes.
3. Add rum and crème de banane or other banana-flavored liqueur and flame.
4. Peel the bananas and cut in half crosswise and lengthwise.
5. Place the hard frozen ice cream on an appropriate dessert dish. Lay the sliced bananas along the side. Top with a dash of cinnamon.
6. Spoon or pour the flaming sauce over each serving and serve immediately.

Note: This preparation is usually prepared before the guest.

Avocado Sauce for Seafood

1 qt

Ingredients

6 large avocados
¼ oz lime juice
2 oz grated onion
6 oz salad oil
salt and white pepper to taste

Procedure

1. Peel, seed, and chop the avocados.
2. Place all the ingredients in the container of an electric blender. Blend at medium speed until mixture is very smooth. Remove from blender.
3. Place in a stainless steel or plastic container. Cover with plastic wrap and refrigerate until ready to use.
4. Serve with sautéed, poached, baked, or broiled fish or shellfish.

Blueberry Whip

25 servings

Ingredients

2 lb 8 oz frozen blueberries (thawed)
8 oz powdered egg whites
8 oz sugar
1 lb 8 oz whipping cream

Procedure

1. Place the egg whites into a small stainless steel mixing bowl. Using the wire whip, whip at high speed until they start to froth.
2. Add the sugar slowly while continuing to whip. Whip until soft, wet peaks are formed.
3. Fold the meringue mixture gently into the blueberries.
4. Place the whipping cream into the stainless steel mixing bowl. Using the wire whip, whip until stiff.
5. Fold the whipped cream gently into the blueberry/meringue mixture. Incorporate thoroughly.
6. Serve in a sherbet glass topped with whipped cream or topping and a red maraschino cherry.

 *Trade Tip*

Blueberries are low in calories, low-fat, a good source of fiber, and high in antioxidants.

Cherries Jubilee

5 servings

Ingredients

1 pt bing cherries and juice
4 oz sugar
2 oz Kirschwasser (cherry-flavored liqueur)
¼ tsp arrowroot starch
5 large scoops vanilla ice cream

Procedure

1. Pour the juice from the cherries into the blazer of a chafing dish. Hold back just enough juice to dissolve the arrowroot starch.
2. Place the blazer of the chafing dish over the flame of the chafing dish. Bring the cherry juice to a boil.
3. Slowly pour the dissolved starch into the boiling juice while stirring rapidly. Cook until the juice thickens slightly.
4. Add the sugar. Reduce heat to simmer.
5. Add the cherries. Stir to incorporate.
6. Heat the liqueur in a separate pan until just warm. Pour the liqueur over the cherry mixture.
7. Ignite the liqueur and pour the flaming sauce over each mound of vanilla ice cream. The flaming should be done before the guest. Serve while still aflame.

Creamy Cranberry Fruit Salad

25 servings

Ingredients

1½ oz plain gelatin
8 oz cold water
1 lb boiling water
3 oz sugar
⅛ oz salt
8 oz mayonnaise
2 oz lemon juice
¼ oz grated lemon rind
2 lb canned whole cranberry sauce
1 peeled, diced orange
1 cored, diced apple
3 oz chopped walnuts

Procedure

1. Place the plain gelatin in a stainless steel bowl, add the cold water, and let soak for 5 min.
2. Add the boiling water, salt, and sugar. Stir until thoroughly dissolved.
3. Add the mayonnaise, lemon juice, and lemon rind. Using a wire whip, whip until all ingredients are incorporated and the mixture is smooth. Place in the refrigerator until it becomes syrupy and starts to thicken.
4. Remove mixture from the refrigerator. Place in the stainless steel mixing bowl and, using the wire whip, whip until light and fluffy. Remove from the mixer.
5. Fold in the cranberry sauce, oranges, apples, and nuts. Fill individual molds and return to the refrigerator until firm.
6. When ready to serve, unmold and serve on crisp salad greens.

Cherry Sauce

2 qt

Ingredients

2 lb cherry juice
8 oz water
3 c drained canned cherries
1 lb 8 oz sugar
¼ oz lemon juice
4 oz butter or margarine
4 oz waxy maize starch (clearjel) or cornstarch
⅛ oz salt

Procedure

1. Place the cherry juice in a sauce pot, place on the range, and bring to a boil.
2. Place the starch in a small stainless steel bowl and add the water slowly while stirring constantly until thoroughly dissolved.
3. Pour the dissolved starch into the boiling juice, whipping vigorously with a wire whip. Cook until thickened and clear.
4. Add the sugar, salt, and lemon juice. Stir thoroughly. Return to a simmer.
5. Remove from the range and place in a stainless steel container.
6. Serve hot or cold over cake, bread pudding, cream puffs, etc.

Cranberry Raisin Sauce

2 qt

Ingredients

2 lb cranberry juice
5 oz raisins
12 oz brown sugar
½ tsp ground allspice
3 oz waxy maize starch (clearjel) or cornstarch
salt to taste

Procedure

1. Place the cranberry juice in a sauce pot. Reserve 1 pt to dissolve the starch. Bring the juice to a boil.
2. Add the raisins, brown sugar, and allspice. Bring to a simmer.
3. Place the starch and the cranberry juice held in reserve in a plastic or stainless steel bowl. Stir until starch is thoroughly dissolved.
4. Pour the dissolved starch slowly into the boiling mixture while stirring constantly with a kitchen spoon. Cook until thickened and clear.
5. Add salt to taste, remove from the range, and pour into a steam table pan. Hold for service.
6. Serve with ham, turkey, or chicken.

Grape Tarts

25 servings

Ingredients

25 baked tart shells
2 qt prepared vanilla pie filling
1 qt seedless grapes
1 qt plain fruit glaze, tinted the color of the grapes

Procedure

1. Place the baked tart shells on a sheet pan.
2. Fill a plastic pastry bag with the vanilla pie filling.
3. Fill each tart shell about ¾ full of vanilla pie filling.
4. Cover the surface of the filling with grapes.
5. Cover the grapes with the tinted fruit glaze.
6. Garnish the edge of the crust around the tart shell with toasted macaroon coconut, chopped nuts, or toasted cake crumbs.

Kiwi Tarts

24 servings

Ingredients

24 baked tart shells
2 qt vanilla pie filling
1 qt plain fruit glaze, tinted green
12 peeled, sliced kiwi
1 lb toasted macaroon coconut

Procedure

1. Place the cream filling in a pastry bag and pipe it into each tart shell until each shell is approximately three-fourths full.
2. Place the sliced kiwi on top of the cream filling. Cover the filling completely.
3. Cover the top of the tart completely with fruit glaze to seal the fruit from the air. Brush the glaze over the top edge of each tart shell.
4. Touch the complete top edge of the tart shells into the toasted macaroon coconut to cover and garnish the edge.
5. Place in a paper baking cup and serve.

Lemon Cream Pudding

25 servings

Ingredients

5 lb milk
8 oz sugar
3½ oz cornstarch
⅛ oz salt
1 lb milk
10 oz eggs
3 oz butter
2 oz lemon juice
¼ oz grated lemon rind

Procedure

1. Place the first amount of milk in the top of a double boiler and heat until scalding hot. Scum will appear on surface.
2. Place the sugar, cornstarch, and salt in a stainless steel bowl. Add the second amount of milk and blend thoroughly.
3. Pour the mixture into the scalding milk while whipping constantly with a wire whip. Cook until the mixture becomes smooth and thick.
4. Place the eggs in a stainless steel bowl and beat slightly. Add a little of the hot mixture to the beaten eggs and mix. Pour into the hot mixture, whipping constantly. Cook for approximately 5 more minutes. Remove from the double boiler.
5. Whip in the butter, lemon juice, and grated lemon.
6. Place into appropriate dessert glasses and cool in the refrigerator.
7. Serve topped with whipped cream or topping and garnish with a red maraschino cherry.

Lemon Sauce

2 qt

Ingredients

4 lb water
2 lb sugar
4 oz cornstarch
⅛ oz salt
4 oz egg yolks
8 oz lemon juice
¼ oz grated lemon rind
6 oz butter or margarine

Procedure

1. Place the water in a sauce pot and bring to a boil.
2. Place the sugar, cornstarch, and salt in a stainless steel bowl. Add a small amount of the boiling water while stirring constantly with a kitchen spoon. Stir until thoroughly dissolved
3. Pour the dissolved starch mixture slowly into the remaining boiling water while at the same time whipping gently with a wire whip. Cook until thickened and clear.
4. Place the egg yolks in a stainless steel bowl. Whip slightly with a wire whip while adding the lemon juice. Add slowly to the thickened mixture while whipping constantly.
5. Remove from the range, add the butter and grated lemon rind, and blend thoroughly.
6. This sauce may be served with any dessert item that can be improved with the addition of lemon flavor.

Key Lime Pie

six 8″ pies

Ingredients

4 lb water
2 lb 4 oz sugar
1 lb beaten pasteurized eggs
12 oz powdered egg whites
6 oz cornstarch
¼ oz salt
10 oz lime juice
1 oz grated lime rind
4 oz butter or margarine
1 lb 4 oz sugar
6 prebaked pie shells

Procedure

1. Place 3 lb of the water in a sauce pot. Add the first amount of sugar and salt and bring to a boil.
2. Place the cornstarch in a stainless steel bowl. Add the remaining water and stir until starch has dissolved.
3. Add the beaten eggs to the dissolved starch. Slowly pour this mixture into the boiling liquid while at the same time whipping rapidly with a wire whip. Cook until mixture is thickened and smooth.
4. Remove from the range and stir in the butter, lime juice, and grated rind. Cool slightly.
5. Place the reconstituted powdered egg whites in the stainless steel mixing bowl. Using the wire whip, whip until whites start to froth.
6. Slowly add the sugar while continuing to whip at high speed. Whip until soft wet peaks have formed.
7. Fold the cooked mixture slowly and gently into the meringue. Fill the prebaked pie shells and refrigerate until set.

Creamy Lime Salad

25 servings

Ingredients

12 oz lime gelatin
1 lb 8 oz hot water
1 lb 8 oz pear juice
¼ oz cider vinegar
⅛ oz salt
1 lb soft cream cheese
⅛ oz ginger
1 lb 8 oz diced canned pears

Procedure

1. Place the gelatin in a stainless steel bowl. Add the hot water and stir until thoroughly dissolved.
2. Add the pear juice, vinegar, and salt. Stir until blended.
3. Fill each of the individual molds one-third full of the prepared gelatin. Place in the refrigerator and let set until firm.
4. Place the remaining gelatin in a shallow pan and chill until slightly thickened. Remove from the refrigerator and place in the stainless steel mixing bowl.
5. Using the wire whip, whip the gelatin until light and fluffy.
6. Add the soft cream cheese and ginger. Continue to whip until the cheese is thoroughly blended into the gelatin.
7. Remove from the mixer and fold in the diced pears.
8. Complete filling the partly-filled molds. Again, place the molds in the refrigerator until firm.
9. When ready to serve, unmold onto crisp salad greens and serve.

Lime Dressing

2 qt

Ingredients

2 qt mayonnaise
4 oz finely-minced onions
2 dashes Tabasco sauce
2 oz lemon juice
2 oz lime juice
¾ oz finely-grated lime peel

Procedure

1. Place all the ingredients into a plastic or stainless steel bowl. Blend thoroughly using a kitchen spoon.
2. Serve with seafood.

Trade Tip

Watermelon, considered one of America's favorite fruits, is actually a vegetable. Watermelons typically range in size from 7 lb to 100 lb.

Watermelon Basket

Cut the top off of a ripe watermelon lengthwise and remove the flesh using a parisienne or melon ball scoop. Discard the seeds. The edges of the cut walls may be scalloped or cut into triangles with a utility knife if desired. A design may also be scraped on the outside surface of the melon with the tines of a dinner fork. Place watermelon balls, sliced fresh peaches, fresh strawberries, pineapple chunks, cantaloupe balls, honeydew melon balls, seedless green grapes, pitted ripe cherries, and blueberries in a stainless steel bowl. Sprinkle lightly with sugar and chill in the refrigerator. When ready to serve, fill the melon cavity with the mixed fruit, sprinkle with champagne or other desired wine or liquor, and serve on a buffet table or to a special party of approximately 12 people.

Melon Surprise

Cut a circular piece off the stem end of a large cantaloupe, honeydew, or casaba melon. Reserve the cut-off piece. Remove the seeds and the center fiber. Scoop out the flesh using a parisienne or melon ball scoop, leaving a wall approximately ½″ thick. Place the melon balls in a stainless steel bowl and add an assortment of diced fresh fruit and berries such as peaches, pineapple, apricots, strawberries, blackberries, seedless grapes, blueberries, and pitted cherries. Pour over this assortment one cup of strawberry or raspberry puree sweetened with sugar and flavored with Kirschwasser (cherry-flavored liqueur) and chill. When ready to serve, sprinkle the interior of the melon shell generously with additional Kirschwasser. Fill with chilled assorted fruit. Replace the circular piece that was cut off the stem end and serve on a buffet table or to a party of two or three people.

Trade Tip

Oranges are the largest citrus crop in the world. They are high in fiber and Vitamin C. One orange provides 130% of daily recommended Vitamin C requirements. Oranges contain no fat, no cholesterol, and no sodium.

Orange Bavarian Cream

25 servings

Ingredients

2 oz plain gelatin
2 lb 8 oz cold water
2 lb 8 oz orange juice
4 oz lemon juice
12 oz sugar
2 lb whipping cream

Procedure

1. Soak the gelatin in the cold water for approximately 5 min. Place in a stainless steel bowl over a pot of boiling water. Stir until thoroughly dissolved. Remove from the range.
2. Add the orange juice, lemon juice, and sugar. Stir until thoroughly blended.
3. Place in the refrigerator and chill until the mixture starts to thicken.
4. Place the cold whipping cream in a cold stainless steel mixing bowl. Using the wire whip, whip at high speed until stiff. Remove from the mixer.
5. Fold the whipped cream into the partly jellied orange mixture. Blend gently, but thoroughly.
6. Place into individual glass dessert dishes and chill until ready to serve.
7. Top with additional whipped cream and an orange segment.

Orange Sauce

1½ qt

Ingredients

1 lb 8 oz water
1 lb 8 oz orange juice
1 lb 8 oz sugar
⅛ oz salt
3 oz cornstarch
¼ oz orange grating
4 oz lemon juice
4 oz butter

Procedure

1. Place the juice and half of the water in a sauce pot. Bring to a boil.
2. Place the sugar, salt, and cornstarch in a stainless steel bowl. Add the remaining water. Stir with a kitchen spoon until the ingredients are thoroughly blended and the dry ingredients are dissolved.
3. Pour slowly into the boiling mixture, whipping constantly with a wire whip. Cook until slightly thickened and clear. Remove from the range.
4. Add the lemon juice, orange gratings, and butter. Stir until incorporated. Pour into a stainless steel container.
5. Serve with items that are complemented by a sweet orange flavor, such as crepes and cake.

Pears in Red Wine

25 servings

Ingredients

25 fresh, ripe pears
12 eating apples
6 oz butter or margarine
8 oz sugar
6 pieces cinnamon stick
6 oz chopped walnuts
2 lb red wine
1 lb 8 oz sugar
1 oz lemon juice

Procedure

1. Peel, core, and mince the apples.
2. Place the butter in a sauce pot and heat.
3. Add the apples, first amount of sugar, and three cinnamon sticks. Cook until apples become soft. Discard the cinnamon.
4. Add the walnuts, remove from the range, and hold for later use.
5. Place the wine, second amount of sugar, the remaining three pieces of cinnamon stick, and lemon juice in a braising pot. Bring to a boil.
6. Peel the pears, place into the wine syrup, and poach covered until the pears are tender. Remove pears but continue cooking the syrup until reduced by about one-third.
7. Put equal amounts of the apple/nut mixture into each serving dish. Place pear on top and ladle the wine syrup over each pear.

Note: If desired, just before serving, pour a small amount of warmed rum or cognac over the pear and ignite at the table.

Pineapple Cream Pudding

25 servings

Ingredients

5 lb milk
8 oz sugar
3 oz cornstarch
¼ oz salt
12 oz milk
8 oz beaten eggs
3 oz butter
1 pt crushed pineapple

Procedure

1. Place the first amount of milk in a double boiler and heat.
2. Place the sugar, cornstarch, salt, and second amount of milk in a stainless steel bowl. Using a wire whip, work until a smooth paste is formed.
3. Add the beaten eggs to the smooth paste mixture and blend.
4. Pour the paste mixture slowly into the scalded milk while whipping vigorously with a wire whip.
5. Cook until thick. Remove from the double boiler.
6. Stir in the pineapple and flavor with vanilla.
7. Place into dessert glasses and chill until ready to serve. Top with whipped cream and a maraschino cherry.

Variation

For a pineapple whip, blend in 1 pt of 36% to 40% whipping cream that has been whipped to a stiff peak. Fold the cream in while the pudding is still slightly warm.

Pineapple Sauce

2 qt

Ingredients

1 qt crushed pineapple
1 lb sugar
8 oz water
1 oz lemon juice
3 oz waxy maize starch (clearjel)
6 oz water
4 oz white corn syrup

Procedure

1. Place the pineapple, sugar, first amount of water, and lemon juice in a sauce pot and simmer for approximately 5 min.
2. Place the starch and second amount of water in a stainless steel bowl. Stir until thoroughly dissolved.
3. Pour the dissolved starch slowly into the boiling mixture while whipping rapidly with a wire whip. Cook until slightly thickened and clear.
4. Add the corn syrup and stir until blended. Remove from the range and place in a steam table pan.
5. Serve with cake, hot cakes, ham, etc.

Plum Sauce

2 qt

Ingredients

4 lb plum syrup from canned plums
3 oz cornstarch or waxy maize starch (clearjel)
2 oz lemon juice
½ tsp almond extract
salt to taste
red color if needed or desired

Procedure

1. Place the plum juice in a saucepan. Reserve only enough to dissolve the starch. Bring to a boil.
2. Place the starch in a plastic or stainless steel bowl. Add the plum juice held in reserve and stir until thoroughly dissolved.
3. Pour the dissolved starch into the boiling juice, whipping rapidly with a wire whip. Cook until thick and smooth.
4. Add the lemon juice and almond extract. Remove from the range and pour into a stainless steel steam table pan.
5. Tint with red color if needed or desired.

Strawberry Shortcake

25 servings

Ingredients

25 shortcake biscuits
3 qt fresh strawberries
1 lb 6 oz sugar
1 qt whipping cream
2 oz powdered sugar

Procedure

1. Cut the shortcake biscuits in half crosswise. Set aside.
2. Cut the berries into halves or quarters.
3. Place the cut berries in a stainless steel or plastic bowl. Add the sugar and place in the refrigerator until thoroughly chilled.
4. Place the cream in a cold stainless steel mixing bowl. Using the wire whip, whip until cream starts to stiffen. Add the sugar and continue to whip until stiff. Place in a pastry bag with a star tube. Refrigerate until ready to use.
5. To set up for service, place the bottom half of the biscuit on a dessert plate or in a bowl. Place approximately 2 oz of the sweetened strawberries on top. On top of the berries, place the top half of the biscuit and cover with 2 more ounces of berries. Pipe a spiral of whipped cream on the berries. Serve immediately.

Strawberry Jubilee

4 servings

Ingredients

1 pt frozen whole strawberries and juice
½ tsp arrowroot starch
1 oz cointreau
2 oz brandy
4 large scoops ice cream

Procedure

1. Strain the juice from the berries into the blazer of a chafing dish. Reserve just enough juice to dissolve the starch.
2. Place the blazer pan over the flame of the chafing dish and bring the juice to a simmer.
3. Dissolve the arrowroot in the juice that was held in reserve. Slowly pour the dissolved starch into the simmering juice while stirring constantly. Cook until slightly clear. Add cointreau. Stir to blend.
4. Place the brandy in a saucepan and heat slightly (do not boil). Pour the warm brandy over the strawberries.
5. Ignite the brandy and pour the flaming mixture over each scoop of hard ice cream.

Note: This preparation should be done before the guest.

Strawberries Romanoff

25 servings

Ingredients

3 qt fresh strawberries
1 lb confectioners' sugar
1 tsp brandy
1 pt whipping cream

Procedure

1. Separate the berries into two equal amounts. Hold one amount for later use. Place the other in a stainless steel or plastic bowl and mash with the tines of a fork.
2. Add half the sugar and the brandy and stir until thoroughly incorporated. Sprinkle the remaining sugar over the berries that were left whole. Refrigerate both mashed and whole berries until chilled.
3. Place the cream into a cold stainless steel mixing bowl. Using the wire whip, whip until thick.
4. Remove the mashed berries from the refrigerator and fold in the whipped cream with a very gentle motion.
5. Fill selected dessert glasses and top with the whole fresh berries.

Glazing Fruit

Glazing fruit can be used in some preparations to improve appearance and in some cases to seal out air so the fruit does not dry out or discolor. Glazing is usually done by coating the fruit with a thick, shiny, sweet liquid. The fruit on tarts, open-face pies, coffee cakes, and tortes are usually glazed. Other methods of glazing may be accomplished by coating the fruit with sugar and browning slightly under a broiler or on the griddle.

Apricot Glaze

5 qt

Ingredients

1 No. 10 can apricot halves
6 lb sugar
2 lb light corn syrup

Procedure

1. Rub apricots through a fairly fine sieve placed in a sauce pot.
2. Add the sugar and bring to a boil. Reduce heat and simmer for approximately 5 min.
3. Stir in the corn syrup. Remove from the heat.
4. Use the glaze hot or warm and apply over the surface of the fruit with a pastry brush.

Fresh Strawberry Glaze

2 qt

Ingredients

2 lb water
1 lb 8 oz sugar
8 oz (1 c) water
6 oz waxy maize starch (clearjel)
1 lb sugar
red color
8 oz light corn syrup
2 oz lemon juice

Procedure

1. Place the 2 lb of water in a sauce pot.
2. Add the first amount of sugar and bring to a boil.
3. Dissolve the starch in the second amount of water. Add to the boiling liquid, mixing rapidly using a wire whip. Cook until the mixture is clear.
4. Add the remaining sugar, corn syrup, and color. Bring back to a boil and remove from the heat.
5. Add the lemon juice and let cool.

Plain Fruit Glaze

2 qt

Ingredients

2 lb water
2 lb 8 oz sugar
4 oz waxy maize starch (clearjel)
8 oz (1 c) water
4 oz light corn syrup
1 oz lemon juice

Procedure

1. Place the first amount of water and the sugar in a sauce pot. Bring to a rapid boil.
2. Dissolve the starch in the second amount of water. Add slowly to the boiling liquid, mixing rapidly with a wire whip. Cook until the mixture becomes clear.
3. Add the corn syrup and bring the mixture back to a boil. Remove the mixture from the heat.
4. Add the lemon juice and color as desired. The color should blend with the fruit being glazed.

Marzipan Fruit

Marzipan fruit is a specialty item that is most popular during the holiday season, but can be used year-round to decorate and add eye appeal to cakes, tortes, and pies. Marzipan is a candy with an almond flavor. It is simple to prepare, but requires patience and some skill in forming, coloring, and glazing the various fruits. The first step in the preparation of marzipan fruit is to prepare the marzipan mixture.

Marzipan

1 qt

Ingredients

1 lb almond paste
1 lb powdered sugar
2 egg whites

Procedure

1. Place the almond paste and sugar in the stainless steel mixing bowl. Blend at low speed using the paddle.
2. Add the amount of egg white needed to obtain a mixture of stiff consistency. Consistency should resemble molding clay. Mix until smooth.
3. Remove the marzipan from the mixer. Place in a plastic or stainless steel container and keep covered with a damp towel. Store in a cool place.

Note: If marzipan is too soft to mold, add sifted powdered sugar. If too dry, add additional egg white.

Coloring the mixture to obtain natural fruit colors can be done using the following method:

1. Place the desired amount of marzipan in the bowl of the electric mixer. Using the paddle, blend the selected paste food color at slow speed until desired shade or tone is reached.
2. Dissolve powdered or paste colors in alcohol and apply to the molded fruit using a brush or atomizer. Alcohol is used because it evaporates and dries quickly.
3. Moisten the molded fruit over steam and apply the powdered colors using a brush or dabbing with a cotton ball.

Forming the fruit into apples, peaches, pears, bananas, etc. is the most difficult part of making marzipan fruit. This step requires patience, skill, and some artistic talent. The forming is done by hand and requires knowledge of the shape and markings of the fruit being copied. The more natural the fruit appears, the more appealing it will be. Always work with clean hands and clean tools. Molding tools made of wood are best. However, other tools may be substituted to mark and help mold the fruit. Small paintbrushes, toothpicks, or toothbrushes are used to mark the fruit with natural marks, streaks, or tone of color (such as the touch of green at the end of the bananas and a streak of brown on a banana, pear, or apple). Cloves can be used for stems and leaves can be molded, or imitation leaves and stems may be purchased from bakery supply houses.

Peach

Color the marzipan yellow. Form the peach by hand. Mark the seam of the peach around the center using the back of a paring knife. Tint with a faint touch of light red on each side. Roll in cornstarch and blow off excess starch. Use clove for the stem. Insert upside-down. Add a leaf if desired.

Banana

Color the marzipan yellow. Form the banana by hand. Touch each end with green color and then with a spot of brown. Draw fine brown lines on the banana to indicate a natural peel.

Pear

Color the marzipan yellow. Form the pear by hand. Tint with a faint touch of light red on one side. Mark with a few brown streaks or spots. Apply a clove stem and a leaf.

Orange

Color the marzipan orange. This can be done by blending red and yellow colors. Form the orange and speckle with a toothbrush or toothpick. Insert a clove right side up for the stem.

Strawberry

Color the marzipan red. Form the strawberry by hand. Speckle with a toothpick. Roll in plain or red sugar. Place a green marzipan or imitation leaf at the top.

Apple

Color marzipan red, green, or yellow, depending on the apple desired. Form the apple by hand. Spot, streak, or spray with various contrasting colors to give a natural look. Use a clove, inserted upside-down, for the stem and add a leaf.

After the fruits are formed and colored, a glaze may be applied to some to give a shine and improve eye appeal. The glaze is usually applied with a brush and may be made by thinning glucose with water or by using a glaze.

Marzipan Glaze

1 qt

Ingredients

1 lb sugar
1 lb glucose
1 lb water
4 oz pure alcohol

Procedure

1. Place the sugar, water, and glucose in a sauce pot. Bring to a boil.
2. Simmer for about 5 min, remove from the fire, and let cool.
3. Add the alcohol and blend. Adding alcohol allows the glaze to dry quicker.

Fruit Garnishes

Fruit garnishes may be simple or complex, depending on the food item being garnished. They are used to improve the appearance and flavor of the food. Some garnishes used with popular foods are:

- *Ham*: glazed pineapple, glazed peach, cinnamon pear, orange or grapefruit basket, twisted orange slice

- *Seafood*: slice, twist, or crown of lemon or lime

- *Roast pork*: baked apple

- *Broiled, sautéed, and fried pork*: cinnamon apple ring, apple fritter, pear with cavity filled with cranberry sauce

- *Roast beef and braised beef*: kumquats, spiced crab apple, spiced peach

- *Roast, broiled, and sautéed chicken*: spiced crab apple, peach half with cavity filled with cranberry sauce, mandarin orange, twisted orange slice

- *Roast turkey*: pear or peach half with cavity filled with whole cranberry sauce, cinnamon pear half, crown of orange, cinnamon apple ring, cranberry orange relish

- *Pork sausage*: cinnamon apple ring, baked apple, apple fritters

Fruit Flavor

Fruit flavor is an important ingredient in many preparations. The flavor may be added in the form of juice, pulp, or chopped fruit. The following are examples of fruits used to improve the flavor of certain preparations.

- Add fresh apples and bananas to a curry sauce.
- Add applesauce to sauerkraut.
- Add diced sautéed apples or canned whole cranberries to bread dressing.
- Add lemon juice to chicken salad.
- Add lemon juice to fish sauce.
- Add diced lemon to mock turtle soup.
- Add diced apples to mulligatawny soup.
- Add lemon juice to borscht.
- Add chopped pineapple to cream cheese for a pineapple-cheese spread.
- Add grape halves and pineapple chunks to a chicken or turkey salad.
- Add julienne orange peel and concentrated orange juice to the liquid when baking rice.
- Scallop sweet potatoes with apples.
- Add slices of orange and lemon to candied sweet potatoes before baking.
- Add orange juice and julienne orange peel to hollandaise sauce to create maltaise sauce.
- Add chopped apples, bananas, oranges, and strawberries to certain cake batters to create special variations.

Vegetable Preparation

While vegetable preparations are sometimes overshadowed by the main entree, they are an important food item used in many preparations. Vegetables may be purchased fresh, frozen, canned, or as dried legumes and are classified as fresh or dried. Fresh vegetables can be eaten when ripe and must be preserved by canning or freezing. Dried vegetables have all their moisture removed and can be stored indefinitely.

Fresh vegetables are classified by their color: green, yellow, red, and white. Green vegetables include peas, green beans, and broccoli. Yellow vegetables include carrots, corn, and squash. Red vegetables include red cabbage and beets. White vegetables include white cabbage, white onions, and turnips. Vegetables are also classified by the part of the plant that is edible.

Some dried vegetables are classified as dried legumes or cereals. Dried legumes include lima beans, lentils, and dried peas. Cereals include rice and barley.

VEGETABLE PREPARATION

Vegetables are an important but sometimes neglected part of any meal. Most food service establishments emphasize the entree or dessert. While these items are important, the vegetables that accompany the entree are essential. Many menus give so little importance to the vegetable that it is simply listed as the vegetable *du jour* (of the day) rather than describing the vegetable preparation. Vegetable preparations sometimes lack variety and are often overcooked. However, with a little creativity, vegetables can be prepared to add variety and complement the entire menu.

VEGETABLE CLASSIFICATIONS

Vegetables are classified as fresh or dried, depending on their nature when ripe. Fresh vegetables are eaten at the time the plants are ripe. A fresh vegetable can be preserved by canning or freezing. Dried vegetables have all their moisture removed and keep indefinitely. Fresh vegetables are classified according to color as well as the portion of the plant that is edible such as leaves, stems, flowers, seeds, bulbs, or roots. Dried vegetables are classified as legumes (beans, peas, etc.) and cereals (rice, barley, etc.).

Green vegetables include peas, string beans, broccoli, lima beans, and others. They obtain their green color from a pigment known as *chlorophyll*. Chlorophyll is easily destroyed by alkalis and acids when heat is applied. Therefore, green vegetables require proper cooking in order to retain their natural green color when they are served. Steam fresh vegetables in a small amount of water, as they contain water soluble vitamins that are easily lost when overcooked. Most green vegetables contain a high percentage of acid. They should be cooked slowly and uncovered so the acid, which would destroy the green color, can escape in the steam.

Yellow vegetables include carrots, yellow turnips, corn, rutabagas, squash, and others. Yellow vegetables do not change color when heat is applied unless they are overcooked. Overcooked vegetables become dull in appearance. Most yellow vegetables should be cooked with a lid to minimize the loss of nutrients. Yellow vegetables contain fat soluble vitamins that are not lost when cooked in water. Exceptions to this are yellow turnips and rutabagas, which possess a strong flavor and should be cooked uncovered.

Red vegetables include red cabbage and beets. Red vegetables react directly opposite to green vegetables when heated. Acids improve the color of red vegetables, while alkalies make them fade and turn a bluish-gray. To improve the natural color of these vegetables when cooking, diluted acids, such as lemon juice, vinegar, or cream of tartar are added. Usually, one tablespoon to each quart of water is sufficient. Beets should be cooked covered to utilize the acid content present. Red cabbage should be cooked partly uncovered to provide an escape for its sulfur content.

White vegetables include white cabbage, white onions, turnips, cauliflower, and others. White vegetables have a tendency to turn yellow when cooked in hard water or if overcooked. To prevent overcooking, cook the vegetables in just enough water to submerge them. Do not cover. Cook until the vegetables are just tender. A small amount of vinegar or lemon juice can be added to improve the appearance and eating qualities of white vegetables.

Dried legumes include lima beans, lentils, peas, kidney beans, navy beans, great northern beans, and others. Dried legumes are cooked by simmering. They are commonly soaked for several hours and then rinsed before cooking.

When preparing dried legumes, soak them overnight to replace the water lost in ripening and drying. The next day, they should be covered with water, using about 1 gal. of water for each pound of legumes, and cooked slowly by simmering until tender. Never allow legumes to be subjected to boiling temperatures because boiling tends to toughen them. Soaking is not required when preparing split peas and lentils.

Grains include rice and barley. Rice and barley should be washed, thoroughly covered with water, and simmered until tender. When preparing rice, the ratio is two parts water to one part rice. For every 2 qt of water, add 1 qt of rice. For cooking barley, a ratio of four to one is preferred. Rice can also be prepared by baking with excellent results. Both rice and barley must be cooked covered.

USA Rice Federation

A rice and bean burrito is a healthy vegetarian alternative to the traditional beef and bean burrito.

VEGETABLE MARKET FORMS

Vegetables may be purchased in the following market forms: fresh, frozen, canned, and dried. The preparation techniques required may vary depending on the market form.

Fresh vegetables should be thoroughly washed before cooking to remove dirt and grit. All blemishes should be removed and the vegetable cut or shaped. Water is brought to a boil separately and poured over the vegetables. The water is brought to a simmer as quickly as possible. The vegetables may also be added to simmering water. When vegetables are slightly tender but still somewhat crisp, they are removed from the heat and placed in another container to cool. Ice may be added to the liquid to shock the vegetables. Shock is a term used for adding ice to food to stop the cooking. The cooked vegetables can be reheated and seasoned when they are to be served. The seasoning varies with different vegetables, but generally consists of salt, pepper, and sugar.

Frozen vegetables are prepared in the same manner as fresh vegetables with the exception of washing. The cooking time is less since all frozen vegetables are *blanched* (partly cooked) before they are frozen. Frozen vegetables may be cooked in their frozen state, but thawing or partial thawing at refrigerator temperatures (34°F to 39°F) results in more uniform cooking. Spinach, kale, and other leaf greens cook more uniformly when completely thawed. Frozen vegetables are extremely popular in all food service establishments because they are convenient to use. Frozen vegetables can be prepared quickly, requiring minimal labor cost. Most frozen vegetables are packed IQF (individually quick frozen) so that the user can prepare any desired amount. Frozen vegetables have better color and texture than canned vegetables.

Canned vegetables are fully cooked and require only heating and seasoning before serving. Canned vegetables should be reheated in their own liquid and seasoned with pepper. In some cases, other items such as butter, bacon grease, onions, ham stock, or fat will improve the taste of the vegetable. Canned vegetables should only be heated in small quantities and never overcooked. Canned vegetables contain salt and are higher in sodium than fresh vegetables.

Dried legumes have excellent keeping qualities because there is no moisture present to breed bacteria, making them easy to store. To prepare legumes, soak in water overnight at room temperature. Soaking is not required for split peas and lentils. Legumes should be simmered, not boiled, to prevent toughening from high cooking temperatures.

VEGETABLE PREPARATION

In most commercial kitchens, vegetables are already cleaned and peeled before use by the chef or cook. Therefore, recipes usually do not call for the vegetables to be cleaned or peeled. When cooking vegetables, certain procedures must be followed to achieve good eating qualities and eye appeal.

Recommended Procedures for Cooking Vegetables

- Use only enough water to cover the vegetable.
- If called for, use simmering water that has been salted.
- Cook in small quantities, if practical.
- Cook only until the vegetable is tender.
- Cook about 1 hr before serving time, if practical.
- After cooking, shock the vegetables with ice and cold running water if they are not to be used immediately.
- Save the liquid the vegetables were cooked in for reheating the vegetables or for use in stocks and sauces.
- Season vegetables to taste just before serving.
- When preparing fresh vegetables, cut to uniform size for proper and even cooking.
- Clean all fresh vegetables thoroughly and store all vegetables properly for best results.

Avoid:

- Letting vegetables soak before cooking, except for some dried legumes.
- Stirring air into the water while cooking.
- Using excessive amounts of water; the result is loss of flavor and food value.
- Overcooking the vegetables; food value, flavor, and appearance will suffer.
- Letting the cooked vegetables stand in hot water after cooking. The vegetables will continue to cook, become extremely soft and lose their natural color.
- Cooking in large quantities; food value will be lost and appearance and flavor will suffer.
- Mixing fresh-cooked vegetables with the old; color, texture, and flavor will be different.
- Thawing frozen vegetables too far in advance of preparing; food value is lost and the chance of spoiling is present.
- Boiling vegetables when cooking. Boiling has a tendency to break up and overcook vegetables.

Vegetable Storage

Fresh vegetables require refrigeration to preserve their appearance and flavor, except for onions and potatoes, which should be stored in a cool, dry place. Vegetables should be placed in the refrigerator in baskets or containers that are vented so the cold, moist air can circulate properly around the vegetables. Peeled and cut vegetables must be refrigerated and sealed from the air by placing them in a plastic bag, covering with water, or treating them with chemical oxidizing agents to prevent discoloration. If sulfides are used, it should be noted on the menu to inform people who are allergic to sulfides. The method used to seal out the air depends on the vegetable.

Frozen vegetables should be stored at temperatures of 0°F to 10°F. For best results, frozen vegetables should be thawed at refrigerator temperatures (34°F to 39°F). Frozen vegetables should never be refrozen after being thawed.

Dried vegetables must be stored in a cool (70°F to 75°F), dry place in cans or bags and placed on shelves off the floor.

Canned vegetables must be stored in a cool, dry place out of sunlight. Canned vegetables should be placed on shelves off the floor. Rotate the stock by moving the old stock forward and the new stock to the rear so the old is used first. If cans appear rusted or punctured, check contents for spoilage. If there is any doubt about the contents, throw it out.

STEAM PRESSURE AND STEAM JACKET COOKING

Vegetables cooked on the range reach a maximum temperature of 212°F. In a *pressure cooker*, which operates on 5 lb to 6 lb of steam pressure, the temperature is 225°F to 230°F. This allows vegetables to be cooked much faster. Since less water is added to the vegetables, pressure cooking reduces the loss of vitamins and minerals and provides a better flavor. Some larger pressure cookers operate on 15 lb of steam pressure, which produces a temperature of approximately 250°F.

Steam jacket cooking is different from steam pressure cooking in that the food is not directly exposed to the steam. The steam flows around the outer jacket of the kettle, providing an equal distribution of heat around the sides and bottom. The food is still cooked in the same amount of water as cooking on the range at 212°F, but the boiling point is reached much faster because the kettle is uniformly surrounded by heat. A great advantage of cooking vegetables in a steam jacket is speed. The quicker cooking time results in more minerals, vitamins, and flavor preserved.

Steam Pressure Cooking Procedures

1. Cut the vegetables to uniform size to promote even cooking.
2. Place the vegetables in a solid basket and add just enough salt water to cover. If using a perforated basket, water is not necessary.
3. Place the basket in the steam cabinet and lock the door.
4. Turn on the steam. A short period of time must be allowed for the steam to build cooking pressure (about 5 lb or 6 lb).
5. Cooking time varies with the vegetable being cooked and the amount of steam pressure.
6. Remove the vegetables when just slightly tender. Season and serve at once, or cool the vegetables by placing them in a bain-marie and placing the container in ice water.

Caution: Do not open the lid of a steam pressure cooker until there is no steam pressure left and the pressure gauge is at zero.

Steam Jacket Cooking Procedures

1. Cut the vegetables to uniform size to promote even cooking.
2. Place the amount of water needed to cover the vegetables in the kettle. Add salt.
3. Bring the water to a rolling boil.
4. Add the vegetables and bring the liquid back to a boil.
5. Reduce the heat and let simmer until the vegetables are slightly tender.
6. Cook the vegetables uncovered to preserve their natural color.
7. Remove the vegetables from the kettle immediately after they are cooked and leave only enough liquid to cover.
8. Season and serve immediately, or cool the vegetables by placing the container in ice water.

VEGETABLE RECIPES

Most vegetable preparations are easy to prepare if basic information about each vegetable is known. The recipes listed include some basic information.

Frozen vegetables can be substituted for most of the preparations calling for fresh vegetables, and vice versa. However, in most cases, the amounts of frozen vegetables used will be less than the fresh vegetables. Fresh vegetables must be thoroughly washed prior to preparation. The edible portion of each vegetable varies. The cooking procedure would also be different in most cases because fresh vegetables are processed and prepared differently from the frozen product. Most vegetable preparations are served with a 3 oz kitchen spoon.

ASPARAGUS RECIPES

Asparagus has been a highly regarded vegetable for thousands of years. It is a perennial plant of the lily family believed to have originated in Asia. During the days of the Roman empire, asparagus grew wild along the coast of the Mediterranean. The banquet-loving Romans soon learned how to cultivate and cook it. After the Romans, the British discovered this tasty, thistle-like vegetable, which they called sparrowgrass.

Asparagus is very popular in the United States and grows in many states. It takes about three years to establish an asparagus bed; but, once established, the plants grow extremely fast. The mature plant produces year after year. Good quality asparagus possesses round, compact tips. The stalks are straight and brittle. If the stems are tough and woody, peel them with a vegetable peeler before cooking. The stems require longer cooking time than the tips, so sometimes the stems are stood up in water and precooked for a short period of time before the complete spear (stem and tip) is cooked. Cooking in a pressure steamer is the preferred method.

When opening a can of asparagus spears, always open the end containing the stems. The can is usually marked to indicate which end is the correct end to open, so read any notice that may appear on the top or bottom of the can. If the top end is opened, the tips are usually mashed and destroyed. Canned asparagus spears can be purchased as white or all green. The white spears are the most expensive, but are considered the best.

Asparagus with Bleu Cheese and Walnuts

25 servings

Ingredients

10 lb fresh asparagus spears
boiling water to cover
2 tsp salt
8 oz melted butter
1 lb bleu cheese
3/4 lb walnuts, chopped

Procedure

1. Cut off the tough ends and peel the remaining stalks slightly with a potato peeler.
2. In simmering pot of water, add salt and asparagus.
3. Cook for 1 min to 2 min until tender.
4. Remove aparagus from pot and drain, toss with butter.
5. Place asparagus on plate, crumble bleu cheese over asparagus and top with chopped walnuts.
6. Serve.
Note: Lemon juice may be added to the butter if desired. Also, frozen asparagus may be substituted for the fresh.

Creamed Asparagus

25 servings

Ingredients

5 lb frozen, cut asparagus
boiling water to cover
2 tsp salt
2 qt cream sauce

Procedure

1. Place the cut asparagus in a saucepan.
2. Add enough boiling water to cover.
3. Add the salt and simmer until the asparagus is tender. Drain thoroughly.
4. Add the cream sauce and blend gently.
5. Check seasoning and serve.
Note: Fresh asparagus may be used if desired. If so, cut off the tough ends and peel the stalks slightly.

Trade Tip

When steaming vegetables, the liquid should never boil dry or touch the food. Make sure the food is at least 1" above the boiling water. The steam should circulate freely in the steamer.

Asparagus Au Gratin

25 servings

Ingredients

5 lb frozen, cut asparagus
boiling water to cover
2 tsp salt
2 qt cream sauce
3 oz Parmesan cheese or 8 oz grated Cheddar cheese
2 oz butter
paprika as needed

Procedure

1. Place the cut asparagus in a saucepan.
2. Cover with boiling water and add the salt.
3. Simmer until the asparagus is just tender, then drain.
4. Place in a baking pan. Pour the cream sauce over the asparagus.
5. Check seasoning. Sprinkle the grated cheese over the top.
6. Dot the butter over the top and sprinkle slightly with paprika.
7. Bake in a preheated oven at 375°F until the cheese melts and the top becomes slightly brown. Serve.

Asparagus Hollandaise

25 servings

Ingredients

10 lb fresh asparagus spears
boiling water to cover
2 tsp salt
2 qt hollandaise sauce

Procedure

1. Cut off the tough ends of the asparagus and peel the remaining stalks slightly with a potato peeler.
2. Place the spears in a baking pan, cover with boiling water, and add the salt.
3. Simmer on the range or cook by steam pressure until the stalks are just tender.
4. Serve three or four spears (depending on size) to each order, half-covered with the hollandaise sauce.

Note: Frozen asparagus spears may also be used.

WAX AND GREEN BEAN RECIPES

Wax beans and green beans (sometimes known as string beans and snap beans) are similar in both cooking and eating qualities. They differ only in color. Some varieties have round pods while others have flat pods; however, the shape of the bean has no bearing on the flavor or tenderness. Beans that are fresh and tender snap readily when they are bent. Fresh beans produce a very good finished product. However, much depends on their freshness at cooking time. After picking, the beans should be refrigerated or cooked immediately for the finest quality.

Canned green beans also produce a good finished product, but they lack color and appearance. Frozen green beans, on the other hand, possess excellent appearance when cooked properly, but lack flavor.

All green bean preparations need help to improve their flavor. Onions can also be an excellent flavor additive. Green beans blend well with other foods, so they are frequently placed on the menu combined with other vegetables and ingredients such as corn, mushrooms, almonds, onions, and spaetzels.

Green Beans Chuck Wagon Style

25 servings

Ingredients

1 No. 10 can green beans, whole or cut
½ No. 10 can whole tomatoes, slightly crushed
1 lb onions, julienne cut
8 oz celery, julienne cut
1 lb bacon, julienne cut
1 pt ham stock
salt and pepper to taste

Procedure

1. Place the julienne cut bacon in a saucepan and cook to a soft crackling.
2. Add the onions and celery and continue to sauté until slightly tender.
3. Add the ham stock and tomatoes and simmer until the celery is tender.
4. Drain the liquid from the can of green beans. Add the beans to the tomato-vegetable mixture and simmer for 5 min.
5. Season with salt and pepper.
6. Serve garnished with chopped parsley.

Green Beans or Wax Beans Lyonnaise

25 servings

Ingredients

1 No. 10 can green beans or wax beans, whole or cut
8 oz onions, julienne cut
6 oz butter
salt and pepper to taste

Procedure

1. Place the butter in a saucepan and heat.
2. Add the julienned onions and sauté until slightly tender.
3. Add the green beans or wax beans (with liquid). Simmer for 5 min.
4. Season with salt and pepper and serve.

Green Beans Amandine

25 servings

Ingredients

5 lb frozen green beans, French cut
boiling water to cover
2 tsp salt
8 oz butter
6 oz almonds, sliced
salt and pepper to taste

Procedure

1. Place the green beans in a saucepan.
2. Pour enough boiling water to cover. Add 2 tsp salt and simmer until the beans are slightly tender.
3. Place the butter in a separate saucepan and melt.
4. Add the sliced almonds and brown until golden.
5. Add the butter-almond mixture to the cooked green beans.
6. Season with salt and pepper and serve.

Green Beans with Mushrooms

25 servings

Ingredients

1 No. 10 can whole green beans
1 lb fresh mushrooms, sliced
8 oz butter
salt and pepper to taste

Procedure

1. Place the butter in a saucepan and heat.
2. Add the mushrooms and sauté until slightly tender.
3. Add the green beans (with liquid). Simmer until the mushrooms are completely tender.
4. Season with salt and pepper and serve.

Green Beans with Pimientos

25 servings

Ingredients

1 No. 10 can wax beans
4 oz butter
4 oz pimientos, diced small
salt and pepper to taste

Procedure

1. Place the wax beans (with liquid) and butter in a saucepan and bring to a simmer.
2. Add the pimientos and stir gently.
3. Season with salt and pepper and serve.

BEET RECIPES

Beets are native to Europe, Africa, and Asia. In early times, they were raised for their leaves rather than the root. Leaves may be simmered or sautéed and can be added separately to many preparations. Small leaves can be added to salads. In fact, the roots only achieved popularity when they became larger and more tasty through cultivation of the plant. They are now in general cultivation chiefly for their succulent roots, which provide not only a tasty table vegetable, but are also a source of sugar. They are available in red, white, and yellow varieties.

Beets of good quality are smooth and free from growth cracks and blemishes. Small beets are more desirable than larger ones mainly because they present a better appearance when served. Fresh beets and canned beets give good results in most beet preparations.

If convenience is a major factor, canned beets should be used. They can be purchased in many forms: sliced, diced, quartered, julienned, and whole. The small, whole, rosebud beets are most desirable. The flavor of orange blends well with and improves the flavor of beets. Orange juice or orange zest (grated orange rind) can be used with excellent results.

Rosebud Beets in Orange Juice

25 servings

Ingredients

1 No. 10 can whole rosebud (small) beets
1 qt orange juice
2 oz sugar
¼ c orange peel, grated
2 oz cornstarch
¼ c cider vinegar

Procedure

1. Drain the liquid from the beets and place in a saucepan.
2. Add half of the orange juice and the sugar, vinegar, and grated orange peel. Bring to a simmer.
3. Dissolve the cornstarch in the remaining orange juice. Pour slowly into the liquid, stirring constantly until the mixture becomes slightly thickened and smooth.
4. Add the drained beets, and again bring to a boil.
5. Check the seasoning and serve.

Hot Pickled Beets

25 servings

Ingredients

1 No. 10 can small beets, whole or sliced, drained
1 qt beet juice or beet juice and water
8 oz onions, sliced into thin rings
⅓ c salad oil
1 tbsp salt
8 whole cloves
1 c sugar
2 bay leaves
3 c cider vinegar

Procedure

1. Place the salad oil in a saucepan and heat.
2. Add the onions and sauté until slightly tender.
3. Add the beet juice, cloves, bay leaves, salt, sugar, and vinegar. Simmer for 15 min and then remove the spices.
4. Pour the hot liquid over the drained beets and bring to a boil again.
5. Adjust the seasoning and serve.

Spiced Beets

25 servings

Ingredients

1 No. 10 can beets, sliced or diced
3 pt beet juice and water
1½ pt cider vinegar
1 c sugar
10 whole cloves
4 whole allspice
4 cinnamon sticks
1 tsp salt

Procedure

1. Place the beet juice and water, vinegar, salt, and sugar in a saucepan and bring to a boil.
2. Add the cloves, allspice, and cinnamon. Simmer for 10 min.
3. Strain the hot liquid over the beets and again bring to a boil.
4. Adjust the seasoning and serve.

Buttered Beets

25 servings

Ingredients

1 No. 10 can whole beets, sliced or diced
6 oz butter
1 tsp sugar
salt and white pepper to taste

Procedure

1. Place the beets and beet juice in a saucepan and heat.
2. Add the butter and sugar and continue to heat.
3. Season with salt and pepper to taste and serve.

BROCCOLI RECIPES

Broccoli is a variety of the cabbage species. It is closely related to cauliflower, but has a small green head rather than a firm white head. For many years it was grown only in Europe, but today it has become a very important crop and popular table vegetable in the United States. The broccoli head consists of green leaves and small green flower buds. The entire broccoli, which consists of the stalk, leaves, and flower bud clusters, is eaten.

The two types of broccoli are *cauliflower broccoli*, which forms a head similar to cauliflower, and *Italian broccoli*, which does not form a head. The cauliflower broccoli is the most popular in the United

States. Broccoli of good quality possesses a compact head with the flower buds unopened. The length of the head and stalk should be about 5″. The color should be a deep green with no yellow showing. (Yellow indicates poor quality.)

Broccoli is one of the more difficult vegetables to cook because of the difference in tenderness between the stem and the tip or flower head. Many cooks solve this problem by peeling the tough and sometimes woody stem with a vegetable peeler so the stem becomes tender when the tip, or flower head, is done. Broccoli is sometimes split lengthwise to

aid in rapid cooking. The less broccoli is handled or disturbed during cooking, the better the results. Cook in a steam table pan (2″ deep stainless steel pan) preferably in a compartment or convection steamer under approximately 5 lb of pressure. Cook until stems are just tender, remove from the steamer, season, and take directly to the steam table. Keep covered with a clean, wet cloth while holding on the steam table. Fresh broccoli is usually preferred, but frozen broccoli spears give excellent results. Raw broccoli separated into flowerets is a popular garnish for the salad bar.

Buttered Fresh Broccoli

25 servings

Ingredients

4 bunches fresh broccoli (approx 12 lb)
boiling water to cover
5 tsp salt
8 oz butter

Procedure

1. Remove any outer leaves and tough stems from the broccoli. Wash thoroughly in cold salt water. Do not bruise the head.
2. If the stalk is tough or woody, split it in half lengthwise and peel with a vegetable peeler.
3. Place the broccoli in a bake pan and cover with boiling water. Add the salt and cover the pan with a wet towel to keep the broccoli submerged in the liquid so it will cook uniformly.
4. Simmer on top of the range until the stalks become tender. Do not overcook or the heads will come apart.
5. Drain the liquid from the broccoli and serve, dressing each portion with melted butter.

Note: Broccoli may also be cooked by steam pressure. The broccoli is prepared in the same manner as for simmering.

Broccoli with Cheese Sauce

25 servings

Ingredients

4 bunches fresh broccoli (approx 12 lb)
boiling water to cover
2 tsp salt
2 qt cheese sauce

Procedure

1. Remove any poor outer leaves and tough stems from the broccoli. Wash thoroughly in cold salt water. Do not bruise the head.
2. Split the stalks halfway up and peel if tough or woody.
3. Place the broccoli in a bake pan and cover with boiling water. Add the salt and cover the pan with a wet towel.
4. Simmer on top of the range until the stalks become tender. Do not overcook or the heads will be destroyed.
5. Remove from the range, drain the liquid from the broccoli, dress each portion with a generous amount of cheese sauce, and serve.

Note: Frozen broccoli spears can be used in place of fresh broccoli. If desired, the broccoli may be cooked in the pressure steamer.

Broccoli Amandine

25 servings

Ingredients

6 lb frozen broccoli spears
boiling water to cover
2 tsp salt
6 oz butter
5 oz almonds, sliced

Procedure

1. Thaw the broccoli spears and place in a bake pan.
2. Cover with boiling water, add the salt, and cover with a wet towel.
3. Simmer on top of the range until the stalks become tender. Do not overcook or the heads will be destroyed.
4. Place the butter in a skillet or saucepan and heat.
5. Add the sliced almonds and sauté until golden brown.
6. Drain the liquid from the broccoli, sprinkle each portion with the toasted almonds, and serve.

Note: Fresh broccoli can be used in place of frozen broccoli. If desired, the broccoli may be cooked in a pressure steamer.

Broccoli Hollandaise

25 servings

Ingredients

6 lb frozen broccoli spears
boiling water to cover
2 tsp salt
2 qt hollandaise sauce

Procedure

1. Thaw the broccoli spears and place in a bake pan.
2. Cover with boiling water, add the salt, and cover the pan with a wet towel to keep the broccoli submerged in the liquid so it will cook uniformly.
3. Simmer on top of the range until the stalks become tender. Do not overcook or the heads will come apart.
4. Drain the liquid from the broccoli and serve, dressing each portion with a generous amount of hollandaise sauce.

Note: Fresh broccoli can be used in place of frozen broccoli. If desired, the broccoli may be cooked in the pressure steamer.

Procter and Gamble Co.

BRUSSELS SPROUTS RECIPES

Brussels sprouts belong to the cabbage family and look like miniature cabbages. They originated in Belgium, near the city of Brussels, from which they took their name. The sprouts grow about 1″ in diameter and are attached to the long stalks of the plant, which sometimes grows 3′ long. Brussels sprouts are available on the market from October through May. A sprout of good quality is firm, compact, and possesses a good green color. Puffy-looking sprouts provide poor eating qualities.

Brussels sprouts are prepared and used in the same manner as cabbage; however, the sprouts possess a superior flavor. Brussels sprouts are very delicate and become difficult to handle when cooked; therefore, care must be taken when seasoning and serving. Because of this delicate nature, best results can be obtained by steaming if a steamer is available. Before removing them from the heat, test for doneness by cutting one of the larger sprouts in half and testing the center. Many times, the outer portion is done but the center remains raw. Remove from the heat when still slightly on the tough side. Never overcook. Cook as close to serving time as possible.

Brussels Sprouts in Sour Cream

25 servings

Ingredients

6 lb fresh brussels sprouts (approx 3 qt)
boiling water to cover
2 tsp salt
4 oz onions, minced
4 oz butter
2 lb sour cream
salt and white pepper to taste

Procedure

1. Remove wilted and discolored outer leaves and trim the stems of the brussels sprouts.
2. Soak in cold water for approximately 30 min and drain.
3. Place in a saucepan and cover with boiling water. Add the 2 tbsp of salt and simmer until tender.
4. Drain off the liquid and hold.
5. Place the butter in a separate saucepan and heat.
6. Add the minced onions and sauté without coloring until tender.
7. Stir in the sour cream and heat slightly while stirring gently.
8. Pour the sour cream mixture over the cooked brussels sprouts and fold gently.
9. Season with salt and white pepper and serve.

Buttered Brussels Sprouts

25 servings

Ingredients

6 lb fresh brussels sprouts (approx 3 qt)
boiling water to cover
2 tsp salt
8 oz butter
salt and pepper to taste

Procedure

1. Remove wilted and discolored outer leaves and trim the stems of the brussels sprouts.
2. Soak in cold salt water for approximately 30 min and drain.
3. Place in a saucepan, cover with boiling water, add the salt, and simmer until tender.
4. Drain off part of the liquid and add the butter.
5. Season with salt and pepper and serve.

Brussels Sprouts Hollandaise

25 servings

Ingredients

6 lb fresh brussels sprouts (approx 3 qt)
boiling water to cover
2 tsp salt
2 qt hollandaise sauce

Procedure

1. Remove wilted and discolored outer leaves and trim the stems of the brussels sprouts.
2. Soak in cold salt water for approximately 30 min and drain.
3. Place in a saucepan and cover with boiling water. Add the 2 tbsp of salt and simmer until tender.
4. Remove from the range and serve, dressing each portion with a generous amount of hollandaise sauce.

Brussels Sprouts Au Gratin

25 servings

Ingredients

5 lb frozen brussels sprouts
boiling water to cover
2 tsp salt
2 qt cream sauce
8 oz American cheese, grated
salt and white pepper to taste
paprika as needed

Procedure

1. Place the partly-thawed sprouts in a saucepan and cover with boiling water. Add the 2 tsp of salt and simmer until tender.
2. Drain the liquid from the sprouts and pour the hot cream sauce over the sprouts. Stir gently so the sprouts will not be broken.
3. Season with salt and white pepper.
4. Turn into a baking pan and sprinkle the cheese over the top. Sprinkle the cheese lightly with paprika.
5. Bake in a preheated oven at 375°F until the cheese melts and becomes slightly brown.
6. Remove and serve.

Note: Fresh brussels sprouts may be used if desired.

CABBAGE RECIPES

Cabbage, cultivated since prehistoric times, has been developed into many varieties, including the common cabbage, brussels sprouts, broccoli, and kale. In cultivation, the common cabbage grows into a head comprised of many leaves. The head may be spherical, conical, or flat-shaped. The leaves may be green or red and wrinkled or smooth, depending on the cabbage being cultivated. The common cabbage is categorized in five classes:

- *Early cabbage* is also known as pointed cabbage because the head comes to a slight point.

- *Danish cabbage* has a firm, solid head and is grown mainly for winter use.

- *Domestic cabbage* consists of both early and late varieties. Its head is not as firm and solid as the Danish cabbage.

- *Red cabbage* has a firm, solid head with reddish-purple leaves. The flavor of the red cabbage is much stronger than that of the other cabbages.

- *Savory* or *curly cabbage* has a very loose head with wrinkled, dark green leaves. It has a flavor that is milder than the other cabbages.

Buttered Cabbage

25 servings

Ingredients

10 lb cabbage
1 tsp salt
boiling ham stock to cover
8 oz butter
salt and pepper to taste

Procedure

1. Remove outer leaves. Trim and wash the cabbage.
2. Cut the heads into 25 wedges. Do not remove the core unless there is an excessive amount on the wedges.
3. Place the wedges in a braiser or deep baking pan. Cover with the boiling ham stock.
4. Add the salt and simmer uncovered until the cabbage is tender but still retains its shape.
5. Drain off a small amount of the liquid. Add the butter.
6. Season with salt and pepper and serve.

Fried Cabbage: Chinese Style

25 servings

Ingredients

10 lb cabbage, trimmed, cored, and shredded coarse
2 c chicken stock
¼ c soy sauce
¾ c salad oil
2 garlic cloves, minced
1 tbsp salt
¼ c sugar

Procedure

1. Prepare the chicken stock.
2. Place the oil and garlic in a braising pot and heat.
3. Add the shredded cabbage and cook for 10 min, stirring occasionally.
4. Add the chicken stock, sugar, salt, and soy sauce. Simmer until the cabbage is tender.
5. Check the seasoning and serve.

Shredded Cabbage

25 servings

Ingredients

10 lb cabbage
boiling water to cover
2 tbsp salt
4 oz onions, sliced thin
8 oz butter (variable)
2 tsp celery seed
salt and pepper to taste

Procedure

1. Remove outer leaves. Trim and wash the cabbage.
2. Cut the heads into six wedges, remove the core, and shred coarsely.
3. Place the cabbage in a sauce pot and cover with boiling water. Add the salt and simmer for 10 min or until slightly tender. Drain well.
4. Place the butter in a large skillet or braiser until melted.
5. Add the onions and sauté until slightly tender.
6. Add the cooked, shredded cabbage and sauté for approximately 5 min longer.
7. Add the celery seed and blend thoroughly.
8. Season with salt and pepper and serve.

CARROT RECIPES

The carrot is a biennial plant that has an orange-yellow tapering root. The early crop of carrots produces the most desirable carrots. They are generally small, mild in flavor, extremely tender, and have a bright color. The late crop of carrots has a more pronounced flavor, a deeper color, and a much coarser texture. They are a very simple vegetable to cook because they do not break easily and very few things affect their bright, attractive color.

Carrots must be peeled before they are cooked. This can be done using one of many methods. They can be scraped raw with the blade of a paring knife or vegetable peeler or they can be placed in a pot, covered with water, brought to a quick boil, removed from the heat, and drained. The skin can then be easily removed by scraping. After peeling, the carrots may be cut into the desired shape and size or they may first be cooked, cooled, and then cut. When cooking, cover with water, add salt and sugar, and simmer until just tender. Let cool in the juice they were cooked in, and use this natural juice when reheating.

Carrots can also be cooked by steaming, but simmering on the range gives best results. Fresh carrots always produce the best finished product. Canned and frozen carrots, however, also give good results if prepared and handled properly and, of course, are most convenient.

Buttered Carrots

25 servings

Ingredients

6 lb fresh carrots, peeled
boiling water to cover
1 tsp salt
6 oz butter
salt and sugar to taste

Procedure

1. Slice the carrots diagonally. Place in a saucepan.
2. Cover with boiling water, add 1 tsp of salt, and simmer until tender.
3. Add the butter and season with sugar and additional salt if needed, then serve.

Note: One No. 10 can of whole or sliced carrots may be used if desired.

Candied Carrots

25 servings

Ingredients

6 lb fresh carrots, peeled
boiling water to cover
1 tbsp salt
8 oz dark brown sugar
4 oz butter

Procedure

1. Cut the carrots into strips 1″ long and ½″ thick.
2. Cover with boiling water, add 1 tbsp of salt and brown sugar, and simmer until tender.
3. Add the butter and serve.

Creamed Carrots

25 servings

Ingredients

6 lb fresh carrots, peeled
boiling water to cover
1 tsp salt
2 qt cream sauce
2 oz butter
salt and white pepper to taste

Procedure

1. Slice or dice the carrots. Place in a saucepan.
2. Cover with boiling water, add 1 tsp of salt, and simmer until tender. Drain thoroughly.
3. Pour the hot cream sauce over the drained carrots, return to the range, and bring to a simmer.
4. Season with salt and white pepper.
5. Add the butter, blend, and then serve.

Carrots Vichy

25 servings

Ingredients

6 lb fresh carrots, peeled
boiling water to cover
1 tsp salt
8 oz butter
1 tbsp sugar
salt and white pepper to taste

Procedure

1. Slice the carrots crosswise fairly thin and place in a saucepan.
2. Cover with boiling water, add 1 tsp of salt, and simmer until tender.
3. Add the butter and sugar.
4. Season with salt and white pepper.
5. Serve garnished with chopped parsley.

CAULIFLOWER RECIPES

Cauliflower is a variety of the common cabbage. It has a white or cream-white head, which is the only part prepared for food. Cauliflower of good quality has a smooth white color with no blemishes on the surface of the head and is heavy for its size. The outer leaves that protect the delicate head should be green, firm, and fresh in appearance. If the leaves should grow up through the head of the cauliflower, this does not indicate a poor-quality cauliflower. It only hinders the appearance.

Before cooking, all leaves, blemishes, and core are removed. The cauliflower is then washed thoroughly. When cooking, the head may be left whole or separated into flowerets. It is usually cooked by steaming or simmering in liquid with milk or lemon juice and salt present to help preserve its white color. When steam is used, color is usually affected and overcooking occurs.

When testing for tenderness, test the solid stem if cooking flowerets. If the whole head is cooked, test the area where the core was removed. Cauliflower, like broccoli, overcooks quickly if one is not alert during the cooking period. Remove from the heat while still slightly tough. After cooking, if not serving immediately, cool rapidly by placing ice over the drained cauliflower and hold in a cool place covered with the liquid in which it was cooked and milk.

Buttered Cauliflower

25 servings

Ingredients

6 lb fresh cauliflower, trimmed
boiling water to cover
2 tbsp salt
1 tsp lemon juice
8 oz butter
salt and white pepper to taste

Procedure

1. Place enough boiling water in a sauce pot to cover the cauliflower. Add 2 tbsp of salt and the lemon juice. Bring to a boil.
2. Add the cauliflower and simmer until the base of the head is tender. Drain and break into segments.
3. Serve with a teaspoon of melted butter over each serving.

Cauliflower with Cheese Sauce

25 servings

Ingredients

6 lb fresh cauliflower, trimmed
2 tbsp salt
1 tsp lemon juice
boiling water to cover
2 qt cheese sauce

Procedure

1. Prepare cheese sauce.
2. Place enough boiling water in a sauce pot to cover the cauliflower. Add 2 tbsp of salt and the lemon juice. Bring to a boil.
3. Add the cauliflower and simmer until the base of the head is tender, then drain thoroughly.
4. Serve, dressing each portion with a generous amount of cheese sauce.

Note: Frozen cauliflower may be used if desired.

Cauliflower Parmesan

25 servings

Ingredients

6 lb fresh cauliflower, trimmed
boiling water to cover
2 tbsp salt
1 tsp lemon juice
6 oz Parmesan cheese
2 oz bread crumbs
6 egg yolks
¼ c flour
3 c milk
salt and white pepper to taste

Procedure

1. Place enough boiling water in a sauce pot to cover the cauliflower. Add 2 tbsp of salt and the lemon juice. Bring to a boil.
2. Add the cauliflower and simmer until the base of the head is tender, then drain thoroughly.
3. Place the cooked cauliflower in a baking pan and break into segments. Combine the Parmesan cheese and bread crumbs and sprinkle over the cauliflower.
4. Break the eggs into a container and beat slightly, blend in the flour, and stir until smooth. Add the milk and blend thoroughly.
5. Pour this mixture over the cooked cauliflower and season with salt and white pepper.
6. Bake in a preheated oven at 350°F for 20 min to 30 min until the surface becomes golden.
7. Remove from the oven and serve.

Cauliflower Creole

25 servings

Ingredients

6 lb fresh cauliflower, trimmed
2 tbsp salt
1 tsp lemon juice
boiling water to cover
2 qt creole sauce
salt and pepper to taste

Procedure

1. Prepare the creole sauce.
2. Place enough boiling water in a sauce pot to cover the cauliflower. Add 2 tbsp of salt and the lemon juice. Bring to a boil.
3. Add the cauliflower and simmer until the base of the head is tender, then drain thoroughly.
4. Break the cooked cauliflower into segments and place in a saucepan.
5. Pour the hot creole sauce over the cauliflower and bring to a boil.
6. Remove from the range, season with salt and pepper, and serve.

Note: Frozen cauliflower may be used if desired.

Cauliflower Au Gratin

25 servings

Ingredients

6 lb frozen cauliflower
1 tbsp salt
1 tsp lemon juice
boiling water to cover
2 qt cream sauce
10 oz Cheddar cheese, grated
salt and white pepper to taste
1 tsp paprika

Procedure

1. Prepare cream sauce.
2. Place the partly-thawed cauliflower in a saucepan and cover with boiling water.
3. Add 1 tbsp of salt and the lemon juice, simmer until the base of the cauliflower is tender, then drain thoroughly.
4. Place the cooked cauliflower in a baking pan, pour over the hot cream sauce, and season with salt and white pepper.
5. Sprinkle on the grated cheese and top with a sprinkle of paprika.

CORN RECIPES

Corn is a cereal grass grown mainly for food and livestock feed. It is native to both North and South America and was the chief source of food for the American Indians. In world grain production, it ranks behind rice and wheat; however, in the United States it is the chief grain crop. The two main types of corn on the market are white, or sweet, corn and yellow corn. Both types of corn are extremely popular in the commercial kitchen. The white corn, however, is generally sweeter and is more tender and superior in flavor to the yellow corn. White corn requires less cooking time than yellow corn.

Corn deteriorates rapidly after it is picked. The sugar in corn begins to lose its sweetness and converts into starch as soon as it is picked, resulting in a loss of flavor and tenderness.

To prepare corn shortly after it has been picked in the field is the ideal way; however, in most cases, this is not possible. Immediately upon receipt, the corn should be shucked and should be kept covered with a damp cloth in the coldest part of the refrigerator. All corn requires little cooking. Heat it thoroughly or bring it to a boil and it is ready to be served. In fact, corn will toughen if overcooked. Use caution if cooking in a compartment steamer. It is easy to overcook.

Corn Marie (Corn and Tomatoes)

25 servings

Ingredients

2½ lb frozen corn, whole kernel
boiling water to cover
1 tsp salt
1 tsp sugar
½ No. 10 can whole tomatoes
2 tbsp cornstarch (variable)
2 tsp sugar
salt and white pepper to taste

Procedure

1. Place the corn in a sauce pot and cover with boiling water.
2. Add 1 tsp of sugar and 1 tsp of salt. Simmer the corn 3 min to 5 min. Remove from the range and drain thoroughly.
3. Place the tomatoes in a saucepan, reserving ½ c of the juice to dissolve the cornstarch, and bring to a boil.
4. Dissolve the cornstarch in the tomato juice and pour slowly into the boiling tomatoes, stirring constantly until slightly thickened and smooth.
5. Add the drained corn and the second teaspoon of sugar and bring to a simmer.
6. Season with salt and white pepper. Remove from the range and serve.

Note: The amount of cornstarch used may vary depending on the desired thickness.

Corn on the Cob

25 servings

Ingredients

25 ears of white or yellow corn
2 tbsp sugar
boiling water to cover
1 qt milk
butter as needed

Procedure

1. Remove the husk and all the corn silk and trim.
2. Place the milk, sugar, and enough boiling water to cover the corn in a sauce or stockpot and bring to a boil.
3. Add the corn and cook slightly, covered, for 4 min to 8 min or until done.
4. Remove from the range and hold the corn in the liquid until ready to serve.
5. Serve one ear of corn with corn holders and a generous portion of butter.

National Broiler Council

Note: If the corn is old, add more sugar to the water and cook for a longer time. Cook all corn as near to serving time as possible.

Corn O'Brien (Mexican)

25 servings

Ingredients

5 lb frozen corn, whole kernel
boiling water to cover
1 tsp salt
1 tbsp sugar
6 oz green pepper, diced
3 oz pimientos, diced
4 oz butter

Procedure

1. Place the corn in a sauce pot and cover with boiling water.
2. Add the sugar and salt. Simmer 3 min to 5 min and remove from the range.
3. Poach the diced green pepper in a separate saucepan until just tender. Drain.
4. Add the green peppers and pimientos to the cooked corn.
5. Add the butter, season with additional salt and sugar, and serve.

Corn Pudding

25 servings

Ingredients

1 No. 10 can cream-style corn
1 qt milk
10 eggs
4 oz butter
2 oz flour, sifted
1 tbsp sugar
1 pt bread crumbs
salt and white pepper to taste

Procedure

1. Place the eggs in a stainless steel container and beat slightly.
2. Blend in the milk and corn.
3. Add the butter and stir.
4. Add flour, sugar, and bread crumbs. Blend thoroughly.
5. Season with salt and white pepper.
6. Place in a buttered baking pan and bake in a preheated oven at 350°F for approximately 45 min or until the pudding becomes slightly solid.
7. Serve 3 oz to each portion.

Trade Tip

To roast vegetables, the natural sugars must be quickly caramelized through high, intense heat. Use fresh, evenly-sliced vegetables. Oil lightly and place in the pan in a single layer, apply herbs, and roast at 450°F, turning once.

Butter Succotash (Corn and Lima Beans)

25 servings

Ingredients

2½ lb frozen corn, whole kernel
2½ lb frozen fordhook lima beans
boiling water to cover
2 tsp salt
1 tbsp sugar
4 oz butter
salt and white pepper to taste

Procedure

1. Place the corn and lima beans in separate saucepans. Cover both with boiling water. Add 1 tsp of salt and the sugar to the corn. Add 1 tsp of salt to the lima beans. Simmer both until tender.
2. Remove both vegetables from the range and combine. Pour off excess liquid.
3. Add the butter, season with salt and white pepper, and serve.

Creamed Corn

25 servings

Ingredients

5 lb frozen corn, whole kernel
boiling water to cover
1 tsp salt
1 tsp sugar
2 qt cream sauce
2 oz butter

Procedure

1. Prepare the cream sauce.
2. Place the corn in a sauce pot and cover with boiling water.
3. Add the sugar and salt and simmer 3 min to 5 min. Remove from the range, then drain thoroughly.
4. Add the cream sauce and blend.
5. Add the butter, season with additional salt and sugar, and serve.

EGGPLANT RECIPES

Common eggplant is a purple-skinned, pear- or egg-shaped vegetable. Many feel that the shape resembles an egg rather than a pear because of its name. The edible flesh has a watery, grayish pulp and a flavor somewhat similar to that of a cooked oyster. Other varieties are available, such as Chinese eggplant, which is long, skinny, and green. Eggplants of good quality are medium-sized and have a rich purple color, a smooth skin, and are firm but light for their size. Eggplants that are picked just before reaching full growth are the best.

Eggplant can be prepared by various cooking methods. It can be fried, sautéed, baked, grilled, or stewed with excellent results. Usually the eggplant is peeled before cooking; however, when baking, the skin may be left on if desired. After peeling, keep the eggplant covered with a damp cloth because when the flesh is exposed to air it discolors rapidly.

Eggplant, like most vegetables, should be prepared as close to serving time as possible. This is especially true of fried or sautéed eggplant.

Scalloped Eggplant and Tomatoes

25 servings

Ingredients

4 lb eggplant, peeled
2 qt canned whole tomatoes
4 oz onions, minced
4 oz butter
2 lb bread, cut into ½″ cubes, then toasted
1½ oz sugar
1 tsp basil
¼ c Parmesan cheese
salt and pepper to taste

Procedure

1. Cut the eggplant into ½″ cubes. Simmer in boiling salt water for 8 min to 10 min, drain thoroughly, and hold.
2. Place the butter in a saucepan and heat.
3. Add the onions and sauté until tender.
4. Add the tomatoes, sugar, and basil. Simmer for 5 min.
5. Stir in the bread cubes and cooked eggplant. Season with salt and pepper.
6. Pour into a lightly greased bake pan and sprinkle with Parmesan cheese.
7. Bake in a preheated oven at 350°F until the top becomes slightly brown and the bread cubes absorb most of the liquid.

Fried Eggplant

25 servings

Ingredients

6 lb eggplant, peeled
1 lb flour
5 eggs
3 c milk
2 lb bread crumbs
2 tsp salt
½ tsp pepper

Procedure

1. Cut the eggplant in half lengthwise. Slice crosswise ¼" to ½" thick.
2. Place the slices in cold salt water to prevent discoloration, then drain well.
3. Season the flour with salt and pepper, add the sliced eggplant, and coat thoroughly.
4. Beat the eggs slightly and add the milk, making an egg wash. Remove the eggplant from the flour and deposit the slices in the egg wash, coating thoroughly.
5. Remove the slices from the egg wash and place in the bread crumbs. Press crumbs on firmly. Fry in deep fat at 350°F until golden brown.
6. Serve two or three slices for each portion.

Eggplant Creole

25 servings

Ingredients

6 lb eggplant, peeled
boiling water to cover
1 tbsp salt
2 qt creole sauce
salt and pepper to taste

Procedure

1. Cut the eggplant into ½" cubes. Place in a sauce pot and cover with boiling water. Add 1 tbsp of salt and simmer for 8 min to 10 min, then drain thoroughly.
2. Pour over the creole sauce, return to the range, and simmer.
3. Season with salt and pepper and serve.

Trade Tip

The visual appearance of vegetables provides a clue to their doneness. When vegetables are done, they will brighten and look waxed.

KOHLRABI RECIPES

Kohlrabi is a variety of cabbage, but it also has an association with the turnip. The edible portion is a swollen, turnip-like green or purple stem that grows above the ground, whereas the turnip grows below the ground. The plant is harvested while the stems are small and tender. Young kohlrabi has the best eating qualities. It should possess a very pale green color and the diameter of the bulb should not be more than 3". If allowed to mature, it develops a woody texture and a very strong flavor. The taste of kohlrabi resembles the combined flavors of turnips and cabbage.

To cook kohlrabi, after peeling and cutting into desired uniform pieces, place in a saucepan, cover with boiling water, add salt, and simmer until just tender. If cut into fairly small pieces, kohlrabi will blend well with peas, green beans, and lima beans.

Creamed Kohlrabi

25 servings

Ingredients

8 lb kohlrabi, peeled
boiling water to cover
2 tbsp salt
2 qt cream sauce
2 oz butter
salt and white pepper to taste

Procedure

1. Cut the peeled kohlrabi into ½" cubes.
2. Place in a saucepan and cover with boiling water.
3. Add the salt and simmer for 30 min or until tender, then drain thoroughly.
4. Pour the hot cream sauce over the cooked kohlrabi and fold gently.
5. Season with salt and white pepper.
6. Blend in the butter and serve.

Buttered Kohlrabi

25 servings

Ingredients

8 lb kohlrabi, peeled
boiling water to cover
2 tbsp salt
8 oz butter
salt and pepper to taste

Procedure

1. Cut the peeled kohlrabi into ¼″ thick slices or dice into ½″ cubes.
2. Place in a saucepan and cover with boiling water.
3. Add the salt and simmer approximately 30 min or until tender.
4. Drain off excess liquid and add the butter.
5. Adjust the seasoning with salt and pepper and serve.

LIMA BEAN RECIPES

Lima beans originated in Peru and were probably named after Peru's principal city, Lima. Francisco Pizarro, Spanish explorer and conqueror of Peru, took the lima bean plant back to Europe, and from there it spread to its worldwide popularity.

The two types of lima beans on the market are the small, or baby limas, and the large, or fordhook limas. The fordhook limas are the most popular because they are plumper, have a more tender skin, possess a superior flavor, and present a more desirable appearance. Lima beans may be purchased frozen, canned, fresh, or dried. Frozen limas are most popular because they are convenient and present an excellent flavor and appearance. Fresh and frozen limas are cooked following the procedure given for green vegetables. Canned limas only require heating and dried limas should be soaked overnight in water before simmering in water until tender. Drain and add fresh water.

Never overcook limas because they become mushy. Ham and butter give extra flavor to lima beans. Sautéed onions provide additional flavor.

Lima Beans with Bacon

25 servings

Ingredients

5 lb frozen fordhook lima beans
boiling water to cover
1 tsp salt
1 lb bacon, diced medium
4 oz onion, minced
1 tbsp chopped chives
salt and white pepper to taste

Procedure

1. Place the lima beans in a saucepan and cover with boiling water.
2. Add the salt and simmer until slightly tender.
3. Drain off any excess liquid.
4. Place the diced bacon in a separate saucepan and cook to a light brown crackling.
5. Add the minced onions and sauté until tender. Do not brown.
6. Pour in the cooked limas and the remaining juice.
7. Add the chives and bring to a simmer.
8. Season with salt and white pepper and serve.

Lima Beans and Mushrooms

25 servings

Ingredients

5 lb frozen fordhook lima beans
boiling water to cover
1 tsp salt
8 oz butter
1 lb mushrooms, sliced
4 oz onion, minced
salt and white pepper to taste

Procedure

1. Place the lima beans in a saucepan and cover with boiling water.
2. Add the salt and simmer until slightly tender.
3. Drain off any excess liquid.
4. Place the butter in a separate saucepan and heat.
5. Add the onions and sauté slightly.
6. Add the mushrooms and continue to sauté until they are tender.
7. Pour in the cooked lima beans and the remaining juice and simmer for 5 min.
8. Season with salt and white pepper and serve.

Buttered Lima Beans

25 servings

Ingredients

5 lb frozen fordhook lima beans
boiling water to cover
1 tsp salt
8 oz butter
salt and white pepper to taste

Procedure

1. Place the lima beans in a saucepan and cover with boiling water.
2. Add the salt and simmer until slightly tender.
3. Drain off any excess liquid and add the butter.
4. Season with salt and white pepper and serve.

Creamed Lima Beans

25 servings

Ingredients

5 lb frozen fordhook lima beans
boiling water to cover
1 tsp salt
2 qt cream sauce
2 oz butter
salt and white pepper to taste

Procedure

1. Place the lima beans in a saucepan and cover with boiling water.
2. Add the salt and simmer until slightly tender. Drain thoroughly.
3. Pour the hot cream sauce over the cooked limas. Stir gently.
4. Add the butter. Season with salt and white pepper and serve.

OKRA RECIPES

Okra is grown and used mainly in the southern states and is also known as gumbo plant. It is a fuzzy, tapered, many-seeded pod vegetable that contains from 5 to 12 sides and grows approximately 3″ long. Okra of good quality has a fresh green color, a plump appearance, and snaps easily when bent. Okra is used in soups and stews and as a vegetable preparation. When used as a vegetable, it is usually blended with tomatoes.

Okra may be purchased fresh, canned, or frozen. Canned and frozen okra are most popular because they are more convenient than the fresh product. When cooking fresh or frozen okra, prepare it like other green vegetables. If it is used in a soup or stew, cook it in those preparations.

Buttered Okra

25 servings

Ingredients

6 lb fresh okra
boiling water to cover
¼ c cider vinegar
1 tbsp salt
8 oz butter
salt and pepper to taste

Procedure

1. Cut off the stems and wash the okra in cold salt water. Cut into ½″ pieces.
2. Place in a sauce pot and cover with boiling water.
3. Add 1 tbsp of salt and the vinegar. Simmer for approximately 20 min or until tender, then drain.
4. Add the butter and stir until thoroughly melted.
5. Season with salt and pepper and serve.

Okra and Tomatoes

25 servings

Ingredients

5 lb frozen okra
4 oz butter
2 qt canned whole tomatoes
8 oz minced onions
salt and pepper to taste

Procedure

1. Place the butter in a saucepan and heat.
2. Add the onions and sauté until slightly tender. Do not brown.
3. Add the tomatoes and bring to a boil.
4. Add the okra and simmer until the okra is tender.
5. Season with salt and pepper and serve.

Fried Okra

25 servings

Ingredients

5 lb frozen okra
1 c flour
½ c cornmeal
½ c cooking oil
salt and pepper to taste

Procedure

1. Cut the stems off the okra. Cut into ½" pieces.
2. Mix the flour and cornmeal. Add salt and pepper. Coat the okra thoroughly in the flour and cornmeal mixture.
3. Sauté rapidly in the cooking oil until golden brown. Do not overcook. Drain and cover.
4. Serve while hot.

ONION RECIPES

Onions are of two basic types: dry onions and green onions. The three types of dry onions are categorized according to color: red, yellow, and white. Red onions are usually used in cooking, especially Italian cuisine. They have also gained some popularity as a salad bar garnish. The yellow onions, Spanish and Bermuda being the two main varieties, contain a mild, sweet flavor and although they are popular in general food preparation, they are best fried, sautéed, or sliced thin and served raw. The medium-sized white and yellow onions are strong in flavor but are preferred for baking and boiling because they retain their shape after cooking and the white color presents a desirable appearance.

Green onion varieties include scallions, shallots, leeks, and chives. Scallions contain a small bulb and a strong flavor. They are usually served as a relish and eaten raw. Shallots have a bulb that consists of several cloves resembling the garlic bulb. They have a very strong flavor and are usually used in the preparation of stews and sauces. They are grown in sandy soil and must be washed well. Leeks have long, flat, wide, green stems and little or no bulb. The stems are the usable part and are desirable because their very delicate flavor is perfect for salads and certain soups. Leeks are also used when decorating other foods for buffet display. They make perfect leaves and stems for flowers and can also be used for borders. Chives are small, green, onion-flavored sprouts that are long and thin. They possess a very desirable mild onion flavor, so they are suitable for dips, salads, and certain entrees and sauce preparations. Onions can be sautéed, broiled, baked, boiled, fried, or grilled with equally fine results. They blend well with other foods and can provide additional flavor for most food preparations.

Onions are one of the most popular vegetables. They are chopped, sliced, diced, ringed, and served in combination with many other vegetable dishes as well as in soups and salads. Onions should be stored in a cool temperature, low light, and well-ventilated area.

French-Fried Onion Rings

25 servings

Ingredients

5 lb Bermuda or Spanish onions, peeled, cut in ¼" slices, separated into rings
ice water to cover
1 lb cake flour

Batter

5 eggs
1 pt milk
3 tsp baking powder
1 lb cake flour, sifted
1 tsp paprika
1 tsp salt

Procedure

As soon as the onion rings are sliced, they should be placed in ice water to prevent them from bleeding (losing water). Keep them in the ice water while preparing the batter.

1. Break the eggs in a stainless steel container. Beat slightly.
2. Add the milk and blend.
3. Combine the flour, paprika, salt, and baking powder and sift.
4. Add the dry ingredient mixture to the milk-egg mixture. Blend thoroughly until the batter is smooth.
5. Remove the onion rings from the ice water and drain thoroughly.
6. Place them in the flour and dip into the batter.
7. Fry in deep fat at 350°F to 360°F until golden brown.
8. Drain and serve five to eight rings per portion.

Creamed Onions
25 servings

Ingredients

1 No. 10 can small whole onions
2 qt cream sauce
2 oz butter
salt and white pepper to taste

Procedure

1. Place the onions and liquid in a saucepan. Heat, then drain thoroughly.
2. Add the hot cream sauce and fold in gently so the onions do not break or become mashed.
3. Add the butter and season with pepper.
4. Serve two or three onions to each portion.

Broiled Onions
25 servings

Ingredients

10 Bermuda or Spanish onions, peeled, cut into five slices each
salt as needed
sugar as needed
paprika as needed

Procedure

1. Place the onion slices on sheet pans. Place in the steamer and steam approximately 7 min to 10 min or until the onions are slightly tender.
2. Remove and sprinkle each onion with salt, sugar, and paprika.
3. Place under the broiler and brown.
4. Serve two slices to each portion.

Buttered Onions
25 servings

Ingredients

6 lb small white onions, peeled
boiling water to cover
1 tbsp salt
8 oz butter
white pepper to taste

Procedure

1. Place the peeled onions in a saucepan and cover with water.
2. Add the salt and simmer for approximately 30 min or until tender.
3. Drain off half the liquid and add the butter.
4. Season with white pepper.
5. Serve two onions to each portion.

National Onion Association

Onions can be sautéed, broiled, baked, fried, or grilled with good results.

PEA RECIPES

Peas are the edible seeds of a pod-bearing vine cultivated to a considerable extent as a field crop in the northern United States and Canada. The plant withstands light frosts and is therefore planted very early in the spring. Peas, along with green beans, are one of the most popular green vegetables. Their popularity is due to a number of reasons. They have excellent appearance when served, a good taste, are easy to prepare and serve, and blend well with certain other vegetables. Because they are easy to serve and satisfy a majority of the people, they are usually the vegetable selected when catering a large party or when preparing the menu for a busy day such as Mother's Day or Easter.

A variety of peas are available on the market. They vary in size, taste, and shape; however, the variety is of little importance unless purchasing canned peas, then the variety is generally stated on the label, such as "sweet," or "early June." Frozen peas, which are the popular choice on today's market, are not labeled but are usually of the same variety because they are always uniform in color, shape, and size.

Peas require very little cooking time. Frozen and fresh peas are cooked using the recommended method for green vegetables. Remove them from the heat when they are still on the firm side. Avoid overcooking. If the cooked peas are to be held for a period of time, chill them quickly by adding ice to the liquid. This helps preserve color and flavor. Canned peas, which lack color, only require heating.

Snow peas, which are tender edible pea pods, are becoming popular even though they are a little more expensive than the average vegetable. They are picked before the peas inside the pod have developed. The pods are approximately 3½″ to 4″ long, curved, have a fairly smooth surface, and are fleshy. They may be purchased fresh or frozen and are prepared the same as other green vegetables. They are often used in Chinese cuisine.

Buttered Peas

25 servings

Ingredients

5 lb frozen peas
boiling water to cover
1 tbsp salt
4 oz butter
1 tbsp sugar

Procedure

1. Thaw, place the peas in a saucepan, and cover with boiling water.
2. Add the salt and simmer until tender.
3. Drain off excess liquid. Add the butter and sugar and serve.

Creamed Peas

25 servings

Ingredients

5 lb frozen peas
boiling water to cover
1 tbsp salt
2 qt cream sauce

Procedure

1. Thaw, place the peas in a saucepan, and cover with boiling water.
2. Add the salt and simmer until tender. Drain thoroughly.
3. Blend in the hot cream sauce. Adjust the seasoning and serve.

Peas and Carrots

25 servings

Ingredients

5 lb frozen peas
boiling water to cover
1 tbsp salt
1 tbsp sugar
2 lb fresh carrots, peeled, diced small
boiling water to cover
4 oz butter

Procedure

1. Thaw, place the peas in a saucepan, and cover with boiling water.
2. Add the salt and sugar. Simmer until tender.
3. Place the diced carrots in a separate saucepan, cover with boiling water, and simmer until tender.
4. Combine the cooked peas and carrots. Drain off any excess liquid.
5. Add the butter, adjust the seasoning, and serve.

Minted Peas

25 servings

Ingredients

5 lb frozen peas
boiling water to cover
1 tbsp salt
1 tbsp sugar
4 oz butter
¼ c chopped mint

Procedure

1. Thaw, place the peas in a saucepan, and cover with boiling water.
2. Add the salt and sugar. Simmer until tender.
3. Drain off any excess liquid.
4. Add the butter and chopped mint.
5. Adjust the seasoning and serve.

Trade Tip

Microwaving is a good choice for preparing vegetables. Because of the short cooking time, few nutrients are lost and colors and flavors are brighter. Chop, slice, or cut vegetables uniformly so they will cook evenly. Place around the edge of a microwave-safe dish for faster cooking.

RICE RECIPES

Rice, which seems to have originated in southeast Asia, is now grown in all parts of the world where warm and moist conditions exist. The seed or grain of the plant is a white grain enclosed by a layer of bran surrounded by a brown husk. Rice marketed as white rice has had the husk and bran removed by special machines. The rice kernel is then polished to improve the appearance.

Rice marketed as brown rice is dried and cleaned with the husk and bran remaining on the kernel.

Since most of the vitamins found in rice are contained in the husk, the brown rice is richer in nutritional value. Three rices on the market are long grain, medium grain, and short grain. They all contain the same food value but differ in size and texture of the grain. Therefore, each requires a slightly different cooking time.

Rice Pilaf

25 servings

Ingredients

1 qt raw rice
2 qt hot chicken stock
6 oz onions, minced
6 oz butter
1 small bay leaf
salt to taste
yellow color as needed, if desired

Procedure

1. Place the butter in a fairly small braising pot and melt.
2. Add the onions and sauté slightly. Do not brown.
3. Add the rice and continue to sauté for 3 min longer.
4. Add the chicken stock and stir.
5. Season with salt, add the bay leaf and yellow color if desired. Stir and bring to a boil.
6. Cover the braising pot and bake in a preheated oven at 400°F for approximately 20 min or until the rice kernels become slightly tender. (Do not stir the rice during the baking period.)
7. Remove from the oven and turn the rice out on a sheet pan. Work in additional butter, remove the bay leaf, and check the seasoning.
8. Place in a bain-marie and serve with a No. 12 dipper.

Note: For rice risotto, add approximately ¼ c of Parmesan cheese when working in the additional butter.

Rice Valencienne

25 servings

Ingredients

1 qt raw rice
2 qt hot chicken stock
4 oz butter
½ c onions, minced
½ c lean ham, minced
3 fresh tomatoes, peeled, diced
1 small bay leaf
¼ tsp thyme
salt and white pepper to taste

Procedure

1. Place the butter in a fairly small braising pot and melt.
2. Add the onions and ham and sauté until the onions are slightly tender.
3. Add the rice and continue to sauté for 3 min longer.
4. Add the chicken stock, tomatoes, thyme, and bay leaf. Stir and bring to a boil.
5. Season with salt and white pepper. Cover the braising pot and bake in a preheated oven at 400°F for approximately 20 min or until the rice kernels become slightly tender. (Do not stir the rice during the baking period.)
6. Remove from the oven and turn the rice out on a sheet pan. Work in additional butter, remove the bay leaf, and check the seasoning.
7. Place in a bain-marie and serve with a No. 12 dipper.

Spanish Rice

25 servings

Ingredients

1 lb raw rice
½ No. 10 can whole tomatoes
1½ qt hot chicken stock
6 oz green peppers, diced small
6 oz celery, diced small
8 oz onions, diced small
2 oz pimientos, diced small
4 oz butter
1 bay leaf
1 tbsp salt
1 tbsp sugar
pepper to taste

Procedure

1. Place the butter in a saucepan and heat.
2. Add the green peppers, celery, and onions and sauté until slightly tender.
3. Add the tomatoes, bay leaf, salt, sugar, and chicken stock. Bring to a boil.
4. Add the rice, cover, and simmer for approximately 20 min or until the rice is tender.
5. Remove the bay leaf and season with pepper.
6. Serve with a No. 12 dipper.

Curried Rice

25 servings

Ingredients

1 qt raw rice
2 qt hot chicken stock
8 oz onions, minced
6 oz apples, minced
8 oz butter
¼ tsp thyme
2 tsp curry powder
salt and white pepper to taste

Procedure

1. Place the butter in a fairly small braising pot and melt.
2. Add the onions and sauté until slightly tender.
3. Add the rice, apples, and curry powder. Continue to sauté 3 more minutes while stirring constantly.
4. Add the thyme and chicken stock and bring to a boil.
5. Season with salt and white pepper. Cover the braiser and bake in a preheated oven at 400°F for approximately 20 min or until the rice is tender.
6. Remove from the oven and turn the rice out on a sheet pan. Work in additional butter and place in a bain-marie.
7. Serve with a No. 12 dipper.

Trade Tip

Rice is one of civilization's most ancient and most important foods. It has been recognized as a food for over 5000 years. Today, this grain helps sustain two-thirds of the world's population.

RUTABAGA RECIPES

The rutabaga is a turnip-like root that grows partly above and partly below the ground. The flesh of the rutabaga is generally yellow in color, although there are some varieties that are white. The rutabaga is similar to the turnip in that it contains about 90% water; however, the flavor is similar to kohlrabi. Rutabagas are usually purchased fresh, and after peeling they are sliced or diced before they are cooked by simmering in water or steaming. Rutabagas, like turnips, are used for vegetable carvings. The firm flesh produces an excellent yellow rose.

Buttered Rutabagas

25 servings

Ingredients

6 lb rutabagas, peeled
boiling water to cover
1 tbsp salt
6 oz butter
1 tsp sugar

Procedure

1. Cut the rutabagas into ½″ cubes.
2. Place in a saucepan and cover with boiling water.
3. Add the salt and sugar and simmer until tender.
4. Drain off any excess liquid and add the butter.
5. Check the seasoning and serve.

Mashed Rutabagas

25 servings

Ingredients

6 lb rutabagas, peeled
boiling water to cover
1 tbsp salt
1 tbsp sugar
6 oz butter
4 oz hot milk or cream
½ tsp white pepper

Procedure

1. Cut the rutabagas into thin slices.
2. Place in a saucepan and cover with boiling water.
3. Add the salt. Simmer until thoroughly cooked and drain.
4. Place in the bowl of the rotary mixer. Using the paddle, mix at second speed until slightly smooth.
5. Add the butter, hot milk or cream, sugar, and pepper. Continue to mix until thoroughly blended and smooth.
6. Remove from the mixer. Place in a bain-marie and serve with a No. 12 dipper.

SAUERKRAUT RECIPES

Unusual as it may seem, kraut supposedly originated in ancient China during the building of the Great Wall. It was included in the workers' food rations to supplement their diet of rice. At that time, the shredded cabbage was fermented in wine. This method was used until about the 16th century, when the people of western Europe found that fermenting cabbage with salt was a far better method. Kraut was introduced to the regions of Germany and northern Europe by the Tartars. It became a favorite of this region and acquired its present name, sauerkraut, which means "sour cabbage," from the Germans.

Sauerkraut is usually not listed on the vegetable menu, but is served with certain meat entrees with which it has been associated for many years. Meat such as spare ribs, pig knuckles, wieners, pork sausage, and bratwurst blend very well with sauerkraut.

Sauerkraut may be purchased in canned or bulk form. The canned kraut is heated in the process of canning but still requires further cooking before it is served. The bulk kraut has only been cured and is in a raw state when purchased. When cooking kraut, add a little caraway seed. Caraway seed is a German favorite and is used in many of their preparations. To reduce the sour taste of the kraut, applesauce and grated raw potatoes may be added.

Sauerkraut Old World Style

25 servings

Ingredients

1 No. 10 can sauerkraut
boiling water to cover
2 tsp salt
1 tsp caraway seed
5 oz applesauce
10 oz onions, julienne cut
1 lb bacon, julienne cut
6 oz raw potatoes, grated
pepper to taste

Procedure

1. Place the julienne cut bacon in a braising pot and sauté until it becomes a light crackling.
2. Add the onions and continue to sauté until slightly tender.
3. Add the caraway seed, sauerkraut, salt, and enough boiling water to cover. Place a lid on the pot.
4. Simmer about 1 hr until the kraut is tender.
5. Add the applesauce and grated raw potatoes. Continue to simmer for 10 min longer.
6. Season with pepper and serve.

Sauerkraut Modern Style

25 servings

Ingredients

1 No. 10 can sauerkraut
boiling water to cover
2 tsp salt
1 tsp caraway seed
1 lb ham hocks
6 oz ham fat or bacon grease
12 oz onions, julienne cut
pepper to taste

Procedure

1. Place the ham fat or bacon grease in a braising pot and heat.
2. Add the onions and sauté until slightly tender.
3. Add the caraway seed, salt, sauerkraut, and enough boiling water to cover. Bring to a boil.
4. Add the ham hocks and press into the center of the kraut. Cover the pot and continue to simmer for approximately 1 hr.
5. Season with pepper and serve.

SPINACH RECIPES

Spinach is the edible young leaves of the spinach plant, which is grown in sandy soil. It grows fairly close to the ground, so when it rains the sandy soil has a tendency to splash on to the leaves and drop into the fairly deep crevices in the leaves. Therefore, spinach must be washed two or three times before using.

The desirable part of the plant is the broad, thick, dark green leaves, which can be used as both a vegetable and a salad green. The undesirable part is the stems attached to the leaves. These stems must be removed before using.

Spinach may be purchased fresh, frozen, or canned. Fresh spinach, which is most plentiful during fall and winter, is usually simmered in water until the leaves are wilted and tender then drained and seasoned. Frozen spinach is best when cooked by steaming. Canned spinach needs only to be heated before serving. Most canned spinach is overcooked and lacks taste and appearance.

Buttered Spinach

25 servings

Ingredients

10 lb fresh spinach
boiling water to cover
2 tsp salt
8 oz butter
salt and pepper to taste

Procedure

1. Wash the spinach thoroughly in cold water at least three times. Remove all stems and any discolored leaves. (Note that the curly leaf spinach is more difficult to clean than the smooth leaf and requires more attention.)
2. Place the spinach leaves in a sauce pot. Cover slightly with boiling water.
3. Add the salt and simmer until the leaves are wilted and tender, then drain.
4. Add the butter. Season with salt and pepper and serve.

Creamed Spinach

25 servings

Ingredients

6 lb frozen spinach, chopped
boiling water to cover
1 tsp salt
1½ qt cream sauce
2 oz butter
salt and white pepper to taste

Procedure

1. Place the partly thawed spinach in a saucepan and cover with boiling water.
2. Add the salt and simmer only until the spinach is tender, then drain thoroughly.
3. Add the hot cream sauce. Stir in gently.
4. Season with salt and white pepper to taste.
5. Add the butter. Blend and serve.

Spinach Country Style

25 servings

Ingredients

6 lb frozen spinach, chopped
boiling water to cover
1 tsp salt
8 oz bacon, diced
4 oz onion, minced
1 pt raw potatoes, diced medium
3 c ham stock (variable)
¼ tsp nutmeg
salt and white pepper to taste

Procedure

1. Prepare the ham stock.
2. Place the partly-thawed spinach in a saucepan and cover with boiling water.
3. Add the salt and simmer only until it is tender, then drain thoroughly.
4. Place the bacon in a separate saucepan and cook until it becomes a crackling.
5. Add the onions and sauté until tender. Do not brown.
6. Add the diced potatoes and ham stock. The amount of ham stock may vary depending on the moisture still present in the spinach. Simmer until the potatoes are tender.
7. Add the spinach and nutmeg, stirring gently so the potatoes do not break. If mixture is too wet, remove some of the liquid.
8. Season with salt and pepper and serve.

Baked Spinach Parmesan

25 servings

Ingredients

6 lb frozen spinach, chopped
boiling water to cover
1 tsp salt
4 oz butter
4 oz onion, minced
1 tsp Worcestershire sauce
6 eggs, slightly beaten
1½ c cracker crumbs (variable)
⅓ c Parmesan cheese
salt and pepper to taste

Procedure

1. Place the partly-thawed spinach in a saucepan and cover with boiling water.
2. Add the salt and only simmer until it is tender, then drain thoroughly.
3. Place the butter in a separate saucepan and heat.
4. Add the minced onions and sauté until slightly tender. Do not brown.
5. Add the cooked spinach, Worcestershire sauce, and Parmesan cheese. Stir until all ingredients are blended. Remove from the range and allow to cool slightly.
6. Add the eggs while stirring constantly.
7. Add the cracker crumbs. The amount may vary, depending on the moisture still present in the spinach. Blend thoroughly.
8. Season with salt and pepper. Place in a buttered baking pan.
9. Bake in a preheated oven at 350°F until the mixture binds and becomes firm.
10. Remove from the oven, cut into squares, and serve with cream sauce accented with additional Parmesan cheese.

SQUASH RECIPES

Squash is the edible fruit of a vine plant belonging to the gourd or cucumber family. The vines, which are similar to pumpkin vines, produce fruits of widely different shapes and sizes. The two classes of squash are summer and winter. The summer squashes, including Italian, patty-pan, and crookneck, are harvested early, before the rind begins to harden. They do not keep well. The winter squashes include Hubbard, acorn, and winter crookneck and have hard rinds and excellent keeping qualities.

Italian squash, also known as zucchini, is one of the most popular squashes used in the commercial kitchen. It is long and narrow with a dark green skin and it somewhat resembles the cucumber. It grows from 3″ to 20″ long, although the very young zucchini, 3″ to 6″ long, possess the best eating qualities. Zucchini is generally very tender and mild in flavor and for this reason can be featured on the menu in a variety of ways.

Patty-pan, also known as scalloped squash, is round and flat with scalloped edges. It ranges from 3″ to 15″ in diameter, has a thin, smooth rind, and a color that may be yellow or white. The flesh of patty-pan squash is watery.

Crookneck squash, so named because of its curved neck, has a thin, yellow, slightly warted skin. The flesh is very tender with a color that varies from yellow to cream.

Hubbard squash is the most popular of the winter squash. It has a globular shape and a hard, warted rind that may be orange, green, or yellow. The green Hubbard is most commonly preferred. The flesh of the Hubbard squash is slightly orange, thick, and fine-grained. Hubbard is the largest squash variety and can grow in sizes exceeding 40 lb.

Acorn squash, also known as Danish squash, is shaped somewhat like an acorn. It has a hard, smooth, dark green rind and a yellow, sweet-flavored flesh.

Winter crookneck squash is similar in almost all respects to the summer crookneck; however, it has a tougher skin and better keeping qualities.

Baked Acorn Squash

25 servings

Ingredients

12 or 13 acorn squash (approx 1¼ lb each)
8 oz butter
1 tbsp salt
6 oz dark brown sugar
water as needed

Procedure

1. Cut the squash in half lengthwise and remove he seeds.
2. Butter the surface of the flesh lightly. Place on a sheet pan skin side down. Add enough water to cover the bottom of the pan about ¼″ deep.
3. Bake in a preheated oven at 350°F for approximately 35 min.
4. Brush the surface of the squash with butter a second time. Sprinkle with salt and brown sugar.
5. Return to the oven and continue to bake until golden brown.
6. Serve one half of the squash to each portion.

Buttered Summer Crookneck Squash

25 servings

Ingredients

8 lb summer crookneck squash
boiling water to cover halfway
2 tsp salt
8 oz butter
salt and white pepper

Procedure

1. Cut off the ends of the squash. Do not peel. Score the squash lengthwise with the tines of a dinner fork.
2. Slice crosswise into ½″ disks. Place in a braising pot.
3. Cover halfway with boiling water, add the salt, cover the braiser, and simmer until the squash is just tender. Drain off half of the liquid.
4. Add the butter. Adjust the seasoning with salt and white pepper and serve.

Note: Young summer squash, with its characteristic soft rind, need not be peeled.

Trade Tip

Pinto beans are the freckled dried beans that are popular in Southwestern cooking, where they are used in chili, refried beans, etc. In the South, pinto beans, cooked with ham hock and onions, are served with corn bread and hot pepper sauce.

TOMATO RECIPES

It is believed that the tomato originated in Peru and was known as *Xitomatles* by the Aztec Indians. This was later modified to the English *tomato*. Tomatoes were first cultivated for their decorative bright red appearance, but were soon found to be edible. It was said to be the belief in colonial New England that the tomato was poison and unfit for consumption. This belief must have persisted for some time because many of the popular preparations that originated in that area do not contain tomatoes. New England baked beans contain maple syrup and no tomatoes. New England and Boston Clam Chowder contain milk and no tomatoes.

Many people consider the tomato a vegetable because they are usually prepared and served as a vegetable. Actually, the tomato is classified as a fruit. There are many varieties of tomatoes, and each differs in plant form, fruit shape, and size. The color is either red or yellow. A tomato of good quality is vine-ripened, firm, well-formed, free of cracks or blemishes, and has a smooth skin and a rich red or sharp yellow color.

During the off season, tomatoes are picked green, packed in wooden boxes, and shipped to market. They ripen in the box without the benefit of sunshine. Although they are wholesome, they lack the color, texture, and flavor of the vine-ripened tomato. They do, of course, make the tomatoes available year-round.

Tomatoes are also sold in puree, sauce, and paste forms. Tomato puree is the pulp of tomatoes cooked down with all the skin, cores, and seeds removed. It is used in stews, gravies, sauces, and soups. Tomato sauce is similar to tomato puree, but the pulp is cooked down to a thicker consistency and is usually flavored with basil or bay leaves. It is used in soup, sauces, and stews. Tomato paste is similar to tomato sauce, but the pulp is cooked down to a very heavy consistency, close to a solid. It is used in meat loaf, soup, pasta, and to thicken stews and sauces.

Stewed Tomatoes

25 servings

Ingredients

1 No. 10 can whole tomatoes
8 oz bread, diced, toasted
6 oz butter

4 oz celery, minced
6 oz onions, minced
2 oz sugar
salt and pepper to taste

Procedure

1. Place the butter in a saucepan and heat.
2. Add the onions and celery. Sauté until tender.
3. Add the tomatoes and sugar. Simmer for 5 min, then remove from the range.
4. Add the toasted bread cubes and season with salt and pepper.
5. Pour into a lightly greased bake pan.
6. Dot with additional butter and bake in a preheated oven at 350°F until the top becomes slightly brown. Serve.

French Fried Tomatoes

25 servings

Ingredients

13 fairly large tomatoes, half-ripe
12 oz flour
2 tsp salt
½ tsp pepper
5 beaten eggs
3 c milk
1 lb bread crumbs (variable)

Procedure

1. Cut each tomato into four thick slices.
2. Season the flour with salt and pepper.
3. Blend the beaten eggs and the milk, making an egg wash. Dip each tomato slice into the egg wash until thoroughly coated.
4. Remove from the egg wash and place in the bread crumbs, pressing slightly.
5. Fry in deep fat at 360°F until golden brown.
6. Serve two slices for each portion.

Baked Tomatoes Italiano

25 servings

Ingredients

25 medium-sized fresh tomatoes, fairly ripe and solid
1 c salad oil
2 tsp sweet basil
1 tsp oregano
salt and pepper to taste

Procedure

1. Remove the stem of each tomato and slice off the bottom.
2. Place in a baking pan bottom side up.
3. Rub each tomato with salad oil.
4. Rub the sweet basil and oregano together and sprinkle over each tomato.
5. Season with salt and pepper.
6. Bake in a preheated oven at 350°F until the tomatoes are just tender.
7. Serve one tomato for each portion.

TURNIP RECIPES

Turnips are a hardy annual or biennial plant belonging to the mustard family and grown for the edible globular white or yellow root they produce. Turnips are native to Europe and some parts of Asia, but they are also cultivated in temperate regions throughout the world.

Turnips of good quality are smooth and firm with very few roots at the base. They are heavy for their size and the tops are green and fresh-looking. The color of the root may be yellow or white, depending on the variety grown. Yellow turnips are stronger in flavor.

Turnips are always cooked by moist heat. Simmering or steaming are the methods used to make the firm texture palatable. Because of their strong flavor, turnips are seldom served by themselves. Usually, they are blended with other foods to limit or control this strong flavor. They are used in stews and ragouts or blended with other vegetables such as peas and green beans in limited quantities. Their firm texture and pure white color make them an excellent choice when carving vegetable flowers. The turnip greens may be simmered and served with the turnips.

Buttered Turnips

25 servings

Ingredients

8 lb white turnips, peeled
boiling water to cover
1 tbsp salt
6 oz butter
1 tsp sugar
½ tsp white pepper

Procedure

1. Dice the turnips into ½″ cubes, place in a saucepan, and cover with boiling water.
2. Add the salt, simmer uncovered until slightly tender, then drain off any excess liquid.
3. Add the butter, sugar, and pepper. Remove from the range and serve.

Creamed Turnips

25 servings

Ingredients

6 lb white turnips, peeled
boiling water to cover
1 tbsp salt
2 qt cream sauce
4 oz butter
salt and white pepper to taste

Procedure

1. Dice the turnips into ½″ cubes, place in a saucepan, and cover with boiling water.
2. Add the salt, simmer uncovered until slightly tender, then drain thoroughly.
3. Add hot cream sauce, stirring gently.
4. Add the butter. Season with salt and white pepper and serve with a No. 12 dipper.

Mashed Yellow Turnips

25 servings

Ingredients

6 lb yellow turnips, peeled
boiling water to cover
2 tsp salt
2 lb potatoes, peeled
boiling water to cover
1 tsp salt
2 tbsp sugar
6 oz butter
salt and white pepper to taste

Procedure

1. Cut the turnips into uniform pieces, place in a saucepan, and cover with boiling water. Add 2 tsp of salt and simmer until tender, then drain thoroughly.
2. Cut the potatoes into uniform pieces, place in a saucepan, and cover with boiling water. Add 1 tsp of salt and simmer until tender, then drain thoroughly.
3. Place the cooked turnips and potatoes in the bowl of the rotary mixer. Using the paddle, mix until smooth.
4. Add the sugar and butter, continuing to mix.
5. Season with salt and pepper and mix.
6. Place in a bain-marie and cover with wax paper.
7. Serve using a No. 12 dipper.

Potato Preparation

Potatoes are one of the most popular food items served in the commercial kitchen. They are served for breakfast, lunch, and dinner and can be prepared in a variety of ways. Baked potatoes are popular when served with fish and beef. French-fried potatoes are popular when served with hamburger, hot dogs, sandwiches, etc.

Potatoes are classified according to their size and shape. The long type includes potatoes such as Idaho potatoes, and the round type includes potatoes such as red or new potatoes. Sweet potatoes are different from white potatoes in both color and taste.

Potatoes are purchased in four market forms: fresh, canned, frozen, and dehydrated. Fresh potatoes are stored in a cool, dry place. They should not be exposed to sunlight in storage, as they will develop chlorophyll, turn green, and create a food hazard. Refrigeration is not required. Fresh potatoes should be purchased in quantities for economy.

Potatoes may be baked, boiled, roasted, steamed, French fried, or sautéed. The cooking method used is determined by the form and type of equipment available.

POTATO PREPARATION

Potato preparations are very important, as very few meals served in the United States are served without a potato preparation. Potatoes are the world's second largest food crop (behind rice). The United States is the world's largest producer of potatoes.

Although the potato is a staple in kitchens across the United States, some people avoid the potato because they believe it is starchy and fattening. This belief has been disproved by nutrition experts. The potato does contain a fairly large percentage of starch. However, the starch found in potatoes is more digestible than other starches. When the potato is properly cooked, the starch particles become very tender. Overcooked or undercooked potatoes can be hard, watery, or soggy and are usually indigestible.

Good-quality potatoes are firm when pressed in the hand. They should be clean and have shallow eyes. When cut, they should display a yellow-white color and moisture should appear on the cut side. Occasionally, potatoes that are starting to sprout or that are partly green are found. These potatoes should *not* be eaten because they contain a small amount of alkaloid poison. Although alkaloid poison is not harmful when consumed in small amounts, it renders the potato unpalatable. Storage in a dark location helps prevent this condition.

POTATO CLASSIFICATIONS

White potatoes are commonly classified into two main categories: the long potato and the round or intermediate potato. The *long potato* becomes mealy (grain-like and easily broken up) when cooked. The long potato, such as the *Idaho potato*, is best for baking, mashing, or French frying. The *round* or *intermediate potato* is hard and stays firmer when cooked. The round or intermediate potato, such as the *red* or *new potato*, is best for boiling, sautéing, or roasting.

Idaho potatoes have excellent eating qualities and a fine white meat. They have a light brown, thin skin and a long shape with shallow eyes. A large percentage of these potatoes are grown in the state of Idaho, hence their name. However, Idaho potatoes are also grown in other states.

Idaho Potato Commission

The Idaho potato is best for baking, mashing, or French frying.

New potatoes or *early crop potatoes* are harvested before reaching full maturity. They have a very thin red skin and contain more water than other potatoes. New potatoes are generally small and uniform in size. New potatoes cannot be stored as long as other potatoes.

Intermediate potatoes, sometimes known as *red potatoes* because they generally have a reddish skin, are very similar to new potatoes in appearance but they are larger. They come on the market in September, between the early and late crop potatoes. They are often used when the preparation calls for a potato that holds together. These potatoes have better storage qualities than new potatoes but do not keep as well as late potatoes.

Old or *late potatoes* are the mature, main crop potatoes. They are harvested in the fall, stored, and used during the year because they have excellent storage qualities. They possess a tough brown skin and less water than the new potato. Late potatoes are more of an all-purpose potato because they can be used in many different ways.

Sweet potatoes are different from the white potato because of the color of their meat and characteristic sweet taste. Sweet potatoes originated in America as the Indians first utilized this root vegetable. Sweet potatoes are sometimes known as *yams*. However, this is an incorrect use of the term. The true yam is similar to but larger than the sweet potato. The yam comes from an entirely different plant. Sweet potatoes are commonly used in the commercial kitchen because they can be prepared in many different ways. They bake, mash, fry, and sauté with excellent results. They complement pork and poultry, providing a new taste experience when served with these items.

POTATO MARKET FORMS

Potatoes may be purchased in four market forms: fresh, canned, frozen, or dehydrated. The market form purchased depends on the need, the time element of preparation, equipment, storage, and cost. Many chefs believe there is no substitute for the fresh potato. However, with new preparation techniques, time- and labor-saving potato forms offer additional options. Frozen, precooked French fries, dehydrated mashed potatoes, and canned, whole, cooked potatoes are used extensively with good results.

Trade Tip

When preparing potato mixtures that will be fried, such as croquette potatoes, sweet potato patties, and potato puffs, if the potato mixture is not stiff or dry enough for successful frying, add dehydrated potato granules to absorb the moisture.

Idaho Potato Commission

Potatoes are available by the bag or carton.

POTATO PREPARATIONS

Potato preparations can be classified as *simple* or *complex*. Potatoes may be fried, French fried, sautéed, mashed, boiled, baked, etc.

Simple Potato Preparations

American fried potatoes (also known as *home fries*): Peel red potatoes and boil until they become tender. Allow to cool overnight in the refrigerator. Slice to a medium thickness and sauté in a steel skillet until golden brown. Serve garnished with chopped parsley. *Note:* New potatoes may be used for home fries.

Sautéed potatoes: Prepare in the same manner as American fries, but slice the potatoes thicker.

Lyonnaise potatoes: Cut onions julienne style and sauté in butter in a steel skillet. Prepare the potatoes in the same manner as American fries. When the potatoes are golden brown, add the sautéed julienned onions and continue to sauté until the onions have blended with the potatoes. Serve garnished with chopped parsley.

German-fried potatoes: Prepare in the same manner as sautéed potatoes.

Hash brown potatoes: Peel red potatoes and boil until they become tender. Allow to cool overnight in the refrigerator. Chop or hash the potatoes into slightly small pieces and sauté in shallow oil until golden brown. Serve garnished with chopped parsley.

Hash brown O'Brien potatoes: Prepare in the same manner as hash brown potatoes, but add small diced pimientos and sautéed small diced green peppers. Serve with chopped parsley.

Hash-in-cream potatoes: Peel red potatoes and boil until they become tender. Allow to cool overnight in the refrigerator. Chop or hash the potatoes into fairly small particles. Prepare a thin cream sauce by adding hot milk to a roux comprised of butter and flour. Add the thin cream sauce to the hashed potatoes. Season with salt and a touch of nutmeg. Serve garnished with a touch of paprika. Cream can be used in place of the cream sauce if a richer product is desired.

Delmonico potatoes: Prepare in the same manner as hash-in-cream potatoes, but add diced, blanched green peppers, diced pimientos, and coarsely-chopped hard-boiled eggs. Place in a bake pan topped with bread crumbs and bake until brown.

Au gratin potatoes: Peel red potatoes and boil until they become tender. Allow to cool overnight in the refrigerator. Chop or hash the potatoes into medium-sized pieces. Prepare a thin cream sauce by adding hot milk to a roux comprised of butter and flour. Season with salt and a touch of nutmeg. Place this mixture in a bake pan and sprinkle the top with grated Cheddar cheese and paprika. Bake in a 350°F oven until the cheese is melted and slightly brown.

Baked potato: Select uniform-sized Idaho potatoes. Wash them thoroughly. Lay on a bake sheet or oven rack and bake at a temperature of 375°F until they become slightly soft when squeezed gently. Baked potatoes can also be wrapped in aluminum or gold foil and baked. However, steam is created inside the foil. When wrapped, the potato will stay hotter after baking; but it is not, strictly speaking, a baked potato because baking is by dry, not moist heat.

POTATO PREPARATIONS

National Potato Promotion Board

HASH BROWNS

National Potato Promotion Board

BAKED

Idaho Potato Commission

FRENCH FRIES

Idaho Potato Commission

MASHED

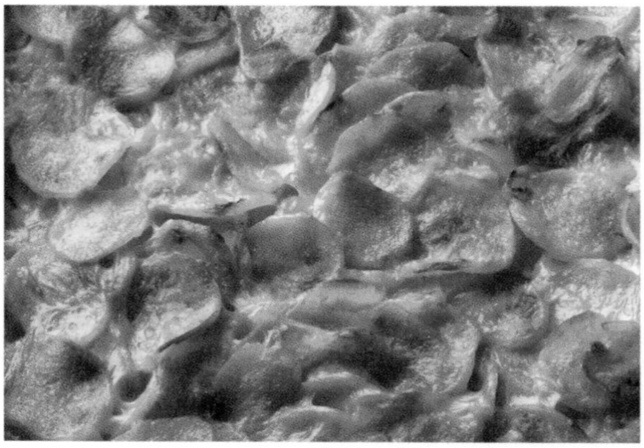

National Potato Promotion Board

SCALLOPED

National Potato Promotion Board

STEAK FRIES

Potatoes are served on the breakfast, luncheon, and dinner menus.

Rissole or *oven brown potatoes*: Place shortening in a roast pan and heat in a 375°F oven until hot. Add the potatoes, sprinkle with paprika, and season with salt and pepper. Return to the oven and roast, turning occasionally, until potatoes become golden brown and tender.

French-fried potatoes: Peel Idaho potatoes and cut about 3″ long and ½″ thick with special cutter or French knife. Place in deep fry baskets and drain thoroughly. Blanch the potatoes in deep fat at a temperature of 325°F until partly done. Do not brown them. Drain and place on sheet pans that have been covered with brown paper and allow to cool. Before serving, fry again in deep fat at a temperature of 350°F until golden brown and crisp. Sprinkle with salt. French-fried potatoes are also known as "pommes frites."

National Potato Promotion Board
French-fried potatoes are available in different shapes.

Long branch potatoes: Follow the method of preparation as for French fries, but cut the potato longer and narrower.

Julienne potatoes: Peel Idaho potatoes and cut into long and very thin strips with a French knife. Drain off any water that may be present and fry the potatoes in deep fat at a temperature of 350°F to 375°F until golden brown and crisp. Sprinkle with salt and serve.

Waffle potatoes: Peel Idaho potatoes and cut with a special waffle cutter. Drain off any water that may be present and fry the potatoes in deep fat at a temperature of 350°F to 375°F until golden brown and crisp. Sprinkle with salt and serve.

Shoestring potatoes: Peel Idaho potatoes and cut with a special cutter that cuts the potatoes into a spring or curl shape. Drain thoroughly and fry in deep fat at a temperature of 350°F to 375°F until golden brown and crisp. Sprinkle with salt and serve.

Riced potatoes: Peel Idaho potatoes, boil or steam them until they are very tender, then drain thoroughly. Force them through a potato ricer and serve sprinkled with melted butter.

New potatoes in cream: Select uniform new potatoes, peel, and cook by boiling or steaming until they are just tender. Drain thoroughly. Prepare a thin cream sauce by adding hot milk to a roux comprised of melted butter and flour. Add the cooked whole potatoes and season with salt and white pepper. Serve garnished with chopped parsley.

Hash lyonnaise: Mince onions and sauté in butter (about ¼ c per quart of potatoes) in a steel skillet until slightly tender. Do not brown. Prepare the potatoes in the same manner as hash brown potatoes, but when the potatoes are golden brown, add the sautéed onions and continue to sauté until the onions have blended with the potatoes. Serve garnished with chopped parsley.

Potatoes fines herbes: Select uniform new potatoes, peel, and cook by boiling or steaming until they are just tender. Drain. Pass them through melted butter, then through fine-chopped herbs. The herb mixture is made by combining parsley, chives, and chervil or tarragon.

Polonaise potatoes: Select uniform new potatoes, peel, and cook by boiling or steaming until they are just tender. Drain. Pass them through melted butter, then through a Polonaise made by combining bread crumbs browned in butter, chopped parsley, and chopped hard-boiled eggs.

Use these ingredients in making the Polonaise (approximately enough to cover 50 potatoes):

1½ lb butter
2 qt fresh bread crumbs
4 hard-boiled eggs, chopped
3 tbsp parsley, chopped

French-fried sweet potatoes: Peel Idaho potatoes and cut about 3″ long and ½″ thick with a special cutter or a French knife. Place in fry baskets and blanch in deep fat at a temperature of 325°F until partly done, then drain. Place on sheet pans that have been covered with brown paper. Allow to cool. Before serving, fry again in deep fat at a temperature of 350°F to 375°F until golden brown.

Swiss potatoes: Peel Idaho potatoes and grate with the medium to large cut of a food grater. Place the shreds of potatoes in cold water and drain thoroughly. Place shortening or butter in a steel skillet and heat.

Add the potato shreds and sauté until they are golden brown and tender. Season with salt and pepper and serve garnished with chopped parsley.

Chateau potatoes: Select medium-sized new potatoes and peel and cut them into the shape of very large Spanish olives. Cook in shortening over a very low flame until they are tender and golden brown. Sprinkle with chopped parsley and serve.

Minute or *cabaret potatoes*: Peel red potatoes, dice to a medium size, drain thoroughly, and blanch in deep fat at 350°F until slightly tender. Finish in a skillet by sautéing them in butter with a small amount of minced garlic.

O'Brien potatoes: Prepare in the same manner as minute potatoes, but omit the garlic and add fine-diced green peppers and pimientos.

Mashed potatoes: Peel Idaho potatoes, steam or boil them in salt water until very tender, and drain thoroughly. Place in the bowl of a mixing machine and mix with the whip or the paddle until fairly smooth. Add hot milk until desired consistency is reached.

Parsley potatoes: Select uniform new potatoes, peel, and cook by steaming or boiling in salt water until they are just tender, then drain thoroughly. Pass them through melted butter and sprinkle with chopped parsley.

Cottage fried potatoes: Select medium-sized red potatoes, peel, and slice very thin. Dry the slices on a cloth and arrange in circles on the bottom of a steel skillet with the potato slices overlapping one another. Reverse each circle until the bottom of the skillet is covered. Proceed in the same manner with a second layer. Cover the potatoes with melted shortening and place in a 400°F oven until the potatoes are tender. Remove from oven, drain off grease. Brown both sides of potatoes on range top. (Potatoes will adhere when cooked.) Tilt onto a platter and garnish with chopped parsley.

Anna potatoes: Prepare in the same manner as cottage fried potatoes, but when the potatoes are partly browned add Parmesan cheese and continue to brown until golden. Tilt onto a platter and serve garnished with Parmesan cheese.

Parisienne potatoes No. 1: Peel Idaho or red potatoes. Cut into round balls the size of a large marble with a parisienne or melon ball scoop. Cook in the steamer until slightly tender. Finish by browning in deep fat at 350°F until golden brown. Sprinkle with melted butter and chopped parsley before serving.

Parisienne potatoes No. 2: Peel Idaho or red potatoes. Cut out into small round balls the size of a large marble with a parisienne or melon ball scoop. Cook in steamer until slightly tender. Sprinkle with melted butter and chopped parsley before serving.

Red skin potatoes: Select medium-sized new potatoes and wash them thoroughly. Place them in a perforated stainless steel pan and cook by steaming until they are just tender. Sprinkle with melted butter or margarine and serve.

Steak fries (sometimes known as *Kentucky fries*): Select uniform-sized Idaho potatoes, wash them thoroughly, and drain. Leaving the skin on, slice them lengthwise approximately ½″ thick with a French or utility knife. Place in the deep fry basket. Drain thoroughly if they were placed in cold water during the cutting period. Blanch the potatoes in deep fat at 325°F until partly done. Do not brown the potatoes. To absorb excess grease, drain and place on sheet pans that have been covered with brown paper and allow to cool. Before serving, fry again in deep fat at a temperature of 350°F until golden brown and crisp.

National Potato Promotion Board

Steak fries are cooked with the skin left on.

Potato rounds: These potatoes are prepared in the same manner as steak fries, but sliced crosswise approximately ½″ thick.

Note: Many potato preparations can be purchased frozen, dehydrated slightly, or completely cooked. If convenience and speed of service is necessary, they will fill these requirements. Commercial sizes are available for the food service.

FRIED AND SAUTÉED POTATO RECIPES

Sweet Potato Amandine

25 servings

Ingredients

1 No. 10 can whole sweet potatoes
1 tsp lemon rind, grated
2 tsp orange rind, grated
½ tsp ground cloves
1 tsp nutmeg
1 tbsp salt
1½ tsp cinnamon
6 egg yolks
½ c brown sugar
1 qt egg wash
½ c dehydrated potatoes (variable)
1 c bread crumbs (variable)
1 c almonds, shaved (variable)

Procedure

1. Drain the sweet potatoes thoroughly. Place on a sheet pan and dry in the oven at 300°F.
2. Place the potatoes, sugar, and all seasoning ingredients in a mixing bowl. Mix using the paddle in the mixing machine until the mixture is fairly smooth and free of lumps.
3. Add the egg yolks and bread crumbs, blend well, then remove from the mixing bowl.
4. Form into miniature sweet potatoes about 1½″ long.
5. Pass through flour, egg wash (6 eggs to 1 qt of milk), and a mixture of bread crumbs and shaved almonds. Press almonds to the potatoes tightly.
6. Fry in deep fat at 350°F until golden brown.
7. Serve two to each order.

Potato Pancakes

25 servings

Ingredients

8 lb red potatoes, peeled
10 oz onions
8 eggs
8 oz cake flour (variable)
1 oz salt
¼ c parsley, chopped
pepper to taste

Procedure

1. Grate or grind the potatoes and onions and pour off all liquid.
2. Beat the eggs slightly and blend into the potato-onion mixture.
3. Add the remaining ingredients and blend well.
4. Cover the bottom of an iron skillet with ¼″ of salad oil or shortening and heat. Fill a kitchen spoon (3 oz) half-full of the potato mixture and deposit the mixture in the shallow grease. Repeat this process until the skillet is filled.
5. Brown one side of each pancake. Turn and brown the other side.
6. Remove the potato pancakes from the skillet and allow to drain.
7. Serve three potato pancakes to each order.

Soufflé Potatoes

25 servings

Ingredients

6 lb small Idaho potatoes, peeled
salt to taste

Procedure

1. Slice the raw potatoes lengthwise about ⅛″ thick on the slicing machine.
2. Soak in very cold water about 1 hr. Drain and dry thoroughly in a towel.
3. Cook in deep fat at 200°F for about 10 min. Remove and cool.
4. Increase temperature of deep fat to 425°F. Add a few potatoes at a time and cook until they puff and become golden brown. Keep potatoes moving while they are frying so they will brown uniformly and puff to their fullest.
5. Sprinkle with salt and serve about 2 oz of the potatoes to each order.

Sweet Potato Patties with Coconut

Ingredients

10 lb fresh sweet potatoes
4 oz brown sugar
½ c bread crumbs (variable)
½ oz salt
1 tbsp cinnamon
1 tsp nutmeg
2 tsp orange rind, grated
2 tsp lemon rind, grated
1 lb 8 oz coconut, shredded or grated
5 egg yolks

Procedure

1. Scrub the potatoes.
2. Place on sheet pans and bake in a preheated oven at 375°F until the potatoes are very tender.
3. Cut potatoes lengthwise, scoop out all the pulp, and discard the skin.
4. Place the pulp in the bowl of the electric mixer. Add the brown sugar, salt, cinnamon, nutmeg, orange rind, lemon rind, egg yolks, and bread crumbs. Mix using the paddle until slightly smooth.
5. Place on a bake pan, cover with wax paper, and refrigerate until firm.
6. Divide into twenty-five 4 oz balls. Form into round, flat patties. Press into the coconut until it adheres to the patties. Place on brown paper on a sheet pan and refrigerate until ready to cook.
7. Sauté each patty in butter until slightly brown on both sides.
8. Arrange on sheet pans and bake in a preheated oven at 350°F for 10 min.
9. Serve one 4 oz patty to each order.

Croquette Potatoes

Ingredients

6 lb Idaho potatoes
½ oz salt
2 oz cornstarch
8 oz egg yolks
pepper to taste
1 qt egg wash
1 pinch of nutmeg, if desired
yellow color to tint, if desired

Procedure

1. Steam or boil potatoes and drain thoroughly.
2. Place on sheet pans and dry in the oven for about 20 min at 275°F to 300°F.
3. Place the potatoes in the mixing bowl and whip smooth using the paddle.
4. Add the egg yolks and cornstarch while mixing at slow speed. Mix thoroughly.
5. Add the salt, pepper and nutmeg, if desired, and mix at slow speed.
6. Tint potatoes with yellow color, if desired, and mix at slow speed.
7. Remove the potato mixture from the mixer and mold into 3 oz portions. Form into desired shape and bread by passing through flour, egg wash (6 eggs to 1 qt of milk), and bread crumbs.
8. Fry in deep fat at 340°F to 345°F until golden brown. Serve one croquette per portion.

Note: If the potato mixture is not stiff or dry enough for successful frying, add dehydrated potato flakes to absorb the moisture.

To prepare potato puffs from this same mixture, use three-fourths croquette mixture to one-fourth pâte de choux. (Ingredients for pâte de choux are given with the recipe for Lorette potatoes.) Blend thoroughly and drop one soup spoonful at a time into deep fat at 350°F.

To prepare potato cheese puffs, add grated Cheddar cheese to the potato puff mixture and fry the same as for potato puffs.

Dauphine Potatoes

Ingredients

8 lb duchess potato mixture
2½ lb pâte de choux mixture
salt and nutmeg to taste

Procedure

1. Prepare the two potato mixtures. (The pâte de choux recipe is given with the recipe for Lorette potatoes.)
2. Mix the duchess potatoes and the pâte de choux mixture at slow speed on the mixing machine.
3. Season with salt and a touch of nutmeg.
4. Place the mixture in a pan and allow to cool.
5. Mold to the shape of corks. Bread by passing them through flour, egg wash (6 eggs to 1 qt of milk), and bread crumbs.
6. Fry in deep fat at 350°F.
7. Serve one potato to each order.

Lorette Potatoes

Ingredients

Lorette Potatoes

8 lb duchess potato mixture
4 lb pâte de choux mixture
salt and white pepper to taste

Procedure

1. Prepare duchess potato mixture.
2. Combine duchess potato mixture and pâte de choux. Blend thoroughly at slow speed using the paddle in the mixing machine.
3. With the pastry bag and star tube, pipe the mixture onto greased paper into 25 large spiral-shaped mounds.
4. Slide the potatoes off the paper into deep fat at 350°F and fry until puffed and golden brown.
5. Drain and serve one mound to each serving.

25 servings

Ingredients

Pâte de Choux Mixture

3 c boiling water
1 c shortening
½ c butter
3 c pastry flour
12 eggs

Procedure

1. Sift together the flour and salt.
2. Combine the shortening, butter, and boiling water in a saucepan.
3. Heat over a low flame until the shortening and butter are melted.
4. Add the flour-salt mixture all at once and stir vigorously over low heat until the mixture forms a ball and leaves the sides of the saucepan. Remove from the heat and allow to cool.
5. Add unbeaten eggs one at a time. Beat gently after each addition until all eggs have been incorporated into the dough.

BAKED POTATO RECIPES

Duchess Potatoes

25 servings

Ingredients

10 lb Idaho potatoes, peeled
4 oz butter
8 egg yolks
1 pinch nutmeg
salt and white pepper to taste
yellow color, if desired

Procedure

1. Cut the peeled potatoes into uniform pieces. Place in a stockpot, cover with water, add salt, and boil until the potatoes are tender. Do not overcook.

2. Drain the potatoes thoroughly. Place in mixing bowl and mix smooth with the paddle.
3. Add the egg yolk and butter and continue to mix.
4. Season with a pinch of nutmeg, salt, and pepper.
5. Add yellow color, if desired.
6. Place the potato mixture in a pastry bag with a star tube and pipe out 25 spiral cone shapes on sheet pans covered with parchment paper.
7. Brush lightly with egg wash or slightly beaten egg whites.
8. Bake in a preheated oven at 400°F to 425°F until potatoes brown slightly.
9. Remove from the oven. Serve one cone per portion.

Special Baked Potatoes

25 servings

Ingredients

25 medium-sized Idaho potatoes
4 oz butter
8 oz bacon, minced
8 oz green pepper, minced
8 oz onions, minced
4 oz pimientos, minced
1 c warm light cream or milk (variable)
salt and white pepper to taste

Procedure

1. Wash potatoes, place on sheet pans, and bake in a preheated oven at 375°F for about 1½ hr or until the potatoes are soft when gently squeezed. Remove from the oven.
2. Cut off the upper portion of the shell lengthwise.
3. Scoop out the pulp of the potato and save the shell. Place the pulp in the mixing bowl and keep hot.
4. Place the bacon in a saucepan and cook until it becomes light brown.

5. Add the green peppers and onions and continue to cook until they become tender. Do not brown.
6. Remove from the heat and add the pimientos.
7. Mix the potato pulp using the paddle on the mixing machine until it is smooth.
8. Add the cooked garnish and butter and continue to mix.
9. Add the warm cream to obtain proper consistency.
10. Mix until thoroughly blended.

11. Season with salt and white pepper and remove from the mixer.
12. Using a pastry bag and star tube, refill the potato shells with the mixture.
13. Sprinkle the top with paprika and additional butter.
14. Bake in a preheated oven at 400°F until the potatoes are heated through and the tops are brown.
15. Serve one potato to each order.
Note: Instead of restuffing the original potato shells, aluminum potato shells may be used.

Sherried Sweet Potatoes

25 servings

Ingredients

25 medium-sized, fresh sweet potatoes
1 lb 8 oz dark brown sugar
8 oz butter
1 qt sherry

Procedure

1. Boil the potatoes until just slightly tender. Run cold water over them and peel.
2. Place the potatoes in a buttered baking pan. Sprinkle the sugar over them.
3. Dot with butter and pour on the sherry.
4. Bake in a preheated oven at 350°F for about 30 min until the potatoes are completely tender.
5. Serve one potato to each order with a small amount of the remaining liquid.

Pommes Elysees

25 servings

Ingredients

10 lb Idaho potatoes, peeled, julienne cut
1 lb mushrooms, sliced, sautéed
1 lb 8 oz cooked ham, julienne cut
1 lb butter
salt and pepper to taste

Procedure

1. Combine the julienned potatoes, sliced mushrooms, and the julienned ham in a mixing container. Season with salt and pepper.
2. Place the butter in a baking pan. Coat the bottom and sides well.
3. Pack the potato mixture into the buttered pan and top with additional pieces of butter. Bake in a preheated oven at 325°F for about 45 min or until the potatoes are tender and the top is a golden brown.
4. Serve 4 oz to 5 oz with a solid kitchen spoon. Garnish with chopped parsley.

Boulangere Potatoes

25 servings

Ingredients

10 lb red potatoes, peeled, cut boat shape (4 or 6 pieces lengthwise)
12 oz carrots, julienne cut
1 lb onions, julienne cut
⅓ c parsley, chopped
1 lb shortening (variable)
salt and pepper to taste

Procedure

1. Place enough shortening in two large steel skillets to cover the bottoms and heat.
2. Add the boat-shaped potatoes and brown slightly. Place in a roast pan or hotel pan. Bake in a preheated oven at 400°F.
3. Sauté the julienned onions and carrots in shortening until slightly tender. Sprinkle over the potatoes when three-quarters done.
4. Continue to roast until the potatoes are completely tender. Remove from the oven and season with salt and pepper.
5. Serve 4 oz per portion with a solid kitchen spoon. Garnish with chopped parsley.

Scalloped Potatoes

25 servings

Ingredients

8 lb potatoes, peeled, sliced ¼″ thick
8 oz butter
6 oz flour
3 qt hot milk
salt and pepper to taste
paprika as needed

Procedure

1. Place the butter in a saucepan and heat.
2. Add the flour, making a roux, and cook slightly.
3. Add the hot milk, whipping rapidly until cream sauce is slightly thickened and smooth. Season with salt and pepper.
4. Place the sliced potatoes in a baking pan and cover with the cream sauce. Sprinkle paprika lightly over the top.
5. Bake in a preheated oven at 350°F until the potatoes are tender and the top is slightly brown.
6. Remove from the oven and serve 4 oz per portion. Serve with a solid kitchen spoon.

National Potato Promotion Board

Italian Potatoes

25 servings

Ingredients

8 lb red potatoes, cooked and diced into ½″ cubes
1 c salad oil
2 lb onions, sliced thin
1 pt stuffed olives, sliced thin crosswise
1 qt chili sauce
1 qt water
salt and pepper to taste

Procedure

1. Spread half of the salad oil in the bottom of a baking pan. Add the diced potatoes and bake in a preheated oven at 450°F until brown. Turn occasionally.
2. Place the remaining oil in a sauce pot and heat.
3. Add the sliced onions and sauté until tender.
4. Add the chili sauce, water, and sliced olives and simmer for 10 min.
5. Pour the sauce over the potatoes and mix gently.
6. Return to the oven and continue to bake for about 15 more minutes or until the potatoes take on a slightly pink color.
7. Serve 4 oz portions with a solid kitchen spoon. Garnish with minced chives.

Scalloped Sweet Potatoes and Apples

25 servings

Ingredients

10 lb sweet potatoes
4 lb tart apples
1 lb dark brown sugar
5 oz butter
½ oz salt
1 pt water

Procedure

1. Boil sweet potatoes until slightly tender, then drain and peel. Cut into ½″ slices.
2. Core the apples and cut into ½″ slices.
3. Arrange potatoes and apples in alternate layers in baking pans.
4. Place the water, brown sugar, salt, and butter in a saucepan. Cook until the sugar is dissolved and mixture is smooth.
5. Pour over the sweet potato and apple slices.

Mushroom Potatoes

25 servings

Ingredients

25 medium-sized Idaho potatoes
2 tbsp salt
1 tsp white pepper
1 qt hot milk (variable)
3 c mushrooms, chopped
2 c butter
2 tbsp fresh lemon juice
50 small mushroom caps

Procedure

1. Wash the Idaho potatoes, place on sheet pans, and bake in a preheated oven at 375°F until done.
2. Remove from the oven, cut the top off the potatoes lengthwise, and scoop out the pulp. Save the potato shells.
3. Place the pulp in a mixing bowl and keep hot.
4. Sauté the chopped mushrooms in 1 c of the butter. When almost done, add 1 tbsp of fresh lemon juice and continue to cook until completely tender.
5. Cook the mushroom caps in the same manner as the chopped mushrooms and keep warm.
6. Add the sautéed chopped mushrooms to the potato pulp in the mixing bowl. Add salt and pepper. Mix in the mixer with the paddle until slightly smooth.
7. Add the hot milk to the potato mixture to obtain proper consistency.
8. Place the mixture in a pastry bag with a star tube and refill the potato shells.
9. Top each potato with two caps of cooked mushrooms, spot with melted butter, and return to the oven.
10. Bake at 375°F until slightly brown. Serve one stuffed potato per portion.

Note: Aluminum potato shells may be used if desired.

Suzette Potatoes

25 servings

Ingredients

25 medium-sized Idaho potatoes
1/3 c chives, minced
8 oz butter
6 egg yolks
1 qt hot light cream (variable)
salt and pepper to taste
Parmesan cheese as needed

Procedure

1. Wash the potatoes, place on sheet pans, and bake in a preheated oven at 375°F until done.
2. Remove from the oven and cut the top off the potatoes lengthwise. Scoop out the pulp, reserving the shell.
3. Place the pulp in a mixing bowl and beat with the paddle until slightly smooth.
4. Add the butter, chives, and egg yolks and continue to mix.
5. Add the hot cream while mixing at slow speed until proper consistency is obtained.
6. Place the mixture in a pastry bag with a star tube and refill the potato shells.
7. Sprinkle Parmesan cheese on each potato.
8. Bake at 375°F until lightly brown. Serve one potato per portion.

Note: Aluminum potato shells may be used if desired.

Princess Potatoes

25 servings

Ingredients

10 lb Idaho potatoes, peeled
8 oz bacon, minced
8 oz green peppers, minced
8 oz onions, minced
4 oz pimientos, minced
8 egg yolks
salt and white pepper to taste

Procedure

1. Cut the peeled potatoes into uniform pieces. Place in a perforated stainless steel pan and steam until they are very tender.
2. Place the minced bacon in a saucepan and cook to a crisp crackling.
3. Add the green peppers and onions and continue to cook until they are tender. Remove from the heat and add the pimientos.
4. Place the cooked potatoes in the stainless steel mixing bowl while they are still hot. Mix at medium speed using the paddle until fairly smooth.
5. Add the cooked bacon and vegetable garnish and continue to mix until well blended.
6. Add the egg yolks slowly while continuing to mix.
7. Season with salt and white pepper.
8. Place the potato mixture in a pastry bag with a star tube and pipe out 25 spiral cone shapes on sheet pans covered with parchment paper.
9. Bake in a preheated oven at 400°F until potatoes are slightly brown.
10. Remove from the oven, sprinkle each potato with melted butter or margarine, and serve.

Delmonico Potatoes

25 servings

Ingredients

8 lb red potatoes, peeled
6 oz green peppers, diced medium
5 oz pimientos, diced medium
6 oz hard-boiled eggs, diced medium
1 gal. medium cream sauce
1 pt coarse bread crumbs
salt and nutmeg to taste
5 oz bacon, diced, cooked to crackling, drained

Procedure

1. Cook potatoes in salt water or steam until slightly tender. Drain and let cool.
2. Place the green peppers in a small sauce pot, cover with water, and simmer until just slightly tender. Drain.
3. Dice the cooled potatoes medium size and place in a stainless steel mixing bowl.
4. Add the green peppers, pimientos, bacon crackling, eggs, and cream sauce. Using a kitchen spoon, mix thoroughly, but gently.
5. Season with salt and a hint of nutmeg.
6. Place in a baking pan and sprinkle bread crumbs over the top.
7. Bake in a preheated oven at 350°F until heated thoroughly and the surface is medium brown.

Potato Skins or Stuffed Potatoes

One of the latest food trends is setting up and presenting the old-fashioned baked potato in unusual and eye-appealing ways. The two most popular methods used are topping the baked potato with various preparations or preparing a potato skin (potato boat or shell), filling it with assorted preparations, and serving it piping hot. These potato skins may be prepared from scratch or purchased frozen in a partly-prepared state. Purchasing the frozen shell will ensure a product that is uniform in shape and size and can be prepared and set up quickly. However, remember that convenience increases food cost. Potato skins may be served as an appetizer or breakfast, luncheon, or dinner entree. Topped baked potatoes are usually featured as a luncheon entree or as an item in a fast food operation.

The fillings that can be used are almost limitless. The following are suggestions for potato skin fillings:

- Assorted seafood, shrimp, or crabmeat Newburg
- Chili or chili con carne topped with Cheddar cheese
- Cream chicken and mushrooms or broccoli
- Chicken a la king
- Creamed ham or dried beef

- Italian meatballs in sauce or meat sauce topped with Provolone cheese
- Scrambled eggs, plain or with ham or crisp bacon
- Chicken or seafood curry
- Taco filling
- Beef stroganoff
- Seafood, shrimp, or crabmeat creole
- Swedish meatballs

The toppings for baked potatoes are limitless, but select items that will blend with the potato pulp. The following are suggestions for topping the baked potato:

- Crisp bacon bits or crackling
- Sour cream and chives
- Mushrooms in sauce
- Fried chicken livers
- Seafood, shrimp, or crabmeat Newburg or creole
- Scallops in white wine sauce
- Creamed chicken with broccoli, avocado, or mushrooms
- Chili and Cheddar cheese
- Italian meat sauce and Provolone cheese
- Sour cream sauce
- Creamed broccoli
- Italian sausage in sauce

Trade Tip

When baking potatoes, wash and dry them thoroughly, then rub each one with salad oil, bacon grease, or melted margarine before placing them in the oven. This will keep the skin soft so the potato skin will not crack or pop open during the baking period. The potato will be easier to cut open when ready to serve.

BOILED POTATO RECIPES

Bouillon Potatoes

25 servings

Ingredients

10 lb red potatoes, peeled, cut boat shape (four to six pieces lengthwise)
4 qt chicken or beef stock
8 oz butter
12 oz onions, julienne cut
8 oz carrots, julienne cut
1 oz parsley, finely chopped
salt and white pepper to taste

Procedure

1. Prepare chicken or beef stock.
2. Place the butter in a stock pot and melt.
3. Add the julienned onions and carrots and sauté without color (do not brown) until slightly tender.
4. Add the stock and bring to boil.
5. Add the potatoes and simmer until the potatoes are just tender. Remove from the range.
6. Add the chopped parsley and season with salt and pepper. Hold in a warm place until served.
7. Serve 4 oz per portion with a pierced kitchen spoon.

Candied Sweet Potatoes

25 servings

Ingredients

Potatoes

10 lb sweet potatoes

Procedure

1. Boil or steam the sweet potatoes until just tender (do not cook completely done).
2. Place in cold water and remove the skins.
3. Remove all discolored blemishes and cut into uniform pieces about 2″ long. Let cool overnight.
4. Remove from the refrigerator and brown slightly in deep fat. Place in a hotel pan and hold until the syrup is prepared.

Ingredients

Syrup

1 pt water
1 lb 8 oz brown sugar
1 lb sugar
2 qt light corn syrup
juice from two lemons
grating (zest) and juice from 4 oranges

Procedure

1. Bring the water to a boil.
2. Add the sugars, stirring until dissolved.
3. Add the corn syrup, lemon juice, orange grating (zest), and orange juice. Bring to a boil, then turn down to simmer for 5 min to 10 min.
4. Pour the syrup over the sweet potatoes and simmer on the range for 5 min.
5. Serve two potato pieces (4 oz) to each order with a pierced or slotted kitchen spoon.

Kartoffel Klosse (Potato Dumplings)

25 servings

Ingredients

8 lb red potatoes, peeled, boiled the day before
14 eggs, beaten slightly
1 lb cornstarch
⅓ c parsley, chopped
1 lb bacon, cooked crisp and minced
1 c onions, sautéed and minced
10 oz flour
1 lb croutons, small cubes
2 lb fresh bread crumbs
1½ lb butter
salt and pepper to taste
2 gal. chicken stock (variable)

Procedure

1. Prepare the chicken stock.
2. Dice the potatoes into very small cubes or chop coarse, and place in a mixing container.
3. Add the eggs, cornstarch, onions, bacon, parsley, croutons, salt, and pepper and mix by hand until thoroughly blended.
4. Form mixture into balls a little larger than a golf ball. Roll each ball in flour.
5. Place the balls into simmering chicken stock and cook for 10 min. Remove using a skimmer.
6. Roll each ball into bread crumbs previously sautéed in butter until golden brown.
7. Serve one ball to each order.

Pasta Preparation

Pasta is a general name for several products made from wheat flour. It is available in many different shapes and sizes. The shape of the pasta used in a preparation is determined by how the sauce used will cling to the pasta. In addition, the shape of the pasta used should complement the appearance of the topping.

All pasta is cooked in boiling water. Oil is added to the boiling water to prevent the pasta from sticking together. The pasta should be checked for doneness frequently as the cooking process nears completion. Pasta should not be overcooked. The noodles must be soft, but must not be mushy.

Pasta is classified into four general categories: long goods, short goods, specialty items, and egg noodles. Long goods are pastas that are long and narrow, such as spaghetti. Short goods are pastas that are short and broad, such as elbow macaroni. Specialty items are pastas that are large and used for special preparations. Egg noodles are pastas that are made with eggs, such as fettuccine.

PASTA PRODUCTS

Pasta products are used in many popular preparations served in food service establishments throughout the world. Some preparations are named after the pasta ingredient used; for example, macaroni and cheese, chicken fettuccine, and lasagna. Pasta products are classified by their size and shape into four general categories:

- *Long goods*: spaghetti, spaghettini, vermicelli, linguine, etc.

- *Short goods*: elbow macaroni, salad macaroni, rigatoni, rotini, etc.

- *Specialty items*: lasagna, manicotti, mostaccioli, jumbo seashells, etc.

- *Egg noodles*: items made with eggs such as fettuccine, twists, large bows, etc.

Pasta products are made from semolina flour, farina flour, wheat flour, or a mixture of these flours. Water and salt are also added. These flours are made from hard wheat, which contains a very high percentage of gluten. *Gluten* is the protein in flour that provides the strength to hold the shape, form, and texture of bakery products when cooked. Pasta products are formed into many different shapes and sizes for variety and function.

Egg noodles are classified as thin, medium, or wide according to width. Egg noodles must contain at least $5\frac{1}{2}\%$ egg solids to meet government regulations. Every pound of noodles must contain approximately the equivalent of two or more eggs.

Pasta Product Forms

There are over 150 pasta shapes and sizes. Each shape is given a different name. Pasta products commonly used in the commercial kitchen include:

Spaghetti: Long, round, solid rods of flour paste. The diameter of the rod is approximately ³/₃₂″. Spaghetti is one of the most popular pastas. It may be used with any sauce.

Spaghettini: Long, round, solid rods of flour paste. Spaghettini is slightly thinner than spaghetti. The diameter of the rod is approximately ¹/₁₆″.

Vermicelli: Long, round, very thin solid rods of flour paste. The diameter of the rod is approximately ¹/₃₂″. Vermicelli is sold in both straight and coiled forms, although the straight is most popular. Vermicelli is used most often in soup.

Macaroni: Straight, round, hollow tubes of flour paste. Macaroni is one of the most popular pastas.

National Pasta Association

Pasta may be purchased in numerous shapes and sizes.

National Cattlemen's Beef Association
ROTINI

National Turkey Federation
LASAGNA

National Cattlemen's Beef Association
VERMICELLI

TORTELLINI

Pasta can be prepared in many different ways.

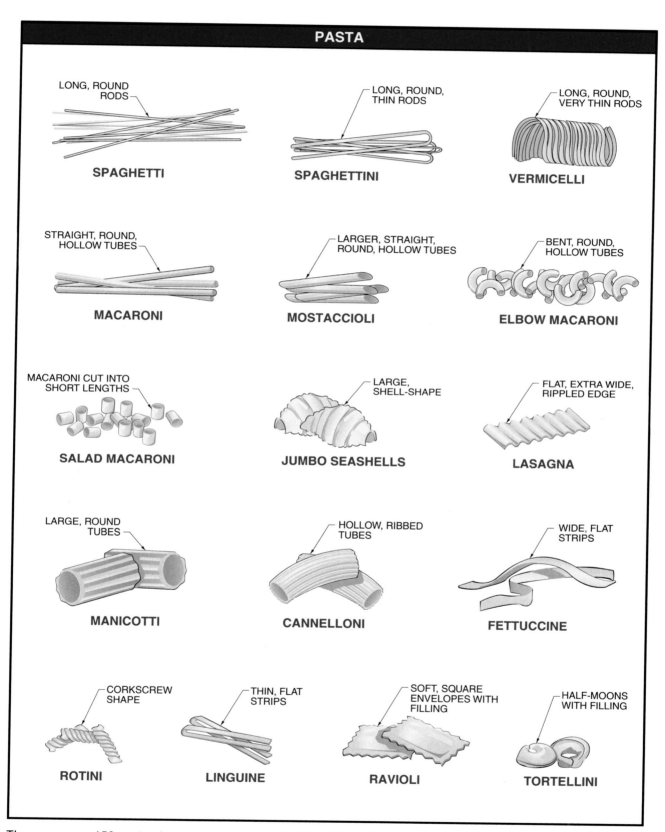

PASTA

SPAGHETTI — LONG, ROUND RODS

SPAGHETTINI — LONG, ROUND, THIN RODS

VERMICELLI — LONG, ROUND, VERY THIN RODS

MACARONI — STRAIGHT, ROUND, HOLLOW TUBES

MOSTACCIOLI — LARGER, STRAIGHT, ROUND, HOLLOW TUBES

ELBOW MACARONI — BENT, ROUND, HOLLOW TUBES

SALAD MACARONI — MACARONI CUT INTO SHORT LENGTHS

JUMBO SEASHELLS — LARGE, SHELL-SHAPE

LASAGNA — FLAT, EXTRA WIDE, RIPPLED EDGE

MANICOTTI — LARGE, ROUND TUBES

CANNELLONI — HOLLOW, RIBBED TUBES

FETTUCCINE — WIDE, FLAT STRIPS

ROTINI — CORKSCREW SHAPE

LINGUINE — THIN, FLAT STRIPS

RAVIOLI — SOFT, SQUARE ENVELOPES WITH FILLING

TORTELLINI — HALF-MOONS WITH FILLING

There are over 150 pasta shapes and sizes, each with a different name.

This highly versatile shape can be topped with any sauce, baked, or put in soups and salads.

Mostaccioli: Straight, round, hollow, ribbed tubes of flour paste. Mostaccioli is similar to macaroni, but with a larger diameter and a diagonal cut.

Elbow macaroni: Bent, round, hollow tubes of flour paste. Elbow macaroni is bent to resemble an elbow.

Salad macaroni: Straight, round, hollow tubes of flour paste cut into short lengths. They are best suited for use in salad production.

Jumbo seashells: Large shell-shaped pasta. Can be served plain or stuffed with meat, poultry, seafood, vegetables, or cheeses.

Lasagna: Flat, extra wide, rippled-edge pasta. Used in the preparation of lasagna.

Manicotti: Large, round tubes of pasta. The diameter of the tube is approximately 1″. Manicotti may be either straight or diagonal cut. They are stuffed with a meat, poultry, or cheese filling and baked in a rich Italian sauce.

Cannelloni: Similar to manicotti, with the differences being the diameter, a ridged outer surface, and the way the ends are cut.

Fettucine: Depending on the manufacturer, it may be a wide, short egg noodle or a slightly slender, long noodle. It is perfect for heavier sauces, such as a white sauce.

Rotini: Shaped like a corkscrew. The twisted shape holds bits of meat, vegetables, and cheeses very well.

Linguine: Long, thin, flat strips of flour paste. Similar to spaghetti, but flat instead of round. This is a good shape for all sauces.

Ravioli: A member of the pasta family, although the dough is soft rather than hard like the majority of pasta products. Ravioli are small, square (approximately 1″ to 1½″) envelopes of soft dough filled with seasoned ground meat, spinach, or cheese. After poaching, they are served with a rich Italian sauce and Parmesan cheese.

Tortellini: A soft dough similar to ravioli dough is used. The filling is the same, but they are formed differently. The very thin rolled dough is cut into two circles and a filling is placed in the center. The circle is folded in half, forming a half moon. The half moon is then formed into a ring by wrapping it around the finger and pressing the ends together.

Uses of Pasta Products

Pasta products are often associated with Italian cuisine because they were popularized by Italians. However, countries all over the world now commonly use pasta products in many recipes. Food service operators have found recipes using pasta products to be an ideal menu item. Most pasta recipes are easy to prepare. Pasta is best when cooked for immediate use, but can be cooked ahead if necessary. Pasta preparations return an excellent profit because the raw food cost is usually low. It is easy to serve and can speed up service.

Cooking Pasta Products

All pasta products are cooked using the same basic procedure. Determine the approximate amount of water needed (1 gal. of water for every pound of pasta). Place the water in a stockpot or steam jacket kettle. Add salt and some olive oil or salad oil to help prevent the pasta from sticking together. Bring the water to a rolling boil. Add the pasta while stirring gently with a paddle. If the pasta is long, such as spaghetti, spaghettini, or vermicelli, it is best not to break it, but to spread it out around the inner wall of the pot while at the same time lifting it gently with the paddle or a kitchen fork. When the water returns to a boil, the lifting and stirring action is only required occasionally during the cooking period.

All pasta should be cooked until it becomes *al dente*, an Italian expression meaning "to the tooth." This refers to the ideal texture of pasta, which should be slightly firm and chewy when bitten. Pasta can be tested by removing a piece from the pot and pressing it between the thumb and forefinger. Spaghetti, spaghettini, and vermicelli can also be tested for doneness by twirling it around the index finger. If it wraps around tightly, it is usually done. Never overcook pasta. Overcooked pasta is very mushy and can lose its shape or fall apart.

A pound of raw pasta yields approximately 3 lb when cooked. Cooking time varies, depending on the shape, size, and quality of the pasta. Always cook pasta uncovered because of the large head of froth that may cause the liquid to boil over. When the minimum cooking time is reached, the pasta should be tested frequently to ensure the proper doneness. See Table I for approximate cooking times.

TABLE I: APPROXIMATE COOKING TIMES FOR PASTA

Pasta	Minutes
Spaghetti	10 – 12
Spaghettini	8 – 10
Vermicelli	5 – 7
Mostaccioli	9 – 11
Elbow macaroni	9 – 12
Salad macaroni	9 – 12
Jumbo seashells	20 – 25
Lasagna	11 – 13
Manicotti	10 – 12
Rigatoni	16 – 18
Rotini	12 – 14
Linguine	9 – 12
Ravioli	12 – 15
Tortellini	10 – 12
Noodles	8 – 14*

*depending on width

Pasta must be processed properly after cooking to prevent sticking. Pour the pasta into a colander. Wash the pasta in cold running water to remove a portion of the starch. Reheat by running hot tap water over the pasta. Shake off excess water and place the pasta in a stainless steel pan. Pour a mixture of half-melted butter or margarine and oil over it. Season with salt and white pepper and toss gently to coat with the oil mixture. Place in a stainless steel steam table pan and place on the steam table. Cover and apply low heat.

For later service, or for use in a cold pasta dish, follow the same cooking procedure, but after draining the cooked pasta in a colander, place it in a plastic or stainless steel container. Cover with cold water and place it in the refrigerator. When it is needed for service, reheat by running hot tap water over it or placing it in a china cap or pasta pan and submerging it in boiling hot water. Pour the heated pasta into a stainless steel mixing bowl, add the oil mixture, season with salt and white pepper, and toss gently. Place in a stainless steel steam table pan and place on the steam table. Cover and apply low heat. If only one order of the cooked pasta is needed, it may be sautéed gently in the oil mixture to reheat.

PASTA SAUCE RECIPES

Tomato Sauce (Marinara)

1½ gal.

Ingredients

6 oz bacon or ham grease
6 oz onions, cut rough
6 oz celery, cut rough
4 oz flour
1 No. 10 can tomato puree
1 pt tomato paste
½ gal. hot ham stock
1 clove minced garlic
1 bay leaf
1 tbsp sweet basil, rubbed
salt and sugar to taste

Procedure

1. Place bacon or ham grease in a sauce pot and heat slightly.
2. Add the garlic, onions, and celery. Sauté until they just start to brown.
3. Add flour, making a roux. Cook 3 min.
4. Add the hot ham stock while whipping rapidly with a wire whip. Bring to a boil.
5. Add the tomato puree and tomato paste while continuing to whip. Bring back to a boil.
6. Add the bay leaf and sweet basil and reduce to a simmer. Let simmer until vegetables are completely cooked.

National Pasta Association

7. Season with salt and a very small amount of sugar to taste. Remove from the range.
8. Strain through a china cap into a 2 gal. stainless steel container and hold for service.

Prosciutto Sauce

1½ gal.

Ingredients

2 lb ground beef
1 lb onions, minced
1 No. 10 can crushed tomatoes
1 pt tomato puree
12 oz prosciutto ham, julienne cut
1 pt dry red wine
2 tsp sugar
1 tsp salt
1 tsp rosemary leaves, crushed fine
1 tsp ground nutmeg
½ tsp black pepper
4 oz Parmesan cheese, grated

Procedure

1. Place the ground beef in a braising pot. Cook until the beef becomes slightly brown. Drain off excess grease.
2. Add the onions, rosemary leaves, and nutmeg. Continue to cook until the onions become slightly tender.
3. Add the tomatoes, tomato puree, red wine, salt, sugar, and pepper. Bring to a boil, then reduce to a simmer. Let simmer until beef is very tender.
4. Add the prosciutto ham and Parmesan cheese. Stir in gently so the pieces will not break. Bring the mixture back to a boil.
5. Remove from the range. Pour into a 2 gal. stainless steel container and hold for service.

Wine Sauce

1½ gal.

Ingredients

1 No. 10 can crushed whole tomatoes
1 pt tomato puree
1 pt dry red wine
2 tsp sugar
1 tsp salt
1 tsp rosemary leaves, crushed fine
½ tsp ground nutmeg
½ tsp black pepper
4 oz Parmesan cheese, grated
1 c olive oil
1 lb onions, minced

Procedure

1. Place the olive oil in a sauce pot and heat slightly.
2. Add the minced onions and sauté until slightly tender. Do not brown.
3. Add all the remaining ingredients except the Parmesan cheese. Bring to a boil.
4. Reduce heat to simmer and simmer for approximately 20 min.
5. Sift in the Parmesan cheese. Continue to simmer for an additional 5 min.
6. Remove from the range and pour into a 2 gal. stainless steel container. Hold for service.

Tetrazzini Sauce

2 gal.

Ingredients

6 lb cooked chicken or turkey, cut into thin strips
8 oz shortening
8 oz butter or margarine
12 oz flour
1 gal. hot chicken stock
8 oz warm cream
4 oz sherry
1 lb fresh mushrooms, sliced, sautéed
salt and white pepper to taste

Procedure

1. Place the shortening and the butter or margarine in a sauce pot and heat until melted.
2. Add the flour, making a roux, and cook the roux just slightly. Stir constantly with a wire whip.
3. Add the hot chicken stock while whipping vigorously with a wire whip until thickened and smooth.
4. Whip in the warm cream and sherry.
5. Add the strips of cooked chicken or turkey and the sautéed mushrooms. Stir in gently with a kitchen spoon.
6. Season with salt and white pepper.
7. Pour into a 3 gal. stainless steel container and hold for service.

Note: Tetrazzini sauce is placed over cooked spaghetti, topped with Parmesan cheese, and browned gently under the broiler before serving.

Clam Sauce

1 gal.

Ingredients

10 oz butter or margarine
8 oz flour
2 qt hot clam juice
½ tsp garlic, minced
6 oz onion, minced
1½ qt light cream
1 pt cooked clams, chopped
4 oz bacon, minced and cooked to a crackling
3 tsp basil
3 tsp oregano
3 tsp rosemary leaves
4 oz Parmesan cheese, grated
8 oz white wine
1 tbsp parsley
salt and white pepper to taste

Procedure

1. Place the clam juice in a sauce pot. Add the herbs.
2. Simmer for approximately 5 min.
3. Remove from the range and strain the liquid through a cheesecloth. Discard the herbs.
4. Place the margarine in a second pot. Heat until the margarine is melted.
5. Add the onions and garlic and sauté until tender. Do not brown.
6. Add the flour, making a roux. Cook the roux just slightly.
7. Add the clam juice and the cream while whipping briskly with a wire whip. Continue to whip until the sauce starts to boil. Reduce to a simmer and simmer approximately 5 min.
8. Add the clams, bacon crackling, cheese, and white wine. Bring to a simmer and remove from the range.
9. Add the parsley and season with salt and white pepper. Pour into a 2 gal. stainless steel container and hold for service.

Note: Clam sauce is usually served over fettuccine or linguine.

POPULAR ITALIAN PASTA ENTREES

Italian Meatballs

120 1 oz meatballs

Ingredients

6 lb ground beef
1 lb onions, minced
1 tbsp garlic, minced
½ c olive oil
8 oz bread crumbs (variable)
8 eggs
½ c Parmesan cheese, grated
¼ c parsley, chopped fine
1 tsp oregano
salt and fresh ground cracked pepper to taste

Procedure

1. Place the olive oil in a sauce pot and heat.
2. Add the minced onions and garlic. Sauté until tender and let cool.
3. Place all the ingredients, including the sautéed onions and garlic, in a large mixing bowl. Mix thoroughly by hand until all ingredients are incorporated.
4. Form into small balls weighing approximately 1 oz each and place on oiled sheet pans.
5. Bake in a preheated oven at 350°F to 375°F until slightly brown and done. Drain off all grease.
6. Take a handful of meatballs and rinse under warm running water to eliminate coagulated blood particles. Place in a stainless steel steam table pan. Continue this process until all the cooked meatballs have been rinsed.
7. Add about ½″ of beef stock to the steam table pan. Cover and reheat in the steam table until ready to serve.

Note: Place the prepared spaghetti in a mound on a dinner plate or in a casserole. Arrange approximately six meatballs on top of the spaghetti. Ladle Italian sauce over the meatballs and serve hot with Parmesan cheese.

Lasagna

Ingredients

1½ gal. tomato sauce
3 lb cooked ground beef, drained
8 lb ricotta, baker's cheese, or dry cottage cheese
4 lb Provolone or Mozzarella cheese
1 lb Parmesan cheese, grated
6 lb lasagna

Procedure

1. Brown the ground beef in a braising pot, pour off all the grease, and add the tomato sauce. Cook until the meat is tender. Set aside for later use.
2. Cook and process the lasagna.
3. Grate the Provolone or Mozzarella cheese by rubbing it across the coarse grid of a box grater.
4. Make up the pans of lasagna by placing a thin layer of meat sauce in the bottom of two 2″ deep × 18″ × 12″ steam table pans. Place two layers of cooked lasagna in opposite directions on top of the meat sauce in each pan. Alternate layers of ricotta, baker's or dry cottage cheese, lasagna, tomato sauce, and Provolone or Mozzarella cheese until the pan is filled and all ingredients are used. Finish with a top layer of lasagna, tomato sauce, and cheese.
5. Bake in a preheated oven at 350°F for approximately 45 min or until heated through and the top is golden brown.
6. Remove from the oven and let set in a warm place for 15 min to 20 min before cutting each lasagna into 24 serving portions.

Pesto Chicken Manicotti

Ingredients

1½ lb manicotti shells
2 lb frozen chopped spinach, thawed and drained
3 lb part-skim ricotta cheese
3 c grated Parmesan cheese, divided
1½ c egg substitute
8 c diced cooked chicken breast
3 tbsp basil
2 tsp pepper
2 qt low-sodium tomato sauce or spaghetti sauce

Procedure

1. Cook pasta according to package directions and drain.
2. Squeeze all water from the spinach.
3. Mix together ricotta cheese, 1½ c Parmesan cheese, and egg substitute. Add all remaining ingredients except sauce.
4. Spoon the mixture into the shells, place in a bake pan, cover with sauce, and sprinkle remaining Parmesan cheese on top.
5. Cover and bake in 350°F oven for 20 min. Remove cover and bake an additional 15 min or until cheese is golden brown.

National Pasta Association

Trade Tip

Dried pasta is packed in bags or boxes. Pasta with a more fragile shape, such as lasagna and manicotti, is often hand-packed to help prevent breakage. Uncooked, dry pasta may be stored for up to one year. Store the pasta in a cool, dry place and use the pasta stored the longest before opening new packages.

Ravioli Squares

To make ravioli squares, divide the dough into eight equal parts (keep dough covered at all times with a moist cloth). Roll out one unit of dough as thin as possible on a floured bench or a piece of heavy canvas until it is about 12″ square. Portion the desired meat filling into 16 mounds using a full teaspoon for each mound or, if using a baker's scale, ¼ oz per mound.

Arrange the mounds of meat on the surface of the rolled-out dough about 1½″ apart. Dip a pastry brush into egg wash (6 eggs to 1 qt of milk) and brush in straight lines between mounds around edges of the dough. Roll out the second unit of dough as thin as the first. Fold in half so it will be easy to lift and unfold over the mounds of filling. Starting in the center, press with fingertips and the sides of the hand around filling and edges to seal the two doughs together.

Use a pizza cutter, pastry wheel, or knife to cut between the mounds. Separate the squares and place on a sheet pan covered with wax paper or freezer paper, wax side up. Form the remaining units of dough. Cover properly and place in the refrigerator if using the next day or freeze for later use.

Setting Up an Order of Ravioli

After the ravioli has been cooked, place the ravioli squares in a steam table pan with about 1″ of chicken stock in the bottom to keep them moist and from sticking to the pan. Six ravioli squares are placed in a flat individual casserole or shirred egg dish for each individual order. They are then topped with a desired sauce such as Italian sauce, meat sauce, or tomato (marinara) sauce. Parmesan or grated Provolone cheese, or both, is sprinkled on top before it is served to the guest.

Trade Tip *Most American-made pasta is produced from durum wheat. The hardness of this particular wheat is excellent for making pasta because it produces a firm product with a consistent cooking quality and a pleasing golden color.*

Provolone Ravioli Gratin

16 servings

Ingredients

2 lb refrigerated ravioli
8 zucchini, sliced ¼″ thick
4 cans cannelini or great northern beans, drained
4 c grated Wisconsin Pepato cheese
2⅔ tbsp coarsely chopped garlic
1 tsp salt
1 tsp pepper
24 slices (1 oz each) Wisconsin Provolone cheese, halved
8 tomatoes, cut into 5 slices each
1 c chopped parsley

Procedure

1. Cook pasta according to package directions and drain.
2. Place zucchini and cannelini into a casserole or gratin dish and top with 2 c Pepato cheese, garlic, salt, and pepper.
3. Alternate layers of Provolone and tomato slices and sprinkle with remaining Pepato cheese and parsley.
4. Bake at 375° for 15 min – 20 min, or until cheese is melted.

Wisconsin Milk Marketing Board

Ravioli Dough

1 qt

Ingredients

2 lb 8 oz bread flour
½ oz salt
6 eggs
1½ oz olive oil
1 lb (1 pt) warm water

Procedure

1. Place the bread flour and salt in the stainless steel mixing bowl. Blend at slow speed using the dough hook.
2. Break the eggs into a stainless steel bowl and beat slightly with a wire whip. Add the warm water and mix until blended with the eggs.
3. Add the egg-water mixture and the olive oil to the blended flour. Mix until a very smooth dough is formed.
4. Turn the dough out of the bowl onto a floured bench. Knead the dough slightly, cover with a cloth, and let rest for 10 min.
5. Form into ravioli squares.

Ravioli Cheese Filling

120 mounds

Ingredients

3 c ricotta or baker's cheese
1 c Parmesan cheese, grated
1 tbsp onions, minced
6 egg yolks, beaten
salt and white pepper to taste

Procedure

1. Place the cheese and onion in a stainless steel mixing bowl. Mix with a kitchen spoon until blended.
2. Add the slightly beaten egg yolks. Mix until incorporated.
3. Season with salt and white pepper.
4. Form ravioli squares.

Note: This cheese filling can also be used in making tortellini or for stuffing manicotti or cannelloni.

Ravioli Meat Filling (Chicken or Turkey)

150 mounds

Ingredients

2 lb cooked chicken or turkey, chopped very fine
8 oz cooked spinach, chopped fine
1 garlic clove, minced
5 eggs
2 oz margarine, melted
2 oz bread crumbs
2 oz Parmesan cheese, grated
salt and pepper to taste

Procedure

1. Place all the ingredients except the eggs in a mixing bowl and blend.
2. Place the eggs in a stainless steel bowl and whip slightly. Add to the mixture. Mix with a kitchen spoon until thoroughly blended.
3. Form ravioli squares.

Ravioli Meat Filling (Beef)

150 mounds

Ingredients

2 lb ground beef
1 garlic clove, minced
4 oz onions, minced
8 oz cooked spinach, chopped fine
5 eggs
2 oz bread crumbs (variable)
2 oz Parmesan cheese, grated
salt and pepper to taste

Procedure

1. Place the ground beef into a sauce pot. Cook over medium heat until thoroughly brown.
2. Add the minced onions and garlic and cook for at least 5 more minutes. Remove from the heat and drain off any liquid.
3. Add the spinach, Parmesan cheese, and bread crumbs. Mix with a kitchen spoon until thoroughly blended.
4. Place the eggs in a stainless steel bowl and whip slightly. Add to the mixture. Mix with a kitchen spoon until thoroughly blended.
5. Season with salt and pepper.
6. Form ravioli squares.

Trade Tip

Pasta is not fattening. According to the United States Department of Agriculture, a ½ c serving of spaghetti contains 99 calories, less than half a gram of fat, and less than 5 mg of sodium.

Tortellini Rings

To make tortellini rings, divide the dough into approximately four equal parts (keep dough covered at all times with a moist cloth). Roll out one unit of dough as thin as possible on a floured bench or a piece of heavy canvas. Using a 2″ biscuit cutter, cut the dough into rounds. Moisten the edge of each round with water or egg wash and place a small ball of filling in the center of each round. Fold the circles in half and press the two edges together until they are tightly sealed. Shape into small rings by stretching the tips of each half circle slightly and wrapping the ring around the index finger. Press the tips together until they are tightly sealed. Form the remaining units of dough. Cover properly and place in the refrigerator if using the next day, or freeze for later use.

After the tortellini has been cooked, place the tortellini rings in a steam table pan with about 1″ of chicken stock in the bottom to keep them moist and from sticking to the pan. Place eight tortellini rings in a flat, individual casserole or shirred egg dish for each order. Top with a desired sauce. Parmesan or grated Provolone cheese, or both, is sprinkled on top before it is served.

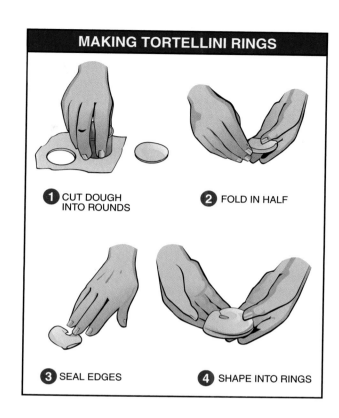

MAKING TORTELLINI RINGS

1. CUT DOUGH INTO ROUNDS
2. FOLD IN HALF
3. SEAL EDGES
4. SHAPE INTO RINGS

Trade Tip

Eight ounces of most uncooked pastas will generally yield 4 c cooked pasta. Noodles will yield less.

Tortellini Dough

1 pt

Ingredients

1 lb 6 oz bread flour
4 eggs
4 egg whites
4 tbsp olive oil
8 oz water (variable)
2 tsp salt

Procedure

1. Place the bread flour and salt in the stainless steel mixing bowl. Blend at slow speed using the dough hook.
2. Place the eggs and egg whites in a stainless steel bowl. Beat slightly. Add all but a small amount of the water. Retain the small amount. Add if the dough is too dry.
3. Add the egg-water mixture and the oil to the blended flour. Mix until a very smooth dough is formed.
4. Turn the dough out of the bowl onto a floured bench. Knead slightly.
5. Form tortellini rings.

Tortellini Meat Filling

150 rings

Ingredients

2 lb 4 oz cooked chicken, chopped fine
10 oz Parmesan cheese, grated
6 egg yolks
¼ tsp lemon rind, grated
¼ tsp ground nutmeg
8 oz spinach, chopped fine
salt and pepper to taste

Procedure

1. Place all the ingredients in a stainless steel mixing bowl. Blend thoroughly by hand.
2. Form tortellini rings.

Carlisle FoodService Products

Pasta salads may be served as a side dish or as an entree.

Pasta Salads

Pasta salads are tasty, inexpensive, colorful, and attractive. They may be prepared in advance and will hold well in the refrigerator. These characteristics qualify them for menu entrees, salad bar, and buffet service. The secret of a good pasta salad is cooking the pasta al dente and combining it with fresh ingredients. These ingredients add flavor and crispness to the salad preparation.

Trade Tip

Cooked pasta may be refrigerated in an airtight container for 3 – 5 days. Add a small amount of oil to keep the pasta from sticking.

Deluxe Macaroni Salad

25 servings

Ingredients

1 lb macaroni, straight or elbow
2 tbsp olive oil
12 oz cooked ham, julienne cut
1 lb celery, diced small
3 oz onion, minced
2 oz green pepper, minced
6 oz Cheddar cheese, shredded or diced small
1 c sweet relish
3 c mayonnaise
2 oz pimientos, minced
2 tbsp parsley, chopped very fine
6 hard-boiled eggs, chopped coarse
salt and white pepper to taste
25 lettuce leaves

Procedure

1. Place a pot containing approximately 2 gal. of water on the range. Add salt and 2 tbsp of olive oil. Bring to a boil.
2. Add the macaroni and cook al dente. Drain and let cool.
3. Add the remaining ingredients except the salt and pepper. Toss gently but thoroughly until mixed.
4. Season with salt and white pepper. Toss a second time.
5. Cover each cold plate with crisp lettuce. Place a mound of salad in the center of each and serve.

Spaghetti Salad

25 servings

Ingredients

1 lb 8 oz spaghetti or vermicelli
2 tbsp olive oil
12 oz Cheddar or longhorn cheese, grated or diced
8 oz relish or sweet pickles, chopped fine
3 c mayonnaise
1 lb celery, julienne cut
8 chopped hard-boiled eggs
6 oz hard salami, julienne cut
salt and white pepper to taste
25 lettuce leaves

Procedure

1. Place a pot of water containing approximately 2 gal. of water on the range. Add salt and 2 tbsp of olive oil. Bring to a boil.
2. Add the spaghetti or vermicelli and cook al dente. Drain and let cool.
3. Add the remaining ingredients except the salt and pepper. Toss gently until thoroughly mixed.
4. Season with salt and white pepper and toss a second time.
5. Cover each cold plate with crisp lettuce. Place a mound of salad in the center of each and serve.

# Minted Pasta and Fruit Salad

Minted Pasta and Fruit Salad

24 servings

Ingredients

1½ lb rotini
12 pears, seeded, cored, and diced
2 tbsp lemon juice
6 c seedless red grapes, cut in half
6 navel oranges, peeled, sliced, and quartered
12 oz Feta cheese, crumbled
3 tbsp chopped fresh oregano or 1 tbsp dried oregano
6 tbsp chopped fresh mint
3 tbsp white wine vinegar
½ c orange juice
6 tbsp vegetable oil
1½ tsp freshly ground black pepper
1½ heads Romaine lettuce, torn into bite-size pieces

National Pasta Association

Procedure

1. Cook pasta according to package directions and drain.
2. In a large bowl, mix pears and lemon juice. Add grapes, oranges, Feta cheese, oregano, mint, and pasta.
3. In a separate bowl, whisk together vinegar, orange juice, oil, and pepper. Add to pasta mixture and mix thoroughly.
4. To serve, place lettuce into salad bowls and spoon pasta mixture on top.

Seashell Salad

25 servings

Ingredients

1 lb 8 oz seashell pasta
2 tbsp olive oil
3 c mayonnaise
¼ c parsley, chopped
¼ c onions, minced
1 tbsp prepared mustard
1 pt celery, minced
6 hard-boiled eggs, chopped
1 lb bacon, diced small, cooked to a crackling
salt and white pepper to taste
25 lettuce leaves

Procedure

1. Place a pot containing approximately 2 gal. of water on the range. Add salt and 2 tbsp of olive oil. Bring to a boil.
2. Add the seashell pasta and cook al dente. Drain and let cool.
3. Add the remaining ingredients except the salt and pepper. Toss gently until thoroughly mixed.
4. Season with salt and white pepper. Toss a second time.
5. Cover each cold plate with crisp lettuce. Place a mound of salad in the center of each and serve.

Salmon Seashell Salad

24 servings

Ingredients

1 lb 12 oz pasta seashells
2 tbsp olive oil
6 hard-boiled eggs, chopped
2 lb salmon, drained, skin and bone removed, meat flaked
1 qt mayonnaise
1 pt celery, minced
½ c green peppers, minced
¼ c minced pimientos
¼ c lemon juice
½ c sweet relish
salt to taste
25 lettuce leaves
1 c grated Parmesan or shredded Provolone cheese

Procedure

1. Place approximately 2 gal. of water in a stockpot. Add the olive oil and a little salt. Bring to a boil.
2. Add the seashells and cook, stirring occasionally until al dente. Drain thoroughly.
3. Place all the ingredients in a large stainless steel mixing bowl. Toss very gently until thoroughly blended.
4. Line each cold salad plate with a leaf of lettuce.
5. Place a mound of salad in the center of each plate.
6. Garnish with Parmesan or shredded Provolone cheese and serve.

Picnic Salad

Ingredients

2 lb elbow macaroni
4 c chopped celery
4 small cucumbers, sliced
24 cherry tomatoes, quartered
1⅓ c radishes, thinly sliced
2 green bell peppers, chopped
1 c green onion, sliced
1 c red onion, chopped
16 hard-boiled eggs, chopped
4 c low-fat mayonnaise
1 c skim milk
4 tbsp prepared mustard
4 tsp salt
1 tsp dill
1 tsp pepper

National Pasta Association

Procedure

1. Cook pasta according to package directions and drain.
2. In a large bowl, combine pasta, celery, cucumber, tomatoes, radishes, bell pepper, green and red onions, and hard-boiled eggs.
3. In a separate bowl, blend mayonnaise, milk, mustard, salt, dill, and pepper. Add to pasta mixture and toss well.
4. Cover and chill thoroughly before serving.

Soups and Stocks

Soups are a liquid food made from stock and nutrients from meat, fish, and vegetables. Stocks are used as a base in the preparation of soups. Stocks are liquids that are used to make soups, sauces, and gravies. Stocks require a great amount of effort to produce a full-flavored base to be used in soups. A stock that is concentrated to one-fourth its volume is a glaze. A stock that is reduced to one-half its volume is a demiglace. Prepared soup bases are available and eliminate the preparation of stock used for a base.

There are many different soups. Soups can be served as an appetizer, as a meal, or as a dessert, depending on the ingredients used. Generally, soups can be classified into four types: clear soups, thick soups, special soups, and cold soups. Clear soups have a clear base and are prepared without starch. Thick soups are soups that are thickened by adding food items containing starch such as rice, potatoes, or barley. Special soups, such as New England clam chowder, are soups that have gained special recognition and an association with a certain locale. Cold soups have a variety of consistencies and are served cold.

SOUPS AND STOCKS

Soup is a liquid food consisting mainly of a broth of meat, seafood, or vegetables. It is an international dish that can be prepared as an appetizer, as a complete meal, or as a dessert, depending upon the ingredients and the size of the portion. In the United States today, soup is most commonly served as an appetizer.

Soups presented as appetizers on the menu are designed to stimulate the appetite for foods to follow. These soups should be light and should not contain food particles that require much chewing. If a heavy soup is served as an appetizer, lighter foods should follow. The rule to follow when serving soup is *heavy soup, light entree*; *light soup, heavy entree*.

When preparing soups, the following procedures should be followed:

- Use a strong, flavorful stock. A soup is only as good as the stock used.

- Cut the garnish small. The garnish should not be filling and should require only minimal chewing.

- Sauté the vegetables slightly when preparing the soup. Sautéd vegetables produce a better flavor.

- For stock based soups, simmer the soup for at least two hours. The soup should be the first item placed on the heat in the morning so it can cook long enough to develop a pronounced flavor.

- Season in moderation. More seasoning can be added when served.

- If flour is added to the vegetables after they are sautéed to make a roux, cook the roux for at least 5 min to avoid a raw flour taste.
- Serve hot soups very hot and cold soups very cold.

Soups are classified differently among chefs. In addition, there are many variations of soups within the various classifications. For simplicity of classification, the four types of soup are clear soups, thick soups, special soups, and cold soups.

Clear Soups

Clear soups include broth (bouillon), consommé, vegetable soup, and borscht. Most clear soups have a clear liquid and are prepared without the use of a starch.

Bouillons are liquids in which meat, vegetables, or seafood has been simmered. They are stronger in flavor and clearer in body than stocks.

Consommés are clarified bouillons or stocks reduced by simmering to increase their richness. Consommés should be clear and should contain a predominant beef or poultry flavor. The word *consommé* comes from the word *consummate,* which means to bring to completion or to make perfect.

Vegetable soups use broth and bouillon and may contain a variety of vegetables. In most cases, vegetable soup is prepared without the use of a starch or thickening agent. If thickening agent is added, it then becomes a thick soup.

Borscht, a soup of Russian origin, is prepared using stock, beet juice, and lemon juice. It is classified as a thin soup because it contains no thickening ingredient. Borscht can be served cold, also classifying it as a cold soup.

Thick Soups

A *thick soup* is a soup that is thickened by adding ingredients to the stock such as potatoes, rice, barley, macaroni, roux, and/or other items containing starch. Thick soups may also be thickened by adding pureed vegetables. Thick soups include cream soups, purees, chowders, and bisques.

Cream soups are soups thickened with a roux and thinned slightly by adding cream or milk. The predominant vegetable or meat ingredient is the name of the soup, such as cream of mushroom, cream of celery, cream of chicken, etc.

Puree soups are thickened by cooking the predominant ingredient, such as split peas, tomatoes, potatoes, dried lima beans, or other vegetables, into a pulp and straining through a fine sieve. Milk or cream should not be added to a puree soup.

Chowders are a cream soup with diced potatoes. They are prepared from fish, shellfish and, in some cases, vegetables. The addition of diced potatoes is the main difference between a cream soup and a chowder. The most popular of all the chowders is clam chowder.

Bisques are similar to chowders but are slightly thicker. They are usually prepared from shellfish and are named according to the shellfish used, such as shrimp bisque, lobster bisque, or crab bisque.

National Fisheries Institute

Seafood soups, stews, and chowders are popular on the luncheon and dinner menus.

Special Soups

Special soups include both thick and thin soups that originated in a certain locale and have retained that association. In some cases, these soups have a great tradition, such as chicken gumbo, which is the traditional soup in the bayous and wetlands of southern Louisiana, and New England clam chowder, which helped the early colonists survive many severe winters. Other famous special soups include minestrone (Italy), English beef broth (England), French onion soup (France), chili bean soup (American Southwest), and gazpacho (Spain).

Cold Soups

Cold soups have become more popular in food service establishments. Cold soups may be classified separately from hot soups or they may be classified with hot soups according to their consistency. Some of the popular cold soups are gazpacho, jellied chicken broth, cold borscht, fruit soups, and vichyssoise.

Garnishes

Most soups are served with a garnish to add eye appeal and/or flavor to the soup. Garnishes most commonly used include vegetables, chopped parsley, minced chives, minced leeks or scallions, croutons (plain or cheese), Parmesan or Romano cheese, meats (finely diced or julienned), sour cream, unsweetened whipped cream, egg dumplings, and marrow dumplings. Garnish may be added to the soup when served or cooked into the soup before served.

Vegetable garnish, which is cooked into the soup, is cut by the soup cook the day before preparation. This is done the day before because soup is the first item placed on the range during the morning preparation. It takes a few hours of simmering to create the desired flavor for most soups. Vegetable garnish is cut small and serves two functions in soup preparation: it adds flavor and color to the soup. Vegetable garnish may be prepared in several ways: *julienne* (cut into long thin strips), *brunoise* (very small dice), *printaniere* (small dice, spring vegetables), and *paysanne* (fine shredded vegetables, in the peasant style).

STOCKS

Stock is a thin liquid that is produced by simmering (cooking in liquid just below the boiling point of 212°F) meat trimmings, meat bones, fish bones, poultry bones, or vegetables. Stock is used in a variety of food preparations. In all preparations, the quality of stock determines the quality of the completed food preparation.

In the past, preparation of stocks was a long process, requiring a pot of stock constantly simmering over heat. The stock was simmered for 12 hr to 24 hr and, in many cases, was reduced to a demiglacé or glacé de Viande (stock reduced to about one-third its original volume). Today, food preparation techniques have reduced the stock simmering time to 4 hr to 6 hr. Reduction of stock to a *demiglacé* or *glacé de Viande* has been replaced by the use of soup bases. Soup bases can be used to enhance the flavors of soups, stocks, and sauces. They are used with good results and save labor and time. Approximately 4 oz of soup base should be mixed with each gallon of water.

Ingredients used in the preparation of stock determine the quality of the finished product. Stocks are commonly prepared using bones to supply the necessary strength and flavor. Some bones produce richer stock than others. For beef, veal, lamb, and brown stocks, shank, knuckle, and neck bones are used. For chicken stock, feet and, if available, the bones of older chickens are used. The bones must be cut into medium-sized pieces for ease in handling and maximum flavor. If a clear stock is required, it may be necessary to blanch the bones. A steam jacket kettle in place of a stock pot will reduce the time required to blanch bones. To blanch bones before starting the stock:

1. Rinse the bones in cold water.
2. Place the bones in a stock pot filled with cold water.
3. Bring to a fast boil.
4. Drain and rinse well. Follow stock recipe.

A number of different stocks are used in the commercial kitchen from which soups, sauces, and gravies are made. Brown stock, beef stock, veal stock, ham stock, chicken stock, and fish stock are most commonly used.

Selecting the stock to use in specific preparations is based on how well the flavors blend with one another. The stock selected should be one that enhances the quality of the other ingredients in a preparation. For example, beef stock enhances the flavor of vegetables when vegetable soup is prepared. Fish stock is used in the preparation of clam chowder. Chicken stock is used in delicately-flavored cream soups. Ham stock is most commonly selected for bean and pea soups.

Gelatin can be added to stock to produce jellied stock for a richer body and flavor. Natural gelatin is extracted from flesh and bones during the cooking process. When cold, the gelatin solidifies for easier handling. Stock will gel if approximately 2 oz of gelatin is added for each gallon of stock. If extra gelatin is required, use plain gelatin and let it soak in a small amount of cold water before stirring it into the hot stock. This allows the gelatin to dissolve more quickly, preventing a lump from forming at the bottom of the pot when the stock is reheated.

Trimmings from tomatoes, celery, onions, leeks, parsley, and other vegetables should be saved for use in stocks. Other products, such as leftover stew, gravies, and sauces, can be used in certain stock preparations. However, ingredients with strong flavors, such as radishes, cabbage, and turnips, should be used with care.

Seasoning Stock

Stocks may require seasoning using selected herbs and/or spices. To season the stock while simmering, a bouquet garni (seasonings placed in a porous cloth bag) or sachet is used. The herbs and/or spices are placed in the bag. The bag is tied with a long string to permit lowering it into the simmering stock. The long string is also used to remove the bag when the correct amount of seasoning is obtained.

SOUP PREPARATION

Soups are a blend of ingredients carefully prepared to achieve the balanced flavor desired. Each ingredient must be properly prepared and added in the correct sequence. The following techniques detail the steps required in preparing soups.

Cutting Raw Vegetable Mirepoix

The raw vegetable mirepoix for soups supplies flavor and enhances the appearance, depending on the soup being prepared. A mirepoix is sometimes referred to as "rough garnish" because the vegetables are cut in a rough or irregular fashion. If the recipe calls for mirepoix, its purpose is to supply only flavor to the soup. After preparation, the vegetables are strained off and discarded. A mirepoix usually consists of carrots, onions, and celery. Peeling the carrots is not necessary. However, the onions should be peeled if the preparation is light in color.

If the soup preparation calls for the garnish to be diced, minced, or julienned it is referred to as matignon. In this instance, the vegetable garnish must be small and uniformly cut. The purpose of matignon is twofold. The matignon must supply both flavor and appearance to the soup preparation, and it will remain in the soup upon completion. If the matignon is cut too large, the soup will become too filling. If the vegetables are cut irregular in size, they will cook unevenly and lack in appearance.

Cutting and Cracking Bones

Soup stocks are prepared from animal bones. In most cases, these bones are fairly large. The bones are cut or cracked into uniform sized pieces to fit into the stock pot and provide flavor and strength to the stock efficiently. The bones are cut with a hand or power meat saw or a cleaver, depending upon the size of the bones. Whatever tool is used, extreme caution must be exercised to prevent injury.

To cut bones with a hand meat saw, the bones should be held firmly on a non-slippery surface. The saw should be pushed and pulled with a slow, easy motion. A sharp saw does not require force when cutting. If force is used, an injury could occur.

Cutting with a power meat saw is more efficient, but it is not always available in the commercial kitchen. Bones should be held firmly in the hand and moved across the rotating blade with a firm, steady motion. Refer to the operator's manual for the proper use of the power meat saw.

The cleaver is used only when a hand meat saw or a power meat saw is not available. The cleaver can be dangerous if used improperly. Bones should be placed on a solid, non-slippery surface. Hold the bones firmly with one hand and make sharp blows with the cleaver. Bones must be placed on a solid surface or the cleaver may spring back rapidly and cause an injury.

Straining Soups and Stocks

All stocks require straining to remove undesirable particles that reduce the appearance and eating qualities of the soup. Some soups, such as cream, puree, broth, bouillon, and consommé, also require straining when completed. Soup is strained using a china cap or a china cap covered with cheesecloth, depending upon the consistency of the item being strained. Thin or clear liquids, such as bouillons, broths, consommés, and most stocks (brown stock is an exception), are strained with a fine-hole china cap that is covered with a cheesecloth.

Thick liquids, such as cream or puree soups, are strained with a china cap upon completion to remove rough garnish or vegetable pulp before serving. When straining, the rough garnish and vegetable pulp are forced down into the tip of the china cap with a ladle to force as much of the flavor as possible into the soup before the vegetable or pulp is discarded. The size of the china cap and cloth used depend on the size of the pot and the consistency of the liquid. An immersion blender may be used to purée the soup, eliminating the need for straining.

Sautéing Vegetable Mirepoix

Sautéing the vegetable mirepoix is an important step that will produce a soup with a superior flavor. Mirepoix usually consists of 50% onions, 25% celery, and 25% carrots, roughly chopped. The vegetables are added to hot melted shortening, butter, or oil, depending on preference and the type of soup being prepared. The mirepoix is sautéed until the onions take on a transparent appearance. The garnish should never be browned unless a caramelized flavor and color are desired, such as when making French onion soup. Otherwise appearance and flavor of the soup will be impaired. Stir the vegetable garnish with a paddle throughout the sautéing period.

Cracking Whole Peppercorns

Peppercorns supply a desirable flavor to many soup and stock preparations. The peppercorns should be cracked before they are added to the soup. When cracked, they release a more potent flavor. Peppercorns can be cracked using a saucepan, a mallet, or a kitchen mortar and pestle. When using a saucepan, peppercorns are placed on a hard surface. Using force, rub across the peppercorns with the bottom of a saucepan. When using a mallet, peppercorns are placed in a kitchen towel and tapped. When using a kitchen mortar and pestle, the peppercorns are cracked with firm, even pressure in the kitchen mortar.

Skimming Fat and Scum from Stock

Skimming the fat and scum from the surface of a stock is an essential step in preparing stocks. The removal of fat and scum produces a clearer stock, which results in a finer soup or sauce. Skimming can be done with a ladle, or a skimmer may be run across the surface of the liquid, removing all the coagulated particles that appear.

To remove any fat that may appear on the surface of stocks, strain the liquid using a cold, wet cheesecloth. Wash the cloth and repeat the process until all signs of fat are removed. Drawing brown paper over the surface of the hot liquid also absorbs the fat. If time is not a factor, refrigerate the stock. The fat, when cold, will solidify for easy removal.

Grinding Foods

Some recipes call for grinding certain ingredients to disperse more flavor particles throughout the soup. Ingredients such as ham fat, cooked giblets, beef, and corn kernels are commonly ground for this purpose. A *power food grinder*, sometimes known as a *hamburger grinder*, is used to grind ingredients. This grinder can be an independent piece of equipment or an attachment on the mixing machine.

The grinder is equipped with chopper plates that fit over its front. The food is passed through the chopper plates with holes that vary in size from very small to very large. The chopper plate used is selected based on the soup being prepared and the hole size designated in the recipe.

Cooling Soups and Stocks

Soups and stocks that are not being used immediately should be quickly cooled to below 40°F to avoid remaining in the temperature danger zone. The most efficient method of cooling is to place the soup or stock in a tub of ice water and stir as necessary with a kitchen spoon or paddle to speed the cooling action. Blast chillers are a faster means of chilling. When the soup or stock is cooled completely, it should be stored in the coldest part of the refrigerator. When preparing the stored soup or stock, bring it to a rapid boil before using it in another preparation or serving.

Trade Tip

Soup and sandwiches may be presented on the buffet menu. Customers can choose from several soups and select bread, meat, cheese, etc. to make their own sandwich. Offer a large variety of "toppings" to go with the soup, such as grated cheese, toasted nuts, etc.

STOCK RECIPES

Brown Stock

Brown stock is prepared by browning then boiling beef and veal bones. Vegetables and seasoning are added for flavor. Brown stock is used in the preparation of soups, sauces, and gravies.

5 gal.

Equipment

- French knife
- Baker's scale
- 10 gal. stock pot
- Spoon measure
- Paddle
- China cap
- Ladle or skimmer
- 5 gal. container
- Meat saw
- Cleaver
- Roast pan
- Gallon measure
- Tub

Ingredients

20 lb beef bones
10 lb veal bones
7 gal. water
1 lb celery
1 lb carrots
3 bay leaves
1 tsp thyme
1 tsp whole black pepper
4 whole cloves
3 garlic cloves
1 qt tomato puree

Procedure

1. Place the bones in a large roast pan and brown thoroughly in a preheated oven at 400°F.
2. When the bones are brown, drain off any grease that may have accumulated in the pan.
3. Add the rough garnish and continue to roast until the garnish is slightly brown.
4. Remove the bones and rough garnish from the pan and place in a large stock pot.
5. Deglaze the roast pan with part of the water.
6. Cover the bones with the remaining water and liquid from deglazing the pan and bring to a boil.
7. Add all remaining ingredients, reduce heat, and simmer for 5 hr to 6 hr. Stir occasionally with the paddle.
8. Strain through a fine china cap into the 5 gal. container. Cool as quickly as possible in a tub of ice cold water.
9. Refrigerate until ready to use.

Preparation

1. Cut the bones with a meat saw or crack with a cleaver into medium-sized pieces.
2. Cut the mirepoix (onions, carrots, and celery) with a French knife into medium-sized pieces.
3. Mince the garlic and crack the peppercorns.

Precautions

- Exercise caution when cutting or chopping the bones.
- Skim fat and scum from the simmering stock frequently using a ladle or skimmer.
- Do not let the liquid boil rapidly. Simmering produces a more flavorful stock.

Beef Stock

Beef stock is prepared by simmering beef or beef bones, vegetables, and seasoning in water to extract all strength and flavor. Beef stock is used in the preparation of soups, sauces, and gravies.

5 gal.

Equipment

- French knife
- Baker's scale
- Spoon measure
- 10 gal. stock pot
- China cap
- Cheesecloth
- Meat saw
- Cleaver
- 5 gal. container
- Gallon measure
- Tub
- Ladle or skimmer

Ingredients

15 lb beef bones
1 lb beef shank
1 lb onions
8 oz celery leaves
8 oz carrots
2 bay leaves
1 tsp whole black pepper
1 tsp thyme
6 gal. water
1 oz parsley stems

Procedure

1. Blanch the bones and meat in the large stock pot using a sufficient amount of boiling water to cover them. Drain and wash thoroughly in cold water.
2. Add the 6 gal. of water, bring to a quick boil, and immediately remove any scum that may appear on the surface. Remove scum with a ladle or skimmer.
3. Add the remaining ingredients and let simmer for 5 hr to 6 hr.
4. Strain through a china cap covered with a cheesecloth to remove all foreign particles. Strain into a 5 gal. container and cool as quickly as possible in a tub of ice cold water.
5. Refrigerate until ready to use.

Preparation

1. Cut the bones with a meat saw or crack with a cleaver into medium-sized pieces.
2. Cut the mirepoix (onions, carrots, and celery) with a French knife into medium-sized pieces.
3. Crack the peppercorns.

Precautions

- Exercise caution when cutting or cracking the bones.
- Do not add the vegetables or seasoning to the stock until all the scum has been removed.
- Do not let the liquid boil vigorously at any time. Simmering produces a clearer, more flavorful stock.

Veal stock is prepared by simmering veal or veal bones, vegetables, and seasoning in water to extract all strength and flavor. Veal stock is used in the preparation of soups and sauces.

Veal Stock

5 gal.

Equipment

- French knife
- Baker's scale
- Spoon measure
- 10 gal. stock pot
- China cap
- Cheesecloth
- Meat saw
- Cleaver
- 5 gal. container
- Gallon measure
- Tub
- Ladle or skimmer

Ingredients

15 lb veal bones
1 lb veal shank
1 lb onions
8 oz celery and celery
    leaves
1 bay leaf
1 tsp thyme
1 tsp whole black pepper
6 gal. water
1 oz parsley stems

Procedure

1. Blanch the bones and meat in the stock pot, using a sufficient amount of boiling water to cover them. Drain and wash thoroughly in cold water.
2. Add the 6 gal. of water, bring to a boil, and remove any scum that may appear.
3. Add the remaining ingredients and let simmer for 5 hr or 6 hr.
4. Strain through a china cap covered with a cheesecloth to remove all foreign particles. Strain into a 5 gal. container and cool as quickly as possible in a tub of ice cold water.
5. Refrigerate until ready to use.

Preparation

1. Cut the bones with a meat saw or crack with a cleaver into medium-sized pieces.
2. Cut the mirepoix (onions and celery) with a French knife into medium-sized pieces.
3. Crack the peppercorns.

Precautions

- Exercise caution when sawing or cracking the bones.
- Do not add the vegetables or seasoning to the stock until all the scum has been removed.
- Do not let the liquid boil rapidly at any time. Simmering produces a clearer, more flavorful stock.

Ham stock is prepared by simmering ham trimmings and bones, vegetables, and seasoning in water to extract the strength and flavor. Ham stock is used in the preparation of soups, vegetables, and sauces.

Ham Stock

5 gal.

Equipment

- 10 gal. stock pot
- China cap
- Cheesecloth
- 5 gal. container
- French knife
- Baker's scale
- Gallon measure
- Ladle
- Tub

Ingredients

16 lb ham trimmings and
    bones
2 lb onions
1 lb celery and celery
    leaves
8 oz carrots
6 gal. water
1 tsp whole cloves
½ tsp garlic

Procedure

1. Place the ham bones and trimmings in a large stock pot. Add the water, bring to a quick boil, and remove any scum that may appear on the surface.
2. Add the remaining ingredients and let simmer for 4 hr to 5 hr, skimming fat and scum from stock frequently.
3. Strain through a china cap covered with a cheesecloth to remove all foreign particles. Strain into the 5 gal. container and cool as quickly as possible in a tub of ice cold water.
4. Refrigerate until ready to use.

Preparation

1. Cut the onions, celery, and carrots rough using a French knife.
2. Mince the garlic.

Precautions

- Do not add the vegetables or seasoning to the stock until all the scum has been removed.
- Do not let the liquid boil rapidly at any time. Simmering produces a more flavorful stock.

Trade Tip

If a dark beef stock is desired for such preparations as beef consommé or French onion soup, before adding the onions to the mirepoix, cut them in half crosswise and burn the exposed flesh on a hot griddle. This action helps darken the stock during the simmering period.

Lamb stock is prepared by simmering lamb or lamb bones, vegetables, and seasoning in water to extract the strength and flavor. Lamb stock is not commonly used in most commercial kitchens; however, occasions do arise when it is needed.

Lamb Stock

5 gal.

Equipment

- French knife
- Baker's scale
- 10 gal. stock pot
- China cap
- Meat saw
- Cleaver
- Spoon measure
- Cheesecloth
- Gallon measure
- 5 gal. container
- Tub

Ingredients

15 lb lamb bones
1 lb lamb shank
1 lb onions
8 oz celery and celery
   leaves
8 oz carrots
1 bay leaf
2 tbsp marjoram
1 tsp whole black peppercorns
6 gal. water

Procedure

1. Blanch the bones and meat in the stock pot using a sufficient amount of boiling water to cover. Drain and wash thoroughly in cold water.
2. Add the 6 gal. of water, bring to a boil, and remove any scum that may appear on the surface with a ladle or skimmer.
3. Add the remaining ingredients and let simmer for 5 hr or 6 hr.
4. Strain through a china cap covered with a cheesecloth to remove all foreign particles. Strain into a 5 gal. container. Cool as quickly as possible in a tub of ice cold water.
5. Refrigerate until ready to use.

Preparation

1. Cut the bones with a meat saw or crack with a cleaver into medium-sized pieces.
2. Cut the mirepoix (onions, carrots, and celery) with a French knife into medium-sized pieces.
3. Crack the peppercorns.

Precautions

- Exercise caution when sawing or cracking the bones.
- Do not add the vegetables or seasoning to the stock until all the scum has been removed.
- Do not let the liquid boil rapidly at any time. Simmering produces a clearer, more flavorful stock.

Chicken stock is prepared by simmering chicken bones, vegetables, and seasoning in water to extract the strength and flavor. Chicken stock is used in the preparation of soups, sauces, and gravies.

Chicken Stock

5 gal.

Equipment

- French knife
- Baker's scale
- Gallon measure
- China cap
- 10 gal. stock pot
- Cheesecloth
- 5 gal. container
- Ladle or skimmer
- Tub

Ingredients

15 lb chicken bones,
   necks, or feet
1 lb onions
1 lb celery
2 oz concentrated chicken
   base
6 gal. water
½ tsp whole black pepper
2 oz parsley stems

Procedure

1. Place the chicken bones in a large stock pot, add the water, bring to a quick boil, and remove at once any scum that may appear on the surface. Remove the scum with a ladle or skimmer.
2. Add the remaining ingredients and let simmer for 5 hr to 6 hr.
3. Strain through a china cap covered with a cheesecloth to remove all foreign matter. Strain into a 5 gal. container and cool as quickly as possible in a tub of ice cold water.
4. Refrigerate until ready to use.

Precautions

- Do not add the vegetables to the stock until all the scum has been removed.
- Do not let the liquid boil rapidly at any time. Simmering produces a clearer, more flavorful stock.

Preparation

1. Cut the celery and onions rough with a French knife.
2. Crack the peppercorns.

Trade Tip

The temperature of the water when starting a stock makes a difference. Starting the stock in cold water produces a stock with superior flavor. A clearer stock is usually obtained by starting with hot water.

Fish Stock

Fish stock is prepared by simmering fish trimmings and bones, vegetables, and seasoning in water to extract the strength and flavor. For best results, the trimmings and bones of lean white meat fish such as cod, haddock, sole, and flounder are preferred. The extraction of fish stock is more rapid than the extraction of meat stocks. Fish stock is used in the preparation of soups and sauces.

5 gal.

Equipment

- French knife
- 10 gal. stock pot
- Spoon measures
- China cap
- Cheesecloth
- 5 gal. container
- Gallon measure
- Baker's scale
- Skimmer or ladle
- Tub

Ingredients

16 lb fish trimmings and bones
2 lb onions
1 lb celery and celery leaves
3 lemons
3 bay leaves
2 oz parsley stems
1 tsp whole black pepper
½ tsp dill weed or seeds
6 gal. water

Procedure

1. Place the fish bones and trimmings in a large stock pot. Add the water, bring to a boil, and remove any scum that may appear on the surface. Remove scum with a ladle or skimmer.
2. Add the remaining ingredients and let simmer for 3 hr to 4 hr.
3. Strain through a china cap covered with a cheesecloth to remove all foreign particles. Strain into a 5 gal. container and cool as quickly as possible in a tub of ice cold water.
4. Refrigerate until ready to use.

Note: White wine may be added to the stock if preferred. Add approximately ⅘ qt to the 5 gal. of stock upon completion for a richer, more flavorful stock.

Preparation

1. Cut the onions and celery rough using a French knife.
2. Crack the peppercorns and cut the lemons into quarters.

Precautions

- Do not add the vegetables or seasoning to the stock until all the scum has been removed.
- Do not let the liquid boil vigorously at any time. Simmering produces a clearer, more flavorful stock.

THIN SOUP RECIPES

Vegetable Soup

Vegetable soup contains a variety of vegetables. It is served in most commercial kitchens as a soup du jour (soup of the day).

3 gal.

Equipment

- 5 gal. stock pot
- Paddle
- Baker's scale
- French knife
- Quart measure
- 3 gal. container

Ingredients

1 lb carrots
1 lb celery
1 lb onions
½ lb cabbage
4 oz shortening
2½ gal. beef stock
½ gal. crushed tomatoes
2 oz salt (variable)
1 oz garlic
1 pinch pepper
1 lb peas
½ lb corn
½ lb lima beans

2. Prepare the beef stock.
3. Mince the garlic.

Procedure

1. Place shortening in stock pot and add the diced onions, celery, carrots, and garlic. Sauté until vegetables are partly done. Do not brown.
2. Add beef stock, bring to a boil, and simmer for 1 hr.
3. Add peas, corn, lima beans, and cabbage. Continue to simmer until all vegetables are tender.
4. Add crushed tomatoes and continue simmering for 5 min.
5. Season with salt and pepper.
6. Pour into 5 gal. container.

Preparation

1. Dice carrots, celery, onions, and cabbage with a French knife.

Precautions

- Use a rich beef stock.
- Cut vegetable garnish as uniformly as possible.

French Onion Soup

French onion soup is a thin soup that originated in France. This soup is extremely popular on the luncheon, dinner, and a la carte menu.

3 gal.

Equipment

- 5 gal. stock pot
- French knife
- Quart measure
- Baker's scale
- Paddle
- 3 gal. container
- Ladle or skimmer

Ingredients

12 oz butter
3 qt onions
1½ gal. beef stock or
   consommé
1½ gal. chicken stock
6 oz sherry (optional)
salt and pepper to taste

Preparation

1. Prepare the beef stock or consommé.
2. Prepare the chicken stock.
3. Cut the onions julienne with a French knife.

Procedure

1. Place the butter in the stock pot and heat at a moderate temperature until melted.
2. Add the onions and sauté until they begin to color.
3. Add the hot stocks and stir with the paddle.
4. Bring back to a boil, reduce heat, and simmer for about 1 hr. Remove scum with a ladle or skimmer.
5. Season with salt and pepper and pour into the 3 gal. container.
6. Add wine if desired. Serve with cheese croutons and grated Parmesan cheese.

National Onion Association

Precautions

- When sautéing the onions, do not burn them. A light brown color is desired.
- Simmer throughout the cooking period so the stocks will not cloud.

Beef Consommé

Beef consommé is a very clear beef liquid. The name is derived from the word consummate, meaning the finest or most perfect. It is one of the most popular soups served and is listed on the menu of most food service establishments. Beef consommé can be served hot or cold. If served cold, plain gelatin must be added.

3 gal.

Equipment

- 5 gal. stock pot with spigot
- Paddle
- Wire whip
- China cap
- Cheesecloth
- Baker's scale
- Quart measure
- 3 gal. container
- Food grinder

Ingredients

3 lb beef shank
4 gal. beef stock
2 lb onions
1 lb celery
8 oz carrots
parsley stems from four
   bunches of parsley
½ tsp thyme
6 cloves
2 bay leaves
2 tsp black peppercorns
1 pt canned whole tomatoes

Preparation

1. Prepare the beef stock. Let cool.

2. Grind the beef shank meat using the coarse chopper plate of the food grinder.
3. Cut the rough garnish (onions, carrots, and celery) with a French knife.
4. Cut the stems from four bunches of parsley.
5. Separate the eggs. The whites and shells will be used in the clarification process. Save the yolks for use in another preparation.
6. Crack the peppercorns.

Procedure

1. Blend the rough vegetable garnish, ground beef, parsley stems, thyme, bay leaves, cloves, peppercorns, eggshells, tomato juice, and whole tomatoes in a large stock pot. Mix well.
2. Beat the egg whites slightly with a wire whip and pour into the stock pot.
3. Add the cold beef stock and stir vigorously with a paddle.
4. Bring to a slow boil, stirring occasionally. Reduce heat to a simmer and allow the coagulated mass to rise to the top of the stock pot, forming a raft (floating mass).

5. Continue to simmer for 2 hr. Do not break or disturb the raft.
6. Remove from the range and strain through a china cap covered with a fine cheesecloth into a 3 gal. container.
7. Serve hot or cold. If serving cold, add 3½ oz of plain gelatin to each gallon of liquid. The gelatin should be soaked in cold water and added to the soup during the second step. The soaking ensures that the gelatin is completely dissolved.

Note: To prepare chicken consommé, use the same recipe but substitute chicken stock for beef stock, omit the tomatoes and tomato juice, and add ½ c of lemon juice.

Precautions

- Use caution when grinding the meat.
- Do not break the raft at any time during the cooking period or when straining.

- When straining, use the spigot (faucet) on the bottom of the stock pot so the raft will not be broken.
- Once the raft has set, do not let the liquid boil.

Note: Consommé is usually served with a garnish. The garnishes used are numerous. Some popular garnishes are:

Celestine: with julienned French pancakes

Brunoise: with assorted vegetables cut in a small dice

Printaniere: with small diced spring vegetables

Royale: with custard cut in a diamond shape

Xavier: with egg drops

Beleview: topped with unsweetened whipped cream and browned under the broiler

Vermicelli: with small pieces of boiled vermicelli

Rice: with boiled rice

Barley: with boiled barley

Marrow Dumpling: with poached marrow dumplings

Trade Tip

Popular luncheon choices include soups and sandwiches. List a variety of soups and sandwiches from which the customer may select. Often, the soup is accompanied by crackers and a pickle spear accompanies the sandwich. Soups and sandwiches are especially appealing in cold weather.

Petite marmite is a soup that is considered a fancy preparation in comparison to most other soups. It is a combination of a rich beef consommé and chicken broth with a garnish of diamond-cut vegetables flowing through its rich liquid. Petite indicates that the vegetables should be cut fairly small. Marmite is the earthen pot in which the soup is served. This soup is a welcome addition to the a la carte menu for special menus on holidays or festive occasions.

Petite Marmite

3 gal.

Equipment

- 5 gal. stock pot
- French knife
- 3 gal. container

Ingredients

- 1½ gal. beef consommé
- 1½ gal. chicken broth or stock
- 2 lb cooked beef
- 2 lb cooked chicken or turkey white meat
- 1 lb carrots
- 1 lb turnips
- 8 oz celery
- salt and white pepper to taste

Preparation

1. Prepare the beef consommé.
2. Prepare a rich, clear chicken broth or stock.
3. Cut the beef and chicken or turkey into a diamond shape with a French knife.

4. Cut the vegetables into a diamond shape with a French knife.

Procedure

1. Combine the beef consommé and chicken broth in a stock pot and bring to a simmer.
2. Add the diamond-cut vegetables and continue to simmer until they are tender.
3. Add the diamond-cut meat and simmer for 10 more minutes.
4. Season with salt and white pepper and pour into a 3 gal. container.
5. Serve in marmite pots topped with toasted cheese.

Precaution

- Do not boil the consommé and chicken broth as they may become cloudy.

Tomato Madrilene

Tomato madrilene is a sparkling clear broth with a tomato flavor. It is clarified by using the same process required for preparing consommé. Tomato madrilene can be served hot or cold. If served cold, plain gelatin must be added.

5 gal.

Equipment

- 10 gal. stock pot with spigot
- French knife
- Spoon measures
- Baker's scale
- Quart measure
- Food grinder
- China cap
- Cheesecloth
- Paddle
- 6 oz ladle
- 5 gal. container

Ingredients

3½ gal. beef or chicken stock
1 No. 10 can whole tomatoes
1 No. 10 can tomato juice
2 qt fresh tomatoes, piece or overripe
2 lb lean beef
2 lb onions
1 lb celery
1 lb carrots
1 pt parsley or parsley stems
1 tsp thyme
4 bay leaves
1 tsp whole cloves
1 tbsp peppercorns
1 qt egg whites

Preparation

1. Prepare the beef or chicken stock. Let cool.
2. Grind the lean beef using the coarse chopper plate on the food grinder.
3. Cut the mirepoix (onions, celery, and carrots) with a French knife.
4. Separate the eggs. The whites and shells will be used in the clarification process. Reserve the yolks for use in another preparation.

5. Crack the peppercorns.

Procedure

1. To make the clarification mixture, blend the mirepoix, ground beef, fresh tomatoes or pieces, parsley or stems, thyme, bay leaves, cloves, peppercorns, eggshells, and whites in a large stock pot. Mix well.
2. Add the remaining ingredients and mix well.
3. Bring to a slow boil, stirring occasionally with a paddle. Reduce heat to a simmer and allow the coagulated mass to rise to the top of the stock pot, forming a raft.
4. Continue to simmer for 2 hr. Do not break or disturb the raft. Remove from the range and strain into a 5 gal. container through a china cap covered with a fine cheesecloth.
5. Add red color as desired because the clarification process destroys most of the tomato color.
6. Serve hot or cold. If serving cold, add 3½ oz of plain gelatin to each gallon of liquid. The gelatin should be soaked in cold water and added to the soup during the second step.

Note: Hot tomato madrilene is generally served with one of the following garnishes: julienne, brunoise, paysanne, or printaniere-cut vegetables, rice, egg drops, cooked pasta, dumplings, etc.

Precautions

- Do not break the raft at any time during the cooking period or when straining. Discard the raft after straining.
- When straining, use spigot in the stock pot.
- Once the raft has set, do not let the liquid boil.

THICK SOUP RECIPES

Puree of Tomato Soup

Puree of tomato soup is prepared from the pulp of tomatoes. This soup is quite popular when served as a soup du jour. It is usually garnished with croutons.

3 gal.

Equipment

- 5 gal. stock pot
- French knife
- Paddle
- China cap
- 6 oz ladle
- 3 gal. container
- Baker's scale
- Quart measure
- Spoon measure
- Wire whip

Ingredients

2 gal. ham stock
1 No. 10 can tomato puree
1 lb onions
1 lb celery
8 oz carrots
8 oz bacon grease
6 oz flour
2 bay leaves
3 oz sugar (variable)
1 tsp basil
½ tsp rosemary leaves
salt and white pepper to taste

Preparation

1. Prepare the ham stock.
2. Cut the mirepoix (onions, carrots, and celery) with a French knife.

Procedure

1. Place the bacon grease in a stock pot and heat.
2. Add the mirepoix and sauté until slightly tender.
3. Add the flour, making a roux, and cook for 5 min.
4. Add the tomato puree and ham stock, whipping vigorously with a wire whip until thickened and smooth.
5. Add the bay leaves, rosemary leaves, basil, and sugar. Simmer for approximately 2 hr. Strain through a fine china cap, forcing through as much of the vegetable flavor as possible with a ladle.

6. Season with salt and white pepper and pour into a 3 gal. container.
7. Serve garnished with croutons.

Precautions

- Do not let the vegetables brown when sautéing.
- While the soup is simmering, stir occasionally with a paddle to avoid sticking.
- Rub basil between the palms to release its flavor before adding to the soup.

Cream of Tomato Soup

Cream of tomato soup is prepared by adding a thin cream sauce to a puree of tomato soup. The result is a soup with a smooth, creamy consistency. This soup is a popular appetizer on both the luncheon and dinner menu.

5 gal.

Equipment

- 10 gal. stock pot
- China cap
- Baker's scale
- Gallon measure
- 6 oz ladle
- French knife
- Paddle
- 5 gal. container for holding soup

Ingredients

3 gal. ham stock
2 gal. tomato puree
1 lb celery
1 lb onions
1 lb carrots
1 lb leeks
2 lb flour
2 lb bacon grease, ham fat, or shortening
2 garlic cloves
3 oz salt
sugar and pepper to taste
3 bay leaves
1 tbsp thyme
1 gal. thin cream sauce
2 tsp baking soda

Procedure

1. Place the grease or shortening in a stock pot, add the mirepoix, and sauté until slightly tender.
2. Add the flour, making a roux. Cook for approximately 5 min.
3. Add the hot ham stock and tomato puree, stirring constantly until slightly thickened and smooth.
4. Add the bay leaves, thyme, and salt. Simmer for approximately 2 hr.
5. Season with sugar and pepper.
6. Add the baking soda and stir well.
7. Strain the soup through a china cap into a 5 gal. container. Using a ladle, force as much of the vegetable pulp and flavor as possible through the china cap.
8. Blend in the hot cream sauce gradually by stirring gently with the paddle.
9. Adjust seasoning and serve with croutons.

Preparation

1. Prepare the ham stock.
2. Cut the mirepoix (onions, celery, carrots, and leeks) into medium-sized pieces with a French knife.
3. Mince the garlic.
4. Prepare the thin cream sauce.

Precautions

- Do not brown the vegetables when sautéing.
- The baking soda is added to keep the soup from curdling when the cream sauce is added. However, if the soup is exposed to excessive heat for too long a period, it still may curdle.

Cream of Cauliflower Soup

Cream of cauliflower is a rich, creamy white soup with a delicate cauliflower flavor. Small pieces of cauliflower flow through the soup and produce a richer flavor and aroma. This hearty preparation is an excellent addition to the fall and winter menu.

3 gal.

Equipment

- French knife
- 5 gal. stock pots (2)
- Kitchen spoon or paddle
- China cap
- 6 oz ladle
- Baker's scale
- Gallon measure
- Quart measure
- Small sauce pot
- Large wire whip
- 3 gal. container

Ingredients

8 lb cauliflower pieces and trimmings
1 lb onions
1 lb celery
3 gal. hot chicken stock
2 oz chicken base
1 lb flour
3 qt milk or light cream
pinch baking soda
1 qt cauliflower, small pieces
salt and white pepper to taste

Preparation

1. Prepare the chicken stock.
2. Cut the onions and celery rough with a French knife.
3. Heat the milk or light cream.
4. Cook the small pieces of cauliflower in a sauce pot.

Procedure

1. Place the margarine in a stockpot and heat until melted.
2. Add the mirepoix and sauté until slightly tender. Do not brown.
3. Add the flour, making a roux. Cook the roux slightly. Do not brown.
4. Add the hot chicken stock and chicken base while whipping rapidly with a large wire whip. Bring to a boil, then reduce to a simmer.

5. Add the cauliflower pieces and trimmings. Continue to simmer until the cauliflower is extremely tender.
6. Remove from the heat and strain through a fine china cap into a separate 5 gal. stockpot. With a ladle, force as much of the vegetable pulp as possible through the china cap.
7. Add a pinch of baking soda and the hot milk or light cream while whipping vigorously with a wire whip.
8. Season with salt and white pepper.
9. Stir in the cooked pieces of cauliflower, pour into a 3 gal. container, and hold for service.

Precautions

- Exercise caution when cutting the vegetables.
- Cook by simmering because more flavor can be extracted from the vegetables.
- Avoid browning the onions and celery when sautéing.
- Exercise caution when straining the soup and forcing the vegetable pulp into the china cap.
- Stir the soup occasionally during the simmering period.

Split Pea Soup

Split pea soup is a thick soup that is cooked until the peas become pureed. It is commonly served as a soup du jour.

5 gal.

Equipment

- 5 gal. and 10 gal. stock pots
- China cap
- Paddle
- French knife
- 6 oz ladle
- Baker's scale
- Spoon measure
- Wire whip
- 5 gal. container
- 1 gal. measure

Ingredients

1 lb of bacon or ham fat
8 lb split peas
1 lb flour
5 gal. ham stock
1 lb ham bones
2 oz salt (variable)
2 lb carrots
2 lb onions
2 lb celery
⅛ oz thyme
1 tsp pepper

Preparation

1. Prepare ham stock.
2. Cut vegetables into medium-sized pieces with a French knife.

Procedure

1. Place peas in 5 gal. stock pot and cover with 3 gal. ham stock. Cook by simmering until peas are well-done.
2. In 10 gal. stock pot, braise the vegetables in bacon or ham fat with a ham bone until tender.
3. Add flour and mix thoroughly, making a roux. Cook slowly without burning for 5 min.

4. Add remaining stock to the roux and vegetable mixture and mix well. Stir with the paddle until smooth, then let simmer.
5. Add cooked split peas, salt, pepper, and thyme to the simmering vegetable mixture. Continue to simmer for an additional hour.
6. Strain the soup through a china cap into a 5 gal. container. Using a ladle, force as much of the vegetable pulp as possible through the china cap.

Precautions

- After peas are added to the soup, stir occasionally to avoid sticking.
- Cook peas until they are pureed.

Curdling in cream soups will result if held too hot for too long a period of time, or if the cream or milk is added incorrectly. To remove curdle, beat cold sweet cream into the soup or whip in some cream or bechamél sauce.

Cream of Corn Soup Washington

Cream of corn soup Washington is a rich cream soup with the flavor of fresh corn. This soup is generally garnished with small diced pieces of pimiento and whole kernel corn. It is served as a soup du jour.

3 gal.

Equipment

- 5 gal. stock pot
- China cap
- French knife
- 6 oz ladle
- 3 gal. container
- Baker's scale
- Wire whip
- Paddle
- Quart measure
- Spoon measures
- Saucepans
- Food grinder

Ingredients

2½ gal. milk
6 oz onion
1 No. 10 can creamed corn
1 lb frozen whole kernel corn
10 oz butter
8 oz flour
2 qt cream
¾ oz. sugar
½ tsp nutmeg
4 oz pimientos
salt to taste

Procedure

1. Place the butter in a stock pot and heat.
2. Add the onions and sauté until slightly tender.
3. Add the flour, making a roux, and cook for 5 min.
4. Add the hot milk, whipping briskly with a wire whip until slightly thick and smooth.
5. Stir in the creamed corn and sugar with a paddle. Simmer for approximately 45 min.
6. Strain through a fine china cap, forcing through as much of the onion and corn flavor as possible with a ladle.
7. Stir in the hot cream and nutmeg and blend thoroughly using a paddle.
8. Stir in the ground whole kernel corn and the pimento.
9. Season with salt and pour into a 3 gal. container.

Preparation

1. Dice the onions with a French knife.
2. Heat the milk and cream separately in a saucepan.
3. Grind the whole kernel corn on the food grinder using the coarse chopper plate.
4. Dice the pimientos small with a French knife.

Precautions

- When heating the butter, do not let it burn.
- When sautéing the onions, do not let them brown.
- While the soup is simmering, stir occaisionally with a paddle to avoid sticking or scorching.
- Do not hold the soup at too high a temperature. It may curdle.

Cream of Celery Soup

Cream of celery soup is a rich cream soup with a strong celery flavor. It is usually served on the menu as a soup du jour.

3 gal.

Equipment

- 5 gal. stock pot
- Paddle
- French knife
- Baker's scale
- Cup measure
- Quart measure
- Cheesecloth
- China cap
- 6 oz ladle
- Wire whip
- 3 gal. container
- 2 saucepans

Ingredients

2 gal. chicken stock
1 gal. milk and cream (half and half)
1 lb butter or shortening
12 oz flour
5 lb celery
1 lb onions
1 pt water
¼ c celery seed
1 bay leaf
salt and white pepper
½ tsp baking soda

Procedure

1. Place the butter or shortening in a stock pot and heat.
2. Add the onions and celery. Sauté until slightly tender.
3. Add the flour, making a roux, and cook for 5 min.
4. Add the hot chicken stock, whipping vigorously with a wire whip until slightly thick and smooth. Simmer for approximately 1½ hr.
5. Strain through a fine china cap into a 3 gal. container. Using a ladle, force as much of the vegetable pulp as possible through the china cap.
6. Add the baking soda and pour in the hot milk and cream, stirring gently with a paddle.
7. Season with salt and white pepper. Adjust the flavor by adding the celery-flavored liquid.
8. Serve garnished with croutons, if desired.

Preparation

1. Cut the onions and celery coarse with a French knife.
2. Heat the milk and cream in a saucepan.
3. Place the celery seed in a saucepan. Add the pint of water and simmer for 5 min. Strain through a cheesecloth. Save liquid to flavor soup.

Precautions

- When sautéing the onions and celery, do not let them brown.
- When cooking the roux, do not let it brown.
- While the soup is simmering, stir occasionally with a paddle to avoid sticking or scorching.
- Do not overheat the soup because it may curdle if held at too hot a temperature.

Potato-Leek Soup

Potato-leek soup is a rich cream soup with a potato-onion flavor. The potatoes are cooked into a puree and the pulp is pressed through a china cap to acquire the rich potato flavor. This soup is usually served on the menu as a soup du jour.

3 gal.

Equipment

- 5 gal. stock pot
- Paddle
- French knife
- Baker's scale
- Gallon measure
- China cap
- Saucepan
- 3 gal. container
- Wire whip

Ingredients

2 gal. chicken stock
1 gal. milk and cream (half and half)
1 lb onions
8 oz celery
6 lb Idaho potatoes
10 oz butter
8 oz flour
1 bunch leeks
½ tsp baking soda
1 bay leaf
salt and white pepper to taste

Procedure

1. Place the butter in a stock pot and heat.
2. Add the onions and celery. Sauté until slightly tender.
3. Add the flour, making a roux. Cook for 5 min.
4. Add the bay leaf and hot chicken stock. Whip briskly with a wire whip until thick and smooth.
5. Add the potatoes and simmer for approximately 2 hr or until the potatoes are well done.
6. Remove from the range and strain through a china cap, forcing as much of the potato pulp through the china cap as possible with a ladle. Return to the range.
7. Add the baking soda and pour in the hot milk and cream slowly, stirring gently with a paddle.
8. Add the fine-diced leeks and stir.
9. Season with salt and white pepper and pour into a 3 gal. container.
10. Serve.

Preparation

1. Cut the mirepoix (onions and celery) with a French knife.
2. Wash the leeks and dice very fine with a French knife.
3. Heat the milk and cream in a saucepan.
4. Peel and slice the potatoes thin.

Precautions

- When sautéing the vegetables, do not let them brown.
- When cooking the roux, do not let it brown.
- While the soup is simmering, stir frequently with a paddle to avoid sticking and scorching.
- The baking soda is added to resist curdling. However, if the soup is held at a very high temperature for too long a period, it could still curdle.

Cream of Mushroom Soup

Cream of mushroom soup is a rich cream soup with a mushroom flavor and small diced pieces of mushrooms. It is served on the menu as a soup du jour.

3 gal.

Equipment

- 5 gal. stock pots (2)
- French knife
- Quart measure
- Baker's scale
- China cap
- 6 oz ladle
- 2 qt and 3 qt saucepans
- Kitchen spoon
- 3 gal. container
- Wire whip

Ingredients

2½ gal. chicken stock
2 qt cream
1 lb butter or shortening
1 lb flour
1 bay leaf
1 lb onions
8 oz celery
8 oz butter
2 lb fresh mushrooms
salt and white pepper to taste
1 tsp baking soda

Preparation

1. Wash and chop the mushrooms fine with a French knife.
2. Cut the onions and celery rough with a French knife.
3. Heat the cream in a saucepan.

Procedure

1. Place the butter or shortening in a stock pot and heat.
2. Add the onions and celery. Sauté until slightly tender.
3. Add the flour, making a roux, and cook for 5 min.
4. Add the hot chicken stock, whipping the mixture vigorously with a wire whip until slightly thickened and smooth.
5. Add the bay leaf and simmer until the vegetables are tender (approximately 1 hr).
6. Strain the soup through a fine hole china cap into the second 5 gal. stock pot. Using a ladle, force as much of the vegetable pulp as possible through the china cap. Return to the range.
7. Place 8 oz of butter in a saucepan and heat.
8. Add the chopped mushrooms and sauté until tender, stirring occasionally with a kitchen spoon. Add to the strained soup and simmer for at least 15 min.
9. Add the baking soda and pour in the cream very slowly while stirring gently with a kitchen spoon.
10. Season with salt and white pepper and pour into a 3 gal. container.

Precautions

- When sautéing the vegetables, do not let them brown.
- Whip vigorously with a wire whip when adding the liquid to the roux to avoid lumps.
- Stir occasionally with a kitchen spoon while the soup is simmering to avoid sticking or scorching.
- Do not overheat the soup because it may curdle.

Bean Soup

Bean soup is a very popular soup in the home as well as in the restaurant. In the restaurant, it is served as a soup du jour.

5 gal.

Equipment

- 10 gal. and 5 gal. stock pots
- Food grinder
- French knife
- Paddle
- Baker's scale
- Quart measure
- 5 gal. container

Ingredients

- 4 lb navy beans
- 4 gal. ham stock
- 1 lb carrots
- 2 lb celery
- 2 lb onions
- 1 lb ham fat
- ½ No. 10 can crushed tomatoes
- 4 oz salt
- 1 lb leeks
- 3 cloves garlic
- 1 lb flour
- ¼ oz pepper
- ⅛ oz nutmeg

Procedure

1. Place the beans and the water in which the beans have been soaked into a 5 gal. pot. Bring to a boil, reduce the flame, and let simmer for about 2 hr or until beans are tender and soft.
2. Place the ground ham fat in a separate 10 gal. stock pot and braise for 5 min. Add diced leeks, carrots, celery, and onions. Continue to braise until vegetables are partly done. Stir occasionally with a paddle.
3. Add flour to take up the fat and create a roux. Cook with other ingredients for 5 min.
4. Add the ham stock, which should already be hot.
5. Bring to a boil. Simmer until all vegetables are tender.
6. Remove the cooked beans from the range and add the beans and 1 gal. of the liquid in which the beans were cooked to the ham stock.
7. Add the crushed tomatoes and seasoning.
8. Continue to boil until the beans and vegetables are very tender, approximately 30 min.
9. Pour into a 5 gal. container.

Precautions

- Do not add tomatoes to soup until beans are thoroughly done.
- Be thorough when picking the beans. Do not overlook stones.
- Be cautious when grinding ham fat.

Preparation

1. The day before preparation, pick over the beans, removing all rocks and foreign matter.
2. Wash and soak beans in water overnight. Do not refrigerate.
3. Dice onions, leeks, celery, and carrots with a French knife.
4. Grind ham fat in food grinder. Chop garlic.

Lentil Soup

Lentil soup is served on the lunch or dinner menu as the soup du jour. Lentils are beans that are small and flat.

5 gal.

Equipment

- 5 gal. and 10 gal. stock pots
- Paddle
- French knife
- Baker's scale
- Quart measure
- 5 gal. container

Ingredients

- 1 lb carrots
- 2 lb onions
- 1 lb celery
- 4 lb lentils
- 5 gal. ham or beef stock
- 1 lb shortening or ham fat
- 1 lb flour
- ½ No. 10 can crushed tomatoes
- salt and pepper to taste

Procedure

1. Cook the lentils separately in 4 gal. of the ham stock in the 5 gal. stock pot until tender (approximately 45 min to 1 hr).
2. Braise vegetables in the shortening in the 10 gal. stock pot. Add flour, making a roux. Mix well and cook for 5 min.
3. Add remaining 1 gal. of beef or ham stock to the roux. Let boil until vegetables are tender.
4. Add the cooked lentils and the liquid. Continue to simmer.
5. Add crushed tomatoes.
6. Season with salt and pepper (and a dash of nutmeg if desired).
7. Pour into a 5 gal. container.
8. Sliced or diced pieces of wieners or franks may be added if desired.

Precautions

- When adding stock, stir until all roux is dissolved.
- Pour in the stock slowly.
- Add tomatoes last. Beans or lentils do not cook well when tomatoes are present.

Preparation

1. Remove all foreign matter from the lentils by picking over them carefully.
2. Wash lentils thoroughly.
3. Dice onions, carrots, and celery with a French knife.

Cheddar Cheese Soup

Cheddar cheese soup is a rich cream soup with a strong Cheddar cheese flavor. This soup is usually served when one wishes to feature a different soup.

3 gal.

Equipment

- 5 gal. stock pot
- French knife
- Wire whip
- Baker's scale
- Spoon measure
- Gallon measure
- 6 oz ladle
- China cap
- Kitchen spoon
- 3 gal. container
- Metal box grater
- Saucepan
- Paddle

Ingredients

2 gal. chicken stock
1 gal. milk and cream
   (half and half)
1 lb carrots
8 oz onions
8 oz celery
1 lb butter
10 oz flour
2 lb sharp Cheddar cheese
2 tbsp Worcestershire
   sauce
1 tbsp paprika
salt and white pepper to
   taste

Procedure

1. Place the butter in a stock pot and heat.
2. Add the mirepoix and sauté until slightly tender.
3. Add the flour and paprika, making a roux, and cook for 5 min.
4. Add the hot chicken stock, whipping vigorously with a wire whip until slightly thick and smooth.
5. Simmer for approximately 1 hr or until all the vegetables are tender.
6. Strain through a fine china cap, forcing through as much of the vegetable flavor as possible using a ladle.
7. Stir in the grated cheese gradually with a paddle.
8. Add the Worcestershire sauce and blend in thoroughly.
9. Add the hot milk and cream and stir in gently with a paddle.
10. Season with salt and white pepper, pour into a 3 gal. container, and serve.

Preparation

1. Dice the onions, carrots, and celery (*mirepoix*) with a French knife.
2. Grate the cheese by rubbing it on the coarse grid of a metal box grater.
3. Heat the milk and cream in a saucepan.

Precautions

- When sautéing the mirepoix, do not let it brown.
- While the soup is simmering, stir occasionally with a paddle to avoid scorching.
- Do not hold the soup at too high a temperature. It may curdle.

Corn Chowder

Corn chowder is a thick soup with a creamy consistency and a rich corn flavor. The soup contains diced potatoes, which are an ingredient in all chowders. Corn chowder is served on the menu as a soup du jour.

3 gal.

Equipment

- 5 gal. stock pot
- French knife
- Paddle
- 6 oz ladle
- China cap
- Baker's scale
- Quart measure
- Saucepan
- 3 gal. container

Ingredients

2 gal. chicken stock
1 No. 10 can creamed
   corn
1 lb butter
10 oz flour
1 lb onions
1 oz celery
3 lb. potatoes
3 lb fresh corn
2 qt cream
salt and white pepper to
   taste

Procedure

1. Place the butter in a stock pot and heat.
2. Add the onions and celery and sauté until partly tender.
3. Add the flour, making a roux, and cook 5 min.
4. Add the hot chicken stock, stirring constantly with a paddle until slightly thick and smooth.
5. Add the creamed corn and simmer, stirring occasionally, for approximately 1 hr until the celery becomes very tender.
6. Remove from the fire and strain through a fine china cap, forcing through as much of the corn pulp as possible with a ladle.
7. Return to the range, add the diced potatoes and fresh corn, and simmer until the potatoes are tender.
8. Add the warm cream slowly, stirring constantly to blend thoroughly.
9. Season with salt and white pepper and pour into a 3 gal. container.
10. Serve.

Preparation

1. Dice the onions, celery, and potatoes with a French knife.
2. Cut the corn off the cob. Cut cobs in half crosswise, stand cob on end, and cut with a downward stroke of French knife.
3. Heat the cream in a saucepan until it is warm.

Precautions

- When sautéing the vegetables, do not let them brown.
- Stir occasionally with a paddle throughout the simmering period to avoid sticking or scorching.
- Do not hold the soup at too high a temperature for a long period of time because it may curdle.

Oxtail Soup

Oxtail soup is a thick soup very similar to English beef broth, but with small diced pieces of oxtail meat added. This soup is fairly popular in England, but not as well accepted in the United States. This soup is usually served as a soup du jour.

3 gal.

Equipment

- 5 gal. stock pots (2)
- Meat saw
- French knife
- Paddle
- Baker's scale
- Quart measure
- Saucepans (2)
- Roast pan
- China cap
- Wire whip
- Spoon measure
- 3 gal. container

Ingredients

2½ gal. beef stock
6 lb oxtail
6 oz uncooked barley
8 oz turnips
2 lb onions
4 oz leeks
8 oz celery
1 lb carrots
6 oz flour
1 pt canned tomatoes
1 c tomato puree
8 oz butter or shortening
2 tbsp Worcestershire
    sauce
salt and pepper to taste

Preparation

1. Cut the oxtail into pieces by cutting through each joint with a French knife. Brown in a preheated oven at 375°F.
2. Place the barley in a saucepan and wash. Cover with water and simmer for approximately 2 hr or until tender, then drain in a china cap and wash with cold water.
3. Dice the onions, leeks, turnips, celery, and carrots with a French knife.
4. Crush the tomatoes by hand.
5. Place the diced turnips in a saucepan and simmer until tender, then drain.

Procedure

1. Place the browned oxtails in a stock pot, cover with the beef stock, and simmer until the meat is tender enough to remove from the bone.
2. Strain the stock through a fine china cap and keep hot. Remove the oxtail meat from the bone and dice into small cubes.
3. Place the butter or shortening in the second stock pot and heat.
4. Add the diced onions, leeks, carrots, and celery and sauté until slightly tender.
5. Add the flour, making a roux, and cook for 5 min.
6. Add the hot beef stock, whipping vigorously with a wire whip until slightly thick and smooth. Simmer until the vegetables are tender.
7. Add the crushed tomatoes, tomato puree, turnips, barley, diced oxtail meat, and Worcestershire sauce. Simmer for an additional 20 min.
8. Season with salt and pepper and pour into a 3 gal. container. Serve.

Precautions

- When sautéing the vegetables, do not let them brown.
- While the soup is simmering, stir occasionally with a paddle to avoid sticking or scorching.

Lobster Bisque

Lobster bisque is a slightly thick, rich cream soup with small pieces of cooked lobster flowing through it to add flavor and color. A small amount of wine is added to enhance the flavor.

3 gal.

Equipment

- 5 gal. stock pot
- French knife
- Baker's scale
- 3 gal. sauce pot
- Gallon measure
- 3 gal. container
- China cap
- Paddle
- Cheesecloth
- Saucepan

Ingredients

4 lb raw lobster tails, cut
    into 3 crosswise pieces
2 gal. water
12 oz onions, diced
1 lemon, sliced
1 bay leaf
6 oz celery, diced
1 lb 4 oz butter
1 lb flour
1 gal. milk and cream
6 oz sherry
1 tbsp paprika
salt and white pepper to
    taste

Preparation

1. Place the water, lemon, celery, onion, and bay leaf in a sauce pot. Simmer for 30 min.
2. Add the lobster tails, which have been cut into 3 pieces. Simmer for 10 min more and remove from the heat.
3. Strain the stock through a fine china cap covered with a fine cheesecloth and hold.
4. Shuck by cracking the lobster shell with the back of a French knife and removing the shell with the hands. Dice the lobster meat with a French knife. Heat the milk and cream in a saucepan.

Procedure

1. Place the butter in a stock pot and heat.
2. Add the diced lobster and sauté slightly.
3. Add the paprika and flour and cook for approximately 3 min.

4. Add the hot stock slowly, stirring continuously with a paddle until slightly thick and smooth. Simmer for 20 min.
5. Pour in the hot milk and cream slowly, stirring gently with a paddle.
6. Add the sherry and blend well. Pour into a 3 gal. container.
7. Serve.

Precautions

- When heating the butter, do not burn.
- When sautéing the lobster, do not brown.
- Do not hold the soup at too high a temperature because it may curdle.

Shrimp Bisque

Shrimp bisque is a slightly thick, rich cream soup with small pieces of cooked shrimp flowing through it to add flavor and color. A small amount of wine is added to enhance the flavor.

3 gal.

Equipment

- 5 gal. stock pot
- French knife
- Baker's scale
- Sauce pot
- Gallon measure
- 3 gal. container
- China cap
- Paddle
- Cheesecloth

Ingredients

3 lb raw shrimp
2 gal. water
12 oz onion, diced
1 lemon, sliced
1 bay leaf
6 oz celery, diced
stems of 1 bunch parsley
1 lb 4 oz butter
1 gal. milk and cream
   (half and half)
6 oz sherry
1 lb flour
1 tbsp paprika
salt and white pepper to taste

Preparation

1. Place the water, lemon, celery, onion, bay leaf, and parsley stems in a sauce pot. Simmer for 30 min.
2. Add the shrimp and continue to simmer for 10 min more, then remove from the heat.
3. Strain the stock through a china cap covered with a fine cheesecloth and hold.

4. Cool the shrimp in cold water. Remove the shell from the meat by pulling it away from the body with the fingers. Devein (remove the sand vein from the back of the shrimp) by taking out the black line with the point of a small knife, a toothpick, or a can opener. Dice the shrimp fine with a French knife.
5. Heat the milk and cream.

Procedure

1. Place the butter in a stock pot and heat.
2. Add the diced shrimp and sauté slightly.
3. Add the paprika and flour and cook for approximately 3 min.
4. Add the hot stock slowly, stirring continuously with a paddle until slightly thick and smooth. Simmer for 20 min.
5. Pour in the hot milk and cream slowly, stirring gently with a paddle.
6. Add the sherry and blend well. Pour into a 3 gal. container.
7. Serve.

Precautions

- When heating the butter, do not burn.
- When sautéing the shrimp, do not brown.
- Do not hold the soup at too high a temperature because it may curdle.

Oyster or Clam Bisque

Oyster or clam bisque is a slightly thick soup prepared with milk and cream, vegetables, and seasoning. Most bisques are prepared from shellfish.

3 gal.

Equipment

- 5 gal. stock pot
- French knife
- Baker's scale
- Gallon measure
- Spoon measures
- Paddle
- 3 gal. container
- 6 qt sauce pot
- China cap
- Wire whip
- 6 qt saucepan

Ingredients

3 qt shucked oysters or clams
1 gal. water
1 gal. milk and cream
1 lb 4 oz butter
8 oz onions
1 lb flour
1 bay leaf
1 tsp paprika
salt and white pepper to taste

Preparation

1. Cut the oysters or clams into quarters with a French knife. Place in a sauce pot, add 1 gal. of water, and simmer for 10 min. Strain the liquid off the cooked oysters or clams by pouring it through a china cap into a separate pan. Chop the oysters or clams very fine and return them to the strained liquid.
2. Mince the onions with a French knife.
3. Heat the milk and cream.

Procedure

1. Place the butter in a stock pot and heat.
2. Add the finely minced onions and sauté until slightly tender. Do not brown.
3. Add the flour and paprika and cook for about 5 min.

4. Add the oysters or clams and oyster or clam liquid, milk, and cream, whipping vigorously with a wire whip until slightly thick and smooth.
5. Add the bay leaf and simmer for 30 min. Remove the bay leaf. Pour the soup into a 3 gal. container.
6. Season with salt and white pepper and serve.

Precautions

- While the bisque is simmering, stir occasionally with a paddle to avoid sticking and scorching.
- When sautéing the onions, do not let them brown.

Seafood Chowder

Seafood chowder is a thick soup similar to clam chowder, but with an assortment of fish and shellfish added instead of only clams. This soup is an excellent choice for the Friday menu.

3 gal.

Equipment

- 5 gal. stock pot
- French knife
- Quart measure
- Baker's scale
- Paddle
- Saucepans (3)
- Cheesecloth
- 3 gal. container

Ingredients

2½ gal. fish stock
1 qt hot milk
1 qt warm cream
2 lb red snapper
2 lb halibut
2 lb shrimp
1 pt canned clams, chopped
12 oz flour
1 lb butter
3 pt potatoes
1 bay leaf
1 lb 8 oz onions
8 oz celery
¼ oz thyme
salt and white pepper to taste

Preparation

1. Place the red snapper and halibut in a saucepan, cover with salt water, and simmer until the fish is cooked. Break a piece of the fish in half. If the flesh is dull in appearance, the fish is done. If it has a slightly glossy appearance, the fish is not fully cooked. When done, drain the liquid from the fish and allow the fish to cool. Remove all skin and bones and flake.

2. Peel the raw shrimp. Remove the shell from the meat by pulling it away from the body with the fingers. Devein (remove sand vein from the back of the shrimp) by taking out the black line with the point of a small knife, toothpick, or can opener. Cut each shrimp into four pieces.
3. Dice the onions, celery, and potatoes with a French knife. Simmer the thyme in water, strain through a cheesecloth and save juice for seasoning.

Procedure

1. Place the butter in a stock pot and heat.
2. Add the diced vegetables and shrimp. Sauté until the vegetables are slightly tender.
3. Add the flour, making a roux, and cook for 5 min.
4. Add the hot fish stock and bay leaf, stirring vigorously with a paddle until slightly thickened and smooth. Simmer until the vegetables are tender.
5. Add the clams, thyme, liquid, and potatoes. Continue to simmer until the potatoes are tender.
6. Add the cooked red snapper and halibut. Simmer for 5 min longer.
7. Pour in the milk and cream slowly, stirring gently with a paddle until well blended.
8. Remove the bay leaf and season with salt and pepper. Pour into a 3 gal. container.

Precautions

- When sautéing the vegetables and shrimp, do not let them brown.
- While the soup is simmering, stir occasionally to avoid sticking or scorching.
- Before adding the cooked fish to the soup, be sure all bones have been removed.

Trade Tip *Chowder literally means the contents of a kettle. While chowder recipes vary from region to region, the most popular are Manhattan clam chowder, which has a tomato base, and New England clam chowder, which has a milk base.*

SPECIAL SOUP RECIPES

Chicken Gumbo Soup

Chicken gumbo soup is a creole soup, very popular in New Orleans and the southern states. It is often served as the soup du jour.

5 gal.

Equipment

- 10 gal. stock pot
- French knife
- Paddle
- Baker's scale
- Quart measure
- 5 gal. container

Ingredients

2 lb celery
1 lb green peppers
2 lb onions
1 lb shortening or chicken fat
5 gal. chicken stock
1 lb rice
2 No. 2 cans okra
1 No. 10 can tomatoes
salt and pepper to taste
chicken base to taste

Preparation

1. Dice onions, celery, and green peppers using a French knife.
2. Crush tomatoes.
3. Wash rice.

Procedure

1. Place shortening or fat in stock pot, add diced vegetables, and sauté slightly.
2. Add chicken stock and rice. Let cook until rice is done.
3. Add crushed tomatoes and let simmer 5 min.
4. Add okra and season with salt and pepper. Add chicken base if stock flavor is lacking.
5. Pour into a 5 gal. container.

Precautions

- Do not overcook the okra.
- Use a rich chicken stock if possible.

Borscht is a thin soup of Russian origin. The main ingredients in the preparation are beets and tomatoes. Borscht can be served hot or cold and is usually garnished with sour cream. It can be served as a soup du jour or as a specialty item.

Borscht

2½ gal.

Equipment

- Quart measure
- French knife
- China cap
- 6 oz ladle
- Spoon measure
- Cup measure
- Pepper mill
- 5 gal. stock pot
- 3 gal. container

Ingredients

1½ qt canned beets
1 pt onions
2 qt canned tomatoes
1½ gal. beef stock
1 pt lemon juice
1 tsp garlic
½ c parsley stems
2 bay leaves
½ c sugar
3 tsp salt
1 tbsp paprika
fresh ground pepper to taste
sour cream as needed

Preparation

1. Dice the onions and canned beets with a French knife. Save the juice from the beets. Squeeze the juice from lemons.
2. Mince the garlic.

Procedure

1. Place the beef stock, tomatoes, onion, and garlic in a stock pot and simmer until the onions are partly done.
2. Add the remaining ingredients and simmer for about 1½ hr.
3. Strain the soup through a china cap into a 3 gal. container. Using a ladle, force as much of the vegetable pulp as possible through the china cap.
4. Serve hot with sour cream on top of each serving or let cool, refrigerate, and serve cold with sour cream on top.

Precaution

- Exercise caution when straining the soup.

Trade Tip

Gumbo is a soup thickened with okra or filé. It contains whatever meats are available (usually sausage, fish, and shellfish) and local vegetables such as onions, tomatoes, peppers, etc. Filé is powdered young leaves of sassafras, which is used as a thickener for soups and stews.

Minestrone Soup

Minestrone soup is an Italian vegetable soup that has become quite popular in the United States since the end of World War II. It is served on the menu as a soup du jour or with Italian entrees.

3 gal.

Equipment

- Stock pot
- Food grinder
- French knife
- Paddle
- Baker's scale
- Spoon measure
- Quart measure
- Cup measure
- 3 gal. container
- China cap
- Sauce pot

Ingredients

2½ gal. beef stock
1 c olive oil
4 oz black-eyed peas
4 oz red beans
1 small can chickpeas (garbanzos)
1 lb 8 oz onions
1 lb celery
1 lb carrots
8 oz green pepper
6 oz cabbage
3 garlic cloves
1 qt canned tomatoes
5 oz salt pork
2 tbsp parsley
1 tsp basil
1 tsp oregano
⅓ c Parmesan cheese
salt and pepper to taste

Preparation

1. Wash the black-eyed peas and red beans. Cover with water and soak overnight. Drain and place them in a sauce pot, cover with salt water, and simmer until tender. Drain a second time through a china cap.
2. Mince onions, celery, green pepper, carrots, cabbage, and garlic with a French knife.
3. Crush the tomatoes by hand.
4. Chop the parsley with a French knife.
5. Grind the salt pork on the food grinder using the very fine chopper plate.

Procedure

1. Place the olive oil in a stock pot and heat.
2. Add the minced vegetables and garlic and sauté until slightly tender.
3. Add the crushed tomatoes, beef stock, basil, and oregano. Simmer for approximately 1 hr or until all the vegetables are tender.
4. Add the black-eyed peas, red beans, and chickpeas and continue to simmer for an additional half hour.
5. Remove from the fire, add the chopped parsley, ground salt pork, and Parmesan cheese and blend thoroughly with a paddle.
6. Season with salt and pepper and pour into a 3 gal. container.
7. Serve by sprinkling Parmesan cheese on top of each serving.

Precautions

- When sautéing the vegetables, do not let them brown.
- When adding the basil and oregano, rub them between your palms to release their flavor.

English Beef Broth

English beef broth is a thick soup with barley as the main ingredient. It is served in restaurants as a soup du jour.

5 gal.

Equipment

- 10 gal. stock pot
- French knife
- Paddle
- Baker's scale
- Quart measure
- 5 gal. container
- Saucepan

Ingredients

2 lb carrots
3 lb onions
2 lb celery
2½ gal. brown stock
1 lb bacon or ham fat
1 lb tomato puree
salt and white pepper to taste
1 lb leeks
2½ gal. ham stock
1 lb flour
1 lb 4 oz barley

Procedure

1. Place bacon or ham fat in a stock pot. Add the diced onions, carrots, celery, and leeks. Sauté until partly done (onions will have a transparent appearance).
2. Add flour to take up the fat, making a roux. Cook for 5 min.
3. Add ham stock, brown stock, and tomato puree and let simmer for 30 min.
4. Cook the barley in a separate saucepan. Cover with a ratio of three parts water to one part barley. Simmer until tender. The barley contains a high percentage of starch, and cooking it in the soup will make the soup too thick and starchy.
5. Add cooked barley to the soup and continue to simmer until all ingredients are tender (approximately 45 min).
6. Season with salt and pepper.
7. Pour into a 5 gal. container.

Precautions

- Cook the barley separately and wash thoroughly after cooking.
- While the soup is cooking, stir occasionally to avoid sticking.
- After the barley is cooked, wash well with water to remove starch before adding it to the soup.

Preparation

1. Wash barley.
2. Dice garnish (carrots, celery, and onions) with a French knife.

New England Clam Chowder

New England clam chowder is similar to Manhattan clam chowder, but the tomatoes are omitted and milk and cream are added. This chowder is popular on the Friday menu.

5 gal.

Equipment

- 10 gal. stock pot
- Paddle
- Baker's scale
- Quart measure
- French knife
- 1 qt saucepan
- Cheesecloth
- 5 gal. container
- 3 gal. sauce pot

Ingredients

3 gal. fish stock and clam juice
2 gal. milk and cream
1 lb 8 oz shortening
1 lb 8 oz flour
2 lb onions
2 lb celery
1 lb green pepper
8 oz leeks
4 lb peeled potatoes
2 qt canned clams
¼ oz. thyme
1 oz garlic
salt and white pepper to taste

Preparation

1. Dice the mirepoix (onions, celery, green peppers, potatoes, and leeks) with a French knife.

2. Simmer the thyme in water for 5 min in a saucepan. Strain through a cheesecloth and save the liquid for seasoning.
3. Heat the milk and cream in a sauce pot.
4. Drain and chop the clams. Mince the garlic.

Procedure

1. Place the shortening in a stock pot and heat.
2. Add the mirepoix and garlic and sauté until slightly tender. Do not brown.
3. Add the flour, making a roux, and cook for 5 min.
4. Add the hot stock and clam juice, whipping to make it smooth. Simmer until the vegetables are just slightly tender.
5. Add the potatoes, clams, and thyme liquid. Continue to simmer until the potatoes are tender.
6. Add the hot milk and cream slowly, stirring gently with a paddle. Bring the soup back to a simmer.
7. Season with salt and pepper and pour into a 5 gal. container.
8. Remove from the range and serve.

Manhattan Clam Chowder

Manhattan clam chowder is also known as Long Island or Philadelphia clam chowder. This chowder differs from the New England chowder in that it contains tomatoes and no milk or cream. A chowder always contains diced potatoes. This chowder is popular on the Friday menu.

5 gal.

Equipment

- French knife
- Baker's scale
- Quart measure
- 10 gal. stock pot
- Paddle
- Saucepan
- Cheesecloth
- 5 gal. container

Ingredients

4½ gal. fish stock and clam juice
2 qt clams, canned, drained, chopped
3 qt potatoes, peeled
2 lb onions
2 lb celery
1 lb green peppers
1 lb leeks
2 lb shortening
½ oz thyme
⅛ oz rosemary leaves
2 oz flour
1 oz garlic
1 No. 10 can whole tomatoes
salt and white pepper to taste

Preparation

1. Dice onions, celery, leeks, potatoes, and green peppers with a French knife.

2. Place the thyme and rosemary leaves in a saucepan. Simmer in water for a few minutes, strain the liquid through a cheesecloth, and save the liquid for later use.
3. Mince the garlic and crush the tomatoes.

Procedure

1. Place the shortening in a large stock pot and heat.
2. Add the diced vegetables and minced garlic. Sauté until the vegetables are slightly tender.
3. Add the flour, making a roux, and cook for 5 min.
4. Add the hot fish stock and clam juice and cook, stirring vigorously with the paddle until slightly thick.
5. Continue to simmer until the vegetables are tender.
6. Add the diced potatoes, clams, tomatoes, and the thyme and rosemary-flavored liquid. Simmer until the potatoes are tender.
7. Remove from the range and season with salt and pepper and pour in a 5 gal. container.
8. Serve hot.

Precautions

- When sautéing vegetables, do not let them brown.
- Stir the soup occasionally with the paddle throughout the cooking period to avoid sticking.

Chili Bean Soup

Chili bean soup is a variation of chili con carne (chili with beans). It is very similar to chili con carne in aroma and taste, but differs in consistency. This soup is usually featured on the luncheon menu.

2 gal.

Equipment

- 5 gal. stockpot
- Baker's scale
- Pint measure
- Gallon measure
- French knife
- Paddle
- 3 gal. stockpot
- China cap
- 3 gal. container

Ingredients

1 lb 8 oz ground beef
12 oz onions
½ c salad oil
¼ oz garlic
1 qt tomato puree
1 gal. water or beef stock
1½ oz chili powder
1/5 c cider vinegar
¼ oz ground cinnamon
1 pinch crushed red peppers
4 bay leaves
¼ oz cumin
½ oz salt
½ oz sugar
1 lb 12 oz red beans, dried or 2 lb canned red beans

Preparation

1. Mince the onions and garlic with a French knife.
2. Place the beans in a 3 gal. stockpot the day before preparation. Cover with four times their amount of water. Let soak at room temperature overnight. The morning of preparation, simmer the beans in the water in which they were soaked. When the beans are tender, remove them from the range, drain off all liquid, and hold.

Procedure

1. Place the ground beef in a stockpot and brown the beef slightly. Drain off excess grease.

2. Add the salad oil, onions, and garlic. Sauté with the beef until tender.
3. Add all the remaining ingredients except the cooked red beans. Bring to a boil, then reduce to a simmer. Simmer for approximately 2 hr.
4. Add the cooked beans and simmer for approximately 15 min more. Remove from the heat.
5. Remove all the bay leaves. Check seasoning. Pour into 3 gal. container and hold for service.

Precautions

- Exercise caution when mincing the onions and garlic.
- Stir the preparation occasionally during the simmering period with a paddle.

COLD SOUP RECIPES

Vichyssoise

Vichyssoise is a rich, creamy, potato soup that is always served cold. It is usually garnished with chopped chives and is an extremely popular appetizer on the warm weather menu.

3 gal.

Equipment

- 5 gal. stock pot
- Gallon measure
- Baker's scale
- French knife
- Paddle
- 3 gal. container
- Vegetable peeler

Ingredients

2 gal. chicken stock
1 gal. light cream
3 lb raw potatoes
1 lb onions
12 oz leeks
2 bay leaves
8 oz celery
4 white peppercorns
salt and white pepper to taste
chives, for garnish

Preparation

1. Mince the chives with a French knife.
2. Dice onions, leeks, and celery with a French knife.
3. Peel the potatoes with a vegetable peeler. Wash and slice the potatoes fairly thin.
4. Crack the peppercorns.

Procedure

1. Place the butter in a stock pot and heat.
2. Add the onions, celery, and leeks and sauté until slightly tender.

3. Add the chicken stock, potatoes, bay leaves, and peppercorns. Simmer until the potatoes are very well done.
4. Strain the mixture through a fine china cap into a 3 gal. container. Using a ladle, force as much of the potato pulp as possible through china cap. Let the mixture cool.

5. Add the cream and blend in thoroughly using a paddle.
6. Season with salt and white pepper.
7. Chill well. Garnish with minced chives and serve ice cold.

Gazpacho

Gazpacho is a cold vegetable soup of Spanish origin. The crisp colorful vegetables blended into the highly seasoned consommé creates a unique preparation that can be featured on the luncheon or a la carte summer menu.

1 gal.

Equipment

- Quart measure
- Cup measure
- Baker's scale
- 2 gal. stainless steel bowl
- French knife
- Kitchen spoon

Ingredients

2 qt tomatoes
1 pt green peppers
1 garlic clove
1 pt onions
½ c pimientos
1 qt beef consommé
½ c salad oil
½ c vinegar, cider
1 c cucumbers
¼ c sugar
¼ oz paprika
⅛ oz (pinch) cumin
salt and fresh ground pepper to taste

Preparation

1. Prepare beef consommé.
2. Mince the garlic, onions, pimientos, green peppers and cucumbers with a French knife.
3. Peel and dice the tomatoes.

Procedure

1. Place all the ingredients except the cucumbers in a stainless steel bowl. Let set at room temperature for approximately 2 hr. Stir frequently.
2. Place in the refrigerator for 2 additional hours.
3. Stir in the cucumbers just before serving. Season with salt and pepper.
4. Serve in cold bouillon cups.

Precaution

- Exercise caution when mincing the vegetables.

Sauces and Gravies

A sauce or gravy is a rich flavored, thickened liquid used to complement another food item. Sauces and gravies enhance the flavor, moistness, and appearance of meats, vegetables, fish, poultry, and desserts. The main difference between sauces and gravies is the flavor. A sauce does not always possess the same flavor as the food item it accompanies. Gravies possess the flavor of the meat with which they are served. The base of a gravy is the meat drippings acquired while roasting the meat.

The sauce or gravy selected should flow over the food item and provide a thin coating that enhances the food item rather than a heavy mass that disguises the food item. In addition, the sauce or gravy should not overpower the flavor of the food item with which it is served.

The use of sauces has declined slightly because of the time and cost involved in preparation. However, sauces offer variety in a menu and contribute to the reputation of the food service establishment.

SAUCES

Sauces are generally classified as warm sauces, cold sauces, and dessert or sweet sauces. *Warm sauces* consist of leading sauces and small sauces. Small sauces are variations of the leading sauces. *Cold sauces* are served cold with both hot and cold foods. *Dessert* or *sweet sauces* contain a high percentage of sugar and are usually served with dessert items. *Butter sauces*, although used frequently in commercial kitchens, are not considered a major sauce category because they are easy to prepare. Some sauces cannot be placed in a specific category. Sauces such as mint sauce and oriental sweet and sour sauce are listed as miscellaneous in the recipes.

Warm Sauces

Warm sauces (leading sauces and small sauces) are the most popular and numerous of the three major sauce categories. Warm sauces can be served with all foods. *Leading sauces*, called *mother sauces* by the famous French Chef Escoffier, are of great importance because they are the basis for all other sauces. The leading sauces are:

- Brown or espagnole sauce
- Cream or béchamel sauce
- Velouté or fricassee sauce
- Hollandaise sauce
- Tomato sauces

Preparing *small sauces* from leading sauce is done by changing ingredient amounts or adding certain ingredients. Adding chopped hard boiled eggs to béchamel or cream sauce yields egg sauce. Adding sautéed onions to brown sauce creates onion sauce.

Most warm sauces are made from stock, which is the basis of many preparations in the commercial kitchen. The quality of the stock used determines the quality of the sauce. Chicken stock is made from chicken bones. Beef stock, sometimes known as white stock, is made from beef bones. Brown stock is made from beef, veal, or pork bones. Fish stock, also known as *fumet*, is made from fish bones and trimmings.

The best stocks are made from young animals. Young animals have more cartilage in their bone structure. Cartilage and connective tissue (collagen) break down during the simmering process to form gelatin to improve the quality of stock. For this reason, most stocks used in sauces are reduced or boiled down to a concentrate. All warm sauces should have the following characteristics:

- The sauce has a slight sheen.
- The consistency is flowing, smooth, and lump-free.
- The taste is velvety.
- The flavor is delicate.
- The starch is completely cooked.
- Brown sauces are a rich brown.
- Velouté sauces are a creamy color.

The thickening agent used in the preparation of warm sauces depends on the sauce and the preference of the cook or chef. In most cases, a roux, whitewash, or cornstarch is used. A roux is considered the best thickening agent because it holds up better under constant heat without breaking back into a liquid. A roux may be a plain roux or a French roux. The difference between the two is in the shortening used. A plain roux is made by blending equal portions by weight of flour and fat. Fats such as shortening, margarine, oil, or rendered animal fat may be used. A French roux is made in the same manner, but butter is used instead of fat, which produces a richer roux.

Roux must always be properly cooked to eliminate the raw flour taste. The amount of cooking time required depends on its intended use. A roux to be used for a white or light sauce is cooked only slightly. A roux to be used in a brown sauce is cooked until it becomes slightly brown. When using a roux as a thickening agent, always add the hot stock to the roux, stirring constantly to eliminate lumps and to take full advantage of the thickening powers of the roux.

Whitewash is a mixture of equal amounts of cornstarch and flour diluted in water. It is poured slowly and stirred into the boiling preparation to be thickened, such as stews, stocks, and fricassees. The amount of thickening used depends on the thickness desired and the amount to be thickened.

Cornstarch is mixed with cold water or stock and poured into the boiling preparation in the same manner as whitewash, stirring constantly while pouring. Cornstarch not only thickens, but also provides a glossy semiclear finish to a product. It is used extensively in thickening sweet sauces. The amount to use depends upon the same conditions required for whitewash.

Cold Sauces

Cold sauces are blended from many different foods, the most popular being mayonnaise. Cold sauces can be served with both hot and cold foods. They are sometimes known as *dressings* since they function as a dressing rather than a sauce when served with foods such as salads. Consequently, mayonnaise is classified as a dressing rather than a cold sauce. The difference between the terms *sauce* and *dressing* is minimal. Sauce usually refers to thickened liquids that enhance the flavor of meats and vegetables. Dressing usually refers to thickened liquids that enhance the flavor of salads. In addition, sauces are usually prepared using a rich stock base. Dressings are usually prepared using a salad oil base.

Trade Tip

The french term for stock is "fond," meaning base or foundation. High-quality stocks are the basis for many soups and sauces.

Butter Sauces

Butter sauces are generally simple to prepare. Most butter sauces are prepared by melting butter in a saucepan and adding other ingredients for flavor, or by placing the butter in a saucepan and heating it until it becomes a medium brown color before the flavoring ingredients are added. Butter sauce increases the flavor, moisture, and appearance of the preparation.

Dessert or Sweet Sauces

Dessert or sweet sauces are usually made from fruit or fruit juice, milk, and/or cream. These sauces contain a high percentage of sugar. Dessert or sweet sauces are commonly served with meats such as ham or duck, breakfast items such as French toast or pancakes, or various desserts. These sauces possess a high sheen since they are usually thickened with high-gloss starches or caramelized sugar. The consistency of dessert or sweet sauces varies from very fluid to very thick. The variation in consistency depends on how the sauce is to be served, the item the sauce is served with, and the thickener used. Dessert or sweet sauces can be served with hot or cold food.

Sauce Preparation Techniques

Adding spices to a sauce requires knowledge of the spice and how it affects the preparation. The seasoning must never overpower the other ingredients used, except in preparations that require a dominating flavor, such as curries. If using whole spices, always remove the spice from the sauce when the desired flavor is obtained. If left in, the spice will continue to disperse flavor. Spices such as paprika, curry, and dry mustard used in a preparation should be worked into the roux or dissolved in liquid for a more uniform distribution and a smoother sauce.

Using onions in a sauce is a common practice because of their desirable flavor. However, the flavor of the onions should not hinder the delicate flavor of the sauce. A milder onion flavor can be obtained by using leeks or chives instead of onions.

Adding whipped or sour cream to a sauce should be done upon completion of the basic preparation. The cream is folded into the sauce using a gentle motion to retain as many air cells as possible to produce a smoother, lighter, and fluffier sauce. Sauces such as bonne-femme, divine, mousseline, and sour cream have cream folded into the basic preparation.

Southern Pride Catfish

A sauce is a rich-flavored, thickened liquid used to complement another food item.

Adding wine to a sauce produces best results if the wine is added at the end of the cooking period just as it is removed from the heat. Some recipes call for wine to be added earlier to reduce and concentrate the flavor. Always follow the recipe instructions carefully when adding wine. Wine contains a high percentage of acid, which breaks down starch. This may require an increase in the thickening agent used.

Browning or glazing a sauce is required when preparing certain preparations such as Coquilles St. Jacques Mornay, Lobster Thermidor, or Fillet de Sole Marquery. Before a sauce browns, unsweetened whipped cream is folded in or Parmesan cheese is sifted into the sauce. After the sauce is placed over the surface of the item, it is browned under a broiler or salamander and served immediately.

Caramelized sugar is called for in certain sauce recipes such as bigarade, brandy, and some fruit sauces. Caramelized sugar supplies a slightly sweet taste and a high sheen. Use a thick bottom pot to caramelize sugar. Place the sugar in the pot and heat until it turns a medium brown color. Do not overcook, as it will produce black jack, which is used to color gravies, stews, and sauces. After the sugar has been browned to the desired color, add the liquid while stirring rapidly with a kitchen spoon. Use caution as a flare-up may occur when liquid is added. Watch carefully to prevent the mixture from boiling over.

Adding a liaison to a sauce at the end of the cooking period increases the flavor and richness of the sauce. The liaison, a blend of egg yolks and cream, is blended in a stainless steel bowl using a wire whip. Part of the hot mixture is blended into the liaison to prevent curdling. Once the liaison is added, heat to a serving temperature, but do not boil.

When holding a sauce for service, dot the sauce with pieces of butter or margarine to prevent a crust from forming. If a crust forms, it will cause lumps when the sauce is stirred and the crust is broken. After the sauce is finished and placed in a bain-marie or steam table pan, spot pieces of butter or margarine on the surface of the hot sauce. As the butter or margarine melts, spread it over the surface using the bottom of a ladle.

When serving a sauce, use a 2 oz or 4 oz ladle. The serving amount may vary depending on the need. The sauce is served with the food item for maximum eye appeal. If the item has an attractive appearance, such as a sautéed or fried item, place it on top of the sauce. If the item is not attractive, such as boiled beef or braised stuffed cabbage, place the sauce over the item.

GRAVIES

Gravies are sauces that have the same flavor as the meat they accompany when served. Gravies are usually prepared from the drippings and juices of roasted meats. The flavor and volume of the drippings and juices can be increased by supplementing with a brown sauce. The brown sauce is prepared by browning and boiling bones of the animal that is being roasted. For example, when preparing a pork roast, pork bones are used.

Adding brown sauce is necessary in order to prepare enough gravy for the amount of meat to be served. Meat drippings and juices will evaporate during the roasting period. Brown sauce also enhances gravies of meats such as pork and veal, which have a very delicate flavor.

The best thickening agent to use in gravy preparation is roux. Salt and pepper are the main seasoning ingredients because they enhance natural flavors. Some gravies may be improved using spices and herbs, but only in moderation. A hint of cloves improves pork flavor, and the herb marjoram enhances lamb gravy.

BROWN OR ESPAGNOLE SAUCE RECIPES

Brown Sauce

This leading sauce is used in the preparation of gravies, small sauces, and certain stews and soups. It is generally prepared in large quantities and kept on hand in the commercial kitchen. Brown sauce is used to supplement natural gravy drippings when preparing gravies and is also used if additional gravy is needed when preparing beef stew.

2 gal.

Equipment

- 3 gal. sauce pot
- French knife
- China cap
- Ladle
- Paddle
- Baker's scale
- Spoon measure
- Quart or gallon measure

Ingredients

2 gal. hot brown stock
1 lb 8 oz onion
1 lb celery
1 lb carrots
1 lb shortening
1 lb bread flour
2 bay leaves
2 tsp thyme
1 c tomato puree
salt and pepper to taste

Procedure

1. Place the shortening in the sauce pot and heat. Add the rough garnish (onions, carrots, and celery) and sauté slightly.
2. Add the flour, making a roux, and cook 5 min.
3. Add the hot brown stock, tomato puree, and seasoning. Bring to a boil and stir with a paddle until thickened and smooth.
4. Continue to simmer for 2 hr, stirring frequently.
5. Strain through a china cap into a stainless steel container. Using a ladle, force as much of the vegetable flavor as possible through the china cap.
6. Use in gravies, beef stews, and small sauces as needed.

Precautions

- Be careful not to scorch the sauce while simmering.
- When sautéing the garnish, do not let it brown.
- Exercise caution when cutting the rough vegetable garnish.

Preparation

1. Cut the vegetables rough with a French knife.
2. Prepare the brown stock.

Madeira Sauce

This sauce is prepared by adding brown sauce to the rich flavor of Madeira wine. Madeira sauce increases the delicate flavors of ham and veal.

1 gal.

Equipment

- Gallon measure
- Pint measure
- 6 qt sauce pot
- China cap
- 1 gal. stainless steel container

Ingredients

- 1 gal. brown sauce
- 1 pt Madeira wine

Procedure

1. Place the prepared brown sauce in a sauce pot and simmer until the sauce is reduced to about three-fourths of its original volume.
2. Add the wine and simmer for 5 min more.
3. Check the seasoning and strain through a fine china cap into a stainless steel container.
4. Serve 2 oz per portion using a ladle. Serve with baked ham, ham steaks, veal cutlets, or veal chops.

Preparation

1. Prepare the brown sauce.

Precaution

- Use caution so the sauce does not scorch while reducing.

Burgundy Sauce

This is a variation of brown sauce. The Burgundy wine gives a nice flavor to the sauce. It is served mostly with beef dishes.

2½ qt

Equipment

- French knife
- Quart measure
- Cup measure
- 4 qt saucepan
- China cap
- Kitchen spoon
- 1 gal. stainless steel container

Ingredients

- 2 qt brown sauce
- 1 c dry Burgundy wine
- ½ c tomato puree
- 2 small garlic cloves

Preparation

1. Mince the garlic with a French knife.
2. Prepare the brown sauce.

Procedure

1. Place all the ingredients in a saucepan and blend.
2. Let simmer for 30 min, stirring occasionally with a kitchen spoon. Strain through a china cap into a stainless steel container.
3. Serve 2 oz per portion using a ladle. Serve with beef dishes.

Precautions

- Do not allow the liquid to scorch.
- Exercise caution when mincing the garlic.

Trade Tip

Reduce brown stock until it is dark and almost the consistency of syrup. Let the stock cool and use a tablespoon or two to flavor sauces. This preparation will keep for several months in the refrigerator.

Bercy Sauce

This is brown sauce flavored with shallots and dry white wine. Bercy sauce is an excellent selection to accompany fish and veal.

1 gal.

Equipment

- French knife
- 6 qt sauce pot
- Kitchen spoon
- Cup measure
- Spoon measure
- Gallon measure
- 1 gal. stainless steel container

Ingredients

1 gal. brown sauce
8 oz butter
12 oz shallots or onions
1 c dry white wine
2 tbsp parsley
1 lemon
salt and pepper to taste

Procedure

1. Place the butter in the sauce pot and melt.
2. Add the minced shallots or onions and sauté without color.
3. Add the dry white wine and simmer until it is reduced to half its original amount.
4. Add the brown sauce and lemon juice. Simmer for 20 min.
5. Remove the sauce from the range and add the chopped parsley. Pour into a stainless steel container.
6. Serve 2 oz per portion using a ladle. Serve with sautéed or broiled fish, veal chops, or sautéed veal cutlets.

Preparation

1. Prepare the brown sauce.
2. Mince the shallots or onions with a French knife.
3. Chop the parsley and squeeze the juice from the lemon.

Precautions

- Exercise caution to avoid scorching throughout the preparation.
- When sautéing shallots or onions, do not let them brown.
- Exercise caution when mincing the shallots or onions.

Chateau Sauce

This sauce is a variation of brown sauce. The dry white wine adds a distinctive flavor. It can be served with sautéed or broiled meat (beef or veal) or sautéed fish entrees.

1 qt

Equipment

- French knife
- 2 qt saucepan
- 1 qt stainless steel
- container
- Cheesecloth
- Baker's scale
- 1 qt measure

Ingredients

8 oz dry white wine
4 shallots
2 pinches thyme
2 bay leaves
1 qt brown sauce
8 oz melted butter
salt and pepper to taste

Procedure

1. Place the dry white wine, chopped shallots, thyme, bay leaves, and salt in a saucepan and bring to a boil.
2. Continue to boil the mixture until it reduces approximately one-half in volume.
3. Add the brown sauce. Continue to boil until the mixture reduces at least one-fourth in volume.
4. Strain through fine cheesecloth into a stainless steel container. Add the melted butter.
5. Serve 2 oz per portion using a ladle. Serve with broiled beef steaks, sautéed veal steak, broiled veal chops, sautéed Dover sole, etc.

Preparation

1. Chop the shallots with a French knife.
2. Prepare the brown sauce.

Precautions

- When melting the butter, do not let it brown or burn.
- Exercise caution when chopping the shallots.

Mushroom Sauce

This is a brown sauce rich with the distinctive flavor of mushrooms and sherry. This sauce is very popular and is generally served with steaks, chops, and loafs.

2½ gal.

Equipment

- 6 qt sauce pot
- Kitchen spoon
- French knife
- Quart measure
- Baker's scale
- 2 gal. stainless steel container

Ingredients

2 gal. brown sauce
1 lb mushrooms
1 lb onions
1 garlic clove
1 qt canned whole tomatoes
8 oz butter
1 c sherry (variable)
salt and pepper to taste

Preparation

1. Slice the mushrooms with a French knife.
2. Prepare the brown sauce.
3. Mince the onions and garlic with a French knife.
4. Crush the tomatoes by hand.

Procedure

1. Place the butter in a sauce pot and melt.

2. Add the minced onions and garlic and sauté slightly without color.
3. Add the brown sauce and bring to a boil. Stir occasionally with a kitchen spoon.
4. Add the crushed tomatoes, salt, and pepper. Simmer for 20 min.
5. Add the sherry to obtain the desired taste.
6. Remove from the heat, check the seasoning, and pour into a stainless steel container.

7. Serve 2 oz per portion using a ladle. Serve with beef steaks, veal chops, meat loaf, etc.

Precautions

- Do not brown the onions and garlic while sautéing.
- Do not allow the brown sauce to thicken too much.
- Exercise caution when mincing the onions and garlic.

Bordelaise Sauce

This is a very rich brown sauce. It is served mostly with steaks.

1 gal.

Equipment

- 6 qt sauce pot
- Kitchen spoon
- French knife
- 1 gal. stainless steel container

Ingredients

1 lb onions
1 garlic clove
1 lb mushrooms
1 gal. brown sauce
1 c Burgundy wine (variable)
½ lb margarine
salt and pepper to taste

Procedure

1. Melt the margarine in a sauce pot. Add onions, garlic, and mushrooms. Sauté until just cooked.
2. Add the brown sauce and cook for 20 min or until vegetables are completely done. Stir occasionally with a kitchen spoon.
3. Remove the sauce from the heat and pour into a stainless steel container. Add salt, pepper, and wine.
4. Serve 2 oz per portion using a ladle. Serve with broiled steaks, roast rib, sirloin of beef, meat loaf, etc.

Preparation

1. Mince the onions and garlic and chop the mushrooms with a French knife.
2. Prepare the brown sauce.

Precautions

- Chop mushrooms fresh. If chopped in advance, they will turn black.
- Use caution when mincing the onions and garlic and chopping the mushrooms.

Sour Cream Sauce

This sauce is a blend of brown sauce and sour cream. It is seasoned mostly with bay leaves. Sour cream sauce is mainly used with Swedish meatballs and beef stroganoff.

1 gal.

Equipment

- French knife
- 6 qt sauce pot
- Quart measure
- Cup measure
- Spoon measure
- Wire whip
- China cap
- 1 gal. stainless steel container
- Kitchen spoon

Ingredients

2 c butter
2 c flour
3 qt brown stock
1 c tomato puree
½ c cider vinegar
1 lb onions
1 qt sour cream
2 tbsp salt
2 bay leaves

Procedure

1. Place the butter in a sauce pot. Add the onions and sauté without color.
2. Add the flour, making a roux, and cook for 5 min.
3. Add the brown stock, tomato puree, vinegar, bay leaves, and salt while whipping constantly with a wire whip. Simmer about 30 min.
4. Add the sour cream by folding in gently with a kitchen spoon. Bring back to a boil, remove from the range, and strain through a china cap into a stainless steel container.
5. Check the seasoning. Serve with Swedish meatballs, beef stroganoff, veal chop stroganoff, etc.

Preparation

1. Prepare the brown stock.
2. Mince the onions with a French knife.

Precautions

- Avoid scorching while simmering the sauce.
- When adding the sour cream, fold in gently.
- After the sour cream is added, bring back to a boil but do not cook for any length of time.

CREAM OR BÉCHAMEL SAUCE RECIPES

Cream or Béchamel Sauce

Cream sauce is a leading sauce made from a roux and hot milk. Several small sauces utilize cream sauce in their preparation.

2 qt

Equipment

- 4 qt sauce pot
- 4 qt saucepan
- Wire whip
- China cap
- Baker's scale
- Spoon measure
- Quart measure
- 1 gal. stainless steel container

Ingredients

Thin

8 oz butter or shortening
2 oz flour
2 qt milk
2 tsp salt

Medium

8 oz butter or shortening
4 oz flour
2 qt milk
2 tsp salt

Thick

4 oz butter or shortening
8 oz flour
2 qt milk
2 tsp salt

Preparation

1. Heat the milk in a saucepan.

Procedure

1. Place the shortening in a sauce pot and melt.
2. Add the flour, making a roux, and cook for 5 min.
3. Add the hot milk, whipping constantly with a wire whip until the desired consistency is reached.
4. Bring to a boil and season with salt. Remove from the heat and strain through a china cap into a stainless steel container. Dot the top of the sauce with butter so it does not form a crust.

Precautions

- When cooking the shortening and flour, be careful not to scorch the mixture.
- Whip constantly when adding the milk to obtain a smooth sauce.

A la King Sauce

This is a cream sauce containing diced mushrooms, green peppers, and pimientos. This sauce is generally associated with poultry; however, it can also be served with ham or sweetbreads to create variety.

1 gal.

Equipment

- Kitchen spoon
- 6 qt sauce pot
- French knife
- Baker's scale
- Cup measure
- Gallon measure
- 1 pt saucepan
- China cap
- 1 gal. stainless steel container.

Ingredients

1 gal. cream sauce
2 oz green peppers
6 oz butter
8 oz mushrooms
4 oz pimientos
½ c sherry

Preparation

1. Prepare the cream sauce.
2. Dice the green peppers, mushrooms, and pimientos with a French knife.

Procedure

1. Place the butter in a sauce pot and melt.
2. Add the mushrooms and sauté until slightly tender.
3. Add the sherry and simmer slightly.
4. Add the prepared cream sauce and continue to simmer, stirring occasionally with a kitchen spoon.
5. Place the green peppers in a small saucepan, cover with water, and poach until tender. Drain and add to the prepared mixture.
6. Add the pimientos and check seasoning.
7. Pour into a stainless steel container.
8. Serve 2 oz per portion using a ladle. Serve with sautéed sweetbreads or turkey steak. The sauce is also used in the preparation of ham, chicken, or turkey a la king.

Precautions

- Do not brown or overcook the mushrooms when sautéing.
- After the cream sauce is added, stir occasionally to avoid sticking or scorching.
- Exercise caution when cutting the garnish.

Newburg Sauce

This popular sauce is a blend of cream sauce, paprika, sherry, and seasoning. It is used in seafood dishes.

1 gal.

Equipment

- Wire whip
- 6 qt sauce pot
- Gallon measure
- Spoon measure
- Cup measure
- Kitchen spoon
- 1 gal. stainless steel container

Preparation

1. Prepare the cream sauce.

Ingredients

1 gal. cream sauce
¼ c butter
2 tbsp paprika
½ c sherry
salt and white pepper to taste

Procedure

1. Place the butter in a sauce pot and melt.
2. Add the paprika and blend it into the butter.
3. Add the sherry and bring the mixture to a simmer.
4. Add the cream sauce, whipping briskly with a wire whip until all ingredients are thoroughly incorporated. Simmer for 5 min. Season with salt and white pepper.
5. Pour into a stainless steel container.
6. Use in the preparation of shrimp, lobster, crabmeat, and seafood Newburg.

Precautions

- Only blend the paprika into the butter, do not cook.
- Avoid scorching at all times.

Cheese Sauce

This sauce is prepared by adding Cheddar cheese and seasoning to cream sauce. It is an excellent sauce to serve with broccoli, asparagus, brussels sprouts, baked potatoes, etc.

2 qt

Equipment

- Wire whip
- Food grinder
- Spoon measure
- Quart measure
- 3 qt saucepan
- China cap
- Kitchen spoon
- 1 gal. stainless steel container

Ingredients

1 lb sharp Cheddar cheese
2 tsp dry mustard
2 tsp paprika
½ c milk
1½ tsp Worcestershire sauce
1½ tsp Tabasco sauce
1¾ qt medium cream sauce
salt to taste

Anchor Food Products, Inc.

Preparation

1. Grind the Cheddar cheese by passing it through a food grinder and using the medium hole chopper plate.
2. Prepare the medium cream sauce.

Procedure

1. Place cheese, mustard, paprika, and milk in the saucepan. Stir with a kitchen spoon.
2. Add 1 c of the white sauce. Heat until cheese is melted, stirring constantly with a kitchen spoon.
3. Add the remainder of the white sauce, the Tabasco sauce, and the Worcestershire sauce. Bring to boil, whipping occasionally with a wire whip.
4. Add salt to taste if necessary.
5. Strain through a china cap into a stainless steel container.
6. Serve 2 oz per portion using a ladle.

Precaution

- When melting the cheese and bringing the sauce to a boil, do not scorch.

Mock Hollandaise Sauce

This is a substitute for the true butter/egg sauce and is generally used to reduce cost. It is a blend of cream sauce, egg yolks, and lemon juice.

2 qt

Equipment

- Quart measure
- Baker's scale
- 4 qt saucepans (2)
- Wire whip
- Stainless steel bowl
- Cup measure

Ingredients

2 qt milk
8 oz butter
6 oz flour
10 egg yolks
¼ c lemon juice
salt and white pepper to taste
yellow color, if desired

Procedure

1. Place the butter in a saucepan and heat.
2. Add the flour, making a roux, and cook for 3 min.
3. Add the hot milk, whipping vigorously with a wire whip. Allow to simmer for 5 min.
4. Place the egg yolks in a stainless steel bowl and beat with a wire whip. Drip in a small amount of the hot cream sauce and blend with the egg yolks. Slowly pour this mixture into the simmering cream sauce, mixing continuously.
5. Add the lemon juice and season with salt and white pepper.
6. Tint with yellow color if desired.
7. Serve the same as true Hollandaise sauce.

Preparation

1. Squeeze juice from lemons.
2. Heat the milk in a saucepan.
3. Separate the eggs. Reserve the whites for later use.

Precautions

- When adding the egg yolks to the hot cream sauce, pour very slowly and whip briskly.
- Exercise caution so as not to scorch the sauce.

Mornay Sauce

This popular sauce is generally used over items that are to be glazed, such as lobster thermidor and Florentine items.

1 gal.

Equipment

- Baker's scale
- Cup measure
- 1½ to 2 gal. sauce pot
- 1 gal. stainless steel container
- Wire whip
- 5 qt saucepan

Ingredients

1 lb 4 oz bread flour
1 lb 4 oz melted butter
4 qt milk
12 egg yolks
½ c light cream
salt to taste
6 oz Parmesan cheese
8 oz cold butter, broken into small pieces

2. Add the hot milk and stir with a wire whip until slightly thickened and smooth.
3. Beat egg yolks and cream with a wire whip.
4. Add slowly to the mixture, whipping constantly with a wire whip.
5. Season with salt and cook for 1 min.
6. Remove from the heat and add the cheese. Pour into a stainless steel container.
7. Add the cold butter, stirring until it is blended into the sauce.
8. Serve approximately 2 oz over items that are to be glazed. Use a kitchen spoon to apply sauce. Use with lobster thermidor, Florentine items, and poached fish.

Preparation

1. Break the eggs and separate the yolks from the whites.
2. Heat the milk in a saucepan.

Procedure

1. Melt the butter in a sauce pot. Add flour, making a roux, and cook for 5 min. Do not brown.

Precautions

- Overheating the sauce may cause it to break and become fluid.
- When adding the egg and cream mixture to the sauce, whip small amounts of sauce into it gradually, making sure the egg does not curdle, then add the mixture to remaining sauce.

Trade Tip

Cream sauces can be enriched before serving by adding egg yolks. Beat some of the hot cream sauce into the yolks before adding to the sauce to prevent curdling. The sauce should be served immediately after the yolks are added.

VELOUTÉ OR FRICASSEE SAUCE RECIPES

Basic Velouté (Fricassee) Sauce

Velouté sauce is one of the five leading sauces used in the commercial kitchen. Fish stock-vin blanc, chicken supreme, or veal or beef stock fricassee are examples of sauces which may be prepared from basic fricassee sauce.

1 gal.

Equipment

- 5 qt saucepan
- China cap
- Cheesecloth
- Wire whip
- Baker's scale
- Quart measure
- 1 gal. stainless steel container

Ingredients

12 oz butter or margarine
12 oz bread flour
1 gal. stock (fish, chicken, veal, or beef)
salt and white pepper to taste

Procedure

1. Heat the butter in a saucepan and add flour. Cook slowly for about 5 min, stirring and forming a roux.
2. Add the stock slowly, whipping constantly until thick and smooth.
3. Season and continue to cook for 20 min. Adjust to desired consistency. If too thick, add more stock; if too thin, add more roux or whitewash.
4. Strain through a china cap into a gallon container. Reserve for use.
5. Serve 2 oz per portion using a ladle.

Precautions

- When cooking the roux, do not scorch or allow to brown.
- Whip constantly when adding the stock to the roux or lumps may form.

Preparation

1. Prepare stock to be used and strain through cheesecloth to remove all scum particles.

Curry Sauce

This sauce is associated with India because that is where curry originated. Curry sauce is used when preparing many meat and seafood dishes, but it is most popular when served with lamb, chicken, shrimp, or lobster.

1 gal.

Equipment

- 6 qt sauce pot
- China cap
- French knife
- Baker's scale
- Ladle
- Spoon measure
- Wire whip
- Kitchen spoon
- Quart measure
- 1 gal. stainless steel container
- 2 qt saucepan

Ingredients

10 oz butter or shortening
8 oz flour
3 qt hot chicken stock
1 qt milk
5 oz onions
½ tsp mace
½ tsp thyme
4 tbsp curry powder
2 bay leaves
1 banana
5 oz pineapple
5 oz apples
salt to taste

4. Dice the onions with a French knife.

Procedure

1. Place the butter in a sauce pot and heat.
2. Add the onions and sauté without browning.
3. Add the flour, making a roux, and cook for 5 min.
4. Add the thyme, mace, bay leaves, and curry powder and blend into the roux using a kitchen spoon.
5. Add the chicken stock and milk, whipping vigorously with a wire whip to avoid lumps. Bring to a boil.
6. Add fruit and simmer for about 1 hr.
7. Season with salt and strain through a fine china cap into a stainless steel container, using a ladle to force as much of the fruit pulp as possible into the sauce.
8. Use in the preparation of the following entree dishes: shrimp curry, lobster curry, lamb curry, chicken curry, and veal curry.

Precautions

- When sautéing, do not brown the onions or the butter.
- Stir occasionally while the sauce is simmering to avoid sticking.
- Exercise caution when dicing the onions and the fruit.

Preparation

1. Prepare the chicken stock.
2. Heat the milk in a saucepan.
3. Dice the fruit with a French knife.

Horseradish Sauce

This sauce is served mostly with boiled meats to increase the flavor of the dish. It is a combination of beef fricassee (velouté sauce) and prepared horseradish.

1¼ gal.

Equipment

- 1½ gal. sauce pot
- Wire whip
- Fine china cap
- Cheesecloth
- 2 gal. stainless steel container
- 1 qt measure
- Baker's scale

Ingredients

1 gal. hot beef stock
10 oz bread flour
10 oz shortening or butter
1 pt horseradish
1 dash Tabasco sauce
salt and pepper to taste

Procedure

1. Place the shortening in a sauce pot and heat.
2. Add the flour, making a roux, and cook slightly.
3. Add the hot beef stock, stirring constantly with a wire whip.
4. Remove from the heat and strain through a fine china cap into a stainless steel container.
5. Add the horseradish and Tabasco sauce.
6. Season with salt and white pepper.
7. Serve 2 oz per portion using a ladle.

Precautions

- Strain the stock through a cheesecloth to eliminate undesirable scum.
- When cooking the roux, do not let it brown or scorch.
- Boil the stock before adding it to the roux.

Preparation

1. Prepare the beef stock. Strain through cheesecloth. Reheat when ready to use.

Poulette Sauce

This sauce contains lightly sautéed onions and mushrooms that are added to a chicken velouté sauce. It is used mainly with chicken or turkey.

1 gal.

Equipment

- French knife
- 5 qt saucepan
- Stainless steel container
- Kitchen spoon
- 1 gal. stainless steel container

Ingredients

1 gal. chicken velouté sauce
2 lb fresh mushrooms
4 oz onions
4 oz butter
yellow color as desired
salt and white pepper to taste

Procedure

1. Place the butter in the saucepan and heat.
2. Add the minced onions and sauté slightly. Do not brown.
3. Add the sliced mushrooms and cook until slightly tender.
4. Stir in the chicken velouté sauce. Continue to simmer the sauce for 20 min, stirring frequently.
5. Season with salt and white pepper. Tint with yellow color.
6. Pour into a 1 gal. stainless steel container.
7. Serve 2 oz per portion using a ladle. Serve with chicken croquettes, stuffed chicken leg, fried turkey wings, etc.

Precautions

- When simmering sauce, do not scorch.
- Use caution when tinting with yellow color.

Preparation

1. Slice the mushrooms and mince the onions with a French knife.
2. Prepare the chicken velouté sauce.

Victoria Sauce

This fish velouté sauce is blended with diced mushrooms and lobster and flavored with white wine. It is generally served with lobster and shrimp preparations.

1 gal.

Equipment

- Quart measure
- Cup measure
- 6 qt saucepan
- French knife
- Kitchen spoon
- 1 gal. stainless steel container

Ingredients

3 qt fish velouté sauce
1 pt lobster
1 c butter
1 c white wine
⅓ c shallots
1 c mushrooms
salt and white pepper to taste

Preparation

1. Prepare the fish velouté sauce.
2. Cook and dice the lobster with a French knife.
3. Dice the mushrooms and mince the shallots with a French knife.

Procedure

1. Place the butter in a saucepan and melt.
2. Add the shallots and sauté slightly. Do not brown.

3. Add the mushrooms and lobster, continuing to sauté.
4. Add the white wine and simmer until it is reduced to one-half its original amount.
5. Add the fish velouté sauce while stirring with a kitchen spoon. Continue to simmer for about 10 min.
6. Pour into a stainless steel container.

7. Serve 3 oz per portion using a ladle. Serve over sautéed shrimp, lobster, king crabmeat, and sole.

Precautions

- When sautéing, do not brown any of the ingredients.
- When simmering, stir occasionally to avoid sticking.

Dugleré Sauce

This is a fish velouté sauce with white wine, shallots, mushrooms, and crushed tomatoes added. It is served with baked and poached fish entrees.

1 gal.

Equipment

- 6 qt sauce pot
- Quart measure
- French knife
- Cup measure
- Spoon measure
- 1 pt saucepan
- Kitchen spoon
- 1 gal. stainless steel container

Ingredients

3 qt fish velouté sauce
1 c white wine
½ c shallots
½ c mushrooms
1 c tomatoes
2 tbsp parsley
¼ c butter
salt and white pepper to taste

Preparation

1. Prepare the fish velouté sauce.
2. Mince the shallots and dice the mushrooms with a French knife.
3. Chop and wash the parsley.
4. Crush the tomatoes by squeezing with the hands.

Procedure

1. Place the wine in a sauce pot, add the shallots, and simmer until the wine is reduced to half its original amount.
2. Add the fish velouté sauce and continue to simmer.
3. Place the butter in a small saucepan, add the diced mushrooms, and sauté slightly.
4. Add the mushrooms to the simmering sauce and stir gently with a kitchen spoon.
5. Add the crushed tomatoes and chopped parsley.
6. Season with salt and white pepper and pour into a stainless steel container.
7. Serve 2 oz per portion using a ladle. Serve over baked fish entrees.

Precaution

- Stir the sauce gently throughout the entire preparation to avoid scorching.

HOLLANDAISE SAUCE RECIPES

Hollandaise Sauce

Hollandaise sauce is one of the leading sauces. It is served mostly with vegetables such as asparagus and broccoli.

1 qt

Equipment

- Stainless steel bowl
- Braiser
- French whip
- 1 qt stainless steel container

Ingredients

8 egg yolks
1 lb butter
½ lemon
salt to taste
Tabasco sauce to taste

Preparation

1. Melt the butter.
2. Break the eggs and separate the whites from the yolks.
3. Squeeze the juice from ½ lemon.
4. Place the water in the braiser and bring to boil.

Procedure

1. Put the egg yolks in the stainless steel bowl, add a few drops of water, and mix well.
2. Put the bowl in hot water with a temperature of 160°F.
3. Beat yolks slowly with a French whip until they foam and tighten.
4. Remove from water and add melted butter very slowly while whipping continuously with a French whip.
5. When all butter is added, forming the emulsion, season with salt, Tabasco sauce, and lemon juice.
6. Serve 2 oz per portion with a kitchen spoon. Serve with cooked vegetables and poached egg dishes or as a base for small sauces.

Precautions

- Do not let the water get too hot or the eggs will cook.
- Whip continuously throughout the entire preparation.

Maximillian Sauce

This sauce is prepared by adding anchovy essence to hollandaise sauce. It is used with poached or baked fish entrees.

1 qt

Equipment

- 2 qt stainless steel bowl
- French knife
- Quart measure
- Wire whip

Ingredients

1 qt hollandaise sauce
2 oz anchovies

Preparation

1. Prepare the hollandaise sauce.
2. Chop the anchovies into a puree with a French knife.

Procedure

1. Place the prepared hollandaise sauce in a stainless steel bowl.
2. Add the pureed anchovies and blend thoroughly with a wire whip.
3. Serve 2 oz per portion using a kitchen spoon. Serve over poached and baked fish entrees.

Precautions

- Exercise caution when chopping the anchovies.
- Keep the sauce warm. Do not keep it hot or it will separate.
- Serve within 2 hr of preparation.

Cherburg Sauce

This is a hollandaise sauce blended with pureed crabmeat to create a sauce that complements certain shrimp and lobster preparations.

1 qt

Equipment

- Quart measure
- Cup measure
- French knife
- Food grinder
- 2 qt stainless steel mixing container
- Kitchen spoon

Ingredients

1 qt hollandaise sauce
1 pimiento
½ c king or blue crabmeat

Preparation

1. Prepare the hollandaise sauce.
2. Puree the pimiento by chopping with a French knife.

3. Puree the crabmeat by running it through the food grinder.

Procedure

1. Place the hollandaise sauce in the stainless steel mixing container.
2. Using a kitchen spoon, blend the pureed crabmeat and pimiento until thoroughly incorporated.
3. Serve 3 oz per portion over sautéed shrimp, lobster, king crabmeat, or flounder.

Precautions

- Serve the sauce within 2 hr after preparation.
- Keep the sauce warm. If the sauce is kept too hot or too cold it may separate.
- Exercise caution when using the food chopper.

Bearnaise Sauce

This sauce is prepared from the leading sauce, hollandaise. Bearnaise sauce, however, substitutes a tarragon vinegar mixture for the lemon juice ordinarily used in hollandaise sauce. Bearnaise sauce is generally served with meats such as steaks and various types of roast.

2 qt

Equipment

- 1 pt saucepans (2)
- Wire whip
- Cheesecloth
- 3 qt stainless steel bowl
- French knife
- Cup measure
- Spoon measure
- Quart measure

Ingredients

4 tbsp shallots
1 tsp peppercorns
1 c tarragon vinegar or
   1 tbsp tarragon leaves
   and 1 c cider vinegar
2 qt hollandaise sauce
1 tsp tarragon leaves or
   parsley

Preparation

1. Mince the shallots with a French knife and crush the peppercorns.
2. Prepare the basic hollandaise sauce, omitting the lemon juice.
3. Chop the tarragon leaves or parsley using a French knife.

Procedure

1. Place the shallots, peppercorns, and tarragon vinegar (or substitute) in small saucepan. Boil slowly until the mixture is reduced by evaporation to half the original amount. Remove from the range and strain the reduced liquid through a cheesecloth into another small pan.

2. Add the desired amount of reduced liquid to the hollandaise sauce. Whip gently with a wire whip until blended.
3. Add a teaspoon of finely chopped tarragon leaves or parsley to the sauce.
4. Serve 3 oz per portion with a kitchen spoon. Serve with roast rib and sirloin of beef, sautéed veal steak, broiled lamb steak, etc.

Precautions

- When reducing the liquid, do not scorch.
- Use caution when preparing the hollandaise sauce so the emulsion does not break.
- Exercise caution when mincing the shallots and chopping the tarragon leaves or parsley.

Choron Sauce

This is a tomato-flavored bearnaise sauce. Choron sauce is a versatile sauce that can be served with steaks, fish, chicken, and eggs with equally good results.

2¼ qt

Equipment

- 3 qt stainless steel container
- Kitchen spoon
- Quart measure
- Cup measure

Ingredients

2 qt bearnaise sauce (omit the chopped tarragon or parsley)
½ c tomato paste or ¾ c tomato puree
salt to taste

Procedure

1. Place the bearnaise sauce in a stainless steel container.
2. Pour the tomato paste or puree slowly into the bearnaise sauce while stirring gently with a kitchen spoon.
3. Season with salt.
4. Serve 2 oz per portion with a kitchen spoon. Serve with broiled steaks, poached sole or salmon, poached eggs, or sautéed or broiled breast of chicken.

Preparation

1. Prepare the bearnaise sauce, omitting the chopped tarragon or parsley.

Precaution

- Stir gently when adding the tomato paste or puree to the bearnaise sauce so it blends in thoroughly and to lessen the chance of the emulsion breaking.

TOMATO SAUCE RECIPES

Tomato Sauce

This is a leading sauce used in all commercial kitchens. It is served with a number of fried items, such as breaded veal cutlets, breaded pork chops, and breaded veal chops.

1½ qt

Equipment

- 2 gal. sauce pot
- Wire whip
- China cap
- French knife
- Kitchen spoon
- 2 gal. stainless steel container

Ingredients

4 oz bacon or ham grease
6 oz onion
6 oz celery
6 oz carrots
5 oz flour
1 No. 10 can tomato puree
½ gal. ham stock
salt and sugar to taste
1 garlic clove
1 bay leaf
1 pinch thyme

Procedure

1. Place the bacon or ham fat in a sauce pot and add the onions, carrots, and celery. Sauté until golden brown.
2. Add the bay leaf, thyme, and garlic.
3. Add the flour, taking up the ham or bacon grease and making a roux. Cook 5 min.
4. Add the tomato puree and ham stock and bring to a boil. Whip occasionally with a wire whip.
5. Let simmer until vegetables are completely cooked.
6. Season with salt and sugar.
7. Strain through a china cap and pour into a stainless steel container.
8. Serve 2 oz per portion using a ladle. Serve with breaded veal chops, pork chops, or pork and veal cutlets.

Precautions

- Do not make the sauce too thick or too heavy with starch.
- Red color may be added to the sauce to increase appearance.

Preparation

1. Cut vegetables rough using a French knife.
2. Prepare the ham stock.

Napolitaine Sauce

This sauce is prepared by adding diced fresh tomatoes, minced ham, and parsley to tomato sauce. This sauce should be served chiefly with pork and veal entrees.

1 gal.

Equipment

- Gallon measure
- Cup measure
- Sauce pot
- Kitchen spoon
- French knife
- Paring knife
- 1 gal. stainless steel container

Ingredients

1 gal. tomato sauce
2 small, firm tomatoes
½ c lean ham
1 small garlic clove
¼ c parsley
4 oz butter

Procedure

1. Place the butter in the sauce pot and melt.
2. Add the garlic and ham and sauté slightly.
3. Add the prepared tomato sauce, bring to a boil, then allow to simmer for 10 min. Stir occasionally with a kitchen spoon.
4. Add the diced tomatoes and simmer for an additional 5 min.
5. Remove from the heat and add the chopped parsley.
6. Adjust the seasoning and pour into a stainless steel container.
7. Serve 2 oz per portion with a ladle. Serve with breaded fried pork chops, veal cutlets, veal chops, etc.

Preparation

1. Prepare the tomato sauce.
2. Mince the ham and garlic with a French knife.
3. Peel and dice the tomatoes with a paring knife and chop the parsley.

Precautions

- Do not brown when sautéing the garlic and ham.
- Use caution throughout the entire preparation so scorching does not occur.
- Exercise caution when dicing the tomatoes and mincing the ham.

Milanaise Sauce

Milanaise sauce is made by adding sliced mushrooms, julienned ham, and tongue to tomato sauce. This sauce blends exceptionally well with spaghetti and can also be served over sautéed pork and veal items.

1 gal.

Equipment

- Gallon measure
- Cup measure
- Baker's scale
- 6 qt saucepan
- French knife
- Kitchen spoon
- 1 gal. stainless steel container

Ingredients

1 gal. tomato sauce
½ c butter
10 oz mushrooms
12 oz ham
12 oz tongue

3. Slice the mushrooms with a French knife.

Procedure

1. Place the butter in a saucepan and heat.
2. Add the mushrooms and sauté until slightly tender.
3. Add the ham, tongue, and tomato sauce. Simmer for 10 min.
4. Check the seasoning and remove from the range. Pour into a stainless steel container.
5. Serve 2 oz per portion with a ladle. Serve over spaghetti, sautéed pork cutlets, or sautéed veal cutlets.

Preparation

1. Julienne the ham and tongue using a French knife.
2. Prepare the tomato sauce.

Precautions

- Exercise caution when julienning the meat.
- Stir through the entire preparation to avoid sticking.

 **Trade Tip** *Always choose the appropriate sauce for the type of service. Use sauces that can be prepared in large quantities and held at the correct temperature for banquets. If food is prepared to order, any type of sauce can be used.*

Creole Sauce

This sauce is suitable for service with many items such as omelets, poached eggs, fish, leftover dishes, meat, poultry, and game. Creole sauce is also known as Spanish sauce.

1 gal.

Equipment

- 6 qt sauce pot
- French knife
- Kitchen spoon
- 1 gal. stainless steel container

Ingredients

- 3 lb celery
- 3 lb onions
- 4 oz flour
- ½ gal. beef stock
- 1 oz salt
- ½ oz pepper
- 1 bay leaf
- 2 lb green peppers
- 1 lb mushrooms
- ½ No. 10 can tomatoes
- ½ No. 10 can tomato puree
- 2 garlic cloves
- 4 oz bacon grease or shortening

Preparation

1. Prepare the beef stock.
2. Julienne the vegetables and chop the garlic with a French knife.
3. Crush the tomatoes by squeezing with the hands.

Procedure

1. Place the bacon grease or shortening in a sauce pot and heat.
2. Add the onions, green peppers, celery, garlic, and mushrooms. Sauté slowly, but do not brown.
3. Add the flour, stir with a kitchen spoon until smooth, and allow to cook for a few minutes.
4. Add the tomatoes, tomato puree, stock, and seasoning.
5. Cook until the vegetables are well done.
6. Verify the seasoning and consistency. Pour into a stainless steel container.
7. Serve 2 oz per portion with a ladle. Serve with meat loaf, Salisbury steak, steamed shrimp, baked fish, etc.

Precautions

- Do not brown the vegetables when sautéing.
- Do not make the sauce too thick.
- Exercise caution when cutting the vegetables.

Italian Sauce

This is a basic Italian sauce that can be converted into meat sauce or served with meatballs.

1½ gal.

Equipment

- Quart measure
- Cup measure
- Spoon measure
- 2 gal. sauce pot
- French knife
- 2 gal. stainless steel container
- Kitchen spoon

Ingredients

- 5 garlic cloves
- 1 c salad or olive oil
- 2 No. 10 cans whole tomatoes
- 6 tbsp parsley
- 1 tbsp sweet basil
- 2 tsp salt
- 2 tsp black pepper
- 3 c tomato paste
- 1 lb Parmesan cheese
- 1 lb onions
- 1 qt tomato puree

Preparation

1. Chop the garlic and parsley and crush the tomatoes.
2. Mince the onions with a French knife.
3. Crush the basil and grate the Parmesan cheese.

Procedure

1. Place the oil in a sauce pot. Add the garlic and onions and sauté.
2. Add the crushed tomatoes, tomato puree, parsley, basil, salt, and pepper and simmer for 30 min, stirring occasionally with a kitchen spoon.
3. Add the oregano and tomato paste. Continue to cook until thick, then remove from the heat.
4. Add the cheese and check the seasoning. Place in a 2 gal. stainless steel container.
5. Serve 2 oz per portion with a ladle. Serve with meatballs and other Italian dishes.

Precautions

- Stir frequently while simmering to avoid scorching.
- Stir vigorously when adding the cheese to avoid lumps.

Trade Tip

Most tomato sauces can be prepared ahead of time and stored in the refrigerator. Tomato sauces also freeze well. For convenience, prepare extra tomato sauce and freeze for later use.

Barbecue Sauce

Barbecue sauce may be used with any barbecue item such as pork, chicken, or beef. It has a tomato base and a very rich, savory taste.

7 gal.

Equipment

- Cup measure
- Spoon measure
- Pint measure
- French knife
- 6 qt sauce pot
- 1 gal. stainless steel container

Ingredients

1 c shortening or oil
2½ c onions
1½ c brown sugar
5 tbsp prepared mustard
1 tsp salt
5 tbsp Worcestershire sauce
5 c catsup
3½ c celery
½ c vinegar

Carlisle FoodService Products

Preparation

1. Mince the onions and celery with a French knife.

Procedure

1. Place the shortening or oil in sauce pot. Add the minced onions and celery and sauté without browning.
2. Add all other ingredients and simmer slowly for about 30 min, stirring occasionally with a kitchen spoon.

3. Remove from the fire and pour into a stainless steel container.
4. Serve 3 oz per portion with a ladle. Serve with barbecued chicken, spare ribs, pork, and beef.

Precautions

- When simmering the sauce, stir frequently to avoid scorching.
- Exercise caution when mincing the onions and celery.

Pizza Sauce

This is a savory tomato sauce. It is used in the preparation of pizza. The sauce is spread over the pizza dough before the toppings are added.

1 gal.

Equipment

- Quart measure
- 6 qt sauce pot
- Spoon measure
- Cup measure
- Kitchen spoon
- French knife
- 1 gal. stainless steel container

Ingredients

3 qt canned tomatoes
1 qt tomato paste
2 c water
3 tsp garlic powder
3 tsp onion powder
3 tsp salt
½ tsp black pepper
2 tbsp basil leaves
2 tbsp oregano
¼ c butter

Procedure

1. Place the tomato pulp, juice, tomato paste, and water in a sauce pot. Bring to a boil.
2. Add the remaining ingredients and stir with a kitchen spoon. Simmer until the mixture reduces slightly and becomes fairly thick.
3. Check the seasoning and pour into a stainless steel container.
4. Use in the preparation of pizza and serve approximately 2 oz with veal chop Italienne or sautéed veal steak.

Precautions

- Stir occasionally while the sauce is simmering to avoid scorching.
- Exercise caution when chopping the tomato pulp.

Preparation

1. Strain tomatoes, reserve juice, and chop the pulp using a French knife.

Chili Sauce

This is a versatile sauce with a very pronounced flavor. It can be used for chili, chili spaghetti, or chili con carne with the addition of ground beef and kidney beans. It can also be used in the preparation of certain hot hors d'oeuvres and chili dogs.

1½ gal.

Equipment

- Baker or portion scale
- Cup measure
- Gallon measure
- Spoon measure
- French knife
- 2 gal. heavy bottom sauce pot or stockpot
- Wire whip
- 2 gal. stainless steel container

Ingredients

- 1 lb onions
- 1 c salad oil
- 2 cloves garlic, chopped
- 1 No. 10 can tomato puree
- ½ c cider vinegar
- ½ c chili powder
- 2 tbsp ground cinnamon
- 1 tbsp red peppers
- 2 tbsp cumin
- 5 bay leaves
- 2 tsp salt

Preparation

1. Mince the onions with a French knife.

2. Crush the red peppers very fine.

Procedure

1. Place the salad oil in a heavy bottom sauce pot or stockpot and heat.
2. Add the minced onions and garlic. Sauté slightly.
3. Add all the remaining ingredients, bring to a simmer, and simmer for approximately 2 hr, stirring with a wire whip to avoid scorching.
4. Remove from the range. Place in a stainless steel container, cool, and refrigerate until ready to use.

Note: When using for chili, chili spaghetti, chili con carne, or hot dogs, add 3 lb to 5 lb cooked ground beef. The amount used is determined by the consistency desired. When using as a hot hors d'oeuvre, just add wiener or meat tidbits to the sauce.

Precaution

- Exercise caution when mincing the onions and garlic.

Trade Tip

Value-added, scratch, and signature sauces and gravies are commercially available as finished (wet) or powdered (dry) sauces and gravies. These products are convenient and add variety to meal preparations, although they can be costly.

COLD SAUCE RECIPES

Cocktail Sauce

Cocktail sauce is a cold tomato sauce. It has a hot, tangy taste and is usually served with hot or cold seafood dishes.

1 gal.

Equipment

- 1 gal. stainless steel container
- Quart measure
- Spoon measure
- Cup measure
- Kitchen spoon
- French knife

Ingredients

- 2 qt catsup
- 1 qt chili sauce
- 1 pt tomato puree
- 2 tbsp lemon juice
- 1 tbsp green pepper (optional)
- 1 c horseradish
- 1 tsp salt
- 2 tbsp Worcestershire sauce
- 1 tsp Tabasco sauce
- 2 tbsp onions

Preparation

1. Mince the onions and green pepper using a French knife.
2. Squeeze the juice from a lemon.

Procedure

1. Place all the ingredients in a stainless steel container and blend thoroughly with a kitchen spoon.
2. Adjust the seasoning if a hotter sauce is desired.
3. Serve 3 oz per portion in a soufflé cup with hot or cold seafood items, such as shrimp, oysters, or crabmeat.

Precaution

- Exercise caution when mincing the onions and green peppers.

Ravigote Sauce

This sauce consists of a mayonnaise base with finely chopped sour gherkins, onions, capers, and tarragon. Dijon mustard is added to give a slightly pungent taste.

2½ qt

Equipment

- 1 gal. stainless steel container
- Kitchen spoon
- Quart measure
- Cup measure
- Spoon measure
- French knife
- Cutting board
- Food chopper or grinder

Ingredients

2 qt mayonnaise
1 c sour gherkins
¼ c onions
¼ c capers
2 tbsp Dijon mustard
1 tbsp tarragon
1 tbsp parsley

Preparation

1. Prepare the mayonnaise.
2. Chop the capers, parsley, and tarragon with a French knife.
3. Grind or chop the sour gherkins fine and drain.
4. Mince the onions with a French knife.

Procedure

1. Place all ingredients in a stainless steel container and blend thoroughly.
2. Adjust the seasoning and serve 2 oz per portion in a soufflé cup with petite salads and fried seafood.

Precautions

- Drain the chopped gherkins and onions thoroughly before adding to the mayonnaise.
- Exercise caution when chopping the capers, parsley, and tarragon.

Tartar Sauce

Tartar sauce consists of a mayonnaise base with minced dill pickles and onions. This sauce can be served with meat or seafood, but most often it is associated with seafood.

2½ qt

Equipment

- 1 gal. stainless steel container
- Cup measure
- Quart measure
- Kitchen spoon
- Food chopper or grinder
- French knife

Ingredients

2 qt mayonnaise
1 pt dill pickles
¼ c parsley
1 c onions

Preparation

1. Prepare the mayonnaise.
2. Chop the parsley with a French knife and wash.
3. Grind or chop the dill pickles very fine and drain.
4. Grind or chop the onions very fine and drain.

Procedure

1. Place all the ingredients in a stainless steel container and blend thoroughly with a kitchen spoon.
2. Adjust the seasoning and serve 2 oz per portion in a soufflé cup with fried seafood, or, in some cases, with meat, such as hamburger.

Precaution

- Drain the chopped onions and pickles fairly dry before adding to the mayonnaise or the sauce will have a fluid consistency.

Dill Sauce

Dill sauce is a tangy cold sauce that stimulates the taste buds when served with cold poached salmon and other seafood items.

1½ qt

Equipment

- French knife
- Spoon measure
- Cup measure
- Quart measure
- 2 qt mixing bowl
- Kitchen spoon
- 2 qt stainless steel container

Ingredients

1 qt sour cream
1 c salad dressing
1 tbsp dry mustard
1 small onion
1 tbsp light brown sugar
2 tbsp dill seed
½ c white vinegar
2 c cucumbers
salt and white pepper to taste

Preparation

1. Mince the onion with a French knife.
2. Chop the cucumbers fine with a French knife.

Procedure

1. Place all the ingredients in a mixing bowl and blend thoroughly using a kitchen spoon.
2. Adjust the seasoning and pour into a stainless steel container.
3. Serve 2 oz per portion in a soufflé cup or ladled over the item. Serve with cold poached salmon or other seafood.

Precaution

- Exercise caution when mincing the onions and chopping the cucumbers.

Chaud-froid Sauce (White)

Chaud-froid, a French term meaning hot-cold, is a jellied sauce used to decorate poultry, fish, and ham for buffet display. It is made by adding plain gelatin and cream to a velouté sauce. The velouté sauce can be made from chicken or fish stock, depending on what the chaud-froid sauce will cover.

1½ gal.

Equipment

- 6 qt sauce pot
- Wire whip
- 1 qt metal bowl
- 2 gal. stainless steel container
- Kitchen spoon
- Quart measure
- Baker's scale

Ingredients

1 gal. velouté sauce
1 qt coffee cream
1 pt hot water
12 oz plain, granulated gelatin
salt and white pepper to taste

Preparation

1. Prepare the velouté sauce.
2. Dissolve the gelatin in the hot water. Place the water in a metal bowl and stir with a kitchen spoon while pouring the gelatin in slowly.

Procedure

1. Place the velouté sauce in a sauce pot and heat.
2. Whip in the dissolved gelatin with a wire whip. Remove from the heat.
3. Add the cream slowly while whipping continuously with a wire whip.
4. Season with salt and white pepper and pour into a stainless steel container. Let cool.
5. The sauce may be colored with various food colors if desired.
6. Use chaud-froid to cover poultry, ham, and fish when decorating these items for buffet display. When applying the chaud-froid sauce, the item being covered should be cold. The sauce should be chilled to the point where it is ready to jell, but still has a flowing consistency. The work should be done in a walk-in refrigerator so the sauce will jell and adhere upon contact to the item being coated. Apply the sauce with a 4 oz to 6 oz ladle.

Note: When covering fish with chaud-froid sauce, mayonnaise may be substituted for the velouté sauce, the gelatin reduced to 9 oz, and the cream omitted.

Precautions

- Be alert when chilling the sauce. If the sauce is too warm, it will not adhere to the item being covered. If the sauce is too cold, it will jell before contact is made. Chill to the point where the sauce will flow, but not run.
- Dissolve the gelatin thoroughly before adding it to the mayonnaise or velouté sauce.

BUTTER SAUCE RECIPES

Lemon Butter Sauce

Lemon butter sauce is a blend of melted butter, lemon juice, and chopped parsley. It is usually served with sautéed or broiled seafood.

1 pt

Equipment

- 1 qt saucepan
- French knife
- Kitchen spoon
- 1 qt stainless steel container

Ingredients

1 lb butter
¼ c lemon juice
2 tbsp parsley

Preparation

1. Chop the parsley with a French knife and wash.
2. Squeeze the juice from the lemons.

Procedure

1. Place the butter in a saucepan and melt.
2. Add the lemon juice and chopped parsley and stir. Pour into a stainless steel container.
3. Serve 1 oz per serving over broiled or sautéed seafood.

Precautions

- Exercise caution when chopping the parsley.
- Do not brown the butter.

Trade Tip

Soften the butter to be used in butter sauces at room temperature until it can be stirred easily with a kitchen spoon. Do not soften butter in the oven, in the microwave, or on the top of the stove.

Meunière Sauce

Meunière sauce is the most popular of the butter sauces. The butter is heated until light brown. The lemon juice and chopped parsley are added and the sauce is served over sautéed seafood.

1 pt

Equipment

- Cup measure
- Baker's scale
- Saucepan
- French knife
- 1 qt stainless steel container
- Kitchen spoon

Ingredients

1 lb butter
2 oz lemon juice
¼ c parsley

Procedure

1. Place the butter in a saucepan and brown slightly.
2. Add the lemon juice and stir with a kitchen spoon.
3. Add the chopped parsley.
4. Serve over sautéed seafood.

Preparation

1. Chop the parsley with a French knife.
2. Squeeze the juice from lemons.

Precautions

- Do not overbrown or burn the butter.
- Exercise caution when chopping the parsley.

Brown Butter Sauce

This sauce adds taste and color to many broiled meats, such as broiled pork chops, liver, veal chops, etc.

1 qt

Equipment

- 2 qt saucepan
- Kitchen spoon
- 1 pt stainless steel container

Ingredients

1 lb butter
1 c brown gravy
2 c consommé

Procedure

1. Place butter in a saucepan and brown on the range (nut brown).
2. Add the brown gravy.
3. Thin with consommé while stirring with a kitchen spoon. Pour into a stainless steel container.
4. Serve 1 oz per portion with broiled meats.

Precautions

- Do not overbrown the butter.
- Exercise caution when adding the gravy to the brown butter; it will flare up. Stir continuously while adding gravy. Exercise caution when adding the gravy to the brown butter; it will flare up. Stir continuously while adding gravy.

Preparation

1. Prepare the consommé.
2. Prepare the brown gravy.

DESSERT OR SWEET SAUCE RECIPES

Cinnamon Sauce

This is a red, cinnamon-flavored dessert sauce. It can be served with any dessert item that is complemented by a cinnamon flavor.

1 qt

Equipment

- 2 qt saucepan
- Kitchen spoon
- Cup measure
- Food grater
- 1 qt stainless steel container

Ingredients

3 c water
2 c sugar
¼ c cornstarch
2 lemons
½ c red cinnamon candy

Procedure

1. Blend the sugar and cornstarch in a saucepan.
2. Pour in the water slowly, blending thoroughly with a kitchen spoon. Bring to a boil on the range.
3. Add the lemon rind, juice, and cinnamon candy.
4. Cook, stirring constantly, until sauce thickens and clears.
5. Serve 2 oz per portion over desserts that are complemented by a cinnamon flavor, such as apple cobbler and apple crisp.

Preparation

1. Grate the lemon peel and squeeze the juice from the lemons.

Precaution

- Stir the sauce occasionally while cooking to avoid scorching.

Custard Sauce

This is a dessert sauce consisting mainly of milk, eggs, and sugar. It is served hot or cold to complement such dessert items as rice pudding or bread pudding.

2 qt

Equipment

- 4 qt saucepan
- Wire whip
- Quart measure
- Cup measure
- Spoon measure
- 1 qt stainless steel bowl
- Kitchen spoon
- 1 gal. stainless steel container

Ingredients

2 qt milk
1 c sugar (first amount)
½ c sugar (second amount)
1½ tsp salt
6 egg yolks
½ c butter
vanilla to taste

Preparation

1. Separate the eggs, saving the whites for use in another preparation.

Procedure

1. Combine the milk and first amount of sugar in a saucepan and bring to a boil.
2. Mix the second amount of sugar, salt, and cornstarch with a little of the mixture using a kitchen spoon in the stainless steel bowl to form a smooth paste. Add the egg yolks and mix smooth. Pour this mixture slowly into the hot sugar and milk mixture, whipping vigorously with a wire whip, and cook until slightly thickened and smooth.
3. Whip in the butter and vanilla. If the color is too pale, adjust by adding yellow color. Pour into a stainless steel container.
4. Serve 2 oz per portion with rice pudding, snow pudding, or bread pudding.

Precautions

- Be alert and stir the milk and sugar mixture while it is on the range or it will boil over.
- Stir the mixture at all times through the cooking period to avoid scorching.
- Pour the cornstarch and egg mixture very slowly and whip vigorously when adding it to the hot milk and sugar mixture.

Vanilla Sauce

Vanilla sauce is a dessert sauce that is generally served with cakes or cobblers.

2 qt

Equipment

- 4 qt saucepans (2)
- Wire whip
- Quart measure
- Baker's scale
- Spoon measure
- 1 gal. stainless steel container
- Kitchen spoon
- 1 qt stainless steel bowl

Ingredients

2 qt water
3 oz cornstarch
2 lb 8 oz sugar
3 oz egg yolks
10 oz butter
1 tsp salt
vanilla to taste

Preparation

1. Boil the water in a saucepan.
2. Separate the eggs. Save the whites for use in another preparation.

Procedure

1. Place the cornstarch, salt, and half the sugar in the saucepan. Mix thoroughly with a kitchen spoon.
2. Add the boiling water while whipping with a wire whip. Cook for about 10 min.
3. Beat the egg yolks with a wire whip in a stainless steel bowl. Blend in the remaining sugar by pouring slowly into the thickened mixture while whipping constantly. Cook 2 min.
4. Remove from the range and whip in the butter and vanilla. Pour into a stainless steel container.
5. Serve 2 oz per portion over cake, bread pudding, brown Betty, or fruit cobblers.

Precautions

- Whip rapidly when adding the egg yolk mixture.
- Use this sauce within 2 hr of preparation.

Trade Tip

To prevent custard sauce from curdling while it cools, strain through a fine sieve into a clean bowl. This also eliminates any egg that may have solidified during cooking.

Foamy Brandy Sauce

This is a cold dessert sauce with a light, smooth, creamy texture. It is an excellent choice for any dessert that can be complemented with a brandy flavor.

2 qt

Equipment

- Baker's or portion scale
- Cup measure
- Pint measure
- Mixing machine
- Wire whip
- 1 gal. stainless steel mixing bowl
- Kitchen spoon

Ingredients

8 oz soft butter
8 oz dark or light brown sugar
1 c egg yolks
1 pt heavy whipping cream
1 c unwhipped cream
1 pt egg whites
10 oz sugar
4 oz brandy

Preparation

1. Separate eggs into yolks and whites.

Procedure

1. Place the butter and brown sugar in the stainless steel mixing bowl. Using the wire whip, beat until creamy.
2. Add the egg yolks and continue to beat until smooth and light. Remove from the mixer, place in a stainless steel mixing bowl, and hold for later use.
3. Place the whipping cream in a clean stainless steel mixing bowl. Using the wire whip, whip at high speed until stiff. Fold gently into the egg mixture.
4. Fold in the unwhipped cream.
5. Place the egg whites in a stainless steel mixing bowl. Using the wire whip, whip at high speed until the whites start to froth. Gradually add the sugar and continue to whip until a stiff meringue has formed. Fold gently into the egg and cream mixture.
6. Add the brandy and stir with a kitchen spoon until it is blended into the sauce.
7. Place in the refrigerator until ready to serve.

Precautions

- Be alert and careful when separating the eggs. If the yolk is left in the white, they do not beat well.
- When whipping the cream, beat only until soft peaks form. Overmixing produces butter.

Chocolate Sauce

This is a dessert sauce with a rich, smooth chocolate flavor. It can be served hot or cold with such items as ice cream and cake or used in making sundaes or parfaits.

2 qt

Equipment

- Quart measure
- Cup measure
- Baker's or portion scale
- Spoon measure
- 2 qt saucepan
- 1 gal. sauce pot
- ½ gal. stainless steel container
- Wire whip
- Small stainless steel bowl

Ingredients

1½ qt water
1½ pt sugar
½ c cornstarch
½ c butter or margarine
8 oz bitter chocolate naps
1 tbsp vanilla

International Ice Cream Association

Preparation

1. Place the water in a 2 qt sauce pot and bring to a boil.
2. Place the chocolate in a small stainless steel bowl, place near the heat, and melt.

Procedure

1. Place the sugar, cornstarch and melted bitter chocolate in a gallon sauce pot.
2. Gradually pour in the boiling water while at the same time whipping rapidly with a wire whip. Bring to a boil. Reduce heat to a simmer and simmer for approximately 2 min to 3 min. Stir constantly with the whip.
3. Remove from the range. Stir in the butter or margarine and vanilla.
4. Serve hot or hold in the refrigerator for serving cold.

Precaution

- Heat chocolate only enough to melt. Too hot a temperature will harm the chocolate.

MISCELLANEOUS SAUCE RECIPES

Oriental Sweet-Sour Sauce

This is a tangy, sweet and sour sauce associated with Oriental cuisine. It is an excellent sauce to serve with sautéed shrimp or with strips of pork.

1 qt

Equipment

- Quart measure
- Cup measure
- 2 qt saucepan
- Wire whip
- Kitchen spoon
- 2 qt stainless steel container
- Small bowl

Ingredients

1 qt water
1 pt sugar
1 c cider vinegar
1 tbsp soy sauce
1 c sweet pickle relish
¼ c cornstarch

Procedure

1. Place the vinegar, sugar, and all but one cup of the water in a saucepan and bring to a boil.
2. Dissolve the cornstarch in the remaining cup of water in a small bowl and add slowly to the boiling mixture, whipping constantly with a wire whip. Cook until smooth and fairly thick.
3. Add the sweet relish and soy sauce and simmer, stirring occasionally with a kitchen spoon.
4. Remove from the range and pour into a stainless steel container.
5. Serve 2 oz per portion with a ladle. Serve with sautéed shrimp, scallops, or strips of pork.

Precaution

- Stir occasionally after the starch is added to avoid scorching.

Mint Sauce

Mint sauce is a sauce with a fresh mint flavor. It is served almost exclusively with roast lamb.

1 gal.

Equipment

- 5 qt saucepan
- Kitchen spoon
- 1 qt measure
- Baker's scale
- 1 gal. stainless steel container

Ingredients

1 gal. water
1 c cider vinegar
6 oz sugar
12 oz fresh mint leaves

Procedure

1. Place the water in the saucepan and bring to a boil.
2. Add the remaining ingredients, simmer for 25 min, and remove from the range.
3. Serve 2 oz per portion, either hot or cold, in a soufflé cup with roast or potted lamb.

Precaution

- Use caution when chopping the mint.

Preparation

1. Chop the mint leaves very fine with a French knife.

GRAVY RECIPES

Roast Beef Gravy

This is a brown gravy with the flavor of roast beef. The flavor is acquired by using the drippings left in the roast pan after the beef is done.

1 gal.

Equipment

- 6 qt sauce pot
- Wire whip
- Quart measure
- Baker's scale
- China cap
- 1 gal. stainless steel container

Ingredients

1 gal. beef or brown stock
12 oz fat (from roast pan) or shortening
10 oz flour
½ c tomato puree
caramel color if desired
salt and pepper to taste

Preparation

1. Prepare the beef or brown stock.
2. Pour drippings from the roast pan in a stainless steel container and deglaze the pan with the stock.

Procedure

1. Place the fat (from the roast pan) or shortening in a sauce pot and heat.
2. Add the flour, making a roux, and cook until it is slightly brown.

3. Add the hot stock, whipping rapidly with a wire whip until slightly thickened and smooth.
4. Add the tomato puree and simmer for 20 min.
5. Season with salt and pepper. Strain through a china cap into a stainless steel container.

6. Serve 2 oz per portion over each order of roast beef.

Precaution

- Stir occasionally while the gravy is simmering to avoid sticking.

Roast Veal Gravy

This is a brown gravy with the flavor of roast veal. The flavor is acquired by using the drippings left in the roast pan after the veal is done.

1 gal.

Equipment

- 6 qt sauce pot
- Wire whip
- Quart measure
- Baker's scale
- China cap
- 1 gal. stainless steel container

Ingredients

1 gal. veal or brown veal stock
12 oz fat (from roast pan) or shortening
10 oz flour
½ c tomato puree
caramel if desired
salt and pepper to taste

Procedure

1. Place the fat (from roast pan) or shortening in a sauce pot and heat.
2. Add the flour, making a roux, and cook until it is slightly brown.
3. Add the hot stock, whipping rapidly with a wire whip until slightly thickened and smooth.
4. Add the tomato puree and simmer for 20 min.
5. Season with salt and pepper. Strain through a china cap into a stainless steel container.
6. Serve 2 oz per portion over each order of roast veal.

Precaution

- Stir occasionally while the gravy is simmering to avoid sticking.

Preparation

1. Prepare the veal or brown veal stock.
2. Pour the drippings from the roast pan into a stainless steel container and deglaze the pan with the stock.

Roast Pork Gravy

This is a brown gravy containing the flavor of roast pork. The flavor is obtained by deglazing the pan after the pork roast is done.

1 gal.

Equipment

- Kitchen spoon
- 6 qt sauce pot
- Baker's scale
- China cap
- 1 gal. stainless steel container
- Gallon measure
- Wire whip
- Cup measure

Ingredients

12 oz fat (from roast pan) or shortening
10 oz flour
1 gal. brown pork stock
½ c tomato puree
salt and pepper to taste
caramel color if desired

Procedure

1. Place the fat in a sauce pot and heat.
2. Add the flour, making a roux, and cook until flour is slightly brown.
3. Add the hot stock, whipping vigorously with a wire whip until slightly thick and smooth.
4. Add the tomato puree and simmer for 20 min.
5. Season with salt and pepper. If the gravy is too light, adjust the color by adding a small amount of caramel color.
6. Strain through a china cap into a stainless steel container.
7. Serve 2 oz per portion over each order of roast pork.

Precautions

- Stock must be hot before adding it to the roux.
- Stir occasionally while the gravy is simmering to avoid scorching.

Preparation

1. Prepare the brown pork stock after the roast is done by deglazing the roast pan with the hot stock.

Roast Lamb Gravy

This is a brown gravy with the flavor of roast lamb. The flavor is acquired by using the drippings left in the roast pan after the roast is done.

1 gal.

Equipment

- 6 qt sauce pot
- Baker's scale
- Spoon measure
- Wire whip
- China cap
- 1 gal. stainless steel container

Ingredients

12 oz fat (from roast pan)
10 oz flour
1 gal. brown lamb stock
2 tbsp marjoram
salt and pepper to taste

Procedure

1. Place the fat in a sauce pot and heat.
2. Add the flour, making a roux, and cook until the flour browns slightly.
3. Add the hot stock, whipping vigorously with a wire whip until the gravy is slightly thick and smooth.
4. Add the marjoram and simmer for 30 min.
5. Season with salt and pepper. Strain through a china cap into a stainless steel container.
6. Serve 2 oz per portion over each order of roast lamb.

Preparation

1. Prepare the brown lamb stock after the roast is done by deglazing the roast pan with the hot stock.

Precautions

- Stock must be hot before adding it to the roux.
- Stir occasionally while the gravy is simmering to avoid scorching.

Giblet Gravy

This gravy can be prepared in two different ways. The cooked, chopped giblets can be added either to a brown or light turkey gravy. Both are excellent to serve with roast turkey. If a light gravy is desired, omit the puree and caramel color. If a plain light or brown gravy is desired, omit the giblets.

1 gal.

Equipment

- 6 qt sauce pot
- Baker's scale
- Gallon measure
- Cup measure
- Wire whip
- China cap
- Kitchen spoon
- 1 gal. stainless steel container
- Food grinder

Ingredients

12 oz chicken or turkey fat
10 oz flour
6 lb giblets
1 gal. chicken or turkey stock (include pan drippings from roast chicken or turkey)
salt and pepper to taste
½ c tomato puree (for brown gravy only)
caramel color as desired (for brown gravy only)

Procedure

1. Place the fat in a sauce pot and heat
2. Add the flour, making a roux, and cook until light brown.
3. Add the stock, whipping vigorously with a wire whip until slightly thick and smooth.
4. If preparing a light turkey gravy, season and strain through a china cap into a stainless steel container and then add the ground giblets. If preparing a brown turkey gravy, add the tomato puree and caramel color and continue to simmer for 15 min. Season and strain through a china cap into a stainless steel container. Add the ground giblets. If a plain light or brown gravy is desired, the ground giblets are omitted.
5. Serve 2 oz per portion with roast chicken or turkey.

Precautions

- Be extremely cautious when using the food grinder.
- When adding the stock to the roux, be sure the stock is hot. Whip vigorously to eliminate lumps.
- Exercise caution when adding the caramel color. A little goes a long way.
- Be sure to include any drippings that are left in the roast pan. Also, deglaze the roast pan to utilize all flavor.

Preparation

1. Prepare the chicken or turkey stock.
2. Boil the giblets and grind on a food grinder using the medium chopper plate.

Trade Tip

Know the thickener you are using. Common thickeners for sauces and gravies include flour, cornstarch, tapioca, and arrowroot. Each has its particular applications. Thickeners have different properties, so an equal substitution of one thickener for another may not produce desired results.

Country Gravy

This gravy, also known as pan gravy, is made with milk and is light brown in color. This gravy is most frequently served with fried chicken or pork chops.

1 gal.

Equipment

- Baker's scale
- Quart measure
- 6 qt sauce pot
- China cap
- 4 qt saucepan
- Kitchen spoon
- Wire whip
- 1 gal. stainless steel container

Ingredients

1 lb chicken or pork fat or butter
1 lb flour
2 qt milk
2 qt chicken or brown pork stock
salt and pepper to taste

Procedure

1. Place the fat or butter in a sauce pot and heat.
2. Add the flour, making a roux, and cook until light brown.
3. Add the hot stock and milk, whipping vigorously with a wire whip until thick and slightly smooth.
4. Season with salt and pepper.
5. Strain through a china cap into a stainless steel container.
6. Serve 2 oz per portion with fried chicken or pork chops.

Preparation

1. Prepare the chicken or brown pork stock.
2. Heat the milk in a saucepan.

Precautions

- Brown the roux slightly, but do not let it burn.
- Exercise caution when straining the gravy.

Beef Preparation

## Chapter 19

National Cattlemen's Beef Association

Beef is the most popular of all edible meats. More beef is consumed in the United States each year than any other type of meat. In recent years, there has been concern over the amount of cholesterol present in beef. High amounts of cholesterol have been associated with some forms of heart disease. Beef preparations today have reduced the amount of cholesterol present. In addition, it has been proven that when properly prepared, beef contains less cholesterol than other popular meats and food items.

Beef is the flesh of steers, heifers, cows, bulls, and stags. The age and sex of these animals have a great effect on the taste and quality of the meat. Meat is graded to ensure a standard level of meat quality. Beef is graded for yield and quality. Yield is the amount of salable meat that can be obtained from a carcass. Quality is determined by the age of the animal, marbling (fat present in the muscle tissue), color, and texture of the meat. The meat grades most commonly used in the commercial kitchen are U.S. prime, U.S. choice, and U.S. good.

CLASSES OF BEEF

The five classes of beef are steers, heifers, cows, stags, and bulls. Two important factors to be considered when beef is graded are quality and yield. The following lists the characteristics of these five classes:

Steer: A steer is a male calf that has been castrated. They produce the best quality beef and a high yield (amount of salable meat). Most steers are graded prime or choice, the two highest grades.

Steers are about 2½ to 3 years old when they are marketed. They weigh approximately 650 lb to 1250 lb, producing sides ranging in weight from approximately 200 lb to 400 lb. Most steers are grain-fed,

which has a great bearing on the quality. Grain-fed animals are superior to grass-fed animals. In recent years, however, less grain has been fed to these animals, producing leaner beef, which in turn is lower in cholesterol.

Heifer: A heifer is a young female that has not borne a calf. Heifers produce high-quality meat and are generally marketed when 2½ to 3 years old. Heifers mature faster than steers, but steers are preferred because they usually produce a higher yield. Most heifers are graded prime or choice.

Cow: A cow is a female that has borne one or more calves. Cow meat generally has an uneven distribution of fat that is usually yellow. Slightly poor quality and yield are also evident in cow meat. This is generally graded good, commercial, or standard. Cows are marketed at an older age because they are kept as long as the calves they bear, or they are used for milk production. Age has an effect on the quality of meat and grading.

Stag: A stag is a male that is castrated after it has become sexually mature. The meat is generally of poor quality, lacking the characteristics necessary to achieve a high grade. Stag meat is rarely used in commercial kitchens. Stags are generally graded commercial, utility, cutter, or canner. Stag meat is commonly used in canned meat products and dried beef.

Bull: A bull is a male that is sexually mature and uncastrated. Bull meat is dark red, characteristic of an older animal. It is commonly used in the making of sausage and dried beef. Bull meat is never used in the commercial kitchen.

BEEF GRADING

Meat grading provides the purchaser with certain standards to follow when purchasing beef or other meats. Grading and inspection is done by the federal government if the packing house is engaged in interstate commerce. If not, this responsibility belongs to the city, municipality, or state in which the plant is located.

Meat is stamped to indicate that it has been inspected and graded. The most important stamp is the *federal inspection stamp*. The federal stamp is round with "U.S. INSP'D & P'S'D" included on the inside. This indicates that the meat is U.S. inspected and that it has passed inspection. It has met minimum government standards and is fit for human consumption. Purple vegetable dye is used for the stamping because it is harmless and does not have to be removed before cooking. The stamp also carries an identifying number, which is the number assigned to the packing plant where the meat is processed.

The *grading stamp* on meat designates the quality of the meat. These stamps are also stamped on the carcasses with purple vegetable dye. The grading is done by the federal meat inspector, who represents the U.S. Department of Agriculture. The inspectors are well-trained experts with ready knowledge of the qualifications each beef grade must possess. Before 1976, grading was based on three factors: finish, quality, and conformation of the animal.

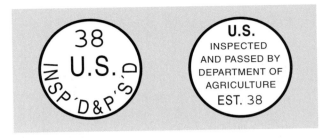

The USDA inspection stamp is required by federal law. This stamp guarantees the meat is fit for human consumption.

Finish refers to the amount and color of the fat on the outside and inside of the carcass. *Quality* refers to the color and texture of the meat, bones, and the marbleization (mixture of fat and lean within the meat). *Conformation* refers to the shape of the beef carcass and does not affect the palatability of the beef. On February 23, 1976, the U.S. Department of Agriculture changed the grading specifications for beef, eliminating conformation as a factor in determining the grade. Beef now is graded for yield and quality.

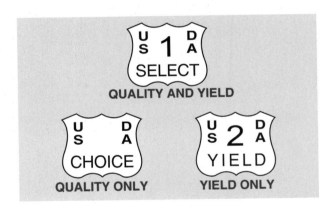

The number in the yield grade stamp indicates how much salable meat can be obtained. Yield grade numbers range from 1 (high) to 5 (low).

Yield grades are numbered 1 to 5 and determine how much salable meat can be obtained from a carcass. The higher the yield number, the more salable meat there is. Quality grading is determined by the age of the animal and the marbling, color, and texture of the meat.

Beef grades, in order of desirability, are prime, choice, select (good), standard, commercial, utility, cutter, and canner. The four grades commonly used in food service establishments are prime, choice, good, and standard.

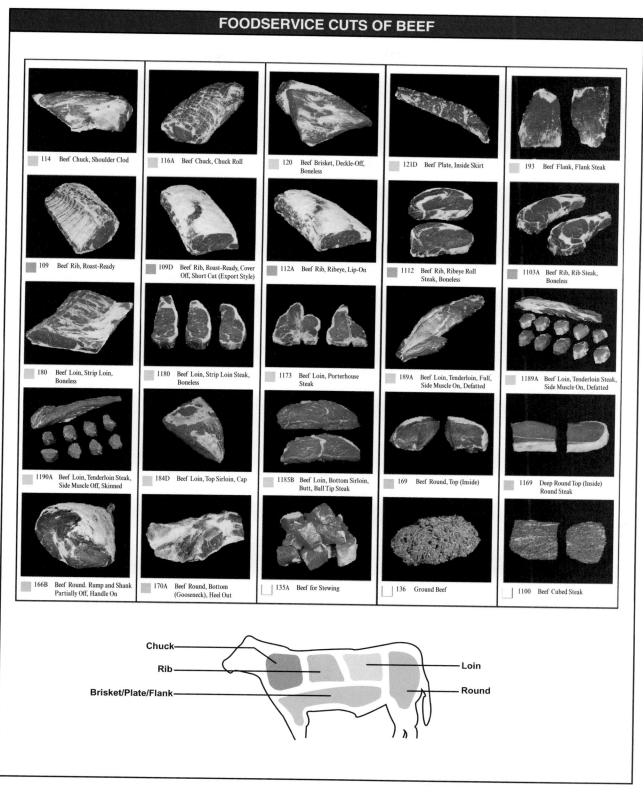

North American Meat Processors Association

NAMP/IMPS (North American Meat Processors Association/Institutional Meat Purchase Specifications) specifies meat cuts by number.

COOKING METHODS FOR BEEF

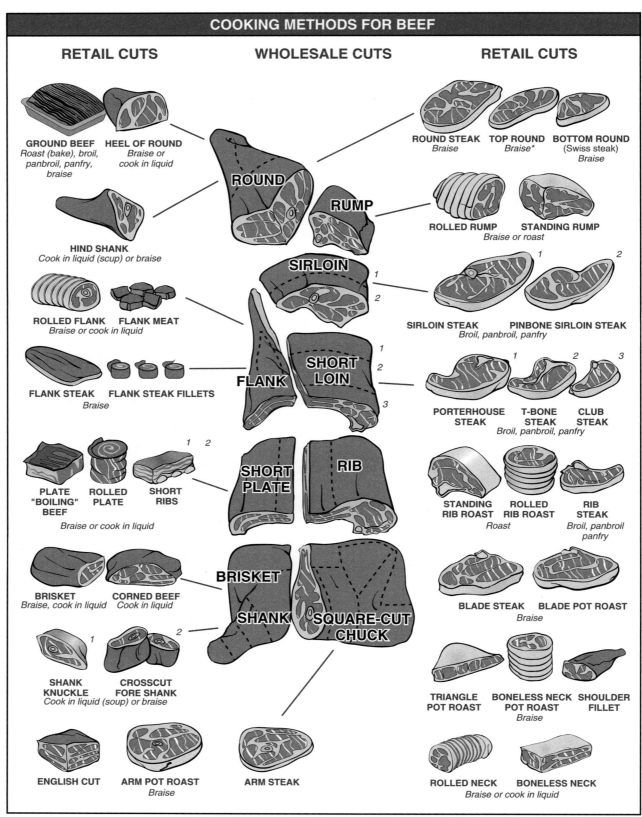

RETAIL CUTS

GROUND BEEF
Roast (bake), broil, panbroil, panfry, braise

HEEL OF ROUND
Braise or cook in liquid

HIND SHANK
Cook in liquid (scup) or braise

ROLLED FLANK   **FLANK MEAT**
Braise or cook in liquid

FLANK STEAK   **FLANK STEAK FILLETS**
Braise

PLATE "BOILING" BEEF   **ROLLED PLATE**   **SHORT RIBS**
Braise or cook in liquid

BRISKET   **CORNED BEEF**
Braise, cook in liquid   *Cook in liquid*

SHANK KNUCKLE   **CROSSCUT FORE SHANK**
Cook in liquid (soup) or braise

ENGLISH CUT   **ARM POT ROAST**   **ARM STEAK**
Braise

WHOLESALE CUTS

ROUND

RUMP

SIRLOIN

FLANK

SHORT LOIN

SHORT PLATE   **RIB**

BRISKET

SHANK   **SQUARE-CUT CHUCK**

RETAIL CUTS

ROUND STEAK   **TOP ROUND**   **BOTTOM ROUND**
Braise   *Braise**   (Swiss steak) *Braise*

ROLLED RUMP   **STANDING RUMP**
Braise or roast

SIRLOIN STEAK   **PINBONE SIRLOIN STEAK**
Broil, panbroil, panfry

PORTERHOUSE STEAK   **T-BONE STEAK**   **CLUB STEAK**
Broil, panbroil, panfry

STANDING RIB ROAST   **ROLLED RIB ROAST**   **RIB STEAK**
Roast   *Broil, panbroil panfry*

BLADE STEAK   **BLADE POT ROAST**
Braise

TRIANGLE POT ROAST   **BONELESS NECK POT ROAST**   **SHOULDER FILLET**
Braise

ROLLED NECK   **BONELESS NECK**
Braise or cook in liquid

*Prime and choice grades may be broiled, panbroiled, or panfried

Commercial kitchens commonly use wholesale cuts for economy and yield.

U.S. prime: The best beef available comes from prize steers and heifers. Of all the beef marketed in the U.S., only 4% is graded prime. Prime beef has a very high fat content, which makes it costly when it is trimmed for cooking. It is generally used in commercial establishments that have an expensive menu.

U.S. choice: This grade is the most popular grade of beef. It has very good fat covering and good marbleization of fat in the lean meat. It is preferred by most commercial establishments because there is less waste than with prime beef.

U.S. select (good): This grade is inexpensive and can produce a fairly good product if cooked with care. Most beef graded good comes from grass-fed steers and heifers. In some cases, corn-fed cows are graded good. This beef has a soft fat covering that is generally yellow. There is just a slight marbleization of fat in the lean meat.

U.S. standard: This grade was created by the U.S. Dept of Agriculture in 1959 and is between the good and commercial grades. Most standard beef is from young steers, heifers, and cows. It has very poor conformation and little fat covering, depending on the age of the animal. It is not handled by many meat purveyors because it is not suitable for use in commercial food service establishments. It is sold mostly in retail outlets.

U.S. commercial: This grade comes mostly from older cows. It is tough meat, and for best results it must be cooked by a lengthy cooking method or treated with a meat tenderizer. Very little of it is used in the commercial kitchen. Commercial meat has poor conformation and, although in some cases the fat covering is fair, it is yellow, indicating an older animal.

U.S. utility: This grade comes mostly from stags, bulls, and older cows. It is seldom, if ever, used in the commercial kitchen. It is tough meat and, like the commercial grade, must be cooked for a lengthy period of time or treated with a tenderizer.

U.S. cutter and *U.S. canner*: These grades of beef are very inferior and very tough. All cutter and canner beef come from bulls and stags. These grades are largely used in canned meat products. The color of cutter and canner beef is dark red to a light brown, and the flesh is soft and watery.

The three beef grades most commonly used in the commercial kitchen are U.S. prime, U.S. choice, and U.S. good.

MARKET FORMS

Beef can be purchased in five different forms: by the side or quarter, wholesale, primal, fabricated, and retail cuts. Retail cuts (cuts found in a supermarket) are of little interest to the commercial chef or cook because the method of cutting beef into units is different in the commercial kitchen than for home use. The main difference between retail and commercial cuts is undesirable fat and bone is left on the retail cuts. In the commercial kitchen, the meat cutter prepares the meat for immediate cooking and discards most of the unnecessary fat and bone.

The meat form best suited for the commercial kitchen is determined by the following:

- meat cutting skill of personnel
- meat cutting equipment available
- working space available
- meat storage space available
- utility of all cuts purchased
- meat preparations served
- overall economy of purchasing meat in a particular form

A *side of beef* is a half of the complete carcass. It ranges in weight from 225 lb to 450 lb, the average weighing approximately 300 lb. A side of beef can be purchased at a cheaper cost per pound than any of the market forms. However, many factors must be considered before purchasing beef in this form. The proper facilities and equipment must be available, and people with knowledge of blocking out a side of beef into individual units are required.

Utilization of cuts and their suitability for use in entree dishes that will sell is another consideration. Whether or not fat or suet can be rendered and used for frying and bones be used for stock is yet another factor. Only the purchaser can decide whether or not purchasing beef by the side is economical. In the past, a large percentage of food service establishments purchased beef by the side. Today, this is no longer true because of the lack of trained personnel for blocking out a side of beef and the emphasis on production.

A *quarter of beef* is a side divided into two parts. The front, or fore part, is the *forequarter*. The back, or hind part, is the *hindquarter*. The side is divided into quarters by cutting between the twelfth and thirteenth ribs, leaving 12 ribs on the forequarter and 1 rib on the hindquarter. The hindquarter contains the most desirable meat. Consequently, there is more demand and a higher price. The forequarter contains less desirable cuts and costs less.

The *forequarter* of beef, when blocked out, consists of five wholesale or primal cuts: the rib, chuck or shoulder clod, brisket, shank, and short plate or navel. Of these, the rib cut is the best because it comes from the part of the animal where the muscles are used the least.

The *hindquarter* consists of four wholesale or primal cuts: the round, flank, rump, and sirloin (sirloin and short loin). Of these, the sirloin, which is sometimes cut in half and known as the sirloin or short loin, is the best. Again, it comes from the part of the animal where the muscles are used the least. The sirloin is the steak meat and is always very much in demand.

The *wholesale* or *primal cuts* are parts of the forequarter and hindquarter of beef. There are five wholesale cuts in the forequarter and four in the hindquarter. This is a very popular way of purchasing beef today because it reduces labor cost and time, requires less refrigeration space, and less expensive butchering equipment is needed.

Fabricated cuts are ready-to-cook meats cut or packaged to certain specifications for quality, size, and weight. Fabricated cuts are the most convenient and popular way of purchasing meat today because they eliminate trimming waste, provide uniform portions, control the cost, eliminate the need for expensive cutting equipment, and cut labor costs. However, before purchasing fabricated cuts, the price per pound must be determined and compared with other types of cuts.

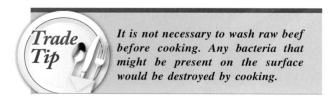

Trade Tip It is not necessary to wash raw beef before cooking. Any bacteria that might be present on the surface would be destroyed by cooking.

WHOLESALE CUTS AND COOKING METHODS

The following is a list of the eight wholesale cuts of beef. Nine cuts are counted if the rump is considered a separate cut. However, when the carcass is blocked out, the rump section usually remains attached to the round and is considered part of the wholesale cut. The rump may be removed later and sold separately.

National Cattlemen's Beef Association

Beef may be as formal as a ribeye roast, as delicious as a grilled steak, or as popular as hamburgers and hot dogs.

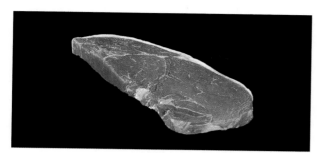

National Cattlemen's Beef Association

Sirloin

Sirloin is the best cut on a side of beef. It is some-times cut into two sections known as the *sirloin* and *short loin*. The sirloin contains two different types of meat: sirloin on the outside and tenderloin on the inside. The European method separates the sirloin into two parts so each cut has only sirloin or only tenderloin. The European method produces such steaks as strip sirloin, New York sirloin, filet mignon, and tenderloin steak. The American method of cutting the loin is to leave the T-bone and tenderloin on the sirloin. Cutting cross sections, starting from the end close to the rib cut, produces club steaks, T-bone steaks, and porterhouse steaks.

Idaho Potato Commission

A T-bone steak with a baked potato is a classical entreé on the dinner menu.

The club steaks are the first three cuts from the sirloin and contain little or no tenderloin. The T-bone contains both sirloin and tenderloin meats, and the porterhouse contains sirloin and the largest amount of tenderloin. Porterhouse is the best of the steaks cut by the American method.

The whole sirloin cut comes from a section of the animal where the muscles are used the least, which produces tender meat. These cuts are cooked by a quick-cooking method such as broiling or sautéing. The sirloin is also roasted in many cases. The tenderloin, when removed from the underside of the whole sirloin cut, may also be roasted. Sirloins graded prime and choice have excellent fat covering and good marbleization, which supplies juices to the meat when cooked.

The tenderness of the sirloin cut can be improved by *aging*. Aging is the process of holding meats at 34°F to 36°F for a certain period of time. The aging time required depends on kind and cut of meat.

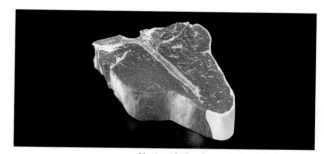

National Cattlemen's Beef Association

Porterhouse

Beef steaks are usually cooked using the broiling method. Beef steaks include cuts such as chateaubri-and, filet mignon, T-bone steak, porterhouse steak, tenderloin steak, club steak, Delmonico or rib steak, and flank steak.

Rib is the best cut from the forequarter of beef and the only cut from the forequarter that can be improved by aging. The rib contains seven of the thirteen rib bones found on a side. These seven ribs make the rib cut the best beef cut for roasting because it forms a natural rack. The rib can be easily iden-tified, not only because of the seven ribs, but also because of the large muscle of meat known as the *ribeye*, which is a continuation of the sirloin. Ribs graded prime, choice, and good generally have ex-cellent fat covering and extremely good marblei-zation, which are the characteristics needed to make an excellent roast. The average weight of a rib is 20 lb to 25 lb, and although there is some waste in trimming for cooking, about 75% to 80% of this cut is usable.

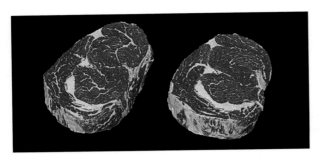

National Cattlemen's Beef Association

Ribeye

In commercial kitchens, the rib is generally roasted and is one of the most popular items to appear on the dinner menu. However, one can also cut steaks from the smaller ribs. These are known as Delmonico or rib steaks. Short ribs are also extracted from the rib cut when the rib is trimmed for roasting. They make excellent short ribs since there is generally a sizable portion of lean meat left on them. Another preparation is to remove the seven rib bones, roll up the meat, and secure it with butcher twine. This roast is known as a rolled rib.

Burgers' Smokehouse

The prime rib roast is the most luxurious of all beef roasts.

Round and rump is the hind leg of the beef that, when blocked out as a commercial cut, includes the rump. When the round cut is boned completely, it consists of five pieces: rump, top round, bottom round, knuckle or tip, and shank meat. When trimmed for cooking, about 70% to 75% of this cut is usable. This cut is versatile in the commercial kitchen and is used in such preparations as Swiss steak, pot roast, sauerbraten, beef rouladen, etc. It is a fairly tough cut of beef because it comes from a section of the animal where the muscles were used extensively and, for this

reason, must be cooked by a lengthy cooking method such as roasting, braising, stewing, or cooking in water. The round with the shank and rump removed also makes a fine roast for buffets and smorgasbords. This roast is referred to as a Chicago round or steamship round roast.

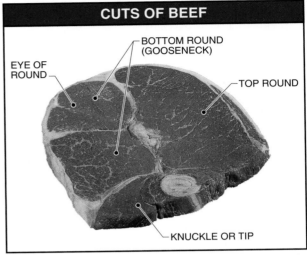

National Cattlemen's Beef Association

A center cut beef round is commonly used to prepare pot roast, Swiss steak, and sauerbraten.

National Cattlemen's Beef Association

Chuck

Chuck is sometimes known as shoulder clod chuck. Chuck includes the section of the forequarter that contains the first five rib bones. It is a fairly tough cut of beef but is quite lean and possesses excellent flavor. When trimmed for cooking, it produces about 75% to 80% usable meat. The chuck, like the round, has become a versatile piece of meat in the commercial kitchen. It is used for such profitable items as meat loaf, beef stew, Salisbury steak, goulash, and ragout. Because the meat is slightly tough, it must be cooked for some length of time or used in items that call for ground beef.

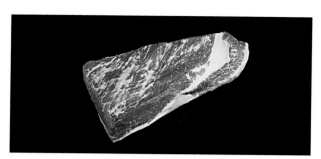

National Cattlemen's Beef Association

Brisket

Brisket contains layers of both lean and fat. It is a thin section of meat with breastbones and short sections of the rib bone present. It has long muscle fibers that run in several directions, making it difficult to slice. The brisket is a tough cut of beef but has excellent flavor when cooked by a long cooking method. In a commercial kitchen, the brisket is used in the preparation of pot roast, sauerbraten, corned beef, and boiled beef.

Shank is the cut from the lower foreleg of the side of beef and contains a high percentage of bone and little meat. It is seldom purchased for use in the commercial kitchen. It is possible to use it as ground beef or in beef stew, but the chuck is far superior for these preparations.

National Cattlemen's Beef Association

Flank

Flank is a section of the hindquarters and is thin and flat. It has long, coarse fibers and a larger percentage of fat to lean. It contains one flank steak, which is used after the fat covering is removed. The steak is oval shaped, thin, boneless, and has long muscle fibers running through it. The flank steak weighs about 1½ lb. It should be scored or cubed before it is cooked. Except for the flank steak, flank is rarely used in commercial kitchens.

Short plate is sometimes known as the navel. It is a thin portion of the forequarter that lies opposite the rib cut. The bones of this cut are the remaining sections of the rib bones. In the commercial kitchen,

short plate is sometimes used in the preparation of short ribs, but the short ribs cut from the end of the rib cut are far superior because they contain more lean meat.

Both raw and cooked beef left from trimming, cutting, boning, or cooking must be utilized if a profit and low food cost are to be maintained.

Uses of Raw Beef

- Grind for chopped steak, Salisbury steak, meat loaf, chili, meat sauce, etc.
- Dice for stew, goulash, pot pie, ragout, a la deutsch, etc.
- Slice for stroganoff, mandarin, pepper steak, etc.
- Cube for cubed steak.

National Cattlemen's Beef Association

Beef may be ground, diced, sliced, or cubed to make a variety of dishes.

Uses of Cooked Beef

- Julienne or dice for salads.
- Chop or grind for barbecue or meat sauce.
- Dice for hash.
- Dice and add to a prepared stew or goulash.
- Slice and add to a prepared stroganoff or beef a la deutsch.
- Grind and use in preparing the meat/rice stuffing for cabbage or green peppers.
- Slice and use in preparing beef mandarin.
- Slice and use for sandwiches and cold plate.
- Grind and use in a deviled beef or barbecue beef spread for canapés or hors d'oeuvres.

National Cattlemen's Beef Association

Ground beef is browned and combined with noodles, frozen vegetables, ginger, green onion, seasonings, and water for Asian Beef and Noodles.

COOKING BEEF		Pan-Broil	Pan-Fry	Broil	Stir-Fry	Roast	Grill	Cook in Liquid	Braise
	Beef Cut								
Chuck	Chuck top blade steak, chuck eye steak, boneless	●	●	●	●		●	●	●
	Chuck shoulder steak, boneless	✻	✻	✻			✻	●	●
	Chuck arm steak							●	●
	Chuck 7-bone steak				✻		✻	●	●
	Chuck pot roast, such as arm, blade shoulder							●	●
Rib	Rib steak, ribeye steak	●	●	●	●		●		
	Rib roast, ribeye roast					●	●		
Short Loin	Porterhouse/T-bone steak	●	●	●			●		
	Tenderloin steak, top loin steak	●	●	●	●		●		
	Tenderloin roast, top loin roast					●	●		
Sirloin	Sirloin steak, tri-tip steak, top sirloin steak, boneless	●	●	●	●		●		
Round	Round tip steak, thin cut	●	●		●				
	Round steak							●	●
	Top round steak	✻	✻	✻	●		✻	●	●
	Eye round steak	✻					✻	●	●
	Eye round roast, round tip roast, top round roast					●		●	●
Other Cuts	Brisket, fresh or corned							●	●
	Skirt steak	✻	✻	✻			✻		●
	Flank steak			✻	●		✻		●

✻ requires marinating

National Cattlemen's Beef Association

VARIETY MEATS

Variety meats are other edible parts of beef besides the wholesale or primal cuts. Variety meats are sometimes known as meat specialties, meat sundries, and glandular meats. Variety meats include brains, liver, tripe, heart, sweetbreads, tongue, kidney, and oxtail. These meat items are often regarded as delicacies.

Liver: Beef liver is the largest and least tender of all the edible livers. It is covered with a very thin membrane that should be removed before slicing. For best results, liver should be partly frozen when it is sliced. Cut at a 45° angle for larger slices. Beef liver is best when broiled or sautéed and should always be cooked medium unless otherwise specified by the guest.

Tongue: Beef tongue is the most popular of all the edible tongues. They may be purchased smoked, pickled, fresh, and corned, but smoked is generally the most popular. Cooking in water is the method by which tongue is always prepared. Feel the tip to test the tenderness of a tongue while boiling. When it is soft, the tongue is done. After it is cooked, it is cooled in cold water, skinned, and stored in the refrigerator. While it is under refrigeration, it is covered with water. After the tongue is cooked and skinned it is served cold or reheated and served hot.

Tripe: This is the muscular inner lining of the stomach of meat animals. The most desirable tripe is known as honeycomb tripe. It is the lining of the second stomach of beef. Tripe may be purchased pickled, fresh, or canned. The fresh tripe is generally cooked at the packing house before it is sold. However, before serving by other methods, it should be further cooked by simmering in water for about 1 hr. Tripe may be fried, creamed, served cold with vinaigrette dressing, or used as an ingredient in Philadelphia pepper pot soup.

Sweetbread: Beef sweetbreads are the thymus glands of the beef. The two types of sweetbreads in a beef animal are the *heart sweetbreads*, which are the best, and the *throat sweetbreads*. Sweetbreads come from calves, veal, and young beef because as the animal matures the thymus gland disappears.

The usual procedure for the preparation of sweetbreads is to blanch them as soon as they are received. The blanching should be done by simmering in a mixture of water, salt, and lemon juice or vinegar for about 10 min. The presence of lemon juice or vinegar keeps the sweetbreads white and firm. After blanching, all membranes should be removed and the sweetbreads kept in water and refrigerated. When the sweetbreads are blanched, they may be prepared for service by utilizing other cooking methods such as sautéing, braising, broiling, or frying.

Brains: These are much like sweetbreads in tenderness and texture. Brains do not keep well, so they should be used as soon as possible after purchasing. When brains are delivered to the commercial kitchen, they are first placed in a solution of cold water, salt, and lemon juice or vinegar, then left to soak for approximately 1 hr. This is done so it will be easier to remove the outer membrane. After the membrane has been removed, the brains are parboiled for about 15 min in another solution of water, vinegar or lemon juice, and salt. The presence of acid in the solution keeps the brains white and firm. The brains are then breaded and fried or floured and sautéed for a more attractive appearance.

Heart: The heart is the toughest of all the variety meats. The heart should be washed thoroughly in warm water and some of the arteries and veins cut away before cooking. Soaking the heart in vinegar improves its tenderness. The heart is prepared by slow moist cooking. Simmering and braising are the normal cooking methods, with a time element of about 3½ hr for proper tenderness. The heart is rarely used in a food service establishment.

Oxtail: This is the tail of the beef animal. It is sometimes known as oxjoint or beef joint. The oxtail has considerable bone but also possesses a good portion of meat and a very fine and rich flavor. Oxtail is most popular when used in stew. The thin end of the tail can be used in oxtail soup. When cutting the tail into sections for cooking, use a French knife and cut at the joints. Do not splinter the bone by using a cleaver.

Trade Tip

When microwaving unequal size pieces of beef, arrange in a dish or on a rack so thick parts are towards the outside of the dish and thin parts are in the center. Cook at medium-high or medium power.

Kidney: Beef kidney is distinguished by the many irregular lobes divided by deep cracks. Their average weight is about 1 lb. Before cooking, all suet and urinary canals must be carefully removed. Beef kidneys are the toughest of the edible animal kidneys and therefore should be cooked by moist heat. Braising or cooking in water are the recommended methods. Kidneys are highly desirable and are used to prepare kidney entrees such as kidney stew, pie, and steak.

Florida Department of Citrus

Shredded grapefruit, flank steak, and broccoli are combined with baby corn and noodles for this taste-tempting delight.

BEEF TERMINOLOGY

Certain terms associated with beef purchasing are commonly used in the commercial kitchen. These terms include the following:

Grain-fed beef is obtained from cattle that were grain-fed for a period of 90 days to a year. The animals producing the best grades of beef, U.S. prime and U.S. choice, are grain-fed in drylots (feeding pens). Most grain-fed beef animals are marketed in April and May.

Grass-fed beef is obtained from cattle that were raised on grass with little or no special feed. Most of the grass-fed animals are marketed during the fall months of the year. Grass-fed beef is generally graded U.S. good or U.S. standard.

Baby beef is a term applied to beef cattle less than 18 months of age. The baby beef carcasses weigh approximately 400 lb to 550 lb. This beef is tender, but it lacks the pronounced flavor of mature beef.

Calf carcasses and *beef dressed veal* are from animals too large to be sold as veal, but not eligible for carcass beef. These carcasses weigh from 150 lb to 375 lb.

Branded beef is beef with a trademark or trade name that is used by some packers to indicate their own grades. These brands are sometimes placed on a product by some packers even though they were already graded by the U.S. Department of Agriculture.

Aging is a term applied to meat held for a period of time under controlled conditions for the purpose of tenderizing and developing a more pronounced flavor. After an animal has been slaughtered, the muscles stiffen. The stiffening of the muscles is known as *rigor mortis*. It is a condition that gradually disappears in 3 or 4 days as the carcass hangs in the refrigerator. This period of hanging causes the muscles to relax or soften.

Once the muscles of the carcass have relaxed or softened, the aging process can begin. Only certain cuts of beef and lamb can be aged because one of the requirements for proper aging is a heavy or thick fat covering to protect the flesh from bacteria and from drying. Veal and pork are young animals that do not require aging. Veal has little or no fat covering, so it could not qualify. The three types of aging are fast aging, dry aging, and vacuum or cryovac aging.

Fast aging is a process of increasing the temperature and humidity in the box in which the aging is taking place to shorten the time it is held. The aging room must be equipped with ultraviolet lights to control bacteria for a successful fast age. This process does not produce a quality age as most fast-aged meat is sold to retail stores.

Dry aging is the process that usually produces the best results because the necessary elements of temperature, air flow, and humidity are monitored and closely controlled. The aging time is longer than for the other methods. The meat may be held for as long as 4 or 6 weeks for extra tenderness and flavor; however, the normal period is 10 days to 2 weeks.

Vacuum or *cryovac aging* is the newest aging process. The carcass is separated into smaller cuts, placed in air- and moisture-proof cryovac (heavy plastic) bags, and kept refrigerated. During aging, the bag protects the meat from bacteria and mold. Juices drawn from the meat during aging accumulate in the bag and help keep the meat moist. This prevents the meat from drying, so weight lost by shrinkage is held to a minimum. The meat is sold and held in the bag until ready to be prepared.

Not all customers like the flavor of aged meat, so if the meat has been aged it is advisable to mention it on the menu. Aging meat is costly for the food service operator because of the cost of storage, weight loss resulting from drying and shrinkage, and excessive trimming resulting from dried and discolored areas on the surface of the meat.

Green meat is meat from animals just slaughtered and before it is hung to relax or soften. This meat is tough and flavorless if it is prepared before ripening. Since it takes several days from slaughter to the kitchen, green meat very seldom causes a problem for the food service operator.

Beef is aged to develop additional tenderness and flavor. It is done commercially under controlled temperatures and humidity.

ROASTED OR BAKED BEEF RECIPES

Roast Tenderloin of Beef

Roast tenderloin of beef is an excellent choice for parties, buffets, or for use on the dinner menu. Beef tenderloin can be roasted in a very short period of time, is easy to carve to order, contains little or no fat when properly trimmed, and blends with many savory sauces.

50 serving

Equipment

- Roast pan
- Kitchen fork
- French knife
- Baker's scale
- Hotel pan
- China cap
- Stainless steel container
- Ladle
- heavy towel

Ingredients

30 lb beef tenderloin (approx. 5 full loins)
8 oz onions
6 oz carrots
8 oz celery
salad oil, as needed
salt and fresh cracked peppercorns, as needed
1 gal. hot beef stock

Preparation

1. Cut the onions, carrots, and celery rough with a French knife.
2. Trim the tenderloin and prepare it for roasting.
3. Preheat the oven to 400°F.
4. Prepare the stock.

Procedure

1. Place the tenderloins in a roast pan. Rub with salad oil and season with salt and cracked peppercorns.
2. Bake in a preheated oven at 400°F until thoroughly brown.
3. Reduce temperature to 350°F. Turn meat over using a kitchen fork on the underside and a heavy towel on the top side. Do not stick the fork into the meat.
4. Add the rough garnish and continue to roast until the desired degree of cooking is obtained: rare, medium, or well-done (time range 1 hr to 1 hr 45 min).
5. Remove roast pan from oven. Remove meat from roast pan and place in a hotel pan. Hold in a warm place for at least 15 min before slicing.
6. In the meantime, pour the fat off the drippings in the roast pan, leaving the rough garnish. Add the hot beef stock. Return the pan to the range and simmer for approximately 5 min.
7. Strain the natural juice (au jus) through a china cap into a stainless steel container. Skim off excess grease with a ladle and use in the preparation of the sauce.
8. Roast tenderloin of beef is sliced with a French knife as ordered. It should be served with a savory sauce such as mushroom, Burgundy, Bercy, bordelaise, chateau, or bearnaise.

Precautions

- If natural drippings evaporate during roasting, add a small amount of water.
- Exercise caution when using the knife.

Au jus should be free of all grease. After the au jus has been prepared, if grease still appears on the surface, strain the liquid a second or third time through a wet cheesecloth. Rinse out the cloth after each straining. If the amount of grease still appears, draw brown paper over the surface of the liquid. The paper will absorb the fat.

Roast sirloin of beef, the sirloin with the tenderloin and all bones removed, makes an excellent roast for parties or buffets or for use on the dinner menu. The meat can be roasted in a short time, is easy to carve, and contains a large portion of lean meat. It is one of the most tender beef cuts available.

Roast Sirloin of Beef

50 servings

Equipment

- Roast pan
- Kitchen fork
- French knife
- Ladle
- Hotel pan
- Baker's scale
- Bake pan
- Cheesecloth

Ingredients

25 lb boneless sirloin of beef
1½ lb onions
½ lb carrots
½ lb celery
1 gal. hot beef stock
salt and pepper to taste

Preparation

1. Cut the onions, carrots, and celery rough with a French knife.
2. Trim the sirloin and prepare meat for roasting.
3. Season the sirloin on the lean side with salt and pepper the day before roasting.
4. Preheat the oven to 400°F.
5. Prepare the beef stock.

Procedure

1. Place the sirloin in the roast pan, fat side down.
2. Place in the 400°F oven until thoroughly brown.
3. Reduce temperature to 350°F. Turn the meat over with a kitchen fork. Add rough garnish and continue to roast until desired degree of cooking is obtained: rare, medium, or well-done.
4. When the sirloin is cooked, remove to a clean bake pan and hold in a warm place. Roast should set at least ½ hr before slicing with a French knife to order.
5. Pour the fat off the drippings in the roast pan. Add the hot beef stock and simmer gently for about 20 min.
6. Strain the natural meat juice (au jus) through a china cap and cheesecloth into a stainless steel container. Adjust seasoning and skim off excess grease before serving.
7. Roast sirloin of beef is sliced with a French knife to order. It may be served with other sauces besides the au jus, for example, bordelaise, bearnaise, mushroom, Burgundy, and Bercy are excellent choices.

Precautions

- If drippings evaporate during roasting, add a small amount of water.
- Exercise caution when using the knife.

The length of time required to roast a rib or standing rib of beef depends upon the weight of the roast, the quality of the beef, and the way the roast is to be finished: rare, medium, or well-done. Low temperature roasting reduces shrinkage and therefore yields more portions. The average weight of a rib of beef is 20 lb to 25 lb and takes from 3 hr to 3½ hr to roast medium rare in the center and medium toward each end. There are always two outside and a few inside well-done cuts.

Roast Rib or Standing Rib of Beef

30 servings

Equipment

- Roast pan
- Kitchen fork
- French knife
- Heavy towel
- Medium size bake pan
- Butcher's twine
- Boning knife
- 1 gal. stainless steel container
- Cheesecloth
- 4 qt sauce pot
- Ladle
- Baker's scale
- Meat saw
- Spoon measure
- Pint measure

Ingredients

20 lb rib of beef
8 oz onions
8 oz celery
8 oz carrots
1 pt water (variable)
salt and pepper to taste

Preparation

1. Prepare the rib of beef for roasting. Cut off the short ribs, remove the blade bone, separate the feather bones from the rib bones (do not remove bones), and tie the rib with butcher's twine.
2. Cut the onions, carrots, and celery rough with a French knife.
3. Preheat the oven to 350°F.

Procedure

1. Place the roast in a roast pan, rib side up.
2. Sprinkle salt and pepper over the ribs and roast in a preheated oven at 350°F until the complete surface of the roast is brown (approximately 1 hr).
3. Remove the roast pan from the oven. With a kitchen fork and heavy towel, lift the rib of beef from the roast pan. Pour off the rendered fat that has collected in the roast pan and discard.
4. Place the rib of beef back into the roast pan, rib side down, and return to the oven.

5. Add the rough garnish and continue roasting until the garnish becomes light brown.
6. Add approximately 1 pt of water to keep vegetables from getting dry and the natural drippings from burning.
7. Reduce oven temperature to 325°F and continue to roast until rare, medium, or done as desired. Total roast time will vary from 3 hr to 3½ hr.
8. Remove the roast from the roast pan, place in a bake pan, and set in a warm place for 1 hr.
9. Pour the drippings and vegetable garnish into a gallon container. Deglaze the roast pan by adding approximately ½ gal. of water and bringing it to a boil on the range to dissolve crusted juices that have dried on the bottom and sides of the pan. Pour this liquid into the same container as the drippings and save for the preparation of au jus, which usually accompanies each order of roast rib of beef.
10. Using a French knife, remove the butcher twine and feather bones from the roast. Save the bones.

11. Stand the roast in the steam table pan by placing the large end of the roast down and the small end up. The beef is now ready to be carved.

Note: The au jus or natural juice that is usually served with each order of roast rib of beef is prepared as follows. Place the liquid and vegetables, saved from the roast pan and all the bones removed from the roast into a sauce pot. Add 1 tbsp of Worcestershire sauce and salt to taste. Simmer for 15 min. Skim off all fat using a ladle and strain through a china cap covered with a cheesecloth into a stainless steel container. Serve 1½ oz with each portion of beef.

Precautions

- Use a low roasting temperature to reduce shrinkage.
- Add water only if drippings evaporate and vegetable garnish becomes dry.
- Tie the rib before roasting so it will hold shape while cooking.
- Exercise caution when removing the roast from the pan and pouring the grease and drippings.

Roast Round of Beef

Roast round of beef is the roasted hind leg of the beef. It produces lean servings and a very rich and flavorful gravy. It is an excellent choice for both luncheon and dinner menus.

50 servings

Equipment

- Baker's scale
- Quart measure
- French knife
- 2 gal. sauce pot
- Wire whip
- Kitchen fork
- China cap
- Ladle
- 2 gal. stainless steel container
- Butcher's twine

Ingredients

25 lb boneless beef round
1 lb onions
½ lb celery
½ lb carrots
1 lb shortening
12 oz flour
1½ gal. beef or brown stock
1 c tomato puree
salt and pepper to taste

Preparation

1. Cut round into 4 lb or 5 lb individual roasts and tie with butcher's twine.
2. Cut the vegetables rough with a French knife.
3. Prepare the beef or brown stock.
4. Season the meat with salt and pepper the day before roasting.
5. Preheat the oven to 375°F.

Procedure

1. Place the meat in the roast pan and roast in a preheated oven at 375°F until meat is thoroughly brown.

2. Add the rough garnish (onions, carrots, and celery) and continue to roast until the vegetables are slightly brown. Reduce the oven temperature to 325°F to 350°F.
3. Add a small amount of beef or brown stock and continue to roast until the meat is done. Approximate time is 2½ hr to 3 hr, depending on the size of the roast and degree of doneness desired.
4. Remove the meat from the roast pan and place in a steam table pan. Keep in a warm place. Add the beef or brown stock to the roast pan to deglaze the pan. Pour into a stainless steel container and hold.
5. Place the shortening in a sauce pot and heat. Add the flour, making a roux, and cook slightly.
6. Add the hot stock and tomato puree, whipping vigorously with a wire whip until slightly thick and smooth. Cook for 15 min. Strain through a fine china cap back into the stainless steel container.
7. Slice the meat across the grain with a French knife or on a slicing machine. Serve covered with gravy.

Precautions

- Turn roast frequently while roasting with a kitchen fork.
- When adding the stock to the roux, whip vigorously with a wire whip to avoid lumping.
- Exercise caution when using the knife.

Trade Tip

All beef found in retail stores is either USDA inspected for wholesomeness or inspected by state systems that have standards equal to the federal government. The "Passed and Inspected by USDA" seal ensures that the beef is wholesome and free of disease.

Meat Loaf

Meat loaf is ground beef bound by soaked bread and eggs. It is an item in which a tough cut of beef can be used and is an excellent item for the luncheon or dinner menu.

50 servings

Equipment

- French knife
- Baker's scale
- Pint measure
- Bake pans
- Mixing machine with grinder attachment
- Skillet
- 4 gal. mixing container or stainless steel mixing bowl
- Full steam table pan

Ingredients

13 lb boned beef chuck
½ oz salt
8 eggs
1 pt milk
1 lb 8 oz trimmed bread
8 oz salad oil
1 lb celery
3 lb onions
¼ oz thyme
¼ oz fresh ground black pepper

Preparation

1. Grind the boned beef chuck on a food grinder using the medium chopper plate.
2. Mince the onions and celery with a French knife.
3. Preheat the oven to 350°F.

Procedure

1. Sauté the onions and celery in a skillet with oil until tender and let cool.
2. Place bread in the large mixing bowl and add milk. Mix thoroughly with hands until smooth.
3. Add eggs, sautéed celery and onions, and seasoning.
4. Add ground beef and mix thoroughly with hands. If the mixture is too moist, adjust consistency by adding bread crumbs or cracker crumbs.
5. Form into 3 lb loaves and place in greased pans, lightly oiling the outside of the loaves.

6. Bake in a preheated oven at 350°F for about 1½ hr. Remove from the oven, let cool, and keep in the refrigerator overnight.
7. Slice the cold meat loaf with a French knife into 5 oz or 6 oz portions and place in a steam table pan. Reheat in a steamer or on the steam table. Serve with an appropriate sauce.

Precautions

- Bake meat loaf the day before serving since it slices better when cold.
- Pack the meat mixture solidly in the pans.
- Exercise caution when using the knife.

Stuffed Cabbage

Slightly poached cabbage leaves are filled with a meat and rice mixture made from cooked meat. This is an excellent item to serve when using cooked meat leftovers. Stuffed cabbage is excellent for the luncheon menu.

50 servings

Equipment

- Roast pan
- French knife
- Large braiser and cover
- Kitchen spoon
- Baker's scale
- Spoon measure
- Food grinder
- 5 gal. stockpot
- Large bake pan
- Full steam table pans (2)

Ingredients

6 large heads of cabbage
11 lb cooked beef or cooked beef and ham (boiled or baked leftovers may be used)
3 lb onions
3 garlic cloves
1 No. 10 can tomatoes
1 qt tomato puree
2½ lb raw rice
2 tbsp chili powder
2 gal. brown gravy
3 tbsp paprika
1 qt water
1½ c salad oil
salt and pepper to taste

Preparation

1. Remove the cores from the cabbage heads and place the cabbage in a stockpot. Cook in boiling water until slightly tender or until leaves can be removed from the heads easily. Cook cabbages on the day they are to be used.
2. Mince the onions and garlic with a French knife.
3. Crush the tomatoes by hand.
4. Grind the cooked meat on food grinder using the coarse chopper plate.
5. Prepare the brown gravy.
6. Preheat the oven to 375°F.

Procedure

1. Place the salad oil in the braiser. Add the minced onions and garlic. Sauté without browning.
2. Add the crushed tomatoes, tomato puree, and water. Let boil until the onions are tender. Stir occasionally with a kitchen spoon.

3. Add the chili powder and paprika and continue to boil.
4. Add the cooked beef or ham and beef and bring back to a boil.
5. Add the rice and season with salt and pepper.
6. Cover the pot and bake in the preheated oven at 375°F until the rice absorbs the liquid and is tender (approximately 45 min).
7. Remove from the oven and check the seasoning. Place in a bake pan to cool. Refrigerate overnight.
8. Remove the meat and rice mixture from the refrigerator. Place cabbage leaves (about two or three leaves for each ball) on a kitchen towel and put a ball of the meat and rice mixture on the cabbage leaves. Repeat this process until all ingredients are used.

9. Place the stuffed cabbages in the roast pan. Bake in the preheated oven at 375°F for about 20 min. Baste with half of the brown gravy. Continue to bake for 30 min more. Remove from the oven. Place in steam table pans.
10. Serve one ball per portion with brown gravy.

Precautions

- When poaching the cabbage, do not overcook.
- Exercise caution when mincing the onions and garlic and grinding the meat.
- Let the meat and rice mixture cool before placing it in the refrigerator.

Trade Tip

Inspection is mandatory, but grading is voluntary. Meat processing plants pay to have their meat graded. USDA-graded beef sold at the retail level is Prime, Choice, and Select. Lower grades (Standard, Commercial, Utility, Cutter, and Canner) are mainly ground or used in processed meat products.

Stuffed Bell Peppers

Stuffed bell peppers are excellent for the luncheon menu and can be a very profitable item when the peppers are in season. These peppers are generally served with creole or tomato sauce.

24 servings

National Cattlemen's Beef Association

Ingredients

2 doz medium green bell peppers

Filling

6 lb lean ground beef
4½ c chopped onion
1½ c uncooked white rice
1 c ketchup
1 tbsp dried oregano leaves
1 tbsp salt
1½ tsp black pepper

Sauce

6 cans (14½ oz) Italian-style stewed tomatoes
⅓ c ketchup
1 tbsp dried oregano leaves

Preparation

1. Combine sauce ingredients.
2. Cut tops off bell peppers and remove seeds.
3. Cook the white rice.

Procedure

1. Combine filling ingredients in large bowl. Mix lightly but thoroughly.
2. Spoon equal amounts of filling into peppers and place in baking dishes.
3. Pour sauce over peppers and cover tightly with aluminum foil.
4. Bake in a 350°F oven for 1½ hr.
5. Top with sauce and serve one pepper per portion.

Precautions

- Exercise caution when using the knife.
- Exercise caution when removing the aluminum foil.

Tacos

Taco shells are filled with meats, cheeses, lettuce, onions, tomatoes, peppers, seasonings, etc. Tacos may be eaten by hand and are very popular on the luncheon menu.

24 serving or 48 tacos

Equipment

- Baker's or portion scale
- Box grater
- Braiser
- French knife
- Sauté pan
- China cap
- Kitchen spoon

Ingredients

4 lb ground beef
3 lb onions
1 oz chili powder
1 oz beef base
½ oz ground black pepper
¼ oz oregano
8 oz mild green chiles (optional)
1 c tomato sauce
¼ c cider vinegar
48 taco shells
12 oz iceberg lettuce
3 lb Monterey Jack cheese
3 lb Cheddar cheese
1 lb 8 oz tomatoes
12 oz scallions

Preparation

1. Place the ground beef in the braiser. Brown the meat thoroughly, drain, cool, and crumble.
2. Shred the lettuce with a French knife.
3. Julienne or dice the tomatoes, mince the onions, chop the chiles, and mince the scallions with a French knife.
4. Shred the cheese on the box grater.

Procedure

1. Combine the browned beef, minced onions, chili powder, beef base, oregano, green chiles, tomato sauce, and vinegar in the braiser. Cook slowly for approximately 10 min, stirring frequently to avoid scorching. Remove from the heat.
2. Fill each taco shell with 2 oz beef mixture, ¼ oz shredded lettuce, 1 oz Monterey Jack cheese, 1 oz Cheddar cheese, ½ oz tomato, and ¼ oz scallions.
3. If desired, garnish with hot peppers. Serve two tacos to each order.

Precautions

- Exercise caution when cutting the vegetables, grating the cheese, and shredding the lettuce.
- Drain the ground beef thoroughly.
- When cooking the beef mixture, stir frequently with a kitchen spoon.

Italian Meatballs

Italian meatballs are a ground beef product. They are highly seasoned and held by eggs and softened bread. They are most often served with an Italian pasta and sauce.

50 servings

Equipment

- Baker's scale
- French knife
- Cup measure
- Spoon measure
- Meat grinder
- 5 gal. mixing container
- Sheet pans
- Kitchen spoon
- Skillet

Ingredients

12 lb boneless beef chuck
3 lb onions
2 tbsp garlic
1 c salad oil
1½ lb bread
2 c milk
8 slightly beaten eggs
½ c parsley
1 c grated Parmesan cheese
1 tbsp oregano
1 tsp basil
salt and pepper to taste

Procedure

1. Sauté the garlic and onions in the salad oil in a skillet.
2. Combine the bread and milk in a mixing container and blend well by stirring with a kitchen spoon.
3. Add the sautéed onions and garlic, the pieces of beef, Parmesan cheese, oregano, basil, salt, and pepper, and blend thoroughly by hand.
4. Pass the mixture through the food grinder twice using the medium hole chopper plate.
5. Add the parsley and beaten eggs and blend thoroughly. Check the seasoning.
6. Form into balls and place on greased sheet pans.
7. Bake in a preheated oven at 350°F until done. Serve with pasta covered with a rich Italian sauce. The number of balls to a serving will depend on the size. Size may range from 1″ to 2″ in diameter.

Preparation

1. Cut the beef chuck into pieces that will pass through the food grinder. Use a French knife.
2. Mince the onions and garlic with a French knife.
3. Cut the bread into cubes with a French knife.
4. Chop the parsley with a French knife.
5. Preheat the oven to 350°F.

Precautions

- When forming the balls, rub hands with a small amount of salad oil for best results.
- Do not let the onions and garlic brown when sautéing.
- After grinding, add bread crumbs to take up the moisture if mixture is too wet or loose.
- Exercise caution when using the knife.

BRAISED AND STEWED BEEF RECIPES

Beef Pot Pie

Beef pot pie is similar to beef stew, but it is served with a flaky pie crust topping. This is an excellent entree for the luncheon menu.

50 servings

Equipment

- Braiser and cover
- French knife
- 4 qt saucepans (2)
- 2 qt saucepan
- Kitchen spoon
- Baker's scale
- Quart measure
- China cap
- Deep steam table pan
- Rolling pin
- Sheet pans

Ingredients

15 lb beef round or chuck
2 gal. hot brown stock
1 pt salad oil
1 lb flour
1 qt tomato puree
3 lb carrots
½ No. 10 can small whole onions
1 No. 10 can small whole potatoes
2½ lb peas
2 lb celery
1 tsp ground thyme
2 bay leaves
salt and pepper to taste
50 baked pastry cutouts

Preparation

1. Cut the beef round or chuck into 1″ cubes with a French knife.
2. Cut the carrots and celery into a medium-sized dice (½″ by ½″) with a French knife.
3. Drain the canned onions and potatoes.
4. Preheat the oven to 350°F.
5. Prepare the brown stock.
6. Prepare the pastry cutouts using pie dough. Roll out the dough with a rolling pin to a thickness of ⅛″. Cut out disks large enough to cover the top of the serving casseroles. Bake the cutouts on sheet pans.

Procedure

1. Place a small amount of salad oil in the braiser and heat. Add the cubes of meat and brown thoroughly.
2. Add the remaining salad oil and flour. Blend well with a kitchen spoon, making a roux, and cook slightly.
3. Add the hot brown stock, tomato puree, bay leaves, and thyme. Stir with a kitchen spoon. Cook until thickened and smooth.
4. Cover and bake in the preheated oven at 350°F for about 2 hr to 2½ hr until meat is tender.
5. Boil all raw vegetables in separate saucepans in salt water until tender. Drain through a china cap.
6. When the meat is tender, remove from the oven. Remove the bay leaves and add all the drained cooked vegetables except the peas.
7. Check the seasoning and place in a deep steam table pan.
8. Serve 6 oz to 8 oz in deep casseroles. Sprinkle cooked peas over each portion and top with a baked pastry cutout.

Precautions

- Do not overcook the meat. It will crumble when served.
- When the flour is added to make the roux, be sure to cook the roux slightly or the pot pie will have a raw flour taste.
- Exercise caution when using the knife.
- Exercise caution when removing the lid from the braiser. Steam will escape.

Hungarian Goulash

Hungarian goulash is a stew served mostly on the luncheon menu. It is highly seasoned with paprika to give it the characteristic flavor associated with Hungarian dishes.

50 servings

Equipment

- Roast pan or braiser with cover
- French knife
- Kitchen spoon
- Baker's scale
- Full steam table pan

Ingredients

18 lb beef chuck or shoulder
1 oz garlic
8 oz flour (variable)
¾ oz chili powder
5 oz paprika
1 lb tomato puree
8 lb or 4 qt water or brown stock
salt and pepper to taste
2 bay leaves
½ oz caraway seed
2 lb onions

Preparation

1. Dice the beef into 1″ cubes with a French knife.
2. Mince the onions and garlic with a French knife.
3. Prepare the brown stock if it is to be used.

Procedure

1. Place the diced beef in roast pan or braiser and brown in the oven.
2. Add the minced onions and garlic and continue to brown for 5 min.
3. Add flour by sprinkling over the beef and stir with a kitchen spoon until the flour is incorporated with the beef.

4. Add paprika and chili powder in the same manner as the flour was added. Cook for 5 min.
5. Add tomato puree and water or brown stock.
6. Season with salt, pepper, bay leaves, and caraway seed.
7. Return to the oven and cook until the meat is very tender.
8. Remove from the oven. Remove bay leaves and place in a steam table pan.
9. Serve 6 oz in an individual casserole with buttered spaetzles or noodles.

Precautions

- While beef is cooking, if too much liquid should disappear, add more brown stock or water.
- While beef is cooking, stir occasionally.
- Exercise caution while using the knife.

Trade Tip

Product dating of beef is not required by federal regulations. However, many stores and processors may voluntarily date packages of raw beef or processed beef products. If a calendar date is shown, there must be a phrase explaining the meaning of the date.

Beef Stroganoff

Beef stroganoff is a braised meat preparation which includes egg noodles and sour cream. It is very popular on the dinner menu.

24 servings

Ingredients

6 lb beef round tip steaks
2 lb wide egg noodles
½ c vegetable oil, divided
6 cloves of garlic
3 lb mushrooms
4 oz brown gravy mix
1½ c sour cream
4 c cold water
1½ tsp salt
1½ tsp black pepper

Preparation

1. Cut beef into thin strips.
2. Slice mushrooms ⅛″ – ¼″ thick.
3. Crush garlic.

Procedure

1. Cook noodles per package directions. Keep warm.
2. In a large skillet, heat ¼ c vegetable oil over medium-high heat until hot. Add beef strips and garlic (½ at a time) and stir-fry 1 min. Do not overcook.
3. Remove from skillet and season with 1½ tsp salt and 1 ½ tsp black pepper.
4. Using the same skillet, cook mushrooms in the remaining ¼ c of vegetable oil for 2 min. Remove from heat.
5. Stir in gravy mix and 4 c cold water. Bring to a boil, reduce heat, and simmer until sauce is thickened, stirring frequently.
6. Return beef and heat through. Serve over noodles with sour cream.

National Cattlemen's Beef Association

Precautions

- Do not overcook the meat.
- Exercise caution when using the knife.

Swiss Steak

Swiss steak is served on both the luncheon and dinner menu. Swiss steak is cooked in its own gravy until it is very tender.

50 servings

Equipment

- Roast pan with cover
- Skillet
- French knife
- Butcher knife
- Kitchen spoon
- 2 gal. sauce pot
- Wire whip
- Full steam table pans

Ingredients

50 round steaks
1 lb onions
2 garlic cloves
12 oz tomato puree
6 qt water or brown stock
3 c salad oil
12 oz bread flour (variable)
salt and pepper to taste

Preparation

1. Cut the steaks into ½" thick (6 oz) portions with a butcher knife.
2. Mince the onions and garlic with a French knife.
3. Prepare the brown stock if it is to be used.
4. Preheat the oven to 350°F.

Procedure

1. Heat the oil in a fry pan or skillet and brown the steaks on both sides.
2. Place the browned steaks in a braiser or roast pan and hold.
3. Place the minced onions and garlic in the skillets where the steaks were browned and sauté.
4. Add flour, mixing a roux, and stir with a kitchen spoon until flour is well blended and lightly browned. Remove this mixture to a sauce pot.
5. Add the hot brown stock (or water) and tomato puree while whipping with a wire whip. Cook until sauce is slightly thickened.
6. Season with salt and pepper.
7. Pour the sauce over the steaks. Cover and bake in a preheated oven at 350°F for about 2 hr or until steaks are very tender.
8. Remove from the oven and place in steam table pans. Serve one steak to each order with 2½ oz to 3 oz of sauce.

Precautions

- Exercise caution when using the knife.
- When sautéing the onions and garlic, do not let them overbrown.
- When testing steaks while they are in the oven, remove the cover with caution to avoid steam burns.
- Do not overcook the steaks or they will fall apart when served.

Braised Flank Steak Polynesian

The steaks are marinated to help tenderize the meat and provide the flavor to both meat and sauce that is associated with Polynesian cuisine. This item can be served on both the luncheon and dinner menu.

50 servings

Equipment

- Large steel skillet
- Kitchen fork
- French knife
- Pint measure
- Cup measure
- Spoon measure
- Braiser
- Large stainless steel container
- Slicing machine
- Large wire whip
- Colander
- Bake pan
- Small stainless steel bowl
- China cap
- 2 gal. stainless steel container
- Full steam table pan

Ingredients

16 flank steaks, approx. 1½ lb each
1½ pt cider vinegar
1½ pt salad oil
1½ c brown sugar
1½ c soy sauce
6 small onions
2 tsp black pepper
2 tsp garlic salt
5 oz cornstarch
1 c pineapple juice
1 c salad oil

Preparation

1. Trim each flank steak by removing all fat and gristle.
2. Peel and slice the onions into thin rings using the slicing machine.
3. Preheat the oven to 350°F.

Procedure

1. Prepare a marinade by combining the onion rings, vinegar, brown sugar, soy sauce, first amount of salad oil (1½ pt), pepper, and garlic salt in a large stainless steel container. Whip the marinade slightly using a large wire whip until the sugar has dissolved.
2. Add the flank steaks. Submerge them in the marinade. Place in the refrigerator to marinate overnight.
3. On the day of preparation, remove the steaks from the refrigerator. Remove the steaks from the marinade. Place them in a colander to drain. Save as much of the marinade as possible.
4. Using the second amount of salad oil (1 c), place a small amount in a large steel skillet. Place on the range and heat.
5. Fill the skillet with flank steaks. Brown one side, then the other. When thoroughly brown, place the steaks in a braiser. Repeat this process until all steaks are browned.
6. Add a small amount of the marinade to the pan. Heat slightly to deglaze the pan, capturing all the meat flavor.

7. Pour all the marinade, including the amount used to deglaze the pan, over the steaks in the braiser.
8. Bake in the preheated oven at 350°F. Cover tightly with a lid and bake until each steak is tender. Test for doneness using a kitchen fork. Insert the fork into the flesh. If it penetrates easily, the meat is usually tender. Remove the braiser from the oven.
9. Remove the steaks from the braiser, place them in a bake pan, and hold in a warm place.
10. Place the cornstarch in a small stainless steel bowl. Add the pineapple juice and dissolve thoroughly.
11. Place the braiser containing the marinade on the range. Bring to a simmer.
12. Slowly pour in the dissolved starch while at the same time whipping vigorously with a large wire whip. Return to a simmer and cook for 2 min to 3 min.
13. Remove from the range and strain through a china cap into a stainless steel container.
14. Slice the flank steaks on a bias with the French knife. Place the slices in a steam table pan. Serve 4 oz to 5 oz per portion with 2½ oz to 3 oz of sauce.

Precautions

- Exercise caution when using the knife and the slicing machine.
- Exercise caution when removing the lid from the braiser.
- Cook meat until just tender. Overcooking causes meat to crumble when sliced.

Baked Stuffed Flank Steak

Baked stuffed flank steak has ground stuffing intertwined with the meat for an interesting and tasty dish served with a savory sauce. This is an excellent choice for both the luncheon and dinner menu.

50 servings

Equipment

- Roast pan
- Stainless steel container
- Boning or utility knife
- French knife
- Butcher's twine
- Kitchen fork
- Bake pan
- Mallet
- Sheet of heavy plastic
- Stainless steel bowl or dish pan
- Cup measure
- Baker's scale
- Small saucepan

Ingredients

10 flank steaks, approx. 1½ lb each
10 lb ground beef
12 oz onions
1 clove garlic
1 c salad oil
1 tbsp thyme
salt and fresh cracked peppercorn to taste
10 eggs
bread crumbs, if needed
¼ c parsley

Preparation

1. Butterfly and flatten each flank steak.
2. Mince the garlic and onions with the French knife. Chop the parsley.
3. Preheat the oven to 350°F.

Procedure

1. Place the ground beef in a stainless steel bowl or dish pan.
2. Place ½ c of the salad oil in a small saucepan and heat. Add the minced onion and garlic. Sauté until tender, let cool, and add to the ground beef.
3. Add the eggs, thyme, parsley, salt, and pepper. Mix thoroughly.
4. If mixture is fairly moist, add some bread crumbs to absorb the moisture. Mix thoroughly a second time.
5. Lay each flattened, butterflied flank steak on a flat surface and distribute the stuffing evenly over each. Use approximately 1 lb stuffing for each steak.
6. Roll each steak into a tight roll. Secure by tying with butcher's twine at three or four places around the stuffed steak.
7. Rub each stuffed steak with the remaining salad oil. Place in a roast pan.
8. Bake in the preheated oven at 350°F until browned.
9. Add approximately 1″ of water to the pan. Cover and continue baking until the meat is tender. Test for doneness using a kitchen fork.
10. Remove from the roast pan. Place in a bake pan and let set for approximately 15 min before slicing.
11. Pour the liquid from the roast pan into a stainless steel container. Reserve for later use in the preparation of a sauce.
12. Baked stuffed flank steak is sliced with a French knife as ordered. It should be served with a savory sauce such as mushroom, bordelaise, chateau, Burgundy, or brown sauce.

Precautions

- Exercise caution when using the knives.
- Turn the meat with a kitchen fork occasionally while roasting.
- Exercise caution when removing the cover from the roast pan. Steam will escape.
- Do not tie the stuffed steak too tightly. Allow for some expansion.

It is safe to freeze beef in its original packaging. If freezing longer than 2 months, overwrap these packages with airtight heavy-duty foil, plastic wrap, or freezer paper, or place the package inside a plastic bag.

Marinating Meat

Marinating is for the purpose of flavoring and seasoning meat, as well as tenderizing by the reaction of the acid in the marinade. The marinade should be prepared and left to stand approximately 2 hr before the meat is added so all flavors have blended. The length of time the meat is left in the marinade is determined by the amount of seasoning and tender-izing desired, the size of the pieces, and preparation. For example, a steak may be marinated a few hours, but a preparation such as sauerbraten takes 3 to 5 days. Many types of marinades are used. Most are highly seasoned and contain wine or vinegar. If wine is used, select a white wine for chicken and veal and a red wine for beef and lamb. It is not necessary to use premium wines when cooking. Common wines will produce the desired taste.

Special Marinade No. 1

1 qt

Ingredients

1 c cider or wine vinegar
1 c salad oil
2 crushed garlic cloves
1 tsp coarsely ground pepper
3 minced onions
½ c brown sugar

Procedure

1. Blend all ingredients in a stainless steel or plastic container.
2. Add steaks and marinate in the refrigerator for approximately 8 hr or until desired tenderness has been achieved. Tougher meats such as flank steak are usually marinated longer than cuts such as sirloin.

Special Marinade No. 2

1½ qt

Ingredients

1 pt salad oil
1 pt red or white wine
1 oz sugar
½ oz salt
1 tsp margarine
1 tsp thyme
8 oz minced onions
¼ oz minced garlic
6 oz lemon juice

Procedure

1. Blend all ingredients in a stainless steel or plastic container.
2. Add steaks and let marinate in the refrigerator for approximately 8 hr or until desired tenderness has been achieved.

Beef Ragout

Ragout is a mixture of well-seasoned meat and vegetables. This beef ragout contains several root vegetables in a thick, highly-seasoned brown sauce. It is served mostly on the luncheon menu.

50 servings

Equipment

- French knife
- Quart measure
- Cup measure
- Braiser
- Baker's scale
- 4 qt saucepans (3)
- 2 qt saucepan
- Deep steam table pans (2)

Ingredients

20 lb lean beef chuck or shoulder
4 c salad oil
1 lb onions
1 lb celery
1½ lb flour
2 gal. hot brown or beef stock
1 pt tomato puree
100 cubes of carrots
100 cubes of rutabaga
100 cubes of potatoes
50 small onions
2½ lb peas
3 No. 2½ cans Italian tomatoes, drained
salt and pepper to taste

Preparation

1. Cut the beef into cubes with a French knife.
2. Cut the carrots, rutabaga, and potatoes into ½″ cubes with a French knife.
3. Dice the celery and onions with a French knife.
4. Prepare the brown or beef stock.
5. Cook the peas in salt water in a saucepan.
6. Preheat the oven to 400°F.

Procedure

1. Place the salad oil in a braiser and heat. Add the beef cubes and sauté until brown.
2. Add the celery and onions and continue to sauté until slightly tender.
3. Add the flour, making a roux, and cook for 5 min.
4. Add the hot stock and stir with a kitchen spoon until thick and smooth.

5. Add the tomato puree, cover the braiser, and cook in the preheated oven at 400°F until meat cubes are tender.
6. Cook the remaining vegetables separately, except tomatoes, in saucepans until just tender, then drain.
7. When the meat is tender, remove from the oven and add the tomatoes and drained vegetables, except the peas, to the ragout. Season with salt and pepper to taste.
8. Return to the oven for ½ hr. Remove and place in deep steam table pans.

9. Serve 6 oz to 8 oz in deep casseroles. Garnish each portion with the cooked peas.

Precautions

- Do not overcook the meat or vegetables.
- Cut vegetables as uniformly as possible so they cook evenly.
- Exercise caution when using the knife.

Trade Tip

All cattle start out eating grass. Three-fourths of them are "finished" (grown to maturity) in feedlots, where they are fed specially formulated feed based on corn or other grains.

Sauerbraten is a sour beef dish that originated in Germany. The beef is marinated in a souring solution for 3 to 5 days. Sauerbraten is always found on the menu of establishments featuring German cuisine.

Sauerbraten

50 servings

Equipment

- Large crock
- French knife
- Kitchen spoon
- Braiser and cover
- 3 gal. sauce pot
- Kitchen fork
- China cap
- Wire whip
- Rolling pin
- Bake pan
- Baker's scale
- Quart measure
- Butcher's twine

Ingredients

25 lb round beef or beef brisket
6 qt cold water
2 qt red wine vinegar
2 lb onions
1 lb carrots
½ lb celery
½ lb brown sugar
4 garlic cloves
2 oz salt
6 bay leaves
1 tsp peppercorns
15 ginger snaps
shortening and flour for roux as needed

Preparation

1. Cut and tie the round of beef into separate roasts, or if using brisket, cut into medium-sized pieces and trim.
2. Cut the onions, carrots, and celery rough with a French knife.
3. Chop the garlic cloves and crush the peppercorns.
4. Crush the ginger snaps with a rolling pin.
5. Preheat the oven to 400°F.

Procedure

1. Place the meat in a large crock or barrel.
2. Cover with a solution of water and red wine.
3. Add the garlic, rough garnish, salt, bay leaves, peppercorns, and brown sugar.
4. Let the meat marinate in the solution for 3 to 5 days in the refrigerator.

5. Remove the meat from the marinade the night before using. Strain marinade through a china cap, saving both liquid and vegetable garnish.
6. Place the meat in a braiser and brown thoroughly in the preheated oven at 400°F.
7. Add the drained vegetable garnish and continue to roast for about 15 min or until the vegetable garnish becomes slightly brown.
8. Add the marinade liquid, cover, and continue to cook until the meat is tender.
9. Remove the meat from the liquid with a kitchen fork. Place in a hotel pan, cover with a damp cloth, and keep warm.
10. Place the shortening in a separate sauce pot. Add flour, making a roux (1 lb shortening, ½ lb flour for each gallon of liquid). Cook roux slightly.
11. Add the liquid in which the meat was cooked to the roux, whipping constantly with a wire whip until slightly thick and smooth.
12. Add the crushed ginger snaps and simmer for about 10 min. Check seasoning, strain through a fine china cap into a stainless steel container, and hold for service.
13. Slice the beef across the grain with a French knife or on the electric slicing machine.
14. Serve with the prepared sauce, accompanied with potato pancakes, potato dumplings, buttered noodles, or spaetzles.

Precautions

- Do not overcook the meat.
- Slice meat across the grain.
- Keep the cooked meat covered with a damp cloth at all times.
- The desired sweet or sour flavor of the sauce can be controlled by varying the amount of the vinegar and brown sugar.
- If a dark sauce is desired, some brown flour may be added when preparing the roux.
- Exercise caution when using the knife.

Pot Roast

Pot roast is cooked using the braising method. Since the browned beef is cooked in its own gravy, most of the flavor is in the sauce or gravy. This is popular on the luncheon or dinner menu.

50 servings

Equipment

- Baker's scale
- Cup measure
- French knife
- Quart measure
- Spoon measure
- Large skillet or frying pans
- China cap
- Braiser or heavy pot
- Sauce pot
- Kitchen spoon
- Kitchen fork
- 2 gal. stainless steel container
- Full steam table pans

Ingredients

25 lb beef brisket or round
2 c salad oil
1 lb onions
½ lb celery
½ lb carrots
1 lb bread flour (all-purpose can be used)
5 qt hot beef or brown stock
1 No. 2½ can crushed tomatoes
1 bay leaf
½ tsp ground thyme

Procedure

1. Place the oil and meat in the braiser. Bake in the preheated oven at 400°F and brown on all sides.
2. Add the onions, celery, and carrots and continue to cook for an additional 15 min.
3. Blend in the flour and cook 10 min. If roux is too thick, add more oil.
4. Add the hot stock and stir with a kitchen spoon until thickened and smooth.
5. Add the tomatoes, juice, and seasoning and mix well.
6. Cover and cook at 400°F for about 2½ hr or until meat is tender. Remove from the oven.
7. Remove the meat from the gravy with a kitchen fork and let the meat cool. Strain the gravy through a china cap into a stainless steel container. Adjust the seasoning.
8. Slice the meat against the grain with a French knife or on the electric slicing machine. Place in a steam table pan and reheat on the steam table or in the steamer.
9. Serve with 2½ oz of gravy. A jardiniere vegetable garnish (carrots, celery, and turnips cut 1″ long by ¼″ thick) is usually served with each order of pot roast.

Preparation

1. Trim the meat. If using round, cut into individual 5 lb roasts and tie.
2. Cut the onions, carrots, and celery rough with a French knife.
3. Prepare the beef or brown stock.
4. Preheat the oven to 400°F.

Precautions

- Exercise caution when cutting the vegetables.
- Simmer the meat rather than boiling so that it will be firmer and easier to slice.
- Use care when straining the gravy to avoid getting burned.

Chinese Pepper Steak

Chinese pepper steak is an American-Chinese preparation. The thin slices of beef are cooked in a sauce highly flavored with green peppers and soy sauce.

50 servings

Equipment

- Baker's scale
- French knife
- Braiser
- Kitchen spoon
- Quart measure
- Deep steam table pan

Ingredients

12 lb beef round
6 oz shortening
5 lb green peppers
3 lb onions
3 lb celery
6 oz pimientos
3 qt beef stock
4 oz soy sauce
3 oz cornstarch (variable)
salt and pepper to taste

Procedure

1. Place the shortening in the braiser and heat. Add the sliced beef and brown.
2. Add the onions, celery, and green peppers. Continue to cook until vegetables are slightly done.
3. Add the beef stock and bring to a boil. Continue to cook until the celery and beef are tender.
4. Add the soy sauce to the cornstarch and dissolve. Pour into the boiling mixture, stirring constantly with a kitchen spoon until thick. Simmer for 5 min.
5. Add the pimientos and season with salt and pepper. Remove from the range and pour into a deep steam table pan.
6. Serve in shallow casseroles with baked or steamed rice.

Preparation

1. Slice the beef round on a bias into thin slices.
2. Prepare the beef stock.
3. Julienne the green peppers, onions, celery, and pimientos with a French knife.

Precautions

- Before adding the cornstarch, be sure the celery and beef are tender.
- Add the dissolved cornstarch to the boiling mixture slowly while stirring constantly.
- Exercise caution when using the knife.

Oxtail Stew

Oxtail stew is a stew made from the tails of beef animals. This mixture of meat and vegetables possesses an excellent flavor. It is commonly served on the luncheon menu.

50 servings

Equipment

- French knife
- Baker's scale
- Quart measure
- Braiser and cover
- Kitchen spoon
- Skimmer
- 4 qt saucepans (3)
- 2 qt saucepan
- Skillet
- Deep steam table pan
- Ladle

Ingredients

30 lb oxtail
1 lb shortening
1 lb 4 oz flour
2 gal. beef or brown stock
1 qt tomato puree
4 garlic cloves
2 lb onions
50 canned, whole, small onions
3 lb carrots
3 lb celery
3 lb turnips
2 bay leaves
3 tbsp salt
1 tsp black pepper
2½ lb peas

Preparation

1. Cut the oxtails at the joints into serving pieces with a French knife. Wash and dry the pieces thoroughly.
2. Cut the celery, carrots, and turnips jardiniere (1" long by ¼" thick) with a French knife.
3. Mince the onions and garlic with a French knife.
4. Prepare the beef or brown stock.
5. Preheat the oven to 400°F.

Procedure

1. Put the shortening in the braiser, place in the preheated 400°F oven, and heat. When the shortening is hot, add the cut sections of oxtail. Season with salt and pepper and brown thoroughly. Turn oxtail with the skimmer.
2. Sprinkle the flour over the brown oxtails and blend thoroughly with a kitchen spoon. Continue to cook for 5 min.
3. Add the minced onions and garlic and cook slightly.
4. Add the beef or brown stock, tomato puree, thyme, and bay leaves. Cover the braiser and cook until the oxtails are tender and the sauce is slightly thick. Remove the bay leaves.
5. In separate saucepans, cook the carrots, celery, peas, and turnips in boiling water until tender. Drain and add all to the stew except the peas.
6. Sauté the whole canned onions in a skillet with a small amount of additional butter or shortening. Add to the stew.
7. Remove the stew from the oven and place in a deep steam table pan.
8. Serve in a deep casserole with an 8 oz ladle. Garnish each serving with the cooked peas.

Precautions

- After the oxtails are washed, be sure they are dried thoroughly before browning.
- When cutting the oxtails, use a French knife and cut at the joints. Do not use a cleaver or the bones will splinter. A power saw may also be used.
- When adding vegetables to the stew, stir gently with a kitchen spoon. Do not break up the vegetables.
- While the oxtails are cooking, skim off the grease frequently with a ladle.

Braised Short Ribs of Beef

The best short ribs are cut from the end of the rib roast. This flavorful cut of beef is made tender by braising. Braised short ribs of beef are an excellent luncheon item.

50 servings

Equipment

- French knife
- Butcher's twine
- Braiser
- China cap
- Ladle
- Kitchen fork
- Baker's scale
- Quart measure
- Spoon measure
- Cup measure
- 1 qt saucepan
- Full size steam table pans (2)
- 2 gal. stainless steel container

Ingredients

50 10 oz short ribs
1½ lb onions
3 garlic cloves
1 tsp thyme
1 tsp sweet basil
6 qt hot beef stock
1 pt tomato puree
2 c salad oil
1 lb bread flour (variable)
salt and pepper to taste

Preparation

1. Trim the ribs, remove the fat, and tie with butcher's twine.
2. Mince the onions and garlic with a French knife.
3. Prepare the beef stock.
4. Preheat the oven to 400°F.

Procedure

1. Place the short ribs in a braiser and pour 1 c of the salad oil over the short ribs. Bake in a preheated oven at 400°F and brown thoroughly.
2. Sauté the minced onions and garlic in a saucepan with 1 c of salad oil, spread evenly over the browned short ribs, and continue to cook for 10 min more.
3. Sprinkle the flour over the short ribs using a kitchen spoon and blend well with the oil. Cook for 10 min.

4. Add the hot beef stock, stirring with a kitchen spoon until slightly thickened.

5. Add the salt, pepper, thyme, basil, and tomato puree and blend well.

6. Cover the braiser and continue to cook until the short ribs are tender, about 2½ hr. Turn the meat occasionally with a kitchen fork.

7. Remove the cooked ribs to steam table pans and cover with a damp cloth. Strain the gravy through a fine china cap into a stainless steel container.

8. Remove excess grease with a ladle. Adjust the seasoning and thickening. Hold for service.

9. When serving, remove the butcher's twine from the short ribs and cover with gravy and a vegetable garnish cut in the jardiniere style (cut 1″ long by ¼″ thick).

Precautions

- Do not overcook the ribs.
- Be sure that the gravy is not too thick. If it is too thick, thin by adding water or beef stock.
- Exercise caution when removing the cover from the braiser. Steam will escape.
- When adding the herbs to the liquid, rub them in the palm of the hand to release the flavor.
- Exercise caution when using the French knife.

Spanish Steak

Spanish steak is baked in a rich tomato and green pepper mixture until it becomes tender. This item is best for luncheon service.

50 servings

Equipment

- French knife
- Quart measure
- Baker's scale
- Cup measure
- Braiser
- Large fry pan
- 1 gal. sauce pot
- Cleaver

Ingredients

14 lb beef round
6 oz bread flour (variable)
3 c salad oil
2 qt beef stock
1 No. 10 can tomatoes
½ No. 10 can tomato puree
2 lb onions
1 lb celery
10 oz green peppers
2 bay leaves
salt and pepper to taste

Preparation

1. Cut the beef round into 4 oz steaks with a French or butcher knife and flatten with a cleaver.
2. Prepare the beef stock.
3. Mince the onions, celery, and green peppers with a French knife. Crush the tomatoes by hand.
4. Preheat the oven to 350°F.

Procedure

1. Place about ⅛″ of oil in a large fry pan and heat. Add the steaks and brown thoroughly.
2. Place the browned steaks in a large braiser.
3. Sauté the onion, celery, and green peppers in the remaining salad oil in a sauce pot. Do not brown.
4. Add the flour, making a roux.
5. Add the beef stock, tomatoes, and tomato puree, stirring constantly with a kitchen spoon until thickened.
6. Pour the sauce over the steaks. Add the bay leaves and season with salt and pepper.
7. Bake in the preheated oven at 350°F for about 2 hr or until steaks are tender.
8. Remove from the oven, remove the bay leaves, and place the steaks and sauce in a steam table pan.
9. Serve one steak per portion covered with the sauce and garnished with chopped parsley.

Precautions

- Do not overcook the steaks or they will crumble when served.
- When the liquid is added to the roux, stir constantly.
- Exercise caution when using the knife.

Sautéed Beef Tenderloin Tips in Mushroom Sauce

Sautéed beef tenderloin tips in mushroom sauce can be prepared very quickly, so it is generally prepared to order. It is an excellent item for the dinner or a la carte menu. The tenderloin tips require very little cooking.

50 servings

Equipment

- Baker's scale
- Quart measure
- French knife
- Frying pan
- Braiser
- Skillet
- Full sized steam table pan

Ingredients

18 lb beef tenderloin tips
4 lb mushrooms
4 oz butter
2 gal. hot brown sauce
1 pt Burgundy wine
1 pt salad oil (variable)

Preparation

1. Slice the beef tenderloin tips on a bias about ¼″ thick with a French knife.

2. Wash the mushrooms and slice fairly thick.
3. Prepare the rich brown sauce. Thicken with a roux.

Procedure

1. Place the butter in a skillet and heat. Add the mushrooms and sauté until tender.
2. Place the brown sauce in the braiser and bring to a boil.
3. Add the sautéed mushrooms and the Burgundy wine. Simmer slowly.
4. Place the salad oil in a large frying pan and heat. Add the tenderloin tips and brown quickly. Drain off all oil.

5. Add the brown tips to the mushroom sauce, bring to a boil, and remove from the heat. Place in a steam table pan.
6. Serve 6 oz in a shallow casserole. Garnish with chopped parsley.

Precautions

- When washing the mushrooms, lift the mushrooms out of the water rather than pouring the water off. All dirt will be removed using this method.
- When sautéing the tenderloin tips, do not overcook. They should be medium done.
- Exercise caution when using the knife.

Beef a la Deutsch

Beef a la Deutsch is an excellent item to feature on the luncheon or dinner menu to use leftover tenderloin tips. These tips remain when tenderloin or filet mignon steaks are cut. The sautéed tips are poached in a very tasty sauce.

25 servings

Equipment

- Quart measure
- Pint measure
- Cup measure
- Baker's scale
- French knife
- Sauce pot
- Skillet

Ingredients

8 lb beef tenderloin tips
1 c salad oil (variable)
3 c thickly sliced mush-rooms
1 ½ c green peppers
1 c claret wine
2 c onions
2 c celery
5 shallots
¾ tsp garlic
3 qt rich brown gravy
1 pt canned tomatoes
salt and pepper to taste
2 bay leaves

Preparation

1. Slice the tenderloin tips on a bias about ¼" thick with a French knife.
2. Prepare the brown gravy.
3. Crush the canned tomatoes by hand.
4. Mince the shallots and garlic with a French knife.
5. Julienne the green peppers, celery, and onions with a French knife.

Procedure

1. Place about ½ c of the salad oil in a sauce pot. Add the onions, shallots, garlic, green pepper, celery, and mushrooms. Sauté until slightly tender, but do not brown.
2. Add the claret wine and simmer for about 15 min. Add the bay leaves.
3. Add the brown gravy and continue to cook until the celery is tender, then remove the bay leaves.
4. Add the crushed tomatoes and continue to simmer.
5. Sauté the tenderloin tips in a skillet in the remaining oil until slightly brown, then add to the sauce and cook until the meat is very tender.
6. Season with salt and pepper, remove from the range, and place in a steam table pan.
7. Serve a 6 oz portion in a shallow casserole with a scoop of baked rice.

Precautions

- Do not overcook the tenderloin or the meat will break apart.
- Use caution when cutting the beef and vegetable garnish.

Beef a la Bourguignonne

Beef a la bourguignonne is a French preparation. It is cubes of beef tenderloin cooked in Burgundy wine. This item is an excellent choice for the dinner menu.

50 servings

Equipment

- Baker's scale
- French knife
- Quart measure
- Braiser
- Kitchen spoon
- 3 qt sauce pot
- Full sized steam table pan

Ingredients

18 lb beef tenderloin
10 oz shortening
4 lb mushrooms
1 lb shallots or green onions
3 oz flour
1½ qt Burgundy wine
salt and pepper to taste

Preparation

1. Wash and slice the mushrooms fairly thick with a French knife.
2. Mince the shallots or green onions with a French knife.
3. Trim and cut the beef tenderloin into 1" cubes with a French knife.

Procedure

1. Place two-thirds of the shortening in the braiser and heat. Add the cubes of beef and brown thoroughly.
2. Place the remaining shortening in a sauce pot and heat. Add the mushrooms and shallots and sauté until tender, then hold for later use.

3. Add the flour to the browned beef cubes and blend thoroughly with a kitchen spoon. Cook for 5 min.
4. Add the wine and sautéed shallots and mushrooms. Blend thoroughly and simmer for about 30 min or until all ingredients are tender.
5. Season with salt and pepper and remove from the range. Place in a steam table pan.
6. Serve 6 oz in a shallow casserole with yellow or wild rice.

Precautions

- Do not overcook the beef tenderloin or it will fall apart when served. Test for doneness by removing a piece of beef with a pierced spoon and pressing with the finger.
- Exercise caution when using the knife.

Garlicky Beef and Pasta

Beef and pasta are popular menu choices. When combined with garlic, green beans, and brown gravy, an appealing and delicious entree is produced.

24 servings

Ingredients

6 lb beef round tip steaks
8 c uncooked rotini
Vegetable cooking spray
1 head of garlic
3 lb frozen green beans
3 c brown beef gravy
3 tsp salt
1½ tsp black pepper
4 tbsp water

Preparation

1. Cut beef into thin strips.
2. Cook rotini per directions.
3. Crush garlic.

Procedure

1. Spray large skillet with vegetable cooking spray and heat over medium-high heat until hot.
2. Add beef and garlic (½ at a time) and stir-fry 1 min. Do not overcook.
3. Remove from skillet and season with 3 tsp of salt and 1½ tsp of black pepper.
4. In same skillet, heat 4 tbsp of water, add green beans, and cook 4 – 5 min, stirring occasionally.

National Cattlemen's Beef Association

5. Stir in gravy and rotini. Return beef to skillet and heat thoroughly.

Precautions

- Exercise caution when using the knife.
- Do not overcook the rotini.

Swiss Steak in Sour Cream

Swiss steak in sour cream is baked in its own sauce until tender. Sour cream and Parmesan cheese are added to give it an unusual flavor and appearance. This is a popular dish on the luncheon menu.

50 servings

Equipment

- French or butcher knife
- Quart measure
- Cup measure
- Spoon measure
- Baker's scale
- Skillet
- Braiser and cover
- Kitchen spoon
- China cap
- Wire whip
- 2 gal. sauce pot
- Full size steam table pan
- 2 gal. stainless steel container
- Cleaver

Ingredients

50 beef round steaks
1 qt salad oil
3 lb onions
1 c Parmesan cheese
4 oz Worcestershire sauce
6 qt brown stock
3 tbsp paprika
2 bay leaves
1 lb sour cream
1 lb bread flour
salt and pepper to taste

Preparation

1. Cut the beef round into 6 oz steaks with a French or butcher knife and flatten slightly with a cleaver.
2. Mince the onions with a French knife.
3. Prepare the brown stock.
4. Preheat the oven to 350°F.

Procedure

1. Cover the bottom of a skillet with the salad oil and heat.
2. Place the steaks in the hot oil and brown both sides. Repeat this process until all steaks are browned. Place the browned steaks in a braiser. Add 1 qt of brown stock to the skillet. Deglaze and save the liquid.
3. Place the remaining oil in a sauce pot and heat. Add the minced onions and sauté until tender. Do not brown.

4. Add the paprika and flour, blending well with a wire whip. Cook slightly.
5. Add the brown stock, whipping constantly with a wire whip until slightly thick and smooth.
6. Add the Worcestershire sauce, Parmesan cheese, bay leaves, salt, and pepper.
7. Pour this sauce over the browned steaks in the braiser.
8. Cover the braiser and bake in the preheated oven at 350°F for about 2 hr to 2½ hr or until the steaks are tender.
9. Remove the steaks from the sauce with a kitchen fork and place in a steam table pan. Keep the steaks covered with a wet towel.
10. Strain the sauce through a china cap into a saucepan. Cook slightly. Stir the sour cream gently into the sauce with a kitchen spoon. Heat and remove from the range.
11. Check the seasoning and pour the sauce into a stainless steel container.
12. Serve one steak per portion. Cover with 2 oz to 2½ oz of sauce.

Precautions

- Do not overcook the steaks.
- When sautéing the onions, do not let them brown.
- Make certain that the liquid in which the steaks were cooked is slightly cooled before adding the sour cream.
- Exercise caution when using the knife.

Beef Mandarin

Beef mandarin is an American dish prepared in the Chinese fashion using crisp vegetables, water chestnuts, and bean sprouts. This is an excellent item to feature on the luncheon menu.

50 servings

Equipment

- Cup measure
- Quart measure
- French knife
- 5 gal. stockpot
- Baker's scale
- Paddle
- Deep steam table pan

Ingredients

15 lb boneless beef chuck
½ c salad oil
5 lb celery
4 lb onions
2 cans drained water chestnuts
14 oz cornstarch
12 oz water
½ c soy sauce
6 qt beef stock
5 whole canned pimientos
1 c tomato puree
2 bay leaves
salt and pepper to taste
1 No. 10 can drained bean sprouts

Preparation

1. Slice the beef chuck with a French knife or on a slicing machine into thin pieces approximately 1″ by 1″ by ¼″.
2. Dice the celery, onions, and pimientos with a French knife.
3. Slice the water chestnuts very thin with a French knife.
4. Prepare the beef stock.

Procedure

1. Place the salad oil in a stockpot, add the sliced beef, and sauté until the meat is brown. Stir frequently with a paddle.
2. Add the celery, onions, and water chestnuts. Continue to sauté for 10 min.
3. Add the beef stock, tomato puree, bay leaves, and soy sauce. Boil until beef is tender.
4. Place the cornstarch in a bowl, add water, and dilute. Add the mixture slowly to the boiling beef, stirring constantly with a paddle.
5. Remove from the heat. Remove bay leaves and add the pimientos and bean sprouts.
6. Season with salt and pepper. Remove from the range and place in a deep steam table pan.
7. Serve 6 oz with rice and fried Chinese noodles.

Precautions

- Pour the diluted starch into the boiling liquid slowly while stirring constantly to avoid lumps.
- Bean sprouts must be added upon completion of cooking so they remain crisp.
- Be sure beef is tender before adding the starch.
- Exercise caution when using the knife.

Trade Tip

There are three safe ways to defrost beef: in the refrigerator, in cold water, and in the microwave. Never defrost on the counter or in other locations. It is safe to cook frozen beef in the oven, on the stove, or grill without defrosting it first. The cooking time may be about 50% longer. Do not cook frozen beef in a slow cooker.

Beef Rouladen

Beef rouladen consists of thin slices of beef round, flattened and spread with a filling. A piece of pickle is added, then it is rolled and baked in the oven. Beef rouladen is an excellent dinner item.

50 servings

Equipment

- Cleaver
- Food grinder
- Mixing container
- Roast pan
- French or butcher knife
- Steel skillet
- Wire whip
- Baker's scale
- Quart measure
- Cup measure
- Spoon measure
- China cap
- Quart bowl
- Stainless steel container
- Kitchen spoon
- Toothpicks

Ingredients

10 lb beef round
1 lb bacon
1 lb lean, raw ham scraps
3 lb raw hamburger
½ c onions
6 eggs
1 qt fine, dry bread crumbs
2 tbsp parsley
50 strips sweet or dill pickles
1 gal. Burgundy sauce

Preparation

1. Trim the beef round, slice the beef with a French or butcher knife into 2 ½ oz to 3 oz square pieces, and flatten with the side of a cleaver until very thin.
2. Combine the bacon and ham scraps. Grind on the food grinder using the medium chopper plate.
3. Chop the parsley with a French knife and wash after chopping.
4. Mince the onions with a French knife and sauté in a skillet with additional shortening.
5. Preheat the oven to 375°F.
6. Prepare the Burgundy sauce.
7. Break the eggs in a bowl and beat slightly with a wire whip.

Procedure

1. Place the bacon, ham, onions, eggs, bread crumbs, hamburger, and parsley in a mixing container. Mix thoroughly by hand.
2. Spread the filling and place a strip of dill or sweet pickle on each piece of meat. Roll and secure with twine or toothpicks. Place in the roast pan.
3. Place the meat rolls in the preheated 375°F oven. Allow to brown, then remove excess grease. Discard grease.
4. Add Burgundy sauce and continue to bake until the meat rolls are tender. Remove from the oven. Remove the toothpicks.
5. Place the meat rolls in a steam table pan. Strain the sauce through a china cap into a stainless steel container.
6. Serve one meat roll covered with Burgundy sauce for each portion.

Precautions

- When placing the meat rolls in the roast pan, place the open side down and keep them close together.
- When baking the meat rolls, baste frequently with the sauce using a kitchen spoon.
- When slicing the beef, have the beef slightly frozen for best results.
- Exercise caution when using the knife.

English Beef Stew

This is a brown stew. It is superior in flavor to a white stew. The dish of about two-thirds cooked beef and one-third cooked vegetables is served mostly as a luncheon entree.

50 servings

Equipment

- French knife
- Braiser and cover
- 6 qt saucepans (2)
- 4 qt saucepan
- Baker's scale
- Quart measure
- China cap

Ingredients

15 lb boneless beef chuck
2 gal. beef stock
1 lb beef fat (suet) or 1 pt salad oil
1 qt tomato puree
½ oz garlic
1 lb flour
3 lb carrots
3 lb celery
½ No. 10 can whole small onions, drained
2½ lb peas
½ No. 10 can cut green beans, drained
¼ oz ground thyme
2 bay leaves
3 oz salt (variable)
pepper to taste

Preparation

1. Cut the beef chuck into 1″ cubes with a French knife.
2. Dice the carrots and celery with a French knife.
3. Mince the garlic and suet with a French knife.
4. Preheat the oven to 375°F.
5. Prepare the beef stock.

Procedure

1. Place the suet in the braiser and render on the range until it starts to become crisp.
2. Add the cubes of beef and brown thoroughly.
3. Add the flour, blend thoroughly with a kitchen spoon, making a roux, and cook slightly.
4. Add the hot beef stock, tomato puree, minced garlic, thyme, and bay leaves. Stir with a kitchen spoon. Cover the braiser, place in the preheated oven at 375°F, and cook for about 2 hr to 2½ hr until the beef is tender.

5. Boil all vegetables in salt water in separate sauce-pans until tender and drain.
6. When the meat is tender, remove it from the oven. Remove the bay leaves and add all the drained, cooked vegetables except the peas.
7. Check the seasoning and place in deep steam table pans.
8. Serve 6 oz in deep casseroles. Garnish the top of each portion with cooked peas.

Precautions

- Do not overcook the beef.
- When adding the flour to make the roux, be sure to cook the roux slightly or the stew will have a raw flour taste.
- Exercise caution when using the knife.

Mediterranean Steak Sandwiches

This contemporary presentation of sirloin steak is an attractive luncheon dish.

24 servings

National Cattlemen's Beef Association

Ingredients

6 lb boneless beef top sirloin steak
24 1″ thick pita breads
1½ c crumbled feta cheese

Seasoning

4 tbsp olive oil
1 head garlic
2 tbsp dried oregano leaves
1½ tsp black pepper
1 tsp salt

Tomato Relish

6 c chopped seeded tomato
1 lb black olives
2 c chopped onion
¾ c Italian dressing

Preparation

Seasoning

1. Crush garlic and mix with remaining ingredients.

Tomato Relish

1. Slice olives and mix with remaining ingredients.

Procedure

1. Press seasoning evenly into both sides of the beef steaks.
2. Place meat on broiler pan and broil 3″ – 4″ from heat. (*Note:* Broil approximately 16 min for medium rare and approximately 21 min for medium. Turn once.)
3. Remove from heat and carve steak into thin slices.
4. Season with 1 tbsp salt and place on pitas.
5. Top with relish and feta cheese.

Precautions

- Do not overcook the meat.
- Exercise caution when using the knife.

Trade Tip

When flattening a boneless piece of meat using a mallet or the flat side of a cleaver, place a heavy piece of plastic over the surface of the meat before pounding. This prevents the mallet or cleaver from sticking to the meat or tearing the flesh.

BROILED BEEF RECIPES

Cheddar Steak

Cheddar steak is made tender by cubing or chipping. It is broiled to the desired degree of doneness and served with a rich Cheddar cheese sauce. This dish is generally served on the luncheon menu.

50 servings

Equipment

- Broiler
- Quart measure
- Kitchen fork
- Spoon measure
- Baker's scale
- Box grater
- 2 qt stainless steel container
- Kitchen spoon
- 2 qt saucepan
- Bake pans (2)
- Mixing bowl

Ingredients

50 4 oz cube or chip steaks (purchase ready to cook)
1 pt salad oil
1 tbsp Worcestershire sauce
1 tsp dry mustard
1 tsp paprika
8 drops Tabasco sauce
1 qt tomato juice
2 lb sharp Cheddar cheese
salt and pepper to taste

Procedure

1. Place the salad oil in a bake pan.
2. Season the steaks with salt and pepper and place in the salad oil.
3. Pat off excess oil and place the steaks on a hot broiler. Broil until medium done and remove from the broiler with a kitchen fork. Place in a bake pan and hold in a warm place.
4. Mix the Worcestershire sauce, Tabasco sauce, dry mustard, and paprika into a paste in a mixing bowl.
5. Place the tomato juice in a saucepan, add the paste mixture, and bring to a boil. Stir occasionally with a kitchen spoon.
6. Add the grated Cheddar cheese and cook until smooth, stirring frequently. Remove from the range and pour into a stainless steel container.
7. Serve the steak on a hot plate covered with the rich Cheddar sauce.

Precautions

- Cook steaks medium for best results.
- Exercise caution when grating the cheese.

Preparation

1. Grate the Cheddar cheese on the coarse grid of a box grater.
2. Preheat the broiler.

Trade Tip

Before grinding beef for Salisbury steak, chop steak, or plain hamburger, add some crushed ice to the meat. Grind the ice with the meat. This results in a juicier product when it is cooked.

Beef Tenderloin en Brochette

Beef tenderloin en brochette is an unusual as well as a very attractive entree. The cubes of beef are alternated with the vegetables on a skewer and cooked as one. This is an excellent dinner item.

24 servings

Equipment

- Baker's scale
- 24 skewers
- French knife
- 1 qt saucepan
- Pastry brush
- Bake pan

Ingredients

8 lb beef tenderloin
24 small tomatoes
96 mushroom caps
96 small onions
8 oz melted butter
4 garlic cloves
salt and pepper to taste

Procedure

1. Arrange four each of 1" steak cubes, mushroom caps, onions, and tomatoes alternately on the skewers.
2. Place the butter in a saucepan and heat. Add the garlic and cook slightly. Remove from the range and hold for later use.
3. Place the oil in a bake pan and marinate the skewered items in the salad oil. Place in the preheated broiler and cook slowly until all items are tender. Turn frequently.
4. Brush on the butter and garlic mixture just before removing from the broiler.
5. Serve at once on toast, white rice, or wild rice. Remove skewers before serving.

Precautions

- Exercise caution when using the knife.
- When broiling, turn the skewered tenderloin every 3 min so it browns more evenly.
- Marinate the item in oil before broiling.

Preparation

1. Trim the tenderloin and cut into 1" cubes with a French knife.
2. Cut the tomatoes into quarters with a French knife.
3. Preheat the broiler.

Salisbury steak is a ground beef product. The ground beef is highly seasoned then pressed into steak form, broiled or sautéed, and served with a flavorful sauce.

Salisbury Steak

50 servings

Equipment

- Baker's scale
- Cup measure
- Spoon measure
- French knife
- 1 qt saucepan
- 5 gal. mixing container
- Wire whip
- 1 qt bowl
- Kitchen spoon
- Full size steam table pan
- Food grinder

Ingredients

14 lb boneless beef chuck
3 lb onions
1 tsp garlic
½ c salad oil
8 eggs
2 lb bread
1½ pt milk
salt and fresh ground pepper to taste

National Cattlemen's Beef Association

Preparation

1. Cut the beef chuck with a French knife into pieces that will pass through the food grinder.
2. Mince the onions and garlic with a French knife.
3. Cut the bread into cubes with a French knife.
4. Break the eggs into a bowl and beat slightly with a wire whip.

Procedure

1. Place the oil in the saucepan and heat. Add the onions and garlic and sauté until tender. Do not brown.
2. Place the bread in the mixing container, add the milk, and mix thoroughly with a kitchen spoon.
3. Add the sautéed vegetables, beef chuck, salt, and pepper, and mix thoroughly by hand.
4. Pass this mixture through the food grinder twice using the medium hole chopper plate.
5. Add the beaten eggs and blend thoroughly. Check the seasoning.
6. Form into 5 oz steaks. Pass through salad oil and broil or sauté. Place in a steam table pan.
7. Serve one steak per portion with an appropriate sauce.

Precautions

- If the mixture is too moist, adjust consistency by adding bread crumbs.
- When forming the steaks, rub hands with a small amount of salad oil.
- When sautéing the onions and garlic, do not brown.
- Exercise caution when using the knife.

BOILED BEEF RECIPES

Boiled fresh brisket of beef is an excellent entree for either the luncheon or dinner menu. It is served with horseradish sauce, boiled cabbage, and boiled potatoes. The potatoes and cabbage are often boiled in the same pot as the brisket for additional flavor or may be boiled in a separate pot.

Boiled Fresh Brisket of Beef

50 servings

Equipment

- 10 gal. stockpot
- French knife
- Kitchen fork
- Ladle
- Steam table pan

Ingredients

17 lb beef brisket
water, enough to cover beef
salt to taste
1 tbsp pickling spices
1 lb onions
½ lb celery
½ lb carrots

Preparation

1. If the meat has excess fat, trim slightly with a French knife.
2. Cut the vegetables rough with a French knife.

Procedure

1. Place the brisket in the stockpot and cover with cold water.
2. Bring to a boil and remove any scum that may appear on the surface.

3. Add the rough garnish (onions, celery, and carrots) for flavor. Reduce heat until liquid simmers.
4. Add the salt and spices and continue to simmer until meat is tender. Remove the meat from the stock, and let cool slightly.
5. Slice the meat against the grain with a French knife. Place in a steam table pan and reheat in the steam table or steamer.
6. Serve 3 oz to 4 oz per portion with horseradish sauce, boiled cabbage, and boiled potatoes.

Note: Save the beef stock for use in the preparation of horseradish sauce or other sauces or soups.

Precautions

- Exercise caution when cutting the vegetables.
- Do not let the liquid boil too fast or the stock will become cloudy.

Roast Beef Hash: Southern Style

This is a profitable item and an attractive dish. Beef or veal leftovers can be used by putting it on the luncheon menu.

50 servings

Equipment

- Large braiser
- French knife
- Deep steam table pan
- Paddle
- Baker's scale

Ingredients

1 lb green peppers
2 lb 10 oz onions
8 oz pimientos
15 lb cooked beef (roasted or boiled leftovers)
9 lb raw potatoes
2 lb 8 oz tomato puree
12 oz salad oil
1 gal. beef stock or brown stock
salt and pepper to taste
1½ tsp nutmeg

Preparation

1. Dice the cooked beef with a French knife.

2. Dice the onions, green peppers, pimientos, and potatoes with a French knife.
3. Prepare the brown stock or beef stock.

Procedure

1. Place the shortening, diced onions, and diced green peppers in the braiser and sauté until partly done.
2. Add the brown stock or beef stock.
3. Add the tomato puree and diced raw potatoes. Cook until potatoes are half done.
4. Add the diced beef and diced pimientos and cook until potatoes are completely done. Stir occasionally with a paddle.
5. Season with salt, pepper, and nutmeg. Remove from the heat and place in a deep steam table pan.
6. Serve 6 oz in shallow casseroles with corn fritters.

Precautions

- Do not overcook the potatoes.
- Exercise caution in dicing the garnish.

BEEF VARIETY MEAT RECIPES

Braised Sweetbreads

Braised sweetbreads are a favorite of many gourmets. The sweetbreads are first blanched then braised with a mirepoix garnish (small diced vegetables). The mirepoix increases the flavor of this popular variety meat.

25 servings

Equipment

- French knife
- Baker's scale
- Quart measure
- 3 gal. sauce pot
- Braiser
- China cap
- Colander
- Kitchen spoon
- 1 gal. stainless steel container

Ingredients

25 pairs heart sweetbreads
1 lb margarine or fat
1 lb carrots
1 lb onions
12 oz celery
2 bay leaves
4 garlic cloves
2 qt brown sauce
salt and pepper to taste

Preparation

1. Place the sweetbreads in a sauce pot, cover with water, and add salt and lemon juice or vinegar. Place

on the range and simmer until partly cooked (approximately 10 min to 15 min). Drain in a colander and let cool. Remove membranes.
2. Dice the carrots, onions, and celery into very small cubes with a French knife. This is the mirepoix. Mince garlic.
3. Prepare the brown sauce.
4. Preheat the oven to 375°F.

Procedure

1. Place the butter in the braiser and melt.
2. Add the mirepoix and spices. Place the sweetbreads over the vegetables. Cook in the preheated oven at 375°F until vegetables are slightly brown.
3. Add the brown sauce. Reduce the oven temperature to 350°F and continue to braise for 35 min, basting frequently with a kitchen spoon.

4. Remove from the oven and strain the juice through a china cap into a stainless steel container.
5. Slice the sweetbreads with a French knife about ½" thick on a bias.
6. Serve a pair of sweetbreads per portion in a shallow casserole with 2 oz of the strained liquid.

Precautions

- Exercise caution when using the French knife.
- Do not let the mirepoix burn during cooking.
- Use caution when slicing the sweetbreads.

Variations

Sautéed sweetbreads: Poach sweetbreads, slice in two, then pass through seasoned flour. Sauté in shortening or butter until golden brown on both sides. Serve with a mushroom, brown, bordelaise, brown butter, or Bercy sauce.

Broiled sweetbreads: Poach sweetbreads, slice in two, and pass through salad oil. Season with salt, pepper, and paprika. Place on the broiler and brown one side, then turn and brown the other side.

Sweetbreads Carolina: Poach sweetbreads, slice in two, then pass through seasoned flour. Sauté in butter or shortening and garnish with julienned ham, browned almonds, and sautéed mushroom caps. Serve topped with a thin sherry cream sauce.

Sweetbread chasseur: Poach sweetbreads, slice in two, then pass through seasoned flour. Sauté in butter or shortening and remove from skillet, let drain, and place in sauce pot. Cover with mushroom sauce. Garnish with julienned ham, turkey, and chopped parsley and serve in a casserole.

Sautéed Beef Liver

Sautéed beef liver is one of the popular variety meats. The liver is sliced, passed through flour, and cooked in shallow grease until golden brown.

50 servings

Equipment

- Butcher or French knife
- Frying pan
- Bake pan
- Kitchen fork
- Baker's scale

Ingredients

16 lb beef liver
1 qt salad oil or melted shortening (variable)
3 lb bread flour
salt and pepper to taste

Preparation

1. Skin the beef liver by hand. Chill or partly freeze the liver and slice on a bias into uniform, fairly thin slices.
2. Place the flour in a bake pan and season with salt and pepper.

Procedure

1. Place about ¼" of oil or shortening in the frying pan and heat.
2. Pass the slices of liver through the seasoned flour.
3. Place the liver in the hot oil or shortening. Brown one side, turn with a kitchen fork, and brown the other side.
4. Remove from the fry pan and let drain.
5. Serve two or three slices per portion with a butter sauce accompanied with cooked onions.

Precautions

- Sauté the liver at a medium temperature or a hard crust will form.
- Liver should be cooked medium for best eating qualities.

Note: Liver may also be broiled with excellent results. If broiled, do not pass through flour. Pass the liver through salad oil before placing it in the broiler.

Veal Preparation

National Cattlemen's Beef Association

Veal is the flesh of young milk-fed beef calves. Veal flesh possesses a very delicate flavor and blends well with other food items and sauces. This tender meat cooks quickly and displays well when served. Veal carcasses average in weight from 60 lb to 160 lb, but the best veal comes from calves that are slaughtered between 6 weeks and 8 weeks of age and weigh about 125 lb. Veal is expensive and limited in supply.

From birth to about 8 weeks old, the calf is fed on mother's milk. At this stage, it is classified as veal. The fat is firm and white and the flesh is light pink. There is no marbling of fat in the lean. At about 8 weeks, the animal is turned out to pasture and fed grass, meal, or hay. This feed causes the fat to turn slightly yellow, and the flesh becomes darker in color. The animal at this stage is a calf.

The stages of development from veal to calf to beef are caused by the natural maturation process and the type of feed. Although there is a significant difference between veal and calf carcasses, both are referred to as veal when placed on the market.

VEAL GRADING

Veal is the meat of young milk-fed beef calves. It has very little fat covering and a high moisture content. Veal is graded by yield and quality. Yield indicates how much usable meat can be obtained from a carcass as determined by U.S. government standards. Quality indicates the color, texture, and firmness of the meat. For veal to receive a high-quality rating, the fat surrounding the areas of the shoulders, rump, and kidneys must be thick and white. The meat must be firm and possess a smooth surface when cut.

The six U.S. Department of Agriculture veal grades are prime, choice, good, commercial, utility, and cull. Of these, only prime, choice, good, and commercial are used by food service.

U.S. prime: Veal of the highest quality. To be graded prime, the animal must be rated superior in yield and quality factors. U.S. prime veal is not available in great quantities.

U.S. choice: High-quality grade of veal used in commercial establishments. Veal graded choice is derived from very compact, thick-fleshed, and fairly plump animals. The bones are small in proportion to the size of the animal.

U.S. good: Economical grade of veal used with good results in some preparations. The veal carcass graded good is thin-fleshed and somewhat slender in appearance. The flesh is slightly soft to the touch, and the cut surface displays some roughness. The bones are large in proportion to the size of the animal.

U.S. standard (commercial): Rarely used in the commercial kitchen except when the preparation is being stewed or braised. Veal carcasses of this grade are thin-fleshed, rough, and sunken in appearance. The flesh is soft and the surface of the flesh is rough when cut.

U.S. utility: This is a poor quality of veal. It is very seldom, if ever, used in the commercial kitchen. The veal carcass of this grade is rough, thin, and sunken in appearance. It has little fat covering and the flesh is very soft and watery. All bones are large in proportion to the weight and size of the animal.

U.S. cull: This is a grade that is never used in the commercial kitchen. It is the lowest quality and yield of all the grades.

Veal can be purchased in six market forms. These are the carcass, side, quarter, wholesale or primal cuts, fabricated, and retail. Retail cuts are rarely used in the commercial kitchen. The cutting of meats for preparation in the commercial kitchen is done much differently than in a retail meat market. The meat form purchased depends on:

- meat cutting skill of personnel
- meat cutting equipment available
- meat storage space available
- utility of all cuts purchased
- meat preparations served
- overall economy of purchasing meat in a particular form

Carcass: A complete animal with the head, hide, and entrails removed. Carcasses on the average weigh from 60 lb to 250 lb, depending on whether they are true veal or calf. The carcass can be purchased at a cheaper cost per pound, but many considerations must first be taken. The most important is the utility of all cuts after the carcass is blocked out.

Side: A side of veal consists of half the carcass split by cutting lengthwise through the spine bone. There are two sides to each carcass, a right side and a left side. Each side weighs from 30 lb to 125 lb, depending on the animal. The side is purchased at a savings to the food service operator. However, one must consider whether all the cuts can be utilized after the veal has been blocked out.

Quarter: The quarter of veal is a side divided into two parts. The fore part is the forequarter and the hind part is the hindquarter. The side is divided into the two quarters by cutting through the twelfth and thirteenth ribs, the thirteenth rib remaining on the hindquarter. The hindquarter contains the most desirable cuts and therefore the purchase price is generally high. The forequarter contains less desirable cuts so the price is usually lower.

Wholesale or *primal cuts*: These are parts of the forequarter and hindquarter of the veal. Four wholesale cuts are in the forequarter and two in the hindquarter. The cuts of the forequarter are the rib, shoulder, shank, and breast. The cuts of the hindquarter are the loin and leg. Wholesale or primal cuts are being replaced by fabricated (ready to cook) forms.

Fabricated: These cuts are purchased ready to cook. This is the most convenient way to purchase veal. The meat can be ordered cut to any desired specification of size and weight. Veal purchased in this form is the most expensive but saves in cutting costs.

National Cattlemen's Beef Association

Veal breast roll on pasta is an attractive and popular choice.

FOODSERVICE CUTS OF VEAL

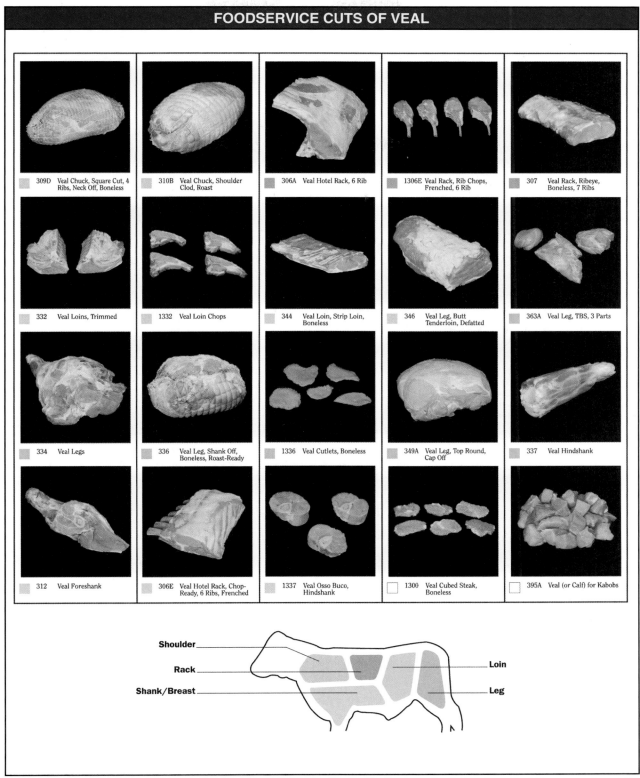

309D Veal Chuck, Square Cut, 4 Ribs, Neck Off, Boneless	
310B Veal Chuck, Shoulder Clod, Roast	
306A Veal Hotel Rack, 6 Rib	
1306E Veal Rack, Rib Chops, Frenched, 6 Rib	
307 Veal Rack, Ribeye, Boneless, 7 Ribs	
332 Veal Loins, Trimmed	
1332 Veal Loin Chops	
344 Veal Loin, Strip Loin, Boneless	
346 Veal Leg, Butt Tenderloin, Defatted	
363A Veal Leg, TBS, 3 Parts	
334 Veal Legs	
336 Veal Leg, Shank Off, Boneless, Roast-Ready	
1336 Veal Cutlets, Boneless	
349A Veal Leg, Top Round, Cap Off	
337 Veal Hindshank	
312 Veal Foreshank	
306E Veal Hotel Rack, Chop-Ready, 6 Ribs, Frenched	
1337 Veal Osso Buco, Hindshank	
1300 Veal Cubed Steak, Boneless	
395A Veal (or Calf) for Kabobs	

Shoulder — Rack — Shank/Breast — Loin — Leg

North American Meat Processors Association

NAMP/IMPS (North American Meat Processors Association/Institutional Meat Purchase Specifications) specifies meat cuts by number.

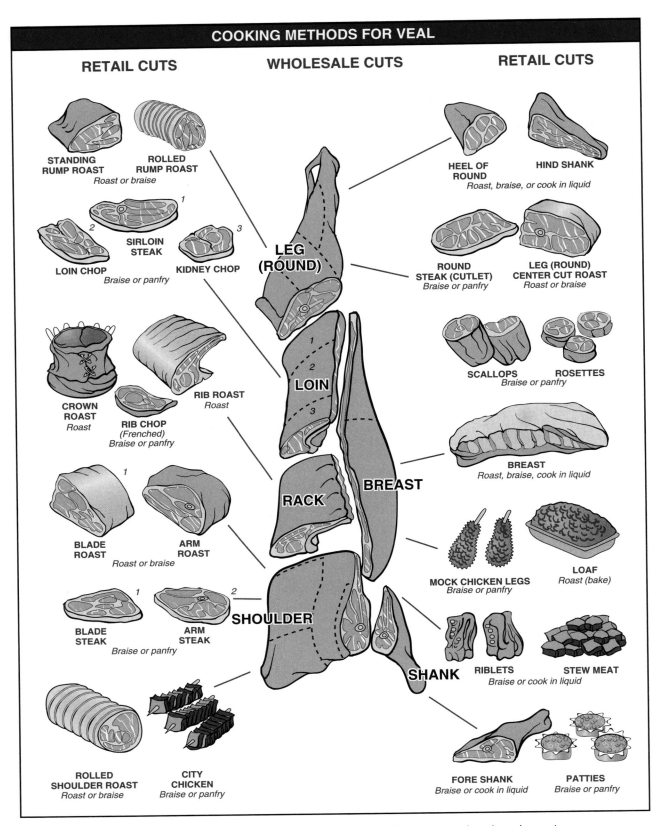

COOKING METHODS FOR VEAL

RETAIL CUTS   WHOLESALE CUTS   RETAIL CUTS

STANDING RUMP ROAST   **ROLLED RUMP ROAST**
Roast or braise

LOIN CHOP   **SIRLOIN STEAK**   **KIDNEY CHOP**
Braise or panfry

CROWN ROAST
Roast

RIB CHOP *(Frenched)*
Braise or panfry

RIB ROAST
Roast

BLADE ROAST   **ARM ROAST**
Roast or braise

BLADE STEAK   **ARM STEAK**
Braise or panfry

ROLLED SHOULDER ROAST
Roast or braise

CITY CHICKEN
Braise or panfry

LEG (ROUND)

LOIN

RACK

BREAST

SHOULDER

SHANK

HEEL OF ROUND   **HIND SHANK**
Roast, braise, or cook in liquid

ROUND STEAK (CUTLET)
Braise or panfry

LEG (ROUND) CENTER CUT ROAST
Roast or braise

SCALLOPS   **ROSETTES**
Braise or panfry

BREAST
Roast, braise, cook in liquid

MOCK CHICKEN LEGS
Braise or panfry

LOAF
Roast (bake)

RIBLETS   **STEW MEAT**
Braise or cook in liquid

FORE SHANK
Braise or cook in liquid

PATTIES
Braise or panfry

Wholesale cuts are less expensive per pound but require meat cutting personnel and equipment.

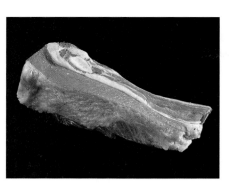

RIBLET

VEAL FOR STEW

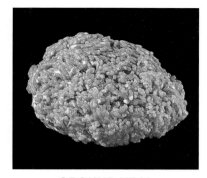

GROUND VEAL

National Cattlemen's Beef Association

Fabricated veal may be purchased ready-to-cook in convenient cuts and sizes.

WHOLESALE CUTS AND COOKING METHODS

The following is a list of the six wholesale or primal cuts of veal, their chief characteristics, and the cooking methods used to convert them into entrees:

National Cattlemen's Beef Association
Leg Cutlet

Leg: The leg is the most desirable of all the veal cuts because of the solid, lean, fine-textured meat it contains. The leg should always be boned by following the muscle structure of the meat so pieces of equal tenderness and less muscle fiber can be removed. The best legs come from veal weighing from 100 lb to 150 lb. Animals of this size provide legs weighing about 20 lb. The leg is good for roasting or for veal steaks, cutlets, or scallopine.

Loin or *saddle*: This veal cut is similar to the sirloin of beef. It contains hip and back bones. The flesh consists of loin eye, tenderloin, and some flank meat, which is generally removed when the loin is trimmed for preparation. Veal loins are sometimes roasted in the commercial kitchen, but in most cases they are converted into veal chops. When a complete loin from the whole veal carcass is unsplit, it is known as a saddle of veal.

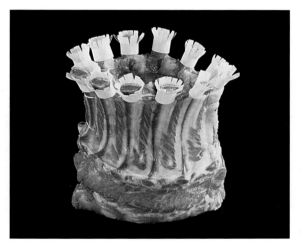

National Cattlemen's Beef Association
Crown

Rib or *rack*: This veal cut is similar to the standing rib roast of beef. It contains seven ribs and a ribeye, which is a solid section of meat. When the complete rib from the veal carcass is unsplit, it is known as a rack of veal. From this rack, a crown roast of veal is generally prepared by forming the ribs into a crown and frenching (removing meat and fat from the bones). The rib cut is sometimes roasted in the commercial kitchen, but in most cases it is converted into veal rib chops, which are best prepared by frying or sautéing.

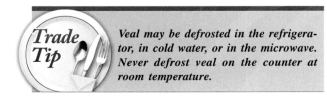

Trade Tip

Veal may be defrosted in the refrigerator, in cold water, or in the microwave. Never defrost veal on the counter at room temperature.

National Cattlemen's Beef Association

Boneless Shoulder

Shoulder: This is the fore section of veal, which contains the blade bone, neck bone, and five rib bones. It produces a fairly high percentage of lean meat. This cut is an excellent choice for preparing such items as veal stew, veal loaf, veal goulash, veal paprika, etc. The shoulder meat displays the best results when it is stewed or braised.

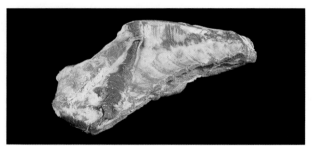

National Cattlemen's Beef Association

Breast

Breast: This is a thin, flat cut containing breast bone and rib ends. It is not a very desirable cut of veal because it contains very little lean meat and some layers of fat. The best way to utilize the breast in the commercial kitchen is to braise or stuff it with a forcemeat (meat stuffing) and bake.

National Cattlemen's Beef Association

Shank

Shank: The shank is another of the undesirable veal cuts. It contains a high percentage of bone and connective tissue with a small amount of lean meat. The full shank is sometimes braised if it is not too large. However, in most cases, the shank is used in stew or ground meat items.

VARIETY MEATS

Veal variety meats are edible parts other than wholesale cuts. Variety meats are highly prized by the restaurant operator and include the sweetbreads, liver, kidneys, brains, and tongue. Veal sweetbreads are white, tender meat prepared in the same manner as beef sweetbreads. Veal liver also has a tender texture and fine flavor. Calf and veal livers are processed and prepared in the same manner as beef liver. Veal kidneys resemble beef kidneys in appearance, but are much more tender. Veal kidneys can be broiled with excellent results, compared to tougher beef kidneys, which must be cooked by moist heat. Veal brains are similar to the brains of other edible animals and are prepared in the same manner. Although veal tongue is not as tough as beef tongue, it is not as popular.

VEAL TERMINOLOGY

The following are terms which are associated with veal:

Veal: Meat of young beef calves fatted on cow's milk.

Calf: Flesh of calves fattened on grass, meal, and hay.

Rack: A complete rib from the whole carcass containing two unsplit ribs.

Saddle: A complete loin from the whole carcass containing two unsplit loins.

Cutlet: A thin, boneless slice of meat.

Scallopine: Small, thin slices of veal, generally leg meat, about equal to a silver dollar in size.

Wiener schnitzel: A Viennese veal steak. A slice of boneless veal that is breaded and fried. It is the same as a veal cutlet.

Baby T-bone steak: A 6 oz to 8 oz steak cut from the loin of veal. It contains loin meat on one side of the small T-bone and tenderloin on the other side. It is similar to the beef T-bone or porterhouse steak, but is much smaller.

ROASTED AND BAKED VEAL RECIPES

Herbed Veal Roast

Herbed veal roast is a good choice for the luncheon or dinner menu. The vegetables accompanying it may be varied depending on what is available.

24 servings

Ingredients

12 lb veal rib roast
3 tbsp chopped fresh sage
6 cloves garlic
1½ tsp cracked black pepper

Apricot-Thyme Chutney

3 tbsp vegetable oil
6 medium onions
18 oz dried apricots
3 c chicken broth
3 tbsp sugar
1¼ tbsp cider vinegar
1½ tsp crushed dried thyme leaves

Preparation

1. Crush the garlic.
2. Slice the onions.
3. Coarsely chop the dried apricots.
4. Heat oven to 325°F.

Procedure

1. Combine sage, garlic, and pepper. Press evenly into surface of veal roast.
2. Place roast, rib ends down, in roasting pan. Insert ovenproof meat thermometer so tip is centered in thickest part of veal, not touching bone or resting in fat.

National Cattlemen's Beef Association

Do not add water or cover. Roast for 22 min – 27 min per pound.

3. Meanwhile, in a large skillet, heat oil over medium heat until hot. Add onions and cook slowly 15 min – 20 min, stirring occasionally. Add remaining chutney ingredients. Cover and simmer 20 min – 25 min.
4. Remove roast when meat thermometer registers 155°F. Let stand for 15 min.
5. Carve into slices and serve with chutney.

Veal Chops Stroganoff

Veal chops stroganoff are sautéed veal chops cut from the rib or loin and baked in a sour cream sauce until tender. This item is a little different and is a good choice for the luncheon or dinner menu.

25 servings

Equipment

- French or butcher knife
- Baker's scale
- Cleaver
- Iron skillet
- Bake pan
- 6 qt sauce pot
- Kitchen fork
- Wire whip
- Braiser and cover
- Stainless steel container
- China cap
- Full size steam table pan

Ingredients

25 5 oz to 6 oz veal chops
1 lb flour
1 lb 8 oz shortening
1 pt melted butter
1 pt flour
3 qt brown stock
1 c tomato puree
½ c vinegar
1 lb onions
1 qt sour cream
1 bay leaf
salt and pepper to taste

Preparation

1. Cut the veal chops with a French or butcher knife from the rib or loin of the veal. Cut 5 oz to 6 oz each, depending on the size desired.
2. Prepare the brown stock.
3. Mince the onions with a French knife.
4. Place the first amount of flour in a bake pan and season with salt and pepper.
5. Preheat the oven to 325°F.

Procedure

1. Place enough shortening in an iron skillet to cover the bottom ¼″ and heat.
2. Pass each veal chop through the seasoned flour and pat off the excess. Place in the hot shortening and sauté until golden brown. Turn with a kitchen fork and brown the second side. Remove, let drain, and place in a braiser.

3. Place the butter in a sauce pot and heat.
4. Add the minced onions and sauté without color.
5. Add the second amount of flour, making a roux, and cook for 5 min.
6. Add the hot brown stock, tomato puree, vinegar, and bay leaf. Whip vigorously with a wire whip until slightly thick and smooth. Let simmer for 30 min.
7. Add the sour cream and bring back to a boil. Remove from the range. Remove the bay leaf, check the seasoning, and pour the sauce over the sautéed chops. Cover the braiser.
8. Bake in a preheated oven at 325°F about 1 hr to 1½ hr until each chop is tender. Remove from the oven.

9. Remove the chops from the sauce and place in a steam table pan. Strain the sauce through a fine china cap into a stainless steel container.
10. Serve one chop per portion, covered with sauce.

Precautions

- Exercise caution when using the knife.
- When sautéing the chops, do not overbrown.
- When sautéing the onions, do not brown.
- Baste the chops frequently with the sauce while baking.

The breast, which contains little meat and many breast bones, is stuffed with a forcemeat and braised to create a very desirable entree out of a slightly undesirable cut of veal. The stuffed breast of veal is served with gravy and noodles.

Stuffed Breast of Veal

25 servings

Equipment

- Boning knife
- Meat saw
- French knife
- Butcher twine
- Large eye needle
- Baker's scale
- Quart measure
- Kitchen fork
- Skillet
- Mixing container
- Roast pan
- Meat grinder

Ingredients

3 5 lb sections (15 lb) veal breast, trimmed
3 lb boneless veal shoulder
3 lb boneless pork shoulder
1½ lb dry bread cubes
1 qt milk (variable)
8 oz onions
6 oz celery
8 oz bread crumbs (variable)
6 egg yolks
8 oz butter
¼ oz sage
2 qt brown gravy
salt and pepper to taste

Procedure

1. Place the dry bread cubes and the milk in a mixing container. Let soak.
2. Sauté the onions and celery in a skillet in the butter and add to mixture.
3. Add the pork, veal, and sage. Mix thoroughly.
4. Grind this mixture twice using the fine chopper plate.
5. Add the slightly beaten egg yolks and season with salt and pepper. Mix thoroughly. If the mixture is too wet, add bread crumbs as needed; if too dry, add more milk.
6. Stuff the forcemeat mixture fairly solid into the pockets cut into the veal breast.
7. Using a large eye needle and butcher twine, sew the opening between the layer of meat and the breastbones. Secure properly so the forcemeat does not come out during the roasting period.
8. Season the stuffed breast with salt and pepper and roast in the preheated 350°F oven until the breasts are thoroughly brown. Turn occasionally with a kitchen fork.
9. Pour the brown gravy over the breasts and continue to braise until the breasts are tender and the forcemeat has become solid.
10. Remove from the oven, place in a steam table pan, and let set in a warm place for 45 min.
11. Slice between the ribs with a French knife. Cut into 8 oz to 10 oz portions. Serve covered with brown gravy and accompanied with buttered noodles.

Preparation

1. Trim the three veal breasts with a French knife to remove the excess fat and bones. Cut a pocket in the breast by slicing with a boning knife between the flesh and the breastbones. Make the opening as large as possible, but do not cut through the flesh at any point.
2. Prepare the brown gravy.
3. Mince the celery and onions with a French knife.
4. Cut the pork and veal shoulders with a French knife into strips that will fit in the grinder.
5. Separate the eggs and beat the yolks slightly. Save the whites for another preparation.
6. Preheat oven to 350°F.

Precautions

- Exercise caution when using the knives.
- Do not break through the flesh when cutting the pocket.
- Use a sharp French knife and apply very little pressure when slicing each order of the veal breast.

Trade Tip *The two basic methods of cooking veal are by dry heat and moist heat. Tender cuts can be prepared by either method. Less tender cuts require moist heat. Less tender cuts can also be marinated and cooked by dry heat.*

City Chicken

City chicken is a mock chicken drumstick prepared by placing cubes of veal or alternating cubes of pork and veal on a wooden skewer. The item is then breaded, fried, and baked until it is tender.

50 servings

Equipment

- 50 wooden skewers
- French knife
- Quart measure
- Baker's scale
- Bake pans (5)
- Quart bowl
- Wire whip
- Full size steam table pan

Ingredients

10 lb boneless veal shoulder
7 lb Boston butt
12 eggs
2 qt milk
3 lb dry bread crumbs
3 lb flour
salt and pepper to taste

Preparation

1. Cut the boneless veal and pork into 1″ cubes with a French knife.
2. Alternate the veal and pork cubes on wooden skewers. Use three cubes of veal and two cubes of pork.
3. Place the flour in a bake pan and season with salt and pepper.
4. Prepare the egg wash. Break the eggs into a bowl and whip slightly with a wire whip. Pour in the milk while continuing to whip. Place in a bake pan.
5. Place bread crumbs in a bake pan.
6. Preheat the deep fat fryer to 325°F and the oven to 300°F.

Procedure

1. Pass each skewer through the flour, egg wash, and bread crumbs. Press the bread crumbs on firmly.
2. Brown lightly in deep fat at 325°F. Place in a bake pan.
3. Bake in the preheated oven at 300°F for about 1½ hr until each cube is very tender. Remove from the oven and place in a steam table pan.
4. Serve one skewer per portion. Serve plain or with brown gravy.

Precautions

- Exercise caution when using the knife.
- Do not overbrown when frying the city chicken in the deep fat.
- Watch carefully while the city chicken is baking in the oven so it does not overbrown or stick to the pan.

Veal Birds

Veal birds are thin slices of veal leg covered with a forcemeat mixture, rolled, and secured with a toothpick. They are browned in hot grease and baked in the oven until tender. This is an excellent choice for the luncheon menu.

50 servings

Equipment

- 50 round toothpicks
- Boning knife
- French knife
- Cleaver
- 1 qt saucepan
- Kitchen spoon
- Ladle
- Meat grinder
- 3 gal. mixing container
- Wire whip
- Quart bowl
- Roast pan
- Steam table pan

Ingredients

50 4 oz slices of veal leg
3 lb boneless veal shoulder
2 lb boneless pork shoulder
1 lb dry bread cubes
1½ pt milk (variable)
6 oz onions
6 oz celery
8 oz bread crumbs (variable)
4 egg yolks
6 oz butter
¼ oz sage
salt and pepper to taste
2 qt brown stock

Preparation

1. Cut thin slices of veal from sections of a boned leg. Cut with a French or butcher knife, against the grain, about ¼″ thick on a bias. Flatten with the side of a cleaver.
2. Mince the onions and celery with a French knife. Sauté in a saucepan in 6 oz of butter.
3. Prepare the brown stock.
4. Cut the boneless pork and veal with a French knife into strips that will fit in the food grinder.
5. Separate the yolks from the whites of the eggs. Save the whites of the eggs for another preparation. Place the egg yolks in a bowl and beat slightly with a wire whip.
6. Cut the bread into cubes with the French knife.
7. Preheat oven to 375°F.

Procedure

1. Place the bread cubes and the milk in a mixing container. Mix with a kitchen spoon until the bread has absorbed the milk.
2. Add the sautéed onions and celery, pork, veal, and sage. Mix thoroughly.
3. Grind the mixture twice on the food grinder using the fine chopper plate.
4. Add the slightly beaten egg yolks and season with salt and pepper. Mix thoroughly by hand. If the mixture is too wet (collapses when formed into a roll), add bread crumbs as needed. If the mixture is too dry (does not hold together), add more milk.
5. Place 2 oz to 3 oz of the stuffing on each flattened thin slice of veal. Roll and secure the ends with a toothpick.

6. Place the veal birds in a roast pan and bake in the preheated oven at 375°F until golden brown.

7. Reduce the oven temperature to 325°F. Pour the brown stock over the birds and continue to bake for 1 more hour or until the birds are tender. Remove from oven and place in a steam table pan. Remove toothpicks.

8. Cover one bird per portion with 2 oz of sauce and serve with buttered noodles.

Precautions

- Exercise caution when slicing and flattening the veal.
- Secure the veal rolls well. If they unroll during the baking period, they are not ready to be served.

FRIED AND SAUTÉED VEAL RECIPES

Veal Chop Sauté Italienne

For veal chop sauté Italienne, the chops are cut from the rib of the veal. A pocket is cut into the meaty side so ham and Swiss cheese can be inserted. The chop is then passed through flour seasoned with Italian herbs and sautéed to a golden brown.

50 servings

Equipment

- French or butcher knife
- Cleaver
- Boning knife
- Bake pans (4)
- Iron skillet
- Kitchen fork
- Baker's scale
- Spoon measure
- Boning knife
- Full size steam table pan
- Quart measure
- Slicing machine

Ingredients

50 6 oz to 8 oz veal chops
50 1 oz to 1½ oz slices of ham
50 1 oz slices of Swiss cheese
3 lb flour (variable)
1 tbsp oregano
1 tbsp basil
salt and pepper to taste
2 qt milk
12 eggs
1 qt salad oil or melted shortening (variable)

beat with a wire whip, and pour in the milk while continuing to whip. Pour into a bake pan.

4. Place the flour in a bake pan and season with the basil, oregano, salt, and pepper.

5. Preheat oven to 300°F.

Procedure

1. Wrap the ham around each slice of Swiss cheese and insert it into the pocket that was cut into each veal chop.

2. Dredge each chop in the seasoned flour. Press firmly with the hand to flatten. Dip into the egg wash and then back into the seasoned flour mixture for the second time. Again, press firmly.

3. Place enough salad oil in the skillet to cover the bottom ¼" and heat.

4. Add the chops and brown both sides slightly. Turn with a kitchen fork. Remove from the skillet and drain.

5. Place the chops in a bake pan flesh side up.

6. Bake in the preheated oven at 300°F for about 1½ hr or until the chops are tender. Remove to a steam table pan.

7. Serve one chop per portion. Place each portion on top of tomato sauce.

Preparation

1. Cut the veal chops from trimmed veal ribs. Cut against the grain ¾" to 1" thick with a French or butcher knife. Cut a deep pocket into the meaty side of each chop with a boning knife.

2. Slice the ham and Swiss cheese about ⅛" thick on a slicing machine.

3. Prepare an egg wash by combining the slightly beaten eggs and the milk. Place the eggs in a bowl,

Precautions

- Exercise caution when using the knives.
- When dredging the chops, press the flour on firmly.
- Exercise caution while baking so the chops do not become too brown.

Product dating is not required for veal products by federal legislation. However, many processors voluntarily date veal products. If a date is shown, there must be a phrase explaining the meaning of the date.

Veal Piccata

These thinly sliced and pounded cutlets can be prepared and cooked very quickly. Serve with light vegetables and a slice of lemon.

24 servings

Ingredients

6 lb veal cutlets
¾ c all-purpose flour
¾ tsp sweet paprika
¾ tsp white pepper
6 tbsp olive oil

Lemon-Caper Sauce

4 c dry white wine
¾ c fresh lemon juice
4 tbsp drained capers
2 tbsp butter

Procedure

1. Pound veal cutlets to ⅛″ thickness if necessary. Combine flour, paprika, white pepper, and 2 tsp salt. Lightly coat cutlets with flour mixture.
2. In large skillet, heat half the oil over medium heat until hot. Add ½ the cutlets. Cook 3 to 4 minutes for medium doneness, turning once. Remove and keep warm. Repeat with remaining oil and cutlets.
3. Add wine and lemon juice to skillet. Cook and stir until browned bits attached to skillet are dissolved and the sauce thickens slightly. Remove from heat. Stir in capers and butter. Spoon over cutlets.

National Cattlemen's Beef Association

Veal Parmesan

This preparation is of Italian origin but was popularized by Americans. A 5 oz or 6 oz veal cutlet is sautéed, topped with a rich Italian sauce and Parmesan and Provolone cheese, placed under the broiler to slightly melt the cheese, and served. It is a popular dinner entree to satisfy those customers who desire Italian cuisine.

50 servings

Equipment

- French or butcher knife
- Cleaver
- Large sauté pan
- Baker's scale
- Gallon measure
- Kitchen fork
- 4 oz ladle
- Bake pan

Ingredients

50 5 oz or 6 oz veal cutlets
1½ gal. Italian sauce
50 thick slices Provolone cheese
1 lb Parmesan cheese (variable)
2 lb shortening (variable)
flour, as needed
salt and pepper to taste

Procedure

1. Place the flour in a bake pan. Pass each cutlet through the flour. Press firmly with palm of the hand so flour adheres to the cutlet.
2. Place enough shortening in the large sauté pan to cover the bottom ¼″. Place on the range and heat.
3. Add the cutlets until the skillet is full. Sauté until one side is brown. Turn with a kitchen fork and brown the second side. Repeat this procedure until all cutlets are sautéed.
4. Place each cutlet on a dinner plate or in a very shallow casserole dish. Top with a 4 oz ladle of Italian sauce, Parmesan cheese, and a slice of Provolone cheese.
5. Place under the broiler to melt the cheese just before serving.

Preparation

1. Cut the veal cutlets from sections of a boned leg. Slice the veal against the grain with a French or butcher knife and flatten slightly with the side of a cleaver.
2. Prepare the Italian sauce.

Precautions

- Exercise caution when using the knife and cleaver.
- Press firmly when passing the cutlets through the seasoned flour so it adheres tightly.
- For best results, sauté at a moderate temperature.

Sautéed Veal Steak

Sautéed veal steaks are similar to veal cutlets, but they are cut slightly thicker and not flattened as much. They are passed through seasoned flour and cooked to a golden brown in shallow grease.

50 servings

Equipment

- Butcher or French knife
- Full size steam table pan
- Baker's scale
- Bake pan
- Iron skillet
- Cleaver

Preparation

1. Cut the 6 oz to 8 oz steaks from the boneless sections of the leg. Cut about ¼″ thick with a butcher or French knife against the grain of the meat. Flatten slightly with the side of a cleaver.

Ingredients

50 6 oz to 8 oz veal steaks
3 lb flour
salt and pepper to taste
shortening as needed

2. Place flour in a bake pan and season with salt and pepper.

Procedure

1. Pass each veal steak through the seasoned flour. Dust off excess.
2. Place the steaks in the hot shortening, brown one side, turn with a kitchen fork, and brown the other side. Remove and let drain. Place in a steam table pan.
3. Serve one steak per portion covered with Bercy, mushroom, Bordelaise, or Madeira sauce.

Precautions

- Exercise caution when using the knife and cleaver.
- Sauté at a moderate temperature.

Veal Cutlet Cordon Bleu

Veal is a delicately flavored meat, so it blends well with other foods. In this case, the veal is blended with ham and Swiss cheese. This combination has caught the fancy of the dining public and is a very popular menu item.

25 servings

Equipment

- French or butcher knife
- Cleaver
- Large iron skillet
- Kitchen fork
- Bake pans (3)
- Quart bowl
- Wire whip
- Mallet
- Full size steam table pan

Ingredients

50 3 oz or 4 oz veal cutlets
25 1 oz slices Swiss cheese
25 1 oz slices boiled ham
1 lb butter
1 lb shortening
8 eggs
1 qt milk
2 lb bread flour
2 lb bread crumbs
salt and pepper to taste

Preparation

1. Cut the veal cutlets from sections of a boned leg. Slice the veal very thin against the grain with a French or butcher knife, and flatten with the side of a cleaver.
2. Place the flour in a bake pan and season with salt and pepper.

3. Prepare the egg wash. Break eggs into a bowl, whip slightly with a wire whip, and pour in the milk while continuing to whip. Place in a bake pan.
4. Place the bread crumbs in a bake pan.

Procedure

1. Place one slice of cheese and one slice of ham on 25 of the cutlets. Cover with the remaining 25 cutlets. Pound the edges of the two cutlets together with a mallet until they adhere to each other.
2. Bread by passing through seasoned flour, egg wash, and bread crumbs.
3. Place half butter and half shortening in a skillet and heat. Sauté the cutlets until they are golden brown on each side. Turn with a kitchen fork. When done, remove and let drain. Place in a steam table pan.
4. Serve with melted butter.

Precautions

- Exercise caution when using the knife and cleaver.
- When flattening the cutlets, do not break the fibers of the meat completely. Tears will develop that will harm the appearance.
- Breading will brown and burn quickly. Exercise caution when sautéing. Sauté at a moderate temperature.
- Press firmly when passing the cutlets through the bread crumbs so they will adhere tightly.

Trade Tip

Veal is classified as red meat, but typical lean meat on a veal carcass has a grayish-pink color. Typical calf carcasses have a grayish-red color of lean meat.

Sautéed Veal Chops

The veal chops, cut from the loin or rib, are passed through seasoned flour and cooked in shallow grease until golden brown. They are served with gravy or sauce.

50 servings

Equipment

- Large iron skillet
- Bake pans (2)
- Butcher knife
- Full size steam table pan
- Kitchen fork
- Baker's scale

Ingredients

50 6 oz or 8 oz veal chops
2 lb shortening (variable)
3 lb flour
salt and pepper to taste

Preparation

1. Cut the veal chops from trimmed loins or ribs of veal with a butcher knife.
2. Place the flour in a bake pan and season with salt and pepper.

Procedure

1. Place shortening in the skillet to cover the bottom about ¼″ deep and heat.
2. Pass each chop through the seasoned flour. Dust off the excess and place in the hot shortening, letting the chop fall away.
3. Sauté one side until golden brown, turn with a kitchen fork, and sauté the other side.
4. Remove the chops from the skillet and let drain. Place in a steam table pan.
5. Serve by placing one chop on top of country gravy, brown sauce, or Bercy sauce.

Precautions

- Exercise caution when cutting the chops.
- Exercise caution when placing the chops in the shortening.
- Use caution when turning the chops. Keep the grease from splashing.
- Do not overbrown the chops.

Trade Tip *All veal is USDA inspected for wholesomeness or inspected by state systems, which have standards equal to the federal government. Each calf and its internal organs are inspected for signs of disease. The "Passed and Inspected by USDA" seal ensures the veal is wholesome and free from disease.*

BRAISED AND STEWED VEAL RECIPES

Fricassee of Veal

Fricassee of veal is a stew consisting of meat and sauce. It is generally served on the luncheon menu with either noodles or baked rice.

50 servings

Equipment

- 5 gal. stockpots (2)
- French knife
- Wire whip
- China cap
- Cheesecloth
- Full size steam table pan
- Ladle
- 3 gal. stainless steel container

Ingredients

18 lb veal shoulder
3 gal. water
2 lb shortening or butter
1½ lb flour
yellow color as desired
salt and white pepper to taste

Preparation

1. Cut the boneless veal shoulder into 1″ cubes with a French knife.

Procedure

1. Place the cubes of veal in the stockpot and cover with water.
2. Bring to a boil and remove any scum that may appear. Continue to simmer until the veal is tender.
3. Remove from the heat and strain the veal stock into a stainless steel container through a china cap covered with a cheesecloth.
4. Make a roux (flour and shortening or butter) in a separate stockpot. Cook for 5 min. Do not brown.
5. Add the strained veal stock, whipping vigorously with a wire whip to make a fricassee sauce. Tint the sauce with yellow color. Season with salt and white pepper.
6. Add the cooked veal to the sauce. Place in a steam table pan.
7. Serve 6 oz to 8 oz with buttered noodles or baked rice.

Precautions

- Exercise caution when using the knife.
- Do not make sauce too thin.
- Do not overcook the veal or it will fall apart when served.
- When adding the veal stock to the roux, whip vigorously.

Hungarian Veal Goulash

Hungarian veal goulash is a stew very similar to beef goulash. The difference is that veal is used and it is generally served with sour cream. It is an excellent luncheon item.

50 servings

Equipment

- French knife
- Baker's scale
- Cup measure
- Quart measure
- Spoon measure
- Braiser and cover
- Kitchen spoon or paddle
- Steel skillet
- Steam table pan

Ingredients

18 lb veal shoulder
3 c salad oil
1 lb flour
½ c paprika
2 tsp caraway seed
2 gal. hot brown stock
1 pt tomato puree
6 lb onions
1 qt sour cream
salt and fresh ground pepper to taste

Procedure

1. Place oil in the braiser. Add the diced veal and brown the meat.
2. Add the flour, paprika, and caraway seed. Continue to cook for about 5 min, stirring frequently with a kitchen spoon.
3. Add the hot brown stock and tomato puree. Stir until thick and smooth.
4. Cover the braiser and bake in the preheated oven at 350°F for 1 hr.
5. While the meat is cooking in the oven, sauté the onions in a skillet in additional oil until tender.
6. After the veal has cooked 1 hr, add the sautéed onions and continue to cook for an additional 15 min or until the meat is tender.
7. Season with salt and pepper. Remove from the oven and place in a steam table pan.
8. Serve 6 oz to 8 oz in a casserole. Top each portion with a spoonful of sour cream.

Preparation

1. Cut the boneless veal shoulder with a French knife into 1″ cubes.
2. Prepare the brown stock.
3. Slice onions thin with a French knife.
4. Preheat oven to 350°F.

Precautions

- Exercise caution when using the knife.
- Veal is a tender meat. Check frequently while cooking so it does not overcook.
- When sautéing the onions, do not burn.

Scallopine of Veal Marsala

Scallopine of veal Marsala consists of thin slices of veal cut from the leg, which are sautéed and poached in Marsala wine to increase the delicate flavor of the veal.

25 servings

Equipment

- French or butcher knife
- Baker's scale
- Quart measure
- Iron skillet
- Cleaver
- Kitchen fork
- China cap
- Bake pan
- Full size steam table pan
- 2 qt stainless steel container

Ingredients

50 2 oz slices of veal leg
1 lb flour (variable)
1 lb butter
1 pt Marsala wine
1 qt brown sauce
8 oz onions
salt and pepper to taste

3. Place the flour in a bake pan and season with salt and pepper.

Procedure

1. Press each scallopine into the seasoned flour.
2. Melt the butter in the skillet. When slightly hot, add the scallopine and brown both sides thoroughly.
3. Pour the wine over the scallopine and simmer gently for about 10 min. Remove the meat with a kitchen fork and place in a steam table pan.
4. Add the sautéed onions and brown sauce to the wine still in the skillet. Simmer for 10 min. Strain through a china cap into a stainless steel container. Check the seasoning.
5. Serve two 2 oz scallopine covered with sauce to each portion.

Preparation

1. Mince the onions with a French knife and sauté in a skillet in additional butter as necessary.
2. Prepare the brown sauce.

Precautions

- Exercise caution when using the knife and cleaver.
- Exercise caution when pouring the wine over the meat as the liquid may flare up.
- Do not overbrown the meat.

Savory Veal Stew

This hearty stew is especially delicious on a cold day. It is colorful and filling.

National Cattlemen's Beef Association

24 servings

Ingredients

8 lb veal for stew
1 c all-purpose flour
½ c olive oil
3 large onions
1 head of garlic
3 cans (13¾ oz – 14½ oz) chicken broth
2 tbsp dried crushed thyme leaves
3 lb baby carrots
3 lb small new red potatoes
3 c frozen peas
1½ tsp salt
1½ tsp black pepper

Preparation

1. Cut the veal into 1″ cubes.
2. Coarsely chop the onions.
3. Crush the garlic.
4. Halve the potatoes.

Procedure

1. In a large bowl, combine veal, flour, salt, and pepper. Toss to coat.
2. In a large pan, heat half the olive oil over medium heat until hot. Add one-half of the veal at a time and brown evenly. Add remaining oil as needed. Remove veal.
3. Add onion and garlic to pan. Cook and stir for 1 min. Gradually stir in broth and thyme. Return veal and bring to a boil. Reduce heat to low, cover, and simmer for 45 min.
4. Add carrots and potatoes. Continue simmering, covered, for 30 min. Skim fat.
5. Add peas and heat through. Serve in individual bowls with a small garnish.

Precautions

- Exercise caution when using the knife.
- Be particular when placing the hot stew in individual bowls.

Veal Scallopine with Mushrooms

Veal scallopine with mushrooms is of Italian origin, as are all scallopine preparations. The small, thin scallopine are sautéed and then baked with the mushrooms and the rich Marsala wine sauce.

50 servings

Equipment

- Boning knife
- Cleaver
- French or butcher knife
- Large skillet
- Bake pans
- Kitchen fork
- Kitchen spoon
- Large braiser
- Quart measure
- Baker's scale
- 2 gal. sauce pot
- Full size steam table pan
- Wire whip

Ingredients

100 2 oz scallopine of veal
3 lb flour
1 qt salad oil (variable)
2 lb fresh mushrooms
3 garlic cloves
8 oz flour (variable)
6 qt brown stock
1 pt Marsala wine
salt and pepper to taste

Preparation

1. Slice the mushrooms thin with a French knife and sauté in a saucepan with part of the salad oil.
2. Mince the garlic with a French knife.
3. Prepare the brown stock and keep hot.
4. Place the first amount of flour in a bake pan and season with salt and pepper.

Procedure

1. Place salad oil in the skillet, covering the bottom about ⅛″ deep, and heat.
2. Dredge (coat) the scallopine in the seasoned flour and shake off excess. Place in the hot oil and sauté until both sides are slightly brown. Turn with a kitchen fork.
3. Remove from the skillet and place, overlapping, in a braiser.
4. Sauté the garlic in the remaining salad oil in the same skillet.
5. Add the second amount of flour, making a roux, and cook slightly. Place the roux in a sauce pot.

6. Add the hot brown stock, whipping continuously with a wire whip until slightly thickened and smooth.
7. Add the sautéed mushrooms and the wine. Season with salt and pepper. Simmer for 10 min.
8. Pour the sauce over the sautéed scallopine and bake in a preheated oven at 350°F for about 20 min. Remove from the oven and place in a steam table pan.

9. Serve two scallopine per portion with a generous amount of sauce. Accompany each portion with a scoop of rice pilaf or risotto.

Precautions

- When adding the brown stock, whip vigorously so lumps do not form.
- Do not overcook the veal. It will fall apart when served.
- Exercise caution when handling the knife and cleaver.

Osso Buco

Osso buco is a preparation of crosscut sections of veal shank cooked by the braising method. This is a luncheon preparation.

24 servings

Ingredients

16 lb veal shanks, crosscut 1½" thick
¾ c olive oil
1 head of garlic
4 cans (14½ oz – 16 oz) Italian-style diced tomatoes
4 c dry white wine
4 tsp crushed dried basil leaves

Gremolata

¼ c chopped fresh parsley
2⅔ tsp freshly shredded lemon peel
2 tsp finely chopped garlic
1 tsp salt

Preparation

1. In a large pan, heat half of the oil over medium heat until hot. Add veal shanks (⅓ at a time) and brown evenly, turning occasionally. Add remaining oil as needed. Remove shanks and season with 1 tsp salt.
2. Add onions, carrots, and garlic to pan. Cook and stir 6 min – 8 min. Add tomatoes, wine, and basil. Return shanks and bring to a boil. Reduce heat to low, cover, and simmer 1½ hr.
3. Meanwhile, combine gremolata ingredients. Set aside.
4. Remove shanks and skim fat. Cook liquid over high heat until slightly thickened, stirring occasionally.

National Cattlemen's Beef Association

5. Spoon sauce over shanks and sprinkle with gremolata. Serve with remaining sauce.

BROILED VEAL RECIPES

Italian Veal Steaks with Marinated Vegetables

Italian veal steaks are broiled and plated with colorful vegetables for an eye-appealing entree.

24 servings

Ingredients

12 veal steaks ¾" thick (approx. 12 lb)
6 cans (14½ oz) artichoke hearts
3 lb medium mushrooms
12 large bell peppers
8 c dried pitted onions

Marinade

4½ c fat-free Italian dressing
1 tbsp crushed red pepper

Preparation

1. Drain and halve the artichoke hearts.
2. Cut and core the bell peppers. Cut into 1" pieces.

Procedure

1. Combine marinade ingredients. Reserve 3 c. Place veal steaks and remaining marinade in a plastic bag. Close the bag securely and marinate in the refrigerator 2 hr or overnight.

2. In a large bowl, combine artichokes, mushrooms, bell peppers, and reserved marinade. Toss to coat. Cover and refrigerate up to 2 hr.

3. Soak 24 10″ bamboo skewers in water for 10 min. Drain. Thread vegetables onto skewers, reserving marinade.

4. Remove steaks from marinade. Discard marinade. Place steaks and vegetable kabobs in broiler pan 4″ from heat. Broil 14 min – 16 min, turning steaks once and vegetables occasionally.

5. Remove vegetables from skewers. Return vegetables to reserved marinade. Add olives and toss. Remove bones and slice veal into thin slices.

6. Serve veal strips with vegetables.

Precautions

- Exercise caution when using the knife.
- Cut vegetables as uniformly as possible so they cook evenly.

National Cattlemen's Beef Association

Trade Tip

To store raw veal, put it in disposable plastic bags to contain leakage. Store the bags in a refrigerator at 40°F for 3 days to 5 days for veal chops and roasts, and 1 day to 2 days for ground veal or stew meat. For longer periods, freeze the veal at 0°F for 4 months to 6 months for veal chops and roasts, and 3 months to 4 months for ground veal or stew meat.

BOILED VEAL RECIPE

Veal Paprika with Sauerkraut

Veal paprika with sauerkraut is a stew. It is a combination of veal cubes cooked in sauerkraut and served with sour cream. This is an excellent item for the luncheon menu.

50 servings

Equipment

- French knife
- Baker's scale
- Spoon measure
- Quart measure
- Braiser and cover
- Kitchen spoon
- Full size steam table pan

Ingredients

15 lb veal shoulder
1 lb butter
6 lb onions
10 lb sauerkraut with juice
3 tbsp salt
5 tbsp paprika
1 tbsp fresh ground pepper
1 qt thick sour cream
3 garlic cloves

Procedure

1. Place the butter in the braiser. Add the onions and garlic and cook slowly until tender.
2. Add the diced veal and cook until the meat is slightly brown. Stir occasionally with a kitchen spoon.
3. Add the sauerkraut, paprika, and seasoning. Cover and simmer gently until the meat is tender. Remove from the range and place in a steam table pan.
4. Serve 6 oz to 8 oz in casseroles. Top each portion with a spoonful of thick sour cream.

Preparation

1. Cut the boneless shoulder of veal into 1″ cubes with a French knife.
2. Slice onions and mince garlic with a French knife.

Precautions

- Exercise caution when using the knife.
- Stir frequently while cooking to avoid scorching.

VEAL VARIETY MEAT RECIPE

Veal Kidney and Brandy Stew

Veal kidney and brandy stew consists of veal kidneys stewed in a rich sauce highly flavored with brandy. Serve with risotto or rice pilaf.

25 servings

Equipment

- French knife
- Braiser
- Cup measure
- Kitchen spoon
- Scissors
- Full size steam table pan
- Quart measure

Ingredients

25 veal kidneys
1 c brandy
1 c onions
1½ c butter
2 lb mushrooms
1 c flour
1 c dry white wine
1 qt brown stock
⅓ c parsley
salt and fresh ground pepper to taste

Procedure

1. Place the butter in the braiser and melt.
2. Add the kidneys and sauté until slightly brown.
3. Add the onions and mushrooms and continue to sauté until tender. Stir frequently with a kitchen spoon.
4. Add the flour and blend thoroughly. Cook for 5 min.
5. Add the wine and brown stock and cook, stirring frequently with a kitchen spoon until thickened.
6. Add the brandy and chopped parsley. Simmer until the kidneys are tender. Remove from the range.
7. Season with salt and freshly ground pepper. Place in a steam table pan.
8. Serve 6 oz to 8 oz in a shallow casserole or shirred egg dish accompanied with risotto or rice pilaf.

Preparation

1. Wash kidneys and remove membranes. Dice into ½" cubes with a French knife. Remove fat and tubes with scissors.
2. Slice the mushrooms with a French knife.
3. Prepare the brown stock.
4. Mince the onions with a French knife.
5. Chop the parsley with a French knife.

Precautions

- Exercise caution when using the knife.
- If the kidneys have a strong odor, soak in salt water for 45 min before cutting.
- Stir the mixture gently when adding the flour and the liquid so the mushrooms do not break.

Pork Preparation

## Chapter 21

National Pork Producers Council

Pork is the meat of hogs usually less than a year old. The best pork on the market comes from hogs 6 to 8 months of age. Pork ranks second to beef in total meat consumption in the United States. In addition, pork is the only popular meat of which all wholesale cuts can be cured (chemically processed). Over two-thirds of all the pork in the U.S. is marketed in the cured form. Pork has a very high fat content. However, if it is properly trimmed and prepared, the fat content is reduced.

Pork is a tender meat because the animal is very young when it is marketed. A sow produces her first litter when she is a year old, so she can still be marketed at a young age. This is not possible with other animals such as cows or sheep. Most hogs are marketed during the fall and winter months.

Pork is a light-colored meat and has a delicate flavor. Pork is always cooked well-done for maximum flavor and to prevent trichinosis from being transmitted to customers.

PORK GRADING

Pork is graded in order of desirability as *U.S. 1*, *U.S. 2*, and *U.S. 3*. The grade of pork is less important than the grade of other animals because all pork is from young animals.

Barrows (male hogs castrated when young) and *gilts* (immature female hogs) are graded U.S. 1. Young sows are graded U.S. 2. Old sows, which have soft fat and oily carcasses, are generally graded U.S. 3. Pork grading, like beef grading, is based on quality and yield. Young hogs usually have a very high proportion of usable meat, so the yield grade is high.

COMMERCIAL CUTS

Pork is commonly marketed as cuts rather than by the quarter, side, or carcass. The quarter, side, and carcass have many extra cuts that are not desirable for use in the commercial kitchen. Only one-third of all the pork marketed is sold as fresh pork. The majority of pork cuts are cured or smoked. All pork cuts can be processed by these two methods. Many pork cuts are more desirable when cured or smoked. Pork is marketed by cuts also because pork spoils more quickly than other edible meats. Too much surplus pork on hand could be very costly for the butcher or restaurant operator.

National Pork Producers Council

Pork is naturally tender and can be prepared many ways.

WHOLESALE CUTS AND COOKING METHODS

Hogs are blocked out for 11 different wholesale cuts. These cuts range from very desirable cuts, which are more expensive, to less desirable cuts, which are less expensive. The following is a list of the wholesale pork cuts.

Loin: The loin consists of the rib end, loin end, and tenderloin. The loin cut extends along the greater part of the backbone from about the third rib, through the rib and loin area. It is considered one of the leanest and most popular pork cuts. Pork loins under 15 lb are considered the best because they are tender and have more flavor.

In the commercial kitchen the tenderloin, the most tender of all pork cuts, is taken from the underside of the loin. It is a fairly long, tapered, narrow strip of lean meat and weighs about 8 oz to 12 oz. Tenderloins are generally saved until enough are available for placement on the menu. Tenderloins can be broiled, sautéed, braised, roasted, or fried, and served in a variety of ways.

The loin is generally separated into the rib end and loin end by cutting through the last two ribs. This leaves one rib on the loin end. All bones are then removed from the cuts except the rib bones,

which are left on the rib end mainly for appearance when the cut is served. These two cuts are generally placed on the menu as roast loin of pork, broiled or sautéed pork chops, or pork cutlets.

Ham: Ham is the thigh and buttock of the hog. It contains a high proportion of lean meat. Ham is marketed in two forms: smoked and cured, and fresh. The most popular form is to cure the ham in a solution of salt, sugar, and sodium nitrate and then smoke it. The skin may be left on or removed. Some hams, such as Virginia and country hams, are cured but not smoked. Hams are marketed in a variety of forms, such as partly cooked, cooked, raw smoked, canned, boneless, tenderized, and shankless. In the commercial kitchen, hams are baked, cut into steaks and broiled or panbroiled, boiled, sliced, and used for sandwiches.

Bacon: Bacon is the cured and smoked belly of the hog. The size of the belly determines the quality of the bacon. Small bellies from very young hogs are the most desirable. High-quality bacon should contain about 40% to 45% fat and 5% rind.

Bacon is a versatile item in the commercial kitchen. It is used for sandwiches, garnishes, appetizers, and entree dishes. It is, of course, a featured item on many breakfast entrees.

Backribs: Backribs are the ribs extending from the backbone. They are separated from the loin and sirloin. Backribs are typically barbecued.

Spareribs: Spareribs are the whole rib section removed from the belly (side) of the hog carcass. They are located on top of the bacon section. Spareribs have little meat, but the meat is tender and has an excellent flavor. The best quality spareribs weigh about 3 lb or less. Spareribs can be served as an appetizer or as an entree and are commonly prepared boiled, barbecued, broiled, and baked.

Boston butt: The Boston butt is a square cut of shoulder located just above the lower half of the shoulder (picnic or callie). It contains the bladebone and a large portion of lean meat. It weighs about 6 lb to 10 lb and contains meat that has a long fiber and a coarse grain. Boston butt is usually sold fresh when the bone is left in. When boneless, the top section is usually cured by smoking and sold as cottage ham. The Boston butt is an excellent cut to choose when a preparation calls for a solid piece of meat that is reasonably priced. The Boston butt can be roasted, boiled, or cut into cutlets and sautéed or fried.

PORK PREPARATIONS

National Pork Producers Council

PORK CHOPS

National Pork Producers Council

ROLLED PORK ROAST

Burgers' Smokehouse

SPIRAL-SLICED COUNTRY HAM

In addition to being popular on the breakfast menu, pork is offered on the luncheon and dinner menus where it may be prepared by a variety of cooking methods.

Picnic or *callie*: This is the lower half of the shoulder of the hog carcass. The picnic ham resembles the full ham in shape, but is smaller and contains more bone and less lean meat. It is marketed in two forms: fresh and smoked. Picnic hams can be purchased at a very low cost per pound, but a large percentage of picnic hams is bone. Fresh picnics are a good choice when preparing such items as chop suey, pork patties, and country sausage. Smoked picnics are a good choice for creamed ham, deviled ham, and ham salad. The average weight of a picnic ham is 3 lb to 6 lb.

Jowls: Jowls, or *jowl bacon*, are the cured and smoked cheek meat of large hogs. They are cured and smoked in the same manner as bacon from the belly and possess a similar flavor. However, the eating qualities are much poorer. The jowl weighs about 2 lb to 4 lb and is used for bacon crackling, seasoning, or for flavoring such items as baked beans and string beans. Jowls can be purchased at a very reasonable price.

Feet: The feet are the cut below the shank. Both front and hind feet are marketed. Feet average from ¾ lb to 1½ lb and are usually sold in a pickled form, although they may also be purchased fresh and smoked.

Pig's feet are commonly prepared as broiled pig's feet or made boneless and served on a cold plate.

Hock: The ham hock is the knee joint of the hog. Hocks are removed from both the front and hind legs. They have little meat, but good flavor. Hocks are purchased fresh or smoked and are popular when cooked with sauerkraut.

Fat back and *clear plate*: These two cuts are very similar, except in some cases the clear plate may contain a few more strips of lean meat running through it. Fat back and clear plate are both fairly solid, rectangular slabs of fat extracted from the surface of the hog's carcass. Lard is usually rendered from these cuts. However, when lean meat is present, it is cured in salt and marketed as salt pork. In the commercial kitchen, salt pork is used in preparations as a flavoring ingredient or it is used to add juice to a preparation. For example, beef tenderloin is often larded. To lard an item, the salt pork is cut into thin strips and drawn, by use of a larding needle, through the meat to increase the juiciness when it is roasted.

Pork variety meats: These meats include liver, brains, and pig tails. However, most variety meats are used in the preparation of various types of sausage. Pork variety meats are not commonly used in the commercial kitchen.

Trade Tip

The safety and wholesomeness of America's pork products is overseen by the Food Safety and Inspection Service of the United States Department of Agriculture. The FSIS-USDA is involved in every aspect of developing wholesome meat products. An FSIS-USDA inspector observes the hog before, during, and after slaughter, and during pork processing.

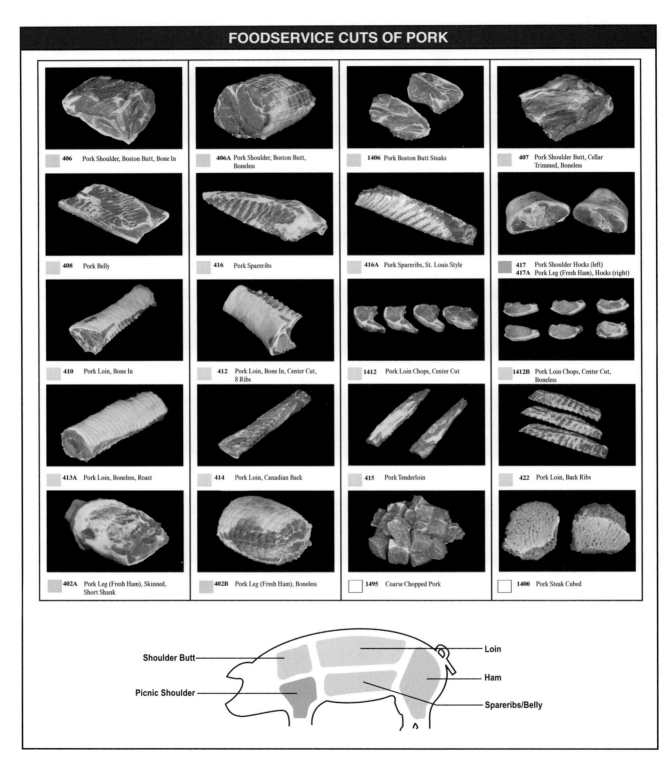

FOODSERVICE CUTS OF PORK

406 Pork Shoulder, Boston Butt, Bone In

406A Pork Shoulder, Boston Butt, Boneless

1406 Pork Boston Butt Steaks

407 Pork Shoulder Butt, Cellar Trimmed, Boneless

408 Pork Belly

416 Pork Spareribs

416A Pork Spareribs, St. Louis Style

417 Pork Shoulder Hocks (left)
417A Pork Leg (Fresh Ham), Hocks (right)

410 Pork Loin, Bone In

412 Pork Loin, Bone In, Center Cut, 8 Ribs

1412 Pork Loin Chops, Center Cut

1412B Pork Loin Chops, Center Cut, Boneless

413A Pork Loin, Boneless, Roast

414 Pork Loin, Canadian Back

415 Pork Tenderloin

422 Pork Loin, Back Ribs

402A Pork Leg (Fresh Ham), Skinned, Short Shank

402B Pork Leg (Fresh Ham), Boneless

1495 Coarse Chopped Pork

1400 Pork Steak Cubed

Shoulder Butt

Picnic Shoulder

Loin

Ham

Spareribs/Belly

North American Meat Processors Association

NAMP/IMPS (North American Meat Processors Association/Institutional Meat Purchase Specifications) specifies meat cuts by number.

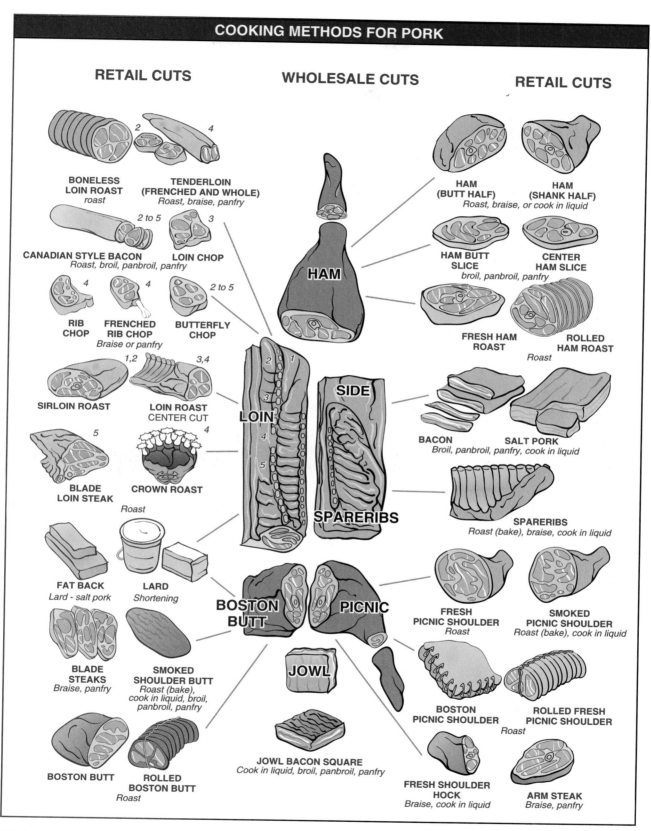

COOKING METHODS FOR PORK

RETAIL CUTS   WHOLESALE CUTS   RETAIL CUTS

BONELESS LOIN ROAST
roast

2   4

TENDERLOIN (FRENCHED AND WHOLE)
Roast, braise, panfry

2 to 5   3

CANADIAN STYLE BACON
Roast, broil, panbroil, panfry

LOIN CHOP

4   4   2 to 5

RIB CHOP   **FRENCHED RIB CHOP**   **BUTTERFLY CHOP**
Braise or panfry

1,2   3,4

SIRLOIN ROAST   **LOIN ROAST CENTER CUT**

5   4

BLADE LOIN STEAK   **CROWN ROAST**
Roast

FAT BACK   **LARD**
Lard - salt pork   *Shortening*

BLADE STEAKS   **SMOKED SHOULDER BUTT**
Braise, panfry   *Roast (bake), cook in liquid, broil, panbroil, panfry*

BOSTON BUTT   **ROLLED BOSTON BUTT**
Roast

HAM

SIDE

LOIN

2   1
3
4
5

SPARERIBS

BOSTON BUTT   **PICNIC**

JOWL

JOWL BACON SQUARE
Cook in liquid, broil, panbroil, panfry

HAM (BUTT HALF)   **HAM (SHANK HALF)**
Roast, braise, or cook in liquid

HAM BUTT SLICE   **CENTER HAM SLICE**
broil, panbroil, panfry

FRESH HAM ROAST   **ROLLED HAM ROAST**
Roast

BACON   **SALT PORK**
Broil, panbroil, panfry, cook in liquid

SPARERIBS
Roast (bake), braise, cook in liquid

FRESH PICNIC SHOULDER   **SMOKED PICNIC SHOULDER**
Roast   *Roast (bake), cook in liquid*

BOSTON PICNIC SHOULDER   **ROLLED FRESH PICNIC SHOULDER**
Roast

FRESH SHOULDER HOCK   **ARM STEAK**
Braise, cook in liquid   *Braise, panfry*

Wholesale (primal) cuts are commonly purchased for use in the commercial kitchen.

PORK TERMINOLOGY

Chitterlings are the large and small intestines of the hog. They are emptied and thoroughly rinsed before boiling or frying.

Canadian bacon is the trimmed, pressed, and smoked boneless loin of pork. The average weight of Canadian bacon is 4 lb to 6 lb. It can be cooked by baking, sautéing, or broiling.

Cottage ham is the smoked, boneless meat extracted from the blade section of the Boston butt. The blade section is the section of meat that is located above the flat bladebone. This section is usually compact and fairly lean. Cottage hams weigh about 1½ lb to 4 lb. For best results, cottage hams should be boiled or steamed.

Head cheese is the jellied, spiced, pressed meat from the hog's head. It is covered with a natural casing and sold as a luncheon meat.

Loin backs are the rib bones removed from the loin. In some cases, loin backs are used in place of spareribs because they contain more meat.

Salt pork is fat back or clear plate that has been cured in salt. It is often used as a seasoning for beans.

PREPARING PORK LOIN

LOIN

A knife is used to separate the rib and loin.

PREPARING PORK TENDERLOIN

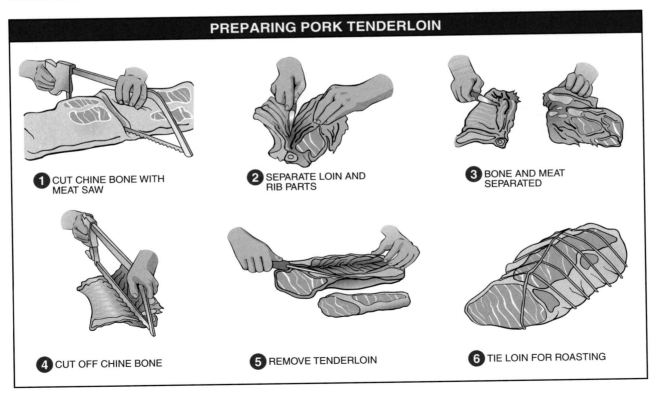

1 CUT CHINE BONE WITH MEAT SAW

2 SEPARATE LOIN AND RIB PARTS

3 BONE AND MEAT SEPARATED

4 CUT OFF CHINE BONE

5 REMOVE TENDERLOIN

6 TIE LOIN FOR ROASTING

The loin is a wholesale cut extending along the backbone of the carcass.

Fresh ham is the unsmoked ham cut from the hind leg of the hog.

Suckling pig is a baby pig 4 to 6 weeks old. A suckling pig weighs from 20 lb to 35 lb dressed. They are purchased with the head on and are priced per pig rather than by the pound. Suckling pigs are roasted whole and are used for ornamental purposes as a centerpiece for banquet tables, etc.

Curing is the salting of an item to retard the action of bacteria and to preserve the meat. Before refrigeration, curing was a major method used to preserve pork.

Virginia ham is a ham that is cured in salt for a period of about 7 weeks. It is rubbed with a mixture of molasses, brown sugar, sodium nitrate, and pepper, and then cured for 2 weeks more. The ham is then hung hock down for a period of 30 days to a year.

Prosciutto ham is very dry and hard. It comes ready to eat and is usually sliced very thin and used in the making of hors d'oeuvres and appetizers. Prosciutto ham originated in Italy.

Burgers' Smokehouse

The lower most wasteful portion of the ribs is removed in St. Louis Style Ribs.

Trade Tip

The National Pork Producers Council offers nutrition information, research results, industry news, recipes, and seasonal menus at www.nppc.org.

ROASTED OR BAKED PORK RECIPES

Roast Pork Loin

Roast pork loin is usually served during the winter months of the year because of its high fat content. An order consists of two slices: one with rib and one with loin. Roast pork can be served on either the luncheon or dinner menu, with or without bread dressing.

50 servings

Equipment

- Wire whip
- Roast pan
- Quart measure
- China cap
- Kitchen fork
- French knife
- Cup measure
- Kitchen spoon
- Baker's scale
- Boning knife
- 2 gal. stainless steel container
- 2 gal. sauce pot
- Full size steam table pan

Ingredients

25 lb pork loin
1 lb onions
½ lb celery
½ lb carrots
10 oz bread flour (variable)
5 qt brown pork stock
1 tsp whole cloves
salt and pepper to taste
⅓ c tomato puree

Preparation

1. Trim the pork loin by separating the rib end from the loin end. Remove the tenderloin with a boning knife. Save the tenderloin for use in another preparation. Remove all bones but the rib bones from both the loin end and the rib end with a boning knife.

2. Season the loin with salt and pepper. Sprinkle on whole cloves.
3. Cut the rough garnish (onions, celery, and carrots) with a French knife.
4. Preheat the oven to 400°F.
5. Prepare the brown pork stock.

Procedure

1. Place the seasoned pork loin in a roast pan, fat side down. Roast in the preheated oven at 400°F.
2. Roast until the meat becomes brown. Remove excess fat from the pan and save for later use in making the gravy. Reduce oven heat to 325°F.
3. Add the rough garnish and continue to roast until the vegetables become light brown. Turn the meat occasionally with a kitchen fork.
4. Add just enough water to cover the bottom of the roast pan. Continue to roast until meat is well done.
5. Remove the meat from the roast pan and place in a steam table pan. Keep covered in a warm place until ready to slice. Pour drippings left in the roast pan into a stainless steel container. Deglaze the roast pan by adding the brown pork stock and bringing it to a boil on the range. Pour this liquid into the stainless steel container.

6. Place the fat that was removed from the pork roast in a sauce pot. Place on the range and heat.
7. Add the flour to the pork fat, making a roux, and cook for 5 min. Stir occasionally with a kitchen spoon.
8. Add the hot stock that was placed in the stainless steel container and whip vigorously with a wire whip until slightly thick and smooth. Add the tomato puree and let simmer for about 30 min.
9. Check color and seasoning. If the gravy is too light in color, add caramel coloring to darken. Strain through a china cap back into the stainless steel container.

10. Slice the pork roast with a French knife. Serve two slices per portion (one rib and one loin) with or without bread dressing. Cover with pork and gravy.

Precautions

- Exercise caution when using the knives.
- Be sure the pork loin is well done before it is removed from the oven. Test by inserting a kitchen fork into the meat. The fork should penetrate easily.
- Do not overcook the pork loin or it will crumble when sliced.
- Use caution when straining the gravy through the china cap; it may splash.

Roast Fresh Ham or Picnic

Fresh ham makes an excellent roast because it contains a large portion of lean meat and enough fat covering for a juicy piece of meat.

50 servings

Equipment

- Roast pan
- French knife
- Boning knife
- Cup measure
- Quart measure
- Baker's scale
- Wire whip
- China cap
- Kitchen fork
- Kitchen spoon
- 2 gal. stainless steel container
- 2 gal. sauce pot
- Butcher's twine
- Full size steam table pan

Ingredients

25 lb fresh ham or picnic
1 lb onions
½ lb celery
½ lb carrots
1 lb bread flour (variable)
5 qt brown pork stock
1 tsp whole cloves
⅓ c tomato puree
salt and pepper to taste
2 lb pork fat

Preparation

1. Remove all the bones from the fresh ham or picnic with a boning knife. Trim off excess fat and tie the meat with butcher's twine.
2. Cut the rough vegetable garnish (onions, celery, and carrots) with a French knife.
3. Preheat the oven to 375°F.
4. Prepare the brown pork stock.

Procedure

1. Place the tied roast in the roast pan, fat side down. Season with salt and pepper and add whole cloves.
2. Roast in the preheated oven at 375°F and brown thoroughly. Remove excess fat and save.

3. Reduce the oven heat to 325°F and add the rough garnish. Continue to roast until the garnish is slightly brown. Turn the roast occasionally with a kitchen fork.
4. Add a small amount of water to cover bottom of roasting pan.
5. Continue to roast until the meat is done, approximately 3 hr, depending on the size of the roast. Remove meat from the roast pan and place in a steam table pan. Cover and keep warm.
6. Pour the drippings left in the roast pan into a stainless steel container. Deglaze the roast pan by adding brown pork stock and bringing it to a boil on the range. Pour into the stainless steel container.
7. Place the pork fat in a sauce pot and heat.
8. Add the flour to the pork fat, making a roux, and cook for about 5 min. Stir occasionally with a kitchen spoon.
9. Add the hot brown pork stock and whip vigorously with a wire whip until the gravy is slightly thick and smooth. Add the tomato puree and let simmer for about 30 min.
10. Check the color and seasoning. If the color is light, darken with caramel coloring. Strain through a fine china cap back into the stainless steel container.
11. Remove the butcher's twine and slice with a French knife or on the slicing machine. Serve 3 oz to 4 oz per portion with the pork gravy.

Precautions

- Use caution when cutting the meat and rough vegetable garnish.
- Be sure the fresh ham is well done before removing it from the oven.
- Do not overcook or the meat will crumble when sliced.
- Use caution when straining the gravy; it may splash.
- If grease forms on the surface of the gravy, dip it off with a ladle before serving.

Trim the exterior fat from pork when roasting. Season the roast with herbs and other seasonings. Place the roast in a shallow roasting pan, uncovered, and bake to an internal temperature of 155°F, or 165°F if stuffed. Allow the pork roast to rest for 10 min to 15 min before slicing.

Ham Loaf

Ham loaf is a ground meat entree. The ground cured ham is mixed with ground fresh pork to help it bind better when baked. This entree is a luncheon item and is generally served with a fruit or tomato sauce.

50 servings

Equipment

- Baker's scale
- 4″ × 9″ loaf pans (5)
- Roast pans (2)
- Quart measure
- Spoon measure
- Meat grinder
- Wire whip
- Mixing container
- 1 gal. stainless steel container
- Kitchen spoon
- Full size steam table pan

Ingredients

5 lb lean fresh picnic
8 lb cured smoked ham
1 lb 8 oz bread crumbs
1 qt milk
14 eggs
1 tsp pepper

Preparation

1. Cut the fresh picnic and smoked ham with a French knife into pieces that will pass through the meat grinder.
2. Grind the meat in a meat grinder using the medium chopper plate.
3. Break the eggs into a stainless steel container and beat with a wire whip.
4. Preheat the oven to 350°F.
5. Coat the inside of the loaf pans with salad oil.

Procedure

1. Place the bread crumbs, milk, beaten eggs, and pepper in a mixing container. Blend thoroughly with a kitchen spoon and let set until the crumbs absorb the liquid.
2. Add the ground meat and mix thoroughly by hand.
3. Pack into the oiled loaf pans, then set the loaf pans in roast pans containing about 1″ of water and place in the oven.
4. Bake in the preheated oven at 350°F for about 2 hr or until the loaf is baked through and is firm.
5. Remove from the oven when done. Let cool and remove the loaf pans, then place in the refrigerator overnight. Slice with a French knife or on a slicing machine into 5 oz to 6 oz portions. Reheat in a steam table pan.
6. Serve one slice per portion with fruit, tomato, or raisin sauce.

Precautions

- Exercise caution when using the knife.
- Always bake the loaf the day before serving. It will be firmer for slicing and serving.
- Pack the ham mixture firmly into the loaf pans. The loaf will bake more solidly.
- Be gentle when handling the slices and reheating them; they may break.

Barbecued Spareribs

This is the most popular method of preparing spareribs. The tangy barbecue sauce is a welcome addition to the delicate flavor of the spareribs. It is a popular item in the commercial kitchen as well as the backyard grill and is served both as an appetizer and an entree.

50 servings

Equipment

- French knife
- Quart measure
- Portion scale
- Sheet pans
- 10 gal. stockpot
- Kitchen fork

Ingredients

50 12 oz pieces pork spareribs
1 pt salad oil (variable)
2 gal. barbecue sauce
paprika as needed

National Pork Producers Council

Preparation

1. Weigh each sparerib section on a portion scale. Cut the trimmed spareribs with a French knife into 12 oz pieces.
2. Prepare barbecue sauce.
3. Preheat oven to 350°F.

Procedure

1. Place the ribs in the stockpot, cover with water, and bring to a boil.

2. Reduce heat by bringing stockpot to the side of the range. Let ribs simmer for about 30 min.
3. Remove ribs from the stockpot and place on sheet pans.
4. Brush oil over each rib and sprinkle with paprika.
5. Brown each side slightly under the broiler. Turn with a kitchen fork. Place in roast pans with the outside of the ribs turned upward.
6. Cover each rib with the barbecue sauce. Bake in the preheated oven at 350°F for about 1 hr or until all ribs are tender.
7. Remove the ribs from the oven. Cut between every other rib and place in a steam table pan.
8. Serve four or five ribs per portion if serving as an entree, two ribs if serving as an appetizer. Accompany each serving with barbecue sauce.

Precautions

- Exercise caution when using the knife.
- When boiling the ribs, do not overcook. Serving portions will be lost and ribs will be difficult to handle.
- When baking the ribs, baste at least twice during the baking period.

Baked sugar cured ham has long been a favorite on both the luncheon and dinner menu. The ham is first boiled and then baked with a covering of honey and sugar to give it its characteristic sweet taste.

Baked Sugar Cured Ham

50 servings

Equipment

- 10 gal. stockpot
- Roast pans
- Kitchen fork
- Boning knife
- French knife

Ingredients

25 lb smoked sugar cured ham
1 qt honey
1 lb brown sugar
2 tsp ground cloves

Preparation

1. Preheat oven to 350°F.

Procedure

1. Place the ham in the stockpot. Cover with hot water and place on the range. Bring to a boil and let simmer for approximately 2 hr.
2. Remove the hams from the water with a kitchen fork. Take off the rind and, using a boning knife, remove the aitchbone, which lies across the upper part of the ham.
3. Trim off some of the excess fat for even shaping and score with a French knife.
4. Place the ham in the roast pans and spread the honey over each ham.
5. Mix the ground cloves with the brown sugar and sprinkle over the ham. Add about ¼″ of water to the bottom of the roast pan and bake in the 350°F oven.
6. Bake until ham is golden brown. Remove and let cool slightly.
7. Place in ham rack and carve to order. Serve 3 oz to 4 oz per portion with cider, raisin, raisin-cranberry, or fruit sauce.

Precautions

- Many types of hams are available on the market. Check to be sure that the ham is sugar cured before boiling.
- Use caution when using the knives.
- Do not overcook the ham while boiling. Remove from the stock pot when the shank bone becomes loose. If the ham is overcooked, it will be difficult to handle, slice, and serve. Overcooking also results in loss of serving portions.

Sweet and Sour Ham Balls

Sweet and sour ham balls are a luncheon item. The ham balls are baked in a sweet and sour sauce to give them a very desirable taste.

50 servings

Equipment

- Bake pans
- Quart measure
- Baker's scale
- French knife
- Cup measure
- Spoon measure
- Mixing container
- 4 qt saucepan
- Kitchen spoon
- Food grinder
- No. 12 ice cream scoop
- Wire whip

Ingredients

25 lb smoked, uncooked ham
8 lb fresh, boneless picnic
2 qt bread crumbs (variable)
10 eggs
1 qt milk
2 lb 8 oz dark brown sugar
1 c onions
1 c celery
1 c butter
1 qt water
1 qt cider vinegar
½ c dry mustard

Preparation

1. Mince the onions and celery with a French knife.
2. Cut the ham and pork into thin strips that will pass through the food grinder.
3. Beat the eggs slightly with a wire whip.
4. Preheat the oven to 325°F.

Procedure

1. Place the bread crumbs in a mixing container. Add the milk and let soak until the crumbs absorb the liquid.
2. Add the ham and pork and mix thoroughly by hand.
3. Grind the mixture twice on the food grinder using the medium chopper plate. Add the beaten eggs and mix thoroughly by hand.

4. With a No. 12 ice cream scoop, form into balls and place in the bake pans.
5. Place the butter in the saucepan and heat. Add the onions and celery and sauté until slightly tender.
6. Add the water, vinegar, brown sugar, and dry mustard. Bring to a boil, stirring constantly with a kitchen spoon.
7. Pour the liquid mixture over the ham balls. Bake in the preheated oven at 325°F until the meat is well done. Baste the balls frequently while baking.

8. Serve two balls per portion topped with the sweet and sour sauce in which the balls were baked. Accompany each order with sautéed apples, noodles, or applesauce.

Precautions

- Use caution when mincing the onions and celery.
- When sautéing the onions and celery, do not brown.
- Cook the ham balls well done.

Trade Tip *Pork provides many of the nutrients necessary for healthy living. It is recommended that we eat two to three 3 oz servings of meat or meat products per day. Pork is a good meat choice because it is high in protein and low in fat when trimmed.*

Baked Stuffed Pork Chops

Baked stuffed pork chops are thick chops cut from the rib or loin end of the pork loin. A pocket is cut into the side of the chop and it is stuffed with a forcemeat mixture. The stuffing adds a nice taste variation to this entree.

50 servings

Equipment

- Boning knife
- French or butcher knife
- Large iron skillet
- 3 gal. stainless steel container
- Roast pan
- Food grinder
- Cup measure
- Baker's scale
- Kitchen fork
- Saucepan
- Ladle
- China cap
- Mixing container
- Wire whip
- Bake pan
- Steam table pan

Ingredients

- 50 pork chops
- 3 lb pork picnic
- 2 lb veal shoulder
- ¾ oz salt
- ¼ oz pepper
- ¼ oz sage
- 1 lb bread, fresh or dried
- 3 c milk
- ⅓ c parsley
- 8 oz onions
- 6 oz celery
- 6 oz shortening or butter
- 8 oz bread crumbs
- 4 egg yolks
- 2 gal. brown sauce
- 3 lb flour
- 1 qt salad oil or melted shortening (variable)
- salt and pepper to taste

Preparation

1. Cut the pork chops from trimmed pork loins. Cut about 1″ thick with a French or butcher knife. Slit the side and cut a pocket in the flesh of each chop with a boning knife. Make the pocket as large as possible without cutting through the flesh on either side of the pocket.
2. Mince the onions and celery with a French knife.
3. Prepare the 2 gal. of brown sauce.
4. Separate the eggs and beat the yolks slightly with a wire whip. (Save the whites for use in another preparation.)
5. Crumble the bread by hand.
6. Cut the picnic and veal shoulder into strips with a French or butcher knife so they will pass through the food grinder.
7. Chop the parsley with a French knife and wash.
8. Put the flour in a bake pan and season with salt and pepper.
9. Preheat the oven to 350°F.

Procedure

1. Place the shortening or butter in the saucepan and heat. Add the onions and celery and sauté until slightly tender. Remove from the range and let cool.
2. Soak the crumbled bread in the milk until the milk is absorbed.
3. Place the pork, veal, salt, pepper, sage, soaked bread, sautéed onions, and celery in a mixing container and mix thoroughly.
4. Grind this mixture twice in the food grinder using the fine chopper plate.
5. Add the slightly beaten egg yolks, bread crumbs, and parsley and mix thoroughly. Check seasoning.
6. Stuff this forcemeat mixture into the pocket that was cut into the side of each chop.
7. Pass each chop through the seasoned flour and pat off the excess.
8. Place salad oil or shortening in the iron skillet and heat. Add the chops and brown slightly on both sides. Turn with a kitchen fork.
9. Line the sautéed chops, cut edge up, in a roast pan.
10. Ladle the hot brown sauce over the chops and bake in the preheated oven at 350°F for approximately 1½ hr or until the chops are tender and well done.
11. Remove from the oven and place the chops in a steam table pan. Strain the sauce in the roast pan into a stainless steel container.
12. Serve one chop per portion with a generous amount of sauce over the top.

Precautions

- Use caution when cutting chops and mincing vegetables.
- When sautéing chops, use caution so the forcemeat will not come out of the chops.
- When baking, baste the chops with the sauce frequently.

Apple-Stuffed Pork Chops

The chops are cut from the rib end of the loin for apple-stuffed pork chops. A pocket is cut into the meaty side of each chop and stuffed with an apple mixture. The apple flavor increases the delicate flavor of the pork.

50 servings

Equipment

- Boning knife
- French or butcher knife
- Mixing container
- Iron skillet
- Spoon measure
- Cup measure
- Quart measure
- Bake pans
- 1 qt saucepans
- Colander
- Kitchen fork
- Steam table pan

Ingredients

50 pork chops, cut from rib end of loin, 1″ thick
1 c onions
1 c celery
6 tbsp bacon grease
1½ qt soft bread crumbs
1 No. 10 can sliced apples
1 c seedless raisins
2 tsp poultry seasoning
1 tsp salt
¼ tsp pepper
1 qt salad oil or melted shortening (variable)

Preparation

1. Cut the chops 1″ thick from the trimmed rib end of a pork loin with a French or butcher knife. Cut a pocket in the meaty side of each chop with a boning knife.
2. Prepare the soft bread crumbs by rubbing fresh bread across the bottom of a colander.
3. Mince the onions and celery with a French knife.
4. Preheat the oven to 350°F.
5. Drain and chop the apples with a French knife.

Procedure

1. Place bacon grease in a saucepan and heat. Add the minced onions and celery. Sauté until slightly tender.
2. Place in a mixing container and add the bread crumbs, apples, raisins, poultry seasoning, salt, and pepper. Toss with the hands gently until thoroughly mixed.
3. Place about ¼ c of stuffing in the pocket of each chop.
4. Brown the chops lightly on both sides in the iron skillet in oil or shortening. Using a kitchen fork, place the chops in bake pans, meaty side up.
5. Bake in the preheated oven at 350°F for about 40 min or until chops are well done and tender. Remove from the oven and place in a steam table pan.
6. Serve one chop per portion with brown sauce.

Precautions

- Exercise caution when using the knives.
- When sautéing chops, do not overbrown.
- Handle chops with care when sautéing so the stuffing does not come out.

Stuffed Spareribs

Stuffed spareribs are spareribs cracked across the middle, stuffed with force-meat, tied, and baked. This is a unique method of serving spareribs on the luncheon menu.

50 servings

Equipment

- French knife
- Baker's scale
- Quart measure
- Wire whip
- Butcher's twine
- Roast pans
- 1 qt saucepans
- Kitchen fork
- Meat grinder
- Full size steam table pan

Ingredients

30 lb spareribs
4 lb fresh, lean pork picnic
2 lb bread, fresh or dried
1½ qt milk (variable)
8 oz onions
8 oz celery
8 oz butter
⅓ c parsley
¼ oz sage
salt and pepper to taste
4 egg yolks
bread crumbs as needed

Preparation

1. Mince the onions and celery and chop the parsley with a French knife.
2. Soak bread in milk until all the milk is absorbed.
3. Cut the fresh picnic into strips for grinding. Use a French knife.
4. Crack all the spareribs with a cleaver lengthwise down the center of each spare rib.
5. Preheat the oven to 375°F.
6. Separate eggs and beat yolks slightly with a wire whip. (Save the whites for use in another preparation.)

Procedure

1. Place the butter in the saucepan and heat. Add the onions and celery and sauté until slightly tender.
2. Place the onions, celery, soaked bread, pork picnic, sage, salt, and pepper into a mixing container and mix thoroughly.
3. Grind this mixture on the meat grinder twice, using the medium chopper plate.
4. Add the beaten egg yolks and chopped parsley and mix thoroughly. If the mixture is too wet, add bread crumbs as needed; if too dry, add more milk.
5. Place the stuffing on one-half of the cracked rib section of each sparerib. Fold over the other half and tie the ribs with butcher's twine to hold in the forcemeat stuffing.

6. Cover the bottom of the roast pans with about ¼″ of water. Place the ribs on a rack in the roast pans. Roast in the preheated oven at 375°F until the ribs become brown on both sides. Turn ribs frequently with a kitchen fork while roasting.

7. Remove from the roast pans and place in a steam table pan.

8. Cut the ribs into 50 equal units with a French knife.

9. Serve one unit per portion with applesauce.

Precautions

- Use caution when mincing the onions and cracking the ribs.
- Tie ribs securely so stuffing does not come out during the roasting. Be gentle when turning the ribs.

Stuffed Fresh Cottage Ham

Stuffed fresh cottage ham is sliced butterfly style, flattened, spread with a forcemeat mixture, rolled, and roasted. It is another profitable addition to the luncheon menu.

50 servings

Equipment

- French knife
- Cleaver
- Roast pan
- 1 qt saucepans
- Mixing container
- Meat grinder
- Wire whip
- Ladle
- Baker's scale
- Cup measure
- Full size steam table pan
- No. 12 scoop

Ingredients

- 16 lb fresh cottage ham
- 3 lb boneless fresh pork picnic
- 2 lb boneless veal shoulder
- 1 lb bread, fresh or dried
- 3 c milk
- 6 oz onions
- 6 oz celery
- 4 egg yolks
- 8 oz bread crumbs (variable)
- 6 oz butter
- ⅓ c parsley
- ¼ oz sage
- salt and pepper to taste
- 2 gal. thin brown sauce

Preparation

1. Slice the fresh cottage ham butterfly-style into 50 units. Cut one thin slice against the grain with a French knife only three-quarters of the way through. Cut the following thin slice all the way through. This results in the butterfly-style cut: two slices joined at the bottom. Spread out the joined slices on a cutting board and flatten with the side of a cleaver.

2. Mince the onions and celery with a French knife.

3. Chop the parsley with a French knife and wash.

4. Cut the pork picnic and veal shoulder into strips that will fit in the grinder.

5. Separate eggs and beat the yolks slightly with a wire whip. (Save the whites for use in another preparation.)

6. Soak the bread in the milk until all the milk is absorbed.

7. Preheat the oven to 350°F.

8. Prepare the brown sauce. Keep hot.

Procedure

1. Place the butter in the saucepan and heat. Add the onions and celery and sauté until slightly tender.

2. Place the onions, celery, soaked bread, pork, veal, sage, and pepper into a mixing container and mix thoroughly.

3. Grind this mixture twice in a meat grinder using the fine chopper plate.

4. Add the beaten egg yolks and chopped parsley and mix thoroughly. If the mixture is too wet, add bread crumbs as needed; if too dry, add more milk.

5. Place the stuffing (meat mixture) on each flattened butterfly slice. Use about 4 oz or a No. 12 scoop full. Roll up each butterfly slice. Place close together in roast pans with the end of each roll on the bottom.

6. Roast in the preheated oven at 350°F until each roll is golden brown. Pour off accumulated grease in bottom of roast pan.

7. Ladle the hot brown sauce over each roll, return to the oven, and continue roasting until the meat is tender and well done. Remove from the oven and place in a steam table pan.

8. Serve one meat roll per portion covered with brown sauce. Accompany with a half baked apple or a canned spiced apple.

Precautions

- Exercise caution when cutting and flattening the pork.
- When flattening the pork, do not pound too hard or the meat will separate along the muscle fibers.
- When the pork rolls are covered with the brown gravy, watch them closely through the remainder of the baking period. They will overbrown quickly.

Trade Tip *Stew dishes with pork can be served immediately after cooking or can be covered, refrigerated overnight, and reheated. The flavor is fuller when the dish is allowed to rest overnight. Also, when chilled, excess fat in the stew will rise to the surface and be easily removed.*

Swedish Meatballs in Sour Cream Sauce

Swedish meatballs are a combination of seven or eight small meatballs and a rich sour cream sauce. The meatballs are light in color and delicate in flavor because they are made from a combination of pork and veal. This profitable luncheon item is generally accompanied with buttered noodles.

50 servings

Equipment

- French knife
- Quart measure
- Baker's scale
- Meat grinder
- Sheet pans
- Meat turner
- Mixing container
- Wire whip
- 1 qt saucepans
- Full sized steam table pan

Ingredients

9 lb boneless, lean fresh pork picnic
9 lb boneless veal shoulder
8 eggs
10 oz milk
1¼ qt bread crumbs
½ pt celery
½ pt onions
8 oz shortening
salt, pepper, and nutmeg to taste
2 gal. sour cream sauce

Preparation

1. Cut the pork and veal into strips that will fit the meat grinder.
2. Mince the onions and celery with a French knife.
3. Grease the sheet pans with salad oil.
4. Prepare 2 gal. of sour cream sauce.
5. Preheat the oven to 350°F.

Procedure

1. Beat the eggs in the mixing container with a wire whip.
2. Add the milk and bread crumbs. Let the mixture stand until all the milk is absorbed into the bread crumbs.
3. Place the shortening in the saucepan and heat. Add the onion and celery, sauté, and let cool.
4. Add the pork, veal, seasoning, onions, and celery to the bread crumb mixture and mix thoroughly.
5. Grind the mixture twice using the fine chopper plate, then mix thoroughly again and check the seasoning.
6. Form into small balls, place on greased sheet pans, and bake in the preheated oven at 350°F until light brown and firm when touched.
7. Remove from the oven, pour off the grease, add water to cover the bottom of the sheet pan, and loosen the meatballs from the bottom of the pan.
8. Using the meat turner, remove meatballs from the sheet pan and place in a steam table pan to keep warm.
9. Serve seven or eight meatballs per portion in a flat casserole covered with the sour cream sauce and accompanied with buttered noodles.

Precautions

- Exercise caution when using the knife and grinding the meat mixture.
- If meat sticks to the hands when forming the meatballs, rub a little salad oil on the hands.
- Do not overbake the meatballs. They will become too brown and too firm and will lack in appearance when served.
- Do not place the meatballs in the sauce before serving. They break up easily.

Ham and Cabbage Rolls

Ham and cabbage rolls are a profitable addition to the luncheon menu because accumulated ham trimmings can be utilized in their preparation. The rolls consist of partly-cooked cabbage leaves rolled around the ham/rice filling.

50 servings

Equipment

- Large braisers and cover
- Bake pans
- 10 gal. stockpot
- French knife
- Baker's scale
- Quart measure
- Spoon and cup measure
- Boning knife
- Meat grinder
- 6 qt sauce pot
- Kitchen spoon
- Full size steam table pans
- Stockpot

Ingredients

6 large heads of cabbage
12 lb cooked, ground ham
2 lb onions
2 garlic cloves
1 No. 10 can tomatoes
1 qt tomato puree
2 lb 8 oz rice
2 tbsp chili powder
3 tbsp paprika
1 qt water
1 c salad oil
2 qt chicken stock
1 qt catsup
2 qt canned tomatoes
salt and pepper to taste

Preparation

1. Remove the cores from the cabbage heads with a boning knife. Place in the stockpot, cover with salt water, and boil until the leaves can be easily removed from the heads.
2. Mince the onions and garlic with a French knife.
3. Crush the tomatoes by hand.
4. With a French knife, cut the ham into pieces that will pass through the meat grinder.
5. Grind the ham in the meat grinder using the coarse chopper plate.
6. Preheat the oven to 400°F.
7. Prepare the chicken stock.

Procedure

1. Place the salad oil in the braiser and heat. Add the onions and garlic and sauté without browning.
2. Add the first amount of crushed tomatoes (No. 10 can), the tomato puree, and the water. Simmer until the onions are tender. Stir occasionally with a kitchen spoon.

3. Add the chili powder and paprika and continue to simmer.
4. Add the ham and rice. Bring back to a boil.
5. Season with salt and pepper, cover the braiser, and bake in the preheated oven at 400°F until the rice absorbs the liquid and is tender.
6. Remove from the oven and check the seasoning. Place in a bake pan to cool, then refrigerate overnight.
7. Remove ham/rice mixture from the refrigerator. Place cabbage leaves (about two leaves for each roll) on a kitchen towel. Place a 2½ oz ball of the ham/rice mixture on the leaves and roll up. Repeat this process until all the cabbage leaves and ham/rice mixture are used. Place the rolls in bake pans.

8. Combine the second amount of crushed tomatoes, catsup, and chicken stock in a sauce pot. Blend thoroughly with a kitchen spoon and pour over the cabbage rolls.
9. Bake in the preheated oven at 350°F for 45 min. Remove from the oven and place in steam table pans.
10. Serve two rolls per portion topped with tomato liquid.

Precautions

- Exercise caution when grinding the ham and mincing the onions and garlic.
- When baking the ham/rice mixture, be cautious in removing the lid.
- Baste the ham and cabbage rolls frequently when baking.

FRIED AND SAUTÉED PORK RECIPES

Fried Pork Chops or Cutlets

For fried pork chops or cutlets, pork chops cut from the rib end or the loin end of the pork loin are passed through seasoned flour and cooked in shallow grease until golden brown. This item is an excellent choice for the luncheon, dinner, or a la carte menu. The cutlet is a thin, boneless, slice of pork loin or Boston butt and is prepared and served in the same manner.

50 servings

Equipment

- French or butcher knife
- Cleaver
- Bake pan
- Kitchen fork
- Iron or steel skillets
- Quart measure
- Spoon measure
- Full size steam table pan

Ingredients

- 50 5 oz pork chops or 50 5 oz pork cutlets
- 3 tbsp salt
- 2 tsp pepper
- 3 lb all-purpose flour
- 2 qt salad oil or melted shortening (variable)

Procedure

1. Place enough shortening or salad oil in the skillet to cover the bottom ¼". Place on the range and heat.
2. Pass each pork chop or cutlet through the seasoned flour and pat off excess.
3. Place the chops or cutlets in the skillet and brown both sides while maintaining a moderate temperature. Turn the meat with a kitchen fork. Cook the pork chops well done.
4. Remove from the skillet and let drain. Place in a steam table pan.
5. Serve one chop or cutlet per portion with country gravy.

Precautions

- Exercise caution when cutting the pork chops or cutlets.
- Do not overbrown the chops or cutlets while frying.
- Do not let the grease in the skillet get too hot.
- Always pat off the excess flour before placing in the grease or it will lie on the bottom of the skillet and burn, causing the item to have poor appearance when served.

Preparation

1. Cut the pork chops or cutlets with a French or butcher knife. Flatten them slightly with the side of a cleaver.
2. Place the flour in a bake pan and season with salt and pepper.

Pork can easily be trimmed to fit desired recipes. For example, cut pork into thin strips for salads and stir-fry dishes; cut pork into cubes for kabobs, stews, and spaghetti sauce; cut pork into thin cutlets for sautéing or braising quickly. Use a sharp knife and a clean cutting board.

Breaded pork chops or cutlets are cut from the trimmed loin end of the full loin of pork. The cutlets are thin, flattened, boneless slices of the loin. Cutlets can also be cut from fresh, boneless Boston butt in the same manner as those cut from the loin. These cuts are coated with a bread or cracker crumb coating and fried to a golden brown. This is a most desirable luncheon entree when served with tomato or cream sauce.

Breaded Pork Chops or Cutlets

50 servings

Equipment

- Baker's scale
- Quart measure
- Spoon measure
- Full size steam table pan
- French or butcher knife
- Cleaver
- Bake pans (3)
- Iron skillet
- 4 qt metal bowl
- Kitchen fork
- Wire whip

Ingredients

50 4 or 5 oz pork chops
  or cutlets
1 lb flour
2 tbsp salt
1 tsp pepper
6 eggs
2 qt milk
2 qt bread or cracker
  crumbs
2 lb shortening (variable)

Preparation

1. Cut the pork chops or cutlets with a French or butcher knife. Flatten them slightly with the side of a cleaver.
2. Beat the eggs with a wire whip. Prepare an egg wash by combining the beaten eggs and milk.

3. Place the flour in a bake pan and season with salt and pepper.
4. Place the bread or cracker crumbs in a bake pan.

Procedure

1. Pass each chop or cutlet through the seasoned flour, coating them completely.
2. Place them in the egg wash, then into the bread or cracker crumbs. Press firmly and shake off excess crumbs.
3. Place the shortening in an iron skillet and heat.
4. Add the chops or cutlets, brown one side, then turn with a kitchen fork and brown the other side. Remove when they become golden brown. Let drain and place in a steam table pan.
5. Serve one chop or cutlet per portion on top of tomato or cream sauce.

Precautions

- Use caution when cutting the chops and cutlets.
- Fry the chops or cutlets slowly so the meat will be well done when the breading is golden brown. An overly high temperature will brown the breading too rapidly.

Ham croquettes are a profitable item when served on the luncheon menu with a cream sauce. This is an excellent way to utilize ham trimmings.

Ham Croquettes

50 servings

Equipment

- Braiser
- Bake pans (4)
- Wire whip
- Deep fat fryer
- Meat grinder
- Kitchen spoon
- No. 20 ice cream scoop
- Cup measure
- Quart measure
- Baker's scale
- French knife
- 8 qt and 3 qt stainless steel bowls
- Oiled brown paper
- Steam table pans

Ingredients

8 lb boiled ham
1 lb celery
1 lb onions
1½ lb melted butter or
  shortening
1½ lb bread flour
2 qt hot milk or ham stock
¼ c prepared mustard
2 tbsp dry mustard
1 c parsley
2 qt milk
12 eggs
2 lb bread flour
3 lb bread crumbs

Preparation

1. Cut the ham with a French knife into strips that will fit into the meat grinder. Grind the ham in the meat grinder using the coarse chopper plate.
2. Mince the onions and celery with a French knife.
3. Chop the parsley with a French knife and wash.

4. Preheat the oven to 350°F.
5. Prepare the ham stock if used.
6. Prepare the egg wash. Place eggs in a stainless steel bowl, beat with a wire whip, and pour in the 2 qt of milk while continuing to whip. On the day the egg wash is to be used, place it in a bake pan.
7. Place the flour and bread crumbs in separate bake pans.

Procedure

1. Place the butter or shortening in a braiser and heat. Add the celery and onions and sauté until tender. Do not brown.
2. Add flour to make a roux and cook slightly. Stir occasionally with a kitchen spoon.
3. Add hot milk or stock and stir until thick and smooth.
4. Blend the dry mustard with the prepared mustard and add to the ham and parsley in a stainless steel bowl.
5. Combine all ingredients, mixing thoroughly with a kitchen spoon in the braiser.
6. Put the mixture in greased bake pans and cover with oiled brown paper.
7. Bake in the preheated oven at 350°F for 45 min.
8. Remove from the oven and let cool. Place in the refrigerator overnight.

9. Portion each croquette with a level No. 20 ice cream scoop.
10. Shape in cones of uniform size.
11. Bread croquettes by passing them through flour, egg wash, and bread crumbs.
12. Fry in deep fat at 350°F until golden brown and place in steam table pans.
13. Serve two croquettes per portion with cream or tomato sauce.

Precautions

- Exercise caution when using the knife and grinding the meat.
- When sautéing the vegetables in butter, be careful not to burn them.
- When breading, use a fairly rich egg wash so the croquettes will hold together better when fried.

Ham turnovers involve a dough mixture and a ham filling mixture. When the turnovers are prepared, the filling is placed inside the dough and the item is deep fried to a golden brown. Many different fillings may be substituted for the ham when preparing this item, but ham and chicken seem to be the most popular. This item should always be served with an appropriate sauce.

Ham Turnovers

50 servings

Equipment

- Baker's scale
- Deep fat fryer
- Rolling pin
- Turnover cutter or 5" round cutter
- 6 qt sauce pot
- Cup measure
- Spoon measure
- Kitchen spoon
- No. 16 ice cream scoop
- French knife
- Stainless dish pan
- Dinner fork
- Quart measure
- Full size steam table pan

Ingredients

Deep Frying Dough

7 lb 8 oz pastry flour
3 lb hydrogenated vegetable shortening
3¾ oz salt
3 lb water

Ham Filling

1 c hydrogenated vegetable shortening
½ c onions
1 qt ham stock
2 qt cooked smoked ham
2 c bread flour
1 c bread crumbs
½ tsp nutmeg
2 tsp salt
1 tsp paprika
½ tsp pepper

Preparation

1. Mince the onions and ham with a French knife.
2. Preheat the deep fry kettle to 375°F.
3. Prepare the ham stock.

Procedure

Deep Frying Dough

1. Place the flour and shortening in the stainless steel dish pan. Cut the shortening into the flour by rubbing together with the palms of the hands until very small lumps are formed.

2. Dissolve the salt in the water by stirring with a kitchen spoon. Add to the flour and shortening and mix into a dough. Cover with a damp towel and refrigerate for later use.

Ham Filling

1. Place the shortening in the sauce pot and heat. Add the onions and sauté without color.
2. Add the flour, making a roux, and stir with a kitchen spoon. Cook until well-blended.
3. Add the hot ham stock gradually, stirring constantly. Cook for about 3 min.
4. Add the ham, bread crumbs, nutmeg, salt, paprika, and pepper and mix thoroughly with a kitchen spoon. Return to the heat, stir until thoroughly heated and mixed, then let cool.
5. Roll out the dough with a rolling pin on a floured bench to about ⅛" thickness, or thinner if desired. Cut out with a 5" round cutter.
6. Dip the ham mixture with a No. 16 ice cream scoop, roll, and shape mixture slightly. Place mixture in the center of the cut out dough.
7. Wash the edges of the dough with cold water and fold the circle of dough in half to form a turnover. Seal the edges completely with the tines of a dinner fork.
8. Pierce the top once or twice with a dinner fork.
9. Fry in deep fat at 375°F for 5 min to 7 min or until golden brown. Place in a steam table pan.
10. Serve one turnover per portion on top of cream sauce, raisin sauce, or other fruit sauce.

Precautions

- Exercise caution when mincing the ham and onions.
- Do not form turnovers until the filling is cool and the dough has been chilled.
- Dust the bench slightly heavy with pastry flour when rolling the dough.
- Secure the edges of the dough tightly with a dinner fork before frying.

When sautéing pork, do not cover the skillet. Cook in a small amount of oil over medium-high heat. Turn the pork occasionally. Sautéing is generally used for more tender cuts of pork. Remember, the thinner the cut, the higher the heat. The thicker the cut, the more moderate the heat.

Orange caramelized onion sauce adds a unique taste to the medallions of pork. The oranges are an attractive and delicious garnish.

Medallions of Pork

50 servings

Ingredients

¼ c water
2 tsp cornstarch
1 tbsp vegetable oil
6 tbsp butter, divided
4 c red onions, sliced ⅛″
salt as needed
black pepper as needed
1¼ c rum
2 tbsp balsamic vinegar
1½ c orange juice
olive oil as needed
24 pork medallions, 3 oz each
36 orange wheels, ¼″ thick
chopped parsley as needed

Florida Department of Citrus

Procedure

Sauce

1. Combine the water and the cornstarch. Mix well and reserve.
2. Combine the vegetable oil and 1 tbsp of butter in a sauté pan over medium heat.
3. When the sauce is boiling, add the onions, ½ tsp salt and ¼ tsp pepper and sauté until golden brown.
4. Deglaze the pan with ½ c rum. Add vinegar and orange juice and bring to a boil.
5. Add the reserved cornstarch, return to a boil, and simmer for 2 min.
6. Remove from heat and hold.

Pork

1. Heat a small amount of olive oil in a hot sauté pan.
2. Sprinkle the pork medallions with salt and pepper.
3. Sauté pork medallions, remove and hold.
4. Deglaze the pan with rum, add the onion sauce, and heat.
5. Place the medallions on serving plates and top with the onion sauce.
6. Garnish with orange wheels and chopped parsley.

BRAISED PORK RECIPES

Pork Chops Creole

Pork chops creole are sautéed pork chops baked in a rich creole or Spanish sauce. This is an excellent choice for the luncheon menu.

50 servings

Equipment

- French or butcher knife
- Kitchen fork
- Bake pans
- Iron skillet
- Baker's scale
- Quart measure

Ingredients

50 5 oz pork chops
2 gal. creole sauce
3 lb flour
salt and pepper to taste
1 qt salad oil or melted shortening (variable)

Preparation

1. Cut the 5 oz pork chops from trimmed pork loin. Cut with a French or butcher knife.
2. Prepare the creole sauce.
3. Place the flour in a bake pan and season with salt and pepper.
4. Preheat the oven to 350°F.

Procedure

1. Place the salad oil or shortening in the iron skillet and heat.
2. Pass each chop through the seasoned flour, pat off excess flour, and place in the hot shortening.
3. Brown one side, turn with a kitchen fork, and brown the other side. Remove from the skillet and drain.
4. Place in bake pans and cover with the creole sauce. Bake in the preheated oven at 350°F until chops are tender.
5. Serve one chop per portion, topped with the creole sauce and accompanied with a mound of rice.

Precautions

- Exercise caution when using the knife.
- Do not overbrown or overcook the chops while sautéing.
- Do not place the chops too close together in the bake pans. They will not bake evenly.
- Use caution when sautéing the chops to avoid burning.

Barbecued Pork Chops

Barbecued pork chops are sautéed and baked in a rich barbecue sauce. They are very popular and are an excellent selection for the luncheon menu.

50 servings

Equipment

- Boning knife
- Baker's scale
- French or butcher knife
- Quart measure
- Iron skillet
- Bake pans
- Kitchen fork

Ingredients

50 5 oz pork chops
3 lb flour
1 qt salad oil or melted shortening (variable)
2 gal. barbecue sauce
salt and pepper to taste

Procedure

1. Place the salad oil or shortening in the iron skillet and heat.
2. Pass each chop through the seasoned flour, pat off excess flour, and place in the hot shortening.
3. Brown one side, turn with a kitchen fork, and brown the other side. Remove from the skillet and let drain.
4. Place in a bake pan and cover with the barbecue sauce. Bake in the preheated oven at 350°F until tender and well done.
5. Serve one chop per portion topped with the barbecue sauce.

Preparation

1. Cut the 5 oz pork chops from trimmed pork loin. Cut with a French knife.
2. Prepare the barbecue sauce.
3. Place the flour in a bake pan and season with salt and pepper.
4. Preheat oven to 350°F.

Precautions

- Exercise caution when using the knife.
- Use caution when sautéing the chops.
- Do not overbrown or overcook the chops while sautéing.
- Do not place the chops too close together in the bake pans. They will not bake evenly.

Pork Chops Hawaiian

Because of its delicate flavor, pork blends well with many fruits. Pork chops are extremely tasty when baked with pineapple. This is an excellent luncheon item.

50 servings

Equipment

- Iron skillet
- French or butcher knife
- Bake pans (2-3)
- Quart measure
- Spoon measure
- Wire whip
- 4 qt saucepan
- Boning knife
- Butcher knife
- Kitchen spoon
- Kitchen fork
- 1 pt stainless steel bowl
- Steam table pan

Ingredients

50 pork chops
50 slices pineapple
2 qt pineapple juice
2 bay leaves
1 qt celery
2 tsp ground cloves
2 garlic cloves
salt and pepper to taste
1 pt salad oil (variable)
⅓ c cornstarch
½ c cold water

Procedure

1. Place the salad oil in the iron skillet and heat. Season the pork chops with salt and pepper and brown on both sides. Turn with a kitchen fork.
2. Remove the chops from the skillet and place them in the bake pans. Pour off the oil left in the skillet.
3. Add the pineapple juice to the skillet and bring to a boil to deglaze the skillet. Pour the liquid over the pork chops.
4. Place a slice of pineapple on top of each chop.
5. Add all remaining ingredients except the cornstarch and water and bake in the preheated oven at 350°F until the chops are well done. Remove from the oven.
6. Pour the juice off the baked chops into a saucepan. Place the chops in a steam table pan and bring the juice to a boil on the range.
7. Dissolve the cornstarch in the cold water in a stainless steel bowl. Pour the starch into the boiling juice while whipping vigorously with a wire whip. Cook until the juice is thickened and clear.
8. Pour the thickened juice over the chops in the steam table pan.
9. Serve one chop per portion with a slice of pineapple and sauce.

Preparation

1. Cut the pork chops about 1″ thick from trimmed pork loins with a French or butcher knife.
2. Mince the garlic and celery with a French knife.
3. Preheat the oven to 350°F.

Precautions

- Use caution when cutting the pork chops and celery.
- When adding the diluted cornstarch to the liquid, stir vigorously to avoid lumps.

Pork Chops Jonathan

Pork chops Jonathan are thick chops cut from the pork loin and baked with apples and apple juice. Apples help increase the delicate flavor of pork. This is an excellent choice for the luncheon menu.

50 servings

Equipment

- Iron skillet
- Quart measure
- French or butcher knife
- Kitchen fork
- Kitchen spoon
- Large braiser and cover
- 1 pt stainless steel bowl
- 2 gal. stainless steel container

Ingredients

50 pork chops
1 No. 10 can sliced apples
2 qt apple juice
½ c lemon juice
½ tsp Tabasco sauce
1 tsp ground cloves
½ c cornstarch
1 c cold water
1 pt salad oil (variable)

Preparation

1. Cut the pork chops about 1″ thick from trimmed pork loins with a French or butcher knife.
2. Preheat oven to 350°F.

Procedure

1. Place the salad oil in the iron skillet and heat. Add the pork chops and brown on both sides. Turn with a kitchen fork.
2. Place the chops in the braiser. Pour off the oil left in the skillet and deglaze the skillet by adding the apple juice and bringing to a boil. Pour this liquid over the pork chops.
3. Cover the pork chops with the sliced apples. Add the cloves, Tabasco sauce, and lemon juice.
4. Cover the braiser and bake in the preheated oven at 350°F until the chops are tender and well done. Remove the chops from the braiser and place in a steam table pan. Cover the chops and keep warm. Place the braiser on the range and bring the apple mixture to a boil.
5. In a stainless steel bowl, dissolve the cornstarch in the cold water. Pour the starch into the boiling mixture while stirring constantly with a kitchen spoon. Cook until mixture is thickened and clear.
6. Check the seasoning and place the thickened apple mixture in a stainless steel container.
7. Serve one chop on top of a small portion of the apple mixture.

Precautions

- Use caution when cutting the chops.
- When adding the diluted cornstarch to the apple mixture, stir constantly until the mixture comes back to a boil to avoid lumps.

Sweet-Sour Pork

Sweet-sour pork is an American version of a famous and popular Chinese preparation. The main difference between the two is the extent to which the vegetables are cooked. The Chinese prefer extra crispness; the American preference is more tenderness. The cubes of pork are cooked and served in a starch-thickened sauce best described as sweet-sour.

50 servings

Equipment

- Quart measure
- Braiser
- Cup measure
- French knife
- Baker's or portion scale
- Kitchen spoon
- Stainless steel bowl

Ingredients

20 lb pork loin or Boston butt
1 c peanut or salad oil (variable)
¾ c soy sauce
1 pt cider vinegar
12 oz dark brown sugar
4 oz onions
½ gal. ham stock
1 qt pineapple juice
8 oz green peppers
8 oz fresh tomatoes
½ oz fresh ginger
1 lb 4 oz canned pineapple chunks, drained
5 oz cornstarch
1 c pineapple juice
1 c sliced onions

Preparation

1. Cut the lean, trimmed pork into 1″ cubes using a French knife.
2. Mince and slice the onions, mince or crush the ginger, dice the green peppers, and cut the tomatoes into wedges with a French knife.

Procedure

1. Place the oil in a large braiser, place on the range, and heat.
2. Add the diced pork and brown thoroughly, stirring frequently with a kitchen spoon.
3. Add the ham stock, soy sauce, vinegar, minced onions, and brown sugar. Simmer until the pork is fairly tender.
4. Add the quart of pineapple juice, ginger, green peppers, and sliced onions. Continue to simmer only until the green peppers are partly cooked. Retain a slight crispness to the peppers.
5. Add the tomatoes and pineapple. Return to a simmer.
6. Place the cornstarch in a stainless steel bowl. Add the cup of pineapple juice and stir until the starch is thoroughly dissolved.

7. Slowly pour the dissolved starch into the simmering liquid, while at the same time stirring rapidly with a kitchen spoon.

8. Bring the sauce to a simmer and simmer until fairly thick and clear.

9. Remove from the range and serve with rice or over fried Chinese noodles.

Precautions

- Exercise caution when cutting the meat and vegetables.
- Do not overcook the vegetables. Retain a slight crispness.
- When adding the starch, pour slowly and stir rapidly to avoid lumps and scorching.

Braised Pork Tenderloin Polynesian

The pork tenderloin is braised in a rich pineapple-flavored sauce often associated with Polynesian cuisine. The sauce is thickened slightly with cornstarch and served with each order of the sliced pork tenderloin. This item can be served on the luncheon or dinner menu. It is an excellent choice for a backyard luau.

50 servings

Equipment

- Quart measure
- Cup measure
- Baker's or portion scale
- Roast pan
- Braiser
- French knife
- Stainless steel bowl
- Kitchen spoon
- Boning knife
- Kitchen fork

Ingredients

18 lb pork tenderloins
2 qt pineapple juice
1 qt water
½ c soy sauce
1 c brown sugar
4 oz green peppers
1 lb 4 oz canned pineapple chunks, drained
4 oz cornstarch
1 c pineapple juice
salad oil as needed

Preparation

1. Trim each tenderloin using a boning knife.
2. Dice the green pepper medium using a French knife.

Procedure

1. Rub each pork tenderloin with salad oil until slightly coated.

2. Place in the roast pan and brown slightly in the preheated oven at 350°F. Turn occasionally with a kitchen fork.
3. Remove from the oven and place in a braiser.
4. Add the first amount of pineapple juice, water, soy sauce, green peppers, and brown sugar. Place on the range and simmer until tenderloins are fairly tender.
5. Remove the tenderloins from the liquid and hold. Add the pineapple chunks to the liquid.
6. Place the cornstarch in a stainless steel bowl. Add the second amount of pineapple juice and stir until the starch is thoroughly dissolved.
7. Slowly pour the dissolved starch into the simmering liquid while at the same time stirring rapidly with a kitchen spoon.
8. Bring the sauce to a simmer and simmer until the sauce is fairly thick and clear.
9. Slice the tenderloins on a bias. Place in a steam table pan. Cover with the thickened sauce and serve.

Precautions

- Exercise caution when dicing the green peppers and slicing the tenderloin.
- When pouring the starch into the liquid, stir continuously or lumps may form.

Honey Style Pork Chops

Honey style pork chops are sautéed lightly, placed in a pan, covered with a honey glaze, and baked until golden and tender. The honey glaze increases the appearance and delicate flavor of the chops.

50 servings

Equipment

- French or butcher knife
- Iron skillet
- 2 qt saucepan
- Wire whip
- Kitchen fork
- Kitchen spoon
- Pint measure
- Baker's scale
- Cleaver
- Bake pans
- Steam table pans

Ingredients

50 5 oz pork chops
1 pt soy sauce
1 pt applesauce
8 oz honey
4 oz sugar
1 oz salt
2 lb shortening (variable)

Preparation

1. Cut the 5 oz pork chops from the pork loin. Trim the pork chops with a French or butcher knife and flatten each chop gently with a cleaver.
2. Preheat the oven to 350°F.

Procedure

1. Place the shortening in an iron skillet and heat. Add the pork chops and brown slightly on both sides. Place in bake pans with a kitchen fork.
2. Combine the remaining ingredients, place in a saucepan, and bring to a gentle boil, whipping slightly with a wire whip.

3. Pour this mixture over the sautéed pork chops. Bake in the preheated oven at 350°F for about 30 min to 45 min or until the chops are well done and tender. Remove from the oven and place in steam table pans.

4. Serve one chop per portion with the honey sauce.

Precautions

- Use caution when cutting and trimming the chops.
- While the chops are baking, turn and brush them often with the honey mixture.

Braised Pork Tenderloin Deluxe

Pork tenderloins that average from 1 lb to 2 lb are browned in hot shortening then cooked in a rich liquid until well done. They are sliced and topped with a rich sauce before serving.

50 servings

Equipment

- Baker's scale
- Quart measure
- Spoon measure
- Cup measure
- Boning knife
- French knife
- Iron or steel skillet
- Large braiser and cover
- 2 gal. sauce pot
- Kitchen spoon
- Kitchen fork
- Full size steam table pan
- 1 gal. stainless steel container
- China cap

Ingredients

25 whole pork tenderloins
1 pt salad oil (variable)
½ c onions
10 oz butter
8 oz flour
4 tbsp dry mustard
¼ tsp black pepper
4 qt hot brown stock
¼ c lemon juice
8 oz Burgundy
2 tbsp sugar
salt to taste

Preparation

1. Trim the pork tenderloin with a boning knife. Remove any membrane or fat.
2. Mince the onions with a French knife.
3. Preheat the oven to 350°F.

4. Prepare the brown stock.

Procedure

1. Brown each tenderloin in a skillet in salad oil. Place them in the braiser using a kitchen fork.
2. Place the butter into the sauce pot and heat. Add the minced onions and sauté without color.
3. Add flour and dry mustard and cook for about 5 min.
4. Add the brown stock, lemon juice, sugar, salt, and wine. Cook, stirring gently with a kitchen spoon until thick and smooth.
5. Pour the sauce over the tenderloins in the braiser, cover, and bake in the preheated oven at 350°F for approximately 1 hr or until the tenderloins are well done.
6. Remove the tenderloins and place them in a steam table pan. Keep covered with a damp cloth.
7. Strain the sauce through a china cap into a stainless steel container. Adjust seasoning and consistency.
8. Slice the pork tenderloin on a bias (slanting) with a French knife and serve 8 oz per portion topped with the rich sauce.

Precautions

- Use caution when trimming and slicing the tenderloins.
- Do not overbrown the tenderloins when sautéing.

Hawaiian Pork

Hawaiian pork is a stew. The pork is cut into strips and cooked with pineapple chunks in a sweet-sour sauce. This item is more suitable for the luncheon menu.

50 servings

Equipment

- Boning knife
- Medium-sized braiser
- French knife
- Cup measure
- Spoon measure
- Quart measure
- Scale
- 2 qt saucepan
- Kitchen spoon
- Full size steam table pan

Ingredients

8 lb pork shoulder
½ c bacon grease
1 c water
½ c cornstarch
1 tbsp salt
1 c dark brown sugar
1 c cider vinegar
1 qt pineapple juice
½ c soy sauce
1 pt green pepper
1 c onion
1 No. 10 can pineapple chunks

Preparation

1. Remove the bones from the pork shoulder with a boning knife. Boil and refrigerate overnight.
2. Cut the cooked pork with a French knife into strips about 3″ long and ½″ square.
3. Julienne the green peppers and onions with a French knife.

Procedure

1. Place the bacon grease in the braiser, heat, add the pork strips, and brown.
2. Add the water and simmer slowly for about 5 min.

3. Dissolve the cornstarch in the pineapple juice. Add the salt, brown sugar, vinegar, and soy sauce and blend thoroughly with a kitchen spoon. Add this mixture to the pork strips, stirring constantly until thick and smooth.
4. Add the onions and pineapple chunks. Cook 10 min or until the onions are tender.
5. Place the green peppers in a saucepan, cover with water, and poach until tender. Drain and add to the pork mixture.

6. Check the seasoning and place in a steam table pan. Serve 4 oz on a mound of baked rice.

Precautions

- Stir the mixture gently when adding the pineapple chunks so they do not break up.
- Use caution when cutting the pork and the vegetables.

BROILED PORK RECIPES

Broiled ham steak Hawaiian is a crosscut section of the ham. The center cuts are best. It is broiled and served with a slice of glazed pineapple and a cherry center.

Boiled Ham Steak Hawaiian

50 servings

Equipment

- French knife
- Small mixing bowl
- Kitchen fork
- Cup measure
- Spoon measure
- Sheet pan
- Bake pan
- Meat saw

Ingredients

50 4 oz or 5 oz ham steaks
50 pineapple slices
50 red maraschino cherries
1 c sugar
½ c brown sugar
1 tsp cinnamon
1 pt salad oil (variable)

Preparation

1. Cut the 4 oz or 5 oz ham steaks with a French knife and hand saw or power saw. Cut approximately ½" thick across the complete width of the ham.
2. Set the broiler at the highest temperature.

Procedure

1. Place the salad oil in a bake pan, marinate each steak in the salad oil, and place fat side out on the hot broiler.
2. Broil one side 2 min to 3 min, turn with a kitchen fork, and broil the other side. Remove from the broiler when done and keep hot.
3. Combine the sugars and cinnamon in a small mixing bowl.
4. Place the pineapple rings on a sheet pan and sprinkle with the sugar/cinnamon mixture.
5. Place pineapple rings under the broiler and glaze.
6. Serve one ham steak per portion with a pineapple slice on top. Garnish the center with a red maraschino cherry and top with brown butter sauce.

Precautions

- Use caution when cutting the steaks.
- When broiling the steaks, do not burn the fat.

Picnic pork is ground and seasoned into a lean pork sausage and formed into 2 oz patties that can be prepared and served on the breakfast or luncheon menu. These patties contain less fat and do not shrink as much as purchased sausage. The patties may be prepared by grilling, baking, or broiling.

Pork Sausage Patties

50 servings

Equipment

- French knife
- Meat grinder
- Mixing container
- Baker's scale
- Measuring spoons
- Boning knife

Ingredients

13 lb picnic pork or callie
1¾ oz salt
¾ oz sugar
1 tbsp fresh ground black pepper
1 ½ tsp rubbed summer savory
1 tsp ginger
1 tsp nutmeg
1 tsp rubbed marjoram
2 ½ tbsp rubbed sage

Preparation

1. Bone the fresh picnic with a boning knife. Using a French knife, cut the meat and fat into strips that will pass through the grinder.

Procedure

1. Place all ingredients in a mixing container and mix thoroughly.
2. Grind the mixture twice in the meat grinder using the medium chopper plate. Check seasoning.
3. Form into 2 oz patties. Place in the refrigerator until ready to use.
4. Prepare by placing on sheet pans and baking, broiling, or grilling.

5. Serve two patties to each order as a breakfast meat or accompanied with applesauce or country gravy on the luncheon menu.

Precautions

- Exercise caution when boning and cutting the fresh picnic or callie and when grinding the meat.
- When grinding the meat, add the meat slowly to the grinder so the grinder does not clog.
- If meat sticks to the hands when forming the patties, coat the hands slightly with salad oil.

The tenderloin is the leanest cut of pork. A 3 oz serving contains 139 calories – comparable to a skinless chicken breast. Other lean cuts of pork include boneless loin roast, boneless sirloin chops, boneless loin chops, and boneless ham.

Broiled Pig's Feet

Broiled pig's feet are split lengthwise, poached, and left to cool. They are then passed through salad oil and bread crumbs and browned under the broiler. This item is served on the luncheon menu with a tart piquant sauce.

50 servings

Equipment

- Bake pans (2)
- 10 gal. stockpot
- Power meat saw
- Sheet pans (3)

Ingredients

50 pig's feet
water as needed
2 tbsp pickling spices
1 gal. piquant sauce
1 qt salad oil (variable)
3 lb bread crumbs (vari-

Preparation

1. Split the pig's feet lengthwise using a power saw.
2. Prepare the piquant sauce.
3. Coat the sheet pans with salad oil.
4. Place the bread crumbs in a bake pan.
5. Preheat the broiler.

Procedure

1. Place the split pig's feet in the stockpot. Add the pickling spices, cover with water, and simmer until tender (about 2 hr to 2½ hr).

2. Remove from the range, let cool, and refrigerate overnight, leaving the split feet in the stock.
3. Remove the split feet from the jellied stock. The stock jells when cold because of the natural gelatin in the pork bones. Place feet in a bake pan with salad oil.
4. Pass the feet through the salad oil then into a pan containing the bread crumbs. Coat thoroughly, pressing firmly, and shake off excess.
5. Place the feet on the oiled sheet pan. Put under a low broiler until brown.
6. Serve two halves per portion topped with piquant sauce.

Precautions

- Watch fingers when splitting the feet with the power saw.
- Exercise caution when browning the feet under the broiler. They sometimes pop, spraying hot liquid.

When stewing pork, the flavor of the stew can be varied considerably by the liquids used (water, broth, wine, beer, etc.) and the different herbs or spices selected. Less tender cuts like pork shoulder are well-suited for stewing.

Ham and Asparagus Rolls Mornay

Ham and asparagus rolls Mornay are a combination of ham and two asparagus spears rolled up in the ham. These rolls are covered with a Mornay sauce and glazed under the broiler.

50 servings

Equipment

- Boning knife
- Broiler
- Casserole dishes (50)
- Ladle
- Stainless steel container
- Bake pans
- Slicing machine
- Quart measure

Ingredients

100 horseshoe slices of cooked ham
200 asparagus spears
2 gal. Mornay sauce
2 qt ham stock

Preparation

1. Prepare ham stock.
2. Bone a cooked ham with a boning knife. Slice the horseshoe side of the ham on a slicing machine approximately ⅛″ thick. Heat the ham slices in the ham stock.
3. Cook asparagus spears if using frozen or fresh asparagus. If using canned asparagus, open and heat.
4. Prepare the Mornay sauce.

Procedure

1. Place two asparagus spears on each slice of ham. Roll and place two rolls in each shallow casserole.
2. Top the rolls with Mornay sauce using a ladle and glaze under the broiler until light brown.
3. Serve at once, two rolls per serving.

Precautions

- Use caution when boning and slicing the ham.
- When glazing the ham rolls, do not overbrown or the item will lack in appearance.

BOILED PORK RECIPES

Pork Chop Suey

Chop suey is a Chinese-American dish that is served in practically all restaurants. It is a very profitable item and can be served on either the luncheon or dinner menu.

4 gal. or 50 servings

Equipment

- French knife
- Paddle
- 5 gal. stockpot
- Small container
- Deep steam table pan
- Quart measure
- Cup measure
- Small braiser
- Kitchen spoon

Ingredients

4 qt fresh pork picnic
3 qt celery
2½ qt onions
1 pt mushrooms
1 c salad oil
½ No. 10 can bean sprouts
½ c soy sauce
2½ c cornstarch
1 lb 8 oz bamboo shoots
2 lb water chestnuts
3 qt water or chicken stock
1 pt water
salt and pepper to taste

Preparation

1. Dice the fresh pork into 1″ cubes with a French knife.
2. Place the cubes of pork in a braiser and brown on the range. Add a small amount of water and cook until tender.
3. Julienne the onions and celery with a French knife.
4. Slice the mushrooms fairly thin with a French knife.
5. Slice the bamboo shoots and water chestnuts thin with a French knife.
6. Prepare the chicken stock if it is to be used.

Procedure

1. Place the salad oil in a stockpot and heat.
2. Add the julienned onions and celery and the sliced mushrooms. Sauté until tender.
3. Add water or stock and the braised pork. Let boil until the vegetables are done but still crisp.
4. Add the soy sauce, bamboo shoots, and water chestnuts. Continue to boil.
5. In a separate container, dilute the cornstarch in water, stir with a kitchen spoon, and add to the chop suey. Stir constantly with a paddle. Cook for 5 min and remove from the heat.
6. Add the bean sprouts. Stir with a paddle.
7. Season with salt and pepper and place in a steam table pan.
8. Serve 6 oz to 8 oz with baked rice and fried Chinese noodles.

Note: Water chestnuts and bamboo shoots may be omitted if desired.

Precautions

- Exercise caution when using the knife.
- Vegetables should be crisp.
- Discontinue cooking when the bean sprouts are added.

Diced Ham and Lima Beans

This item is a combination of cooked dried lima beans and diced cooked ham. It is suitable for cafeteria service. If placed on the regular luncheon menu, it should be served in a casserole.

50 servings

Equipment

- French knife
- 10 gal. stockpot
- Baker's scale
- Kitchen spoon
- 1 gal. measure
- Paddle
- 1 qt saucepans
- Deep steam table pan

Ingredients

- 8 lb dried lima beans
- 10 lb boiled ham
- 1 lb 8 oz onions
- 1 lb salt pork or jowl bacon
- 3 gal. ham stock or 6 oz ham base in 3 gal. water
- salt and pepper to taste

Procedure

1. Place the stockpot containing the soaked lima beans and ham stock on the range and bring to a boil. Place on the side of the range away from the heat and continue to simmer.
2. Cook the salt pork or jowl bacon in a saucepan until it becomes a light brown crackling. Stir occasionally with a kitchen spoon.
3. Add the minced onions and cook until tender. Add to the boiling beans, continuing to simmer until beans are tender. Total cooking time is about 2 hr.
4. Add the diced ham and check seasoning. Stir with a paddle. Place in a deep steam table pan.
5. Serve 6 oz to 8 oz per portion with a ladle into casseroles. Accompany with a slice of Boston brown bread.

Preparation

1. Dice the cooked ham into ½" cubes with a French knife.
2. Prepare the ham stock.
3. Clean and soak the beans in the ham stock overnight. Do not refrigerate.
4. Mince the onions and dice the boiled ham and the salt pork or jowl bacon with a French knife.

Precautions

- Use caution when dicing ham and mincing onions.
- Do not overcook the beans. They will become mushy and appearance will be lacking.
- When sautéing the onions, do not brown.

Using Leftover Pork

Both raw and cooked pork left from trimming, cutting, boning, and cooking must be utilized in some way if profits and low food cost percentages are to be maintained. Some suggestions are as follows:

Cooked Pork
- Dice and use in chop suey or egg rolls.
- Dice and use in hash.
- Dice and use in barbecue.
- Julienne or dice and use in salads.
- Slice and use on a cold meat plate.
- Grind and use in preparing the meat/rice stuffing for green peppers and cabbage.
- Julienne and use in chow mein or stir fry.

Ham
- Dice or julienne and use in salads.
- Dice and cook with dried lima beans.
- Mince and use to flavor beans or bean soups.
- Grind and use for stuffing ham turnovers.

Raw Pork
- Grind for chili, meat sauce, ham loaf, sweet-sour ham balls, stuffing (forcemeat) for pork chops and cutlets, Swedish meatballs, pork sausage, etc.
- Dice for chop suey, Hawaiian pork, etc.

Lamb Preparation

Lamb is the flesh of immature sheep and is the least popular of the meats commonly used in the commercial kitchen. Mutton, which is rarely used in the commercial kitchen, is the flesh of mature sheep. Mutton has a strong flavor and odor that many people find undesirable.

The age and diet of the lamb affect the quality of meat used for preparation. Lamb is marketed as genuine spring lamb, spring lamb, or yearling lamb. Genuine spring lamb is marketed at about 3 months to 5 months old. Spring lamb is marketed at about 5 months to 10 months old. Yearling lamb is marketed at about 12 months to 20 months old. The flesh of lamb feeding on milk is pale pink and tender. As the lamb feeds on grass and grain, the flesh becomes darker and tougher.

Lamb is graded as U.S. prime, U.S. choice, U.S. good, U.S. utility, and U.S. cull. U.S. prime, U.S. choice, and U.S. good are the only grades commonly used in the commercial kitchen. Lamb grading is not as important as beef grading because the lambs are young when marketed. The most popular preparations using lamb are lamb chops and roast lamb.

LAMB PREPARATION

Lamb is the flesh of immature sheep, both male and female, approximately 12 months old. *Mutton* is the flesh of mature sheep, 20 months or older. Lamb is tender and delicate in flavor. Three types of lamb are available on the market: genuine spring lamb, spring lamb, and yearling lamb. *Genuine spring lamb* is available on the market from April to July and is considered the best lamb. Genuine spring lambs are marketed when they are three months to five months old and are fattened primarily on their mother's milk. They are also classified as *milk lambs* and *Easter lambs*.

Spring lambs are marketed during fall and winter and are fattened primarily on grass and grain. Spring lambs are usually 5 months to 10 months old when slaughtered and shipped to market.

Yearling lambs are about 12 months to 20 months old when marketed. Yearling lambs are generally too young to be sold as mutton and too old to be sold as lamb. Yearling lambs are considered the most economical lamb to use. However, special preparation may be required to achieve the desired taste.

When a lamb is born, it feeds on its mother's milk until it is about five months old. It is at this point that it will produce the finest eating qualities. The flesh is a pale pink and has a smooth grain. The bones are soft, porous, and have a reddish tinge and the fat is firm. After five months, the lamb is sent out to pasture. The lamb then feeds on grass and grain and starts to mature. As the lamb ages, changes

in the character, color, and consistency of the flesh, bones, and fat of the animal occur. The animal passes from lamb to yearling to mutton. The flesh becomes darker, and the bones whiter and harder. The fat becomes soft and slightly greasy. When the animal reaches the yearling and mutton stage, the edible flesh is dry and strong in flavor.

LAMB GRADING

Lamb is graded using the same criteria as beef. Grading is based on quality and yield. Like pork and veal, the grading of lamb is not as important as the grading of beef because the animal is young and usually tender. The five grades of lamb, in order of quality, are U.S. prime, U.S. choice, U.S. good, U.S. utility, and U.S. cull. U.S. prime, U.S. choice, and U.S. good are the only grades used commonly in the kitchen.

U. S. Prime is the highest quality of lamb. To be stamped prime, the animal must be three months to five months old and still feeding on its mother's milk. The U.S. prime lamb carcass is compact and has plump legs. The back is wide and thick and the neck short and thick. The interior displays pale pink flesh with a smooth grain, soft and porous bones, and the fat around the kidneys is white and firm.

U. S. Choice is the most popular grade of lamb used in food service establishments. It is high-quality lamb and has excellent eating qualities. To be graded choice, the lamb carcass must have a slightly compact body, the legs must be short and plump, and the back slightly wide and thick. The neck must be slightly short and thick. The interior of the animal has a pink flesh just slightly darker than the prime grade. The flesh grain is smooth, the bones soft and porous, and there is a fairly generous amount of white fat around the kidney.

U.S. good is the most economical grade of lamb. It is only used in certain preparations with good results. The U.S. good lamb is slightly rangy and bony. It has slightly thin legs, a fairly narrow, thin back, and a slightly long, narrow neck. The interior of the carcass shows a dark pink flesh with a slightly rough grain. The bones are just slightly hard and lighter in color with little or no pink tinge visible. There is less fat, with the fat around the kidney not as abundant, and the kidney may be slightly exposed.

U.S. utility is rarely used in the commercial kitchen. It is very rangy and bony with thin, moderately tapered legs. The U.S. utility lamb has a narrow, slightly sunken back and a long, thin neck. The interior of the carcass has a dark red flesh with a fairly rough grain. Little or no fat is present except a small amount slightly covering the kidney.

U.S. cull is the poorest grade of lamb and is never used in the commercial kitchen. It is extremely rangy and bony. The legs are extremely thin, the back very sunken and thin, and the neck extremely thin and long. There is no evidence of interior fat and the flesh is dark red with a soft, watery texture.

COMMERCIAL CUTS

Lamb may be purchased in four commercial cut forms: carcass, saddle (foresaddle and hindsaddle), wholesale or primal cuts, or fabricated. The form best suited for any particular use depends on many factors:

- meat cutting skill of personnel
- meat cutting equipment available
- working space available
- meat storage space available
- utility of all cuts purchased
- meat preparations served
- overall economy of purchasing meat in a particular form

The *carcass* is the complete animal with head, hide, and entrails removed. A typical lamb carcass averages from 30 lb to 75 lb, with the majority being 35 lb to 45 lb. The carcass can be purchased at a cheaper cost per pound because less labor in handling is required. However, with greater emphasis on productivity and standardization, more food service establishments are turning to fabricated cuts.

Saddle cuts are not converted into sides as beef is by splitting or sawing lengthwise through the spine of the animal. Saddle cuts are separated into saddles by cutting across the complete carcass between the twelfth and thirteenth ribs. The fore portion of the lamb becomes the *foresaddle*, the hind portion becomes the *hindsaddle*. The foresaddle includes the first through the twelfth ribs, the thirteenth rib remaining part of the hindsaddle. The hindsaddle, which includes the loin and leg cuts, is generally more popular than the foresaddle. The rib cut, which is part of the foresaddle, is the most popular prepared lamb cut. Other parts of the foresaddle are not quite as popular. As with the carcass, the saddle cut is being replaced by more convenient fabricated cuts.

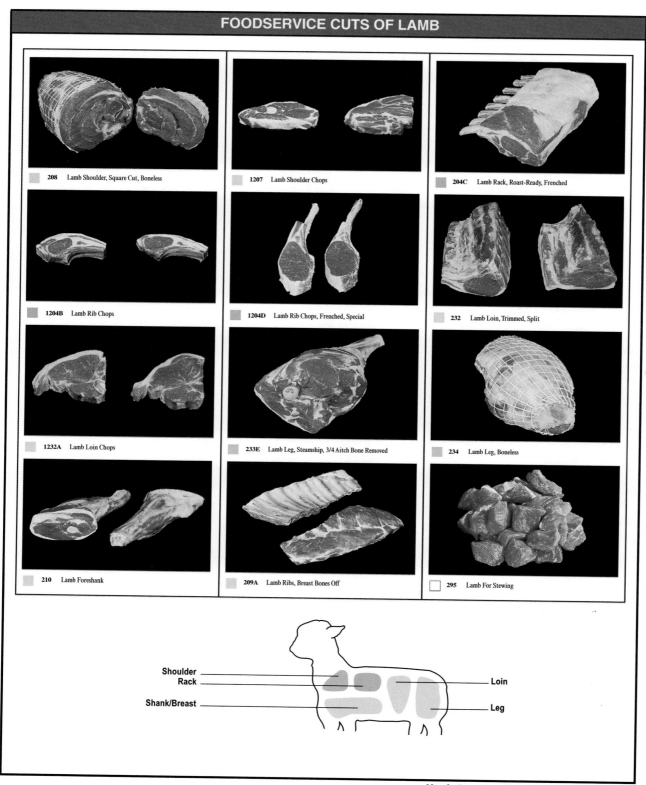

FOODSERVICE CUTS OF LAMB

208   Lamb Shoulder, Square Cut, Boneless

1207   Lamb Shoulder Chops

204C   Lamb Rack, Roast-Ready, Frenched

1204B   Lamb Rib Chops

1204D   Lamb Rib Chops, Frenched, Special

232   Lamb Loin, Trimmed, Split

1232A   Lamb Loin Chops

233E   Lamb Leg, Steamship, 3/4 Aitch Bone Removed

234   Lamb Leg, Boneless

210   Lamb Foreshank

209A   Lamb Ribs, Breast Bones Off

295   Lamb For Stewing

Shoulder
Rack
Shank/Breast
Loin
Leg

North American Meat Processors Association

NAMP/IMPS (North American Meat Processors Association/Institutional Meat Purchase Specifications) specifies meat cuts by number.

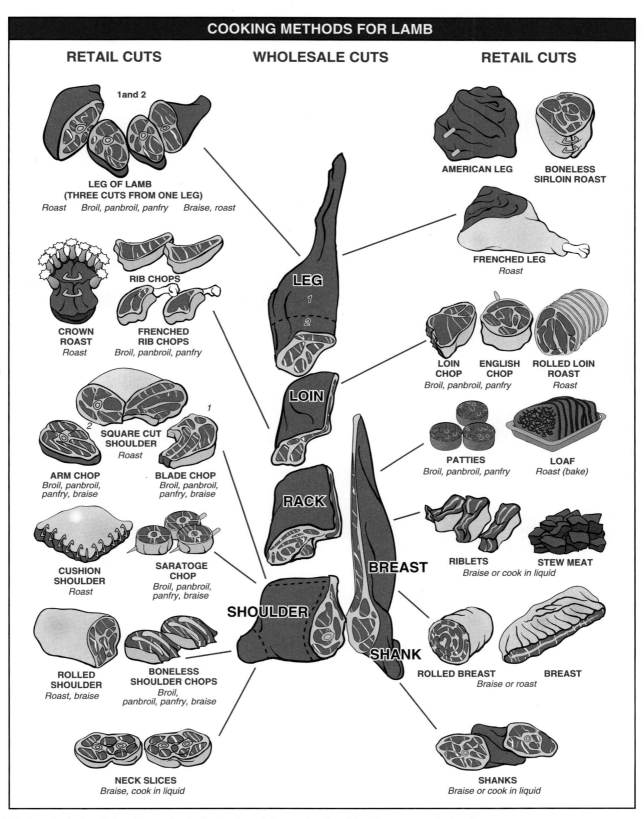

COOKING METHODS FOR LAMB

RETAIL CUTS   WHOLESALE CUTS   RETAIL CUTS

1and 2

**LEG OF LAMB
(THREE CUTS FROM ONE LEG)**
Roast   Broil, panbroil, panfry   Braise, roast

RIB CHOPS

**CROWN
ROAST**
Roast

**FRENCHED
RIB CHOPS**
Broil, panbroil, panfry

**SQUARE CUT
SHOULDER**
Roast

ARM CHOP
*Broil, panbroil,
panfry, braise*

BLADE CHOP
*Broil, panbroil,
panfry, braise*

**CUSHION
SHOULDER**
Roast

**SARATOGE
CHOP**
*Broil, panbroil,
panfry, braise*

**ROLLED
SHOULDER**
Roast, braise

**BONELESS
SHOULDER CHOPS**
*Broil,
panbroil, panfry, braise*

NECK SLICES
Braise, cook in liquid

LEG
1
2

LOIN

RACK

SHOULDER

BREAST

SHANK

AMERICAN LEG

**BONELESS
SIRLOIN ROAST**

FRENCHED LEG
Roast

**LOIN
CHOP**

**ENGLISH
CHOP**

**ROLLED LOIN
ROAST**
Roast

Broil, panbroil, panfry

PATTIES
Broil, panbroil, panfry

LOAF
Roast (bake)

RIBLETS

STEW MEAT
Braise or cook in liquid

ROLLED BREAST

BREAST
Braise or roast

SHANKS
Braise or cook in liquid

Wholesale (primal) lamb cuts include the leg, loin, rack, shoulder, breast, and shank.

Wholesale or *primal* cuts are parts of the foresaddle and hindsaddle of the lamb carcass. The six lamb wholesale or primal cuts are the shoulder, shank, rib, loin, breast, and leg. The cuts of the foresaddle include the shoulder, rib, shank, and breast. The cuts of the hindsaddle include the loin and leg. Purchasing lamb by the wholesale cut is the most popular form after fabricated cuts.

Fabricated cuts are purchased ready to cook. Fabricated cuts are cut to a uniform size and weight. They are easy to store and reduce labor cost at the food service establishment. However, as with most convenient foods, fabricated cuts are more expensive than other cuts. Fabricated cuts are the most popular cuts of lamb used in commercial food preparation.

WHOLESALE OR PRIMAL CUTS: COOKING METHODS

The six wholesale or primal lamb cuts are still used frequently in certain preparations in the commercial kitchen. The most common preparations include rib or rack, loin or saddle, leg, shoulder, breast, and shank.

The *rib* or *rack* is the most popular of all the lamb cuts. It contains seven ribs and a rib eye muscle of solid lean meat. This cut lies between the shoulder and the loin. The rib or rack gained its popularity through the chops it provides. When two ribs are joined or unsplit, they are known as a *rack of lamb*. This rack is generally used to prepare a very popular menu item, roast crown of lamb. The rib of lamb is sometimes roasted in the commercial kitchen, but in most cases, it is cut into chops.

The *loin* or *saddle* cut is located between the rib and leg cuts. It contains a loin eye, a small T-shaped bone, a tenderloin, and a small amount of flank meat, which is removed when the loin is trimmed. When a complete loin from the whole lamb carcass is unsplit, the cut is known as a *saddle of lamb*. It is from the saddle that English lamb chops are cut. English lamb chops are 2″ thick cuts taken along the entire length of the unsplit loin. The loin is roasted on certain occasions in the commercial kitchen, but in most cases it is converted into chops or English lamb chops.

The *leg* is the hind leg of the lamb. It contains a *shank bone* and an *aitchbone*. The aitchbone is the buttock or rump bone and lies at the top of the leg.

The flesh is solid, lean, and fine-textured. The leg averages four pounds to nine pounds. It is popular in the commercial kitchen when cut into lamb steaks or when boned and roasted.

The *shoulder* is the largest and thickest part of the foresaddle. It contains five rib bones and a high percentage of lean meat. It is used in the commercial kitchen in the preparation of lamb stew, patties, fricassee, and curry.

The *breast* is a thin, flat cut containing the breastbone and the tips of 12 ribs. It has alternating layers of fat and lean. It is not a very popular cut of lamb, but is used in the commercial kitchen for such preparations as stuffed breast of lamb and lamb riblets.

The *shank* is the forelegs of the animal. It contains a large portion of bone and connective tissue with little lean meat. To be utilized in the commercial kitchen, the shanks are generally braised and served with jardiniere-cut vegetables. The shank meat is sometimes ground and formed into lamb patties.

Burgers' Smokehouse

The rack of lamb is the most popular lamb cut.

VARIETY MEATS

Lamb variety meats include kidneys, liver, brains, sweetbreads, and tongue. Lamb variety meats are not as popular as the other edible animals. The kidneys, liver, brains, and sweetbreads are all processed and prepared in the same manner as those of beef, veal, and pork. The lamb tongue, if not utilized in some kind of sausage, can be pickled and placed on the market as pickled lamb's tongue.

LAMB TERMINOLOGY

Terms are used within the culinary trade pertaining to lamb, allowing better communication in the field. These terms include:

Lamb: The flesh of immature sheep.

Mutton: The flesh of mature sheep.

Yearling: The flesh of lambs 12 months to 20 months old.

Fell: The thin, paper-like covering over the outside of a lamb's carcass.

English lamb chop: Several 2″ thick cuts taken along the entire length of the unsplit lamb loin.

Frenched: Generally applied to chops. It means the meat and fat are removed from the end of the rib bones. The ribs of a crown roast are frenched and sometimes the leg bone of a roast leg of lamb is frenched.

Cull: The poorest grade of lamb.

Crown roast: Prepared from the unsplit rack. The rib ends are frenched and the ribs are formed into a crown.

Riblets: Rectangular strips of meat, each containing part of a rib bone. Cut from the breast.

Hotel rack: The unsplit rib section of the carcass.

Mock duck: Made from shoulder and shank cuts, shaped like a duck and roasted.

Hothouse lamb: Lamb produced under artificial conditions. It is available on the market from January to March, but supplies are small. Hothouse lamb is generally graded choice or good.

Double lamb chop: One rib chop cut to a thickness equal to two rib chops.

Americans consume approximately seven pounds of lamb per person each year. It is for this reason that lamb entrees do not appear often on the menus of food service establishments. With the exceptions of lamb chops and roast lamb, chefs and managers frequently prefer to use more popular meats. However, there are many lamb entrees that offer additional variety.

ROASTED LAMB RECIPES

The rack, one of the more popular cuts of lamb, is roasted medium or rare and served with a rich au jus. Roasting gives the cut a little more versatility. It can be presented to the guest sliced into individual servings or carved at the table for an interesting display. This entree is usually prepared for a small group.

Roast Rack of Lamb

16 servings

Equipment

- Roast pan
- Meat saw or cleaver
- Boning knife
- French knife
- China cap
- Kitchen fork
- Cheesecloth
- Quart measure
- Wire whip

Ingredients

2 complete lamb racks
salt and pepper as needed
2 garlic cloves
1 qt veal stock
marjoram, as needed

Preparation

1. Prepare the two racks for roasting by cutting down both sides of the feather bones all the way to the chine bone, using a boning knife. Stand the rack up and, with a cleaver, chop through the rib bones on both sides where they join the chine bone. This will separate the rack into two rib sections. Each rib section will contain eight rib bones. Cut off the ends of the rib bones if they appear too long. Using a boning knife, trim off the excess fat. French the ends of the rib bones by removing the fat and meat until the eight bones are exposed.
2. Preheat the oven to 400°F.
3. Mince the garlic using a French knife.

Procedure

1. Place the bones that were removed from the rack in the bottom of the roast pan. Place the trimmed racks, fat side up, on top of the bones. Season with salt, pepper, and marjoram.
2. Roast in the preheated oven at 400°F for approximately 30 min to 45 min or until desired doneness is achieved (usually medium or rare).
3. Remove from the oven. Using a kitchen fork, remove the lamb from the roast pan and place in a bake pan. Hold in a warm place. Leave the bones in the roast pan.
4. Pour the fat from the roast pan and set the pan on the range. Add the garlic. Cook just slightly.
5. Add the veal stock and let the liquid simmer in the pan to deglaze the pan. Remove from the range and strain the liquid through a china cap covered with cheesecloth to eliminate all foreign matter. Season the au jus to taste.
6. Using a French knife, cut between the ribs and serve two ribs to each order with a portion of au jus.

Precautions

- Exercise extreme caution when cutting and chopping the racks.
- Be alert when pouring the grease from the roast pan and straining the au jus.

Roast Leg of Lamb

For roast leg of lamb, the leg is boned and roasted at a moderate temperature. It is sliced and served with its natural gravy and mint sauce. It is an excellent entree for the dinner menu.

50 servings

Equipment

- Baker's scale
- Boning knife
- Roast pan
- 2 gal. sauce pot
- French knife
- Butcher twine
- Kitchen fork
- Bake pan
- Wire whip
- 2 gal. stainless steel container
- China cap
- Spoon measure
- Cup measure

Ingredients

5 lb to 6 lb legs of lamb (2)
12 oz onions
8 oz carrots
8 oz celery
½ tsp garlic
½ c salad oil
1 tsp rosemary leaves
1 tbsp marjoram
1 lb shortening
5 oz flour
1½ gal. brown stock
1 c tomato puree
caramel color as needed
salt and pepper to taste

Preparation

1. Bone the leg of lamb by removing the aitch, leg, and shank bones with a boning knife. Roll the boneless meat and tie with butcher twine.
2. Mince the garlic with a French knife.
3. Cut the vegetables rough with a French knife.
4. Prepare the brown stock.
5. Preheat oven to 375°F.

Procedure

1. Place the tied, boneless legs of lamb in the roast pan. Rub the lamb with salad oil and season with salt, pepper, and marjoram.
2. Roast in the preheated oven at 375°F until the roasts become brown. Add the garlic, rosemary leaves, and rough garnish (onions, carrots, and celery). Reduce oven temperature to 325°F and continue to roast until the meat is done (approximate total time 2 hr to 2½ hr).
3. Remove the lamb from the roast pan with a kitchen fork. Place in a bake pan, remove twine, and hold. Keep warm.
4. Pour the brown stock into the roast pan and bring to a boil on the range to deglaze the roast pan.
5. Pour the liquid from the roast pan into a stainless steel container and keep hot.
6. Place the shortening in a sauce pot and heat.
7. Add the flour, making a roux, and cook for 2 min to 3 min. Add the hot brown stock and tomato puree, whipping vigorously with a wire whip until slightly thickened and smooth. Let simmer for 15 min to 20 min.
8. Check seasoning. If the gravy is too light, add caramel color to darken. Strain through a china cap back into the stainless steel container.
9. Slice the roast lamb against the grain on a slicing machine or with a French knife. Serve approximately 3 oz per portion, covered with the brown gravy and accompanied with a soufflé cup of mint sauce or mint jelly.

Precautions

- Do not overcook the meat. It will be difficult to slice and will lack appearance when served.
- While the gravy is simmering, stir occasionally to avoid scorching.
- If the gravy needs to be darkened, add caramel coloring with caution.

BRAISED OR STEWED LAMB RECIPES

Barbecued Lamb Riblets

The riblets are similar to short ribs of beef. They are cut from the breast of lamb, browned in the oven, and baked until tender in a rich, tasty barbecue sauce.

25 servings

Equipment

- Large braiser and cover
- Kitchen fork
- Quart measure
- Kitchen spoon
- Baker's scale
- Steam table pan
- Hand meat saw or power saw

Ingredients

14 lb lamb breast
1 c salad oil
2 gal. barbecue sauce
salt and pepper to taste

Preparation

1. Cut the lamb breast into riblets. Cut lengthwise with a hand meat saw or power saw about 1½" to 2" wide the full length of the breast. Then cut crosswise into 3" lengths.
2. Prepare the barbecue sauce.
3. Preheat oven to 375°F.

Procedure

1. Place the salad oil in the braiser and heat.
2. Add the riblets and brown thoroughly. Turn occasionally with a kitchen fork.
3. Pour the barbecued sauce over the riblets. Cover the braiser and bake in the preheated oven at 375°F until the riblets are tender.

4. Remove from the oven and check the seasoning. Place in a steam table pan.
5. Serve two riblets per portion accompanied with baked rice.

Precautions

- While baking the riblets in the sauce, stir frequently with a kitchen spoon to avoid scorching or sticking.
- Do not overcook the riblets. They will be lacking in appearance when served.

The breast is a slightly undesirable cut of lamb, since it contains only thin layers of lean meat and many breast bones. It is transformed into a desirable menu item by stuffing with a forcemeat mixture and braising in the oven.

Braised Stuffed Breast of Lamb

25 servings

Equipment

- Boning knife
- Full size steam table
- French knife
- Butcher twine
- Large eye needle
- Baker's scale
- Quart measure
- Kitchen fork
- Kitchen spoon
- Skillet
- Wire whip
- China cap
- 3 gal. mixing container
- Roast pan
- Meat grinder
- 1 gal. sauce pot
- 1 gal. stainless steel container

Ingredients

- 5 lb breasts of lamb (4)
- 5 lb boneless lamb shoulder
- 1½ lb dry bread cubes
- 1 qt milk (variable)
- 8 oz onions
- 6 oz celery
- 8 oz bread crumbs (variable)
- 6 egg yolks
- 8 oz butter
- ¼ oz sage
- 6 oz shortening
- 4 oz flour
- 2 qt brown stock
- salt and pepper to taste

Preparation

1. Trim the four lamb breasts of excess fat and cartilage with a French and boning knife. Cut a pocket in each breast by slicing with a boning knife between the flesh and the breast bones. Make the opening as large as possible, but do not cut through the flesh at any point.
2. Prepare the brown stock.
3. Mince the onions and celery with a French knife.
4. Cut the lamb shoulder into strips that will fit into the grinder. Cut with a French knife.
5. Separate the eggs and beat the yolks slightly with a wire whip. Save the whites for use in another preparation.
6. Preheat oven to 350°F.

Procedure

1. Place the dry bread cubes and the milk in a mixing container and mix with a kitchen spoon until the bread has absorbed the milk.
2. Sauté the onions and celery in butter in a skillet. Add to the mixture.
3. Add the strips of lamb shoulder and sage. Mix thoroughly by hand.
4. Grind this mixture twice in a meat grinder using the fine chopper plate.
5. Add the slightly beaten egg yolks, season with salt and pepper, and mix thoroughly. If the mixture is too wet, add the bread crumbs as needed. If the mixture is too dry, add more milk.
6. Stuff and pack the forcemeat mixture fairly solid into the pockets cut into the lamb breast.
7. Using a large eye needle and butcher twine, sew the opening between the layer of meat and the breast bones. Secure properly so the forcemeat will not come out during the roasting period.
8. Season the stuffed breasts with salt and pepper and place in a roast pan. Brown thoroughly in the preheated oven at 350°F. Turn occasionally with a kitchen fork.
9. Pour the brown stock over the breasts and continue to braise until the breasts are tender and the forcemeat has become solid.
10. Remove from the oven and place in a steam table pan. Let set in a warm place for 45 min.
11. Place the shortening in a sauce pot and heat.
12. Add the flour, making a roux. Cook for 5 min.
13. Add the brown stock in which the breasts were braised, whipping vigorously with a wire whip until thickened and smooth. Strain through a fine china cap into a stainless steel container.
14. Slice the stuffed breasts to order with a French knife. Cut six portions from each breast. Serve over buttered noodles accompanied with brown sauce.

Precautions

- Do not break through the flesh when cutting the pocket.
- Do not stick a fork into the meat during the roasting period.
- Use a sharp French knife and apply very little pressure when slicing each order of the lamb breast.

Trade Tip

When preparing a lamb stew or roasting a leg, shoulder, or rack of lamb, season with the herb marjoram. Marjoram is of the mint family; mint improves the flavor of lamb. Many lamb entrees are served with some kind of mint preparation.

French Lamb Stew

French lamb stew is a brown stew containing meat that is browned before it is stewed to bring out a very rich flavor. The vegetables are cooked separately and blended into the stew when the lamb becomes tender. Like most stews, it is more desirable on the luncheon rather than the dinner menu.

50 servings

Equipment

- French knife
- Large braiser and cover
- 4 qt saucepans (2)
- 2 qt saucepan
- Baker's scale
- Kitchen spoon
- Spoon measure
- Quart measure
- China cap
- Steam table pan

Ingredients

18 lb boneless lamb shoulder
2½ gal. brown stock
3 c salad oil
½ No. 10 can whole tomatoes
2 garlic cloves
1 lb 4 oz flour
3 lb carrots
2 lb celery
½ No. 10 can whole small onions
2½ lb frozen peas
½ No. 10 can cut green beans
8 oz onions
1 bay leaf
1 tbsp marjoram
salt and pepper to taste

Preparation

1. Cut the boneless lamb shoulder into 1″ cubes with a French knife.
2. Dice the carrots with a French knife.
3. Cut the celery ½″ wide on a bias (slant) with a French knife.
4. Mince the onions and garlic with a French knife.
5. Crush the tomatoes by hand.
6. Preheat oven to 375°F.
7. Prepare the brown stock.

Procedure

1. Place the salad oil in a braiser and heat.
2. Add the cubes of lamb and brown thoroughly. Stir with a kitchen spoon.
3. Add the minced onions and garlic and continue to cook for 5 min.
4. Add the flour and blend thoroughly with a kitchen spoon, making a roux. Cook for 5 min more.
5. Add the brown stock, marjoram, and bay leaf and stir. Cover braiser and cook in the preheated oven at 375°F for approximately 1½ hr or until the lamb cubes are tender.
6. Boil all the raw vegetables in separate saucepans in salt water until tender. Drain through a china cap.
7. When the lamb is tender, remove from the oven. Remove the bay leaf and add the crushed tomatoes and all the drained, cooked vegetables except the peas.
8. Serve 6 oz to 8 oz in casseroles topped with the green peas.

Precautions

- Do not overcook the cubes of lamb.
- When adding the flour to make the roux, be sure to cook slightly or the stew will have a flour taste.
- Drain all vegetables thoroughly before adding to the stew.

Sour Cream Lamb Stew

For sour cream lamb stew, cubes of lamb shoulder are gently browned and then baked in a sour cream sauce with mushrooms. This item is similar to stroganoff preparations and is a welcome addition to the luncheon menu when served with buttered noodles or baked rice.

50 servings

Equipment

- French knife
- Baker's scale
- Quart measure
- Large braiser and cover
- Kitchen spoon
- 5 gal. sauce pot
- Cup measure
- Deep steam table pan

Ingredients

18 lb boneless lamb shoulder
2½ gal. brown stock
2 lb shortening
1 lb 8 oz flour
1 c tomato puree
1 c vinegar
2 lb onions
2 bay leaves
½ c chives
1 lb mushrooms
1½ qt sour cream
1 pt salad oil
salt and pepper to taste

Preparation

1. Cut the boneless lamb shoulder into 1″ cubes with a French knife.
2. Prepare the brown stock.
3. Mince the onions with a French knife.
4. Slice the mushrooms thick and chop the chives with a French knife.
5. Preheat oven to 350°F.

Procedure

1. Place the salad oil in a large braiser and heat.
2. Add the cubes of lamb and brown thoroughly. Stir occasionally with a kitchen spoon.
3. Add the minced onions and garlic. Continue to cook for 5 min more. Remove from the range and hold.
4. Place the shortening in a sauce pot and heat.
5. Add the mushrooms and sauté slightly.

6. Add the flour, making a roux, and cook for 5 min.
7. Add the hot brown stock, tomato puree, vinegar, and bay leaves. Stir gently with a kitchen spoon until the sauce becomes slightly thick and smooth.
8. Pour the sauce over the browned cubes of meat, cover the braiser, and bake in the preheated oven at 350°F until the meat becomes tender.
9. Remove from the oven and remove the bay leaves. Stir in the sour cream with a kitchen spoon and season with salt and pepper.
10. Add the chives and blend thoroughly. Place in a deep steam table pan.

11. Serve in casseroles accompanied with buttered noodles and baked rice.

Precautions

- When adding the liquid to the roux, stir gently so the mushrooms will not be broken.
- When the stew is cooking in the oven, stir occasionally.
- Exercise caution when removing the cover from the braiser to minimize the amount of steam escaping.
- Do not overcook the cubes of lamb. They will lack in appearance when served.

Braised Lamb Shanks Jardiniere

Braised lamb shanks jardiniere consist of small lamb shanks cooked using the braising method and served with jardiniere vegetables. One whole lamb shank is served to each order. This is a fairly popular luncheon item.

25 servings

Equipment

- Large braiser and cover
- Kitchen fork
- Wire whip
- French knife
- 4 qt saucepans (2)
- 1 qt saucepan
- Baker's scale
- 2 gal. sauce pot
- Quart measure
- Cup measure
- Boning knife
- Deep steam table pan
- China cap

Ingredients

25 lamb shanks
1½ gal. brown stock
1 lb shortening
12 oz flour
3 lb carrots
2 lb turnips
1 lb celery
1 c tomato puree
2½ lb frozen peas
1 tsp marjoram
1 bay leaf
½ No. 10 can whole onions
1 c salad oil
1 c onion
2 garlic cloves
salt and pepper to taste

6. Heat the canned onions and drain.
7. Mince the onions and garlic with a French knife.
8. Preheat oven to 375°F.

Procedure

1. Place the lamb shanks in a large braiser. Pour the salad oil over them and brown thoroughly in the preheated oven at 375°F.
2. Sprinkle the minced onion and garlic over the shanks and continue to roast for 10 min more. Turn occasionally with a kitchen fork.
3. Add the brown stock, tomato puree, bay leaf, and marjoram. Cover the braiser and reduce the oven temperature to 350°F. Continue to cook until the shanks are tender. Remove from the oven.
4. Boil the carrots, celery, and turnips in salt water in separate saucepans. Drain and hold.
5. Remove the shanks from the braiser and place in a deep steam table pan. Cover and keep warm.
6. Place the shortening in a sauce pot and heat.
7. Add flour, making a roux, and cook for 5 min.
8. Add the hot brown stock in which the shanks were cooked, whipping vigorously with a wire whip until slightly thickened. Simmer for 15 min. Strain through a china cap back over the shanks.
9. Serve one shank per portion, topped with gravy, jardiniere vegetables, and peas.

Preparation

1. Trim the lamb shanks with a boning knife. Use the section of meat between the first and second joints.
2. Prepare the brown stock.
3. Cut the carrots and turnips jardiniere style (1″ by ¼″) with a French knife.
4. Slice the celery on a bias ¼″ thick with a French knife.
5. Cook the frozen peas in boiling salt water in a saucepan.

Precautions

- Do not overcook the shanks because the meat will fall away from the bone.
- Stir occasionally while the sauce is simmering to avoid scorching.

Trade Tip

When broiling French lamb chops, place the chops on the broiler so the exposed rib bone is facing the open end of the broiler. That is, toward the cook, so they will not receive excessive heat and burn. "French" means to remove meat and fat from bones a little distance from the end.

Potted Leg of Lamb

Potted leg of lamb is cooked by the braising method. The leg is browned thoroughly in the oven, placed in a stock pot, covered with a liquid, and simmered until tender. This is an excellent luncheon or dinner entree.

50 servings

Equipment

- Butcher twine
- French knife
- Boning knife
- Roast pan
- 10 gal. stock pot
- Wire whip
- China cap
- 5 gal. sauce pot
- 2 gal. stainless steel container
- Bake pan
- Kitchen fork
- Baker's scale

Ingredients

5 lb to 6 lb legs of lamb (3)
1 lb onions
½ lb carrots
½ lb celery
2 gal. brown stock
1 lb 4 oz shortening
1 lb flour
1 c tomato puree
1 tsp marjoram
1 bay leaf
salt and pepper to taste

Preparation

1. Bone the leg of lamb by removing the aitchbone and the leg and shank bones with a boning knife. Roll the boneless meat and tie with butcher twine.
2. Cut the celery, carrots, and onions rough with a French knife.
3. Prepare the brown stock.
4. Preheat the oven to 375°F.

Procedure

1. Place the boned, tied legs of lamb in a roast pan. Brown in the preheated oven at 375°F.
2. Add the rough garnish (onions, carrots, and celery). Continue to roast until the garnish is slightly brown.
3. Remove the meat from the oven and place in a stock pot. Deglaze the roast pan with the brown stock. Pour over the meat.
4. Add the tomato puree, marjoram, thyme, and bay leaf. Place on the range and let simmer until the meat is tender. Remove the meat from the liquid with a kitchen fork, place in a bake pan, cover with a wet towel, and keep warm.
5. Place the shortening in the sauce pot and heat.
6. Add the flour, making a roux. Cook 10 min.
7. Pour the brown stock into the roux, whipping constantly with a wire whip until thickened and smooth.
8. Simmer the gravy for 15 min and strain through a china cap into a stainless steel container.
9. Slice the potted lamb across the grain on a slicing machine or with a French knife. Serve 2½ oz to 3 oz per portion with the rich gravy. Mint sauce or mint jelly should accompany each portion.

Precautions

- Exercise caution when boning the leg of lamb.
- Do not overcook the lamb.

Curried Lamb

Curried lamb is a stew. Cooked cubes of lamb are blended into a rich curry sauce and generally served with rice and chutney. It is an excellent choice for the luncheon menu.

50 servings

Equipment

- French knife
- Wire whip
- 5 gal. sauce pot
- 5 gal. stock pot
- Kitchen spoon
- China cap
- Deep steam table pan
- Cheesecloth
- Bake pan
- 3 gal. stainless steel container
- Quart measure
- Baker's scale
- Paddle
- Skimmer

Ingredients

18 lb boneless lamb shoulder
2½ gal. water
2 lb butter or shortening
1 lb 8 oz flour
⅓ c curry powder
2 qt tart apples
2 lb onions
½ tsp ground cloves
½ tsp nutmeg
2 bay leaves
1 tsp marjoram
salt and white pepper to taste

Preparation

1. Cut the boneless lamb shoulder into 1" cubes with a French knife.
2. Dice the tart apples and onions small with a French knife.

Procedure

1. Place the meat in a stock pot, cover with the water, and bring to a boil. Remove any scum that may appear on the surface with a skimmer.
2. Add the bay leaves and marjoram and let simmer until the cubes of lamb are tender. Remove from the heat.
3. Strain the stock through a china cap covered with a cheesecloth into a stainless steel container. Keep hot. Place the cooked cubes of meat in a pan, cover with a wet towel, and keep warm.
4. Place the butter in a large sauce pot and melt.
5. Add the onions and sauté until slightly tender. Add the flour and curry powder. Blend thoroughly with a kitchen spoon and cook for 5 min more.
6. Add the hot stock, whipping vigorously with a wire whip until thickened and slightly smooth.
7. Add the apples, nutmeg, and cloves. Simmer for about 20 min to 30 min. Stir constantly.
8. Strain through a china cap into the stock pot. Add the cooked cubes of lamb and blend into the sauce with a paddle.

9. Check the seasoning and consistency. Place in a deep steam table pan.
10. Serve into shallow casseroles with baked rice and chutney.

Precautions

- Do not overcook the lamb cubes or their appearance will be affected.
- Keep the cooked lamb cubes covered with a damp towel at all times or they will dry out and discolor.
- When sautéing the onions, do not let them brown.

Irish Stew

Irish stew is a white or boiled stew. All the ingredients are boiled, which makes it easy to digest. Irish stew is commonly served with dumplings.

50 servings

Equipment

- French knife
- 5 gal. stock pot
- Skimmer
- 3 qt saucepans (3)
- China cap
- Cheesecloth
- Wire whip
- Paddle
- Baker's scale
- 5 gal. sauce pot
- Full size steam table pan

Ingredients

18 lb boneless lamb shoulder
2½ gal. water
½ No. 10 can whole onions
2 qt canned whole potatoes
1 qt turnips
1 qt carrots
1 lb 8 oz butter or shortening
1 lb 4 oz flour
½ c leeks
1 tbsp marjoram
2½ lb frozen peas
salt and white pepper to taste

Procedure

1. Place the cubes of lamb in a stock pot. Cover with the water and bring to a boil. Skim off any scum that may appear on the surface.
2. Add the marjoram, reduce the heat, and simmer until the lamb is slightly tender.
3. Cook the carrots and turnips in boiling salt water in separate saucepans. Drain and add the liquid from the carrots to the stock in which the meat is cooking. Discard the liquid in which the turnips were cooked.
4. Place the butter or shortening in a sauce pot and heat.
5. Add the flour and stir with a wire whip, making a roux. Cook slightly but do not brown.
6. After the meat is cooked, strain the stock through a china cap covered with a cheesecloth into the roux, whipping vigorously with a wire whip until thickened and smooth.
7. Add the cooked cubes of lamb, potatoes, leeks, carrots, onions, and turnips. Simmer for 5 min to 10 min, stirring occasionally with a paddle.
8. Season with salt and white pepper. Remove from the range and place in a deep steam table pan.
9. Serve in a slightly deep casserole with dumplings and garnish with the cooked peas.

Preparation

1. Cut the boneless lamb shoulder into 1" cubes with a French knife.
2. Mince the leeks with a French knife.
3. Dice the turnips and carrots into ½" cubes with a French knife.
4. Drain the canned onions and potatoes.
5. Cook the peas in boiling salt water in a saucepan.

Precautions

- When the thickened sauce is simmering, stir occasionally to avoid scorching or sticking.
- Do not let the roux brown while cooking.

Fricassee of Lamb

Fricassee of lamb is a stew consisting of boiled cubes of lamb and fricassee sauce. It is generally served on the luncheon menu with buttered noodles or baked rice.

50 servings

Equipment

- 5 gal. stock pots
- French knife
- Wire whip
- China cap
- Cheesecloth
- 5 gal. sauce pot
- Kitchen spoon
- Baker's scale
- Skimmer
- 3 gal. stainless steel container
- Bake pan
- Deep steam table pan

Ingredients

17 lb boneless lamb shoulder
2½ gal. water
2 lb shortening or butter
1½ lb flour
yellow color as desired
salt and white pepper to taste
⅓ c fresh mint

Preparation

1. Cut the boneless lamb shoulder into 1" cubes with a French knife.
2. Chop the mint with a French knife.

Procedure

1. Place the cubes of lamb in a stock pot and cover with water. Bring to a boil and remove any scum that may appear on the surface. Continue to simmer until the lamb cubes are tender.
2. Remove from the heat and strain the lamb stock through a china cap covered with cheesecloth into a stainless steel container. Keep hot. Place the cooked lamb in a bake pan and cover with a wet towel.

3. In a separate sauce pot, make a roux (2 lb shortening or butter and 1½ lb flour). Cook 5 min.
4. Add the strained lamb stock, whipping vigorously with a wire whip until thickened and smooth.
5. Tint the sauce with yellow color and season with salt and white pepper.
6. Add the cooked lamb cubes and the chopped mint to the sauce. Stir with a kitchen spoon. Place in a deep steam table pan.

7. Serve in shallow casseroles with buttered noodles or baked rice.

Precautions

- Do not make the sauce too thin.
- Do not add too much yellow coloring.

Dublin Style Lamb Stew

Dublin style lamb stew is a white lamb stew with all the items boiled. This stew was made popular in Dublin, Ireland.

50 servings

Equipment

- French knife
- 5 gal. stock pot
- 5 gal. sauce pot
- Skimmer
- Wire whip
- 3 gal. stainless steel container
- China cap
- Cheesecloth
- Baker's scale
- Gallon measure
- Spoon measure
- Deep steam table pan

Ingredients

- 18 lb boneless lamb shoulder
- 10 lb new potatoes
- 6 lb onions
- 3 bunches leeks
- 2½ gal. water
- 2 oz salt
- 1 lb 4 oz butter
- 10 oz flour
- 1 tbsp marjoram
- 2 bay leaves
- salt and pepper to taste

Preparation

1. Cut the boneless lamb shoulder into 1" cubes with a French knife.
2. Slice the new potatoes slightly thick with a French knife.
3. Julienne the onions and leeks with a French knife.

Procedure

1. Place the lamb cubes in the stock pot and cover with water. Bring to a boil and skim off any scum that may appear. Add the salt, marjoram, and bay leaves.

2. Reduce to a simmer and cook until the meat just starts to become tender. Remove from the range and strain the stock through a china cap covered with a cheesecloth into a stainless steel container. Keep both meat and stock warm.
3. Melt the butter in a sauce pot. Add the onions and sauté without color until slightly tender.
4. Stir in the flour with a wire whip, making a roux. Cook slightly.
5. Add the hot stock, whipping vigorously with a wire whip until slightly thick.
6. Add the potatoes and leeks. Simmer until the potatoes start to become tender.
7. Add the cooked cubes of lamb and continue to simmer for about 10 min more.
8. Season with salt and pepper. Remove from the range and place in a deep steam table pan.
9. Serve into a casserole and garnish with dumplings and chopped parsley.

Precautions

- Do not overcook the lamb or the sliced potatoes. The stew will lack appearance when served.
- Do not brown the onions when sautéing.
- When holding the cooked cubes of meat for later use, keep covered with a wet towel.

BROILED LAMB RECIPES

Broiled Lamb Chops

Broiled lamb chops are a very popular entree on the dinner menu. The lamb chops can be cut from the rib or the loin. Chops cut from the rib are more popular because they can be frenched (the meat and fat is cut away from the end of the rib bone) to present a more desirable appearance when served.

25 servings

Equipment

- French or butcher knife
- Boning knife
- Kitchen fork
- Bake pan
- Cleaver

Ingredients

5 oz lamb chops (50)
1 pt salad oil (variable)
salt and pepper to taste

Preparation

1. Cut the 5 oz lamb chops from trimmed loin or ribs with a French or butcher knife. Cut against the grain of the meat. If using ribs, cut between the rib bones. If using loins, cut until the knife blade hits bone, then chop with a cleaver. Chops may also be cut on a power saw. Rib chops may be frenched by cutting away all meat and fat from the rib bones with a boning knife.
2. Pour the salad oil in a bake pan.
3. Preheat the broiler.

Procedure

1. Pass each chop through the salad oil. Place on a hot broiler with the frenched ribs turned away from the heat. Season with salt and pepper.
2. Brown one side, turn, sticking the fork into the fat, and brown the second side.
3. Remove from the broiler when the desired degree of doneness is obtained.
4. Serve two chops per portion with a paper frill on the end of each frenched chop. Accompany with mint jelly and garnish with watercress.

Precautions

- During the broiling period, turn the heat to the highest point. Adjust cooking temperature by moving the rack.
- Protect the frenched ribs from the fire. The exposed ribs will burn quickly.

Lamb Chop Mix Grill

Lamb chop mix grill is a very popular dinner entree. It is a combination of one broiled trenched lamb chop and five other items that combine well with lamb; for example, bacon, grilled tomato, sautéed mushroom cap, link sausage, and toast.

25 servings

Equipment

- Boning knife
- French or butcher knife
- Sheet pans
- Bake pans
- Kitchen fork
- Quart measure
- Skillet

Ingredients

5 oz or 6 oz lamb chops (25)
50 little pig sausages
50 slices of bacon
25 large mushroom caps
25 pieces of toast
25 halves of fresh tomatoes
1 c bread crumbs
1 qt salad oil (variable)
salt, pepper, and basil to taste

Preparation

1. Cut the lamb chops from trimmed ribs with a French or butcher knife. Cut against the grain of the meat between the rib bones. French by cutting away all meat and fat from the end of the rib bones with a boning knife.
2. Preheat oven to 350°F.
3. Line the little pig sausages on a sheet pan. Place in a 350°F oven and bake until done. Drain and keep warm.
4. Line the bacon slices on a sheet pan, overlapping slightly. Place in the preheated oven at 350°F and bake until medium. Remove and let drain. Keep slightly warm.
5. Sauté the mushrooms in butter or salad oil in a skillet. Drain and keep warm.
6. Preheat the broiler.
7. Wash the tomatoes and remove the stems. Cut the tomatoes in half crosswise with a French knife and place in a bake pan, cut side up. Rub each tomato with salad oil and season with salt, pepper, and basil. Sprinkle bread crumbs on the top of each tomato and brown slightly under the broiler. Finish by baking in the oven at 325°F until the tomato slices are fairly soft.
8. Place remaining salad oil in a bake pan.
9. Toast the bread and cut into triangles with a French knife.

Procedure

1. Pass each lamb chop through the salad oil. Shake off excess oil and place on a hot broiler with the frenched ribs turned away from the heat. Season with salt and pepper.
2. Brown one side, turn, sticking the fork into the fat, and brown the second side.
3. Remove from the broiler when the desired degree of doneness is obtained.
4. Serve on a hot plate with the lamb chop placed on top of the toast and surrounded with a half broiled tomato, two little pig sausages, two slices of bacon, and a cooked mushroom cap. Drip melted butter over the lamb chop, place a paper frill on the end of each frenched chop, and garnish with parsley or watercress.

Precautions

- Protect the frenched ribs from the heat. The exposed ribs will burn quickly.
- During the broiling period, turn the heat to the highest point. Adjust the cooking temperature by moving the rack.

Broiled Lamb Steak

Lamb steaks are crosscut sections of the leg of lamb with the leg bone left in. Approximately ten 6 oz to 8 oz steaks can be cut from the average leg. The steaks are passed through salad oil, seasoned, and broiled.

25 servings

Equipment

- Butcher knife
- Boning knife
- Bake pan
- Kitchen fork
- Meat saw

Ingredients

6 to 8 oz lamb steaks (25)
1 qt salad oil (variable)
salt and pepper to taste

Preparation

1. Remove tail and aitchbones from the legs of lamb with a boning knife. Cut the leg across the grain into steaks with a butcher knife and a hand meat saw or power saw.
2. Place the salad oil in a bake pan.
3. Preheat the broiler.

Procedure

1. Pass each steak through the salad oil and shake off the excess.
2. Place steaks on a hot broiler, fat side out. Season with salt and pepper.
3. Brown one side. Turn with a kitchen fork and brown the other side.
4. Serve one steak per portion with mint jelly. Garnish with watercress.

Precautions

- If meat sticks to the broiler, loosen gently so it does not tear.
- Do not burn the outside fat.

Shish Kebab

Shish kebab is a combination of lamb cubes and three or four vegetables placed on a skewer alternately, cooked by the broiling method, and served on rice. It is a popular and unusual item for the dinner menu.

25 servings

Equipment

- French knife
- 25 metal skewers
- 1 gal. stainless steel container
- Hotel pans
- Quart measure
- Spoon measures
- Boning knife
- Kitchen spoon
- Pastry brush

Ingredients

Shish Kebab

8 lb lamb leg
50 pieces small tomatoes
50 canned mushroom caps
50 pieces canned, small, whole onions

Marinade

1 qt salad oil
1 pt olive oil
1 c wine vinegar
5 tbsp lemon juice
2 garlic cloves
2 tsp fresh ground pepper
2 tbsp salt
½ tsp thyme
½ tsp marjoram
½ tsp basil
½ tsp oregano

Preparation

1. Bone the legs of lamb with a boning knife and cut into 1″ cubes with a French knife.
2. Mince the garlic with a French knife.

3. Squeeze the juice from the lemons.
4. Blend all the marinade ingredients with a kitchen spoon in a stainless steel container. Add the lamb cubes and let marinate overnight.
5. Preheat broiler.
6. Cut the thick slices of tomato with a French knife just before using.

Procedure

1. Place the items on the metal skewers alternately, including two slices of tomatoes, two mushroom caps, two whole onions, and five cubes of lamb.
2. Place the shish kebabs in a bake pan. Pour the marinade over them and marinate until ready to broil.
3. Drain the shish kebabs thoroughly.
4. Place under the broiler and broil for approximately 15 min under a low fire. Brush frequently with the marinade, turning as needed.
5. Serve immediately on a bed of baked rice or pineapple rice.

Precautions

- Cook the shish kebabs medium unless requested otherwise.
- Avoid burning the shish kebabs while broiling. Turn frequently.
- Be careful when turning the broiling shish kebabs. The metal skewers get quite hot.

Lamb and Mushrooms en Brochette

Lamb and mushrooms en brochette is cubes of lamb shoulder, mushrooms, and bacon alternated on a skewer. It is cooked by the broiling method and served with rice.

25 servings

Equipment

- 25 metal skewers
- French knife
- Bake pan
- Kitchen fork
- Slicing machine

Ingredients

8 lb boneless lamb shoulder
2 lb bacon
75 mushroom caps
1 qt salad oil (variable)
salt and pepper to taste

Preparation

1. Cut the boneless lamb shoulder into 1″ cubes with a French knife.
2. Slice the bacon slightly thick on a slicing machine and cut into 1″ pieces with a French knife.
3. Pick the stems from the mushroom caps.
4. Place the salad oil in a bake pan.

Procedure

1. Place three cubes of lamb, three mushroom caps, and three pieces of bacon on the skewer alternately.
2. Place the brochettes in salad oil and season with salt and pepper.
3. Place on the broiler and broil slowly for about 15 min until the cubes of lamb are done, turning occasionally with a kitchen fork.
4. Serve at once with mint jelly and garnish with watercress and a twisted slice of orange.

Precautions

- Broil slowly and turn occasionally to prevent burning.
- Keep broiler flame slightly low while broiling.

BOILED LAMB RECIPE

Boiled Lamb with Dill Sauce

Boiled lamb with dill sauce is lamb shoulder cut into cubes, boiled, and served in a rich dill sauce. The sauce is prepared by using the stock in which the lamb was boiled.

25 servings

Equipment

- French knife
- 5 gal. sauce pots (2)
- Wire whip
- China cap
- Cup measure
- Baker's scale
- Cheesecloth
- 3 gal. stainless steel container
- Bake pan
- Kitchen spoon
- Deep steam table pan
- Skimmer

Ingredients

8 lb boneless lamb shoulder
2 gal. water
1 lb 4 oz butter
1 lb flour
1/4 c sugar
1/3 c vinegar
8 white peppercorns
1/2 c dill seed
8 oz onion
salt and white pepper to taste

Procedure

1. Place the cubes of lamb in a large sauce pot. Cover with water. Add a small amount of salt and bring to a boil.
2. Remove any scum that may appear on the surface of the liquid with a skimmer. Add the sugar, vinegar, peppercorns, onions, and dill seed. Simmer until the meat is tender, about 1 hr to 1 1/2 hr.
3. Melt the butter in a separate sauce pot. Add the flour and blend thoroughly to form a roux. Cook for 5 min.
4. Strain the stock the lamb was cooked in through a china cap covered with a cheesecloth. Strain into a stainless steel container. Wash the meat cubes in warm water and place in a bake pan. Cover with a damp towel and keep warm.
5. Pour the stock into the roux, whipping vigorously with a wire whip until thickened and smooth. Let simmer 10 min. Strain a second time through a fine china cap. Add the cooked meat cubes, stir with a kitchen spoon, and place in a deep steam table pan.
6. Serve in shallow casseroles with buttered noodles or baked rice.

Preparation

1. Cut the boneless lamb shoulder into 1" cubes with a French knife.
2. Cut onions rough with a French knife.

Precautions

- Do not overcook the cubes of lamb. The appearance will be lacking.
- Stir frequently while the sauce is simmering to avoid scorching.

Poultry Preparation

The National Chicken Council

Poultry is extremely popular on the menu. Poultry refers to domestic edible birds. The flesh is the muscle tissue of the bird. Different parts of the bird have different types of meat. Light meat is meat that comes from the breast and wings of the bird. Dark meat is meat that comes from the legs and thighs of the bird.

The cooking method used for poultry is determined by the cut or part of the bird. Whole poultry requires more cooking time to produce evenly cooked meat. Poultry that has been cut into parts cooks more quickly. In addition, the age of the bird and the type of muscle also determine the cooking technique used.

All poultry must be inspected by the United States Department of Agriculture (USDA) for wholesomeness (fit for human consumption). In addition, a U.S. grading stamp may also be attached or stamped on the poultry or packaging material to indicate the shape, distribution of fat, condition of skin, and general appearance of the bird. The most popular poultry meats used are chicken and turkey. Chicken is the most popular poultry meat prepared in the commercial kitchen.

POULTRY

Poultry is the classification of all domestic edible birds for human consumption. Poultry meats have always been popular edible meats in both the domestic kitchen and the commercial kitchen because they are low in cost and can be prepared utilizing most cooking methods. Poultry meats are quite tender and easy to digest when cooked properly. Poultry is popular among diet-conscious people seeking a meat that is low in fat, calories, and cholesterol. However, the skin must be removed before cooking if low fat and cholesterol are a major concern. The two most common poultry meats are chicken and turkey.

Chicken

Chicken is the most popular poultry meat served in the commercial kitchen. Chicken can be prepared in a variety of ways using quick or slow cooking techniques. It can be cooked whole or in parts. This allows great flexibility for the chef or cook. Chickens are classified by age and weight. Different chicken types require different cooking techniques.

Fryers and *broilers*: Chickens of either sex under 16 weeks of age. They have a very tender flesh and a flexible skin. The most common size of fryer found in food service operations is $2\frac{1}{2}$ to 3 lb.

POULTRY PREPARATIONS

GRILLED CHICKEN

National Broiler Council
CHICKEN SALAD

ROAST DUCK

Poultry meats are low in cost and can be prepared utilizing most cooking methods.

Roasters: Chickens of either sex averaging in age from 5 months to 9 months old. They have tender meat and flexible skin. Roasters average in weight from 3 lb to 5 lb.

Hens (fowl): Sometimes known as *stewing chickens*. They are mature female chickens that have laid eggs for one or more seasons and are usually over 10 months of age. The flesh and skin are tough, which requires cooking in moist heat in order to make the meat tender enough to use in such preparations as chicken a la king, salads, and other dishes. Hens average in weight from 4 lb to 6 lb. Hens are excellent for preparing chicken stock. In addition, the fat derived from the hen is used in many preparations.

Stags: Mature male chickens with a fairly tough meat and skin. They are usually over 10 months old and weigh from 2 lb to 6 lb. Stags are rarely used in a commercial kitchen. If necessary, stags are cooked by moist heat for a long period of time in order to make the flesh tender.

Cocks or old roosters: Mature male chickens over one year old with coarse skin and tough, dark flesh. They average in weight from 2 lb to 6 lb. Cocks or old roosters are rarely found on the market.

Capons: Castrated young male chickens 8 months to 10 months old. Capons are specially fattened to produce a large, well-formed breast with flesh that is more tender and better flavored than the average

chicken. Once the bird is castrated, the tenderness of the flesh is affected very little as the bird ages. The average weight of a capon is 5 lb to 8 lb.

Turkey

Turkeys are native to America and are bred as lightweights and heavyweights. The lightweights are bred for fast growth and a more marketable size. The heavyweights are bred for the larger-sized turkeys that are used mostly in food service establishments. The larger turkeys produce more meat in proportion to bone and sell at a lower cost per pound than the lightweights.

Baby turkeys: Lightweights under 16 weeks old. They are very tender with a soft flexible skin. Baby turkeys can be roasted, fried, or broiled. They average in weight from 4 lb to 8 lb.

Young hens: Female turkeys, usually less than one year old, with a soft, tender meat and flexible skin and breastbone. Young lightweight hens average in weight from 6 lb to 10 lb. Young heavyweight hens average 12 lb to 16 lb. Young hens are best when roasted or boiled.

Old hens: Mature female turkeys over one year old with flesh and skin that has become slightly tough and a hardened breastbone. (In poultry, the harder the breastbone is, the older the bird is.) Old lightweight hens average in weight from 6 lb to 10 lb. Old heavyweight hens average from 12 lb to 16 lb. It is best to boil old hens until the flesh becomes tender.

Young toms: Male turkeys, usually less than a year old, with tender meat, flexible skin, and breastbone. Young lightweight toms average in weight from 12 lb to 16 lb. Young heavyweight toms average from 18 lb to 30 lb. Young tom turkeys are usually roasted. However, the breast meat can also be cut into steaks and sautéed or broiled. In addition, young tom turkeys can also be boiled and used for sandwiches, salads, and entree items. Young tom turkeys are the most popular turkey used in food service establishments.

Old toms: Mature male turkeys over a year old with toughened flesh and a hardened breastbone. Old lightweight tom turkeys average in weight from 12 lb to 16 lb. Old heavyweight toms average from 18 lb to 30 lb. It is best to boil old tom turkey meat to make it tender enough for such preparations as chicken hash, pot pies, fricassee, and tetrazzini.

USA Rice Federation

Rice and chicken are often paired in popular dishes.

GAME BIRDS

Game birds are wild birds that are less commonly used than chicken or turkey in food service establishments. Game birds, because of their scarcity, are high priced. They are prepared like domestic poultry, but require a slightly different preparation to help preserve the true game flavor. Game birds are usually aged or ripened for a short period of time in the open air. The term *high* is used to refer to birds that have been ripened for 1 or 2 weeks. Game birds lack a sufficient amount of fat covering when aged longer. When a bird is high, the tail feathers pull out easily. Game birds include ducks, Cornish hens, geese, grouse, guineas, partridges, pheasants, quail, and squabs. Many game birds are commercially raised today. They are available whole or in fabricated cuts.

Ducks

The meat of the duck is all dark and provides less meat in proportion to bone than other birds. The terms *Long Island* and *Western*, which are usually associated with the marketing of ducklings, refer to the way they were grown and fattened. Long Island ducklings are specially-fattened young ducks grown on Long Island duck farms. They are force-fed on special grain and marketed when they weigh from 4 lb to 6 lb. Ducks of this style are now being produced in various parts of the country, but still carry the Long Island name. Western ducklings are young ducks that are not force-fed or specially fattened. Their meat is not as tender and desirable as that of the Long Island style.

Broilers or *fryers*: Ducks of either sex, usually less than 8 weeks of age. They have a tender meat, a soft bill, and a soft windpipe. They weigh about 3 lb.

Roaster ducklings: Ducks of either sex, usually less than 16 weeks of age. They possess a tender meat and a bill and windpipe that is just starting to harden. Roaster ducks weigh about 4 lb.

Trade Tip

Although turkey may be prepared using dry heat cooking or moist heat cooking, the most common method is roasting. Dry heat cooking methods include baking, roasting, broiling, grilling, pan-frying, deep-fat frying, and sautéing. Moist heat cooking methods include stewing, simmering, braising, and steaming.

Mature ducks: Mature ducks of either sex, usually over 6 months of age. The flesh is fairly tough and the bill and windpipe have hardened. The average weight of mature ducks is 4 lb.

Cornish Hens

Cornish hens resemble the chicken in appearance, but contain all white meat. They are small in size, and usually a whole bird will supply only one or two servings. The breast is large and the flesh is fine-grained and tender.

Geese

Geese, like ducks, contain all dark meat. However, unlike ducks, geese contain a very high percentage of fat. Geese weighing less than 11 lb are considered light and geese over 12 lb are considered heavy. They are classed into two groups: young and mature (or old) geese.

Young geese: Geese of either sex, usually less than 6 months old. They have a tender flesh and a windpipe that is easily dented. Young geese weigh from 4 lb to 10 lb.

Mature geese: Geese of either sex over 6 months old. They have less tender flesh and a hardened windpipe. Mature geese average from 10 lb to 18 lb and are rarely used in quality food establishments.

Grouse

Grouse resemble small domestic fowl in appearance, but have thicker and stronger legs. There are over 40 species of grouse found in North America. The most common grouse are the ruffled grouse, the sage grouse, and the blue or dusty grouse. All species have a fairly long feathered tail, a medium-sized wingspan, and a short thick bill. The grouse has a dark meat that is universally recognized by gourmets and connoisseurs as one of the finest. The grouse can be prepared by using different methods. However, large grouse are best when roasted. Smaller grouse may be sautéed to produce their best eating qualities. The flesh of the female bird is usually superior in flavor to that of the male.

Guineas

Guineas are related to the pheasant and have been domesticated in most parts of the world. Guineas are an agile, colorful bird with a flesh that is darker than chicken and has a flavor of wild game. Guineas are available as young or mature guineas.

Young guineas: Guineas of either sex that have tender flesh and an average weight from 1 lb to 1½ lb. Young guineas are best when roasted.

Mature guineas: Old guineas of either sex that have tough flesh and average from 1 lb to 2 lb. Mature guineas are never used in food service establishments.

Partridges

Partridges are smaller than the pheasant and usually provide only enough meat to serve two people. The meat is white and must be cooked slightly on the done side to develop the desired succulent gamey flavor. Partridges can be broiled or sautéed but, like pheasants, the most popular method is roasting.

Pheasants

Pheasants are a fairly large, long-tailed bird with a dark, rich, gamey-tasting meat. Pheasant is prepared by roasting or braising. However, the most popular method of preparation is to stuff the bird with wild rice, roast, and serve while still rare. To produce the best eating qualities, a pheasant should be left to hang and ripen slightly before it is plucked and cooked.

Quail

The quail is similar to the partridge, with short legs and neck. The quail has white meat and should be prepared in the same manner as partridges. The meat of most game birds is best when cooked slightly rare; however, the white-meat birds seem to have a more desirable flavor when cooked slightly longer.

Squabs

Squabs are very young pigeons of either sex that have never flown. They are specially fed to produce meat that is extra tender and light in color. Squabs are marketed when they are 3 weeks to 4 weeks old and weighing from 6 oz to 14 oz. Squabs are expensive and are found only on higher-priced menus. Squabs are most commonly prepared using the roasting method. Squabs can also be sautéed or broiled.

POULTRY GRADING

The U.S. inspection stamp is required for all ready-to-cook poultry products. In addition, this stamp must be present if the poultry is intended for interstate commerce. The U.S. inspection stamp is a guarantee by the United States Department of Agriculture that the meat is wholesome, that it was processed under proper sanitary conditions, and that it was inspected by trained personnel to make sure it is fit for human consumption.

The USDA inspection stamp is required by law and guarantees wholesomeness.

In addition to the inspection stamp, a U.S. grading stamp may also be stamped or clipped on the poultry or packaging material to indicate the quality of the bird. Grading is based on shape, distribution of fat, condition of the skin, and general appearance of the bird. The U.S. government grades include U.S. Grade A, U.S. Grade B, and U.S. Grade C.

The USDA grade stamp indicates the quality of poultry.

Poultry is also graded and inspected by some states to give further assurance that the meat is wholesome. Most food service establishments use only Grade A birds because they yield a greater amount of meat per pound than the other grades.

The major responsibilities of USDA-licensed graders are:

- Grading products for class, quality, quantity, and/or condition.
- Making condition inspections of poultry containers and transportation vehicles.
- Making weight tests when requested.
- Supervising the grading of all authorized personnel under the grader's jurisdiction and check-grading all products graded by them.
- Observing the packing and marking of products to be officially identified.
- Issuing grading certificates or other documents required by the military, other government agencies, or institutional buyers.
- Completing all reports incidental to the grading service.
- Keeping files and indexes up to date.
- Keeping official marking devices, certificates, official memoranda, and any other assigned equipment in custody at all times.
- Promptly reporting any irregularities to the appropriate supervisor.

MARKET FORMS

Poultry can be purchased in two forms: whole or cut. The whole form is the form usually used in the commercial kitchen. Poultry is available fresh or frozen. The feathers, head, neck, and feet are removed and the bird is eviscerated (the entrails and viscera are removed). Fabricated cuts are available in various sizes.

Always wash poultry thoroughly before preparing. It can transmit salmonella, listeria, and other bacteria. Food poisoning and gastrointestinal inflammation can be caused by salmonella bacteria. Remember, *always* wash poultry thoroughly before preparing. After handling poultry, *always* wash hands, tools, and equipment to prevent possible contamination of other foods.

It is best to use fresh poultry within 24 hr of receiving. If the poultry is not used within 3 days, wash it thoroughly in salt water, drain, wrap properly in freezer paper, and freeze.

Cut poultry is birds that have been cut into several pieces. They are usually divided into eight pieces, which includes two breasts, two wings, two legs, and two thighs. However, chickens are usually cut from whole birds in the commercial kitchen. Ground turkey and ground chicken are also available.

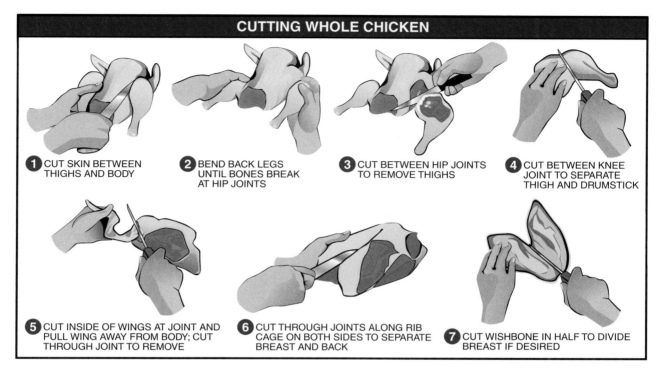

CUTTING WHOLE CHICKEN

1 CUT SKIN BETWEEN THIGHS AND BODY

2 BEND BACK LEGS UNTIL BONES BREAK AT HIP JOINTS

3 CUT BETWEEN HIP JOINTS TO REMOVE THIGHS

4 CUT BETWEEN KNEE JOINT TO SEPARATE THIGH AND DRUMSTICK

5 CUT INSIDE OF WINGS AT JOINT AND PULL WING AWAY FROM BODY; CUT THROUGH JOINT TO REMOVE

6 CUT THROUGH JOINTS ALONG RIB CAGE ON BOTH SIDES TO SEPARATE BREAST AND BACK

7 CUT WISHBONE IN HALF TO DIVIDE BREAST IF DESIRED

Chicken is cut into parts for different preparations and convenience.

The *giblets* are the heart, gizzard, and liver of the poultry. These edible internal parts are utilized in different preparations. Giblets are usually wrapped in paper inside the butt or neck cavity of the dressed (trimmed and cleaned) bird. The most desirable giblet is the liver. It has excellent eating qualities and can be prepared in a variety of ways. Chicken livers are a very popular menu item. The gizzard and heart can be utilized in soups, gravies, and luncheon entrees.

STORING POULTRY

All poultry perishes quite rapidly. Spoiled poultry develops an odor immediately recognizable as unsafe. Fresh poultry should be refrigerated as soon as it is received. If possible, poultry should be packed in crushed ice when it is placed in the lowest part of the refrigerator. Placement on the bottom shelf prevents uncooked poultry from touching or dripping on other foods. Frozen poultry, such as turkey, should be kept frozen until a day or two before using. It should be thawed in the refrigerator or in cold running water. Spoiled poultry is dangerous. Always follow the tip "*when in doubt, throw it out.*" Fresh poultry should be purchased the day before use to eliminate a long holding period and the chance of spoiling.

COOKING POULTRY

Cooking procedures for poultry are similar to those used for other meats. Tougher meats are cooked using lengthy cooking methods and more tender meats are cooked using quicker methods. When cooking older and tougher chickens in liquid, add 1 tbsp of lemon juice or vinegar to the water in which it is being simmered. The meat will become whiter and more tender.

Regardless of the cooking method, poultry should always be cooked to a minimum internal temperature of 165°F, except in the case of some wild game. Larger poultry should be cooked slowly to reduce shrinkage and to retain moisture. Smaller poultry should be cooked at temperatures of 375°F or 400°F to prevent drying out while cooking. In the commercial kitchen, it is a practice to stuff small birds, such as the Cornish hen and squab, but not larger birds. If stuffing is to be served with the large birds, it must be prepared and baked separately for sanitation and safety. This saves time and makes serving easier. This also leaves the carcasses in a better condition for use in making stock.

To improve the flavor of strong-flavored poultry, cover with cold water, add a teaspoon of baking soda, and let set for at least 1 hr before cooking. To improve

the taste and tenderness of all poultry, rub the inside and outside of the bird with lemon juice just before cooking.

Boning Chicken Legs

When boning chicken legs for the purpose of stuffing or serving bone-free in a cacciatore or a la Marengo preparation, start boning on the inside of the leg. Working the boning or utility knife under the bone, free the leg bone and then the thigh bone by cutting away from the joint that joins the two together. When this is accomplished, free the flesh around the joint and remove both bones.

If the leg is to be stuffed, flatten the surface by placing a slightly heavy piece of plastic over the surface and pounding with a mallet until the desired thinness or diameter is achieved. The plastic makes this task easier because the mallet will not stick or tear the flesh.

After the stuffing (rice, forcemeat, duxelle, etc.) has been applied and the boned flesh rolled up, place the roll in an aluminum potato shell with the seam end down. This method produces a more plump and uniform stuffed leg and prevents the flesh from un-rolling during the baking period. Before baking, coat each stuffed leg with salad oil to provide moisture and to keep the skin covering the flesh from blister-ing and cracking during the baking period.

Boning a Chicken Breast

For best results when boning a raw chicken breast, use a very sharp boning or utility knife and use short quick strokes with the tip of the blade when cutting. The first cut should be made at the joint where the wing joins the breastbone. Continue cutting, follow-ing the bones of the rib cage. Stay next to the rib cage. Remove half of the complete breast. Turn to the opposite side and repeat the process. When the boning of the two half breasts are completed, cut the wings at the second joint, leaving the inner portion of the wing attached to the breast. The outer portions are kept to be used as hors d'oeuvres or in the prepa-ration of chicken stock.

After the breast has been boned, if it is to be stuffed (kiev, cordon bleu, a la Suisse, etc.), proceed to flatten the surface using the same method for flattening a boned chicken leg. Boning a whole chicken breast af-ter it has been roasted is a much simpler task.

National Broiler Council

Chicken parts are classified as light meat or dark meat. Light meat includes breasts and wings. Dark meat includes legs and thighs.

When the roasted breast is removed from the oven, let it cool slightly or until it is completely cold. The length of cooling time is usually determined by the time the item must be ready for service. Take the breast in both hands with the thumbs pressed against each side of the wishbone and the four fingers of each hand pressed against the flesh on each side of the breast. Apply slight pressure to the wishbone with the thumbs while at the same time moving the hands outward. Separate the two breast portions by tearing the flesh away from the rib cage. Any bones that remain on each half breast can be easily removed by hand. Leave the inner portion of the wing attached to the breast. Usually, paper frills or stockings are placed on these wing bones when the item is served. If time is a factor and the breast must be boned im-mediately upon removal from the oven, submerge your hands in ice water for a few moments to numb them against the hot flesh.

Trade Tip

Insert a thermometer into the thickest part of the meat. Poultry needs to be at 165°F to be completely cooked. Also, check the juices that run out of the meat. The juices should run clear, not pink.

Chopped cooked chicken or turkey may be used to prepare salad and meatloaf.

Cutting Turkey Steaks

Turkey steaks are best obtained from a tom turkey because of the amount of breast meat. The breast is separated from the legs and the wings are removed from the breast by cutting at the second joint. Both wings and legs are held for other preparations. Using a boning or utility knife, remove all the skin from the breast. Lay the breast on its side and cut the steaks using a French knife or a slicer. Cut the steaks into approximately 6 oz portions. Flatten the steaks slightly by covering with a piece of plastic and striking with a mallet or the flat side of a cleaver. Do not strike a hard blow because the turkey steaks are very tender and tear easily.

The best method of cooking turkey steaks is sautéing. Before sautéing, press the turkey steak into seasoned flour. If a crispier surface is desired, pass the turkey steak through seasoned flour, a rich egg wash, and back into the seasoned flour. Serve with a poulette or supreme sauce.

National Turkey Federation

Turkey cutlets may be grilled and served with colorful vegetables.

Trade Tip

The National Turkey Federation is the national advocate for all segments of the turkey industry. Their website, www.eatturkey.com, offers nutritional information and other information about high-quality turkey products.

ROASTED AND BAKED POULTRY RECIPES

Roast Turkey

Roast turkey is an American tradition dating back to the first Thanksgiving. In recent years, this holiday treat has become a year-round favorite. The bird is roasted to a golden doneness, carved, and served with dressing and giblet gravy.

50 servings

Equipment

- Roast pan
- Kitchen fork
- Kitchen spoon
- French knife
- Baker's scale
- 1 gal. stainless steel container
- China cap

Ingredients

25 lb turkey, dressed
12 oz onions
8 oz celery
6 oz carrots
10 oz ham or bacon grease
salt and pepper to taste

Preparation

1. Lock the turkey's wings by bending under the body. Season the inside of the bird with salt and pepper.
2. Cut the rough garnish (onions, carrots, and celery cut into medium-sized pieces) with a French knife.
3. Preheat oven to 325°F.

Procedure

1. Rub the ham or bacon fat over the surface of the turkey. Place it in a roast pan breast up.
2. Place the turkey in the preheated oven at 325°F and brown the complete surface by occasionally turning it from side to side using a kitchen fork. When the bird is completely browned, turn to the original position, breast up.
3. Roast, basting frequently using a kitchen spoon, for approximately 1½ hr. Add the rough garnish to the pan and continue roasting for 2 hr more or until the bird is done.
4. Remove from the oven and place the turkey in a clean pan. Strain the drippings through a china cap into a stainless steel container. Deglaze the roast pan and pour this liquid into the container.
5. Prepare the giblet gravy.
6. Prepare the bread dressing.
7. To serve roast turkey, a dipper of bread dressing is placed on the plate, dark meat is arranged on top of the dressing, and two to three slices of white meat are placed over the dark meat. Each portion is covered with giblet gravy.

Note: To test the turkey for doneness, a fork is inserted into the thigh and twisted slightly to cause liquid to flow. If the liquid is white, the turkey is done. If the liquid is red or pink, more cooking is required. This part of the bird is selected for testing because it is usually the part that requires the longest cooking.

Note: In most food service establishments, turkeys are usually roasted the day before serving, sliced cold, lined in pans with dark meat on one side and white meat on the other, reheated, portioned, and served.

Note: To cut roasting time approximately in half and still retain much of the natural turkey juices, use the following procedure when preparing roast turkey. Separate the legs from the breast. Cut the wings off at the sec-

ond joint. Save the wings for a separate preparation. Season the inside of the breast with salt and pepper, and rub the outside with bacon grease or salad oil. The bacon grease or salad oil keeps the turkey skin from cracking and helps the breast brown more uniformly. Place the breast in a roast pan and proceed to roast at 350°F until the breast is done. Roasting time depends upon the size of the breast, but usually with the breast of a tom turkey, roasting time is less than 2½ hr.

Note: The legs are boned in the same manner as in boning chicken legs. However, a turkey leg contains more and tougher sinews. Be sure all are removed. When all bones are removed, roll and tie the boneless meat. Season, rub with bacon grease or salad oil, and roast separately from the breast at 350°F until done. Roasting time is usually less than 2 hr, depending on the size of the rolls. When the turkey breast is roasted separately from the legs, overcooking should not occur if the cook is alert. When roasting the whole bird, the breast is usually overcooked and dried out by the time the legs, with the bones in, are done.

Precautions

- Exercise caution when handling the knife.
- Even though the turkey is greased before heat is applied, it still may stick to the pan if not turned occasionally during the first hour of the roasting period.
- If the turkey drippings evaporate during the roasting period, a small amount of water may be added to the pan.
- If the bird begins to overbrown during the roasting period, cover with oiled brown paper or aluminum foil. However, do not cover too completely or steam will be created. Proper roasting is by dry heat.

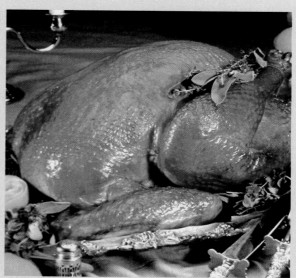

National Turkey Federation

Roast Chicken

Roast chicken uses select young tender birds weighing 3 lb to 5 lb and under 9 months of age. The roasting is done at a fairly high temperature to improve surface browning. The chicken is roasted on a bed of mirepoix (small diced vegetables) to improve the flavor of both the chicken and the drippings. A half chicken is usually served to each order on either the luncheon or dinner menu.

24 servings

Equipment

- French knife
- Roast pan
- Baker's scale
- Kitchen fork
- Saucepan
- Full size steam table pan
- China cap

Ingredients

12 3 lb roasting chickens
10 oz onions
8 oz celery
6 oz carrots
10 oz butter or shortening, melted
salt and pepper to taste

National Broiler Council

Preparation

1. Season the inside of each chicken with salt and pepper and lock the wings by bending under the body.
2. Cut the mirepoix (onions, carrots, and celery cut into a small dice).
3. Preheat the oven to 375°F.

Procedure

1. Place the mirepoix in the bottom of the roast pan.
2. Rub or brush the surface of each chicken with melted butter or shortening. Place in the roast pan breast up.
3. Roast in the preheated oven at 375°F for approximately 1 hr or until done, turning the birds with a kitchen fork from side to side during the roasting period for more uniform roasting and browning.
4. Remove from the oven and place the chickens in a steam table pan. Keep warm.
5. Strain drippings through a china cap. Deglaze pan and save both liquids for use in the gravy.
6. Prepare the giblet gravy.
7. Disjoint the chickens with a French knife.
8. Serve a half chicken per portion with a scoop of bread dressing covered with giblet gravy.

Precautions

- Exercise caution when handling the knife.
- During the roasting period, check the mirepoix occasionally to prevent burning. If necessary, a small amount of water may be added.
- Do not overcook the chickens. Appearance will be lacking.
- Turn the chicken occasionally during the roasting period with a kitchen fork.

Roast Breast of Chicken Virginia

For roast breast of chicken Virginia, the whole chicken breast is roasted and separated into two bone-free halves. Each half is placed on top of Virginia or regular smoked ham and served with a rich chicken sauce. This item is popular on the dinner menu.

24 servings

Equipment

- Roast pan
- French knife
- Quart measure
- Kitchen fork
- Sheet pans (2)
- Slicing machine
- Boning knife

Ingredients

12 whole chicken breasts
1 c salad oil
24 slices cooked Virginia or smoked ham
2 qt chicken velouté sauce
salt and pepper to taste

Preparation

1. Cut the wings off the breast by cutting through the second joint with a French knife, leaving on the section of wing bone that is attached to the breast.
2. Slice the ham fairly thick on a slicing machine. Place on sheet pans and heat in the oven or under the broiler.
3. Prepare the chicken velouté sauce.
4. Preheat the oven to 375°F.

Procedure

1. Season the inside of each whole chicken breast with salt and pepper.
2. Rub each breast with salad oil and oil the roast pan slightly.
3. Place the whole chicken breasts in the roast pan with the neck cavities facing up.
4. Roast in the preheated oven at 375°F for approximately 1 hr to 1½ hr or until done. Remove with a kitchen fork and let cool slightly.
5. Cut the joint at the base of the wing bone, on each side of the breast, with a boning knife. Free the meat from the rib cage by working the hands along the rib cage, separating the whole breast into two boneless halves.
6. Place each boneless half on a hot slice of Virginia or smoked ham. Ladle over the chicken velouté sauce and serve.

Precautions

- Exercise caution when using the French knife and the boning knife.
- Do not overcook the breast.
- If the chicken velouté sauce is too thick, it may be thinned slightly by adding a small amount of hot cream or chicken stock.

Refrigerate raw chicken promptly. Never leave it on the countertop at room temperature. The chicken should be left in the original wrappings and placed in the coldest part of the refrigerator. If it is not to be used within two days, freeze the chicken.

For roast duck, the duck is roasted at a fairly high temperature to extract excessive fat from the bird and to improve surface crispness. It is roasted on a bed of small diced vegetables (mirepoix). This improves the flavor of both the duck and its drippings. Roast duck is a favorite on the dinner menu.

Roast Duck

24 servings

Equipment

- Roast pan
- French knife
- China cap
- Kitchen fork
- Bake pan
- 6 qt sauce pot
- Wire whip
- Stainless steel container
- Quart measure
- Baker's scale

Ingredients

- 6 4 to 6 lb ducklings
- 6 oz carrots
- 4 oz celery
- 8 oz onions
- 8 oz duck fat
- 6 oz flour
- 3 qt brown stock
- 1 c orange juice
- salt and pepper as needed

Preparation

1. Wash and clean the ducks. Season the insides with salt and pepper.
2. Dice the onions, carrots, and celery with a French knife.
3. Prepare the brown stock.
4. Preheat the oven to 375°F.

Procedure

1. Place the mirepoix on the bottom of a roast pan.
2. Place the ducks on top of the mirepoix, breasts up.
3. Roast in the preheated oven at 375°F for approximately 2 hr or until done. Pour off excess grease and turn the ducks at intervals with a kitchen fork. Save the excess grease.
4. Remove from the oven, place the ducks in a clean pan, set in a warm place, and hold.
5. Pour the duck grease in the sauce pot and heat. Add the flour, making a roux, and cook for 5 min.
6. Pour the brown stock in the roast pan. Place on the range and bring to a boil to deglaze the pan.
7. Pour the stock from the roast pan into the roux, whipping vigorously with a wire whip until slightly thickened.
8. Season with salt and pepper and strain through a fine china cap into a stainless steel container.
9. Add the orange juice. Stir to blend.
10. Cut each duck into eight equal pieces and serve a piece of breast and a piece of leg per portion. Garnish with a slice or twist or orange.

Precautions

- Exercise caution when handling the knife.
- Turn the ducks occasionally with a kitchen fork during the roasting period for thorough browning and even cooking.
- Pour off excess grease during the roasting period.

For baked stuffed chicken leg, all bones are removed from the chicken leg. It is stuffed with forcemeat (ground meat mixture) and baked in the oven until tender. Baked stuffed chicken leg is served with an appropriate sauce on the luncheon menu.

Baked Stuffed Chicken Leg

25 servings

Equipment

- French knife
- 25 aluminum potato shells
- Food grinder
- Kitchen fork
- Full size steam table pan
- Baker's scale
- Cup measure
- Mallet
- Skillet
- Mixing container
- Sheet pans (2)
- Pastry brush
- Boning knife
- Towel

Preparation

1. Remove all bones from the chicken legs with a boning knife. Using a mallet, flatten the boneless legs slightly with the skin side down.
2. With a French knife, cut the boneless pork and veal into strips that will fit into the food grinder.
3. Soak the bread in the milk.
4. Mince the onions and celery. Sauté in the butter in a skillet.
5. Separate the eggs and save the whites for use in another preparation.

Ingredients

- 25 chicken legs
- 1 lb 8 oz fresh boneless pork picnic
- 1 lb boneless veal shoulder
- 1 lb fresh or dried bread
- 3 c milk
- 6 oz onions
- 6 oz celery
- 4 egg yolks
- 8 oz dried bread crumbs
- 6 oz butter
- 1/3 c parsley
- 1/4 oz sage
- 1/2 c salad oil
- salt and pepper to taste

6. Chop the parsley with a French knife. Wash in a towel and wring dry.
7. Preheat the oven to 350°F.

Procedure

1. Place the pork, veal, softened bread, sautéed onion and celery, and sage in a mixing container. Mix thoroughly by hand.
2. Grind twice on the food grinder using the fine chopper plate.
3. Add the egg yolks, chopped parsley, and bread crumbs. Mix thoroughly by hand until well-blended.
4. Season with salt and pepper and mix again.
5. Place a fairly generous amount of the forcemeat mixture on the meat side of each boned chicken leg, roll up, and place each leg in an aluminum potato shell with the end of the roll facing down.
6. Place on sheet pans. Brush each stuffed leg with salad oil.
7. Bake in the preheated oven at 350°F until the tops are slightly brown and the meat is completely done (approximately 1 1/2 hr, depending on the size of the legs being used).
8. Take the legs from the oven and remove from the aluminum shells. Place the legs in a steam table pan.
9. Serve one stuffed leg portion, covered with poulette or barbecue sauce.

Precaution

- Exercise caution when handling the knives.

Cheddar chicken macadamia is a preparation that brings together the succulent flavor of white meat chicken, the sharp flavor of Cheddar cheese, and the crunch of nuts. This unusual combination is a taste treat that will provide the opportunity to present an unusual and different menu item.

Cheddar Chicken Macadamia

24 servings

Equipment

- Baker's or portion scale
- Box grater
- Bake pans
- Sauce pot
- Utility knife
- Mallet
- Steam table pans
- Wire whip
- Small sheet pan
- Small piece of heavy plastic
- Cutting board
- Stainless steel bowl
- French knife

Ingredients

- 24 chicken breast halves, 6 oz portions (twelve 3 1/2 to 4 lb chickens)
- 4 oz butter
- 4 oz flour
- 2 lb (1 qt) hot chicken stock
- 8 oz warm heavy cream
- 4 oz sherry
- salt to taste
- 2 lb Cheddar cheese
- 12 oz macadamia nuts

Preparation

1. Bone the chicken breasts with the utility knife. Start at the joint where the wing is connected to the breast. Follow the breastbone until the breast is boned. Follow this procedure for each side of the complete breast. A complete breast provides two serving portions (breast halves).
2. Shred the Cheddar cheese using the large hole grid of a box grater.
3. Chop the macadamia nuts, place on a small sheet pan, and toast.
4. Preheat the oven to 350°F.

Procedure

1. Place each boned chicken breast on a cutting board, cover with a piece of plastic, and pound using a mallet until thin enough to roll.

2. Combine the shredded cheese and nuts in a stainless steel mixing bowl. Mix thoroughly.
3. Sprinkle the cheese mixture fairly heavy over the surface of each flattened chicken breast. Roll up the breast, tucking in the sides, and place seam side down in bake pans.
4. Place the butter in a sauce pot and melt.
5. Add the flour, making a roux. Cook just slightly.
6. Add the hot chicken stock, whipping vigorously with a wire whip. Simmer until the sauce is smooth and thickened. Move to the side of the range and whip in the warm cream and sherry. Season with salt and remove from the range.
7. Bake the stuffed breasts in the preheated oven at 350°F for approximately 15 min. Remove from the oven and pour off excess liquid.
8. Cover with half the sauce and return to the oven until done. Reserve the remaining sauce for service.
9. Remove from the oven. Combine the sauce in the bake pan with the amount held in reserve.
10. Serve one stuffed breast topped with sauce to each order. Garnish with additional chopped nuts and chopped parsley if desired.

Precautions

- Exercise caution when boning the breasts.
- When flattening the breasts, strike gently with the mallet. The tender white meat will flatten easily.
- Be alert when toasting the nuts.
- Whip vigorously when adding the stock to the roux and pouring in the cream.

Chicken in citrus sauce is low in sodium and calories but high in taste, and with the chicken skin removed it is also low in cholesterol. It is usually placed on the luncheon menu garnished with an assortment of citrus fruit.

Chicken in Citrus Sauce

24 servings

Equipment

- Box grater
- Wire whip
- Utility knife
- Sheet pans
- Convection oven
- Large braising pot with lid
- Kitchen fork
- Kitchen spoon
- Stainless steel steam table pan
- Stainless steel bowl

Ingredients

12 fryer chickens
1 c salad oil (variable)
2 oz ginger root
4 oz onions
2 qt orange juice
1 pt lemon juice
1 oz fresh orange rind
1 oz fresh lemon rind
6 oz brown sugar
4 oz cornstarch

Preparation

1. Grate the ginger root and orange and lemon rind using the medium grid of a box grater.
2. Disjoint and remove the skin of the chicken using a French knife and a utility knife.
3. Mince the onions with a French knife.

Procedure

1. Rub each piece of chicken with salad oil. Place on a sheet pan and brown just slightly in the oven at 400°F.
2. Remove from the oven and place in a large braising pot. Sprinkle the ginger root, onions, and grated orange and lemon rind over the chicken.
3. Add the lemon juice and all but 1 pt of the orange juice and brown sugar.
4. Cover and bake in a preheated oven at 375°F or until the chicken is done (approximately 30 min).
5. Remove from the oven. Remove the chicken from the braiser and place in the stainless steel steam table pan. Keep in a warm place.
6. Place the braiser containing the liquid on the range. Bring to a simmer.
7. Dissolve the cornstarch in the pint of orange juice held in reserve. Pour the dissolved starch slowly into the simmering liquid, whipping rapidly with a wire whip. Cook until slightly thickened and clear. Remove sauce from the range.
8. Pour the thickened sauce over the chicken and serve a half chicken to each order covered with sauce.
9. Garnish each serving with a slice of orange and lemon.

Precautions

- Exercise caution when disjointing and skinning the chicken.
- Exercise caution when mincing the onions and grating the fruit rind.
- Work the whip vigorously when pouring in the dissolved cornstarch.

Trade Tip

Marinades enhance the flavor and tenderness of poultry. Long marinating times are not required because chicken is very tender. About 1 c of marinade is enough for six chicken breasts. Marinade in which raw chicken has been soaking should never be used on cooked chicken.

FRIED AND SAUTÉED POULTRY RECIPES

Chicken Maryland

This chicken is prepared in the style that originated in Maryland. The chicken is disjointed, breaded, and fried in deep fat until it is golden brown. It is served with cream sauce, topped with two strips of bacon and two golden brown corn fritters. Chicken Maryland is an excellent choice for the dinner menu.

24 servings

Equipment

- French knife
- Stainless steel bowl
- Wire whip
- Quart measure
- Baker's scale
- Bake pans
- Ladle
- Deep fat fryer
- Sheet pans (2)
- Full size steam table pan

Ingredients

12 2½ lb chickens
1 lb bread flour
egg wash (1 qt milk and 6 eggs)
2 lb bread crumbs (variable)
2 qt cream sauce
48 bacon slices
48 corn fritters
salt and pepper as needed

Preparation

1. Clean and disjoint the chickens with a French knife. When properly disjointed, a whole chicken will yield two wings, two breasts, two thighs, and two legs.
2. Prepare the egg wash. Whip 6 eggs and 1 qt of milk with a wire whip.
3. Place the flour and bread crumbs in bake pans. Season the flour with salt and pepper.
4. Cook the strips of bacon on sheet pans in the oven until fairly crisp.
5. Prepare the corn fritters.
6. Prepare the cream sauce.
7. Preheat deep fat fryer to 325°F.

Procedure

1. Bread the disjointed chickens by passing them through flour, egg wash, and bread crumbs. Pat off excess crumbs.
2. Place the chickens in fry baskets and fry in deep fat at 325°F until golden brown and completely done. Remove and let drain. Place in a steam table pan.
3. To serve, ladle cream sauce on a plate, place a half chicken on top of the sauce, crisscross two strips of bacon over the chicken, and add two corn fritters. Garnish with a sprig of parsley.

Precautions

- Exercise caution when disjointing the chicken.
- Press the bread crumbs on firmly so they will not come off when fried.
- Fry the chicken slowly or the breading will brown before the chicken is done.

BREADING AND FRYING CHICKEN

1 COAT CHICKEN PIECES IN FLOUR

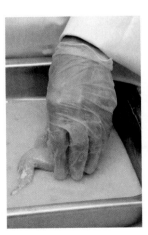
2 DIP IN EGG WASH

3 COAT WITH BREAD CRUMBS

4 FRY IN DEEP FAT

For chicken a la kiev, the chicken breast is completely boned and an herb-flavored butter is rolled into its center. The breast is breaded and fried to a golden brown. The herb-flavored butter adds an unforgettable flavor to the meat. Chicken a la kiev is featured on the dinner menu.

Chicken a la Kiev

25 servings

Equipment

- Boning knife
- Mixing machine
- Baker's scale
- French knife
- Pepper mill
- Bake pans (3)
- Stainless steel bowl
- Wire whip
- Mallet
- Quart measure
- Iron skillet
- Kitchen fork
- Full size steam table pan

Ingredients

25 chicken breast halves (from 3 to 3½ lb chickens)

3 lb butter

2 garlic cloves

½ oz chives

¼ oz marjoram

salt and fresh ground pepper to taste

1 lb flour

egg wash (8 eggs and 1 qt of milk)

1 lb 8 oz bread crumbs (variable)

3 lb shortening for frying (variable)

National Broiler Council

Preparation

1. Cut off the wing tips of each chicken breast, leaving the small wing bone attached to the breast to act as a handle. Starting at the joint where the wing is connected to the breast, follow the breastbones using the tip of a boning knife until the breast is boned.
2. Mince the garlic and chives with a French knife.
3. Grind six peppercorns in the pepper mill.
4. Prepare the rich egg wash. Whip 8 eggs and 1 qt milk with a wire whip. Place in a bake pan.
5. Place the flour and bread crumbs in separate bake pans.

Procedure

1. Place the butter in the mixing machine. Mix at slow speed using the paddle until the butter reaches a plastic consistency.
2. Add the minced garlic and chives. Rub in the marjoram by hand and season with salt and fresh ground pepper. Mix until well blended.
3. Remove the butter mixture from the mixer and place in the refrigerator until it becomes slightly firm.
4. Flatten the boneless chicken breasts, skin side down, using a mallet.
5. Place a finger of rolled cold herb butter in the center of each breast. Roll and fold in the end. Place in the freezer until the butter is very firm.
6. Bread each chicken breast by passing it through flour, egg wash, and bread crumbs. Pat off the excess crumbs.
7. Fry in an iron skillet in fairly deep fat until the breasts are golden brown and completely cooked.
8. Remove from the skillet with a kitchen fork and let drain. Place in a steam table pan.
9. Serve one half chicken breast per portion. Place a paper frill (stocking) on the wing bone and serve with a poulette or velouté sauce.

Precautions

- Exercise caution when using the boning knife and French knife.
- Hit gently when flattening the boneless breasts. The tender white meat flattens easily.
- Do not overmix the butter mixture. It will become too soft.
- Before frying the breasts, chill them in the freezer for a short time so the butter mixture becomes hard.
- Exercise caution when frying. Do not overbrown the breasts.

Trade Tip

When grilling chicken, cook the chicken until the juices run clear. The meat, either white or dark, should reach a minimum internal temperature of 165°F on a meat thermometer.

Chicken Croquettes

Chicken croquettes are a ground meat preparation. This is an item in which chicken and turkey trimmings can be utilized. Chicken croquettes are shaped into cones, breaded, fried in deep fat, and served with an appropriate sauce.

50 servings

Equipment

- Large braising pot
- French knife
- Deep fat fryer
- Bake pans (4)
- Paddle
- Baker's scale
- Food grinder
- Wire whip
- Stainless steel bowl
- Full size steam table pan

Ingredients

9 lb boiled turkey or chicken
1 lb 8 oz shortening or chicken fat
1 lb 4 oz flour
1 lb onions
2 qt hot chicken stock
½ c parsley
6 eggs
1 tsp nutmeg
salt and pepper to taste
bread crumbs as needed
12 eggs
2 qt milk
3 lb flour
3 lb bread crumbs

Procedure

1. Place the shortening or chicken fat in a braising pot and heat.
2. Add onions and sauté without browning.
3. Add the flour, making a roux. Cook for approximately 5 min, stirring occasionally with a paddle.
4. Add the chicken stock, whipping vigorously with a wire whip until thickened and smooth. Cook slightly.
5. Add the ground chicken or turkey, nutmeg, and chopped parsley. Mix thoroughly with a paddle.
6. Season with salt and pepper and remove from the fire.
7. Stir in the six beaten eggs and check consistency. If mixture is too wet, add bread crumbs to absorb some of the moisture.
8. Turn the mixture out into a bake pan and let cool. Refrigerate overnight.
9. Remove from the refrigerator. Form into 100 2 oz croquettes. Bread each croquette by passing through flour, egg wash, and bread crumbs.
10. Fry in deep fat at 350°F until golden brown. Place in a steam table pan.
11. Serve two croquettes per portion, accompanied with cream, poulette, or fricassee sauce.

Preparation

1. Grind the chicken or turkey on the food grinder using the medium chopper plate.
2. Mince the onions and chop the parsley with a French knife.
3. Break the six eggs into a container and beat slightly with a wire whip.
4. Place the 3 lb of flour and the 3 lb of bread crumbs in separate bake pans.
5. Prepare the egg wash. Place the 12 eggs in a stainless steel bowl and beat with a wire whip. Add the 2 qt of milk while continuing to beat. Place the egg wash in a bake pan.

Precautions

- Exercise caution when using the knife.
- Cook the roux to avoid a raw flour taste in the croquette mixture.
- When adding the six beaten eggs to the hot mixture, work the mixture vigorously with a paddle to avoid scrambling the eggs.
- Exercise caution when cooking the croquette mixture. It will scorch easily.

Fried Boneless Turkey Wings

For fried boneless turkey wings, the wings are saved until enough have accumulated to place them on the luncheon menu. The wings are simmered until they are tender. The two wing bones are removed, and the boneless wings are left to cool in the refrigerator overnight. The next day they are breaded, fried to a golden brown in deep fat, and served with poulette or chicken velouté sauce.

25 servings

Equipment

- Cleaver
- Bake pans (3)
- Deep fat fryer
- French knife
- 10 gal. stockpot
- Stainless steel container
- Wire whip
- Baker's scale
- Skimmer
- China cap

Ingredients

50 turkey wings
water as needed to cover wings
8 oz onions
4 oz carrots
4 oz celery
egg wash (9 eggs to 1½ qt of milk)
2 lb flour
3 lb dry bread crumbs
salt and pepper as needed

Preparation

1. Trim excess skin off the sides of the turkey wings with a French knife. Chop off the bone tips on the wing ends with a cleaver.
2. Cut the rough garnish (onions, carrots, and celery) with a French knife.
3. Preheat the deep fat fryer to 350°F.
4. Prepare the egg wash. Whip 9 eggs and 1½ qt milk with a wire whip and place in a bake pan.
5. Place the flour in a bake pan and season with salt and pepper.
6. Place the bread crumbs in a bake pan.

Procedure

1. Place the turkey wings in a stockpot and cover with water. Bring to a boil and skim off any scum that may appear on the surface.
2. Add the rough vegetable garnish and simmer until the wings are tender.
3. Strain off the stock through a china cap and save for use in another preparation. Pull the bones from the wings, leaving the meat in one piece. Let them cool and place them in the refrigerator overnight.
4. Remove from the refrigerator and bread each wing by passing it through flour, egg wash, and bread crumbs.
5. Pat off excess bread crumbs and fry in deep fat at 350°F until golden brown.
6. Serve two wings per portion with poulette or velouté sauce.

Precautions

- Exercise caution when using the knife.
- Do not overcook the wings. They will not hold together when the bones are removed.
- When pulling the bones from the wings, pull gently so the meat stays in one piece.
- Exercise caution while frying the wings. Do not over-brown.

Fried chicken is the most popular method of preparing chicken. There are many different methods used to fry chicken, but the country style method seems to be the most popular. The chicken is passed through a mixture of seasoned flour and fried to a golden brown in fairly shallow grease. It is served with country gravy and can be featured on any menu with assured results.

Country Style Fried Chicken

24 servings

Equipment

- French knife
- Iron skillet
- Kitchen fork
- Bake pan
- Baker's scale
- Full size steam table pan

Ingredients

6 2½ lb fryer chickens
1 lb flour (variable)
salt and pepper, as needed
2 lb shortening (variable)

National Broiler Council

Preparation

1. Clean and disjoint the chickens with a French knife.
2. Place the flour in a bake pan and add salt and pepper.

Procedure

1. Place enough shortening in the iron skillet to cover the bottom ½″ deep and heat.
2. Coat the pieces of chicken in the seasoned flour.
3. Shake off excess flour and place the chicken pieces in the hot grease.
4. Brown one side to a golden brown. Turn with a kitchen fork and brown the other side. It takes approximately 20 min to fry a chicken.
5. When the chicken is done, remove from the grease and let drain. Place in a steam table pan.
6. Serve two pieces of chicken per portion. Accompany each order with country gravy.

Precautions

- Exercise caution when disjointing the chickens.
- While the chickens are frying, be alert for popping grease caused when moisture under the skin comes in contact with the grease.
- Fry to a golden brown. Do not overbrown or burn.
- A lid may be placed on the skillet when starting to fry to speed the cooking time.

Chicken Cacciatore

For chicken cacciatore, the chicken is sautéed to a golden brown, placed in a baking pan, covered with a rich Italian-type tomato sauce, and baked until tender. This Italian dish is popular on both the luncheon and dinner menus.

48 servings

Equipment

- French knife
- Large braising pot and cover
- Kitchen fork
- Bake pan
- Iron skillets
- Baker's scale
- Kitchen spoon
- Quart measure
- 3 gal. sauce pot
- Spoon measure
- Deep steam table pan

Ingredients

12 2½ lb chickens
2 lb flour, seasoned with salt and pepper
1½ qt salad oil (variable)
2 lb 8 oz mushrooms
2 lb onions
6 garlic cloves
5 qt canned whole tomatoes and juice
2 qt tomato puree
1 pt Marsala wine
2 tsp crushed basil
2 tsp crushed oregano
¼ oz chives
salt and pepper to taste

Preparation

1. Clean and disjoint the chicken with a French knife.
2. Place the flour in a bake pan and season with salt and pepper.
3. Cut the mushrooms into a medium dice with a French knife.
4. Mince the onions, garlic, and chives with a French knife.
5. Crush the tomatoes by hand.
6. Preheat oven to 325°F.

Procedure

1. Place enough salad oil in an iron skillet to cover the bottom approximately ¼" deep.
2. Pass each piece of chicken through the seasoned flour and shake off excess. Place in the hot oil and sauté until golden brown. Remove with a kitchen fork and let the pieces drain.
3. Line the sautéed chicken in a large braising pot and hold.
4. Cover the bottom of a sauce pot with the oil used to sauté the chickens.
5. Add the onions, mushrooms, and garlic, and sauté until slightly tender. Do not brown. Stir occasionally with a kitchen spoon.
6. Add the wine and simmer for 5 min.
7. Add the tomatoes and juice, tomato puree, oregano, basil, and chives and season with salt and pepper. Simmer for 5 min, stirring constantly with a kitchen spoon.
8. Pour the sauce over the sautéed chicken. Cover the braising pot and bake in the preheated oven at 325°F for approximately 45 min or until chicken is tender.
9. Remove from the oven and check the seasoning. Place in a deep steam table pan.
10. Serve two pieces of chicken per portion. Cover with sauce and garnish each portion with chopped parsley.

Precautions

- Exercise caution when disjointing the chickens.
- When sautéing the vegetables, do not let them brown.
- Do not overbake the chicken or it will fall away from the bone.

Chicken Paprika

For chicken paprika, disjointed chicken is sautéed to a golden brown and baked in a rich paprika sauce. This item is most popular when served with rice on the luncheon menu.

24 servings

Equipment

- French knife
- Large iron skillet
- Large braising pot and cover
- Baker's scale
- Kitchen fork
- Quart measure
- Bake pan
- Kitchen spoon
- Full size steam table pan

Ingredients

6 2½ lb chickens
1 lb flour
1 qt salad oil (variable)
12 oz onions
1 garlic clove
1½ oz paprika
4 oz green peppers
2 qt chicken stock
6 oz tomato paste
3 oz flour
salt and pepper to taste

Preparation

1. Clean and disjoint the chickens with a French knife.
2. Mince the onions, garlic, and green peppers with a French knife.
3. Prepare the chicken stock.
4. Preheat the oven to 375°F.
5. Place the pound of flour in a bake pan and season with salt and pepper.

Procedure

1. Dredge (coat with flour) the chicken in the seasoned flour.
2. Place the oil in an iron skillet and heat.
3. Add the pieces of chicken and sauté until golden brown.

4. Remove the chicken from the skillet and place in the braising pot.
5. Pour most of the oil from the skillet, leaving only enough to sauté the vegetables.
6. Add the onions, garlic, and green pepper. Sauté until tender.
7. Add 3 oz flour, stir with a kitchen spoon into the sautéed vegetables, and cook slightly.
8. Add the paprika and stir until thoroughly blended.
9. Add the hot chicken stock and tomato paste, stirring constantly with a kitchen spoon until the mixture comes to a boil. Season with salt and pepper.
10. Simmer for 5 min and pour over the sautéed chicken in the braising pot. Cover the pot and place in the preheated oven at 375°F.

11. Bake for approximately 20 min to 30 min or until the chicken is tender.
12. Remove from the oven and place in a steam table pan.
13. Serve two pieces of chicken per portion with a generous amount of sauce. Accompany each order with baked rice or spaetzles.

Precautions

- Exercise caution when using the knife.
- When sautéing the chicken, do not overbrown.
- When sautéing the vegetables, do not let them become brown.
- Do not overcook the chicken or the meat will fall from the bones.

Trade Tip

Chicken may be safely thawed in cold running water. Place chicken in its original wrap or a water-tight plastic bag in cold water. A whole chicken will thaw in approximately 2 hours. Do not thaw chicken in standing water, as this can encourage bacteria growth.

Chicken Marengo

For chicken Marengo, disjointed, sautéed chicken is baked in a rich sauce with mushrooms and served with sliced ripe and green olives. Chicken Marengo is usually served on the dinner menu.

24 servings

Equipment

- French knife
- Kitchen fork
- Iron skillet
- Medium size braising pot and cover
- Olive pitter
- Quart measure
- Baker's scale
- Full size steam table pan

Ingredients

6 2½ lb chickens
1 qt salad oil (variable)
1 gal. brown sauce
1 pt sherry
1 garlic clove
6 oz onions
2 lb mushrooms
1 pt canned whole tomatoes
3 oz ripe olives
3 oz green olives
salt and pepper to taste

Procedure

1. Season the disjointed chicken with salt and pepper.
2. Place the oil in an iron skillet and heat.
3. Add the pieces of chicken and sauté until golden brown on both sides. Turn with a kitchen fork.
4. Remove from the skillet and place in the braising pot.
5. Pour most of the oil from the skillet, leaving only enough to sauté the vegetables.
6. Add the onions, mushrooms, and garlic. Sauté in the oil until slightly tender.
7. Add the crushed tomatoes and sherry. Bring to a boil and pour this mixture over the sautéed chicken.
8. Add the brown sauce. Cover the braising pot and bake in the preheated oven at 350°F until the chickens are tender (approximately 30 min to 40 min).
9. Remove from the oven, add the sliced olives, and season with salt and pepper. Place in a steam table pan.
10. Serve two pieces of chicken per portion with a generous portion of the sauce in which the chicken was baked.

Preparation

1. Clean and disjoint the chickens with a French knife.
2. Prepare the brown sauce.
3. Mince the onions and garlic with a French knife.
4. Slice the mushrooms fairly thick with a French knife.
5. Crush the tomatoes by hand.
6. Pit the olives with an olive pitter. Slice thin with a French knife.
7. Preheat the oven to 350°F.

Precautions

- Exercise caution when disjointing the chicken.
- Do not overbake the chicken or the meat will fall from the bones.

The breast should be boneless, making it easier to cut and consume. It is then passed through flour, dipped in an egg mixture containing Parmesan cheese, and sautéed to a golden brown. This item is appropriate for both the luncheon and dinner menu.

Sautéed Chicken Breast Parmesan

24 servings

Equipment

- Small bake pan
- Stainless steel bowl
- Boning or utility knife
- Mallet
- Kitchen fork
- Sauté pan
- Baker's or portion scale
- Steam table pan
- Small wire whip

Ingredients

24 chicken breast halves, 5 to 6 oz portions (twelve 3½ to 4 lb chickens)
8 oz flour (variable)
10 eggs
8 oz Parmesan cheese
4 oz milk
10 oz margarine or shortening
salt and white pepper to taste

Preparation

1. Cut off the wing tips of each chicken breast, leaving the small wing bone attached to the breast to act as a handle. Starting at the joint where the wing is connected to the breast, follow the breastbone using the tip of a boning or utility knife until the breast is boned. Follow this procedure for each side of the complete breast. A complete breast will provide two serving portions.
2. Melt the margarine or shortening.
3. Flatten each chicken breast slightly with a mallet.

Procedure

1. Place the flour in a small bake pan and season with salt and white pepper.

2. Beat the eggs in a stainless steel bowl using a wire whip. Add the Parmesan cheese and milk. Continue to whip until incorporated.
3. Place the sauté pan on the range. Add enough melted margarine or shortening to completely cover the bottom of the pan.
4. Dip each chicken breast in the flour and press flour on the breast firmly. Then dip the chicken breast into the cheese batter and coat the complete surface.
5. Place the coated breast in the hot grease and sauté until the bottom is golden brown. Turn, using a kitchen fork, and brown the second side. Cook at a fairly low temperature until the breasts are done.
6. Remove the pan from the range and place the breasts in a steam table pan until ready to serve.
7. Repeat the sautéing procedure until all the breasts are cooked.
8. Serve plain or with an appropriate sauce.

Precautions

- Exercise caution when boning the chicken breast.
- Hit gently when flattening the boneless breast. The tender white meat flattens easily.
- Be alert when sautéing chicken breasts. They may contain excess moisture.
- For less cholesterol, remove the skin from the breast before sautéing.

Note: Boneless chicken breast can be purchased on the market in various portion sizes, but will cost more.

In traditional chicken cordon bleu, the chicken breast is boned and stuffed with ham and Gruyère cheese, breaded, and fried to a golden brown. In this variation, the breading is omitted and an herbed crust is substituted.

New Chicken le Cordon Bleu

24 servings

Equipment

- Large skillet or sauté pan
- Parchment paper
- Stainless steel bowl
- Wire whip
- Baker's or portion scale
- Food tongs
- Bake pans (2)
- Mallet
- Slicing machine

Ingredients

24 chicken breast halves (twelve 3½ to 4 lb chickens)
24 1 oz slices smoky, dry cooked ham
24 1 oz slices Gruyère cheese
2 c fresh parsley, chopped
1½ c fresh rosemary, chopped
1½ c fresh sage, chopped
4 oz olive oil
salt and freshly ground black pepper to taste

Preparation

1. Flatten each boneless chicken breast by placing it skin side down, covering with a piece of parchment paper and striking lightly with the mallet until they are ¼" thick and 5" in diameter.
2. Slice the Gruyère cheese and ham into 2" x 4" pieces.

Procedure

1. Season each breast with salt and pepper.

2. Place a slice of ham and cheese horizontally along the bottom of each flattened breast. Fold in the two sides and roll the breast to secure the ham and cheese inside.
3. Mix the parsley, rosemary, and sage together in a wide shallow bowl.
4. One at a time, put a filled chicken breast in the herb mixture, press to adhere as many herbs as possible, then turn and coat the other side with herbs. Coat all the breasts with herbs and set them on a plate until ready to cook. Season the herbed, stuffed breasst with salt and pepper. Coated breasts can be covered with plastic wrap and stored in the refrigerator for up to 24 hours.
5. Heat the oil in a large skillet or sauté pan over medium-high heat. When the oil is hot, carefully lower the chicken into the pan, reduce the heat to medium, and cook, uncovered, until the underside is a deep brown color, about 5 to 6 min. Turn the chicken and cook the other side until well browned and cooked though, another 5 to 6 min.
6. Serve one breast per portion. Serve with a poulette, velouté, or supreme sauce, if desired.

Precautions

- Exercise caution when slicing the ham and cheese.
- Hit gently when flattening the boneless breast. The tender white meat will flatten easily.
- Exercise caution when pan frying the breast. Do not let it become too brown.

The National Chicken Council

BROILED POULTRY RECIPES

Chicken Livers en Brochette

For chicken livers en brochette, chicken livers, bacon, and mushroom caps are partly cooked are alternated on a skewer. This is passed through salad oil and placed under the broiler to complete the cooking. Chicken livers en brochette are usually served on toast covered with a butter sauce. They are featured on the dinner or a la carte menu.

25 servings

Equipment

- 25 metal skewers
- French knife
- Bake pan
- Sheet pans (2)
- Broiler
- Skillets (2)
- Kitchen fork
- Pint measure

Ingredients

100 whole chicken livers
100 mushroom caps
50 strips of bacon
1 pt salad oil (variable)
8 oz butter
25 pieces of bread

Preparation

1. Clean the chicken livers and sauté until they are half done in part of the salad oil in a skillet.
2. Clean and sauté the mushroom caps until partly done in a skillet in butter.
3. Line the strips of bacon on a sheet pan, overlapping slightly. Place in the oven and cook until medium. Remove and cool. Cut in half with a French knife.
4. Preheat the broiler. Adjust the heat fairly low.
5. Toast the bread and cut into triangular halves.
6. Place the remaining salad oil in a bake pan.

Procedure

1. Alternate chicken livers, half strips of bacon, and mushrooms on skewers. Use four of each.
2. Marinate in salad oil, then place on sheet pans.
3. Place the sheet pans under the broiler approximately 6″ from the heat. Cook until all items are completely done. Turn the chicken livers occasionally with a kitchen fork.
4. Place the skewered items across two half pieces of toast. Remove the skewer and serve with a small amount of melted butter ladled over each portion. Serve at once.

Precautions

- Exercise caution when sautéing the chicken livers. The oil may pop if moisture is present in the livers.
- Do not overcook any of the items. They should just be partly cooked.
- Broil the skewered items slowly.
- Be careful in turning the skewers. The metal becomes hot.

Chicken teriyaki is a Polynesian preparation. The chicken is marinated in the teriyaki mixture for a period of 1 or 2 days, drained, and baked. The same marinade that is used to prepare teriyaki sauce is served with each order of chicken teriyaki.

Chicken Teriyaki

25 servings

Equipment

- Portion or baker's scale
- Spoon measure
- Quart measure
- Kitchen fork
- Small stainless steel bowl
- Wire whip
- Sauce pot
- China cap
- Large stainless steel container
- Sheet pan
- French knife
- Silicon paper
- Steam table pan

Ingredients

25 6 oz boneless chicken breasts
1 pt soy sauce
2 cans (46 oz) pineapple juice
½ oz fresh ginger
½ oz fresh garlic
¼ oz black pepper
1 c salad oil
3 oz brown sugar
2 oz cornstarch

Preparation

1. Peel and mince the fresh ginger and garlic.
2. Preheat the oven to 350°F.

Procedure

1. One or two days before preparation, place the soy sauce, pineapple juice, ginger, garlic, pepper, salad oil, and brown sugar in a large stainless steel container. Blend thoroughly using a wire whip.
2. Add the boneless chicken breasts to one-half the marinade. Place in the refrigerator and marinate for one or two days.
3. On the day of preparation, remove the chicken breasts from the marinade and drain thoroughly.
4. Place the remaining half of the marinade in a sauce pot, reserving 1 c for dissolving the starch. Place the marinade on the range and simmer for approximately 1 hr until it has reduced slightly.
5. Place the chicken breasts on a sheet pan covered with silicon paper. Bake in the preheated oven at 350°F until tender. Remove and let cool slightly. *Note:* The chicken breasts may be cooked by broiling, if desired.
6. Place the cornstarch in a small stainless steel bowl. Add the reserved marinade and stir with a wire whip until thoroughly dissolved.
7. Pour the dissolved starch slowly into the simmering marinade while whipping vigorously with a wire whip. Cook until clear and slightly thickened.
8. Remove the skin from each baked chicken breast and place in a steam table pan.
9. Strain the thickened sauce through a china cap over the chicken breasts.
10. Place on the steam table and serve one breast with sauce to each order with baked white or blended rice.

Precautions

- Exercise caution when mincing the ginger and garlic.
- Bake the chicken until it is just done. Do not overbake.
- When adding the starch to the simmering liquid, whip vigorously to avoid lumps.

BOILED OR STEWED POULTRY RECIPES

Curried chicken consists of cubes of cooked chicken or turkey placed in a rich creamy curry sauce and served in a casserole, usually accompanied with rice or chutney. Curried chicken can be featured on either the luncheon or dinner menu.

Curried Chicken

25 servings

Equipment

- French knife
- 3 gal. sauce pots (2)
- Wire whip
- Baker's scale
- Quart measure
- Kitchen spoon
- 2 qt saucepan
- Deep steam table pan
- China cap

Ingredients

6 lb chicken or turkey
12 lb onions
8 oz apples
12 oz butter
10 oz flour
1 oz curry powder
3 qt chicken stock
1 qt single cream
1 banana
salt to taste

Preparation

1. Boil, cool, and dice the chicken or turkey with a French knife into ½″ cubes.
2. Prepare the chicken stock. Keep hot.
3. Heat the cream in a saucepan.
4. Cut the onions and apples into a ½″ dice and slice the banana.

Procedure

1. Place butter in a sauce pot, add the onions, and sauté until they are slightly tender. Do not brown.

2. Add the flour, making a roux, and cook for 5 min. Stir occasionally with a kitchen spoon.
3. Add the curry powder and blend into the roux.
4. Add the hot stock and cream gradually, whipping vigorously with a wire whip until thickened and smooth.
5. Add the apples and the banana and let the sauce simmer until the apples are thoroughly cooked.
6. Strain the sauce through a fine china cap into another sauce pot.
7. Heat the chicken or turkey meat. Add to the sauce and stir in gently with a spoon.

8. Season with salt and place in a deep steam table pan.
9. Serve 6 oz to 8 oz with baked rice and chutney.

Precautions

- Exercise caution when using the knife.
- Stir the sauce occasionally while it is simmering to Avoid scorching.
- Stir the cooked meat into the sauce gently to avoid breaking.

Chicken a la King

Chicken a la king is a colorful entree consisting of chunks of cooked chicken or turkey, green peppers, pimientos, and mushrooms flowing through a rich, flavorful cream sauce. Chicken a la king is an appropriate item for the luncheon or dinner menu, as well as the a la carte menu or buffet.

50 servings

Equipment

- 5 gal. sauce pot
- French knife
- Kitchen spoon
- Wire whip
- 4 qt saucepan
- 2 qt saucepans (3)
- Quart measure
- Baker's scale
- China cap
- Deep steam table pan

Ingredients

10 lb chicken or turkey
1 lb green peppers
8 oz pimientos
2 lb mushrooms
2 lb shortening or butter
1 lb 10 oz flour
3 qt chicken stock
3 qt milk
1 qt light cream
1 pt sherry
salt to taste
yellow color as desired

Preparation

1. Boil the chicken or turkey and let cool. Dice into 1" cubes with the French knife and heat.
2. Dice the green peppers, mushrooms, and pimientos into a ½" dice with a French knife.
3. Cook the diced green peppers in salt water in a saucepan until tender. Drain through a china cap and hold.
4. Sauté the mushrooms in additional butter in a saucepan until slightly tender.
5. Prepare the chicken stock.
6. Heat the milk and cream in saucepans.

Procedure

1. Place the shortening or butter in the sauce pot and heat.
2. Add the flour, making a roux. Cook for 5 min. Stir with a kitchen spoon.
3. Add the chicken stock, whipping vigorously with a wire whip until thickened and smooth.
4. Add the hot milk and cream, continuing to whip until the sauce is smooth.

5. Add the sherry and tint the sauce with yellow color.
6. Add the cooked green peppers, mushrooms, pimientos, and turkey or chicken.
7. Serve 6 oz to 8 oz over toast or a patty shell. For the buffet, serve in a chafing dish.

Precautions

- Exercise caution when using the knife.
- Cook the roux for at least 5 min to avoid a raw flour taste in the sauce.
- When combining the cooked meat and the vegetable garnish with the sauce, blend very gently with a kitchen spoon to prevent the pieces from breaking up in the sauce.
- Exercise extreme caution when adding the yellow color. More can always be added.

Chicken Pot Pie

Chicken pot pie consists of fairly large chunks of chicken and cooked, assorted vegetables. These are placed in deep individual casseroles, covered with chicken velouté sauce, topped with a prebaked pie crust disk, and served on the luncheon menu.

25 servings

Equipment

- French knife
- 4 qt saucepans (2)
- 2 qt saucepan
- Baker's scale
- 25 individual deep casseroles
- Steel skillet
- Ladle
- Bake pans

Ingredients

6 lb boiled chicken or turkey
25 canned small onions
50 carrot pieces
50 canned small potatoes
50 celery pieces
25 mushroom caps
1 gal. chicken velouté sauce
25 pie crust disks
6 oz butter or margarine
1½ lb frozen peas

Preparation

1. Dice the boiled turkey meat fairly large with a French knife.
2. Prepare the chicken velouté sauce.
3. Sauté the mushrooms in the butter.
4. Cut the raw vegetables with a French knife.
5. Drain the onions and potatoes.
6. Cook the raw vegetables in separate saucepans in boiling salt water until tender and drain.
7. Prepare the pie crust disk, cutting each disk the same shape but slightly larger than the top of the casserole. The disk will shrink slightly when baked.

Procedure

1. Place approximately 3 oz to 4 oz chicken meat, two carrot pieces, two celery pieces, two potatoes, one onion, one mushroom cap, and some peas in each individual casserole.
2. Ladle the velouté sauce over the meat and vegetables in each casserole and top with a prebaked pie crust disk.
3. Place the casserole in bake pans and add enough water to surround the bottom half of the casseroles with water.
4. Place the bake pans on the side of the range and keep hot until ready to serve.

Precautions

- Exercise caution when using the knife.
- Handle the baked pie dough disks gently. They break easily.
- Exercise caution when ladling the sauce over the meat and vegetables. Ladle with a smooth, easy motion.

Chicken Tetrazzini

Chicken tetrazzini is a combination of cream chicken and mushrooms placed over cooked spaghetti, sprinkled with Parmesan cheese, and browned gently under the broiler. It was supposedly a favorite of the famous Italian opera singer Luisa Tetrazzini and was named in her honor.

25 servings

Equipment

- French knife
- Kitchen spoon
- 3 gal. sauce pots (2)
- Saucepans
- Colander
- Baker's scale
- Quart measure
- 25 individual casseroles
- Wire whip

Ingredients

6 lb chicken or turkey
2 lb mushrooms
1 lb shortening or butter
12 oz flour
1 gal. chicken stock
1 pt cream
½ c sherry
1 gal. thin spaghetti (variable)
1 pt Parmesan cheese
salt and pepper to taste

Preparation

1. Boil the chicken or turkey until tender and let cool. Remove the meat from the bones and cut into 2″ × ½″ × ¼″ strips with the French knife.
2. Boil the spaghetti in salt water in a sauce pot. Drain in a colander and rinse in cold water, then reheat in warm water. Season lightly with salt and white pepper and hold.
3. Prepare the chicken stock.
4. Warm the cream in a saucepan.
5. Slice the mushrooms and sauté slightly in additional butter in a saucepan.
6. Preheat the broiler.

Procedure

1. Place the butter or shortening in a sauce pot and heat.
2. Add the flour, making a roux. Cook for 5 min. Stir with a kitchen spoon.
3. Add the hot chicken stock, whipping vigorously with a wire whip until thickened and smooth.
4. Whip in the warm cream and sherry.
5. Add the cooked strips of chicken or turkey and the cooked mushrooms. Stir in gently with a kitchen spoon so the meat does not break.
6. Season with salt and white pepper.

7. Arrange spaghetti in the bottom of each casserole. Place the chicken mixture over the spaghetti and cover completely. Sprinkle with grated Parmesan cheese and brown slightly under the broiler.
8. Serve at once.

Precautions

- Exercise caution when using the knife.
- When adding the liquid to the roux, whip vigorously to produce a smooth sauce.
- Add the cream slowly to the sauce while whipping briskly.
- Be alert when browning each order under the broiler.

Chicken chow mein is a Chinese-American dish that appears frequently on the luncheon menu of commercial food establishments. This mixture of crisp Chinese vegetables and cooked chicken or turkey is blended in a rich, thickened, highly-seasoned sauce.

Chicken Chow Mein

50 servings

Equipment

- 10 gal. stockpot
- French knife
- Paddle
- Stainless steel container
- Deep steam table pan
- Baker's scale
- Quart measure
- 50 individual casseroles
- China cap

Ingredients

- 8 lb chicken or turkey
- 1 pt salad oil (variable)
- 6 lb onions
- 6 lb celery
- 2 lb mushrooms
- 1 No. 2½ can water chestnuts
- 1 No. 2½ can bamboo shoots
- 1 gal. chicken stock
- 4 oz soy sauce
- 3 qt canned bean sprouts
- 14 oz cornstarch (variable)
- 1 pt cold water
- salt and pepper to taste

Procedure

1. Place the salad oil in the stockpot and heat.
2. Add the onions and celery and sauté for 5 min. Stir with a paddle.
3. Add the mushrooms and continue to sauté until vegetables are partly done.
4. Add the bamboo shoots, water chestnuts, soy sauce, and chicken stock. Simmer until the celery is done but retains a crisp texture.
5. Dilute the cornstarch in the cold water. Add to the simmering mixture, stirring vigorously with a paddle until it is thickened and smooth.
6. Add the turkey and stir in gently. Remove from the range.
7. Add the bean sprouts and stir in gently. Season with salt and pepper. Place in a deep steam table pan.
8. Serve 6 oz to 8 oz into casseroles with baked rice, fried noodles, or both.

Preparation

1. Boil, remove bones by hand, and cut the chicken or turkey into ½″ diagonal pieces with a French knife.
2. Cut the onions and celery with a French knife.
3. Slice the mushrooms, bamboo shoots, and water chestnuts with a French knife.
4. Prepare a rich chicken stock.
5. Drain the bean sprouts in a china cap.

Precautions

- Exercise caution when using the knife.
- Add the bean sprouts after all cooking is completed or they will lose their crispness.
- Stir constantly and vigorously when adding the cornstarch.
- Do not overcook the vegetables. Attempt to retain a slight crispness.

Trade Tip

Cook turkey progressively, in batches, so that it can be served as soon as possible. Slice and hold only the amount that can be served within a 20 min to 30 min period.

DRESSING RECIPE

Bread dressing is a highly-seasoned bread mixture that is used extensively with poultry preparations. Through the years, it has become heavily associated with roast turkey. There are many variations of this basic bread dressing, such as oyster, raisin, giblet, shrimp, and chicken liver stuffing.

Bread Dressing

50 servings (No. 16 scoop)

Equipment

- Baker's scale
- French knife
- Skillet
- Large round-bottom bowl
- Spoon measure
- Braising pot or roast pan
- Sheet pans
- Apple corer

Ingredients

8 lb dry bread (2 days old)
1 lb celery
3 lb onions
1 lb fresh apples
1 lb margarine or bacon grease
4 tbsp sage
3 tbsp poultry seasoning
½ c parsley
2 gal. chicken stock (variable)
salt and pepper to taste

Preparation

1. Cut the bread into cubes, place on sheet pans, and toast in the oven.
2. Dice the celery fine and mince the onions with a French knife.
3. Core and slice the apples into fairly small pieces with a French knife.
4. Chop the parsley with a French knife.
5. Prepare a rich chicken stock or combine approximately 2 oz to 3 oz of chicken soup base to 1 gal. of water.
6. Grease a large braising pot or roast pan.
7. Preheat oven to 375°F.

Procedure

1. Place the margarine or bacon grease in a skillet and heat.
2. Add the celery and onions. Sauté until slightly tender.
3. Add the apples and continue to sauté until tender. Remove from the heat and cool slightly.
4. Place the toasted bread, parsley, sautéed onions, celery, apples, chicken stock, sage, and poultry seasoning in the large round-bottom bowl. Mix thoroughly using the hands. The mixture should be soft and slightly wet. If it is stiff and dry, add more stock.
5. Season with salt and pepper and place in the greased braiser or roast pan.
6. Bake in the preheated oven at 375°F for approximately 1½ hr to 2 hr until the stuffing is hot throughout and golden brown on the surface. Serve with poultry and other meats.

Variations (Add ingredients before baking.)

Raisin stuffing: Add 1 lb raisins that have been soaked in warm water and drained.
Giblet stuffing: Add 2 lb boiled, chopped giblets.
Oyster stuffing: Add 1 to 1½ qt oysters that have been sautéed slightly in butter.
Shrimp stuffing: Add 2 lb cooked, diced shrimp.
Chicken liver stuffing: Add 3 lb diced chicken livers that have been sautéed in butter.

Precautions

- Exercise caution when using the knife.
- Exercise caution when toasting the bread cubes to avoid burning.
- Blend all ingredients before placing the stuffing in the prepared pan.
- Bake the dressing thoroughly. If a cold center remains after baking, fermentation could start.

Uses of Leftover Chicken or Turkey

The following are suggestions for using leftover cooked chicken or turkey:

- Dice and use in chicken or turkey a la king, creamed chicken, chicken salad, etc.
- Mince and use in various soups and ravioli fillings.
- Grind and use in croquette mixture, stuffing for turnovers, and sandwich and canapé spreads.
- Julienne and use in salads or for stuffing pita bread.
- Slice and use in sandwiches.

Fish and Shellfish Preparation

Chapter 24

Fish and shellfish have been an important source of food since prehistoric times. Today, fish and shellfish are becoming even more popular as people are more diet-conscious. Modern advancements in catching and processing have permitted new markets for fish and shellfish. The popularity of fish and shellfish has increased to such an extent that most food service establishments feature three or four choices on their menus.

The main difference between fish and shellfish is that fish have an internal skeleton with a backbone, while shellfish have an external skeleton with no backbone. Both fish and shellfish have flesh that is naturally tender. Compared to the flesh of animals, fish and shellfish require very little cooking. Overcooking, which is a common mistake, toughens or dries out the preparation.

Fresh fish and shellfish are preferred for most preparations. However, the distance from the supply often requires that frozen or other market forms be used.

FISH CLASSIFICATION

Fish are classified as freshwater fish and saltwater fish based on their natural habitat. The natural habitat of a fish also affects the characteristics of the edible flesh. An active fish that comes from a running stream possesses a superior flavor to fish from a less active body of water.

Fish are further divided into types based on fat content as fat fish and lean fish. Fat fish contain as much as 20% fat or more, while lean fish contain as little as 0.5% fat. Table I lists the fat content of common saltwater and freshwater fish. The fat content of fish varies with the type of fish and season of the year. The fat content of a fish produces a difference in flavor and affects the cooking method required. Fat fish, such as mackerel and salmon, produce superior eating qualities if baked or broiled because their natural fat prevents them from drying during the cooking process. Lean fish, such as haddock and cod, are usually best when steamed or poached because the flesh is firm and holds together during the cooking process. Both fat and lean fish can be sautéed or fried with excellent results.

Exceptions can be made to cooking procedures for fat and lean fish if allowances are made for the fat content. For example, lean fish can be baked or broiled if basted during the cooking process. Fat fish may also be steamed or poached with special care in cooking and handling the fish.

Unlike the flesh of animals, fish flesh has very little connective tissue. This results in naturally tender flesh. Fish requires very little heat for cooking compared to other food items. The most common mistake made in cooking fish is overcooking. Cook just enough to enable the flesh to begin to flake easily from the bones. Allow for the continued cooking that takes place after the fish is withdrawn from the heat.

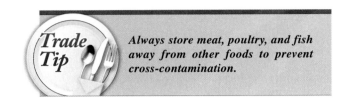

Trade Tip

Always store meat, poultry, and fish away from other foods to prevent cross-contamination.

TABLE I: COMMON FISH AND SHELLFISH			
Species	**Fat or Lean**	**Market Range of Round Fish**	**Market Forms**
Saltwater Fish			
Bluefish	Lean	1 lb to 7 lb	Whole and drawn
Butterfish	Fat	¼ lb to 1 lb	Whole and dressed
Codfish	Lean	3 lb to 20 lb	Drawn, dressed, steaks, and fillets
Croaker	Lean	½ lb to 2½ lb	Whole, dressed, and fillets
English and Dover sole	Lean	1 lb to 4 lb	Whole, drawn, dressed, and fillets
Flounder	Lean	¼ lb to 5 lb	Whole, dressed, and fillets
Haddock	Lean	1½ lb to 7 lb	Drawn and fillets
Hake	Lean	2 lb to 5 lb	Whole, drawn, dressed, and fillets
Halibut	Lean	8 lb to 75 lb	Dressed and steaks
Mullet	Lean	½ lb to 3 lb	Whole
Pompano	Fat	¾ lb to 1½ lb	Whole, drawn, dressed, and fillets
Salmon	Fat	3 lb to 30 lb	Drawn, dressed, steaks, and fillets
Sea bass	Lean	¼ lb to 4 lb	Whole, dressed, and fillets
Shad	Fat	1½ lb to 7 lb	Whole, drawn, and fillets
Snapper, red	Lean	2 lb to 15 lb	Drawn, dressed, steaks, and fillets
Spanish mackerel	Fat	1 lb to 4 lb	Whole, drawn, and dressed
Striped bass	Lean	1 lb to 10 lb	Whole, drawn, and dressed
Freshwater Fish			
Brook trout	Fat	¼ lb to 1 lb	Whole, dressed, and fillets
Catfish	Fat	1 lb to 10 lb	Whole, dressed, and skinned
Lake trout	Fat	1½ lb to 10 lb	Drawn, dressed, and fillets
Northern pike	Lean	2 lb to 12 lb	Whole, dressed, and fillets
Rainbow trout	Fat	1½ lb to 2½ lb	Whole, drawn, and dressed
Smelt	Lean	⅓ lb to ½ lb	Whole and drawn
Whitefish	Fat	2 lb to 6 lb	Whole, drawn, dressed, and fillets
Walleyed pike	Lean	2 lb to 5 lb	Whole, dressed, and fillets
Yellow perch	Lean	½ lb to 1 lb	Whole and fillets
Shellfish			
Clams	Lean		In the shell, shucked
Crabs	Lean		Live, cooked meat
Lobsters	Lean		Live, cooked meat
Oysters	Lean		In the shell, shucked
Scallops	Lean		Shucked
Shrimp	Lean		Headless, cooked meat

SHELLFISH CLASSIFICATION

Shellfish is a general name for seafood having shells. Shellfish are different from fish in that shellfish have an external skeleton with no backbone. Fish have an internal skeleton with a backbone on the inside. Shellfish are classified as *crustaceans* and *mollusks*. Crustaceans have a jointed body, a number of jointed legs, and a hard shell. Crustaceans include lobster, shrimp, crayfish, and crab. Mollusks have very soft bodies covered with a hard-hinged shell. Mollusks include oysters, clams, mussels, snails, and scallops.

Most shellfish are commonly used in preparations in the commercial kitchen except snails and mussels. Snails are often served in food service establishments that feature French cuisine. Mussels are more commonly prepared near the source of supply because of limited storage life.

National Fisheries Institute

Mussels are shellfish of the mollusk variety. Mollusks have very soft bodies covered with a hard hinged shell.

MARKET FORMS

Fish and shellfish may be purchased in various forms. Each form has certain advantages and disadvantages in cost, convenience, and labor. The more preparation that is done before delivery to the commercial kitchen, the higher the cost per pound will be. The market form best suited for the food service establishment depends on the type of seafood and the preparation, storage facilities, availability of skilled labor, and equipment. Common market forms of fish and shellfish are listed in Table II.

TABLE II: MARKET FORMS: FISH AND SHELLFISH	
Fish	**Shellfish**
Whole or round Drawn Dressed or pan-dressed Steaks Single fillets Butterfly fillets Sticks	Live Shucked Headless Cooked meat

Fish Market Forms

Whole or *round fish* are those marketed the way they are taken from the water. Whole fish must be scaled and eviscerated (entrails removed) before cooking. The head, tail, and fins must be removed. The fish is then split, filleted, or cut into serving portions. Fish purchased whole costs less per pound but requires more preparation.

Drawn fish have only the entrails removed and must be scaled. Drawn fish can be prepared whole. The head, tail, and fins must be removed. The fish is then split, filleted, or cut into serving portions.

Dressed or *pan-dressed fish* are scaled and eviscerated. The head, tail, and fins are also removed. Preparation for cooking requires the fish to be split, filleted, or cut into serving portions, except when preparing the fish in its whole form. Smaller fish are usually prepared whole. Dressed fish are a very popular market form used in the commercial kitchen.

Steaks are cross-sectional slices of the larger sizes of dressed fish. They are ready to cook when purchased. Generally the only bone present is a small section of the backbone.

Single fillets are the flesh of the sides of the fish cut lengthwise away from the backbone. There are two single fillets to each fish. Fillets are practically boneless and require no preparation for cooking. Single fillets can be purchased with the skin on or removed. Skin that is left on is already scaled.

Butterfly fillets are two single fillets held together by the uncut belly of the fish.

Sticks are pieces of fish cut lengthwise or crosswise from single fillets into portions of uniform length and thickness.

MARKET FORMS - FISH

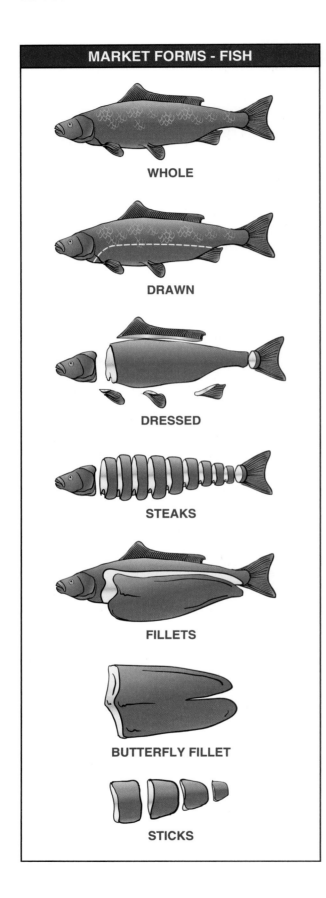

WHOLE

DRAWN

DRESSED

STEAKS

FILLETS

BUTTERFLY FILLET

STICKS

Southern Pride Catfish

Fillets are practically boneless and require no additional preparation for cooking.

Shellfish Market Forms

Live: Shellfish such as crabs, lobsters, clams, oysters, and mussels should be alive if purchased in the shell.

Shucked: Shucked shellfish are marketed with their shell removed. Shucked oysters, scallops, and clams are marketed fresh or frozen. Some shucked shellfish, such as shrimp and oysters, may be purchased canned.

Headless: This form applies to shrimp and sometimes warm water lobsters, which are marketed frozen with the head and thorax removed. Headless shrimp can also be canned.

PURCHASING FISH

Fresh fish: The amount of fresh fish available depends on the time of the year. For economy, it is best to purchase when fresh fish is most abundant. Consult the local fish supplier for price, availability, and quality of fresh fish available throughout the year.

Fresh fish is not federally inspected as meat and poultry are. This requires careful inspection by the purchaser. When purchasing fresh fish, consider all parts of the fish to determine freshness. Refer to Table III for fish condition.

Fresh fish directly out of the water and into the pan gives the best results; however, this is seldom possible. To preserve as much freshness as possible, proper storage is necessary. Fresh fish is best stored a maximum of 1 to 2 days. If the fish is not cooked immediately, it should be packed in ice and placed in the coldest part of the refrigerator as soon as possible. The ice helps hold the proper temperature, keeps the fish moist, and reduces bruising. The entrails of the fish should be removed before storing. Store the fish away from other food items in the refrigerator to prevent the odor from affecting other foods.

The quantity of fresh fish required is determined by the number of people being served, the portion size, and the market form. Portion allowances commonly used are:

- *Sticks, steaks, and fillets*: Use ⅓ lb per person.
- *Dressed fish*: Use ½ lb per person.
- *Drawn fish*: Use ¾ lb per person.
- *Whole or round fish*: Use 1 lb per person.

Frozen fish: More frozen fish is served than fresh fish. Frozen fish allows a greater variety of fish available year-round. If handled and prepared properly, frozen fish is comparable to fresh fish. Thaw frozen fish close to cooking time, allowing time for trimming, cleaning, and breading. Frozen fish should be thawed overnight in the refrigerator at a temperature of 38°F to 40°F for best results. If speed is required, thaw under cold running water or in a microwave oven. Never thaw frozen fish at room temperature. Once frozen fish is thawed, it must be used. Never refreeze frozen fish after thawing. Smaller frozen fillets and portion-cut fish that are breaded and pan-ready can be cooked while still frozen. For best results, reduce the cooking temperature and cook longer.

Frozen fish should be delivered frozen solid and stored in the freezer at 0° to 10°F immediately. Rotate the stock received, using the oldest fish first. The fish is kept frozen until just prior to using. Fish stored in the freezer must be wrapped properly with special freezer paper to prevent freezer burn.

The portion allowance is the same for frozen fish as it is for fresh fish. However, because frozen fish cannot be purchased whole or in the round, allow ⅓ lb to ½ lb of the edible part per person.

Many types of convenience fish forms are available on the market. Fish may be purchased stuffed, breaded, or topped with assorted foods. However, convenience is expensive and must be considered when determining cost. A combination of letters that describe the process used may appear on packages of convenience fish:

- IQF: individually quick frozen.
- PV: peeled and veined. This term is used with shrimp.

Canned fish is seldom used in the commercial kitchen. Cans should be checked for signs of damage or bulging. Cans that are opened must be properly covered and refrigerated. Canned salmon, tuna, anchovies, and sardines are the most common canned fish.

Specialty fish items are the result of developments in the methods of preserving fish products and include smoked, salted, and pickled fish. Popular smoked fish are cod, haddock (finnan haddie), salmon, sturgeon, and herring. Popular salted fish are cod, mackerel, and hake. Popular pickled fish are salmon and herring. Specialty fish items are commonly used in hors d'oeuvres and canapés.

PURCHASING SHELLFISH

Lobsters, like most shellfish, must be kept alive until they are ready to be cooked. Lobsters are purchased through a local dealer or direct from a supplier. Lobsters are shipped by rail or air express in containers filled with seaweed. When received, the lobsters are carefully inspected. Lobsters that are alive have a tightly curled tail. Lobsters that are dying are cooked immediately to save as much meat as possible. Lobster meat has a high moisture content. When a lobster dies, the meat evaporates quickly. Lobsters are graded in four sizes, as shown in Table IV. Cooked lobster meat is picked from the shell and marketed in frozen and canned form. However, production is limited, and lobster meat is not always available in these forms.

TABLE III: FISH CONDITION		
	Desirable	**Undesirable**
Eyes	Bright and clear	Dull and clouded
Gills	Bright red, free from slime	Dull gray or brown
Flesh	Firm, elastic, not separating from bones	Soft, dents when pressed, separating from bones
Scales	Bright, adhere tightly to the skin	Lack of sheen, loose
Odor	Fresh, free from objectionable odors	Strong, objectionable

TABLE IV: LOBSTER GRADES	
Chicken	¾ to 1 lb
Quarters	1¼ lb
Large	1½ to 2¼ lb
Jumbo	Over 2½ lb

PREPARING FISH

1 WASH THE FISH; SCALE THE FISH BY SCRAPING IT WITH THE KNIFE ALMOST VERTICAL

1 CUT DRESSED FISH IN 1″ SECTIONS FOR STEAKS

CUTTING INTO STEAKS

2 CUT THE ENTIRE LENGTH OF THE BELLY FROM THE VENT TO THE HEAD; REMOVE THE INTESTINES

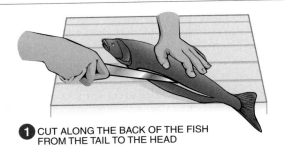

1 CUT ALONG THE BACK OF THE FISH FROM THE TAIL TO THE HEAD

3 REMOVE THE HEAD AND PECTORAL FINS BY CUTTING BEHIND THE COLLARBONE

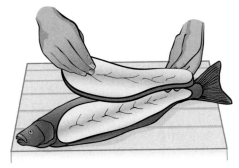

2 TURN THE KNIFE HORIZONTALLY AND CUT THE FLESH AWAY FROM THE BACKBONE AND RIB BONES

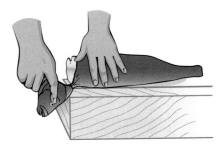

4 PLACE THE FISH ON THE EDGE OF THE CUTTING BOARD AND SNAP THE BACKBONE TO REMOVE THE HEAD; CUT THE REMAINING FISH

DRESSING

3 REMOVE THE FILLET; TURN OVER AND REPEAT THE FILLETING OPERATION

FILLETING

Rock lobsters are also known as spiny lobsters or langoustes. They are warm water relatives of the northern lobster but have no claws. Only the tail is marketed as lobster tail. The flesh is similar to that of the Maine lobster, but is less flavorful.

Crayfish or *crawfish* come from fresh water and are smaller in size than lobsters. Crayfish are commonly used in southern style and French preparations.

Shrimp are always marketed in headless form. The tail is the only edible part. The body and thorax are not edible and are removed and discarded. Shrimp are graded and sold according to size. Shrimp size is determined by the number of headless shrimp to the pound. The count, or number of shrimp per pound, is listed in Table V.

TABLE V: SHRIMP GRADES	
Jumbo	25 or fewer per lb
Large	25 to 30 per lb
Medium	30 to 42 per lb
Small	42 or more per lb

Most shrimp used in the commercial kitchen are purchased in the frozen form and nearly all are sold IQF (individually quick frozen). Shrimp may be purchased in four market forms:

- *Fresh shrimp* are commonly used if near a source of supply. Because shrimp spoils rapidly, fresh shrimp are rarely used away from the coasts because of transportation costs.

- *Frozen shrimp* are packed mainly in 5 lb blocks for sale to food service establishments. Frozen shrimp can be purchased as green shrimp (uncooked), either peeled or unpeeled; cooked and peeled; or peeled, cleaned, and breaded. Frozen shrimp is sold by the pound. The cost of frozen shrimp is based on how much preparation has been done before purchase.

- *Cooked shrimp* may be purchased either peeled and cleaned or in the shell. Cooked shrimp are sold by the pound and are commonly found near the source of supply.

- *Canned shrimp* are available packed in brine or dry and sold in various size cans. Canned shrimp are seldom used in the commercial kitchen.

Carlisle FoodService Products

Shrimp size is determined by the number of headless shrimp to the pound.

Oysters and *clams* are typically purchased in three forms: live in the shell, shucked, or canned. Oysters also can be purchased pasteurized and individually quick frozen (IQF). Oysters and clams live in the shell are sold by the dozen, bushel, or barrel. They must be alive when purchased, which is indicated by a tightly closed shell. If the shell is open and does not close when handled, the oyster or clam is dead and no longer fit for human consumption. Shucked oysters and clams that have been removed from their shell are usually packaged fresh in gallon containers. Frozen oysters and clams are used if fresh ones are unavailable.

Carlisle FoodService Products

Fresh oysters and clams must be alive when purchased, or be purchased frozen.

Trade Tip

Mussels, clams, and oysters should be purchased alive in the shell. The shells should be closed when purchased. The shells may be open naturally but will close tightly when tapped, indicating that they are alive. Discard any dead ones. Shell stock tags should be kept on file for 90 days after the last shellfish was used.

Fresh shucked oysters and clams are packed in metal containers and must be kept refrigerated and packed in ice at all times to prevent spoilage. If handled in the proper manner, oysters and clams remain fresh for about 1 week. Shucked oysters and clams are graded by size based on the number per gallon. Clams are graded large, medium, and small, but oysters are graded according to federal standards as shown in Table VI. Canned oysters and clams are rarely used in the commercial kitchen.

TABLE VI: OYSTER GRADES	
Extra large or counts	160 or less per gal.
Large or extra select	161 to 210 per gal.
Medium or select	211 to 300 per gal.
Small or standard	301 to 500 per gal.
Very small	Over 500 per gal.

Crabs are purchased in three forms: live, cooked meat (fresh or frozen), and canned. On the coasts, crabs are sold alive and must be kept alive until they are cooked. This applies to both the hardshell crab and the softshell crab (molting blue crabs). Hardshell crabs are not sold alive away from the coast because they do not ship well. However, softshell crabs are packed in seaweed and shipped alive throughout the country. Softshell crabs may be purchased through a local dealer or a direct supplier. Cooked crabmeat may be purchased in the shell, fresh or frozen, or as fresh, cooked meat. The fresh, cooked meat is packed in several grades, including:

- *Lump meat*: Solid lump of white meat from the body of the crab.
- *Flake meat*: Small pieces of white meat from the remaining parts of the body.
- *Lump and flake meat*: A combination of lump and flake meat.
- *Claw meat*: Meat from the claws, which has a brownish tint.

Fresh, cooked crabmeat can also be purchased in frozen form. Crabmeat is purchased frozen if it is not going to be used immediately. Fresh, cooked crabmeat should be kept packed in ice and refrigerated until it is used.

Canned cooked crabmeat is available in various size cans and is commonly used in commercial kitchens. The biggest advantage of canned cooked crabmeat is its shelf life. In addition, it can be used interchangeably with cooked frozen crabmeat.

- *Synthetic crabmeat*: A crabmeat product that looks, cooks, and tastes like crabmeat. Synthetic crabmeat is made from a mixture of pollock fish, snow crabmeat, turbot fish, wheat starch, egg whites, vegetable protein, and other ingredients. It is low in calories, sodium, fat, and cholesterol and high in protein. It is marketed precooked and frozen to protect its flavor and can be purchased as legs, chunk meat, or flake meat.

Scallops: Scallops are always marketed in shucked form either fresh or frozen. Fresh scallops can be purchased by the gallon or the pound. Frozen scallops are usually purchased in 5 lb blocks.

POPULAR FRESHWATER FISH

Whitefish is considered the king of the freshwater fish. They are taken from northern lakes and Canada. Whitefish average in weight from 2 lb to 6 lb. Whitefish weighing 2 lb to 4 lb are considered the best. The flesh is white with a flaky grain. Whitefish is a member of the salmon family and is on the market year round. May, June, July, and August provide the largest catch, resulting in the best price. The whitefish is a fatty fish with a black and white skin, a small short head, and a deep forked tail. The fish is very popular in the commercial kitchen and is best when broiled or sautéed.

Lake trout is the largest of all trout and is taken from the Great Lakes with the exception of Lake Erie. Lake Michigan provides the majority of the total catch. Lake trout is most plentiful on the market from May to October. The fish has a dark to pale gray skin that is covered with white spots, and a fairly large head. The flesh may be red, pink, or white depending on the lake from which it was taken. The average size of this trout is about 10 lb. Lake trout weighing 4 lb to 10 lb with pink flesh are considered the best. The lake trout is a fatty fish with a very delicate and desirable flavor that is as highly prized as the whitefish. The lake trout is popular in the commercial kitchen and is best when broiled or sautéed.

Brook trout is a medium-fat fish and one of the finest eating fish available, provided it is taken from ice cold water. Brook trout have silver-gray, slightly speckled skin, and a square, slightly forked tail. They average in weight from $\frac{1}{4}$ lb to 1 lb. A trout weighing 8 oz to 10 oz provides a generous serving. The popularity of brook trout has increased greatly. To meet this increased demand, trout farms have been established throughout the United States.

Rainbow trout is considered a freshwater fish; however, like the salmon, it passes freely from the ocean to fresh water to spawn. Most trout prefer to remain in one general location, but this is not true of the rainbow trout. It migrates to lakes, streams, and oceans, preferring a cold environment. Rainbow trout vary greatly in color depending on the waters they inhabit. They have a characteristic marking of a purplish red band that extends along their sides from the head to the tail. It is because of this band that the name rainbow was derived. The average weight of a rainbow trout is about 2 lb, but the size of the trout depends to a great degree on the size of the body of water from which it is taken. The larger the body of water is, the larger the fish is. Rainbow trout is considered a fat fish and are commonly used in the commercial kitchen. Rainbow trout is best when sautéed or broiled.

Yellow perch are taken from the Great Lakes and northern Canada. The name of the fish varies depending on its size. The small sizes are known as lake perch, the larger ones as Lake Erie perch, and the extra large ones as jumbo or English perch. The skin of the yellow perch is dark olive green on the back, merging into a golden yellow on the sides and becoming lighter as it extends to the belly. Its sides are marked with six to eight dark broad vertical bands that run from the back to just above the belly. Yellow perch average about 12″ in length and weigh about 1 lb. Yellow perch are available on the market all year, but are most abundant from April to November. It is a lean fish and is commonly used in the commercial kitchen. Yellow perch are usually sautéed, fried, or broiled.

Walleyed pike, also known as jack salmon, is a lean fish and a favorite sport fish because of its willingness to strike at any kind of lure. The walleyed pike inhabits many rivers, lakes, and streams in most states except those of the far west and extreme south. The color of the walleye varies with its environment, but it is usually a dark olive green on the back, shading into a light yellow on the sides and belly. Walleye have exceptionally large shiny eyes. Walleyed pike vary considerably in size depending on where they are caught. The average weight is between 2 lb and 5 lb. The flesh is slightly fine-grained and has an excellent flavor, but many small bones are present. Walleyed pike is very popular in food service establishments. It is best when fried or sautéed.

Northern pike, also known as pickerel, is an enormous eater. It consumes a daily portion of food equal to one-fifth its own weight. Northern pike is very much like the muskellunge, but is not as large. It has a long lean body, a long, broad, flat snout, and broad bands of sharp teeth on both jaws. It is a lean fish with a very firm, flaky flesh that is dry with many bones present. The northern pike average from 2 lb to 4 lb, but 10 lb to 15 lb is not rare. The northern pike inhabits most of the cold, fresh waters of the world, and in North America it is found mainly in the Canadian lakes. Northern pike is available all year but is most plentiful in June. The eating qualities are best during the cold months of the year. Northern pike is a popular menu item whether fried, sautéed, broiled, or baked.

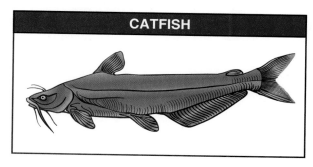

CATFISH

Catfish and *bullhead* (a similar species) generally have a forked or deeply notched tail. Both are fat fish and provide a firm, flaky meat with excellent eating qualities. The skin of the catfish varies in color depending on its environment, but it usually has a brown or black tone. The catfish or bullhead does not have scales, and the skin adheres tightly to the flesh, making cleaning difficult. Catfish are most plentiful from March to October and average in size from ¾ lb to 40 lb. In the commercial kitchen, the ¾ lb size is preferred and placed on the menu either fried or sautéed. Most catfish now on the market are farm-raised and have a better taste than those taken from a natural habitat. Breaded catfish is available as breaded fillets, breaded whole, breaded strips, and breaded nuggets, reducing preparation time in the commercial kitchen.

Smelts are classified as freshwater or saltwater fish. The largest catch of freshwater smelts comes from Lake Michigan. Saltwater smelts come from the Atlantic coast, from New York to Canada, with Canada producing the largest percentage. Smelts taken from the coldest water are the finest. The smelt is a small, lean fish with a very slender body, a long pointed head, a large mouth, and a deeply forked tail. It has an olive to dark green color along the top, blending into a lighter shade with a silver cast along its sides. The belly is silver and the fins are slightly speckled with tiny spots. Smelts are of the salmon family and reach a size of about 10″ long and weigh as much as 1 lb, although most are much smaller. Most smelts used in the commercial kitchen average about 6 to 8 per pound. Smelts are marketed in the round, and once the entrails are removed, the whole fish is prepared by frying or sautéing.

Southern Pride Catfish

Fried catfish is a very popular entreé.

Frog legs produce a delicious meat. Although frogs are amphibians, frog legs are classified as a seafood. The best frog legs come from the bullfrog, which is raised on frog farms. Bullfrogs produce a large white meat leg. Only the hind legs of the frog are marketed.

The common frog, an uncultivated product, produces a dark meat. The green grass frog, which is very small, produces a very sweet meat leg and is the cheapest to purchase, but its small size keeps it from becoming popular.

Frog legs are on the market all year, but are most plentiful from April to October. In the commercial kitchen, a large percentage of the frog legs used come from India or Japan. The most desirable legs average 2 or 3 pairs to the pound. Frog legs are best when fried or sautéed.

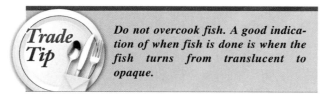

Trade Tip *Do not overcook fish. A good indication of when fish is done is when the fish turns from translucent to opaque.*

POPULAR SALTWATER FISH

Haddock is very similar to the codfish, but the meat is slightly darker and fibrous. However, haddock is still considered a white-meat fish and has a firm flesh with an excellent flavor. When haddock is smoked, it is known as finnan haddie, a product that is quite popular in the commercial kitchen. The quickest way to distinguish a haddock from a codfish is by the black lateral line and the dusky blotch on each side over the pectoral fin just below the lateral. These dusky blotches are sometimes referred to as the devil's mark. The average weight of a haddock is 4 lb. Haddock is available on the market all year, but reaches its peak in the spring. The largest catch of haddock comes from the waters off the New England coast. The flesh of the haddock is lean and dry. It is best to serve this fish with a sauce. Steaming, baking, and broiling are the best cooking methods to use.

Codfish is a very well-known and important fish in the United States. The codfish is very abundant throughout the year off the coast of Newfoundland and Massachusetts. The skin is brownish gray on the back and upper sides and dirty white on the lower sides and belly. Most of the skin is spotted with brown specks. The young codfish are known as scrod. Scrod average from 1 lb to 1½ lb and are very much in demand in the commercial kitchen. The mature codfish weighs 10 lb to 25 lb. Codfish for use in the commercial kitchen is usually purchased in fillet form with all skin and bones removed. The meat of the codfish is lean, white, and dry, with a flaky grain. When cooking, it is best to poach, steam, or bake codfish. A sauce should always be served with codfish because of its dryness. The scrod is more moist and is delicious when broiled or sautéed.

Alaska Seafood Marketing Institute

Codfish for use in the commercial kitchen is usually purchased in fillet form with all skin and bones removed.

FLOUNDER

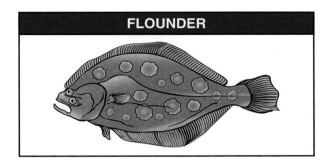

Flounder is of the flat fish family, which also includes turbot, sole, and halibut. All these flat, lean fish are very popular in the commercial kitchen. They are widely distributed geographically and are comprised of many hundreds of species. The flat fish are very easy to distinguish from other fish because their bodies are flat and, except in the very young fish, the color and both eyes are only on one side of the body. The very young flat fish have eyes and color on both sides, but as they mature the color leaves one side and it becomes white. The eye on one side moves to a position just above the eye on the other side. The mouth in some species becomes distorted.

The largest catch of flounder comes from the waters off the New England coast. There are five different species sent to market from this region. Many names may be applied to these five species, but the most commonly used are winter flounder, sand dab, yellowtail, lemon sole, and gray sole.

The *winter flounder*, or common flounder, is noted for its excellent flavor and thick, meaty fillets. This flounder averages about 1 lb. *Sand dab*, also known as windowpane flounder, is a left-handed flounder because its color and eyes are on the left side, whereas with most flounders the eyes and color are on the right side. Sand dab is an excellent pan fish. It has bone-free fillets and a sweet-tasting, oil-free meat. The sand dab weighs from 1 lb to 2 lb. *Yellowtail* is so named because its blind side is a lemon yellow rather than the usual white. It has a fairly thin body and is considered a less desirable food fish. However, it has a good flavor and is marketed in large quantities. *Lemon sole* is also known as George's Bank flounder because this is where most of them are caught. Lemon sole is very similar to winter flounder but is large in size, averaging over 2 lb. It has excellent eating qualities and is highly regarded by gourmets. *Gray sole* is the largest of the winter flounders. It has an excellent flavor. Gray sole averages about 4 lb. Any of these five species produce their best eating qualities when sautéed, broiled, or fried.

Halibut is of the flat fish family and resembles a giant flounder. It has the usual blind and colored side. Both eyes are on the colored side. Halibut is found in both the Atlantic and Pacific Oceans. The Pacific halibut is more slender than the Atlantic halibut, but otherwise they are alike. Both provide a fleshy, lean, fine-flavored white meat. Halibut is on the market all year, with the Pacific Ocean producing the largest catch.

Alaska Seafood Marketing Institute

Halibut can be fried, broiled, steamed, poached, sautéed, grilled, or baked with equally excellent results.

Halibut is a very large fish and only swordfish, tuna, and some sharks reach a larger size. The female halibut is usually larger than the male. The female, when full-grown, weighs about 100 lb to 150 lb. The male averages 50 lb to 150 lb. The flesh of halibut weighing under 100 lb has a richer flavor than halibut weighing over 100 lb. Halibut purchased for use in the commercial kitchen average from 25 lb to 70 lb.

The very young halibut, weighing 4 lb to 12 lb, is known as chicken halibut and is preferred by many because of its very fine eating qualities. Halibut is one of the most popular fish used in the commercial kitchen because of its versatility. It can be fried, broiled, steamed, poached, sautéed, grilled, or baked with equally excellent results. Since halibut has a fairly dry texture, it should always be served with some type of sauce to add moisture.

English and *Dover sole* are very similar fish. They are right-handed flounders averaging about 10″ long with brown to pale brown skin. They are most plentiful in the Pacific Ocean and have become popular in the commercial kitchen. The flesh of the English sole is considered superior to the Dover sole, but both are lean and tasty when sautéed and served a la meunière or amandine.

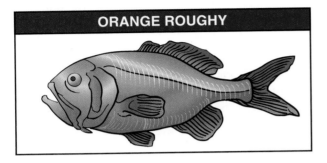

ORANGE ROUGHY

Orange roughy is abundant in the south Pacific Ocean in the area of New Zealand. Most of the orange roughy appearing on menus in the United States is imported from New Zealand. Orange roughy is so named because of its orange, light gray colored skin, which has a rough appearance. The edible flesh is white and possesses a lean, sweet, delicate flavor. This fish is available year-round and can be prepared in various ways. It can be baked, stuffed and baked, broiled, sautéed, or fried with excellent results.

Sea bass are also known as black will, blackfish, rock bass, and black bass. They have mottled black skin interspersed with white markings. They are caught in both the Atlantic and Pacific Oceans; however, the Atlantic sea bass is the leanest and is therefore considered the best. The sea bass is lean, has a white, flavorful meat, and averages from ¼ lb to 4 lb. Sea bass weighing ¾ lb to 1 lb are considered the best and are usually cleaned and sautéed, broiled, or baked whole in commercial kitchens. Sea bass are available on the market year-round but are most plentiful during the winter months.

Striped bass are native to the Atlantic coast, but in the years 1879-1881, they were brought from New Jersey and placed in the San Francisco Bay. Striped bass are also known as rock bass, white bass, striper, or rockfish. They are a lean fish and have skin that is slightly brownish-green on the upper sides, shades to a silver green on the sides, and light silver on the belly. The sides are marked with seven to eight well-defined dark stripes running from the head to the tail. It is because of these stripes that the most common of its many names is striped bass.

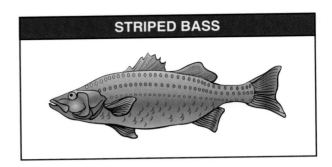

STRIPED BASS

The size of striped bass varies greatly, depending on the locality of the catch. An estimated average is 1 lb to 10 lb. Striped bass are plentiful on the market during May and June, but are available all year. Striped bass are a popular fish in the commercial kitchen and are best when sautéed, baked, or broiled.

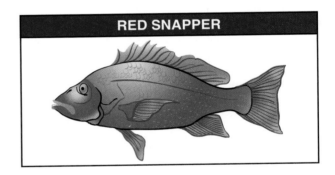

RED SNAPPER

Red snapper is from southern waters. It is a lean fish with a juicy, fine-flavored white meat that is held in very high esteem by gourmets. Red snapper is caught from the coast of New York to Brazil, but the largest catch is made in the Gulf of Mexico. Red snapper is so named because the fish have a deep red-colored skin and red fins. The red color shades slightly on the belly and around the throat. It is one of the most attractively-colored fish taken from the coastal waters. The red snapper averages about 7 lb, but 15 lb to 25 lb is not uncommon. The hard, tough bones of the red snapper make it a difficult fish to fillet. This tasty saltwater fish makes a popular menu entree when baked, broiled, steamed, or poached.

Bluefish has a blue-green color above, becomes lighter along the sides, and is silver on the belly. The flesh, which is lean, sweet, and delicate in flavor, has a slight blue tone. Bluefish are abundant in the waters of the Atlantic Ocean from Florida to Maine. They are available on the market all year, but are most abundant in the New York markets from May to October. The average weight of a bluefish is 2 lb.

Bluefish that weigh between 4 lb to 5 lb are best to use. Generally speaking, the heavier the fish, the finer the quality. Bluefish is an excellent eating fish and is best broiled, sautéed, or baked.

Pompano is regarded as the most choice of all the saltwater fish. It has a firm, flaky white flesh and is usually expensive. Pompano is abundant in the south Atlantic and the Gulf of Mexico. The largest percentage of the catch is brought to the Florida markets. The average size is 2 lb, but the best for use in the commercial kitchen are fish averaging ¾ lb to 1½ lb. Pompano is a fatty fish and has skin that is covered with very smooth scales and color that is blue above but silvery with a light golden tone below. Pompano produces the best eating qualities when sautéed, broiled, or baked.

Spanish mackerel, next to the pompano, is considered the best-flavored saltwater fish. The flesh is firm, slightly dark in color, and rich in flavor. The skin is a dark bluish-brown on black with golden spots both above and below the lateral line. The belly is a silver shade. The mackerel is a long streamlined fish, tapering toward the rear. Spanish mackerel is a fatty fish that ranges from Massachusetts to Brazil and is available on the market year-round. It measures from 14″ to 18″ long and weighs from 1 lb to 2½ lb. Spanish mackerel is a very popular menu item and, for best results, it should be cooked by broiling or baking.

Shark comes from all the oceans of the world. There are approximately 250 species of shark, but only a few supply quality meat. The mako shark supplies most of the quality shark meat for the U.S. market. It is caught throughout the oceans of the world, but most of the domestic supply is caught along the Atlantic coast. The blacktip shark is the premier species found in the Gulf of Mexico and Florida waters. It has a snowy white meat and a quality taste, but it is somewhat drier than mako. Thresher sharks, landed on the West Coast, have a coarser-textured meat.

Shad, like the salmon, is a fat fish and comes in from the sea to freshwater streams to spawn. The shad resembles the whitefish to a degree because it has slightly similar markings and skin color; however, these are the only resemblances. The shad is valuable not only for its edible flesh, but also because of the valuable roe that are obtained from the female. The roe are considered a delicacy and are highly prized by gourmets. Shad can be found in both the Atlantic and Pacific Oceans and is most plentiful during March, April, and May. Shad averages about 4 lb but weighs as much as 12 lb in some cases. The valuable roe weigh ¼ lb to 1 lb per pair. Shad is best when broiled or sautéed.

Redfish are members of the drum family and are sometimes known as red drum or channel bass. They inhabit the Gulf of Mexico and southern Atlantic waters. This fish is easily recognized by its characteristic coppery skin and one or more black spots at the base of the tail. Redfish can grow to as much as 80 lb, but the smaller fish are preferred by the food service operator. Quality redfish weigh under 12 lb. The fish produces a thick, firm textured, meaty fillet that produces best results when baked, broiled, or blackened. Commercial fisheries in Florida and Louisiana process several million pounds of redfish a year. However, even this large catch does not meet the demand, so some redfish are imported from Mexico. The popular preparation, blackened redfish, has increased the demand for the redfish, resulting in increased costs and decreased availability.

Butterfish, also commonly known as dollarfish, is abundant in the middle and north Atlantic during the summer months and migrates to other waters during the fall and winter months. The butterfish averages from ¼ lb to 1 lb and measures approximately 4″ long. The largest on record is 9″ long. They are a fat fish and contain a high percentage of oil. Butterfish is excellent when pan-fried and is considered the best pan fish from the Atlantic waters. They have round, firm bodies, a deep-forked tail, a single, long, thin dorsal (back) fin, and a small head. Most of the catch is marketed in whole, drawn, or dressed form; however, some are smoked. Butterfish may also be broiled with excellent results.

Trade Tip

Whole fresh fish are sold in the same form as they are taken from the water. When purchased, they should have bright, clear, and shiny eyes. Scales should be shiny and cling tightly to the skin. The gills should be bright pink or red.

Croaker, also known as crocus and hardhead, acquired its most common name from the unusual croaking sound made by both the male and female. The croaker is a lean fish and averages from ½ lb to 2½ lb. Croakers are most plentiful from March to October but are available all year. Chesapeake Bay produces the largest catch; however, they are also taken from other areas of the middle and south Atlantic Ocean. The croaker has a brassy color above the lateral line and a lighter color below with irregular, pale, vertical bars running the length of the fish. The upper portion of the fish is spotted with irregular dark brown spots. They have two dorsal fins. The first one is high and the second is low. The tail is concave and the head is fairly large for the size of the fish. They are marketed whole, dressed, or in fillets. Although they are inexpensive, they have a fairly good eating quality when fried or broiled.

Hake is of the codfish family, but is inferior in food value. The flesh is lean but darker and more fibrous than the codfish. The two major species of hake are squirrel hake and white hake. They are not separated when marketed. The squirrel hake averages 1 lb to 4 lb and measures approximately 1′ to 2′ long. The white hake usually runs a little larger, weighing 2 lb to 5 lb and measuring 1½′ to 3′ long. They have a slender body with two sets of dorsal fins. The first is short and the second is long. They also have a forked tail and a pointed snout. The largest catch comes from the north and middle Atlantic during September and October when the large schools in which they usually travel come in to shore.

On the market they can be purchased whole, drawn, dressed, or in fresh or smoked fillets. Sometimes they are substituted for haddock since the flesh and eating qualities are somewhat similar. Although hake is not a popular fish in the commercial kitchen, it can be made desirable if baked or poached and served with an appropriate sauce, such as creole or dugleré.

Monkfish is also known as goosefish, bullmouth, devilfish, frogfish, and lotte (its French name). The tail section is the edible part. In Europe, the head is sometimes retained because it produces an excellent soup stock. At one time, this fish was unheard of on the domestic market. It was popular in Europe, so all the domestic catch was exported to Europe. Exports have dropped dramatically in response to the growing popularity of monkfish in the United States. The fish is caught along the Atlantic coast from Newfoundland to South Carolina; however, the largest catch is off the coast of the New England states.

The flesh of the monkfish is white with a sweet, firm texture, and when cooked, it is similar to lobster meat. In fact, it is referred to as the "poor person's lobster" because of the firm texture of the flesh and also because, like the lobster, it is a bottom dweller. Monkfish is a fairly large fish that can grow to 50 lb. The primary market forms are whole tails and fillets. It can be cooked by a variety of methods, including broiling, baking, sautéing, and poaching. For best results, serve with an appropriate sauce.

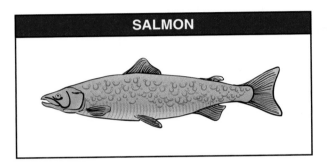

SALMON

Salmon is available commercially in six varieties. These include the Pacific (Alaska) salmon, which are king, silver, red, chum, pink; and the Atlantic salmon. Farm-raised salmon is also commercially available. After the salmon lays its eggs, it dies. This is characteristic of all Pacific salmon, but not true of the Atlantic salmon.

Salmon steaks are often broiled or grilled.

Alaska produces over 90% of the total U.S. salmon production. Salmon contains fish oils, referred to as omega 3, which are highly polyunsaturated. Omega 3 oils offer protective effects against heart disease, inflammatory processes, and certain cancers.

Salmon is graded according to variety and color of the flesh. A large percentage of salmon is marketed fresh and frozen. Frozen salmon is held at 32°F until it is flash frozen. It is protected from dehydration by glazing (covering with water to form a protective sheet of ice) which results in a product that tastes as fresh as the day it left the water. Approximately 30% to 35% of Alaska salmon is canned. There is no waste in canned salmon and nothing is added except a pinch of salt for flavor.

King (Chinook) is rated the finest of the Alaska salmon. It has a deep red flesh and a superb flavor. The high oil content means chinook steaks or fillets can be baked, grilled, sauteed, or poached, and served with little additional flavoring. The chinook is the largest of the Pacific salmon and averages about 20 lb.

Sockeye (Red) is rated second best of the Pacific salmon. Like the chinook, it has a deep red flesh and relatively high fat content. It is excellent for broiling, grilling, baking, steaming, poaching, or sautéing. The sockeye averages 3 lb to 8 lb; however, 12 lb is not uncommon.

Coho (Silver) is considered the third best of the Alaska salmon. It has a deep pink flesh and a good-rated flavor. Silvers have a lower oil content than other salmon, so be careful not to overcook. It is highly desirable for table use and smoking. The coho averages 10 lb, but 15 lb to 20 lb in not uncommon.

Pink (Humpback) has the lightest flesh of all salmon. It is so named because of the noticeable hump that develops in front of the dorsal fin on the male salmon during the spawning season. The humpback is the smallest of the Pacific salmon and is rated the fourth best. It has a soft, pink flesh and a lower fat content. Be sure not to overcook this salmon. It averages 3 lb.

Chum (Keta) is classified as the poorest of the Pacific coast salmon. It has a pale yellow, soft flesh containing a moderate amount of oil. It is used in dishes where price/value is an important criterion. The chum salmon averages 8 lb.

Atlantic (Kennebec) differs from the Pacific coast varieties in that it rarely dies after spawning. It returns to the sea and lives and spawns a second time. The Kennebec comes mainly from the north Atlantic and the rivers and streams along the coast of Maine, Nova Scotia, Quebec, Labrador, and New Brunswick. The flesh of the Kennebec is medium pink, and the eating qualities are considered extremely good. It averages from 10 lb to 15 lb; however, over 20 lb is not uncommon.

Turbot is a large, flat fish that for years has been a very popular food fish in Europe, and in recent years has gained popularity in the United States. Turbot is found in shallow waters from the Mediterranean Sea and northward along the European Atlantic coast. The English Channel provides a large percentage of the catch.

If the tail is eliminated, the turbot has an almost circular shape. Being a member of the flat fish family, it has a flattened body with one side dark and the other white. The dark side is brownish with very small lumps on the skin. Both eyes are on the dark side. The body is approximately 2′ long and weighs an average of 10 lb. However, some have been known to attain a weight of 40 lb. It has a habit of lying on the ocean floor with its blind side partially buried in the sand.

The flesh of the turbot is white, lean, and mild in flavor. It can be cooked using a variety of cooking methods; however, sautéing, frying, and steaming give best results. If baked, serve with a light sauce.

POPULAR SHELLFISH

Shrimp is the most popular of the shellfish family. They have a tender white meat with a distinctive flavor. Shrimp are also known as *prawns* and, although there may be a very slight difference between the two, they are both marketed as shrimp in the United States. Only when the domestic product is shipped to England is the word *prawns* used.

There are four types of shrimp taken from domestic coastal waters: the *common* or *white shrimp*, which has a greenish gray color when caught; the *brown* or *Brazilian shrimp*, which is brownish red in color; the *pink* or *coral shrimp*, which has a medium or deep pink color; and the *Alaska* and *California shrimp*, which vary in color and are small in size. Although these four types of shrimp vary in color when caught, they differ very little in appearance when cooked. Only the tail section of the shrimp is edible. The whole shrimp may be sold fresh near the source of supply, but the majority of the catch is processed by removing the head and the thorax (body), frozen in 5 lb blocks or consumer-sized packages, and shipped throughout the country.

Shrimp, like most seafood, perishes rapidly, so it must be cooked, frozen, or packed in ice and refrigerated immediately after processing. Shrimp is sold according to size or grade. The size or grade of the shrimp is important to the food service operator from the standpoint of time and cost. The jumbo and large shrimp cost the most, but take less time to peel and clean. The smaller shrimp cost less, but take longer to peel and clean because there are more of them.

All shrimp have the same distinctive flavor and food value. Boiling is the most common method of cooking shrimp, although frying is another extremely popular method. Shrimp is a versatile item in the commercial kitchen and is featured on the menu in the form of an appetizer, entree, or salad.

Lobster, known as the king of the shellfish, is the largest of the shellfish group. It has a sweet-tasting, white-meat flesh that is highly prized as food. The

PREPARING BOILED SHRIMP

1 BOIL IN WATER

2 PEEL SHELL

3 DEVEIN WITH KNIFE

two types of lobsters available on the market are the cold water lobster, coming mainly from the North Atlantic, and the spiny lobster, which is nearly worldwide in its distribution, coming from the warmer waters of the Atlantic, Pacific, and Indian Oceans.

The *cold water lobster* has a dark bluish-green shell, two large heavy claws, medium-sized antennae, and four slender legs on each side of the body. The whole lobster is edible except for a small section of membranes located around the eye and the shell. Cold water lobsters are sold alive and must be kept alive up to the time of cooking. They are available on the market all year; however, they are most plentiful during the summer months when they come closer to shore. Lobsters vary in size and are graded according to size.

Spiny lobsters, or *rock lobsters*, have many prominent spines on their bodies and legs, very long slender antennae, no claws, and five very slender legs on each side of the body. Only the tail section is edible, and the flesh is coarser in texture and not as delicate in flavor as the cold water lobster.

The degree of smoothness of the spiny's shell and the way the shell is spotted or marked depend on the part of the world from which it comes. For instance, spiny lobsters from Florida and Cuba have slightly smooth shells with large yellowish spots on the brownish-green colored tail section. South Africa and New Zealand spiny lobsters have rough shells with no spots on the brownish-maroon colored tail sections. The tail section of the spiny lobster, which weighs from 4 oz to 1 lb, is frozen and shipped to market.

The shell of the lobster turns red when cooked. Lobster, regardless of the type, is an extremely popular item in the commercial kitchen and is placed on the menu in a variety of preparations. When preparing lobsters, the best cooking methods are broiling, boiling, or steaming.

Crayfish resembles and is related to the lobster. It is smaller than the lobster and lives in freshwater rivers and streams in temperate climates. Crayfish has a flavor similar to shrimp and is used most often in southern cooking and French cuisine. The European crayfish is highly prized as a food, although it contains little meat, being only 3″ to 4″ long.

Oysters have been a delicacy for a long time. There are three important species of oysters. The *Eastern oyster*, which comprises about 89% of all the oyster production in the United States, is found along the Atlantic and Gulf Coast from Massachusetts to Texas. The *Olympia oyster*, which is quite small, is found on the Pacific coast from Washington to Mexico. The *Japanese oyster*, which is quite large, is found on the Pacific coast. This oyster was introduced from Japan in the year 1902 and today is cultivated in large quantities.

Today, most oysters are cultivated in beds that require much care and attention if they are to continue to produce. Along the Atlantic coast, Chesapeake Bay is one of the biggest oyster-producing areas. Most states along the eastern seaboard produce oysters, except the states of Maine and New Hampshire, where the oyster beds were destroyed years ago.

Oysters are at their best from September to April, but they are available year-round. Oysters of good quality are plump, well-shaped, and surrounded by a clear, jelly-like semi-liquid. Oysters must have a tightly closed shell to be of good quality. If the shell is open, the oyster is dead and not edible. Oysters have a special appeal to cooks and chefs in the commercial kitchen not only because of their delicious flavor but also because of the ease with which they can be prepared and served. Oysters may be eaten raw or prepared by poaching or frying. Regardless of the cooking method used, the secret to proper oyster cooking is to apply just enough heat to heat them through, leaving them plump and tender. Avoid overcooking.

Scallops are the large adductor muscle that opens and closes the scallop shell. The muscle is a solid section of cream-colored flesh that is very lean, juicy, and possesses a sweet delicate flavor. The two types of scallops on the market are bay scallops and sea scallops. The *bay scallops* are taken from shallow waters and are fairly small. They are considered to have the best flavor and usually are higher in price than the sea scallops. The *sea scallops* come from deep waters, are larger in size, and have a coarser texture. Scallops are available on the market all year, but are best from April to October. The Atlantic coast is the largest producer of scallops with small quantities coming from the Gulf of Mexico. The scallop shell is shaped like a fan and, when polished, displays many interesting colors that are pleasing to the eye. Many food service operators use scallop shells as a serving dish when featuring certain seafood appetizers and entrees.

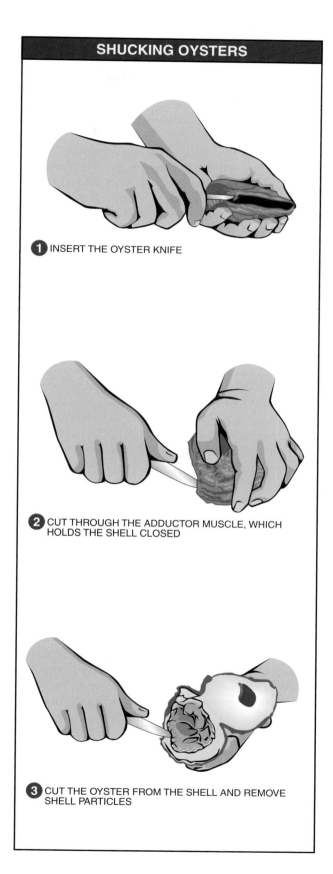

SHUCKING OYSTERS

1 INSERT THE OYSTER KNIFE

2 CUT THROUGH THE ADDUCTOR MUSCLE, WHICH HOLDS THE SHELL CLOSED

3 CUT THE OYSTER FROM THE SHELL AND REMOVE SHELL PARTICLES

Color is the best way to judge the quality of scallops. The best quality has a cream color. If they are white, it indicates they have been packed in ice water and the flavor has been impaired. If they possess a brownish color, it indicates they are slightly old. Scallops are a popular seafood in the commercial kitchen and are best when fried or sautéed. However, they may be poached and broiled with fairly good results.

Clams are available in several species that are used for food. The varieties on the market depend on the source of supply because the varieties from the East Coast differ from those from the West Coast.

The Atlantic coast produces three important species: the soft, hard, and surf clam. The *soft clam,* also known as the long-neck clam, is taken from Cape Cod north to the Arctic Ocean. This clam is popular in the New England area. The *hard clam* is found in abundance south of Cape Cod. Hard clams are sometimes known as *quahaug* by the people of New England. This is an old Indian name for the hardshell clams. The small-sized hard clams are known as littlenecks and cherrystones and are usually served raw on the half shell in the same manner as oysters. The larger hard clams are known as chowders and are used mainly in soups and chowders. The hard clam has a stronger flavor than the soft or surf clam. *Surf clams* have a sweet flavor and are not as important or desirable as the other two mainly because they are usually gritty. They are sometimes used in chowders and soup, but are used mainly for the production of clam juice and broth.

The Pacific coast produces four important species of clams: butter, razor, littleneck, and pismo clams. The *butter clam* is a hardshell clam possessing a very desirable flavor. The *razor clam* is so named because of its sharp razor-edged shell. The *Pacific littleneck* is a different species from the Atlantic littleneck and to a degree lacks the flavor of the Atlantic variety. The *pismo clam* comes from a coastal area in California (Pismo Beach) made famous by these delicious clams.

Clams are prepared in many different ways in food service establishments, the most popular of which is clam chowder. The chowder has been and will continue to be the most famous of the many clam preparations. Clams can also be fried and steamed with excellent results.

Trade Tip *Always wash hands thoroughly with hot, soapy water before and after handling raw seafood.*

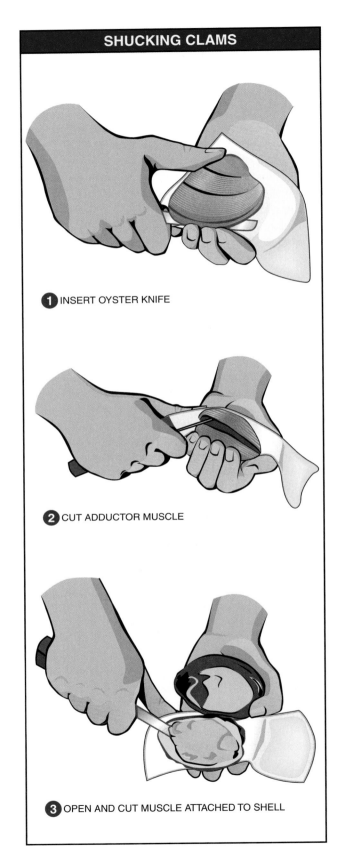

SHUCKING CLAMS

1 INSERT OYSTER KNIFE

2 CUT ADDUCTOR MUSCLE

3 OPEN AND CUT MUSCLE ATTACHED TO SHELL

Trade Tip

Oysters fed with a very thin flour batter are easily opened. Spread a very thin liquid flour mixture over the amount of oysters to be opened. The oysters will eat so much of it that on the next day their shells will have opened about $\frac{1}{4}''$. Taste will not be affected. Oatmeal sprinkled over oysters has nearly the same effect.

Crabs have become a very popular shellfish because of their tender, juicy, sweet-tasting meat that can be converted into many menu items. Four principal types of crabs taken from the waters of the Atlantic and Pacific Oceans are available on the market. These include the blue crab, Dungeness crab, king crab, and rock crab.

The *blue crab* comes from the Atlantic coast and comprises about three-fourths of all the crabs marketed in the United States. It measures about 5″ across its shell and weighs approximately 5 oz. When the blue crab molts (sheds its hard shell), it is marketed as a softshell crab. The molting season is in the spring of the year. This is when the softshell crab is available on the market. Softshell crabs are handled and packed with special care so they arrive at their destination alive. They must be alive up to the time of cooking, unless they are cleaned and quick-frozen. The hardshell blue crab is marketed alive within comparatively short distances but can not be obtained live at points inland because they do not ship well. The majority of the hardshell blue crabs are marketed as frozen or canned cooked meat.

Dungeness crabs are found on the Pacific coast from Alaska to Mexico. They are larger than blue crabs, weighing approximately 1¾ lb to 4 lb. The meat has a pinkish tinge and a very desirable sweet taste. The bulk of the catch is marketed as frozen and canned cooked meat.

King crabs come from the north Pacific off the coast of Alaska. They are the largest of the crab family, weighing from 6 lb to 20 lb and measuring as much as 6″ from the tip of one leg to the tip of the opposite leg. The meat has a pinkish tinge similar to that of the Dungeness crab. It is marketed as frozen cooked meat, frozen cooked in shell, and canned cooked meat.

Rock crabs are taken from the coastal waters of California and New England. They are small in size, weighing approximately 4 oz to 8 oz and measuring about 3″ across the shell. The meat is brownish in color and considered to be inferior to the white meat of the blue crab. Rock crabs are marketed live, but the bulk of the catch is sold as fresh cooked meat and canned meat.

FISH AND SHELLFISH RECIPES

In serving fish and other seafood, the importance of an attractive and appetizing garnish should not be overlooked. A dash of color or a touch of garnish can turn a plain dish into a highly satisfying one. Some of the most common garnishes, with their suggested methods of preparation, are listed in Table VII. The sauces served with seafood preparations also contribute to the success of the recipe. The method of preparation determines the kind of sauce to be used. Table VIII lists suggested sauce selections for fish and shellfish.

TABLE VII: GARNISHES FOR FISH	
Beets	Cooked whole or slices
Carrots	Tops, sticks, curls, or shredded
Celery	Tops, hearts, sticks, or curls
Chives	Chopped
Cucumbers	Slices or sticks
Dill	Sprigs or chopped
Green or red peppers	Sticks or rings
Hard-boiled eggs	Slices, wedges, deviled, or grated yolks
Lemons or limes	Slices, twists, or wedges
Lettuce	Leaves or shredded
Mint	Sprigs or chopped
Nut meats	Toasted whole, halved, slivered, or chopped
Olives	Whole, sliced, or chopped
Oranges	Slices, twists, or wedges
Paprika	Sprinkled sparingly
Parsley	Sprigs or chopped
Pickles	Whole, sliced, or chopped
Radishes	Whole, sliced, or roses
Watercress	Sprigs or chopped

TABLE VIII: RECOMMENDED SAUCES FOR FISH AND SEAFOOD

Cooking Method	Recommended Sauces	Other Sauces
Sautéed	Meunière, bercy, chateau, lemon butter, choron*, cherburg*	Colbert, anchovy butter, oriental sweet-sour, dill, homard
Broiled	Lemon butter, bercy, choron*, maximillian*, cherburg*	Anchovy butter, colbert, chateau, dill*, divine
Baked	Maximillian, cherburg, choron, dugleré, victoria, newburg	Divine, figaro, creole, homard, caper, cardinal, mustard
Poached, Boiled, or Steamed	Mornay, maximillian, cherburg, choron, creole, velouté (vin blanc), dugleré, victoria, curry, newburg	Mousseline, divine, figaro, bonne femme, homard, caper, cardinal, mustard, egg, dill
Fried	Cocktail, tartar	Dill

* Denotes sauces that should be served on the side of the prepared item. The sauce may be served in a gooseneck container, soufflé cup, or other appropriate container.

FRIED, SAUTÉED, AND GRILLED FISH RECIPES

Deep-Fried Fish Fillets

Most lean fish, such as halibut, haddock, and flounder, is best for frying. Fish must be breaded before frying to acquire the crisp golden brown coating. Deep-fried fish is most suitable when speed of service and preparation is stressed.

25 servings

Equipment

- Deep fat fryer
- Quart measure
- French knife
- Wire whip
- Bake pans (3)
- 2 qt stainless steel container
- Fry baskets
- Baker's scale

Ingredients

25 5 oz fish fillets
6 eggs
1 qt milk
1 lb flour
1 lb 8 oz bread crumbs
salt and pepper to taste

Procedure

1. Place the flour in a bake pan and season with salt and pepper.
2. Add the fish and coat thoroughly.
3. Pour the egg wash into the second bake pan. Remove each portion of fish from the flour and place in the egg wash.
4. Place the bread crumbs in the third bake pan. Dip each piece of fish into the bread crumbs and press on the crumbs thoroughly. Shake off excess.
5. Place in fry basket and fry in deep fat until golden brown and done. Let drain.
6. Serve garnished with a wedge or slice of lemon and tartar or fish sauce.

Preparation

1. Prepare an egg wash. Break the eggs into a stainless steel container. Beat slightly with a wire whip. Pour in the milk and blend with the eggs.
2. Preheat the deep fat fryer to 350°F.

Precautions

- Use caution when handling the knife.
- Do not fry too many orders at one time. The temperature of the fat will reduce too much and the pieces will not cook evenly.
- Do not overcook. Most fried fish is done when it starts to float on the surface.

Trade Tip

To pan fry fish, coat with milk, coat the fish with seasoned flour, bread crumbs, cracker crumbs, or cornmeal. Fry in a skillet with preheated butter and oil for 10 minutes to 12 minutes per inch, turning once during cooking. Remove the fish from the skillet and allow to drain before serving.

Salmon Croquettes

The drained canned salmon is bound together in a thick cream sauce, cooled, and shaped into croquettes. They are breaded and deep fried to a golden brown and served with an appropriate sauce. Salmon croquettes make an excellent luncheon entree.

25 servings

Equipment

- French knife
- Baker's scale
- 2 gal. sauce pot
- Deep fat fryer
- Fry baskets
- Bake pans (3)
- Kitchen spoon
- Wire whip
- 2 qt stainless steel container
- 2 qt saucepan

Ingredients

Croquettes

4 lb canned salmon
10 oz onions
10 oz butter or shortening
10 oz bread flour
1 qt milk
½ c parsley
salt and pepper to taste
6 egg yolks

Breading

1 lb flour
2 lb bread crumbs
1½ qt egg wash (9 eggs to 1½ qt milk)

Preparation

1. Open the canned salmon. Drain and remove the bones.
2. Mince the onions with a French knife.
3. Heat the milk in a pan.
4. Prepare an egg wash. Break the nine eggs into a stainless steel container, beat with a wire whip, pour in the 1½ qt of milk, and blend with the eggs.
5. Chop the parsley with a French knife.
6. Preheat the deep fat fryer to 350°F.

Procedure

1. Place the butter or shortening in a sauce pot and heat.
2. Add the onions and sauté slightly. Do not brown.
3. Add the flour, making a roux, and cook slowly for 5 min.
4. Add the hot milk, stirring constantly with a kitchen spoon until thickened and smooth. Simmer for about 5 min.
5. Add the salmon and chopped parsley. Blend thoroughly with a kitchen spoon.
6. Season with salt and pepper. Remove from the range. Let cool slightly.
7. Beat the egg yolks in a stainless steel container with a wire whip. Pour very slowly into the croquette mixture while stirring rapidly with a kitchen spoon.
8. Turn the mixture into a shallow pan. Cover and refrigerate overnight.
9. Remove the mixture from the refrigerator and shape into uniform croquettes.
10. Bread by passing each croquette through flour, egg wash, and bread crumbs.
11. Fry in deep fat at 350°F until golden brown.
12. Serve one or two croquettes, depending on desired size, to each order with an appropriate sauce.

Precautions

- Use caution when handling the knife.
- Stir the mixture occasionally while it is cooking to avoid sticking or scorching.
- Pour the beaten egg yolks very slowly into the mixture to avoid scrambling.
- When frying, do not overfill the fry baskets. The croquettes will not cook evenly and the temperature of the grease will reduce rapidly.
- Do not overbrown the croquettes.

Fried Seafood Platter

This is a combination dish consisting of fish and shellfish that are breaded and deep fried to a crisp golden brown. Fried seafood platters are usually served with tartar or cocktail sauce on the luncheon, dinner, or a la carte menu.

10 servings

Equipment

- French knife
- Bake pans (3)
- Wire whip
- Quart measure
- Baker's scale
- Fry baskets
- 2 qt stainless steel container

Ingredients

30 pieces raw shrimp, peeled and deveined
30 scallops
30 oysters
10 3 oz sticks fillet of sole (halibut)
9 eggs
1½ qt milk
1 lb bread flour
2 lb bread crumbs (variable)
salt and pepper to taste

Preparation

1. Cut the fish with a French knife and prepare both the fish and shellfish for breading.
2. Prepare an egg wash. Break eggs into a stainless steel container, beat slightly with a wire whip, pour in the milk, and blend with the eggs.
3. Set the deep fryer at 350°F.

Procedure

1. Place the flour in a bake pan and season with salt and pepper.
2. Add the assorted pieces of seafood and coat thoroughly.

3. Pour the egg wash into the second bake pan. Remove the seafood from the flour and place in the egg wash.
4. Place the bread crumbs in the third bake pan. Remove the seafood from the egg wash. Place in the bread crumbs, press the crumbs on firmly, and shake off the excess.
5. Place in fry baskets and fry until golden brown. Let drain.
6. For each portion, serve three shrimp, three scallops, three oysters, and one fish stick. Serve with tartar or cocktail sauce and a slice or wedge of lemon.

Precautions

- Press the breading on firmly so the breading does not fall off while frying.
- Do not overbrown or the servings will lack in appearance.
- Do not fry too many pieces at one time. The temperature of the fat will reduce too quickly and frying will not be uniform.

Sautéed Fish

Small, whole fish or fillets are best for sautéing. The fish is seasoned and coated with flour before it is placed in the hot grease. It is cooked until golden brown on both sides. This is an excellent method for preparing fish when speed of service is required.

25 servings

Equipment

- French knife
- Quart measure
- Spoon measure
- Skillet
- Kitchen fork
- Bake pan

Ingredients

25 small, whole, dressed fish or 5 oz fillets or steaks
1 qt flour
2 tsp paprika
salt and pepper to taste

Preparation

1. If cutting 5 oz fillets or steaks, use a French knife.

Procedure

1. Place the flour, paprika, salt, and pepper in a bake pan and mix by hand.
2. Pass each portion of fish through the flour mixture, and press firmly so flour will adhere to the fish.
3. Place enough shortening in the skillet to cover the bottom of the pan, about ¼" deep, and heat.
4. Add fish and sauté until golden brown on each side. Turn with a kitchen fork. Remove and let drain.
5. Serve with meunière sauce, butter, or a sauce that will complement the type of fish being sautéed. Sautéed fish is best when served with a thin, light sauce. Garnish each serving with a wedge or slice of lemon.

Precautions

- The fat should be hot before placing the fish in the pan. This will prevent sticking.
- If skin is left on the fish, place skin side up. Exercise caution when turning the fish in the pan to avoid burning self.
- Sauté at a moderate temperature. The temperature is regulated by moving the pan toward or away from the heat.

Blackened Redfish

This is one of the most popular Cajun preparations. The fish is coated with melted butter or margarine, pressed into the hot Cajun spice mix to cover the surface of the fish, and blackened (burnt) in a very hot skillet.

25 servings

Equipment

- French knife
- Kitchen fork or offset spatula
- Iron skillet
- Small saucepan
- Bake pans (2)
- Steam table pan

Ingredients

25 6 oz portions redfish
Cajun spice blend as needed
1 lb melted butter or margarine (variable)

Preparation

1. Place the Cajun spice blend in a bake pan.
2. Melt the butter or margarine in a saucepan and place in a bake pan.

Procedure

1. Dip the fish in the melted butter or margarine.
2. Remove from the melted butter or margarine and place the flesh side of the fish in the Cajun spice blend. Press so the spices adhere to the surface of the flesh.
3. Place the skillet on the range and heat. Add the fish by placing the coated surface on the hot metal.
4. Sear the coated surface of the fish until it is quite black. Remove from the skillet using an offset spatula.
5. Place in a steam table pan containing just a small amount of butter or margarine, and finish in a preheated oven at 350°F if the fish requires more cooking.
6. Place the pan in the steam table and serve garnished with lemon.

Precautions

- Exercise caution when cutting the fish.
- Be sure the kitchen is properly ventilated when blackening the fish. Smoke will accumulate.
- *Note*: Other fish such as halibut, orange roughy, and codfish can be blackened with excellent results.

This preparation gained popularity during World War II in England. The batter-fried fish was served in a newspaper with French-fried potatoes. It was known as fish and chips. After the war, its popularity spread to the United States. Select a lean fish such as codfish, haddock, halibut, sole, or orange roughy.

Batter-Fried Fish

25 servings

Equipment

- Deep fat fryer
- French knife
- Wire whip
- Baker's or portion scale
- Large fry basket
- Large stainless steel bowl
- Bake pan

Ingredients

25 5 oz fish fillets
½ gal. batter (variable)
1 lb flour or cornmeal

Preparation

1. Prepare the batter.
2. Place the flour in a bake pan.
3. Preheat deep fat fryer to 350°F.
4. Place fry basket in hot grease.

Procedure

1. Pass each piece of fish through the flour (or cornmeal). As it is removed, pat off excess.
2. Place a few pieces of the fish in the batter at a time.
3. Remove one piece of fish at a time from the batter using the thumb and index finger. As the fish is lifted from the batter, cup the hand to catch any dripping batter.
4. Gently drop the fish into a fry basket that has already been lowered into the hot grease.

Southern Pride Catfish

5. After the fish has fried for about 30 sec to 40 sec, shake the basket so the fish will come to the surface.
6. Fry until golden brown and done. Let drain.
7. Repeat this process until all the fish has been fried.
8. Serve garnished with a wedge of lemon and tartar, dill, or cocktail sauce.

Precautions

- Use caution when cutting the fish.
- Consistency of batter may need adjusting.
- Be alert when dropping the fish into the hot grease.
- Drain the fish thoroughly before serving.

No oil or fat are added to this light yet rich twist on quesadillas. The combination of fresh citrus sections and Havarti cheese with smoked salmon slices is as eye-appealing as it is mouth-watering.

Smoked Salmon Quesdilla with Citrus

8 appetizer servings

Ingredients

8 oz smoked salmon, thinly sliced
4 6″ flour tortillas
1 c orange sections
1 c grapefruit sections
1 ½ c Havarti cheese, grated
1 bunch scallions, cross-cut
dill sprigs and orange sections for garnish

Procedure

1. Place Havarti cheese on two tortillas and arrange grapefruit and orange sections on top of the cheese.
2. Sprinkle on chopped dill and cover with remaining tortillas.
3. Grill on a medium-hot grill for 1 to 1½ min per side.

Florida Department of Citrus

4. Remove from grill and arrange smoked salmon on top of quesadilla. Cut into pie-shaped pieces.
5. Garnish with citrus sections, cross-cut scallions, and dill sprigs.

BROILED AND BAKED FISH

Broiling is an excellent method for cooking fish. Fat fish broil the best, but many lean fish are cooked by this method with good results. All fish should be broiled to order. A thin sauce such as butter sauce, lemon butter, or anchovy butter is usually served with broiled fish.

Broiled Fish

25 servings

Equipment

- Bake pan
- Cup measure
- French knife

Ingredients

- 25 5 oz fish steaks or fillets
- 1 c salad oil
- ¼ c paprika
- salt and pepper to taste

Preparation

1. Light broiler and turn heat to highest point.

Procedure

1. Place the salad oil in a bake pan.
2. Place each piece of fish in the oil.
3. Season with salt and pepper and sprinkle the paprika on the flesh side of the fish.
4. Remove the fish from the oil and place on a hot broiler rack, skin side up if skin is left on.
5. Broil 6″ from the heat for approximately 5 min on each side and until the flesh side is slightly brown.
6. Remove and serve with a butter sauce and a wedge or slice of lemon.

Southern Pride Catfish

Precautions

- Use caution when handling the knife.
- Exercise caution when turning the fish on the broiler as some fish break easily.
- Do not overcook the fish.

Broiled Fillet of Sole English Style

The halibut used is cut into 3 oz sticks. The sticks are passed through salad oil and bread crumbs and broiled under very low heat. Two sticks are served to each order with a generous amount of butter sauce.

25 servings

Equipment

- French knife
- Sheet pans (2)
- Bake pan
- Meat turner or kitchen fork
- Broiler
- Pint measure
- Baker's scale

Ingredients

- 9½ lb halibut, boned and skinned
- 1 pt salad oil (variable)
- 2 lb bread crumbs
- salt and pepper to taste

Preparation

1. Light the broiler and turn the heat low.
2. Cut the halibut into 3 oz sticks with a French knife.

Procedure

1. Coat the sheet pans with salad oil.
2. Roll each piece of fish in the salad oil, coating thoroughly. Season with salt and pepper.
3. Place the bread crumbs in a bake pan. Remove the fish from the salad oil and roll it in the bread crumbs. Press the crumbs on the fish firmly.
4. Return the fish to the oil-covered pans, rolling each piece in the oil a second time to moisten the crumbs slightly.
5. Place the pans under the broiler and broil very slowly until crumbs become light brown on top.
6. Remove the pans from the broiler and place on top of the range for a few minutes to cook the bottom of the fish.
7. Remove from the range and remove fish from the pan with a fork or meat turner.
8. Serve two sticks to each order with a generous amount of butter sauce. Garnish with a wedge or slice of lemon.

Precautions

- Exercise caution when broiling to prevent overbrowning the breading.
- Do not overcook.

Baked Seafood Casserole

The assorted fish and shellfish selected is placed in a rich cream sauce, portioned into individual casseroles, covered with cheese, and baked until golden brown. This eye-appealing entree is an ideal selection for the luncheon or dinner menu.

25 servings

Equipment

- 2 gal. sauce pot
- Baker's scale
- French knife
- Kitchen spoon
- Cup measure
- Pint measure
- Spoon measure
- Shallow casseroles (25)
- Box grater

Ingredients

- 3 lb shrimp, deveined, cut in half
- 2 lb king crabmeat
- 3 lb codfish
- 4 oz butter
- ½ c onions
- 2 small bay leaves
- 1 lb mushrooms
- 8 oz sherry
- 2½ qt cream sauce, medium consistency
- 4 tbsp chives
- 1 pt Parmesan or Cheddar cheese
- salt and white pepper to taste

Preparation

1. Cook, peel, and flake the fish and shellfish into uniform pieces with a French knife.
2. Prepare the cream sauce.
3. Mince the onions and chop the chives with a French knife.
4. Dice the mushrooms with a French knife.
5. Grate the cheese on a box grater.
6. Light the broiler and set heat low.

Procedure

1. Place the butter in a sauce pot and heat.
2. Add the onions and mushrooms and sauté until slightly tender.
3. Add the bay leaves, shrimp, crabmeat, codfish, and sherry. Cover the pot and cook on low heat for 3 min.
4. Add the cream sauce and stir with a kitchen spoon. Season with salt and white pepper. Simmer for 5 min.
5. Add the chives and blend thoroughly with a kitchen spoon. Simmer for 3 min. Remove from the range and remove the bay leaves.
6. Place a 6 oz to 8 oz ladled portion in each individual shallow casserole. Sprinkle the Parmesan or Cheddar cheese over the top.
7. Dust the top of each casserole lightly with paprika and cook slowly under the broiler until golden brown.
8. Serve at once garnished with a sprig of parsley.

Precautions

- Use caution when handling the knife.
- Stir in the cream sauce gently so the seafood does not break into small pieces.
- Do not let the onions and mushrooms brown when sautéing.
- While the mixture is simmering, stir occasionally to avoid sticking.

Stuffed Fillet of Flounder

The fillets are rolled around a crabmeat stuffing, baked, and served with a lemon butter sauce. This is a good selection for adding variety to the seafood choices on the dinner menu.

25 servings

Equipment

- Baker's scale
- French knife
- 2 qt and 4 qt saucepans
- Kitchen spoon
- Measuring spoons
- Quart measure
- Bake pan
- Toothpicks

Ingredients

- 9 lb fillet of flounder
- 4 lb frozen king crabmeat
- 12 oz celery
- 12 oz onions
- 6 oz green pepper
- 12 oz butter
- 4 oz flour
- 1 qt milk
- 4 tbsp Worcestershire sauce
- ¼ tsp Tabasco sauce
- 3 tbsp prepared mustard
- 10 oz bread crumbs
- 4 egg yolks
- salt and pepper to taste

Preparation

1. Cut the flounder into 5 oz or 6 oz fillets with a French knife.
2. Thaw the frozen crabmeat.
3. Mince the onions, celery, and green pepper with a French knife.
4. Heat the milk in a saucepan.
5. Separate the eggs and save the whites for use in another preparation.
6. Preheat the oven to 375°F.

Procedure

1. Place the butter in a saucepan and melt.
2. Add the onions, celery, and green peppers. Sauté until slightly tender.
3. Add the flour and continue to cook for 5 min longer.
4. Add the hot milk and stir constantly with a kitchen spoon until mixture thickens.

5. Add the Worcestershire sauce, Tabasco sauce, mustard, and crabmeat. Mix thoroughly with a kitchen spoon. Cook for 5 min and remove from the heat.
6. Stir in the egg yolks and the bread crumbs with a kitchen spoon. Season with salt and pepper. Let cool slightly.
7. Place a portion of the stuffing on each flounder fillet, roll up, and secure with a toothpick.
8. Place the rolled fillets in a bake pan. Bake in a preheated oven at 350°F for approximately 30 min. After the first 10 min of the baking period, add a small amount of melted butter and water to the pan to prevent sticking.

9. Remove from the oven, remove toothpicks, and serve with lemon butter sauce, cardinal sauce, or dugleré sauce.

Precautions

- Use caution when handling the knife.
- Do not let the vegetables brown when sautéing.
- Secure the rolled fillets properly so they do not unroll during the baking period.

Baked Stuffed Orange Roughy

The fillets are flattened slightly by butterflying the thick part of the fillets, covering with a sheet of plastic, and tapping gently with a mallet. The seafood stuffing is placed on one end of the flattened fillets and rolled so the stuffing is in the center of the rolled fillets. This fish is then baked and served with an appropriate sauce.

25 servings

Equipment

- Baker's scale
- French knife
- Wire whip
- Kitchen spoon
- Bake pan
- Mallet
- French knife
- Spoon measure
- Utility knife
- 1 gal. sauce pot
- 2 qt sauce pot
- Kitchen spoon

Ingredients

25 fillets of orange roughy, approx 5 oz each
5 lb assorted cooked seafood
10 oz celery
10 oz onions
12 oz margarine
8 oz flour
1 qt fish stock
2 tbsp prepared mustard
8 eggs
6 oz bread crumbs (variable)
2 tbsp Worcestershire sauce
salt and white pepper to taste

Procedure

1. Place the butter in a sauce pot and melt.
2. Add the onions and celery. Sauté slightly; do not brown.
3. Add the flour and cook for approximately 3 min.
4. Add the hot fish stock while whipping rapidly with a wire whip. Cook until mixture thickens.
5. Add Worcestershire sauce, mustard, and cooked seafood. Mix thoroughly with a kitchen spoon. Remove from the range.
6. Stir in the eggs and bread crumbs with a kitchen spoon. Season with salt and white pepper. Let cool slightly.
7. Place approximately 3 oz of the stuffing on each orange roughy fillet, roll up tightly, and place in a bake pan with the end of each roll facing downward.
8. Bake in a preheated oven at 350°F for approximately 30 min.
9. Serve with an appropriate sauce.

Precautions

- Use caution when mincing the onions and celery and when butterflying and flattening the fish fillets.
- Place the rolled fillets in the bake pan properly or they will unroll during the baking period.
- Do not brown the vegetables when sautéing.

Preparation

1. Mince the onions and celery with a French knife.
2. Butterfly and flatten the fish fillets slightly.
3. Heat the fish stock.
4. Select and cook seafood.

Trade Tip

Fresh fish may be purchased drawn with only the entrails removed or dressed with the entrails, head, tail, and fins removed. Use fresh fish within a day or two of purchase or freeze it. Cook fresh shellfish within one or two days. Cook live lobsters and crabs the day of purchase.

The steaks are cut into 6 oz portions from the mako shark, which supplies the best shark steaks. A pocket is cut into the side of each steak to hold the stuffing. The steak is baked and served with an appropriate sauce. The stuffing, comprised of assorted seafood, creates an element of surprise when the guest cuts into the steak.

Stuffed Shark Steak

25 servings

Equipment

- Baker's or portion scale
- French knife
- Utility knife
- Kitchen spoon
- Sheet pan
- 2 qt sauce pot
- Parchment paper

Ingredients

25 shark steaks, 6 oz portions, cut to a thickness of 1"
8 oz margarine
6 oz flour
½ c white wine
1½ c fish stock, hot
4 oz onions
1 tbsp parsley
3 lb assorted seafood
3 egg yolks
3 oz bread crumbs
salt and white pepper to taste
salad oil as needed
paprika as needed

Preparation

1. Cut a large pocket into the side of each shark steak so that only a small incision appears in the side of the steak.
2. Mince the onions with a French knife.
3. Chop the parsley with a French knife.
4. Separate the eggs. Save whites for another preparation.
5. Select and cook seafood.
6. Preheat oven to 350°F.

Procedure

1. Place the margarine in a sauce pot and melt.
2. Add the onions and sauté slightly; do not brown.
3. Add the flour, making a roux. Cook slightly.
4. Add the hot stock while stirring rapidly with a kitchen spoon. Cook until mixture thickens.
5. Add the cooked seafood and white wine. Mix thoroughly with a kitchen spoon. Remove from the range.
6. Stir in the egg yolks, bread crumbs, and chopped parsley. Season with salt and white pepper. Let mixture cool.
7. Insert approximately 2 oz of the stuffing into the pocket cut into the shark steaks.
8. Place the stuffed steaks on sheet pans coated with salad oil. Sprinkle paprika lightly on the surface of each steak.
9. Bake in a preheated oven at 350°F until just done.
10. Serve with an appropriate sauce.

Precautions

- Exercise caution when cutting pockets into the steaks.
- Do not let the onions brown when sautéing.
- Bake the steaks until they are just done. Never overcook seafood.
- Shark meat can be dry. Serve with a sauce.

POACHED FISH RECIPES

Boiling, poaching, and steaming are similar. The difference lies in the amount of liquid used and the cooking temperature. Since poaching is done at a simmering temperature (200°F) and slow cooking produces best results with fish, this method is recommended. Fish should be poached in a court bouillon. Court bouillon is a liquid consisting of celery, onions, carrots, water, vinegar or lemon juice, salt, and spices if desired. Most poached fish are served with some type of thickened sauce.

Poached Fish

25 servings

Equipment

- Quart measure
- Baker's scale
- French knife
- Deep bake pan
- Saucepan
- China cap

Ingredients

12 lb dressed fish
3 lemons
12 oz celery
8 oz onions
10 oz carrots
2 oz salt
1 gal. water
1 tsp whole peppercorns
butter as needed

Preparation

1. Dice the onions, carrots, and celery with a French knife.
2. Slice the lemons with a French knife.

Procedure

1. Place the water, celery, onions, carrots, salt, peppercorns, and lemons in a saucepan. Simmer for 30 min.
2. Strain the liquid through a china cap to make the court bouillon.
3. Grease the bottom and sides of a deep bake pan. Line the fish in the pan. This fish may be whole, steaks, or fillets.
4. Pour the court bouillon over the fish, place on the range, and continue to poach (simmer) until done.
5. Serve with a sauce prepared from the court bouillon or a desired sauce that will complement the fish being poached.

Precautions

- Use caution when handling the knife.
- For best results, start poaching fairly large fish in cold court bouillon and small fish in hot court bouillon.
- Always coat the bottom of the poaching pan with butter or oil to prevent sticking.
- Avoid overcooking or the fish will break and become difficult to serve.

Poached Halibut Dugleré

The halibut is poached in a court bouillon and served covered with a generous amount of dugleré sauce. This entree is an excellent choice for either the luncheon or dinner menu.

25 servings

Equipment

- Deep bake pans
- French knife
- China cap
- 6 qt saucepan
- Baker's scale
- Quart measure
- Spoon measure

Ingredients

9 lb halibut fillets
2 lemons
8 oz celery
6 oz onions
6 oz carrots
1 oz salt
3 qt water
½ tsp peppercorns
butter as needed
3 qt dugleré sauce

Preparation

1. Dice the onions, carrots, and celery with a French knife.
2. Slice the lemons with a French knife.
3. Grease the bake pans with butter.

4. Cut the halibut fillets into 5½ oz to 6 oz portions with a French knife.
5. Prepare the dugleré sauce.

Procedure

1. Place the water, celery, onions, carrots, salt, peppercorns, and lemons in a saucepan. Simmer for 30 min.
2. Strain the liquid through a china cap. This is the court bouillon in which the fish will be poached.
3. Line the halibut fillets in the greased bake pans. Pour the court bouillon over the fish and poach (simmer) on the range until done. (Approximate cooking time is 10 min.)
4. Serve each halibut fillet with a generous amount of dugleré sauce.

Precautions

- Use caution when handling the knife.
- Cook slowly so the fish does not toughen or break when poaching.

FRIED SHELLFISH RECIPES

Fried Softshell Crabs

The molting blue crabs with their soft shell are dressed for cooking, breaded, and deep fried to a crisp golden brown. Fried softshell crabs are popular on the luncheon and dinner menu during the spring season. They are usually accompanied with tartar or fish sauce.

10 servings

Equipment

- Quart measure
- Baker's scale
- Deep fat fryer
- Fry baskets
- 2 qt stainless steel container
- Wire whip
- Bake pans (3)

Ingredients

30 softshell crabs
6 eggs
1 qt milk
1 lb flour
2 lb bread crumbs
salt and pepper to taste

Procedure

1. Place the flour in a bake pan and season with salt and pepper.
2. Add the dressed crabs and coat thoroughly.
3. Pour the egg wash into the second bake pan. Remove the crabs from the flour and place in the egg wash.
4. Place the bread crumbs in the third bake pan. Remove each crab from the egg wash and place in the bread crumbs, press the crumbs on firmly, and shake off the excess.
5. Place in fry baskets and fry until golden brown. Let drain.
6. Serve three softshell crabs per portion with a soufflé cup filled with tartar or fish sauce.

Precautions

- Press the breading on each crab firmly so it does not fall off while frying the crabs.
- Handle the crabs gently; they are very fragile.
- Do not overbrown or the servings will lack in appearance.
- Do not fry too many crabs at one time. The temperature of the fat drops too quickly and frying will not be uniform.

Preparation

1. Prepare an egg wash. Break the eggs into a stainless steel container. Beat slightly with a wire whip, pour in the milk, and blend with the eggs.
2. Set deep fat fryer at 350°F.
3. Dress the crabs by removing the face, just in back of the eyes, the apron on the underside of the crab, and the entrails under the pointed tip on each side of the soft shell.

Fried Oysters or Clams

These two favorite shellfish are breaded and deep fried to a golden brown. Generally, six or eight oysters or clams are served to each order, depending on the size. Tartar sauce or cocktail sauce usually accompanies each order.

25 servings

Equipment

- Baker's scale
- Bake pans (3)
- Quart measure
- Wire whip
- Deep fat fryer
- Fry baskets
- 2 qt stainless steel container
- Colander

Ingredients

150 large oysters or clams
9 eggs
1½ qt milk
1 lb flour
2 lb 8 oz bread crumbs

Preparation

1. Prepare an egg wash. Break the eggs into a stainless steel container, beat slightly with a wire whip, pour in the milk, and blend.
2. Preheat the deep fat fryer to 350°F.
3. Place the oysters or clams in a colander and drain thoroughly.

Procedure

1. Place the flour and 8 oz of the bread crumbs in a bake pan, season with salt and pepper, and mix together.
2. Add the oysters or clams and coat thoroughly.
3. Pour the egg wash into the second bake pan. Remove the oysters or clams from the flour/bread crumb mixture and place in the egg wash.
4. Place the remaining bread crumbs in the third bake pan. Dip each oyster or clam in the bread crumbs, press the crumbs on firmly, and shake off excess.
5. Place in fry baskets and fry in deep fat until golden brown. Let drain.
6. Serve six pieces per portion with tartar or cocktail sauce. Garnish with a wedge or slice of lemon.

Precautions

- Press the breading on each oyster or clam firmly so it does not fall off while frying.
- Do not fry too many pieces at one time. The temperature of the fat will drop too quickly and frying will not be uniform.
- Do not overbrown or the servings will lack in appearance.

Fried Scallops or Shrimp

Frying is a popular and quick method of preparing these two favorite shellfish. They should be cooked to order and approximately six to eight pieces served per portion. These crisp, golden brown shellfish are usually served with tartar or fish sauce.

25 servings

Equipment

- Baker's scale
- Bake pans (3)
- Quart measure
- Wire whip
- Deep fat fryer
- Fry baskets
- 2 qt stainless steel container
- Shrimp peeler
- Paring knife

Ingredients

150 large shrimp or scallops
9 eggs
1½ qt milk
1 lb flour
2 lb bread crumbs
salt and pepper to taste

Preparation

1. Prepare an egg wash. Break the eggs into a stainless steel container, beat slightly with a wire whip, pour in the milk, and blend.
2. Preheat deep fat fryer to 350°F.
3. If frying shrimp, peel the raw shrimp by hand or with a plastic shrimp peeler. Remove the mud vein by scraping the back of each shrimp with the tip of a paring knife. Cut halfway through each shrimp lengthwise using a paring knife (butterfly style). Flatten the shrimp with the heel of the hand.

Procedure

1. Place the flour in a bake pan and season with salt and pepper.
2. Add the shellfish and coat thoroughly.
3. Pour the egg wash into the second bake pan. Remove each piece of shellfish from the flour and place in the egg wash.
4. Place the bread crumbs in the third bake pan. Dip each piece of shellfish in the bread crumbs and press the crumbs on firmly. Shake off excess.
5. Place in fry baskets and fry until golden brown and done. Let drain.
6. Serve six pieces per portion with tartar or fish sauce. Garnish with a wedge or slice of lemon.

Precautions

- Exercise care when handling the knife.
- Do not fry too many pieces at one time. The temperature of the fat will drop too quickly and the pieces will not cook evenly.
- Do not overbrown or the servings will lack in appearance.

Lobster and Shrimp Croquettes

Lobster and shrimp croquettes are extremely tasty. After the croquettes are formed and breaded, they are deep fried to a golden brown and served on the dinner menu with a rich, tasty sauce.

25 servings

Equipment

- French knife
- Baker's scale
- Cup measure
- 2 gal. sauce pot
- Deep fat fryer
- Fry baskets
- Bake pans (3)
- Kitchen spoon
- Wire whip
- Stainless steel container
- Spoon measures
- 2 qt saucepan

Ingredients

Mixture

3 lb shrimp
2 lb lobster meat
6 oz butter
6 oz flour
1 qt milk
1/3 c brandy
1 tbsp lemon juice
2 tsp dry mustard
1 tbsp chives
8 egg yolks
salt and pepper to taste

Breading

1 lb flour
2 lb bread crumbs
1½ qt egg wash (9 eggs to 1½ qt milk)

Preparation

1. Cook, peel, clean, and dice the lobster and shrimp fairly small.
2. Heat the milk in a saucepan.
3. Chop the chives with a French knife.
4. Separate the eggs. Save the whites for use in another preparation.
5. Prepare an egg wash. Break the eggs into a stainless steel container, beat slightly with a wire whip, pour in the milk, and blend.

6. Preheat the deep fat fryer to 350°F.
7. Place the flour and the bread crumbs in separate bake pans.

Procedure

1. Place the butter in a sauce pot and heat.
2. Add the flour, making a roux, and cook for approximately 5 min. Stir with a kitchen spoon.
3. Pour in the hot milk gradually, whipping briskly with a wire whip until very thick and smooth.
4. Add the lobster, shrimp, brandy, lemon juice, dry mustard, and chives. Stir in gently with a kitchen spoon so the seafood does not break.
5. Season with salt and pepper. Cook over low heat 5 min.
6. Beat the egg yolks with a wire whip in a stainless steel container and pour very slowly into the hot mixture while stirring rapidly with a kitchen spoon.
7. Turn the mixture into a bake pan, cover, and refrigerate overnight.
8. Remove the mixture from the refrigerator. Shape into 50 uniform croquettes.
9. Bread by passing each croquette through flour, egg wash, and bread crumbs.
10. Fry in deep fat at 350°F until brown.
11. Serve two croquettes per portion with an appropriate sauce.

Precautions

- Use caution when handling the knife.
- Stir the mixture occasionally to avoid sticking or scorching.
- Pour the beaten egg yolks very slowly into the hot mixture to avoid scrambling.
- Do not overbrown or the serving will lack in appearance.

POACHED OR STEAMED (IN SAUCE) SHELLFISH RECIPES

Lobster or Shrimp Curry

The cooked lobster or shrimp is placed in a rich curry sauce and generally served in a casserole accompanied with baked or boiled rice.

25 servings

Equipment

- 2 gal. sauce pot
- Kitchen spoon
- Baker's scale
- Quart measure
- French knife

Ingredients

8 lb lobster or shrimp
3 qt curry sauce

Preparation

1. Cook the lobster or shrimp by steaming or boiling. Remove the shells by hand.
2. Dice the lobster or shrimp into uniform pieces with a French knife.

3. Prepare the curry sauce.

Procedure

1. Place the prepared curry sauce in a sauce pot and bring to a simmer.
2. Add the cooked lobster or shrimp, blend into the sauce gently with a kitchen spoon, and bring back to a simmer. Remove from the heat.
3. Dish with a 6 oz ladle into shallow casseroles. Serve with boiled or baked rice.

Precautions

- Use caution when handling the knife.
- When adding the shellfish to the sauce, stir in gently so the meat does not break.

Creamed Lobster

The cooked pieces of lobster are added to a rich cream sauce and served in a casserole with baked yellow rice. This item is an excellent choice for the luncheon, dinner, or a la carte menu.

25 servings

Equipment

- 2 gal. sauce pot
- Kitchen spoon
- Baker's scale
- Quart measure
- French knife

Ingredients

- 8 lb lobster meat
- 3 qt cream sauce
- salt and white pepper to taste

Preparation

1. Cook the whole lobsters or lobster tails in boiling salt water (1 tbsp per gallon of water) until the lobster shell turns red, or steam in a steam pressure cooker at 5 lb of pressure for approximately 10 min to 12 min or at 15 lb of pressure for approximately 3 min to 5 min. The exact time depends on the size of the whole lobsters or lobster tails.
2. Remove the meat from the shell and dice with a French knife. (To remove the lobster meat from the shell, cut through the underside of the shell with heavy scissors, insert the fingers between the shell and the lobster meat, and push the meat out of the shell.)
3. Prepare a rich cream sauce.

Procedure

1. Place the prepared cream sauce in a sauce pot and bring to a simmer.
2. Add the cooked lobster and blend into the sauce gently with a kitchen spoon. Bring back to a simmer and season with salt and white pepper. Remove from the heat.
3. Dish with a 6 oz ladle into shallow casseroles. Serve with baked yellow rice.

Precautions

- Use caution when handling the knife.
- When adding the lobster to the sauce, stir in gently so it does not break.

Lobster Thermidor

The northern cold water lobster is boiled in salt water or steamed, cooled, and split in half lengthwise. The meat is removed from the shell, diced, placed in a rich sauce, and placed back into the lobster shell. It is then covered with a Mornay sauce and glazed lightly under the broiler. Lobster thermidor is served on the dinner menu.

12 servings

Equipment

- 5 gal. stockpot and cover
- 3 gal. sauce pot
- Cup measure
- Spoon measure
- Quart measure
- Kitchen spoon
- French knife
- Sheet pan
- Mallet

Ingredients

- 12 1½ lb to 2 lb northern cold water lobsters
- 1 qt Newburg sauce
- 2 tbsp chives
- ½ c mushrooms
- 4 oz butter
- 1 qt Mornay sauce

Preparation

1. Place the lobsters in boiling salt water (allow 1 tbsp of salt for each gallon of water). Cover the pot and boil for 20 min. Drain and let cool.
2. Prepare the Newburg sauce.
3. Mince the chives with a French knife.
4. Dice the mushrooms fairly small with a French knife.
5. Prepare the Mornay sauce.

Procedure

1. Place the butter in a sauce pot and melt.
2. Add the diced mushrooms and sauté until slightly tender.
3. Add the Newburg sauce and bring to a simmer.
4. Stir in the chives with a kitchen spoon. Place on the side of the range and hold.
5. Place each lobster on its back and split in half lengthwise using a French knife. Remove the intestines and sack near the head. Remove all the meat from the body, saving the shells. Break the claws using a mallet, and pick out all the meat. Discard the shells of the claws.
6. Cut the lobster meat into a fairly small dice and fold into the Newburg sauce gently with a kitchen spoon.
7. Refill the lobster shells with a generous amount of the mixture. Cover the top with Mornay sauce.
8. Place on sheet pans and glaze slightly under the broiler.
9. Serve two lobster halves per portion. Garnish with a wedge of lemon and a sprig of parsley.

Precautions

- Exercise caution when splitting the lobsters.
- Do not let the mushrooms brown while sautéing.
- Cover the top of the filled lobster shells completely with the Mornay sauce.
- Do not burn the Mornay sauce while glazing.

Lobster, shrimp, or crabmeat Newburg is a cream dish colored slightly with paprika and highly seasoned with sherry. The Newburg is named according to the seafood used. Newburgs are generally served in a chafing dish over toast or in a patty shell.

Lobster, Shrimp, or Crabmeat Newburg

50 servings

Equipment

- French knife
- Baker's scale
- Quart measure
- Spoon measure
- 5 gal. sauce pot
- Kitchen spoon

Ingredients

10 lb lobster, shrimp, or crabmeat, steamed or boiled
1 lb butter
2 gal. medium cream sauce
juice of 1 lemon
6 oz dry sherry
3 tbsp monosodium glutamate
5 tbsp paprika
salt and white pepper to taste

Procedure

1. Place the butter in the sauce pot and melt. Add paprika and heat slowly.
2. Add the seafood and heat slowly.
3. Add the cream sauce and blend thoroughly with a kitchen spoon. Bring to a boil.
4. Add the sherry, lemon juice, and monosodium glutamate. Blend with a kitchen spoon.
5. Season to taste with salt and white pepper. Remove from the range.
6. Serve with a 6 oz ladle into shallow casseroles. Serve with toast points.

Precautions

- Use caution when handling the knife.
- Do not burn the paprika when heating in the butter.
- Do not break the seafood into small pieces when stirring.
- Use white pepper when seasoning any light or cream dish. Black pepper will ruin the appearance.

Preparation

1. Cut the shrimp, lobster, or crabmeat into ½″ pieces with a French knife.
2. Prepare cream sauce.

This has been a favorite for many years. It is a mixture of rich plump oysters and bread or cracker crumbs baked together with cream. The simple, easy-to-cook preparation is ideal for the luncheon menu.

Scalloped Oysters

25 servings

Equipment

- Casseroles (25)
- Quart measure
- Baker's scale
- Spoon measures

Ingredients

150 shucked oysters and their liquid
2 qt cracker or bread crumbs, very coarse
1 qt light cream (variable)
1 pt oyster liquid (variable)
12 oz butter
2 tbsp Worcestershire sauce
salt and white pepper to taste

3. Preheat the oven to 350°F.

Procedure

1. Place a layer of crumbs in each casserole.
2. Cover the crumbs with six oysters.
3. Add more crumbs to slightly cover the oysters.
4. Mix the cream, oyster liquid, and Worcestershire sauce. Season with salt and white pepper.
5. Pour enough liquid over each casserole to moisten the crumbs.
6. Dot the top of each casserole with butter.
7. Bake in a preheated oven at 350°F until thoroughly heated and the edges of the oysters ruffle.
8. Serve at once.

Preparation

1. Brush melted butter on the bottom and sides of each casserole.
2. Prepare the bread or cracker crumbs.

Precaution

- Do not overcook the oysters as they will become tough.

In this preparation, the cooked seafood is placed in a creole sauce and is served in a casserole with baked rice. Usually, a combination of shrimp, fish, and crabmeat is used. However, just fish may be used. The preparation is also made with shrimp or lobster. This item is an excellent choice for the luncheon, dinner, or a la carte menu.

Shrimp, Lobster, or Seafood Creole

25 servings

Equipment

- Baker's scale
- 2 gal. sauce pot
- Kitchen spoon
- Quart measure

Ingredients

8 lb shrimp, lobster, or other seafood
3 qt creole sauce
salt and pepper to taste

Preparation

1. Cook seafood by steaming or boiling. Remove shells or skin of the fish or shellfish by hand.
2. Prepare the creole sauce.

Procedure

1. Place the prepared creole sauce in a sauce pot and bring to a simmer.
2. Add the cooked shrimp, lobster, or seafood. Blend into the sauce gently with a kitchen spoon, and bring back to a simmer. Season with salt and pepper.
3. Dish with a 6 oz ladle into shallow casseroles. Serve with baked rice.

Precaution

- When adding the shellfish or seafood to the sauce, stir in gently so the meat does not break.

BAKED AND BROILED SHELLFISH RECIPES

The rich, sweet-tasting blue crabmeat is seasoned and flavored to acquire a slightly tangy taste. It is bound together with a thick cream sauce, packed in an aluminum crab shell, and baked until it becomes slightly brown. Crabs are placed on the luncheon or dinner menu. To cut food cost, use half crabmeat and half poached codfish.

Deviled Crabs

25 servings

Equipment

- 2 gal. sauce pot
- Baker's scale
- French knife
- Aluminum crab shells (25)
- Kitchen spoon
- Paring knife
- Quart measure
- Spoon measure
- Bake pan
- Sheet pans (2)
- 2 qt saucepan

Ingredients

6 lb crabmeat, blue or king
12 oz butter or shortening
12 oz flour
12 oz onions
1½ qt milk
3 tbsp prepared mustard
1 tbsp Worcestershire sauce
¼ tsp Tabasco sauce
2 tbsp lemon juice
1 c sherry
salt and pepper to taste
paprika as needed

Preparation

1. Mince the onions with a French knife.
2. Heat the milk in a saucepan.
3. Cook the crabmeat by steaming or boiling. Remove the meat from the shell by hand. Flake the meat by hand.
4. Preheat the oven to 350°F.

Procedure

1. Place the butter or shortening in a sauce pot and melt.
2. Add the onions and sauté slightly. Do not brown.
3. Add the flour, making a roux, and cook for 5 min. Stir with a kitchen spoon.
4. Add the hot milk, stirring constantly with a kitchen spoon until thickened.
5. Stir in the crabmeat gently with a kitchen spoon.
6. Add the mustard, Tabasco sauce, wine, and Worcestershire sauce. Stir to blend thoroughly.
7. Season with salt and pepper. If the mixture is too wet, add bread crumbs to stiffen.
8. Pour into a bake pan and cover with oiled brown paper. Cool and refrigerate overnight.
9. Remove from the refrigerator. Pack each shell with a generous amount of the crabmeat mixture. Score the top of the mixture in each shell with a paring knife.
10. Sprinkle paprika over the top of each shell and dot slightly with additional melted butter.
11. Place on sheet pans and bake in a preheated oven at 350°F until hot and slightly brown.
12. Serve one shell per portion with tartar or fish sauce.

Precautions

- Use caution when handling the knife.
- While the mixture is cooking, stir occasionally with a kitchen spoon to avoid sticking or scorching.
- Do not let the onions brown when sautéing.

Grilled Shrimp with Gorgonzola Dip

Grilled shrimp is a very popular dish that may be served as an appetizer or as an entree. Large shrimp are used for this dish.

8 appetizer servings

Wisconsin Milk Marketing Board

Ingredients

Dip

½ c cottage cheese
6 oz plain yogurt
1 c milk
¾ c gorgonzola cheese, crumbled
pepper to taste

Marinade

2 tbsp brown sugar
2 tbsp chopped onion
2 tbsp vinegar
2 tbsp catsup
1 tbsp Worcestershire sauce
4 tsp hot pepper sauce
1 clove garlic, minced
2 tbsp water

Shrimp

1 ½ lb shrimp, peeled and deveined

Procedure

Dip

1. Combine all ingredients and mix in blender until slightly chunky.
2. Pour dip into serving dish and chill.

Marinade

1. Combine all ingredients and mix in blender until smooth.
2. Pour marinade over shelled and deveined shrimp; cover and refrigerate 1 hr to 2 hr.

Shrimp

1. Drain shrimp and save marinade.
2. Grill shrimp over hot coals for 3 min to 5 min, turning once and basting with marinade.
3. Remove shrimp from grill and serve immediately with dip.

Broiled Shrimp Scampi

This preparation is associated with the flavor of garlic. Large shrimp are butterflied, saturated with garlic-flavored butter, and broiled until done. This entree adds variety to the seafood choice on the dinner menu.

12 servings

Equipment

- Paring knife
- Half sheet pans
- Saucepan
- French knife
- Ladle
- Aluminum foil

Ingredients

72 large shrimp
1 lb butter
½ oz garlic
1 oz lemon juice
½ oz parsley

Preparation

1. Squeeze the juice from the lemons and remove the seeds.
2. Chop the parsley and garlic fine.
3. Butterfly the shrimp. Pull off the legs and peel the shrimp by hand. Leave the tail section attached. Using a paring knife, make a shallow slit down the back of the shrimp and remove the dark vein that lies just below the surface. Make a second cut down the back of the shrimp approximately three-quarters of the way through its body. Fold the two halves outward butterfly style. Pound slightly with the fist to flatten the butterfly cut.
4. Light the broiler. Place broiler rack in low position. Adjust a medium flame.

Procedure

1. Place the butterflied shrimp in half sheet pan with the cut side down and tails curled up.
2. Place the butter in a saucepan and melt.
3. Add the garlic and lemon juice and cook slightly.
4. Pour the garlic butter over the shrimp. Cover the tails with a strip of aluminum foil. Place under the broiler and cook slowly until shrimp is done. Remove from the broiler.
5. Add the chopped parsley to the garlic butter.
6. Serve six shrimp to each order topped with a small amount of garlic butter.

Precautions

- Exercise caution when butterflying the shrimp.
- Avoid burning the tails when broiling the shrimp.

Broiled Lobster

Whether cold or warm water lobsters are used, this is an extremely popular shellfish entree. It is generally served with melted butter on the dinner or a la carte menu. Broiled lobster is always cooked to order.

1 serving

Equipment

- French knife
- Kitchen fork
- Broiler
- Bake pan
- Small container
- Cup measure
- Spoon measure

Ingredients

1 cold water lobster or two 3 oz or 4 oz spiny lobster tails
½ c salad oil
paprika as needed
2 tbsp dry bread crumbs
1 tsp melted butter

Preparation

- Use the directions accordingly, whether for cold water lobsters or spiny lobster tails.

Cold Water Lobster

1. Using a French knife, chop off the claws and legs. Crack the shell of the claws with the back of the French knife. Save the legs. Insert the blade of the French knife into the back of the lobster shell at the point where the tail section joins the body. Split the tail section then the body section in half lengthwise. Remove the stomach that lies just behind the head and the intestinal vein. The cold water lobster is now ready for broiling.

Spiny Lobster Tails

1. Using a French knife, insert the blade into the back of the lobster shell in the center of the tail. Split the lower section and then the upper section in half lengthwise. The spiny lobster tail is now ready for broiling.

2. Melt the butter.

Procedure

1. Place the lobster halves (and claws if cold water lobster is used) in a bake pan. Pour the salad oil over the exposed flesh.
2. Season with salt and sprinkle with paprika.
3. For cold water lobster, blend the melted butter into the bread crumbs and fill the cavity left by removing the stomach with this mixture. Place a few of the legs on top of the stuffing.
4. Place the lobster halves (and claws if cold water lobster is used) on the broiler flesh side up. Broil under low heat until the flesh side browns slightly and the shell becomes red.
5. Remove the lobster halves (and claws if cold water lobster is used) with a kitchen fork and place in a shallow pan.
6. Place in the oven or the hot chamber above the broiler until lobster is completely done. The cooking time depends on the size of the lobster or lobster tail. Total cooking time of 15 min to 20 min is usually sufficient.
7. Serve on a platter with a small cup of melted butter and a wedge or slice of lemon.

Precautions

- Exercise caution when splitting the lobsters or cracking the claws.
- Broil the lobster slowly. Do not overbrown.

Baked Stuffed Shrimp

Large shrimp are butterflied and flattened slightly. A mound of assorted seafood stuffing is placed on top of the shrimp, coated lightly with fine bread crumbs, and baked. This is a popular addition to the dinner menu. It is served with tartar or dill sauce.

12 servings

Equipment

- French knife
- Paring knife
- Sauce pot
- Baker's or portion scale
- Kitchen spoon
- No. 30 scoop
- Sheet pan

Ingredients

48 large shrimp
3 lb assorted seafood
4 oz onions
5 oz margarine
5 oz flour
12 oz hot fish stock
4 oz white wine
1 oz parsley
5 oz eggs
5 oz bread crumbs
1 oz prepared mustard
salt and white pepper to taste

Preparation

1. Butterfly the shrimp. Pull off the legs and peel the shrimp by hand. Leave the tail section attached. Using a paring knife, make a shallow slit down the back of each shrimp and remove the dark vein that lies just below the surface. Make a second cut down the back of each shrimp approximately three-quarters of the way through the body. Fold the two halves outward, butterfly style. Pound slightly with the fist to flatten the butterfly cut.
2. Mince the onions and chop the parsley using a French knife.
3. Prepare the fish stock.
4. Preheat the oven to 350°F.
5. Select, cook, and chop the assorted seafood.
6. Line sheet pan with parchment paper.

Procedure

1. Place the margarine in a sauce pot and melt.
2. Add the onions and sauté without color.
3. Add the flour, making a roux, and cook just slightly.
4. Add the hot fish stock. Work the mixture rapidly with a kitchen spoon until it becomes very thick. Remove from the range.
5. Add the cooked assorted seafood, white wine, prepared mustard, parsley, eggs, and bread crumbs. Mix with a kitchen spoon until thoroughly blended.
6. On each flattened butterflied shrimp, place a No. 30 scoop full of the seafood stuffing. Dust lightly with ad-

ditional fine bread crumbs and place on a sheet pan covered with parchment paper. Add salt and white pepper to taste.
7. Bake in a preheated oven at 350°F until golden brown.
8. Serve four shrimp to each order with tartar or dill sauce.

Precautions

- Exercise caution when butterflying the shrimp.
- When sautéing the onions, do not let them brown or the appearance of the stuffing will be affected. Never overcook seafood. Cook only until done.

Trade Tip — To remove the meat from the leg of a king crab, place the leg in hot water until the shell softens. Use scissors to cut down the side of the softened shell to expose the meat. This allows the meat to be removed easily.

Grilled Shrimp with Grapefruit Corn Relish

The grapefruit provides nice color and a different taste to the grilled shrimp. This dish has a contemporary flair and gives a healthy boost to the menu.

24 servings

Ingredients

1 c olive oil
4 cloves garlic, minced
4½ tbsp cilantro, chopped
2½ c grapefruit juice
48 jumbo shrimp, peeled and deveined
2 tbsp vegetable oil
1 c onion, minced
⅜ c poblano peppers, roasted, seeded, peeled, and diced
⅓ c red peppers, roasted and diced
2 c grapefruit sections
salt and pepper to taste

Florida Department of Citrus

Procedure

Marinade

1. Combine olive oil, 2 cloves garlic, cilantro, and 2½ c grapefruit juice.
2. Pour marinade over shrimp. Refrigerate for 1 hr – 2 hr.

Relish

1. Heat vegetable oil in a sauté pan. Add corn, onion, and remaining garlic and cook until softened but not brown.
2. Add poblano peppers, red peppers, and corn and sauté until heated through.
3. Remove relish from heat and cool slightly. Stir in grapefruit sections and the remaining cilantro. Add salt and pepper to taste.

Shrimp

1. Grill shrimp over hot coals for about 2 min per side.
2. Mound 3 tbsp of relish in the center of a serving plate. Arrange the shrimp around the relish.

Quickbread Preparation

Quickbreads are easy to prepare compared to other baked goods. Although they are easy to prepare, quickbreads are very important to the success of the food service establishment. Customers want and expect high-quality quickbreads.

Quickbreads are so named because of the quick-acting leavening agent used in their preparation. This allows the quickbread mixture to be taken directly from the mixer, made up, and baked without waiting for the dough to rise. There are many types of quickbreads, each requiring specific recipes. All quickbread recipes should result in a product that supplies tenderness, moisture, leavening, flavor, and body.

Common quickbread preparations include biscuits, muffins, corn bread, and quick loaf breads. Quickbreads can be served at any meal and are chosen for a particular meal based on the menu. Ingredients such as nuts, fruits, vegetables, whole wheat flour, rye flour, and cornmeal are commonly added, which makes quickbreads one of the most versatile preparations in the commercial kitchen. These ingredients also add nutritional value to the breads.

QUICKBREAD PREPARATION

A *quickbread* is a baked product that is quick to mix; uses quick-acting leavening agents such as baking powder, baking soda, or steam; and bakes in a short period of time. Quickbread is usually made with six basic ingredients: flour, sugar, salt, oil or shortening, milk, and eggs. Menu items such as biscuits, muffins, corn bread, and corn sticks can be served as quickbreads. Pancakes and waffles are also quickbreads.

In many cases, food service establishments have built their reputation on the quickbreads they feature on their menu. Quickbreads that are placed before the guest freshly baked and still warm are a great addition to any meal.

Quickbreads have acquired their name because they are made with a quick-acting leavening agent, such as baking powder or baking soda, instead of the slower-acting yeast. Quickbreads can be prepared in a comparatively short time since the mixture is taken directly from the mixing machine, made up, and baked. There is no waiting period of one or more hours for gases to develop to leaven the dough. *Leavening* is a chemical, biological, or physical reaction that causes baked products to rise, increase in volume, and become light. Successful quickbreads require high-quality ingredients and good recipes.

Biscuits

Biscuits can be made using many different recipes. A *biscuit* is a light, layered quickbread made with baking powder or baking soda. Eggs are not commonly used in biscuits and shortening is used instead of oil. However, all contain the same basic ingredients with amounts and procedures varying. Basic ingredients in biscuits include sugar, baking powder, salt, flour, shortening, and milk. Eggs and butter may be added to improve the tenderness and richness of the biscuits. The flavoring may also be varied to create different products, such as orange biscuits or cheese biscuits.

Two basic types of biscuits are the cake biscuit and the flaky biscuit. The difference between the two types is how the ingredients are mixed. The *cake biscuit* is mixed to a fairly smooth dough. The *flaky biscuit* contains a higher percentage of shortening and is mixed like a pie crust. The shortening is cut into the dry ingredients and the liquid is added slowly and blended gently until a dough is formed. Making the flaky dough requires more skill because if the dough is overhandled, it will toughen.

CUTTING BISCUITS

1 BISCUIT DOUGH IS ROLLED TO A ¾" THICKNESS; DIP BISCUIT CUTTER IN FLOUR AND CUT BISCUITS

2 PLACE BISCUITS CLOSE TOGETHER ON SHEET PAN

Cutting is one way to form biscuits. They may also be dropped from a spoon.

Shortening is considered the most important ingredient in preparing biscuits because it supplies tenderness. Baking powder and baking soda are the quick-acting leavening agents that cause the dough to rise when liquid is added and heat is applied. Flour supplies body, form, and texture to the biscuit. Milk provides moisture, regulates the consistency of the dough, develops the flour, and causes the baking powder to generate its gas. Salt brings out the flavor and taste of the other ingredients. Sugar supplies sweetness, helps retain moisture, and helps provide the golden brown color desired.

Muffins

Muffins are a popular item that can be served on the breakfast, luncheon, and dinner menus. A *muffin* is a quickbread that is made with eggs and either oil, butter, or margarine. It is shaped like a cupcake and has a pebbled top and a coarse inner texture. Muffins, like biscuits, should always be served fresh from the oven. This can be accomplished relatively easily as most muffin recipes call for a baking period of only 15 min to 20 min.

There are many different muffin recipes. The amount of ingredients varies, but the method of mixing is basically the same. The sugar, shortening, and salt are creamed together; the eggs are blended in; and the dry ingredients are sifted together and added alternately with the milk. All the mixing is done at slow or second speed with the paddle. The reaction of each ingredient within the batter is similar to that of the biscuit mix.

Proper mixing procedure is the key to a successful muffin. A common fault in mixing muffin batter is mixing the batter to a point where the gluten within the flour becomes tough. *Gluten* is the protein contained in flours. It provides structure for baked products. To prevent the gluten from becoming tough, the dry and liquid ingredients should be mixed together at slow speed with the paddle. Mix just enough to moisten, leaving the batter with a slightly rough appearance. The batter should not be smooth.

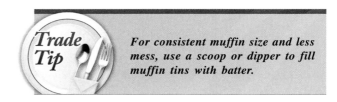

Trade Tip

For consistent muffin size and less mess, use a scoop or dipper to fill muffin tins with batter.

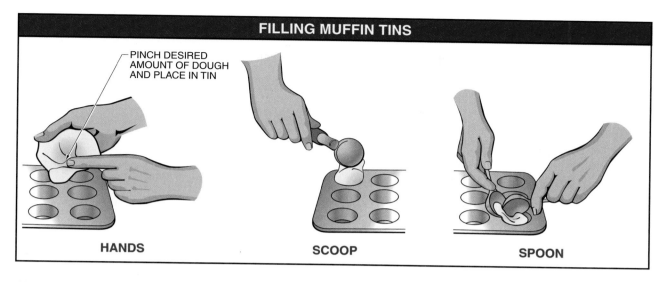

FILLING MUFFIN TINS

PINCH DESIRED AMOUNT OF DOUGH AND PLACE IN TIN

HANDS

SCOOP

SPOON

Corn Bread and Corn Sticks

Corn bread or *corn sticks* are quickbreads made from a batter containing flour and cornmeal as well as eggs and oil. Corn bread and corn sticks are generally prepared using the same batter mix. In some cases, the consistency of the batter may be a little heavier when preparing corn sticks. The batter is made heavier by eliminating a small amount of the liquid. The recipes for preparing corn bread and corn sticks also vary in the mixing method. A recipe may call for a mixing method similar to muffins or for a three-stage method of blending an egg-milk mixture into the dry ingredients and stirring in melted shortening.

Whichever method is used, the mixing should be done at slow or second speed using the paddle. The liquid must be added slowly because cornmeal does not absorb liquid very quickly. If the liquid is added too fast, lumps will form. Avoid overmixing by mixing the batter until fairly smooth.

Quick Loaf Breads

A *quick loaf bread* is a quickbread that uses a batter similar to muffin batter. The shape is changed by baking the batter in a loaf pan. Quick loaf breads are so named because they contain a quick-acting leavening agent such as baking powder. Baking powder starts to rise as soon as heat is applied. These breads provide variety in appearance from the usually served muffin. They can be prepared using different fruits and nuts to produce a variety of flavors. Although quick loaf breads are similar to muffins in eating qualities and texture, they tend to stay fresh longer because of their size.

Quick loaf breads can be baked in a variety of loaf pans to create loaves of various sizes. Since the loaf is a fairly large unit, it can be frozen after baking with excellent results. If the loaves are not served immediately, they should be refrigerated after baking. This allows them to be sliced easier and thinner without crumbling. To extend the freshness of a quick loaf bread, line each pan with parchment paper for storage, or wrap the loaf tight when cool.

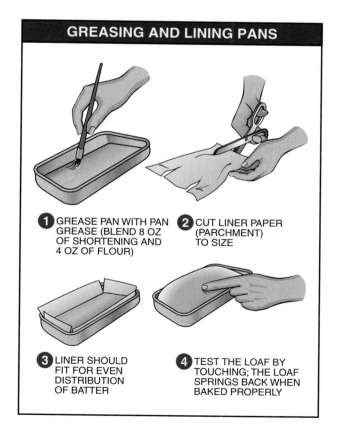

GREASING AND LINING PANS

1 GREASE PAN WITH PAN GREASE (BLEND 8 OZ OF SHORTENING AND 4 OZ OF FLOUR)

2 CUT LINER PAPER (PARCHMENT) TO SIZE

3 LINER SHOULD FIT FOR EVEN DISTRIBUTION OF BATTER

4 TEST THE LOAF BY TOUCHING; THE LOAF SPRINGS BACK WHEN BAKED PROPERLY

QUICKBREAD RECIPES

Baking Powder (Flaky) Biscuits

6 oz

Ingredients

1 lb 4 oz cake flour
1 lb 4 oz bread flour
4 oz granulated sugar
¾ oz salt
2¼ oz baking powder
1 lb 8 oz skim milk
1 lb hydrogenated vegetable shortening

Procedure

1. Place the flours, sugar, salt, and baking powder in a large stainless steel mixing bowl. Blend by hand.
2. Add the shortening and cut in by hand until a fine crumb (small lumps) is formed.
3. Add the milk and mix gently by hand until all the ingredients are moistened and a dough is formed.
4. Turn the dough out of the mixing container onto a floured bench. Cover with a cloth and let rest 10 min.
5. Roll the dough out to a ¾" thickness with a rolling pin.
6. Cut with a small biscuit cutter (2" to 2½") and place on lightly greased or parchment paper-covered sheet pans. Place fairly close together.

7. Brush the top of each biscuit with a rich egg wash (blend 2 eggs and 1 c milk). Let rest 10 min.
8. Bake in a preheated oven at 425°F until golden brown.

Variations

Cheese Biscuits: Add 6 oz of grated Cheddar cheese to the mixture.
Raisin Biscuits: Add 1 lb 4 oz of raisins to the mixture.

Trade Tip

Rolling biscuit dough on a pastry cloth prevents sticking to the surface when cutting and placing the biscuits on baking sheets. Placing biscuits close together on baking pans makes the sides soft and moist; spreading them apart results in crispier sides.

Fried Biscuits

8 doz

Ingredients

12 oz hydrogenated vegetable shortening
4 oz sugar
1 oz salt
4 oz dry milk
8 oz eggs
2 lb 8 oz water (variable)
1 lb 8 oz cake flour
2 lb 8 oz bread flour
2 oz baking powder
4 oz yeast
2 lb raisins (soaked in warm water for 5 min, drained, and dried)

Procedure

1. Combine the flours, baking powder, salt, and sugar. Sift into the stainless steel mixing bowl.

2. Add the shortening and mix slightly at slow speed using the dough hook.
3. Dissolve the dry milk and yeast in the water.
4. Add the eggs to the mixture and whip slightly with a wire whip.
5. Gradually pour the liquid mixture into the flour-shortening mixture while mixing at slow speed.
6. Continue to mix at slow speed until a fairly smooth dough is formed. Remove from the mixer, place dough on a floured bench, and let rest for 10 min.
7. Roll the dough out into a rectangle approximately ½" thick.
8. Cover half the dough with the presoaked raisins. Fold the uncovered half of the dough over the raisins and roll the dough out a second time to a thickness of approximately ½".
9. Cut with a biscuit cutter approximately 3" in diameter.
10. Fry at 375°F in deep fryer until golden brown and done throughout.
11. Dust with powdered sugar.

Blueberry Biscuits

8 doz

Ingredients

2 lb bread flour
1 lb pastry flour
4 oz baking powder
1 lb hydrogenated vegetable shortening
4 oz dry milk
1 qt (2 lb) water
1 oz salt
6 oz sugar
2 lb blueberries

Procedure

1. Place the flours, sugar, salt, and baking powder in a large stainless steel mixing bowl. Blend by hand.
2. Add the shortening and cut in by hand until a crumb mixture is formed.
3. In a separate stainless steel container, dissolve the dry milk in the water. Add this mixture to the crumb mixture and mix gently by hand until all ingredients are moistened and a dough is formed.
4. Turn the dough out of the mixing container onto a floured bench, cover with a cloth, and let rest 10 min.
5. Using a dusted rolling pin, roll the dough out into a rectangle approximately 20″ × 30″.
6. Spread the blueberries over half the rolled-out dough.
7. Fold the uncovered half of the dough over the blueberries.
8. Roll the dough a second time to ¾″ thickness.
9. Cut with a small biscuit cutter (2″ to 2½″) and place on parchment paper-covered sheet pans. Place biscuits fairly close together.
10. Brush the top of each biscuit with a rich egg wash (blend 2 eggs and 1 c milk). Let rest 5 min.
11. Bake in a preheated oven at 400°F until golden brown.

Buttermilk Biscuits

6 doz

Ingredients

5 lb 10 oz all-purpose flour
2¼ oz baking powder
½ oz baking soda
¾ oz salt
1½ oz sugar
1 lb 11 oz butter
3 lb 6 oz buttermilk

Procedure

1. Sift the dry ingredients. Place in a mixing container.
2. Add the butter and cut in by hand until a fine crumb is formed.
3. Stir in the buttermilk gradually with a kitchen spoon until all the ingredients are moistened and a dough is formed.
4. Place the dough in the refrigerator until thoroughly chilled (approximately 45 min).
5. Turn the dough out of the mixing container onto a floured bench. Roll out with a rolling pin to a ¾″ thickness.
6. Cut with a small biscuit cutter (2″ to 2½″) and place on greased and floured or parchment paper-covered sheet pans.
7. Brush the top of each biscuit with melted butter.
8. Bake in a preheated oven at 425°F until golden brown.

Cake Biscuits

6 doz

Ingredients

8 oz granulated sugar
½ oz salt
5 oz dry milk
8 oz hydrogenated vegetable shortening, butter, or margarine
2 lb cold water
1 lb cake flour
2 lb 4 oz bread flour
3 oz baking powder

Procedure

1. Place the sugar, salt, dry milk, and shortening in a stainless steel mixing bowl. Blend with a kitchen spoon to a soft paste.
2. Add the cold water and stir with a kitchen spoon.
3. Sift the cake flour, bread flour, and baking powder. Add to the mixture and mix by hand to a smooth dough.
4. Turn the dough out of the mixing container onto a floured bench. Cover with a cloth and let rest 10 min.
5. Roll the dough out to a ¾″ thickness with a rolling pin.
6. Cut with a small biscuit cutter (2″ to 2½″) and place on lightly greased or parchment paper-covered sheet pans. Place fairly close together.
7. Brush the top of each biscuit with a rich egg wash (blend 2 eggs and 1 c milk). Let rest 5 min.
8. Bake in a preheated oven at 425°F until golden brown.

Golden Rich Biscuits

10 doz

Ingredients

1 lb 8 oz butter
8 oz granulated sugar
8 oz egg yolks
2 lb 4 oz cold milk
2 lb 4 oz bread flour
2 lb 4 oz cake flour
1 oz salt
3¾ oz baking powder

Procedure

1. Place the butter and sugar in a mixing container. Cream with a kitchen spoon until a soft paste is formed.
2. Add the egg yolks. Blend with a kitchen spoon until mixture is slightly smooth.
3. Add the cold milk and stir gently with a kitchen spoon.
4. Sift the cake flour, bread flour, salt, and baking powder. Add to the mixture and mix by hand to a slightly smooth dough.
5. Place the dough in the refrigerator until thoroughly chilled (approximately 45 min).
6. Turn the dough out on a floured bench. Roll out with a rolling pin to a ¾" thickness.
7. Cut with a small biscuit cutter (2" to 2½") and place on greased and floured or parchment paper-covered sheet pans.
8. Brush the top of each biscuit with a rich egg wash (blend 2 eggs and 1 c milk).
9. Bake in a preheated oven at 450°F until golden brown.

Variations

Cheese Biscuits: Add 10 oz of grated Cheddar cheese to the mixture.
Raisin Biscuits: Add 1 lb 12 oz of raisins to the mixture.

Basic Muffin Mix

12 doz

Ingredients

3 lb granulated sugar
2 lb hydrogenated vegetable shortening
1½ oz salt
2 lb eggs
3 lb 8 oz cake flour
1 lb bread flour
2½ oz baking powder
2 lb skim milk (variable)

Procedure

1. Place the sugar, shortening, and salt in a stainless steel mixing bowl. Cream 3 min to 5 min at second speed using the paddle.
2. Add the eggs gradually while continuing to mix at second speed. Mix about 2 min.
3. Add the flour, baking powder, and about two-thirds of the milk. Mix smooth.
4. Add the remaining milk and mix for 1 min more. Remove from the mixer.
5. Fill greased muffin tins or muffin tins lined with paper baking cups two-thirds full of batter.
6. Bake in a preheated oven at 400°F until golden brown.

Variations

Add the following ingredients to each pound of muffin mix for varieties.
Corn Muffins: 4 oz yellow cornmeal and 2 oz skim milk
Date and Walnut Muffins: 2 oz chopped dates and 2 oz chopped walnuts
Marmalade Muffins: 4 oz marmalade
All-Bran Muffins: 4 oz all-bran and 2 oz skim milk
Bacon Muffins: 1 oz chopped fried bacon
Molasses Muffins: 2 oz molasses
Cinnamon Muffins: 4 oz raisins and ½ tsp cinnamon
Apricot Muffins: 4 oz chopped apricots
Honey Whole Wheat Muffins: 2 oz whole wheat flour, 2 oz honey, and 2 oz milk
Banana Muffins: 4 oz well-chopped bananas
Pineapple Muffins: 4 oz chopped pineapple

Trade Tip

To keep a biscuit cutter from sticking when cutting biscuits, dip the biscuit cutter in flour after each cut. Discard the remaining flour after use as it may become lumpy.

Molasses Muffins

6 doz

Ingredients

1 lb 2 oz whole wheat flour
1 lb 2 oz cake flour
6 oz molasses
7 oz brown sugar
5 oz hydrogenated vegetable shortening
2 lb 8 oz milk (variable)
4 oz eggs
¾ oz salt
1 oz baking powder
½ oz baking soda

Procedure

1. Using the paddle, cream the sugar, molasses, shortening, and eggs in a stainless steel mixing bowl.
2. Add the milk while mixing at slow speed.
3. In a separate stainless steel bowl, blend all the dry ingredients. Add to the mixture in the bowl. Continue to mix at slow speed until a fairly smooth batter is formed. Remove from the mixer.
4. Fill greased muffin tins or muffin tins lined with paper baking cups two-thirds full of batter.
5. Bake in a preheated oven at 375°F until golden brown.

Blueberry Muffins

9 doz

Ingredients

2 lb 8 oz cake flour
1 lb 4 oz hydrogenated vegetable shortening
2 lb 8 oz granulated sugar
1½ oz salt
8 oz honey
½ oz baking soda
½ oz baking powder
1 lb 4 oz buttermilk
1 lb 8 oz eggs
2 lb blueberries, fresh or frozen

Procedure

1. Place the flour and shortening in a stainless steel mixing bowl. Mix from 3 min to 5 min at slow speed using the paddle. Scrape down the bowl at least once.
2. Add the sugar, salt, honey, baking soda, buttermilk, and baking powder. Mix for 3 min to 5 min at second speed. Scrape down at least once.
3. Add half of the eggs and mix smooth at second speed. Scrape down and mix smooth again.
4. Add the remaining eggs and continue mixing at second speed for a total of 3 min to 5 min. Scrape down again to ensure a smooth batter.
5. Drain the blueberries thoroughly. Sprinkle them with flour to absorb excess moisture and fold into the batter gently with a kitchen spoon.
6. Fill greased muffin tins or muffin tins lined with paper baking cups two-thirds full of batter.
7. Bake in a preheated oven at 385°F until golden brown.

Bran Muffins

9 doz

Ingredients

2 lb granulated sugar
1 lb hydrogenated vegetable shortening
¾ oz salt
2 lb eggs
3 lb milk
1 lb bran
3 lb bread flour
3 oz baking powder
8 oz honey
8 oz molasses

Procedure

1. Place the sugar, shortening, and salt in a stainless steel mixing bowl. Cream 3 min to 5 min at slow speed using the paddle.
2. Add the eggs gradually while continuing to mix at slow speed.
3. Add the milk and the bran. Mix until thoroughly blended.
4. Sift the flour and baking powder. Add to the mixture. Continue to mix until a batter is formed.
5. Add the honey and molasses and blend into the batter.
6. Fill greased muffin tins or muffin tins lined with paper baking cups two-thirds full of batter.
7. Bake in a preheated oven at 400°F to 425°F until done.

Variations

Raisin Bran Muffins: Add 1 lb of raisins that have been soaked in hot water for a few minutes then drained thoroughly.
Banana Bran Muffins: Add 1 lb of fresh bananas that have been chopped fairly fine.
Apple Bran Muffins: Add 1 lb 4 oz of canned apples that have been chopped fairly fine.

Golden Corn Bread

10 pans

Ingredients

1 lb 10 oz yellow cornmeal
1 lb pastry flour
2 lb 12 oz bread flour
4½ oz baking powder
1½ oz salt
2 lb 10 oz granulated sugar
⅛ oz nutmeg
4 lb skim milk (variable)
1 lb 9 oz melted hydrogenated vegetable shortening
1 lb eggs

Procedure

1. Place the dry ingredients in a stainless steel mixing bowl. Blend at slow speed using the paddle.
2. Add the milk gradually, continuing to mix at slow speed until the mixture is smooth.
3. Add the melted shortening and mix smooth at slow speed.
4. Add the eggs gradually and mix at slow speed until the batter is smooth. Remove from the mixer.
5. Place approximately 24 oz of batter into each greased pan (9″ × 9″ × 1½″).
6. Bake in a preheated oven at 375°F to 400°F until golden brown.

Spicy Apple Muffins

6 doz

Ingredients

1 lb 8 oz sugar
1 lb hydrogenated vegetable shortening
½ oz salt
1 lb eggs
1 lb 12 oz cake flour
8 oz bread flour
1½ oz baking powder
¼ oz cinnamon
¼ oz nutmeg
¼ oz ginger
¼ oz mace
1 lb skim milk (variable)
1 lb 12 oz apples, peeled, cored, and diced fine

Procedure

1. Place the sugar, shortening, and salt in a stainless steel mixing bowl. Cream at second speed using the paddle until smooth.
2. Add the eggs gradually while continuing to mix at second speed. Mix about 2 min.
3. Add the flours, baking powder, and spices with two-thirds of the milk. Mix smooth at second speed.
4. Add the remaining milk and mix smooth a second time at second speed. Remove from the mixer.
5. Fold in the diced apples with a kitchen spoon.
6. Fill greased muffin tins or muffin tins lined with paper baking cups two-thirds full of batter.
7. Bake in a preheated oven at 400°F until golden brown.

Whole Wheat Muffins

7 doz

Ingredients

12 oz granulated sugar
½ oz salt
5 oz dry milk
12 oz hydrogenated vegetable shortening
6 oz molasses
¼ oz cinnamon
12 oz eggs
2 lb water
¼ oz baking soda
12 oz bread flour
12 oz cake flour
12 oz whole wheat flour
2 oz baking powder

Procedure

1. Place the sugar, salt, dry milk, shortening, molasses, and cinnamon in a stainless steel mixing bowl. Cream at slow speed using the paddle until soft and smooth.
2. Add the eggs gradually while continuing to mix at slow speed.
3. Dissolve the soda in the water and sift the three flours and baking powder.

4. Add the water and sifted flours alternately to the mixture. Mix at slow speed until smooth.
5. Fill greased muffin tins or muffin tins lined with paper baking cups two-thirds full of batter.
6. Bake in a preheated oven at 400°F until done.

Variation

Raisin Whole Wheat Muffins: Add 1 lb of raisins that have been soaked in hot water for a few minutes then drained thoroughly.

Southern Spoon Bread

fifty 2½ oz spoonfuls

Ingredients

2 lb white cornmeal
4 lb boiling water
¾ oz salt
4 lb milk
8 oz butter
8 oz egg yolks
1 oz baking powder
12 oz egg whites

Procedure

1. Place the cornmeal and salt in a mixing container. Pour in the boiling water and mix with a kitchen spoon until smooth. Let set until the mixture cools.
2. Blend in the milk by stirring with a kitchen spoon.
3. Place the egg yolks and butter in a stainless steel mixing bowl. Mix at second speed using the paddle until creamy.
4. Add the cornmeal mixture gradually, mixing at second speed.
5. Add the baking powder and continue to mix at second speed until a smooth batter is formed. Remove the batter from the mixer.
6. Beat the egg whites separately with a wire whip until they form fairly stiff peaks.
7. Fold the beaten egg whites into the batter with a kitchen spoon.
8. Pour the batter into buttered baking 9″ × 13″ pans. Fill the pans about one-third full.
9. Bake in a preheated oven at 350°F until the batter sets (approximately 45 min).

Popovers

4 doz

Ingredients

2 oz granulated sugar
2 lb milk
1 lb eggs
1 oz salt
1 lb 8 oz bread flour

Procedure

1. Place the sugar, milk, eggs, and salt in a stainless steel mixing bowl. Beat with a wire whip at high speed until well blended.
2. Add the flour and mix with the paddle at slow speed until a smooth batter is formed.
3. Fill greased muffin tins or muffin tins lined with paper baking cups three-fourths full of batter. For best results, fill every other cup so the batter will have room to pop over.
4. Bake in a preheated oven at 400°F until golden brown and popped over.

Orange Quick Loaf Bread

ten 1 lb loaves

Ingredients

1 lb 14 oz cake flour
10 oz bread flour
2 lb sugar
12 oz hydrogenated vegetable shortening
1 oz salt
½ oz baking soda
¾ oz baking powder
8 oz chopped walnuts (optional)
1 lb 10 oz buttermilk
1 lb 6 oz oranges, ground medium, drained
8 oz white corn syrup
12 oz eggs

Procedure

1. Place the first eight ingredients in a stainless steel mixing bowl. Mix at slow speed using the paddle for approximately 2 min.
2. In a separate stainless steel bowl, blend the buttermilk, ground oranges, corn syrup, and eggs.
3. Add half of the blended liquid mixture to the dry mix in the stainless steel mixing bowl. Mix at slow speed until smooth. Scrape down the bowl and paddle and mix smooth a second time.
4. Add the remaining liquid mixture and continue to mix at slow speed until fairly smooth.
5. Increase machine speed to medium and mix an additional 2 min. Remove from the mixer.
6. Place approximately 1 lb of batter in each prepared paper-lined 4″ × 8″ × 2½″ loaf pan.
7. Bake in a preheated oven at 375°F until done.

Banana Quick Loaf Bread

ten 1 lb loaves

Ingredients

1 lb 14 oz cake flour
10 oz bread flour
2 lb sugar
12 oz hydrogenated vegetable shortening
1 oz salt
½ oz baking soda
¾ oz baking powder
1 lb 2 oz buttermilk
8 oz white corn syrup
2 lb ripe bananas, crushed
12 oz eggs

Procedure

1. Place the flours, sugar, shortening, salt, baking soda, and baking powder in a stainless steel mixing bowl. Mix at slow speed using the paddle for approximately 2 min.
2. In a separate stainless steel bowl, blend the buttermilk, corn syrup, eggs, and crushed bananas.
3. Add half of the blended liquid mixture to the dry mix in the mixing bowl. Mix smooth at slow speed. Scrape down the bowl.
4. Add the remaining liquid mixture and mix smooth at slow speed a second time.
5. Mix at medium speed an additional 2 min. Remove from the mixer.
6. Place approximately 1 lb of batter in each prepared paper-lined 4″ × 8″ × 2½″ loaf pan.
7. Bake in a preheated oven at 375°F until done.

Cranberry Quick Loaf Bread

ten 1 lb loaves

Ingredients

2 lb sugar
1 lb 8 oz cake flour
12 oz bread flour
1 lb 2 oz hydrogenated vegetable shortening
12 oz eggs
12 oz milk
3 lb cranberry sauce
¾ oz baking soda
¼ oz ground cinnamon
¼ oz ground cloves
1 oz salt
1 lb 8 oz chopped cherries

Procedure

1. Place the sugar, shortening, and salt in a stainless steel mixing bowl. Mix at slow speed using the paddle until light and creamy.
2. Gradually add the eggs and the cranberry sauce while continuing to mix at slow speed.
3. Dissolve the baking soda in the milk and gradually add to the mixture in the bowl. Continue to mix at slow speed.
4. Sift the flours and spices. Mix until batter is smooth, scraping down the bowl and paddle at least once.
5. Add the chopped cherries and mix until blended. Remove from the mixer.
6. Place approximately 1 lb of batter in each prepared paper-lined 4″ × 8″ × 2½″ loaf pan.
7. Bake in a preheated oven at 350°F until done.

Ginger-Pineapple Quick Loaf Bread

thirteen 1 lb loaves

Ingredients

3 lb 2 oz pastry flour
1 lb 12 oz sugar
1 lb 4 oz hydrogenated vegetable shortening
2 oz baking powder
½ oz baking soda
½ oz salt
5 oz dry milk
½ oz ground ginger
¼ oz nutmeg
¼ oz cinnamon
12 oz chopped pecans
3 lb drained crushed pineapple
14 oz pineapple juice
2 lb eggs

Procedure

1. Place the first eleven ingredients in a stainless steel mixing bowl. Mix at slow speed using the paddle for approximately 2 min.
2. In a separate stainless steel bowl, blend the crushed pineapple, pineapple juice, and eggs.
3. Add half of the blended liquid mixture to the dry mixture in the stainless steel mixing bowl. Mix at slow speed until fairly smooth. Scrape down the bowl and paddle and mix smooth a second time.
4. Add the remaining liquid mixture. Continue to mix at slow speed until smooth.
5. Increase machine speed to medium and mix an additional 2 min. Remove from the mixer.
6. Place approximately 1 lb of batter in each prepared paper-lined 4″ × 8″ × 2½″ loaf pan.
7. Bake in a preheated oven at 375°F until done.

Cookie Preparation

Cookies are a popular dessert item that can be very profitable for a food service establishment. They can be served alone or with other food items. The ingredients used in cookies are similar to the ingredients used in cakes. However, cookie doughs and batters generally have a higher fat and lower moisture content than cake batters. In addition, cookies are usually baked at higher temperatures for less time than cakes.

Cookies can be classified as soft or crisp cookies, depending upon the texture. Methods commonly used to prepare cookies include the icebox, rolled, bagged, bar, sheet, and drop method. The method used is determined by the consistency of the dough or batter.

Although many cookie recipes are available, all recipes utilize one of three basic mixing methods: single stage, creaming, or sponge method. Overmixing of any cookie batter or dough results in a coarse, hard-to-handle product. Undermixing causes lumps and excessive spreading when baking.

COOKIE PREPARATION

Cookies are a very important and profitable item prepared in the commercial kitchen. Cookies are commonly served alone or with ice creams, sherbets, puddings, fruit cups, and on buffet tables. In each case, cookies can add eye appeal and a finishing touch to a well-prepared meal.

Cookies are classified as soft cookies or crisp cookies according to their texture. *Soft cookies* are cookies prepared from dough that contains a great deal of moisture. *Crisp cookies* are cookies prepared from dough that contains a high percentage of sugar. These two groups of cookies can be further divided into six different types depending on the preparation method used:

- *Icebox method*: Cookies are prepared from a stiff, fairly dry dough. The dough is scaled into units of 1 lb to $1\frac{1}{2}$ lb, rolled into round strips approximately 16″ long, wrapped in parchment paper, and refrigerated overnight. The next day, the dough is sliced into units approximately $\frac{1}{4}$″ thick, placed on sheet pans covered with parchment paper, and baked. The thinner the slice, the crisper the cookie becomes.

- *Rolled method*: Cookies are prepared from a stiff, dry dough. The dough is refrigerated until thoroughly chilled and rolled out on a floured piece of canvas until about $\frac{1}{8}$″ thick. Cookies are cut into desired shapes and sizes with a cookie cutter, placed on sheet pans covered with parchment paper, and baked.

- *Pressed method*: Cookies are prepared from a moist, soft dough. The dough is placed in a cookie press or a pastry bag containing a pastry tube of desired shape and size, squeezed or piped onto sheet pans covered with parchment paper, and baked.

- *Bar method*: Cookies are prepared from a stiff, fairly dry dough. The dough is scaled into 1 lb units, refrigerated until thoroughly chilled, and rolled into round strips the length of a sheet pan. Three strips are placed on each parchment-lined pan, leaving a space between each. The strips are flattened by pressing with the hands, brushed with egg wash, baked, and cut into bars.

- *Sheet method*: Cookies are prepared from a moist, soft batter. The batter is spread over the surface of a parchment-lined sheet pan, brushed with egg wash and sometimes sprinkled with nuts, baked, and cut into square or oblong units.

- *Drop method*: Cookies are prepared from a moist, soft batter. The batter should be at room temperature and is dropped by spoonfuls onto parchment-covered sheet pans. The amount dropped should be as uniform as possible and about the size of a quarter for even baking.

MIXING METHODS

Although many types of cookies are on the market, all cookie doughs and batters are usually mixed using one of three methods:

- The *single-stage method* is performed by placing all the ingredients in the mixing bowl at one time and mixing until the ingredients are blended to a smooth dough. All mixing is done at slow speed.

- The *creaming method* is performed by creaming together the shortening or butter, sugar, salt, and spices until light and fluffy; adding the eggs and liquid if any is required; and sifting in the flour and leavening agent. All mixing is done at slow speed.

- The *sponge* or *whipping method* is named after the method used when preparing a sponge cake. The eggs are whipped at high speed with the sugar until either light and fluffy (soft peaks) when whipping egg whites, or until slightly thickened and lemon-colored when whipping whole eggs or yolks. The remaining ingredients are usually folded into the beaten egg mixture in a very gentle motion to preserve as many air

cells as possible. This method produces cookies that have a light and aerated texture.

Regardless of the mixing method used, never overmix a cookie dough or batter. This results in a finished product that is coarse and difficult to handle. The sides and bottom of the mixing bowl should be scraped down with a plastic scraper at least once or twice during the mixing period to ensure a lump-free batter.

Rules to Follow When Preparing Cookies

- Always use the highest grade of ingredients for best results.
- Follow mixing instructions carefully. Overmixing makes the batter tough, and the cookies do not spread properly when baked. Undermixing results in too much spreading.

TABLE I: DEFECTS AND CAUSES OF FAULTY COOKIES	
Defects	**Causes**
Lack of spread	Too fine a granulation of sugar Adding all sugar at one time Excessive mixing, causing toughening of the flour structure or breakdown of the sugar crystals, or both Too acid a dough condition Too hot an oven
Excess spread	Excessive sugar Too soft a batter consistency Excessive pan grease Too low an oven temperature Excessive or improper type shortening Too alkaline a batter
Fall during baking	Excessive leavening Too soft a batter Weak flour Improper size
Tough cookies	Insufficient shortening Overdeveloped batter Flour too strong
Sticks to pans	Too soft flour Excessive egg content Too slack a batter Unclean pans Sugar spots in dough Improper metal used in pan construction
Greenish cast or dull dark color	Excess bicarbonate of soda
Black spots and harsh crumb	Excessive ammonia
Loss of flavor	Overbaking Too alkaline a dough

- Weigh all ingredients carefully. Too much or too little of any ingredient results in a finished product of poor quality. Use recipes that give weights rather than measures. Weights are more exact than measures.

- Form cookies in uniform size so they bake evenly.

- Bake cookies on pans covered with parchment paper.

- Bake according to recipe instructions. Check the baking progress periodically for necessary adjustments. Baking time for cookies varies depending on size, thickness of dough, and quality of ingredients.

- Double pan if cookies are getting too much bottom heat. This can be detected if the edges of the cookies brown rapidly. To double pan, an extra sheet pan is placed under the original pan to make a false bottom.

- Check the baked cookies for defects.

- Store cookies properly:
  - *Crisp cookies*: Place in a tin container with a loose-fitting top. Store in a dry place. Place in oven at 225°F for 5 min before serving.
  - *Soft cookies*: Place in an airtight tin container. A slice of bread can be added, which will help to keep cookies fresher.

COOKIE RECIPES

The following cookie recipes are for some of the popular types of cookies served at food service establishments. The mixing method (single-stage, creaming, or sponge) is given for each recipe. Each of these cookie recipes contains the list of ingredients by weight rather than measures. Use a scale to measure each ingredient very carefully.

SINGLE STAGE METHOD RECIPES

Chocolate Chip Cookies

These are prepared from a soft, moist dough. The chocolate chips flowing through the batter create eye appeal and supply a rich chocolate flavor to the finished product. This cookie is mixed using the single-stage method.

13 doz

Equipment

- Mixing machine and paddle
- Sheet pans
- Parchment paper
- Plastic scraper
- Pastry bag and large plain tube
- Baker's scale
- French knife

Ingredients

1 lb 8 oz sugar
1 lb hydrogenated vegetable shortening
½ oz salt
¼ oz soda
1 lb 8 oz pastry flour
8 oz pecans
4 oz water (variable)
8 oz eggs
vanilla to taste
1 lb 8 oz chocolate chips or pieces

Preparation

1. Cover sheet pans with parchment paper.
2. Chop pecans fairly fine with a French knife.

Procedure

1. Scale all the ingredients in the stainless steel mixing bowl.
2. Mix at medium speed using the paddle until the dough is smooth. Scrape down sides and bottom of bowl with a plastic scraper at least once during the mixing period.
3. Place the batter in a pastry bag with a large hole plain tube. Squeeze out quarter-sized cookies (approximately 1″ diameter) onto parchment-covered sheet pans.
4. Bake in a preheated oven at 375°F until light.
5. Remove from the oven and let cool.

Precautions

- Use caution when chopping the pecans.
- Do not overmix the dough.
- Do not form a cookie larger than a quarter.
- Do not brown the cookies when baking. Bake lightly.
- Do not remove the cookies from the sheet pan until they are cold.

Fruit Bars

These are a soft, tender, chewy cookie. They are prepared from a slightly soft dough, which can prove difficult to form if the raisins are not dried properly. The dough is mixed using the single-stage method.

20 doz

Equipment

- Baker's scale
- Mixing machine and paddle
- Sheet pans
- Parchment paper
- Plastic scraper
- Saucepan
- Towel
- French knife
- Flour sifter

Ingredients

2 lb sugar
1 lb dark brown sugar
1 lb hydrogenated vegetable shortening
¼ oz baking soda
1 lb eggs
½ oz cinnamon
3 lb pastry flour
3 lb 8 oz raisins
½ oz salt

Preparation

1. Soak the raisins in warm water in a saucepan for 15 min, drain thoroughly, and dry in a towel.
2. Line the sheet pans with parchment paper.

Procedure

1. Place all the ingredients in the stainless steel mixing bowl.
2. Mix at slow speed using the paddle until all ingredients are thoroughly blended. Scrape down sides and bottom of bowl with a plastic scraper at least once during the mixing period.
3. Scale the dough into 1½ lb units and refrigerate until thoroughly chilled.
4. Remove one unit at a time from the refrigerator, place on a floured bench, and form into a roll.
5. Place four rolls across the width of each sheet pan. Flatten each roll with the hands.
6. Brush with egg wash (2 eggs and 1 c milk blended together) or slightly-beaten egg whites.
7. Bake in a preheated oven at 360°F until fairly brown. Let cool.
8. Cut with a French knife into bars approximately 1″ × 3″.
9. Dust with sifted powdered sugar.

Precautions

- Do not overmix the dough.
- Dry the raisins thoroughly before adding to the other ingredients.
- Do not overbake.
- Do not cut the cookies until they have cooled.

FORMING FRUIT BARS

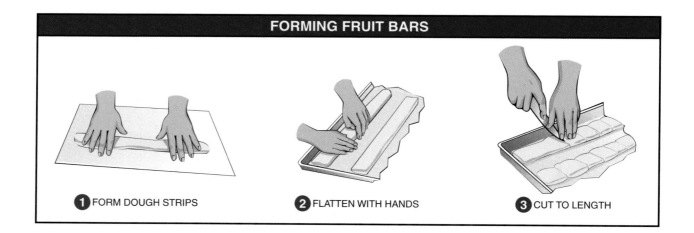

1 FORM DOUGH STRIPS   **2** FLATTEN WITH HANDS   **3** CUT TO LENGTH

Trade Tip

If a cookie dough does not hold together or is difficult to roll, divide the dough into small units and work some egg white into each unit of dough. The added egg white also improves the baking condition.

Fruit Tea Cookies

These are a type of drop cookie. They are formed by depositing small amounts of batter on sheet pans. Fruit tea cookies are mixed using the single-stage method.

18 doz

Equipment

- Baker's scale
- Sheet pans
- Pastry bag and large plain tube
- Kitchen spoon
- Mixing machine and paddle
- French knife
- Parchment paper
- Plastic scraper

Ingredients

1 lb 6 oz hydrogenated vegetable shortening
1 lb 6 oz powdered sugar
2 lb 8 oz pastry flour
2 oz milk (variable)
6 oz raisins
2 oz pecans
2 oz pineapple
2 oz peaches
8 oz eggs
½ oz salt
¼ oz baking soda
vanilla to taste

Preparation

1. Chop the raisins, pecans, pineapple, and peaches fairly fine with a French knife.
2. Cover the sheet pans with parchment paper.

Procedure

1. Place all the ingredients in the stainless steel mixing bowl.
2. Mix at medium speed using the paddle until a smooth batter is formed. Scrape down the sides and bottom of the bowl with a plastic scraper at least once during the mixing period.
3. Remove from the mixer and, using a kitchen spoon, place the dough into a pastry bag with a large plain tube. Squeeze out quarter-sized cookies (approximately 1″ diameter) onto parchment-covered sheet pans.
4. Bake in a preheated oven at 375°F until very light brown.
5. Remove from the oven and let cool.

Precautions

- Do not overmix the dough. Mix only until dough is smooth.
- Do not form a cookie larger than a quarter.
- Do not overbake the cookies. Bake only until light brown.
- It may be necessary to increase or decrease the moisture slightly by adding or leaving out milk to produce a dough best suited to individual use.
- Do not remove the cookies from the sheet pans until they are cold.

Brown Sugar Cookies

These are prepared from a fairly stiff, dry dough with a high percentage of brown sugar present to add flavor and color to the finished product. Brown sugar cookies are mixed using the single-stage method.

14 doz

Equipment

- Mixing machine and paddle
- Plastic scraper
- Baker's scale
- Parchment paper
- Cookie stamp
- Sheet pans

Ingredients

3 lb 2 oz brown sugar
2 lb 4 oz hydrogenated vegetable shortening
1 oz salt
½ oz baking soda
4 lb 8 oz pastry flour
1 lb eggs
vanilla to taste

Preparation

1. Line the sheet pans with parchment paper.

Procedure

1. Place all the ingredients in the stainless steel mixing bowl.
2. Mix at medium speed using the paddle until a smooth dough is formed (approximately 2 min). Scrape down the sides and bottom of the bowl with a plastic scraper at least once during the mixing period.
3. Remove from the mixer and scale the dough into 12 oz pieces (larger if desired). Refrigerate until dough is chilled.
4. Remove one unit of dough at a time from the refrigerator and roll by hand into round strips 16″ long.
5. Cut into 12 equal pieces and place on the prepared sheet pans.
6. Flatten each cookie by hand. Use a cookie stamp to produce an embossed effect or cut with a scallop-edged cookie cutter if desired.
7. Bake in a preheated oven at 375°F until slightly brown.

Precautions

- Do not overmix the dough. Mix only until dough is smooth.
- Do not overbake the cookies. Bake only until they are slightly brown. If the bottom is getting too much heat, double pan.
- Let the cookies cool thoroughly before removing them from the sheet pan.

Bon Bon Cookies

These are prepared from a fairly stiff, dry dough. The dough will form and roll well with this consistency. After baking, the center of each cookie is spotted with various colors of fondant icing. This dough can also be used for cut-out cookies. It is mixed using the single-stage method.

18 doz

Equipment

- Mixing machine and paddle
- Baker's scale
- Sheet pans
- French knife
- Plastic scraper
- Parchment paper

Ingredients

10 oz powdered sugar
1 lb hydrogenated vegetable shortening
4 oz butter
¼ oz salt
2 oz dry milk
2 lb 8 oz cake flour
3 eggs
6 oz water (variable)
pinch of baking soda
flavor to taste

Preparation

1. Line the sheet pans with parchment paper.

Procedure

1. Place all the ingredients in the stainless steel mixing bowl.
2. Mix at medium speed using the paddle until a smooth dough is formed (about 2 min). Scrape down the sides and bottom of the bowl with a plastic scraper at least once during the mixing period.
3. Remove from the mixer and scale the dough into 12 oz units.
4. Mold and roll the dough by hand into round strips approximately 16″ long. Roll each strip in chopped nuts, chocolate shot, or colored sugar. Refrigerate until firm.
5. Remove from the refrigerator and cut into 36 equal pieces.
6. Place on sheet pans covered with parchment paper and press with index finger to form an indentation for adding the fondant icing.

Procter and Gamble Co.

7. Bake in a preheated oven at 375°F until just slightly brown.
8. Remove from oven, let cool, and fill indented center with fondant icing of various pastel colors.

Precautions

- Do not overmix the dough. Mix only until dough is smooth.
- Let the cookies cool thoroughly before decorating with the fondant icing.

Honey Coconut Cookies

These are prepared from a stiff, fairly dry dough that contains the flavor of honey and coconut. They are mixed using the single-stage method.

10 doz

Equipment

- Baker's scale
- Mixing machine and paddle
- Plastic scraper
- Parchment paper
- Cookie stamp
- Sheet pans
- French knife

Ingredients

1 lb sugar
14 oz hydrogenated vegetable shortening
¼ oz salt
2 lb pastry flour
¼ oz baking soda
4 oz macaroon coconut
1 lb honey
2 oz water (variable)
vanilla to taste

Preparation

1. Line the sheet pans with parchment paper.

Procedure

1. Place all the ingredients in the stainless steel mixing bowl.
2. Mix at medium speed using the paddle until a smooth dough is formed (approximately 2 min to 3 min). Scrape down the sides and bottom of the bowl with a plastic scraper at least once during the mixing period.
3. Remove the dough from the mixing machine and divide into 1 lb units. Mold and roll by hand into round strips approximately 16″ long.

4. Cut each round strip into 24 equal pieces, place on a prepared sheet pan, and flatten slightly by hand or use a cookie stamp or other design to produce an embossed effect.
5. Bake in a preheated oven at 375°F until golden brown.

Precautions

- Do not overbake. Cookies continue to bake when removed from the oven.
- Cut the cookies as uniformly as possible so they bake evenly.
- Let the cookies cool thoroughly before removing from the sheet pan.

Schoolhouse Cookies

These are prepared from a fairly stiff, dry dough with ground raisins and nuts added to create a flavor combination. They are mixed using the single-stage method.

5 doz

Equipment

- Baker's scale
- Mixing machine and paddle
- Grinding attachment and medium chopper plate
- Parchment paper
- Sheet pans
- Plastic scraper
- French knife

Ingredients

1 lb 2 oz sugar
8 oz hydrogenated vegetable shortening
½ oz salt
¼ oz baking soda
1 pinch mace
4 oz eggs
4 oz raisins
2 oz pecans
1 lb 6 oz pastry flour
2 oz water

Preparation

1. Line the sheet pans with parchment paper.
2. Grind the raisins and pecans separately in the food grinder using the medium chopper plate.

Procedure

1. Place all the ingredients in the stainless steel mixing bowl.
2. Start mixing at slow speed using the paddle. When ingredients are slightly blended, increase machine speed to medium.
3. Mix until a fairly smooth dough is formed. Scrape down the sides and bottom of bowl with a plastic scraper at least once during the mixing period.
4. Remove from the mixer and scale the dough into 12 oz units. Roll each unit by hand into round, 12" long, rope-like strips.
5. Cut each roll into 12 equal pieces using a French knife. Place on prepared sheet pans and flatten into an oblong shape.
6. Bake in a preheated oven at 375°F until slightly brown.

Precautions

- Do not overbake the cookies. Bake only until they are slightly brown. Double pan if necessary.
- Exercise caution when cutting the cookies.
- Let the cookies cool before removing them from the sheet pan.

Peanut Butter Chocolate Chip Cookies

Peanut butter and chocolate have been associated in a number of successful candy bars. This preparation brings them together in the form of a very tasty, successful cookie. They are mixed using the single-stage method.

14 doz

Equipment

- Baker's scale
- Mixing machine and paddle
- Plastic scraper
- Sheet pans
- Parchment paper

Ingredients

1 lb 8 oz sugar
12 oz hydrogenated vegetable shortening
¼ oz salt
10 oz eggs
1 lb 8 oz peanut butter
1 lb 8 oz cake flour
½ oz dry milk
3 oz water
¼ oz baking soda
1 lb chocolate chips

Preparation

1. Cover the sheet pans with parchment paper.
2. Preheat oven to 375°F.

Procedure

1. Scale all the ingredients into the stainless steel mixing bowl. Using the paddle, mix at medium speed until a fairly smooth dough has formed.
2. Remove from the mixer. Place the dough on a floured bench. Scale the dough into 1 lb units.
3. Roll each unit by hand into a round strip approximately 18" long and 1" thick. Cut into 24 equal pieces.
4. Place each cookie on the prepared sheet pans. Flatten gently by hand.
5. Bake in a preheated oven at 375°F for approximately 10 min.
6. Remove from the oven and let cookies cool thoroughly before handling.

Precautions

- Do not overmix the dough. Mix only until dough is smooth.
- Do not remove the cookies from the sheet pans until they have cooled.

Macaroon Bars

These are a chewy type of cookie because of the presence of moist macaroon coconut. The soft, easy-to-form dough is mixed using the single-stage method.

10 doz

Equipment

- Sheet pans
- Parchment paper
- Mixing machine and paddle
- Plastic scraper
- Utility knife
- Baker's scale

Ingredients

1 lb 6 oz brown sugar
1 lb hydrogenated vegetable shortening
½ oz salt
¾ oz dry milk
½ oz baking powder
1 lb 14 oz pastry flour
8 oz macaroon coconut
4 oz honey or invert syrup
6 oz water (variable)

Preparation

1. Line the sheet pans with parchment paper.
2. Preheat the oven to 375°F.

Procedure

1. Scale all the ingredients into the stainless steel mixing bowl. Using the paddle, mix at medium speed until a smooth dough is formed (approximately 2 min).
2. Remove from the mixer and turn the dough out onto a floured bench.
3. Scale the dough into 1½ lb units. Refrigerate until thoroughly chilled.
4. Remove one unit at a time from the refrigerator, place on a floured bench, and form into a roll.
5. Place three rolls across the width of each sheet pan. Flatten each roll with the hands and cut into bars using a utility knife.
6. Bake in a preheated oven at 375°F until fairly brown. Remove from the oven and let cool.

Precautions

- Do not overmix the dough.
- Do not overbake.
- Do not move the cookies until they are thoroughly cool and set.

Note: Add 8 oz of chocolate pieces to prepare chocolate macaroon bars.

Ginger Cookies

These are prepared from a stiff, fairly dry dough that is seasoned with ginger to create a spicy ginger taste. Ginger cookies are mixed using the single-stage method.

8 doz

Equipment

- Mixing machine and paddle
- Plastic scraper
- Baker's scale
- Parchment paper
- Cookie stamp
- Sheet pans
- French knife

Ingredients

1 lb sugar
8 oz brown sugar
8 oz hydrogenated vegetable shortening
¼ oz baking soda
¼ oz salt
1 lb 8 oz cake flour
¼ oz ginger
1½ oz molasses
8 oz eggs
3 oz water

Preparation

1. Line the sheet pans with parchment paper.

Procedure

1. Place all the ingredients in the stainless steel mixing bowl.
2. Mix at medium speed using the paddle until a smooth dough is formed (approximately 2 min). Scrape down sides and bottom of bowl with a plastic scraper at least once during the mixing period.
3. Remove from the mixer and scale the dough into 1 lb pieces. Refrigerate until dough is chilled.
4. Remove one unit of dough at a time from the refrigerator and roll by hand into round strips 16″ long.
5. Cut with a French knife into 24 equal pieces and place on the prepared sheet pans.
6. Flatten each cookie by hand and use a cookie stamp to produce an embossed effect if desired.
7. Bake in a preheated oven at 375°F until slightly brown.

Precaution

- Do not overbake the cookies. Bake only until they are slightly brown. Double pan if necessary.

Oatmeal-Raisin Cookies

These are prepared from a stiff, fairly dry dough with oatmeal and ground raisins being the two main ingredients. Oatmeal-raisin cookies are mixed using the single-stage method.

12 doz

Equipment

- Sheet pans
- Mixing machine with paddle and grinder attachments
- Plastic scraper
- Parchment paper
- Cookie stamp
- Baker's scale
- French knife

Ingredients

1 lb 12 oz sugar
13 oz hydrogenated vegetable shortening
½ oz baking soda
½ oz salt
1 pinch cinnamon
vanilla to taste
10 oz whole oatmeal
4 oz raisins
1 lb 10 oz cake flour
8 oz water (variable)

Preparation

1. Line the sheet pans with parchment paper.
2. Grind the raisins in the food grinder using the medium chopper plate.

Procedure

1. Place all the ingredients in the stainless steel mixing bowl.
2. Mix at medium speed using the paddle until a smooth dough is formed (approximately 2 min). Scrape down the sides and bottom of the bowl with a plastic scraper at least once during the mixing period.
3. Remove from the mixer and scale the dough into 1 lb pieces. Refrigerate until the dough is chilled.
4. Remove one unit of dough at a time from the refrigerator and roll by hand into round strips 16″ long.
5. Cut into 24 equal pieces with a French knife and place on the prepared sheet pans.
6. Flatten each cookie by hand and use a cookie stamp to produce an embossed effect if desired.
7. Bake in a preheated oven at 350°F until slightly brown.

Precautions

- Do not overmix the dough. Mix only until dough is smooth.
- Do not overbake the cookies. Bake only until they are slightly brown. Double pan if necessary.
- Let the cookies cool before removing them from the pan.

Fudge Cookies

These are prepared from a fairly stiff dough that is very rich with chocolate flavor. Fudge cookies are mixed using the single-stage method.

14 doz

Equipment

- Mixing machine and paddle
- Plastic scraper
- Baker's scale
- Sheet pans
- Parchment paper
- Pastry bag

Ingredients

2 lb sugar
1 lb 8 oz hydrogenated vegetable shortening
¾ oz salt
2 lb 4 oz cake flour
6 oz cocoa
1½ oz baking powder
8 oz eggs
8 oz skim milk (variable)

Procter and Gamble Co.

Preparation

1. Line the sheet pans with parchment paper.

Procedure

1. Place all the ingredients in the stainless steel mixing bowl.
2. Mix at medium speed using the paddle until a smooth dough is formed (approximately 2 min). Scrape down the sides and bottom of the bowl with plastic scraper.
3. Remove from the mixer and scale the dough into 16 oz pieces. Mold and roll by hand into round strips about 16″ long.
4. Cut into 24 equal pieces with a French knife and place them on prepared sheet pans.
5. Flatten each cookie by hand and use a cookie stamp to produce an embossed effect. Top with fruit or nuts if desired.
6. Bake in a preheated oven at 375°F until cookies start to brown.

Note: They may also be formed by placing the batter in a pastry bag with a medium-sized star or plain tip tube, squeezing out onto the prepared sheet pans in pieces about the size of a quarter, then decorating the tops with decorettes and baking.

Precautions

- When baking, avoid strong bottom heat.
- Do not overbake. Cookies continue to bake when removed from the oven.
- Form the cookies as uniformly as possible so they bake evenly.
- Let the cookies cool thoroughly before removing from the sheet pan.

Sugar Cookies

These are prepared from a fairly stiff, dry dough with a high percentage of sugar present to create a brittle cookie. Sugar cookies are mixed using the single-stage method.

14 doz

Equipment

- Sheet pans
- Mixing machine and paddle
- Plastic scraper
- Baker's scale
- Parchment paper
- Cookie stamp
- French knife

Ingredients

2 lb sugar
1 lb 8 oz hydrogenated vegetable shortening
¾ oz salt
⅛ oz mace
2 lb 12 oz cake flour
1½ oz baking powder
8 oz eggs
8 oz skim milk (variable)

Preparation

1. Line the sheet pans with parchment paper.

Procedure

1. Place all the ingredients in the stainless steel mixing bowl.
2. Mix at medium speed using the paddle until a smooth dough is formed (approximately 2 min). Scrape down the sides and bottom of the bowl with a plastic scraper at least once during the mixing period.
3. Remove from the mixer and scale the dough into 1 lb pieces. Refrigerate until dough is chilled.
4. Remove one unit of dough at a time from the refrigerator and roll by hand into 16″ long, round strips.
5. Cut into 24 equal pieces with a French knife and place on the prepared sheet pans.
6. Flatten each cookie by hand and use a cookie stamp to produce an embossed effect.
7. Bake in a preheated oven at 375°F until a very light brown.

Precautions

- Do not overmix the dough. Mix only until dough is smooth.
- Do not overbake the cookies. Bake only until they are slightly brown. If the bottom is getting too much heat, double pan.
- Let the cookies cool thoroughly before removing them from the sheet pan.

Icebox Cookies (Plain)

These are prepared from a stiff, fairly dry dough that is formed into rolls and refrigerated for at least 6 hr. Fruits, nuts, chocolate, and spices can be added to create variety. This recipe is mixed using the single-stage method.

12 doz

Equipment

- Baker's scale
- Sheet pans
- Parchment paper
- Mixing machine and paddle
- French knife
- Plastic scraper
- Wax paper
- Pastry brush

Ingredients

1 lb 12 oz powdered sugar (4X)
1 lb 8 oz hydrogenated vegetable shortening
¾ oz salt
2 lb pastry flour
8 oz eggs
vanilla to taste

Preparation

1. Line the sheet pans with parchment paper.

Procedure

1. Place all the ingredients in the stainless steel mixing bowl.
2. Mix at medium speed using the paddle until a smooth dough is formed (approximately 2 min). Scrape down sides and bottom of bowl with a plastic scraper at least once during the mixing period.
3. Remove dough from mixing machine, divide into 1 lb units, and roll into units about 18″ long. Roll in ground nut meats, macaroon coconut, or colored sugars.
4. Wrap each roll in wax paper and place on sheet pans. Place in the refrigerator for at least 6 hr.
5. Remove from the refrigerator, unwrap, and slice each roll into ½″ slices with a French knife. Place on the prepared cookie sheets.
6. Bake in a preheated oven at 375°F until the cookies become light brown. Avoid too much bottom heat.

Note: Wash with egg wash before baking, or brush with glaze after baking. (Glaze: Add 1 lb water to 2 lb glucose and bring to a boil.) Cookies may also be decorated with icing after baking.

Variations

Fruit cookies: Add 1 lb chopped fruits to 6 lb cookie dough.
Nut cookies: Add 1 lb chopped nuts to 6 lb cookie dough.
Molasses-Spice cookies: Add 3 oz spice combination, 4 oz molasses, and 2 oz flour to 6 lb cookie dough. (*Suggested spice combination*: 8 oz cinnamon, 3 oz mace, 1 oz ginger, 1 oz allspice, and 3 oz nutmeg. Blend together.)

Precautions

- Do not overmix the dough. Mix only until a smooth dough is formed.
- Do not overbake the cookies. Avoid too much bottom heat.
- Handle the cookies carefully after baking. Let them cool thoroughly before removing them from the sheet pans.

CREAMING METHOD RECIPES

Almond Toffee Bars

These are prepared from a soft, moist batter of spreading consistency. The basic dough is spread on sheet pans and baked, then spread with a thin layer of chocolate and topped with nuts. Almond bars are mixed using the creaming method.

4½ doz

Equipment

- Sheet pan
- Baker's scale
- French knife
- Plastic scraper
- Mixing machine
- Double boiler
- Spatula

Ingredients

1 lb butter
1 lb brown sugar
1½ oz egg yolks
1 lb 3 oz pastry flour
1 lb semi-sweet chocolate
4 oz sliced almonds
vanilla to taste

Preparation

1. Grease the bottom and sides of the sheet pan with additional butter.
2. Melt the chocolate in a double boiler.
3. Separate the egg yolks from the whites. Save whites for use in another preparation.

Procedure

1. Place the butter in the stainless steel mixing bowl. Mix at slow speed using the paddle until the butter is creamed.
2. Add the brown sugar and continue creaming until the mixture is light and fluffy.
3. Add the egg yolks and vanilla. Increase the speed of the machine to high and beat well.
4. Reduce the speed of the mixer to slow, add the flour, and mix until well-blended. Scrape down the bowl at least once with a plastic scraper while mixing in this stage.
5. Spread the mixture on the greased sheet pan as evenly as possible with a spatula.
6. Bake in a preheated oven at 325°F for about 20 min or until the batter is set.
7. Remove from the oven and spread a thin layer of melted chocolate over the cookie layer with a spatula while it is still warm.
8. Sprinkle with almonds. Cut into 2″ squares with a French knife while still slightly warm.
9. Let cool before removing the cookies.

Precautions

- Do not overbake the cookie mixture. Bake only until set.
- Spread the cookie mixture on the sheet pan as evenly as possible so it will bake uniformly.
- Melt the chocolate in a warm place or in the top of a double boiler.

Short Paste Cookies

These are prepared from a smooth, stiff, fairly dry dough containing a large percentage of shortening. The finished product is crisp and extremely tender. They are mixed using the creaming method.

10 doz

Equipment

- Baker's scale
- Sheet pans
- Parchment paper
- Rolling pin
- Cookie cutters
- Plastic scraper
- Pastry brush
- Mixing machine and paddle

Ingredients

1 lb hydrogenated vegetable shortening
8 oz butter
1 lb sugar
½ oz salt
4 oz eggs
4 oz milk
2 lb 8 oz pastry flour
vanilla to taste

Preparation

1. Line sheet pans with parchment paper.

Procedure

1. Place the sugar, shortening, butter, and salt in the stainless steel mixing bowl. Cream at slow speed using the paddle. Scrape down the bowl with a plastic scraper.
2. Add the eggs gradually while continuing to mix at slow speed until thoroughly blended with other ingredients.
3. Add the milk and mix at slow speed until thoroughly blended.
4. Add the vanilla and flour while continuing to mix at slow speed. Mix until dough is smooth (approximately 2 min). Scrape down the bowl.
5. Remove the dough from the mixer and divide into 1 lb units. Refrigerate until chilled.
6. Remove one unit of dough from the refrigerator at a time. Roll out about ⅛″ thick on a floured piece of canvas. Cut out cookies with cookie cutters into various shapes and place on prepared sheet pans. Brush with egg wash and decorate. Bake in a preheated oven at 375°F until light brown.

Precautions

- Do not overmix the dough. Mix only until smooth.
- Roll only one unit of dough at a time. Chilled dough is easier to work with.
- Do not overbake. Bake only until cookies are light brown.

French Macaroons

These are prepared from a smooth, medium stiff, rich dough containing a large percentage of almond paste to provide the characteristic almond flavor. They are mixed using the creaming method.

5 doz

Equipment

- Sheet pans
- Parchment paper
- Pastry bag and star tube
- Baker's scale
- Mixing machine
- Plastic scraper
- Kitchen spoon

Ingredients

2 lb almond or macaroon paste
1 lb sugar
8 oz powdered sugar
⅛ oz salt
10 oz egg whites (variable)

Preparation

1. Line the sheet pans with parchment paper.

Procedure

1. Place the almond or macaroon paste, sugar, salt, and powdered sugar in the stainless steel mixing bowl. Mix at slow speed using the paddle until slightly blended. Scrape down the bowl with a plastic scraper.
2. Add the egg whites gradually while continuing to mix at slow speed until a medium stiff dough is formed.

The amount of egg whites added will vary depending on the dryness of the paste.

3. Remove the dough from the mixer. Using a kitchen spoon, place the dough in a pastry bag with a medium star tube and squeeze out onto prepared sheet pans. Form into various shapes and decorate with fruit or nuts.
4. Place on a rack and allow to dry for several hours. This will cause a crust to form, which is desirable for this kind of cookie.
5. Bake in a preheated oven at 375°F until light brown. Brush with a glaze as soon as they are removed from the oven. (Glaze: 1½ lb glucose and 8 oz water. Combine the two ingredients and bring to a boil, stirring occasionally with a kitchen spoon.)

Precautions

- Add the egg whites only until the desired dough consistency is reached.
- Do not overbake the cookies. Bake until they become light brown.
- When glazing the warm cookies, brush lightly so the cookies do not break.

Gingerbread Cookies

These are prepared from a smooth, stiff, fairly soft dough that is highly seasoned with ginger. The dough can be formed into bars or other shapes. They are mixed using the creaming method.

16 doz

Equipment

- Sheet pans
- Parchment paper
- Baker's scale
- Mixing machine and paddle
- French knife or cookie cutter
- Rolling pin
- Plastic scraper

Ingredients

1 lb hydrogenated vegetable shortening
2 lb sugar
1 lb brown sugar
¾ oz baking soda
3 oz dark molasses
½ oz salt
1 lb eggs
3 lb pastry flour
¼ oz ginger
¼ oz cinnamon

Preparation

1. Line the sheet pans with parchment paper.

Procedure

1. Place the shortening, sugar, brown sugar, baking soda, salt, and molasses in the stainless steel mixing bowl. Cream at slow speed using the paddle until smooth. Scrape down the bowl with the plastic scraper.
2. Add the eggs and continue to cream at slow speed until light and smooth.
3. Sift the flour, ginger, and cinnamon. Add to the mixture slowly. Mix at slow speed until the dough is very smooth. Scrape down the bowl with a plastic scraper at least once during this mixing stage.
4. Remove the dough from the mixer and divide into 1½ lb units. Refrigerate until chilled.
5. Remove one unit of dough from the refrigerator at a time. Roll out about ¼" thick on a floured piece of canvas.
6. Cut into bars with a French knife or cut out with a cookie cutter.
7. Place on prepared sheet pans and bake in a preheated oven at 400°F until they brown slightly.
8. Remove from the oven, let cool, and decorate with icing.

Precautions

- Chill the dough thoroughly before rolling.
- Roll the dough on a floured piece of canvas to avoid sticking.
- Do not overbake. Bake until light brown.
- Let cookies cool before removing them from the sheet pan.

Danish Butter Cookies

These are prepared from a soft, smooth, moist dough that is very rich in butter content. The cookies are formed by using a star tube in a pastry bag. They are mixed using the creaming method.

16 doz

Equipment

- Sheet pans
- Parchment paper
- Flour sifter
- Pastry bag and large star tube
- Baker's scale
- Plastic scraper
- Mixing machine and paddle
- Kitchen spoon

Ingredients

1 lb 8 oz sugar
12 oz hydrogenated vegetable shortening
12 oz butter
¼ oz salt
2 lb 6 oz cake flour
1 oz dry milk
⅛ oz baking powder
5 oz eggs
4 oz water
vanilla to taste

Preparation

1. Line the sheet pans with parchment paper.

Procedure

1. Place the sugar, shortening, butter, salt, vanilla, and dry milk in the stainless steel mixing bowl. Cream at slow speed using the paddle until light and fluffy. Scrape down the bowl with a plastic scraper.
2. Add the eggs and continue to mix at slow speed until well incorporated (approximately 3 min).
3. Add the water and blend in thoroughly, while continuing to mix at slow speed.
4. Sift the baking powder and flour, add to the mixture, and mix at slow speed until a smooth batter is formed. Scrape down the bowl with a plastic scraper.
5. Remove from the mixer and use a kitchen spoon to place the batter in a pastry bag with a fairly large star tube.
6. Squeeze out onto the prepared sheet pans. Form the cookies about the size of a quarter.
7. Bake in a preheated oven at 375°F until they start to brown.

Note: They can be decorated with chopped nuts, colored sugars, or by placing a cherry, raisin, icing, cinnamon candy, or pecan in the center of each cookie.

Precautions

- Do not overmix the batter. Mix only until smooth.
- Form the cookies as uniformly as possible so they bake evenly.
- Do not overbake. Remove from the oven when cookies become slightly brown. Avoid too much bottom heat.

Vanilla Wafers

These are prepared from a soft, smooth, moist dough flavored with vanilla. They are mixed using the creaming method.

10 doz

Equipment

- Baker's scale
- Sheet pans
- Parchment paper
- Plastic scraper
- Mixing machine and paddle
- Pastry bag and plain tube
- Kitchen spoon

Ingredients

1 lb sugar
10 oz butter
6 oz hydrogenated vegetable shortening
¼ oz salt
vanilla to taste
12 oz eggs
1 lb 6 oz pastry flour

Preparation

1. Line the sheet pans with parchment paper.

Procedure

1. Place the sugar, butter, shortening, salt, and vanilla in the stainless steel mixing bowl. Cream at slow speed using the paddle until smooth. Scrape down the bowl with a plastic scraper.
2. Add the flour gradually and mix at slow speed until smooth. Scrape down the bowl again while mixing in this stage.
3. Remove from the mixing machine and, using a kitchen spoon, place the batter in a pastry bag with a plain tube.
4. Squeeze out onto the prepared sheet pans. Form the cookies about the size of a quarter.
5. Bake in a preheated oven at 375°F until the edges start to brown.

Precautions

- Do not overmix the batter. Mix only until smooth.
- Form the cookies as uniformly as possible so they bake evenly.
- Do not overbake the cookies. Remove from the oven when the edges start to brown.
- Let the cookies cool before removing them from the sheet pan.

Coconut Drop Cookies

These are formed by depositing small mounds of batter on prepared sheet pans. This cookie will stay moist for a fairly long time if stored under proper conditions. The batter is mixed using the creaming method.

17 doz

Equipment

- Baker's scale
- Pastry bag
- Plastic scraper
- Mixing machine and paddle
- Kitchen spoon
- Sheet pans
- Parchment paper
- Flour sifter
- Spoon measures
- French knife

Ingredients

10 oz hydrogenated vegetable shortening
1 lb 2 oz sugar
1 tsp lemon juice
2 tsp vanilla
8 oz eggs
1 lb milk
1 lb 8 oz flour
1 oz baking powder
1 tsp salt
1 lb 8 oz coconut

Preparation

1. Chop the coconut fairly fine using a French knife.
2. Squeeze the juice from one lemon.
3. Line the sheet pans with parchment paper.

Procedure

1. Place the shortening, sugar, lemon juice, and vanilla in the stainless steel mixing bowl. Cream at slow speed using the paddle until light and fluffy. Scrape down the bowl with a plastic scraper.
2. Gradually add the eggs and continue to mix at slow speed until blended.
3. Add the milk and blend in thoroughly while continuing to mix at slow speed.
4. Sift the flour, baking powder, and salt. Add to the mixture with the coconut. Mix at slow speed until a fairly smooth batter is formed. Scrape down the bowl with a plastic scraper.
5. Remove from the mixer and place the batter in a pastry bag with an open tip approximately the size of a quarter.
6. Spot approximately ½ oz of dough into small mounds approximately 1½" apart on sheet pans covered with parchment paper.
7. Bake in a preheated oven at 375°F until golden brown.

Precautions

- Exercise caution when chopping the coconut.
- Add the milk gradually, while at the same time mixing at slow speed to avoid splashing.
- Do not overbake the cookies. Bake on the light side.
- Do not remove cookies from the sheet pan until they are cold.

Peanut Butter Cookies

These are prepared from a smooth, stiff, fairly soft dough that is very rich with peanut flavor. This dough has a consistency that is easy to roll and cut. It is mixed using the creaming method.

10 doz

Equipment

- Mixing machine and paddle
- Baker's scale
- Plastic scraper
- Sheet pans
- Parchment paper
- Rolling pin
- Pastry brush
- Hotel pan
- Cookie cutter or knife

Ingredients

8 oz peanut butter
1 lb 2 oz sugar
3 eggs
¼ oz baking soda
6 oz light cream
1 lb 5 oz cake flour
½ oz cream of tartar

Preparation

1. Line the sheet pans with parchment paper.

Procedure

1. Place the peanut butter and sugar in the stainless steel mixing bowl. Cream at slow speed using the paddle, add the eggs, and continue to mix.
2. Dissolve the baking soda in the cream. Add slowly to the peanut butter mixture while continuing to mix at slow speed.
3. Sift together the flour and cream of tartar, add to the mixture, and mix until a smooth dough has formed.
4. Remove the dough from the mixing bowl, place in a pan, and refrigerate for at least 1 hr.
5. Roll out until approximately ¼" thick. Brush the surface lightly with beaten eggs. Sprinkle the moistened surface with sugar.
6. Cut the dough into desired shapes and place on the prepared sheet pans.
7. Bake in a preheated oven at 375°F until light brown.

Precautions

- Let the cookies cool thoroughly before removing from the pan.
- Do not overbake.

SPONGE AND WHIPPING METHOD RECIPES

Chocolate Brownies

These are a rich, fudgy sheet cookie. The batter is moist and smooth with a slightly runny consistency. Nuts are generally added to the batter to enhance the eating qualities. They are mixed using the sponge or whipping method.

8 doz

Equipment

- Baker's scale
- Sheet pans
- Spatula
- Mixing machine and wire whip attachment
- French knife
- Plastic scraper
- Kitchen spoon
- 1 qt saucepan
- Kitchen spoon

Ingredients

1 lb 8 oz butter
1 lb bittersweet chocolate
3 lb sugar
1 lb 4 oz eggs
1 lb cake flour
1 lb pecans
vanilla to taste

Preparation

1. Chop the pecans fairly fine with a French knife.
2. Melt the chocolate and butter together in a saucepan.
3. Grease sheet pan slightly.

Procedure

1. Place the eggs and sugar in the stainless steel mixing bowl. Beat for approximately 10 min at high speed using the wire whip until the eggs become a lemon color.
2. Reduce the mixing speed to slow and pour in the melted butter/chocolate mixture. Mix until thoroughly incorporated. Scrape down the bowl with a plastic scraper.
3. Remove from the mixing machine and fold in the sifted flour with a kitchen spoon.
4. Add the vanilla and fold in the chopped pecans with a kitchen spoon.
5. Pour the batter onto a greased sheet pan. Spread evenly with a spatula.
6. Bake in a preheated oven at 350°F until slightly firm to the touch.
7. Remove from the oven and let cool. Cut into 2" squares with a French knife and remove from the pan with a spatula.

Precautions

- Do not overbake.
- Do not cut the cookies until they have cooled slightly.

Lady Fingers

These are prepared from a very light, fluffy sponge batter. The secret of a successful preparation lies in beating the eggs and sugar to a proper degree of stiffness and folding the flour in gently so as not to break the air cells. Lady fingers are used to decorate or set up special pastry items or as a special cookie. They are mixed using the sponge or whipping method.

9 doz

Equipment

- Baker's scale
- Sheet pans
- Parchment paper
- Pastry bag and large hole plain tube
- Mixing machine and wire whip attachment
- Plastic scraper
- Kitchen spoon
- Flour sifter

Ingredients

12 oz eggs
12 oz egg yolks
1 lb 8 oz sugar
2 oz glucose
¼ oz salt
¼ oz vanilla
1 lb 10 oz pastry flour
powdered sugar (10X), as
  needed

Preparation

1. Line the sheet pans with parchment paper.

Procedure

1. Place the eggs, egg yolks, sugar, salt, and glucose in the stainless steel mixing bowl. Beat using the wire whip until slightly thick and lemon-colored. Scrape down the bowl with a plastic scraper. Remove from the mixer.
2. Add the vanilla and fold in gently using a large kitchen spoon.
3. Sift the flour and fold in gently using a large kitchen spoon.

4. Place the mixture in a pastry bag with a large hole plain tube. Press out strips 2½″ long and ½″ wide on the prepared sheet pans.
5. Dust the tops of the lady fingers with sifted powdered sugar. Remove the excess sugar from the sheet pan.
6. Bake in a preheated oven at 400°F until very light brown. Let cool.
7. Remove the fingers from the paper and press two fingers together, forming a sandwich.

Precautions

- When folding the flour, incorporate thoroughly but do not overmix.
- After dusting the tops of the fingers with powdered sugar, bake immediately.
- Do not overbake the fingers. Bake only until they start to brown.
- If a problem develops when removing the fingers from the paper, moisten the back of the paper with water.

Nut Finger Wafers

Nut finger wafers are prepared from a soft, moist, smooth batter that is formed into finger shapes using a pastry bag and plain tube. They are mixed using the sponge or whipping method.

12 doz

Equipment

- Sheet pans
- Parchment paper
- Baker's scale
- Mixing machine and wire whip attachment
- Pastry bag and plain tube
- Plastic scraper
- Kitchen spoon
- French knife

Ingredients

1 lb 8 oz egg whites
1 lb 8 oz granulated sugar
1 lb 8 oz powdered sugar
1 lb 6 oz pecans or walnuts
4 oz cornstarch
¼ oz cinnamon

Preparation

1. Chop the pecans or walnuts very fine with a French knife.
2. Line the sheet pans with parchment paper.

Procedure

1. Place the egg whites in a mixing bowl. Beat at high speed using the wire whip until they start foaming.
2. Add the granulated sugar gradually while continuing to mix at high speed until the mixture forms a soft peak. Remove from the mixer. Scrape the bowl with a plastic scraper.
3. Blend the powdered sugar, cornstarch, cinnamon, and nuts. Fold into the meringue mixture with a kitchen spoon.
4. Place the mixture in a pastry bag with a plain tube. Squeeze out in finger shapes on to the prepared sheet pans.
5. Bake in a preheated oven at 275°F until slightly brown.

Precautions

- Do not overwhip the meringue. Whip only until soft peaks form.
- Form the cookies as uniformly as possible so they bake evenly.
- Do not overbake the cookies. Remove from the oven when they start to brown slightly.
- Let the cookies cool before removing them from the oven.

Trade Tip

A little corn syrup or glucose added to cookie dough, especially in hot weather, makes the dough roll easier and does not affect the quality of the finished product.

Rolls, Breads, and Sweet Doughs

Rolls, breads, and sweet doughs are versatile and can be served at any meal. The dough generally contains flour, liquid, shortening, sugar, eggs, salt, and yeast.

Flour supplies strength to the dough and acts as an absorbing agent. Liquid, usually milk, supplies moisture and helps form the gluten. Shortening supplies tenderness and improves the keeping qualities of the dough. Eggs supply structure to the dough and add color. Sugar supplies sweetness and acts as a stimulant to the yeast. Salt brings out the flavor and taste in the dough.

Proofing allows yeast to increase the dough volume to the required size before baking rolls, breads, and sweet doughs. Yeast growth is controlled mostly by temperature.

Fillings and toppings provide additional taste and eye appeal. Fillings are added prior to baking and toppings are applied prior to or after baking.

ROLLS, BREADS, AND SWEET DOUGHS

Rolls, breads, and sweet doughs are yeast-dough products commonly prepared in the food service industry. Variations of yeast-dough products, often classified as homemade, may add distinction and become a house specialty. In addition, baked homemade specialties often help sales of other preparations on the menu.

Rolls, breads, or sweet doughs can be served at any meal. Because of their popularity, it is important that the chef is familiar with the ingredients in dough and its reaction during the mixing, proofing, and baking periods.

Most dough recipes consist of flour (bread or pastry), liquid (dry milk and water or milk), shortening, sugar, eggs, salt, and yeast. Sweet dough usually has a spice, such as mace, and a flavor, such as vanilla, added to it. Each of these ingredients is important in producing successful rolls, breads, and sweet dough varieties.

Flour is one of the most important ingredients used in the preparation of rolls, breads, and sweet doughs. Wheat, from which flour is made, is the only grain that contains a high percentage of the protein *gluten*.

Gluten must be present in the flour used for successful rolls, breads, or sweet doughs. When gluten

507

and flour are mixed with water, the mixture gives the dough the strength to hold the gases produced by the yeast. Flour, when mixing and baking rolls, breads, and sweet doughs:

- supplies strength to the dough
- supplies structure to the baked product
- supplies nutritional value
- acts as an absorbing agent

Milk used in the preparation of rolls, breads, and sweet doughs may be used in liquid or dry form. The milk is usually specified as whole or skim milk in the recipe. Dry milk must be reconstituted in water before or during the mixing period. Milk, when used in doughs:

- improves the texture of the dough
- supplies moisture
- causes the gluten to form
- adds food value
- improves the flavor

Shortening when used in doughs:

- supplies richness and tenderness to baked products
- improves eating qualities
- improves grain and texture of baked products
- develops flaky layers in puff and Danish pastry
- improves keeping qualities of baked products

Flour, liquid, shortening, sugar, eggs, salt, and yeast are mixed for rolls, breads, and sweet doughs.

Most yeast-dough recipes call for hydrogenated shortening because it produces the best results.

Sugar, in granulated or syrup form, is usually used in yeast-dough recipes. Sugar, when used in doughs:

- supplies the necessary sweetness
- serves as a form of food to stimulate the growth of yeast
- supplies color to the baked product
- supplies moisture and helps prolong freshness
- helps provide a good grain and texture to the baked product

Eggs in yeast-dough products are used in whole form (yolk and whites). Both fresh or frozen eggs produce good results. Eggs, when used in doughs:

- supply the dough with added color
- add flavor to the baked product
- supply structure to the dough
- increase the volume of the dough
- improve the grain and texture of the product

Salt is used to season yeast-dough products. Salt, when used in doughs:

- improves the taste
- controls the yeast
- adds seasoning

Yeast, the leavening agent used in yeast-dough recipes, is available as *compressed yeast* and *dry yeast* (most of the moisture is removed). The yeast called for in the recipe is the one that should be used. However, if necessary, dry yeast may be substituted for compressed yeast. In this case only 40% of dry yeast by weight is used. For example, if the recipe calls for 1 lb (16 oz) of compressed yeast, then 6.4 oz are used (16 × 40% = 6.4 oz). The remaining 60% (9.6 oz) is made up of water (16 × 60% = 9.6 oz). If the dry yeast is purchased in small 1/4 oz packages (equivalent to 2/3 oz of compressed yeast), three packages must be used for every 2 oz of compressed yeast called for in the recipe.

In the preparation of yeast-dough products, the action of yeast must be carefully controlled. The amount of salt used in the recipe controls the yeast to some degree, but the greatest controlling factor is the temperature of the dough:

- Storage stage: 30°F to 40°F
- Slow action: 60°F to 75°F
- Normal action: 80°F to 85°F
- Fast action: 90°F to 100°F

- Action stops: 140°F

The yeast starts to grow when it is mixed with flour, water, and sugar. Yeast, as a microscopic plant, multiplies rapidly and produces carbon dioxide gas, which causes the dough to rise.

Yeast, when used in doughs:

- increases the volume
- improves the flavor
- adds grain and texture

DOUGH PRODUCTION

Steps to follow when producing dough for rolls, breads, and sweet doughs include:

1. *Scale* all ingredients correctly. A baker's scale is normally used.
2. *Mix* to develop the dough. An electric mixer is normally used.
3. *Knead* the dough. Work it smooth and force out all the air.
4. *Proof* the dough. Place in a lightly greased container and let rise to double in bulk.
5. *Punch* the dough by pressing it back to its original size. Place on a floured bench.
6. *Knead* a second time to remove all air.
7. *Scale* the dough into individual units.
8. *Make up* into desired shapes and sizes.
9. *Pan*: Place the units on prepared pans allowing space for proofing.
10. *Pan proof*: Let each unit rise to double in bulk. This is usually done in a proofing cabinet under proper moisture and temperature conditions (high moisture content and a temperature of 85°F to 90°F).
11. *Bake* at required temperatures until golden brown and done.
12. Store until needed.

ROLL DOUGH RECIPES

Soft Dinner Roll Dough No. 1

This is a white, smooth-textured dough that is easy to work with and can be used to prepare many different varieties of rolls. Soft dinner roll dough produces a baked product with excellent eating qualities.

16 doz rolls

Equipment

- Mixing machine and dough hook
- Baker's scale
- Plastic scraper
- Dough cutter (metal scraper may be used)
- 5 gal. stockpot
- Sheet pans
- Parchment paper
- Dough thermometer
- Proofing cabinet
- Kitchen spoon
- 1 gal. stainless steel container

Ingredients

- 1 lb sugar
- 1 lb 4 oz hydrogenated shortening
- 8 oz dry milk
- 2 oz salt
- 6 oz eggs
- 6 oz compressed yeast
- 4 lb cold water
- 7 lb bread flour

Procedure

1. Dissolve the yeast in the water. Place in a stainless steel container and stir with a kitchen spoon.
2. Place all the ingredients, including the dissolved yeast, in the stainless steel mixing bowl. Mix at medium speed using the dough hook until the gluten develops and the dough leaves the sides of the bowl and clings to the dough hook. Check the dough temperature with a dough thermometer. Temperature should be approximately 80°F.
3. Remove dough from the mixing bowl using a plastic scraper. Place on a floured bench and knead.
4. Place the dough in a greased container. Let rise until double in bulk.
5. Turn out on a floured bench and knead again. Cut the dough into 1¼ oz units using a dough cutter.
6. Form into rolls of desired shape. Dip in egg wash and sesame or poppy seeds if desired. Place the rolls 1″ apart on the parchment-covered sheet pans.
7. Proof in proofing cabinet until double in bulk.
8. Bake in a preheated oven at 375°F until golden brown.

Preparation

1. Prepare an egg wash (4 eggs to 1 pt of milk).
2. Cover sheet pans with parchment paper.
3. Preheat oven to 375°F.
4. Grease the inside of a 5 gal. stockpot.

Precautions

- Use cold water to control the yeast while mixing.
- Scale all ingredients. Double-check all weights.
- For best results, the dough temperature should be approximately 80°F when it is removed from the mixer.

Soft Dinner Roll Dough No. 2

This is a white, smooth-textured dough that is easy to work with. It can be used to create a variety of soft roll products that have excellent eye appeal and eating qualities.

16 doz rolls

Equipment

- Mixing machine
- Baker's scale
- Plastic scraper
- Dough cutter (metal scraper may be used)
- 5 gal. stockpot
- Sheet pans
- Parchment paper
- Dough thermometer
- Proofing cabinet
- Kitchen spoon
- 1 gal. stainless steel container

Ingredients

- 1 lb 4 oz sugar
- 1 lb 4 oz hydrogenated shortening
- 2 oz salt
- 6 oz dry milk
- 6 oz eggs
- 7 lb 8 oz bread flour
- 4 lb water
- 10 oz compressed yeast

Preparation

1. Prepare an egg wash (4 eggs to 1 pt of milk).
2. Cover sheet pans with parchment paper.
3. Preheat oven to 375°F.
4. Grease the inside of a 5 gal. stockpot.

Procedure

1. Dissolve the yeast in the water. Place in a stainless steel container and stir with a kitchen spoon.
2. Place all the ingredients, including the dissolved yeast, in the stainless steel mixing bowl. Mix at medium speed using the dough hook until the gluten develops and the dough leaves the sides of the bowl and clings to the dough hook. Check the dough temperature with a dough thermometer. Temperature should be approximately 80°F.
3. Remove dough from mixing bowl using a plastic scraper. Place on a floured bench and knead.
4. Place the dough in a greased container. Let rise until double in bulk.
5. Turn out on a floured bench and knead again. Cut the dough into 1¼ oz units with a dough cutter.
6. Form into rolls of desired shape. Dip in egg wash and sesame or poppy seed if desired. Place the rolls 1" apart on the parchment-covered sheet pans.
7. Proof in the proofing cabinet until double in bulk.
8. Bake in a preheated oven at 375°F until golden brown.

Precautions

- Use cold water to control the yeast while mixing.
- Scale all ingredients. Double-check all weights.
- For best results, the dough temperature should be approximately 80°F when it is removed from the mixer.
- Do not overproof the rolls.

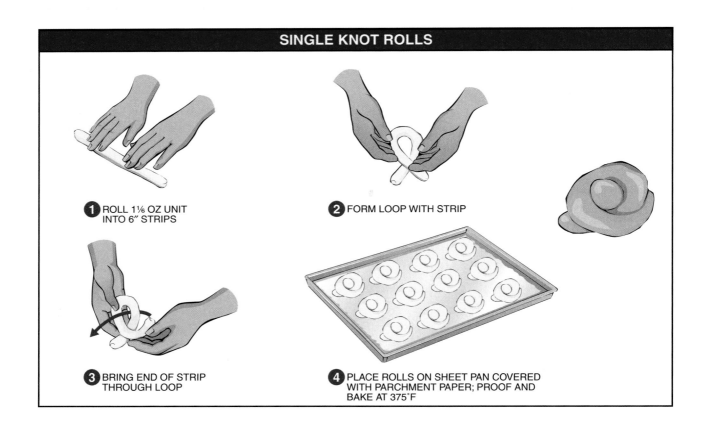

SINGLE KNOT ROLLS

1 ROLL 1⅛ OZ UNIT INTO 6" STRIPS

2 FORM LOOP WITH STRIP

3 BRING END OF STRIP THROUGH LOOP

4 PLACE ROLLS ON SHEET PAN COVERED WITH PARCHMENT PAPER; PROOF AND BAKE AT 375°F

SPLIT ROLLS

1 ROLL OUT 2 OZ UNIT, CUT INTO TWO PARTS AND FORM INTO BALLS

2 PLACE TWO BALLS FOR EACH ROLL INTO LIGHTLY GREASED MUFFIN TIN

3 PROOF THE ROLLS UNTIL DOUBLE IN BULK; BAKE AT 375°

CLOVERLEAF ROLLS

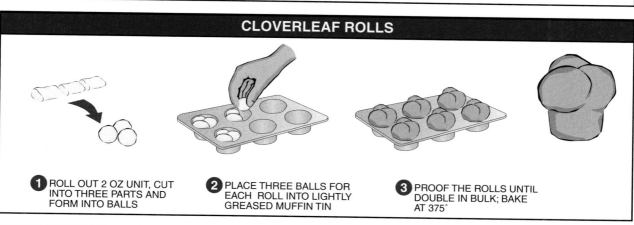

1 ROLL OUT 2 OZ UNIT, CUT INTO THREE PARTS AND FORM INTO BALLS

2 PLACE THREE BALLS FOR EACH ROLL INTO LIGHTLY GREASED MUFFIN TIN

3 PROOF THE ROLLS UNTIL DOUBLE IN BULK; BAKE AT 375°

FIGURE 8 OR TWIST ROLLS

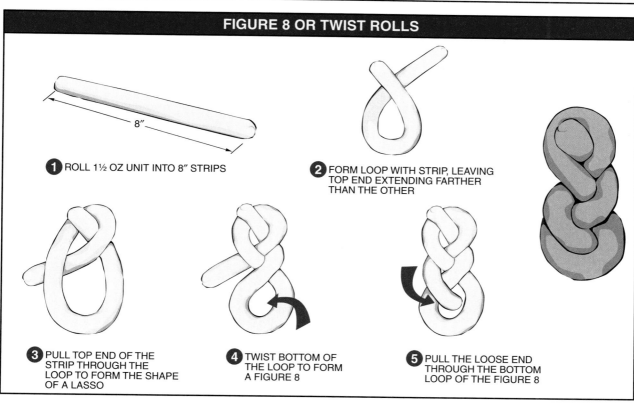

1 ROLL 1½ OZ UNIT INTO 8" STRIPS

2 FORM LOOP WITH STRIP, LEAVING TOP END EXTENDING FARTHER THAN THE OTHER

3 PULL TOP END OF THE STRIP THROUGH THE LOOP TO FORM THE SHAPE OF A LASSO

4 TWIST BOTTOM OF THE LOOP TO FORM A FIGURE 8

5 PULL THE LOOSE END THROUGH THE BOTTOM LOOP OF THE FIGURE 8

Soft Rye Dough

This dough can be used to prepare many different varieties of rolls. The dough is easy to work and the baked product is of excellent quality.

12 doz rolls

Equipment

- Mixing machine and dough hook
- Baker's scale
- Plastic scraper
- Dough cutter (metal scraper may be used)
- 5 gal. stockpot
- Sheet pans
- Parchment paper
- Dough thermometer
- Proofing cabinet
- Kitchen spoons
- 1 gal. stainless steel container

Ingredients

6 lb 6 oz bread flour
1 lb 4 oz dark rye flour
6 oz compressed yeast
1¾ oz salt
5 oz dry milk
1 lb hydrogenated shortening
1 lb sugar
1½ oz malt
4 lb 8 oz water (variable)
6 oz caraway seed
Note: If rye blend flour is used, combine the bread and rye flour to acquire the proper amount.

Procedure

1. Dissolve the yeast in the water. Place in a stainless steel container and stir with a kitchen spoon.
2. Place all the ingredients, including the dissolved yeast, in the stainless steel mixing bowl. Mix at medium speed using the dough hook until the dough has developed. Check dough temperature with a dough thermometer. Temperature should be 80°F.
3. Remove the dough from the mixing bowl using a plastic scraper. Place on a floured bench and knead.
4. Place the dough in a greased container, ferment for 1½ hr, punch, and allow to rest for 30 min.
5. Take to the bench and knead a second time. Scale into 1½ oz units. Cut the units with a dough cutter.
6. Form into rolls of desired shape and dip in egg wash. Place the rolls on the parchment-covered sheet pans.
7. Proof in the proofing cabinet until double in bulk.
8. Bake in a preheated oven at 400°F until golden brown.

Preparation

1. Prepare an egg wash (4 eggs to 1 pt of milk).
2. Cover sheet pans with parchment paper.
3. Preheat oven to 400°F.
4. Grease the inside of a 5 gal. stockpot.

Precautions

- Use cold water to control the yeast while mixing.
- Scale all ingredients correctly. Double-check all weights.
- For best results, the dough temperature should be approximately 80°F when it is removed from the mixer.
- Do not overproof the rolls.

Hard Roll Dough

This dough will produce rolls with a crisp, slightly hard crust. In order to produce a hard roll of good quality, steam must be injected into the oven for the first 10 min of the baking period.

12 doz rolls

Equipment

- Mixing machine and dough hook
- Baker's scale
- Sheet pans
- Plastic scraper
- Dough cutter (metal scraper may be used)
- Dough thermometer
- Proofing cabinet
- 5 gal. stockpot
- 1 gal. stainless steel container
- Kitchen spoon

Ingredients

7 lb 8 oz bread flour
3 oz salt
3½ oz sugar
3 oz shortening
3 oz egg whites
4 lb 8 oz water (variable)
4½ oz compressed yeast

Preparation

1. Sprinkle cornmeal on sheet pans.
2. Preheat oven to 400°F. Inject steam by turning on the steam valve (oven must be equipped to produce steam).
3. Grease the inside of a 5 gal. stockpot.

Procedure

1. Dissolve the yeast in the water. Place in a stainless steel container and stir with a kitchen spoon.

2. Place all the ingredients, including the dissolved yeast, in the stainless steel mixing bowl. Mix at medium speed using a dough hook for approximately 12 min until the dough is developed. Check dough temperature with a dough thermometer. Temperature should be approximately 80°F.

3. Remove the dough from the mixing bowl using a plastic scraper. Place on a floured bench and knead.

4. Place the dough in a greased container. Let rise for approximately 1 hr. Punch down and knead again.

5. Cut the dough with a dough cutter into 1½ oz units. Form into rolls of desired shape. Place the rolls on the prepared sheet pans.

6. Proof in the proofing cabinet until double in bulk.

7. Bake in a preheated oven at 400°F until golden brown. Have steam in the oven for the first 10 min of the baking period.

Precautions

- For best results, take the dough from the mixer at a temperature of 75°F to 80°F.
- Scale all ingredients. Double-check all weights.
- Exercise caution when opening the oven.
- Do not overproof the rolls.

Trade Tip — When a recipe for rolls or bread calls for steam in the oven and there is no steam attachment on the oven, a similar condition may be created by spraying a thin mist of water over the bottom and walls of the oven with a simple water sprayer. This should be done just before the item is placed in the oven.

Bagels

Bagels are of Jewish origin and are a popular item among all nationalities. In appearance, the bagel resembles a doughnut, but the taste and texture are very different. Bagels have a tough, heavy texture that is brought about by poaching the bagel slightly in water before it is baked.

20 bagels

Equipment

- Cloth
- Baker's scale
- Skimmer
- Mixing machine and dough hook
- Sheet pans
- Doughnut cutter
- Rolling pin
- Large braising pot
- Small saucepan
- Dough thermometer
- Plastic scraper

Ingredients

2 lb 8 oz bread flour (high gluten content)
1 lb 2 oz water (variable)
½ oz yeast
3 oz sugar
½ oz salt
1¼ oz hydrogenated shortening
2 oz eggs

Preparation

1. Fill the braising pot three-fourths full of water. Place on the range and bring to a simmer.

2. Melt the shortening in a small saucepan.

3. Preheat oven to 450°F.

4. Grease sheet pans lightly.

Procedure

1. Scale all ingredients into the mixing bowl. Mix at medium speed using the dough hook until a medium-firm dough is formed.

2. Remove dough from mixer at approximately 80°F and place on a floured bench.

3. Cover the dough with a cloth. Let it rest for approximately 30 min and then knead.

4. Using a rolling pin, roll out the dough on a floured bench to a thickness of ½″.

5. Cut out the bagels with the desired size doughnut cutter. Let bagels rest for 15 min.

6. Drop the bagels into the simmering water. When the bagels come to the surface, remove from the water using a skimmer.

7. Place the bagels on greased sheet pans. Bake in a preheated oven at 450°F until medium brown.

8. Remove from the oven and let cool before serving.

Precautions

- When cutting the bagels, dip the cutter in flour so it does not stick to the dough.
- Use caution when poaching the bagels.

Croissants

Croissants are rich, flaky rolls in a crescent shape. The dough is given three rolls and three folds each time it is rolled. This dough has exceptionally fine eating qualities. The dough may also be used to form butter flake rolls.

8 doz croissants

Equipment

- Baker's scale
- Mixing machine and dough hook
- Plastic scraper
- Rolling pin
- Dough thermometer
- Pastry wheel
- Muffin tins or sheet pans
- Proofing cabinet
- 1 gal. stainless steel container
- Wire whip
- Kitchen spoon

Ingredients

9 oz sugar
1¼ oz salt
8 oz butter-flavored vegetable shortening
4 oz dry milk
10 oz egg yolks
4 oz compressed yeast
1 lb 12 oz water, 80°F
4 lb bread flour (variable)
1 lb butter-flavored vegetable shortening (roll-in)

Preparation

1. Grease muffin tins or cover sheet pans with parchment paper. The pan used depends on the type of roll being made.
2. Preheat oven to 400°F.

Procedure

1. Place the egg yolks in a stainless steel container, beat slightly with a wire whip, add the water, and blend together.
2. Dissolve the yeast in the water/egg yolk mixture by stirring with a kitchen spoon.
3. Place all the dry ingredients, including the first amount of butter-flavored vegetable shortening, in the stainless steel mixing bowl.
4. Add the liquid mixture while mixing at slow speed using the dough hook. Increase mixing speed to medium and mix until a smooth dough is formed. Scrape down the bowl at least once with a plastic scraper during the mixing period.
5. Remove the dough from the mixer at approximately 75°F. Check the temperature with a dough thermometer. Place the dough on a floured bench, knead slightly, and let it rest for 40 min.
6. Roll the dough with a rolling pin into an oblong shape ½" thick. Cover two-thirds of the dough with the roll-in shortening or butter. Fold the uncovered third of the dough toward the center and fold the other third over it toward the center.
7. Roll dough again into a ½" thick oblong shape and fold as before. Repeat this process for a total of three rolls, with three folds to each roll.
8. Place the dough in the freezer for several hours. Return to the bench and prepare the croissants.

Precaution

- The shortening or butter to be rolled in must be slightly soft. If it is too hard, it will break through the dough, causing an inferior product.

French Croissants

8 doz croissants

Roll the dough fairly thin into a rectangular shape. Cut with a pastry wheel into strips about 4" wide, then cut into 4" squares. Cut the squares into 2 triangular shapes. Wash with a mixture of half egg and half milk. Roll or twist the dough and shape into a crescent. Brush the top with the egg wash and sprinkle with poppy seed or sesame seed. Place on the sheet pans, give a three-fourth proof, and bake in a preheated oven at 400°F until golden brown. Remove from the oven and brush with melted butter.

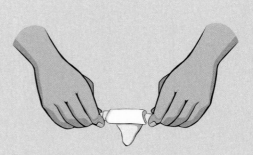

1 CUT INTO TRIANGULAR SHAPES AND ROLL FROM THE WIDE END TO THE POINT

Trade Tip

If there is doubt whether yeast is still active and can be used, drop some of the yeast in a container of warm water. If it rises to the surface, it is still usable and will function properly in a dough.

BREAD RECIPES

Golden Miniature Bread Loaves Dough

These are made from a golden, smooth-textured dough. They have excellent eating qualities and are commonly served before and during dinner.

40 miniature loaves

Equipment

- Mixing machine and dough hook
- Baker's scale
- Rolling pin
- Miniature loaf pans
- Plastic scraper
- Proofing cabinet
- Saucepan
- Kitchen spoon
- 5 gal. stockpot
- Dough cutter (metal scraper)

Ingredients

1 lb sugar
1 lb golden butter-flavored shortening
12 oz eggs
5 oz salt
8 oz dry milk
10 oz yeast
5 lb water
9 lb bread flour
1 lb pastry flour

Preparation

1. Grease miniature loaf pans lightly with shortening.
2. Grease the inside of a 5 gal. stockpot.
3. Preheat oven to 400°F.
4. Set proofing cabinet for proper heat and humidity.
5. Melt butter or margarine in a saucepan.

Procedure

1. Place 2 lb (1 qt) of the water in a saucepan and heat until lukewarm. Add the yeast and dissolve by stirring with a kitchen spoon.
2. Place all the ingredients, including the dissolved yeast, in the stainless steel mixing bowl. Mix at medium speed using the dough hook until the gluten develops and the dough leaves the sides of the bowl and clings to the dough hook.
3. Remove dough from mixing bowl using a plastic scraper. Place on a floured bench and knead.
4. Place the dough in a greased container. Let the dough ferment approximately 1½ hr or for one full rise. Punch and allow dough to relax 20 min before making up. If desired, after kneading, the dough can be placed in the coldest part of the refrigerator and held for makeup the next day.
5. Turn the dough out on a floured bench and knead a second time. Cut the dough into 7 oz units using a dough cutter.
6. Using a rolling pin, roll out each unit of dough approximately 5" wide and 6" to 7" long. Roll by hand and place seam down in the greased miniature loaf pans.
7. Proof in the proofing cabinet until double in bulk.
8. Bake in a preheated oven at 400°F for 30 min to 35 min or until golden brown.
9. Brush each loaf with melted butter or margarine halfway through the baking period. Repeat this process when the miniature loaves are removed from the oven.

Note: If larger loaves are desired, use large loaf pans and scale each unit of dough 1 lb 2 oz.

Precautions

- Scale all ingredients. Double-check all weights.
- Do not overproof or underproof the loaves.
- Exercise caution when brushing the loaves with butter or margarine as the pans are very hot.

Whole Wheat Dough

Whole wheat dough can be used to prepare a variety of whole wheat rolls or bread. The dough is easy to work and the finished baked product has excellent eating qualities.

9 doz rolls or 22 miniature loaves

Equipment

- Mixing machine and dough hook
- Baker's scale
- Sheet pans
- Plastic scraper
- Dough cutter (metal scraper)
- Dough thermometer
- Proofing cabinet
- 5 gal. stockpot
- Saucepan
- Kitchen spoon

Ingredients

5 lb whole wheat flour
8 oz hydrogenated shortening
2 oz salt
8 oz sugar
4 oz dry milk
4 oz yeast
3 lb 6 oz water

Preparation

1. Prepare an egg wash (4 eggs to 1 pt of milk).
2. Cover sheet pans with parchment paper.
3. Preheat oven to 400°F.
4. Grease the inside of a 5 gal. stockpot.
5. Set proofing cabinet for proper heat and humidity.

Procedure

1. Place 1 lb (1 pt) of the water in a saucepan and heat until lukewarm. Add the yeast and dissolve by stirring with a kitchen spoon.
2. Place all the ingredients, including the dissolved yeast, in the stainless steel mixing bowl. Mix at medium speed using the dough hook for approximately 12 min or until the dough develops. The desired dough temperature is 80°F.
3. Remove dough from mixing bowl using a plastic scraper. Place on a floured bench and knead.
4. Place the dough in a greased container. Let the dough ferment 1 hr or for one full rise. Punch the dough down and take to the bench for makeup.

5. Turn the dough out on a floured bench and knead a second time. If making rolls, cut the dough into 1½ oz units using a dough cutter. If making miniature loaves, cut into 7 oz units.
6. Form into rolls of desired shape or roll out and form miniature loaves. Dip rolls into egg wash and then into sesame seeds. Place the rolls on the parchment-covered sheet pans. If forming miniature loaves, place them in lightly greased miniature loaf pans.
7. Proof in the proofing cabinet until double in bulk.
8. Bake in a preheated oven at 400°F until golden brown.

Precautions

- Scale all ingredients. Double-check all weights.
- Make sure the water for dissolving the yeast is just lukewarm. If it is too hot, it will stop the action of the yeast.
- For best results, the dough temperature should be approximately 80°F when the dough is removed from the mixer.
- Do not overproof the rolls.

FORMING LOAVES OF BREAD

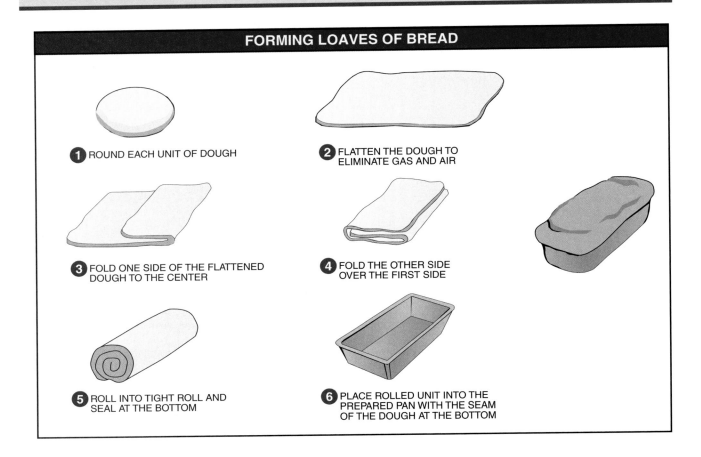

1 ROUND EACH UNIT OF DOUGH

2 FLATTEN THE DOUGH TO ELIMINATE GAS AND AIR

3 FOLD ONE SIDE OF THE FLATTENED DOUGH TO THE CENTER

4 FOLD THE OTHER SIDE OVER THE FIRST SIDE

5 ROLL INTO TIGHT ROLL AND SEAL AT THE BOTTOM

6 PLACE ROLLED UNIT INTO THE PREPARED PAN WITH THE SEAM OF THE DOUGH AT THE BOTTOM

Vienna Bread Dough

This dough produces a loaf with a crisp, slightly hard crust. It can be baked without steam and still produce a quality product. The dough can be formed into regular or miniature loaves.

16 regular loaves or 32 miniature loaves

Equipment

- Baker's scale
- Mixing machine and dough hook
- Sheet pans
- Kitchen spoon
- Dough cutter (metal scraper may be used)
- Parchment paper
- Proofing cabinet
- 5 gal. stockpot
- Stainless steel bowl

Ingredients

4 lb water
4 oz compressed yeast
2 oz sugar
2 oz salt
7 lb bread flour
2 oz dry milk
3 oz egg whites
6 oz hydrogenated shortening

Preparation

1. Cover pans with parchment paper. Dust surface of paper with yellow cornmeal.
2. Preheat oven to 375°F.
3. Lightly grease the inside of a 5 gal. stockpot.
4. Set proofing cabinet for proper heat and humidity.

Procedure

1. Place the yeast in a stainless steel bowl. Add part of the water and stir with a kitchen spoon until thoroughly dissolved.
2. Place all ingredients, including the dissolved yeast, in the stainless steel mixing bowl. Mix at slow speed using the dough hook until all ingredients are blended.
3. Increase machine speed to medium and continue to mix until the dough has developed.
4. Remove the dough from the mixing bowl. Place on a floured bench and knead slightly.
5. Place the dough in a greased container and ferment for approximately 1 hr to 1½ hr. Punch down and knead a second time.
6. Cut the dough into 12 oz units with a dough cutter for a regular loaf or into 6 oz units for a miniature loaf.
7. On a slightly floured bench, roll each unit of dough into a fairly long, slender loaf shape. Place on the prepared sheet pans.
8. Proof in the proofing cabinet until double in bulk.
9. Bake in a preheated oven at 375°F until golden brown and surface is crisp.

Precautions

- Scale all ingredients. Double-check all weights.
- Do not overproof the loaves.
- Have the water fairly cold to control the yeast.

Double-Rich Bread Dough

This is a white, smooth-textured dough that is easy to form into loaves. The dough produces a tender, rich-tasting, golden crust loaf of bread.

10 loaves

Equipment

- Baker's scale
- Mixing machine and dough hook
- Bread pans
- Kitchen spoon
- Dough cutter (metal scraper may be used)
- Parchment paper
- Proofing cabinet
- 5 gal. stockpot
- Stainless steel bowl

Ingredients

- 8 oz sugar
- 8 oz hydrogenated shortening
- 2 oz salt
- 6 oz dry milk
- 1 oz malt powder
- 4 lb water
- 2½ oz compressed yeast
- 6 lb 4 oz bread flour

Preparation

1. Grease bread pans lightly.
2. Preheat oven to 375°F.
3. Lightly grease the inside of a 5 gal. stockpot.
4. Set proofing cabinet for proper heat and humidity.

Procedure

1. Place the sugar, shortening, salt, dry milk, and malt powder in the stainless steel mixing bowl. Mix at slow speed using the dough hook until thoroughly blended.
2. Place the yeast in a stainless steel bowl. Add the water and stir with a kitchen spoon until thoroughly dissolved. Add this mixture to the blended ingredients in the mixing bowl.
3. Add the flour and mix at slow speed until all ingredients are blended.
4. Increase machine speed to medium and continue to mix until the dough has developed.
5. Remove the dough from the mixing bowl. Place on a floured bench and knead slightly.
6. Place the dough in a greased container and let rise until double in bulk. Punch down and let rest 20 min.
7. Turn the dough out of the greased container onto a floured bench. With a dough cutter, cut the dough into 1 lb 4 oz units.
8. Round each unit of dough into a fairly tight ball. Cover with a cloth and let rest 10 min.
9. Form into loaves and place in the prepared loaf pans.
10. Place the loaf pans in the proofing cabinet and proof until they double in bulk and the pan is almost full.
11. Bake in a preheated oven at 375°F until golden brown and firm to the touch.

Precautions

- Scale all ingredients. Double-check all weights.
- Do not overproof the loaves.
- Have the water fairly cold to control the yeast.

Trade Tip

Steam is necessary during the proofing of yeast goods and, if possible, apply a little during the baking period. Applying steam keeps the baked goods soft and fresh longer and makes them more palatable.

SWEET DOUGH RECIPES

Virginia Pastry Dough

Virginia pastry dough is easy to make, easy to work with, and easy to sell. It is an unusual yeast-dough mixture because it requires very little fermentation. The dough can be mixed and made up without waiting for gases to react.

depends upon product made

Equipment

- Baker's scale
- Mixing machine and sweet dough paddle
- Plastic scraper
- Rolling pin
- Baking pans
- 1 gal. stainless steel container
- Wire whip
- Pastry wheel
- Proofing cabinet
- Dough thermometer
- Kitchen spoon
- Pastry brush
- Boning knife

Ingredients

1 lb 8 oz pastry flour
12 oz bread flour
4 oz dry milk
1 lb 12 oz water
1 lb eggs
6 oz compressed yeast
8 oz sugar
2 lb 8 oz emulsified vegetable shortening
1½ oz salt
mace to taste
vanilla to taste
2 lb 4 oz bread flour

Procedure

1. Place the eggs in a stainless steel container. Beat slightly with a wire whip. Add the water and blend.
2. Dissolve the yeast in the water/egg mixture by stirring with a kitchen spoon. Place in the stainless steel mixing bowl.
3. Add the first amount of bread flour, pastry flour, and dry milk. Mix at medium speed using the sweet dough paddle 2 min to 3 min.
4. Add the sugar, shortening, salt, mace, vanilla, and the second amount of bread flour. Mix at low speed for about 2 min until all the ingredients are thoroughly blended. Scrape down the bowl at least once with a plastic scraper.
5. Remove the dough from the mixer at 65°F or below. Check temperature with a dough thermometer. Take directly to the bench and make into the desired product.

Preparation

1. Prepare the baking pans needed for product selected to be made.
2. Preheat oven to required temperature.

Precautions

- The dough will be easier to handle and will produce the best results if the temperature of the dough is kept on the cool side (under 65°F).
- Do not overmix the dough.

Breakfast Cake

15 units

Roll out an 8 oz to 10 oz piece of the dough with a rolling pin to cover the bottom of an 8″ cake pan, building the edges slightly. Place 12 oz of desired fruit filling in the center. Proof in a proofing cabinet for about ½ hr. Bake in a preheated oven at 380°F for 20 min. After removing from the oven, cool and wash over the filling with a light glaze if desired. Edges may be iced with roll icing.

Shortcake Biscuits

115 units

Roll out the dough with a rolling pin to ½″ thickness. Cut out with a 2″ to 2½″ round cutter. Using a pastry brush, wash the tops with skim milk and turn upside down on sugar. Place right side up on bun pans. Proof in a proofing cabinet for ½ hr. Bake in a preheated oven at 400°F for approximately 10 min. Each biscuit should be scaled at 1½ oz.

Fruit Crisp

80 units

Roll out the dough with a rolling pin to about ⅛″ thickness. Cut into 4″ squares with a pastry wheel. Place desired fruit filling in the center of each square. Using a pastry brush, wash edges with skim milk. Fold to form either triangles or rectangles and seal the edges. Turn upside down on sugar. Pierce the center of each crisp to allow steam to escape from the filling. Proof about ½ hr. Bake in a preheated oven at 380°F to 400°F for approximately 10 min. A flat or roll icing may be applied on unsugared crisps after baking.

Turnabouts

70 units

Roll out the dough with a rolling pin to about ¼″ thickness. Cut into 4″ squares with a pastry wheel. Fold each corner to the center, wash with skim milk using a pastry brush, and turn upside down on sugar. Place right side up on bun pans. Proof in a proofing cabinet for about ½ hr. Bake in a preheated oven at 380°F to 400°F. After baking, spot the center of each piece with a small amount of jam.

Concertinas

40 units

Roll out the dough with a rolling pin to about ⅛″ thickness. Spread the surface with coffee cake filling and roll up as for cinnamon rolls. Flatten slightly by hand and slice with a boning knife into units 3″ to 4″ in length. Make three cuts in each piece about three-fourths the way through. Using a pastry brush, wash with skim milk and invert on sugar. Place right side up on bun pans. Proof in a proofing cabinet about ½ hr. Bake in a preheated oven at 380°F to 400°F.

It is traditional to serve hot cross buns during the Lenten season, and usually that is the only time of the year they appear on the menu. The custom of serving them during the Lenten season started in England many years ago. The buns are prepared from a sweet dough containing assorted fruit and raisins. They are marked with an icing cross after baking.

Hot Cross Buns

11 doz

Equipment

- Mixing machine and sweet dough paddle
- Baker's scale
- Plastic scraper
- Sheet pans
- Parchment paper
- Proofing cabinet
- Dough thermometer
- 5 gal. stockpot
- Dough cutter (metal scraper may be used)

Ingredients

14 oz sugar
2 oz malt powder
1 lb emulsified shortening
1 oz salt
12 oz eggs
2 lb (1 qt) milk (variable)
8 oz compressed yeast
vanilla to taste
3 lb bread flour
1 lb 8 oz cake flour
1 lb soaked, dried raisins
1 lb currants
4 oz citron
4 oz orange peel
4 oz chopped cherries
mace to taste

Procedure

1. Place the sugar, malt, mace, shortening, and salt in the stainless steel mixing bowl. Cream at slow speed using the sweet dough paddle until smooth and light.
2. Gradually add the eggs while continuing to mix at slow speed.
3. Dissolve the yeast in the milk. Add the vanilla. Pour slowly into the mixing bowl and continue to mix at slow speed.
4. Add the flours. When mixed to a dough, increase machine speed to medium. Mix until dough is fairly smooth.
5. Add the raisins, currants, citrons, orange peel, and cherries. Mix until well-blended. Remove dough from mixer at about 80°F. Test with dough thermometer.
6. Remove dough from mixing bowl using a plastic scraper. Place on a floured bench and knead.
7. Place the dough in a greased container. Give a three-fourths to full rise and then punch down.
8. Turn out onto a floured bench and knead again. Cut the dough into 1½ oz units using a dough cutter.
9. Roll each unit of dough into a firm ball. Place on parchment-covered sheet pans approximately 1″ apart.
10. Place in the proofing cabinet. Proof until rolls are almost double in bulk.
11. Bake in a preheated oven at 375°F until golden brown.
12. Remove from the oven and let cool. Form an icing cross on each bun.

Precautions

- For best results, remove dough from mixer at approximately 80°F.
- Cover the dough when proofing in greased container.
- Do not overproof the dough.
- Rolls must be cooled thoroughly before icing.

Trade Tip

Adding cream cheese to certain types of yeast doughs will improve the taste and flavor of the dough. If this is done, slightly increase the amount of yeast and decrease the amount of shortening.

A dough cutter is used to cut sweet rolls.

Special Hot Cross Bun Icing

1 qt

Ingredients

5 lb powdered sugar
½ oz salt
6 oz emulsified shortening
¾ oz plain gelatin
12 oz water (variable)

Procedure

1. Place the sugar, salt, and shortening in the stainless steel mixing bowl. Cream at slow speed using the icing whip.
2. Dissolve gelatin in the water. Add slowly to mixture and mix at slow speed.
3. When incorporated, increase machine speed to medium and mix until icing is smooth.

Precaution

- Dissolve gelatin thoroughly in water before adding.

Sweet Dough

This is a rich, flavorful yeast dough. It has a golden yellow color and can be used to produce such popular baked products as sweet rolls, coffee cakes, pecan rolls, and cinnamon buns.

depends upon product made

Equipment

- Baker's scale
- Mixing machine and sweet dough paddle
- Plastic and metal scraper
- Rolling pin
- Baking pans
- Pastry wheel
- 5 gal. stockpot
- Proofing cabinet
- Dough thermometer
- Wire whip
- Kitchen spoon
- 1 gal. stainless steel container
- Pastry brush

Ingredients

1 lb sugar
1 lb golden butter-flavored shortening
1 oz salt
3 lb bread flour
1 lb 8 oz pastry flour
12 oz eggs
4 oz dry milk
2 lb water (variable)
8 oz compressed yeast
mace to taste
vanilla to taste

Preparation

1. Prepare the baking pans for the product selected.
2. Preheat the oven to the required temperature.
3. Grease the 5 gal. stockpot.

Procedure

1. Place all the dry ingredients, including the shortening, in the stainless steel mixing bowl.
2. Place the eggs in a stainless steel container. Beat slightly with a wire whip. Add the water and blend. Add the yeast and stir with a kitchen spoon until thoroughly dissolved.
3. Mix at slow speed using the sweet dough paddle while adding the liquid mixture. Increase speed to medium and mix until a smooth dough is formed. Scrape down the bowl with a plastic scraper at least once during the mixing period.
4. Add the vanilla and mix at slow speed. Check dough temperature with a dough thermometer. Temperature should be approximately 78°F to 85°F.
5. Turn the dough out on a floured bench and knead until all air is worked out and the dough is smooth.
6. Place in a greased container and proof until double in bulk.
7. Turn out onto a floured bench and knead again.
8. Make into desired product.

Precautions

- Remove the dough from the mixer at 78°F to 85°F for best results.
- In making larger batches of sweet dough, if a longer time is required on the bench, the amount of yeast used may have to be reduced.
- Do not overproof the dough.

FORMING SWEET ROLLS

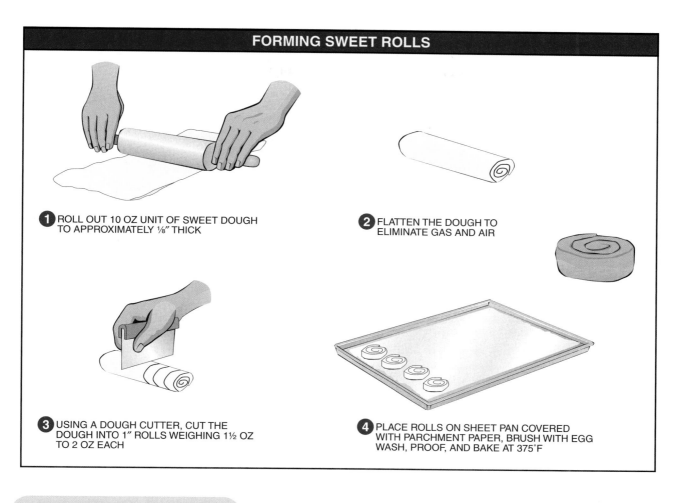

1 ROLL OUT 10 OZ UNIT OF SWEET DOUGH TO APPROXIMATELY ⅛″ THICK

2 FLATTEN THE DOUGH TO ELIMINATE GAS AND AIR

3 USING A DOUGH CUTTER, CUT THE DOUGH INTO 1″ ROLLS WEIGHING 1½ OZ TO 2 OZ EACH

4 PLACE ROLLS ON SHEET PAN COVERED WITH PARCHMENT PAPER, BRUSH WITH EGG WASH, PROOF, AND BAKE AT 375°F

Sweet Rolls

80 units

Using a rolling pin, roll out 10 oz of dough to a thickness of about ⅛″ and about 8″ wide. Spread with desired filling or brush dough slightly with water and sprinkle with a mixture of cinnamon and sugar. Roll up the dough and cut with a dough cutter into 1″ rolls weighing from 1½ oz to 2 oz each. Brush with egg wash using a pastry brush and proof in a proofing cabinet until double in bulk. Bake in a preheated oven at 375°F until golden brown. Remove from the oven and brush with a glaze consisting of 1 lb glucose and ½ lb water brought to a boil. Let rolls cool and ice with roll icing.

Trade Tip

The American Institute of Baking in Manhattan, KS offers a variety of services related to baking. Visit their web site at www.aibonline.org.

Coffee Cake

12 to 14 units

Scale the dough into 10 oz or 12 oz units. Roll out with a rolling pin to a thickness of ¼". Spread with desired filling, roll up the dough, and form into desired shapes. Proof in proofing cabinet until double in bulk. Bake at 375°F until golden brown. Remove from the oven and brush with a glaze. Let the coffee cakes cool and ice with roll icing.

Cinnamon Sugar Buns

80 units

Scale enough dough to cover a sheet pan at a thickness of ½". Dock the dough by tapping it with a docker or the tines of a dinner fork. Brush the top of the dough slightly with water and sprinkle on a mixture of cinnamon and sugar (1 part cinnamon to 3 parts sugar). Proof in the proofing cabinet until double in bulk. Bake in a preheated oven at 375°F until golden brown. Remove from the oven, let cool, and cut with a pastry cutter into 3" squares.

Pecan Rolls

100 units

Prepare a caramel pan smear by blending:

4 lb dark brown sugar
1 lb 8 oz butter-flavored liquid shortening
½ oz salt
12 oz honey or glucose
8 oz milk

Place the pan smear in a pastry bag and cover the bottom of muffin tins about ¼" deep with the smear. Sprinkle a few chopped pecans over the pan smear in each tin. Grease the sides of each tin lightly. Scale the sweet dough into 1½ oz units, shape by hand, and place the dough in the muffin tins. Place in the proofing cabinet and proof until double in bulk. Bake in a preheated oven at 375°F until golden brown. Remove from the oven and turn the pans upside down on a sheet pan. Remove the muffin tins. Let the rolls cool.

Danish Pastry Dough

This is a rich, tender, flaky dough that produces baked products with exceptional eating qualities. It is prepared by mixing a basic sweet dough and rolling in extra shortening to create the tender flaky layers, which are the chief characteristics of a good quality Danish. The dough can be used to make a variety of products such as rolls, coffee cakes, and specialty items.

depends upon product made

Equipment

- Mixing machine and sweet dough paddle
- Baker's scale
- Plastic and metal scraper
- Rolling pin
- Dough thermometer
- Pastry brush
- Proofing cabinet
- Baking pans
- 1 gal. stainless steel container
- Wire whip
- Kitchen spoon
- Bench brush

Ingredients

12 oz sugar
12 oz butter-flavored shortening
1¾ oz salt
3 lb bread flour
1 lb 8 oz pastry flour
1 lb eggs
4 oz dry milk
2 lb water (variable)
8 oz compressed yeast
flavor to taste
2 lb 8 oz butter-flavored shortening (roll-in)

Preparation

1. Prepare the pans selected for baking.
2. Preheat oven to 390°F.

Procedure

1. Place all the dry ingredients, including the shortening, in the stainless steel mixing bowl.
2. Place the eggs in the stainless steel container, beat slightly with a wire whip, add the water, and blend. Add the yeast and stir with a kitchen spoon until thoroughly dissolved. Add the flavoring.
3. Mix at slow speed using the sweet dough paddle while adding the liquid mixture. Increase speed to medium and mix for approximately 4 min to 5 min. Scrape down the bowl with a plastic scraper at least once during the mixing period.
4. Remove the dough from the mixer at 70°F to 75°F. Check with a dough thermometer. Place on a sheet pan (roll out with a rolling pin until the dough fills the pan) and allow it to rest in the refrigerator for approximately 30 min.
5. Bring the dough from the refrigerator and place it on a floured bench. Roll the dough with a rolling pin into an oblong shape ½" thick. Cover two-thirds of the dough with the roll-in butter-flavored vegetable shortening. Fold the uncovered third of the dough toward the center and fold the other third over it toward the center. Brush off any excess flour with a bench brush. Roll dough again into ½" thick oblong shape and fold as before. Place the dough on a sheet pan and retard in the refrigerator for 20 min.
6. Repeat step 5 for a total of three rolls with three folds to each roll. Let the dough rest in the retarder for 20 min between each roll.
7. Return the dough to the bench and make into desired product.

Precautions

- Bring the dough from the mixer at 70°F to 75°F for best results.
- Scale all ingredients. Double-check all weights.
- When rolling in the shortening, dust off any flour that may be present on the surface of the dough before folding.
- The shortening to be rolled in must be slightly soft. If it is too hard, it will break through the dough, causing an inferior product.

- Danish pastry dough must be properly proofed to ensure good results. Test the dough for proper proofing by using a finger to make an indentation in the dough. If the dough closes around the indentation, it is properly proofed. If the dough does not close around the indentation, continue proofing. If the dough is overproofed, it will collapse slightly when indented.
- Place units to be baked on parchment paper to prevent sticking.

Danish Rolls

100 units

Roll out 10 oz of dough with a rolling pin to a thickness of about 1/8″ and about 8″ wide. Spread with desired filling. Roll up the dough and cut with a metal scraper into 1″ rolls weighing from 1½ oz to 2 oz each. Brush with an egg wash using a pastry brush. Proof until double in bulk. Bake in a preheated oven at 390°F until golden brown. Remove from the oven and brush with a glaze consisting of 1 lb glucose and ½ lb water brought to a boil. Let the rolls cool and ice with roll icing.

Fruit-Filled Coffee Cake

16 units

Scale the dough into 10 oz or 12 oz units. Roll out with a rolling pin to about 1/8″ thick, 14″ long, and 7″ to 8″ wide. Spread fruit filling on half of the dough. Wash edges with water and fold unfilled portion over. Seal edges securely. Using a rolling pin, press down to indent surface slightly. Place unit in a prepared square cake pan. Dock the surface of each unit with a docker or with the tines of a fork. Brush with egg wash, top with streusel topping, and proof until double in bulk. Bake in a preheated oven at 390°F until golden brown. Remove from the oven, let cool, and sprinkle with roll icing.

Spiral Coffee Cake

20 units

Scale the dough into 10 oz units. Roll out with a rolling pin to about 1/4″ thick, 20″ long, and 3″ wide. Spread the desired filling over half of the dough. Wash edges with water, fold unfilled portion over, and seal edges securely. Twist by rolling the dough on the bench with the palms of the hands. Coil the twisted dough strip flat on the bench and seal the outside end. Place coffee cakes on sheet pans covered with parchment paper or in 8″ prepared cake pans. Brush with egg wash using a pastry brush. Proof in proofing cabinet until double in bulk. Bake in a preheated oven at 390°F until golden brown. Remove from the oven, brush with a hot corn syrup (1 lb glucose and ½ lb of water brought to a boil), and sprinkle with roll icing.

Honey Fruit Coffee Cake

16 units

Scale the dough into 10 oz to 12 oz units. Roll out with a rolling pin to a thickness of about 1/8″ and about 8″ wide. Spread honey fruit filling on dough and roll two sides toward the center until about 1″ apart. With scissors, cut each side of the roll about 1″ apart. Fill the center with desired filling. Brush with egg wash using a pastry brush and proof until double in bulk. Bake in a preheated oven at 390°F until golden brown. Remove from the oven, brush with a hot corn syrup glaze (1 lb glucose and ½ lb water brought to a boil), and sprinkle with roll icing.

Fruit Cluster Coffee Cake

16 units

Scale the dough into 10 oz to 12 oz units. Roll out with a rolling pin to a thickness of about 1/8″ and about 8″ wide. Spread desired fruit filling on dough and roll up as for sweet rolls. Cut rolls with scissors about every inch, spreading cuts in alternate directions. Brush with egg wash and proof in proofing cabinet until double in bulk. Bake in a preheated oven at 390°F until golden brown. Remove from the oven, brush with a hot corn syrup glaze (1 lb glucose and ½ lb water brought to a boil), and sprinkle with roll icing.

Confection Roll

4 rolls

Scale the dough into 3 lb units. Roll out with a rolling pin to about ⅛″ thickness and about 12″ wide. Brush with egg wash using a pastry brush and spread confection roll filling on dough. Roll up as for sweet rolls and flatten by hand until about 4″ wide. Proof in proofing cabinet until double in bulk. Bake in a preheated oven at 390°F until golden brown. Remove from the oven, brush with a hot corn syrup glaze (1 lb glucose and ½ lb water brought to a boil), and sprinkle with roll icing. Cut into serving units.

DANISH AND SWEET DOUGH FILLINGS AND TOPPINGS

Streusel Topping

3 qt

Ingredients

1 lb 4 oz bread flour
1 lb butter-flavored liquid shortening
1 lb 8 oz sugar
½ oz salt
1 lb 4 oz bread flour

Procedure

1. Place the flour and shortening in a mixing container. Blend by hand.
2. Add the sugar, salt, and bread flour. Rub by hand to a streusel or medium-sized crumb.

Variations

Cinnamon streusel: Add ½ oz cinnamon per pound of flour.

Nut streusel: Add 4 oz ground nutmeats per pound of flour.

Crunch streusel: Add 4 oz macaroon crunch per pound of flour.

Chocolate streusel: In the first mixing stage, add 4 oz melted chocolate per pound of flour.

Orange streusel: Add ½ oz of orange gratings per pound of flour.

Cinnamon Nut Topping

1 qt

Ingredients

1 lb sugar
1 oz cinnamon
2 oz ground nuts

Procedure

1. Place all the ingredients in a mixing container and mix by hand until thoroughly blended.

Almond Paste Filling

1 qt

Ingredients

1 lb almond paste
1 lb sugar
4 oz eggs
2 lb cake crumbs
1 oz dry milk
12 oz water (variable)
½ oz salt

Procedure

1. Place the almond paste and sugar in the stainless steel mixing bowl. Cream at slow speed using the paddle.
2. Add the eggs and continue to cream at slow speed.
3. Add the cake crumbs, dry milk, water, and salt. Mix at medium speed until thoroughly blended.

Honey Fruit Filling

1½ qt

Ingredients

1 lb 8 oz brown sugar
1 lb 8 oz golden butter-flavored shortening
1 lb honey
1 oz salt
1 lb 8 oz chopped fruits
8 oz cake flour

Procedure

1. Place the brown sugar, shortening, honey, and salt in the stainless steel mixing bowl. Cream together at slow speed using the paddle until thoroughly blended.
2. Add the chopped fruits and mix at medium speed.
3. Add the flour and continue to mix until smooth.

Fruit Filling

1½ qt

Ingredients

2 lb cake crumbs
1 lb raisins
8 oz chopped pecans
8 oz chopped maraschino cherries
1 oz cinnamon
2 oz dry milk
14 oz water (variable)

Procedure

1. Place the cake crumbs, raisins, nuts, cherries, and cinnamon in the stainless steel mixing bowl. Mix together at medium speed using the paddle. Dissolve the dry milk in the water and add to the mixture to obtain proper consistency.

Cream Filling

1 qt

Ingredients

2 lb powdered sugar
6 oz golden butter-flavored shortening
6 oz butter-flavored liquid shortening or butter
8 oz eggs
14 oz cake flour

Procedure

1. Place the powdered sugar and both shortenings in the stainless steel mixing bowl. Cream at slow speed using the paddle until thoroughly blended.
2. Add the eggs slowly and continue to cream at slow speed.
3. Add the flour and mix at slow speed until smooth.

Aloha Filling

1½ gal.

Ingredients

6 lb 11 oz canned crushed pineapple
2 ground whole oranges
2 lb water
2 lb 12 oz sugar
2 lb water
8 oz modified starch
½ oz salt
yellow color as needed

Procedure

1. Place the pineapple, ground oranges, water, and sugar in a sauce pot. Bring to a boil.
2. Dissolve the starch in the second amount of water. Add slowly to the mixture, stirring constantly with a kitchen spoon. Cook until thick and clear.
3. Remove from the heat and blend in the salt and yellow color. Let cool before using.

Orange Currant Filling

3 qt

Ingredients

1 lb 8 oz water
12 oz sugar
8 oz ground whole oranges
1½ oz lemon juice
2 oz butter-flavored liquid shortening
1 oz emulsified shortening
8 oz water
4 oz cornstarch
6 oz egg yolks

Procedure

1. Place the water, sugar, ground whole oranges, lemon juice, and both shortenings in a sauce pot. Bring to a rolling boil.
2. Mix the water, cornstarch, and egg yolks thoroughly. Add slowly to the boiling mixture while stirring constantly with a kitchen spoon.
3. Cook until thick. Let cool and use. Apply by spreading.

Almond Filling

1 qt

Ingredients

8 oz ground whole oranges
5 lb sugar
8 oz currants

Procedure

1. Place all the ingredients in the stainless steel mixing bowl. Mix at slow speed using the paddle until thoroughly blended. Apply by sprinkling.

Applesauce-Fruit Filling

1½ qt

Ingredients

1 lb 8 oz chopped pecans
2 lb cake crumbs
1 lb brown sugar
1 lb applesauce

Procedure

1. Place all the ingredients in the stainless steel mixing bowl. Mix the filling at medium speed using the paddle until thoroughly blended.

Date-Nut Filling

2 qt

Ingredients

2 lb 8 oz pitted dates
8 oz brown sugar
1 lb 8 oz water
1 lb chopped pecans

Procedure

1. Place the dates, brown sugar, and water in a sauce pot. Bring to a boil and continue to cook for 5 min.
2. Remove from the fire and stir in the pecans with a kitchen spoon.
3. Let cool before using.

Hazel Nut Filling

1½ qt

Ingredients

1 lb roasted ground filberts
½ oz cinnamon
2 lb sugar
6 oz eggs
3 lb cake crumbs
2 oz dry milk
14 oz water

Procedure

1. Place all the ingredients in the stainless steel mixing bowl. Mix at medium speed using the paddle until thoroughly blended.

Butter Topping

3 qt

Ingredients

4 lb powdered sugar (4, 6, or 10X)
2 lb cake crumbs
1 lb butter
1 lb golden butter-flavored shortening
10 oz eggs
½ oz vanilla

Procedure

1. Place the sugar, cake crumbs, butter, and golden butter-flavored shortening in the stainless steel mixing bowl. Mix at medium speed using the paddle until thoroughly blended.
2. Add the eggs and vanilla and continue mixing until smooth.

Almond Brittle Topping or Filling

1 qt

Ingredients

2 lb sugar
1 lb water
8 oz glucose
12 oz chopped almonds

Procedure

1. Place the sugar, water, and glucose in a sauce pot. Boil to 275°F.
2. Add the chopped almonds and boil to 300°F.
3. Pour into bun pans in thin sheets. Cool and break into fine pieces with a rolling pin.

Orange Filling

2 qt

Ingredients

1 lb ground almonds or almond paste
1 lb emulsified shortening
1 lb sugar
2 oz eggs
4 oz cake flour

Procedure

1. Place all the ingredients in the stainless steel mixing bowl. Mix the filling at medium speed using the paddle until thoroughly blended.

Confection Roll Filling

3 qt

Ingredients

2 lb 8 oz dark brown sugar
1 lb butter-flavored liquid shortening
2½ oz cinnamon
2 lb 8 oz sugar
1 lb 4 oz coarsely ground nutmeats
1 lb 4 oz cake crumbs
½ oz salt

Procedure

1. Place all the ingredients in the stainless steel mixing bowl. Mix at medium speed using the paddle until thoroughly blended.

Pie Doughs and Fillings

Pie is the most popular dessert served in food service establishments. The two basic types of pies are single crust and double crust. The single crust pie consists of one crust on the bottom. The double crust pie consists of two crusts, one on the bottom and one on the top. In both types of pies, the most important characteristic is the tenderness of the crust. The tenderness of the crust often determines how the pie is received. The filling may be of excellent quality, but a tough crust may cause rejection of the whole preparation.

Pie doughs vary in preparation techniques but have very similar ingredients. The mixing process of pie doughs is a critical procedure that is learned with experience.

Pie fillings are divided into four types: fruit, cream, chiffon, and soft fillings. The most popular type of pie filling is fruit. If economically feasible, fresh fruit is the best choice for fruit fillings.

PIE DOUGH TYPES

Most pie doughs have similar ingredients but differ in how the flour and shortening are mixed together and in the amount of liquid added to the dough. The best method of learning how pie dough is properly mixed is by rubbing (using the hands) for close control of the mixing process. The ingredients of a pie dough are rubbed (cut in) with the palm of the hands or pastry blender until the proper consistency is obtained. The amount of rubbing required for proper consistency is a skill acquired through experience. A mixing machine is commonly used for large quantities of pie dough. However, the mixing machine can easily overmix the dough. In addition, the heat in the dough when using the mixing machine can cause the shortening to break down.

Pie doughs are classified into three crust types: mealy, short flake, and long flake. The *mealy crust* absorbs the least amount of liquid because the flour and shortening are rubbed together until the flour is completely covered with shortening. The flour is then unable to absorb a large amount of liquid. The *short flake crust*, which is the most common type used, absorbs a slightly larger amount of liquid because the flour and shortening are only rubbed until no flour spots are evident. The flour is not coated to the degree of the mealy type. The flour is then able to absorb slightly more liquid. The *long flake crust* absorbs the greatest amount of liquid because the flour and shortening are rubbed together less than the mealy or short flake crust. The flour and shortening are rubbed together very lightly, leaving the shortening in chunks about the size of the tip of the little finger.

The mealy crust and the short flake crust are handled the same way after mixing. The doughs are refrigerated about 45 min to an hour until they are firm enough to roll with ease. The long flake crust must be refrigerated for a longer period of time. Usually several hours or overnight is required. If not refrigerated long enough, the dough will be soft and difficult to roll out.

The pie dough used is determined by the type of filling that will be added. The correct amount and mixing of ingredients produces a good pie crust. Table I lists some possible causes and remedies for faulty pies.

PIE DOUGH INGREDIENTS

The ingredients used in most pie dough recipes are flour, shortening, liquid (water or milk), salt, and sugar. Each ingredient plays an important part in the finished product. Pastry flour, milled from soft winter wheat, contains the ideal gluten content for pie dough and produces the best results. If pastry flour is not available, blend 60% cake flour and 40% bread flour. Sift completely to prevent lumping. Flour with too high or too low a gluten content results in tough or sticky dough. The flour should always be sifted when preparing pie dough because the soft pastry flour has a tendency to pack and form lumps. These lumps do not absorb the liquid as readily, which leads to overmixing. Overmixing results in a tough crust. Pastry flour is also used for dusting the bench when the pie dough is rolled to prevent toughness.

TABLE I: CAUSES AND REMEDIES FOR FAULTY PIES

Nature of Problem	Possible Causes	Possible Remedies
Excessive shrinkage of crusts	Not enough shortening Too much water Dough worked too much Flour too strong	Increase the shortening Cut quantity of water Do not overmix Use a weaker flour or increase shortening content
Crust not flaky	Dough mixed too warm Shortening too soft Rubbing flour and fat too much	Have water cold Have shortening at right temperature Do not rub too much
Bottom crust soaks too much juice	Insufficient baking Crust too rich Too cool an oven	Bake longer Reduce amount of shortening More bottom heat
Tough crust	Flour too strong Dough overmixed Overworking the dough Too much water	Increase the shortening Just incorporate the ingredients Work dough as little as possible Reduce amount of water
Soggy crust	Not enough bottom heat Oven too hot Having filling hot	Regulate oven correctly Regulate oven correctly Use only cold filling
Fruit boils out	Oven too cold Fruit slightly sour No holes in top crust Crust not properly sealed	Regulate oven temperature Use more sugar Have a few openings in top crust Seal bottom and top crust on edges
Custard pies curdle	Overbaked	Take out of oven as soon as set
Blisters on pumpkin pies	Oven too hot Too long baking	Regulate oven temperature Take out of oven as soon as set
Bleeding of meringue	Moisture in egg whites Poor egg whites Grease in egg whites	Use a stabilizer in the meringue Check egg whites for body Be sure equipment is free from grease

CHERRY

PUMPKIN

PEAR CHOCOLATE

Pies can be made with a variety of crusts and fillings.

Chilling the flour in the refrigerator before mixing is a method used by some bakers to keep the dough below 70°F during the mixing period. This method is especially useful if mixing is done in a hot environment.

The *shortening* or *fat* used in pie dough may be lard, hydrogenated vegetable shortening, or butter. A high-quality lard is required. However, most lards impart a flavor that may be objectionable if the filling does not cover this taste. Hydrogenated vegetable shortening is most commonly used because it has no taste and has a plastic consistency that is an ideal feature when cutting the flour into the fat. If butter is used to improve the flavor of the dough, it should be blended with hydrogenated vegetable shortening using one-third butter to every two-thirds shortening. This blend must be chilled in the refrigerator and allowed to harden slightly before it is cut into the flour because in mixing the butter and shortening together, the butter tends to soften. The use of butter increases the cost of the product. The increased cost may not pay, as the flavor of the butter may be overpowered by the filling.

The *liquid* used in preparation of pie dough may be water or milk, depending on the recipe being used. Milk produces a richer dough and a better colored crust. If dry milk is used, it must be dissolved in water before it is added to the flour/shortening mixture. Both the water and the milk must be cold. This keeps the fat particles hard and prevents the dough from becoming too soft. The amount of liquid required in the recipe depends on the type of pie dough being prepared. If the mealy crust is used, less liquid is required.

Salt brings out the flavors of all the ingredients used in the dough. The salt is dissolved in the liquid to ensure better distribution and prevent burnt spots.

Sugar adds sweetness and color to the baked crust. The form used may be granulated, syrup, or dextrose, depending on what is called for in the recipe. The sugar should be dissolved in the liquid to ensure complete distribution.

PIE DOUGH PREPARATION

Pie dough must be mixed properly to ensure good results. Most faults in the preparation of pie doughs develop when the dough is being mixed. The flour is sifted into a large round bowl and the shortening or fat is added and cut or rubbed into the flour by hand. The degree of rubbing is determined by the type of dough. The salt, sugar, and cold liquid are blended together in a bain-marie until the salt and sugar are thoroughly dissolved. The liquid mixture is poured over the flour/shortening mixture. The two mixtures should be mixed only until the liquid is absorbed by the flour.

Too many times, if the dough appears sticky at this point, extra flour is added. If the dough appears stiff, extra liquid is added. This unbalances the recipe and causes overmixing, which results in toughness. This is the most common mistake made when preparing a pie dough.

The amount of liquid added is the most important factor in producing a successful dough. After the dough is mixed, it is placed in a pan, covered with a damp cloth, and refrigerated until it is firm enough to be rolled. When the dough is firm, remove it from the refrigerator and scale it into 8 oz units. Return the 8 oz units to the refrigerator to be kept firm until ready to roll.

When rolling the dough, work with one unit of dough at a time. An 8 oz unit provides enough dough for one bottom or one top crust for an 8″ or 9″ pie. The rolling is done on a bench dusted with pastry flour. The amount of flour used to dust depends on the consistency of the dough. In some cases, bakers roll the pie dough on a floured piece of canvas. This keeps the dough from sticking to the bench. After the dough is rolled and the bottom or top crust is formed, any remaining scraps are pressed together and are used for the crust still to be made.

The flakiness of pie dough depends on the ratio of shortening to flour. The higher the percentage of shortening to flour by weight, the more tender the crust will be. Percentages of ingredients used in the average pie dough recipe are:

- pastry flour – 100%
- shortening 60% – 75%
- salt 2% – 3%
- water 25% – 35%
- sugar 1% – 2%
- dry milk 1% – 2%

Flour is the main ingredient of pie dough and represents 100% because other ingredients used are based on the amount of flour. To check the percentages of ingredients used in a pie dough recipe, divide the weight of the flour into the weight of each ingredient. For example:

Golden Pie Dough

twenty-two 8″ double-crust pies

Ingredients

10 lb pastry flour
7 lb 4 oz golden-colored shortening
5 oz salt
2½ oz dry milk
2½ oz sugar
3 lb 8 oz water

- Convert 10 lb flour to 160 oz.

  (10 lb × 16 oz = 160 oz)
- Convert 7 lb 4 oz shortening to 116 oz.

  (7 × 16 oz = 112 oz + 4 = 116 oz)

- Divide 116 oz by 160 oz.

$$(\frac{116\,oz}{160\,oz} = .725 \text{ or } 72.5\% \text{ shortening})$$

Continue dividing the amount of each ingredient. The results are:

- Pastry flour – 100%
- Golden-colored shortening – 72.5%
- Salt – 3%
- Dry milk – 1.5%
- Sugar – 1.5%
- Water – 35%

The ingredients for golden pie dough are consistent with the suggested percentages and produce an excellent pie dough.

Specialty Pie Crust

Specialty pie crust may be prepared by adding cheese, spices, ground pecans, filberts, almonds, or other products to the standard pie dough. The ingredient added usually replaces up to 20% of the flour, except in the case of spices.

PIE FILLINGS

Pie fillings must meet the same high standards as the crust for the best results. The pie filling must be thickened to the right consistency and flavored and seasoned properly. The appearance of the filling is also important. The fillings for dessert pies are usually divided into four types: fruit, cream, chiffon, and soft fillings. Other fillings, such as ice cream and Nesselrode (type of fruit filling flavored with rum) are used; however, these pies are considered specialty pies.

Filling the pie shell with a uniform amount of filling is an essential step to control cost, produce a consistent product, and establish good baking procedures. To accomplish this, determine the proper amount of filling required for each pie by weight. Place the prepared pie shell on the twin platform baker's scale and balance the scale. Set the scale for the required amount of filling. Add the filling until the scale balances a second time. Follow the same procedure for each pie.

Fruit Fillings

The most popular type of pie filling is fruit filling. The fruit used may be fresh, dried, frozen, or canned. Each type of fruit is treated differently when prepared for a pie filling. The filling recipe used should state whether the fruit is fresh, dried, frozen, or canned.

Fresh fruit pie fillings are less popular than in the past because canned and frozen fruits are more convenient. Fresh fruit requires more preparation time than canned or frozen fruit. However, fresh fruits that are in season can be purchased cheaper than at any other time of the year. It may be as economical to use fresh fruit at this time. If economically feasible, fresh fruit is the best choice for achieving a good fruit filling flavor.

Washington Apple Commission

Fruit fillings for pie may be fresh, dried, frozen, or canned. Fresh apples were used for this pie.

The amount of water and sugar to use in a fresh fruit filling is based on the amount of fresh fruit used, the type of fruit, and its natural sweetness. Usually, 65% to 70% water based on the weight of the fresh fruit being used is sufficient. For example, if 10 lb of fruit is used, 6½ lb to 7 lb of water should be used. The amount of sugar required is determined by the type of fruit used and its natural sweetness.

Dried fruit, such as apricots, apples, and raisins, is occasionally used for pie fillings. Dried fruit has most of its natural liquid removed and must be soaked in water to restore the natural moisture. In some cases, the liquid and fruit may be brought to a boil. The dried fruit is then soaked as it cools. This boiling method restores moisture and causes the fruit to become soft and plump. After soaking, the liquid is drained from the fruit, thickened, flavored, and poured back over the fruit.

Frozen fruit is the most common type of fruit used in pie fillings. Frozen fruit has the advantages of fresh fruit and is available year-round. The fruit is frozen as soon as possible after picking, either in raw form or slightly parboiled (partly cooked). The fruit is then packed in cans with liquid, sugar, and, in some cases, additional color. Frozen fruit is commonly on the market in 30 lb cans. In some cases, smaller amounts (6½ lb and 10 lb cans) are available. Frozen fruit must be completely defrosted before it is used for a pie filling. The best method of defrosting is to place the unopened can in the refrigerator. Complete defrosting usually takes about one day. To speed the defrosting, the opened fruit container can be set in hot water, but one must use caution by constantly stirring to be sure the fruit is completely defrosted before using. After defrosting, the juice is drained from the fruit, thickened, and flavored. The fruit must be completely defrosted or it will bleed (continue to release juice) and cause the filling to separate.

Canned fruit is commonly used in pie fillings because it is available year-round and the cans (No. 10) are easy to store. In general, canned fruit can be purchased in a water or syrup pack and a solid pack. The water or syrup pack contains less fruit and a higher percentage of juice and sugar than the solid pack. The solid pack is used for pie fillings because it has more fruit and a lower sugar content. This permits more sugar to be added after the juice is thickened. Better results can be obtained if sugar is added after the juice is thickened. Three accepted methods of preparing fruit fillings are the drained fruit method, the fruit and juice method, and the old style method.

Drained Fruit Method. This method is recommended when preparing cherry, blueberry, peach, apricot, and blackberry pie filling.

1. Drain juice from the fruit and place the juice on the range to boil.
2. Dissolve starch in cold water and pour slowly into the boiling juice while stirring constantly.
3. Bring the juice back to a boil and cook until clear.
4. Add sugar, salt, spices (if used), and lemon juice. Stir until thoroughly blended.
5. Add additional color and stir.

6. Pour the thickened syrup over the drained fruit and stir gently so the fruit is not mashed or broken.

7. Cool slightly and pour the filling into unbaked pie shells.

Fruit and Juice Method. This method is recommended when preparing pineapple, apple, and cranberry-apple pie filling.

1. Place the fruit and juice in a pot with the desired or required spices. Place on the range and bring to a boil.

2. Dissolve starch in cold water and pour slowly into the boiling fruit and juice mixture while stirring constantly.

3. Bring the mixture back to a boil and cook until clear.

4. Add the sugar, salt, and color (if desired), and stir until thoroughly blended.

5. Cool slightly and pour the filling into unbaked pie shells.

Old Style Method. This method is best when used with fresh fruit such as apples, peaches, and apricot. However, it is not a popular method because the consistency of the juice cannot be controlled.

1. Mix the fruit (generally fresh fruit) with a mixture of flour, spices, and sugar.

2. Place this mixture in unbaked pie shells.

3. Dot the top of the fruit mixture with butter.

4. Cover the top of the pie with a sheet or strips of pie dough and bake.

Cream Fillings

Cream fillings are simple to prepare, but care must be taken to achieve a smooth, full-flavored filling. One of the most common mistakes made in preparing this type of filling is undercooking the flour or starch, which results in the finished product having a raw flour or starch taste. Another mistake is not beating vigorously enough once the starch or flour starts to thicken. This causes the filling to become lumpy. The most popular cream pies are chocolate, vanilla, coconut, butterscotch, and banana. After the filling has been prepared, it is placed in a prebaked pie shell and topped with meringue or some other type of cream topping. The usual steps taken in preparing cream fillings are:

1. Place milk in the top of a double boiler, holding back approximately 1 qt (depending on amount of filling being made) to liquefy the dry ingredients. Heat the milk.

2. Beat the eggs in a separate container and add the sugar, salt, and starch or flour. Pour in the remaining milk while stirring constantly until a thin paste forms.

3. Pour the thin paste into the scalding milk, whipping constantly until the mixture thickens and becomes smooth.

4. Cook until all traces of starch are removed. Remove from the heat.

5. Stir in the flavoring and add the required amount of butter or shortening.

6. Pour into prebaked pie shells and let cool.

7. Top with meringue or whipped topping.

Chiffon Fillings

Chiffon fillings are light, fluffy fillings prepared by folding (blending one mixture over another) together a fruit or cream pie filling with a meringue. In most cases, a small amount of plain gelatin is added to the fruit or cream filling to help the chiffon filling set when cooled. The steps in preparing chiffon pie fillings are:

1. Prepare a cream filling or a fruit filling using the fruit and juice method, but chop the fruit instead of leaving it whole.

2. Soak plain gelatin in cold water and add it to the hot cream or fruit filling, stirring until it is thoroughly dissolved. Place the filling in a fairly shallow pan and let cool.

3. Refrigerate until the filling begins to set.

4. Prepare a meringue by whipping powdered egg whites and sugar together to stiff peaks.

5. Fold the meringue into the jellied fruit or cream mixture gently, preserving as many of the air cells as possible.

6. Pour the chiffon filling in prebaked pie shells. Refrigerate until set.

7. Top with whipped cream or whipped topping.

Soft Fillings

Soft fillings are fillings that are uncooked and baked in an unbaked pie crust. These are the most difficult pies to make. The difficulty lies in baking the filling and crust to the proper degree without overbaking or underbaking one or the other. Soft fillings are used in such popular pies as pumpkin, custard, and pecan. To avoid problems that develop when baking soft pies, follow these guidelines:

1. Roll the pie dough on graham cracker crumbs instead of flour to help eliminate the soggy crust that sometimes develops.

2. Use precooked or pregelatinized starch instead of cornstarch to bind the filling. The starches give the filling more body and prevent separation.

3. Egg white stabilizer ($\frac{1}{4}$ oz per quart of filling) or tapioca flour (1 oz per quart of milk) may also be used to bind the filling and improve the appearance of the finished product.

4. Bake soft pies at 400°F for at least the first 10 min to 15 min of the baking period. The temperature may be reduced after this time. Remove the pie from the oven as soon as the filling sets.

5. Fill the pie shell only half full of the fillings and bake for a few minutes before filling the shell completely. This produces, in most cases, a more uniformly baked product.

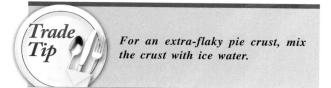

Trade Tip

For an extra-flaky pie crust, mix the crust with ice water.

THICKENING AGENTS USED IN PIE FILLINGS

Starches and flours are used to thicken pie fillings. Starches are used more often than flour because they produce a better sheen and do not discolor or become heavy. Starches used may be corn, tapioca, rice, or a blended product such as waxy maize starch (clear-jel) or modified starch. This starch is a blend of certain starches and vegetable gum that produces a finished product with a high sheen. It gelatinizes quickly when cooked and offsets the action of fruit acids. In addition, these products maintain fruit fla-

vor and color, develop a smooth consistency, and do not cloud when refrigerated.

The amount of starch used in a recipe depends on the jelling quality of the starch and the acidity of the fruit and juice being thickened. More thickener is required for fruits and juices with high acidity. Usually, 3 oz to 5 oz of starch for each quart of liquid is required. The flour or starch being added to a filling should be diluted in cold water or juice before it is poured into the boiling liquid. When added to the boiling liquid, whip vigorously to assure a smooth, creamy filling. A starch or flour begins to swell at approximately 160°F to 170°F. The swelling is complete when the temperature reaches 200°F to 205°F.

New starches have been developed that thicken without cooking. These *pregelatinized starches* thicken when blended with the sugar and added to the liquid. They react quickly without heat because the starch has been precooked and does not require additional heat to absorb liquid and gelatinize. When using this product, always follow the manufacturer's recommended instructions.

Russell Harrington Cutlery, Inc.

Apple pie is the most popular of all pies.

PIE DOUGH RECIPES

The preparation of sure-fire method pie dough is slightly different from the preparation of most pie doughs. The recipe calls for two amounts of flour instead of the usual one. The second amount of flour is mixed into the liquid ingredient before it is added to the flour/shortening mixture. This recipe leaves little chance for failure, hence its name.

Never-Fail Method Pie Dough

twelve 8" double-crust pies

Equipment

- 5 gal. large round bowl
- Plastic scraper
- Baker's scale
- 1 gal. stainless steel container
- Wire whip
- Sheet pan

Ingredients

5 lb pastry flour
4 lb 8 oz shortening
2½ oz salt
1 lb pastry flour
1 lb 8 oz water

Preparation

1. Chill the water.

Procedure

1. Place the first amount of flour in the large round bowl.
2. Add the shortening and rub together by hand until the mixture has formed into small lumps.

3. Place the salt and the second amount of flour in a stainless steel container. Add the cold water and whip with a wire whip until the mixture is smooth.
4. Pour the liquid mixture over the flour/shortening mixture and mix gently by hand until the liquid is absorbed by the flour.
5. Place the dough on a sheet pan. Scrape the bowl clean with a plastic scraper and refrigerate the dough for approximately 1 hr until it becomes very firm to the touch.
6. Remove from the refrigerator, scale into 8 oz units, and refrigerate again until ready to roll.

Precautions

- When blending the salt, flour, and water, mix until all ingredients are blended before adding to the flour/shortening mixture.
- Do not overmix the dough.
- Chill the dough thoroughly before using.

Pie dough (short flake type) is the most common pie dough made. The shortening is cut or rubbed into the flour until small lumps are formed and no flour spots are evident. This dough is versatile and can be used in any preparation calling for a pie dough or crust.

Pie Dough (Short Flake Type)

twenty-two 8" double-crust pies

Equipment

- 5 gal. large round bowl
- Plastic scraper
- Baker's scale
- 1 gal. stainless steel container
- Kitchen spoon
- Sheet pan

Ingredients

10 lb pastry flour
7 lb 8 oz hydrogenated vegetable shortening
5½ oz salt
3 lb water
10 oz sugar or corn sugar solids

Preparation

1. Chill the water.

Procedure

1. Place the flour and shortening in a large round bowl. Rub together by hand until small lumps are formed and no raw flour spots (lumps of dry flour) are evident.

2. Place the water, salt, and sugar or corn sugar solids (dextrose mixture in powdered form) in a stainless steel container. Mix with a kitchen spoon until the sugar and salt are dissolved.
3. Pour the liquid mixture over the flour/shortening mixture and mix gently by hand until the liquid is absorbed by the flour.
4. Place the dough on a sheet pan. Scrape the bowl clean with a plastic scraper and refrigerate the dough for approximately 1 hr until it becomes very firm to the touch.
5. Remove from the refrigerator, scale into 8 oz units, and refrigerate again until ready to roll.

Precautions

- When cutting or rubbing the shortening into the flour, be sure all the flour is worked in before adding the liquid.
- Be sure the salt and sugar are thoroughly dissolved in the water before adding to the flour/shortening mixture.
- Do not overmix the dough.
- Chill the dough before using.

Pie Dough (Mealy Type)

To prepare this pie dough, the flour and shortening are rubbed together thoroughly. All flour particles are thoroughly coated with shortening, thus preventing the flour from absorbing much moisture. Therefore, less water is called for in the mealy recipes. The mealy dough is easy to work with.

twenty 8″ double-crust pies

Equipment

- 5 gal. large round bowl
- Plastic scraper
- Baker's scale
- 1 gal. stainless steel container
- Kitchen spoon
- Sheet pan

Ingredients

10 lb pastry flour
7 lb 8 oz hydrogenated vegetable shortening
5 oz salt
2 lb 8 oz water
10 oz sugar or corn sugar solids

Preparation

1. Chill the water.

Procedure

1. Place the flour and shortening in a large round bowl. Rub together by hand until the flour is completely covered with shortening and the mixture is mealy.

2. Place the water, salt, and sugar or corn sugar solids (dextrose mixture in powdered form) in a stainless steel container. Mix with a kitchen spoon until the sugar and salt are dissolved.
3. Pour the liquid mixture over the flour/shortening mixture and mix gently by hand until the liquid is absorbed by the flour.
4. Place the dough on a sheet pan. Scrape the bowl clean with a plastic scraper and refrigerate until it becomes firm enough to roll.
5. Remove from the refrigerator, scale into 8 oz units, and refrigerate again until ready to roll.

Precautions

- Work the flour/shortening mixture together thoroughly.
- Work all flour into the shortening. No raw flour should be present.
- Be sure the salt and sugar are thoroughly dissolved in the water before adding to the flour/shortening mixture.
- Do not overmix the dough.
- Chill the dough before using.

Trade Tip

Store flour in a dry place between 50°F and 60°F. Humidity and temperature affect the ability of flour to absorb moisture. If the kitchen or work area is extremely hot or humid, chill the flour in the refrigerator. If the flour is lumpy, sift it before use. Weigh and measure ingredients carefully. Too much flour in a recipe will cause a tough pastry.

Pie Dough (Long Flake Type)

This pie dough is usually very tender. Shortening spots are visible throughout the dough and therefore should be used only for top crust or prebaked pie shells.

twenty-three 8″ double-crust pies

Equipment

- 5 gal. large round bowl
- Plastic scraper
- Baker's scale
- 1 gal. stainless steel container
- Kitchen spoon
- Sheet pan

Ingredients

10 lb pastry flour
7 lb 8 oz hydrogenated vegetable shortening
6 oz salt
5 lb water
10 oz sugar or corn sugar solids

Preparation

1. Chill the water.

Procedure

1. Place the flour and shortening in a large round bowl. Rub together lightly by hand, leaving the shortening in fairly large chunks about the size of the tip of the little finger.

2. Place the water, salt, and sugar or corn sugar solids (dextrose mixture in powdered form) in a stainless steel container. Mix with a kitchen spoon until the sugar and salt are dissolved.
3. Pour the liquid mixture over the flour/shortening mixture and mix gently by hand until the liquid is absorbed by the flour.
4. Place the dough on a sheet pan. Scrape the bowl clean with a plastic scraper and refrigerate until it becomes very firm to the touch.
5. Remove from the refrigerator, scale into 8 oz units, and refrigerate again until ready to roll.

Precautions

- Do not overwork the flour and shortening during the rubbing stage or the long flake dough cannot be obtained.
- Work all flour into the shortening. No raw flour should be present.
- Be sure the salt and sugar are thoroughly dissolved in the water before adding to the flour/shortening mixture.
- Do not overmix the dough.
- Chill the dough before using.

Pie Dough for Frying

Pie dough for frying is prepared the same way as the regular pie dough recipes. However, less shortening is used because the dough is fried in deep fat and absorbs fat during the frying period. This dough is used to prepare southern fried pies.

60 pies

Equipment

- Large round bowl
- Plastic scraper
- Baker's scale
- 1 gal. stainless steel bowl
- Kitchen spoon
- Sheet pan
- 5½" round cutter
- Deep fat fryer
- Pastry brush
- Dinner fork

Ingredients

5 lb pastry flour
2 lb shortening
2½ oz salt
2 lb water

Preparation

1. Chill the water.

Procedure

1. Place the flour and shortening in a large round bowl. Rub together by hand until the flour is completely covered with shortening and no raw flour spots are present.

2. Place the water and salt in a stainless steel container. Stir with a kitchen spoon until the salt is dissolved.
3. Pour the salt water over the flour/shortening mixture and mix gently by hand until the liquid is absorbed by the flour.
4. Place the dough on a sheet pan. Scrape the bowl clean with a plastic scraper and refrigerate until it becomes very firm to the touch.

Make up: Roll out the dough on a floured bench to about ⅛" thickness. Cut out with a 5½" round cutter (about 2 oz of dough). Use about 2 oz of filling. Place the filling in the center of the rolled out dough, wash edges with cold water using a pastry brush, and shape to form a turnover (fold the circle in half). Seal the edges securely and pierce the top twice with the tines of a dinner fork. Fry in deep fat for 5 min to 7 min at 375°F. Let drain and cool. Ice with roll icing.

Precautions

- Dissolve the salt in the water thoroughly or small burnt spots will appear on the finished product.
- Do not overmix the dough.

Golden Pie Dough

This recipe is prepared using golden shortening. This shortening has a rich golden color and a butter flavor. It produces a crust that is flaky, extremely tender, very rich in color, and a flavor that suggests that all butter was used.

twelve 8" double-crust pies

Equipment

- 5 gal. large round bowl
- Plastic scraper
- Baker's scale
- 1 gal. stainless steel container
- Wire whip
- Sheet pan

Ingredients

5 lb pastry flour
3 lb 10 oz golden shortening
2½ oz salt
1½ oz sugar
1½ oz dry milk
1 lb 12 oz water

Preparation

1. Chill water until ice cold.
2. Weigh the shortening, place on wax paper, place in a pan, and refrigerate until cold.

Procedure

1. Place the flour and golden shortening in a large round bowl. Rub together using the palms of the hand (cut in) until fairly small chunks develop. Chunks should be a little larger than a grain of rice.

2. Place the salt, sugar, and dry milk in a stainless steel container. Add the cold water slowly while whipping vigorously with a wire whip. Whip until the dry ingredients are thoroughly dissolved in the water.
3. Pour the liquid mixture over the flour/shortening mixture and mix very gently by hand only until the two mixtures are incorporated.
4. Place the dough on a sheet pan. Scrape the bowl clean with a plastic scraper, cover with a damp cloth or plastic wrap, and refrigerate until it becomes firm enough to roll.
5. Remove from the refrigerator, scale into 8 oz units, and refrigerate again until ready to roll.

Precautions

- Chill both water and shortening for best results.
- Work all flour into the shortening. No raw flour should be present.
- Be sure the salt, sugar, and dry milk are thoroughly dissolved in the water before adding it to the flour/shortening mixture.
- Do not overmix the dough.
- Chill the dough before using.

FRUIT FILLING RECIPES

Pineapple Pie Filling

For pineapple pie filling, canned crushed pineapple and juice are thickened with starch, flavored, and placed into unbaked pie shells. Pineapple pie filling produces pies with eye and taste appeal that are a welcome addition to any menu.

six 8"pies

Equipment

- Baker's scale
- 2 gal. sauce pot
- Kitchen spoon
- Small stainless steel bowl
- Quart measure
- Colander

Ingredients

6 lb 8 oz canned pineapple (No. 10 can)
2 lb pineapple juice and water
1 lb 8 oz sugar
¼ oz salt
8 oz water
4½ oz modified starch or cornstarch
6 oz corn syrup
yellow color as needed

Procedure

1. Place the crushed pineapple, sugar, salt, and juice in a sauce pot. Bring to a boil.
2. Dissolve the starch in 8 oz of water in a stainless steel bowl. Pour slowly into the boiling pineapple mixture, stirring until thickened and clear.
3. Simmer the mixture for approximately 2 min, then remove from the heat.
4. Stir in the corn syrup and tint with yellow color. Let the mixture cool.
5. Fill unbaked pie shells.
6. Bake in a preheated oven at 400°F to 425°F.

Precautions

- Pour the dissolved starch slowly into the boiling mixture and stir constantly with a kitchen spoon to avoid lumps. When simmering the thickened mixture, stir occasionally to avoid sticking and scorching.
- Exercise caution when adding the yellow color. A little color enhances the appearance; too much hinders.

Preparation

1. Open the No. 10 can and drain the juice from the pineapple through a colander. Save the juice and add water to equal 2 lb (1 qt).

Spiced Peach Pie Filling

For this filling, peach juice is boiled with vinegar and spices to bring forth a pungent, spicy flavor. The juice is then thickened and folded into the sliced peaches. This desirable, spicy filling is generally used in double-crust pies.

six 8″ pies

Equipment

- Baker's scale
- 2 gal. sauce pot
- China cap
- Wire whip
- Colander
- Small stainless steel bowl
- Quart measure
- Kitchen spoon
- 1 gal. stainless steel container

Ingredients

6 lb 8 oz sliced peaches (No. 10 can)
2 lb peach juice and water
⅛ oz whole cloves
¼ oz cinnamon stick
8 oz cider vinegar
8 oz water
3½ oz modified starch or cornstarch
1 lb 8 oz sugar
¼ oz salt

2. Simmer for approximately 20 min, then strain through a china cap into a stainless steel container.
3. Return the strained juice to the sauce pot and bring back to a boil.
4. Dissolve the starch in 8 oz of water in a stainless steel bowl. Pour slowly into the boiling liquid while whipping briskly with a wire whip.
5. Bring mixture back to a boil. Cook until thickened and clear.
6. Add the sugar and salt. Stir with a kitchen spoon until thoroughly dissolved.
7. Add the drained peaches, gently folding them into the thickened juice with a kitchen spoon to avoid breaking the fruit. Let the mixture cool.
8. Fill unbaked pie shells.
9. Bake in a preheated oven at 425°F.

Preparation

1. Open the No. 10 can and drain the juice from the peaches by placing the peaches in a colander. Save the juice and add water to equal 2 lb (1 qt).

Procedure

1. Place the peach juice, cinnamon stick, cloves, and vinegar in a sauce pot. Bring to a boil.

Precautions

- Pour the dissolved starch slowly into the boiling mixture and whip briskly to avoid lumps.
- Once the starch is added, stir the mixture continuously to avoid scorching.

Canned cherries, which produce an excellent tasting pie filling but lack color, are used. Canned cherries, unlike frozen cherries, do not retain their natural color. Red color is added to improve the appearance. This filling can be used for making double-crust pies or single-crust pies with a streusel or whipped cream topping.

Cherry Pie Filling (Canned Cherries)

six 8″ pies

Equipment

- baker's scale
- 2 gal. sauce pot
- Wire whip
- Kitchen spoon
- Colander
- Small stainless steel bowl
- Cup measure
- Paring knife
- Quart measure

Ingredients

6 lb 8 oz canned cherries (No. 10 can)
2 lb cherry juice and water
4 oz modified starch or cornstarch
1 lb 8 oz sugar
juice of 1 lemon
red color as desired

Procedure

1. Place the cherry juice in a sauce pot, reserving 1 c for dissolving the starch. Bring to a boil.
2. In a stainless steel bowl, dissolve the starch in the cherry juice held in reserve. Pour slowly into the boiling juice, whipping vigorously with a wire whip.
3. Bring the mixture back to a boil and cook until thickened and clear.
4. Add the sugar and whip until dissolved.
5. Add the lemon juice and red color. Stir with a kitchen spoon until blended into the thickened juice. Remove from the range.
6. Add the drained cherries, gently folding them into the thickened juice with a kitchen spoon to avoid crushing the fruit. Let cool.
7. Fill unbaked pie shells.
8. Bake in a preheated oven at 400°F to 425°F.

Preparation

1. Open the No. 10 can and drain the juice from the cherries by placing the cherries in a colander. Save the juice and add water to equal 2 lb (1 qt).
2. Cut the lemon in half with a paring knife and squeeze the juice from the lemon by hand.

Precautions

- Drain the cherries thoroughly.
- Pour the dissolved starch slowly into the boiling mixture and whip vigorously to avoid lumps.
- Once the starch is added, stir the mixture continuously to avoid scorching.

Frozen cherries, which produce a rich, red, natural-looking filling, are used. The cherries are thoroughly thawed before being used. This filling can be used for making double-crust pies or single-crust pies with a streusel or whipped topping.

Cherry Pie Filling (Frozen Cherries)

ten 8″ pies

Equipment

- Baker's scale
- 2 gal. sauce pot
- Wire whip
- Kitchen spoon
- Colander
- Small stainless steel bowl
- Pint measure
- Quart measure

Ingredients

10 lb frozen cherries
4 lb cherry juice and water
8 oz modified starch or cornstarch
2 lb sugar
1 lb corn syrup

Procedure

1. Place all but 1 lb (1 pt) of the cherry juice in a sauce pot and bring to a boil.
2. In a stainless steel bowl, dissolve the starch in the remaining 1 lb (1 pt) of cherry juice. Pour slowly into the boiling mixture while whipping vigorously with a wire whip.
3. Bring the mixture back to a boil and cook until thickened and clear.
4. Add the corn syrup and sugar and stir with a kitchen spoon until thoroughly blended. Remove from the heat.
5. Add the cherries, gently folding them into the thickened juice with a kitchen spoon to avoid breaking or crushing the fruit. Let cool.
6. Fill unbaked pie shells.
7. Bake in a preheated oven at 400°F to 425°F.

Preparation

1. Thaw the frozen cherries.
2. Drain the juice from the cherries by placing them in a colander. Save the juice and add water to the juice if needed to equal 4 lb (2 qt).

Precautions

- Drain the cherries thoroughly.
- Pour the dissolved starch slowly into the boiling mixture and whip vigorously to avoid lumps.

Blueberry Pie Filling (Frozen Blueberries)

The blueberry juice is thickened with starch, flavored, and poured over the blueberries. This method is used to avoid crushing the plump, tender berries.

nine 8″ pies

Equipment

- Baker's scale
- 3 gal. sauce pot
- Wire whip
- Kitchen spoon
- Colander
- Small stainless steel bowl
- Quart measure
- Paring knife

Ingredients

9 lb frozen blueberries
2 lb blueberry juice and water
1 lb sugar
¼ oz salt
¼ oz cinnamon
8 oz water
4½ oz modified starch or cornstarch
2 lb 8 oz sugar
½ oz lemon juice

Procedure

1. Place the blueberry juice or the water/juice mixture, the first amount of sugar, cinnamon, and salt in a sauce pot. Bring to a boil.
2. In a stainless steel bowl, dissolve the starch in 8 oz of water. Pour very slowly into the boiling mixture while whipping briskly with a wire whip.
3. Bring the mixture back to a boil and cook until thickened and clear.
4. Add the second amount of sugar and lemon juice. Stir with a kitchen spoon until thoroughly blended. Remove from the heat.
5. Add the blueberries, gently folding them into the thickened juice with a kitchen spoon to avoid crushing. Let the mixture cool.
6. Fill unbaked pie shells.
7. Bake in a preheated oven at 400°F to 425°F.

Precautions

- Drain the blueberries thoroughly.
- Pour the dissolved starch slowly into the boiling mixture and whip briskly to avoid lumps.
- Once the starch is added, stir the mixture continuously to avoid lumps.

Preparation

1. Thaw the frozen blueberries.
2. Drain the juice from the berries by placing them in a colander. Save the juice and add water to equal 2 lb (1 qt).
3. Cut the lemon in half with a paring knife and squeeze the juice from the lemon by hand.

Peach Pie Filling (Canned Peaches)

For peach pie filling, the peach juice is thickened with starch, flavored, and poured back over the sliced, canned peaches. This filling is usually used in double-crust pies.

nine 8″ pies

Equipment

- Baker's scale
- 2 gal. sauce pot
- Wire whip
- Kitchen spoon
- Colander
- Small stainless steel bowl
- Quart measure

Ingredients

6 lb 8 oz sliced peaches (No. 10 can)
2 lb peach juice and water
12 oz sugar
¼ oz salt
8 oz water
4 oz modified starch or cornstarch
1 lb sugar
yellow color as needed

Procedure

1. Place the peach juice and water, the first amount of sugar, and salt in a sauce pot. Bring to a boil.
2. In a stainless steel bowl, dissolve the starch in the 8 oz of water. Pour slowly into the boiling mixture while whipping vigorously with a wire whip.
3. Bring the mixture back to a boil and cook until thickened and clear.
4. Add the second amount of sugar. Stir with a kitchen spoon until thoroughly dissolved.
5. Add yellow color as needed. Remove from the heat.
6. Add the drained peaches, gently folding them into the thickened juice with a kitchen spoon to avoid breaking or crushing the fruit. Let cool.
7. Fill unbaked pie shells.
8. Bake in a preheated oven at 400°F to 425°F.

Precautions

- Pour the dissolved starch slowly into the boiling mixture and whip vigorously to avoid lumps.
- Once the starch has been added, stir the mixture continuously to avoid scorching.

Preparation

1. Open the No. 10 can and drain the juice from the peaches by placing them in a colander. Save the juice and add water to the juice to equal 2 lb (1 qt).

Apple Pie Filling (Canned Apples)

Canned apples are cooked slightly with water or apple juice, seasoned, and thickened with starch. Apple pies are the all-American favorite and can be made up in single-crust or double-crust varieties.

six to seven 8″ pies

Equipment

- Baker's scale
- 2 gal. sauce pot
- Kitchen spoon
- Small stainless steel bowl

Ingredients

7 lb canned apples (No. 10 can)
1 lb 8 oz water or apple juice
1 lb 4 oz sugar
¼ oz salt
¼ oz cinnamon
⅛ oz nutmeg
3 oz butter
8 oz water
3 oz modified starch or cornstarch

Procedure

1. Place the apples, sugar, salt, spices, and the first amount of water or juice in a sauce pot. Bring to a boil.

2. In a stainless steel bowl, dissolve the starch in the second amount of water. Pour slowly into the boiling mixture, stirring constantly with a kitchen spoon until thickened.
3. Simmer the mixture approximately 2 min. Remove from the heat.
4. Stir in the butter with a kitchen spoon. Let the mixture cool.
5. Fill unbaked pie shells.
6. Bake in a preheated oven at 400°F to 425°F.

Note: For a Dutch apple filling, add 8 oz raisins in Step 1.

Precautions

- When adding the dissolved starch to the boiling mass, stir gently so the fruit does not break.
- When simmering the thickened mixture, stir occasionally to avoid sticking and scorching.

Raisin Pie Filling

This filling can be used to add variety to the dessert menu during the fall and winter months. It can be set up as a single- or double-crust pie.

eight 8″ pies

Equipment

- Baker's scale
- 2 gal. sauce pot
- Stainless steel bowl
- Kitchen spoon

Ingredients

6 lb raisins
1 lb sugar
1 lb light brown sugar
8 lb water
½ oz salt
¼ oz cinnamon
¼ oz lemon juice
4 oz modified starch
2 oz butter or margarine
8 oz water

Procedure

1. Place the raisins, sugar, first amount of water, salt, lemon juice, and cinnamon in a heavy bottom sauce pot. Bring to a boil.
2. In a stainless steel bowl, place the starch and second amount of water. Stir with a kitchen spoon until starch is thoroughly dissolved.

3. Pour the dissolved starch slowly into the boiling raisin mixture while stirring rapidly with a kitchen spoon.
4. Bring the mixture back to a boil while continuing to stir. Cook until mixture is thick and clear.
5. Remove from the fire. Stir in the butter or margarine. Let the filling cool slightly.
6. Pour into unbaked pie shells and top as desired with a double crust, lattice top, or crumb (streusel) topping.
7. Bake in a preheated oven at 400°F until crust is golden.

Precautions

- When the mixture in Step 1 is brought to a boil, proceed with Step 2 immediately or the raisins may overcook and lose their shape.
- When adding the diluted starch to the boiling mixture, stir gently to avoid breaking and mashing the raisins.
- Check pies from time to time during the baking period. Rotate pies in the oven at least once during the baking period.

Trade Tip

When baking fruit pies, a slightly raw or soggy bottom crust sometimes occurs. Some bakers prebake the pie shell slightly before adding the fruit filling. Others sprinkle crushed cornflakes over the bottom pie shell before filling. Crushed cornflakes absorb excess liquid and do not affect the taste of the filling.

This is a sweet, clear, semi-liquid that is used to cover fresh or canned fruit when preparing tarts (individual type of small pie) or open-face pies. The glaze protects the fruit from the air and increases the appearance of the fruit. This glaze can be used to cover fresh fruit or canned fruit.

Fruit Glaze

2 qt

Equipment

- 1 gal. sauce pot
- Wire whip
- Baker's scale
- Paring knife
- Strainer
- Stainless steel bowl

Ingredients

2 lb water
2 lb 8 oz sugar
8 oz water
4 oz modified starch
4 oz corn syrup
1 oz lemon juice
food color as desired

Preparation

1. Cut the lemon in half with a paring knife and squeeze the juice from the lemon by hand. Strain in a strainer.

Procedure

1. Place the first amount of water and the sugar in a sauce pot and bring to a boil.
2. In a stainless steel bowl, dissolve the starch in the second amount of water. Pour slowly into the boiling liquid, whipping vigorously with a wire whip.
3. Cook until thickened and clear.
4. Add the corn syrup and lemon juice and bring the mixture back to a boil. Remove from the heat.
5. Color as desired. The glaze should be colored to blend with the fruit being used.

Note: To set up a fresh strawberry pie, spread 8 oz glaze on the bottom of an 8″ prebaked pie shell. Place 12 oz cleaned strawberries over the glaze. Spread another 8 oz to 10 oz glaze over the strawberries and refrigerate the pie for about 1 hr. Cover the top with whipped cream or topping and serve.

Precautions

- Whip constantly when adding the dissolved starch to the boiling liquid to avoid lumps.
- Once the starch is added, whip or stir continuously to avoid scorching.
- Exercise caution when adding the color. A little color enhances the appearance; too much hinders.

Fresh apples are peeled, cored, and sliced. They are placed in unbaked pie shells, sprinkled with a mixture of starch, sugar, and seasoning, covered with a thin sheet of pie dough, and baked. The juice from the apples thickens during the baking period.

Apple Pie Filling (Fresh Apples)

five to six 8″ pies

Equipment

- Apple corer
- Vegetable peeler
- Paring knife
- Baker's scale
- Containers for mixing (2)
- Pastry brush

Ingredients

10 lb fresh apples
1 lb sugar
2 oz lemon juice
1 lb sugar
1/8 oz nutmeg
1/4 oz cinnamon
1/4 oz salt
5 oz cornstarch
butter as needed

Preparation

1. Core the apples with an apple corer, peel with a vegetable peeler, and slice with a paring knife. Use Winesap or Roman Beauty apples for best results.
2. Squeeze the juice from the lemons.

Procedure

1. Place the apples in a fairly large mixing container. Add the first amount of sugar and the lemon juice, toss together very gently by hand, and let set for approximately 1 hr.
2. Place the second amount of sugar, salt, nutmeg, cinnamon, and cornstarch in a separate container. Mix together thoroughly by hand.
3. Sprinkle the seasoning mixture over the bottom of each unbaked pie shell. Fill the shells with the sliced apples and sprinkle a generous amount of the seasoning mixture over the top of the sliced apple.
4. Dot the top of each filled pie with butter and cover with pie dough. Secure the top layer of dough to the bottom by fluting (forming grooves) the edges of the pies.
5. Brush the top of each pie with melted butter or egg wash (egg and milk) using a pastry brush. Bake in a preheated oven at 425°F to 450°F.

Note: The juice that cooks out of the apples activates the cornstarch and as the pies bake, the filling thickens.

Precautions

- Exercise caution when coring, peeling, and slicing apples.
- Toss the apples with the sugar and lemon juice very gently to avoid breaking the apples.
- Keep the apples covered during the setting period with a damp cloth. They may turn brown if uncovered.

CREAM FILLING RECIPES

Vanilla Pie Filling

This basic vanilla pie filling can be used to prepare many different types of pies, including banana cream, apricot cream, and coconut cream pies.

twelve to fourteen 8″ pies

Equipment

- Baker's scale
- Double boiler
- Wire whip
- Large stainless steel bowl
- Kitchen spoon

Ingredients

12 lb milk
4 lb sugar
1 lb cornstarch
¼ oz salt
2 lb eggs
6 oz butter
vanilla to taste

Procedure

1. Place 10 lb (5 qt) of the milk in the top of a double boiler. Cover and heat until scalding hot (a film will form on the surface of the milk).
2. Place the cornstarch, sugar, salt, and eggs in a large stainless steel bowl. Blend thoroughly with a wire whip.
3. Add the remaining 2 lb (1 qt) of milk and blend until a paste is formed. Pour the paste mixture into the scalding milk, whipping briskly with a wire whip.
4. Continue to cook and whip the mixture until it becomes quite stiff. Remove from the heat.
5. Stir in the butter and vanilla with a kitchen spoon until thoroughly blended.
6. Pour into prebaked pie shells, let cool, top with a meringue, and brown in the oven.

Variations

Banana cream pie: Cover the bottom of a prebaked pie shell about half deep with vanilla pie filling. Slice one banana crosswise and line the slices on the surface of the pie filling. Cover the banana slices with enough pie filling to fill the pie shell. Let the filling cool, cover with a meringue topping, and bake in the oven until the meringue browns.

Apricot cream pie: Add 6 lb 8 oz (one No. 10 can) of drained chopped apricots to the vanilla pie filling. Fold the chopped apricots into the filling and pour into prebaked pie shells. Let the filling cool, cover with a meringue topping, and bake in the oven until the meringue browns.

Coconut cream pie: Stir 1 lb 8 oz of macaroon coconut into the vanilla pie filling and pour into prebaked pie shells. Let the filling cool, cover with a meringue topping, and bake in the oven until the meringue browns.

Precaution

- Whip constantly and briskly when adding the starch mixture to the scalding milk to avoid lumps.

Chocolate Macaroon Pie Filling

Macaroon coconut is blended into a chocolate pie filling which is extremely rich and slightly expensive to prepare, but the eating qualities are outstanding.

seven to eight 8″ pies

Equipment

- 3 gal. sauce pot
- Baker's scale
- Wire whip
- Kitchen spoon
- 1 pt saucepan
- 1 gal. stainless steel container

Ingredients

3 lb 2 oz sugar
¼ oz salt
8 oz dry milk
3 lb 12 oz water
1 lb 4 oz water
6½ oz modified starch or cornstarch
12 oz egg yolks
10 oz bitter chocolate
5 oz shortening
10 oz macaroon coconut
vanilla to taste

Procedure

1. Place the sugar, salt, dry milk, and first amount of water in a sauce pot. Bring to a boil.
2. In a stainless steel container, dissolve the starch in the second amount of water. Blend in the egg yolks using a wire whip. Pour slowly into the boiling mixture while whipping constantly with a wire whip.
3. Cook until the mixture becomes thick.
4. Stir in the melted chocolate, shortening, macaroon coconut, and vanilla with a kitchen spoon. Remove from the heat.
5. Pour into prebaked pie shells, let cool, and top with meringue.
6. Sprinkle additional macaroon coconut over the meringue and brown in the oven.

Note: To prepare a plain chocolate pie, omit the macaroon coconut.

Preparation

1. Place the chocolate in a saucepan set near the heat or over boiling water and melt.

Precautions

- Exercise caution when bringing the ingredients in step 1 to a boil. Do not let the mixture boil over.
- Whip constantly while adding the starch and egg mixture.
- Stir constantly while cooking the filling to avoid scorching.

Lemon Pie Filling

This rich lemon cream filling can be used in a variety of ways, including the very popular lemon meringue pie.

eight to nine 8" pies

Equipment

- Baker's scale
- Food grater
- Paring knife
- Wire whip
- 3 gal. sauce pot
- Stainless steel bowl
- 1 pt saucepan
- Strainer or china cap
- Kitchen spoon

Ingredients

4 lb water
3 lb 6 oz sugar
½ oz salt
3 oz lemon gratings
1 lb water
8 oz cornstarch
12 oz pasturized eggs
1 lb 6 oz lemon juice
4 oz butter or shortening
yellow color as needed

Preparation

1. Using a food grater, grate the rinds of fresh lemons.
2. Cut the lemons in half with a paring knife and squeeze the juice out by hand. Strain in a strainer or china cap.
3. Place the butter or shortening in a saucepan and melt.

Procedure

1. Place first amount of water, sugar, salt, and lemon grating in a sauce pot. Bring to a boil.
2. In a stainless steel bowl, whip the egg slightly with a wire whip. Add the starch and the second amount of water. Stir with a kitchen spoon until the starch is dissolved.
3. Pour the starch/egg mixture slowly into the boiling liquid, whipping vigorously with a wire whip until thickened and clear.
4. Add the lemon juice and melted butter or shortening. Stir with a kitchen spoon until thoroughly blended. Remove from the heat and tint with yellow color if desired.
5. Pour the filling into prebaked pie shells. Let cool and top with whipped topping or meringue.

Precautions

- When adding the starch mixture to the boiling liquid, whip vigorously to avoid lumps.
- Once the starch mixture is added, stir or whip constantly to avoid scorching.

Cream Filling

This is a rich, creamy egg/milk filling that can be used in the preparation of many different desserts.

5½ qt

Equipment

- Double boiler
- Wire whip
- Stainless steel bowl
- Baker's scale
- Kitchen spoon
- Quart measure
- 2 gal. stainless steel container

Ingredients

9 lb milk
2 lb 6 oz sugar
1 lb cake flour
¾ oz salt
1 lb 12 oz eggs
1 oz vanilla
yellow color as needed

Procedure

1. Place 8 lb (4 qt) of the milk in the top of a double boiler. Heat until scalding hot.
2. Combine the dry ingredients and blend with the beaten eggs. Add the remaining milk and stir with a kitchen spoon until a smooth paste is formed.
3. Pour the paste mixture slowly into the scalding milk, whipping vigorously with a wire whip until thickened and smooth.
4. Cook in the double boiler for approximately 15 min, whipping constantly.
5. Tint with yellow color, add the vanilla, and remove from the heat.
6. Place in a stainless steel container, let cool, and refrigerate until ready to use.

Preparation

1. Break the eggs into a stainless steel bowl and beat slightly with a wire whip.
2. Place water in the bottom of a double boiler. Bring to a boil.

Precautions

- Whip vigorously and constantly when adding the paste mixture to the scalding milk.
- Exercise caution when adding the yellow color. A little color enhances the appearance; too much hinders.

Chocolate Cream Pie Filling

This is a rich, smooth, creamy chocolate filling that can be topped with whipped cream or topping or meringue when served. Like all cream pies, it is a single-crust pie.

nine to ten 8″ pies

Equipment

- Baker's scale
- Double boiler
- Wire whip
- Stainless steel bowls (2)
- Box grater
- Sifter

Ingredients

8 lb milk
3 lb 12 oz sugar
1 lb egg yolks
1 lb 8 oz eggs
10 oz sifted cocoa
14 oz cornstarch
2 lb milk
¼ oz salt
6 oz margarine
solid sweet chocolate as needed to garnish

Procter and Gamble Co.

Preparation

1. Separate the eggs from the yolks.
2. Set up the double boiler and place on the range to heat.
3. Grate the sweetened chocolate for garnish.

Procedure

1. Place the 8 lb (4 qt) of milk and sugar in the top of a double boiler and heat.
2. In a stainless steel bowl, dissolve the cornstarch in the 2 lb (1 qt) of milk.
3. In the second stainless steel bowl, blend the egg yolks, eggs, cocoa, and salt. Slowly pour in the dissolved starch and mix until all ingredients are thoroughly blended.
4. Slowly pour the blended mixture into the hot milk while whipping rapidly with a wire whip. Cook until the mixture is thick and smooth.
5. Remove from the heat and whip in the margarine.
6. Fill prebaked pie shells, let cool, top with meringue, and brown slightly in the oven or top with whipped cream or topping. Sprinkle with grated chocolate.

Precautions

- Whip constantly and briskly when adding the starch/egg mixture to the scalding milk to avoid lumps.
- Exercise caution when removing the thickened filling from the double boiler. Steam will escape.

Butterscotch Cream Pie Filling

The brown sugar is cooked slowly in the melted butter or margarine to produce a butterscotch flavor.

ten 8″ pies

Equipment

- Baker's scale
- Thick bottom sauce pot
- Wire whip
- Stainless steel bowls (2)
- Double boiler
- Kitchen spoon

Ingredients

1 lb 8 oz butter or margarine, or a blend of both
4 lb 4 oz dark brown sugar
½ oz salt
9 lb milk
10 oz cornstarch
3 lb milk
1 lb 8 oz beaten eggs
10 oz bread flour
vanilla to taste

Procedure

1. Place the butter or margarine in a thick-bottom sauce pot and melt over low heat.
2. Add the brown sugar and salt. Cook for approximately 20 min until well blended, stirring occasionally with a kitchen spoon.
3. Slowly add the first amount of milk, stirring constantly. The sugar and butter mixture will crystallize if the milk is added too fast. Bring this mixture to a quick boil.
4. Remove from the range and place in a double boiler or steam jacket kettle if one is available. This step is taken to avoid scorching.
5. In the second stainless steel bowl, place the cornstarch, flour, and second amount of milk. Mix using a wire whip until a smooth paste is formed.

6. Blend in the beaten eggs. Add a little of the hot milk mixture while whipping gently with a wire whip. This is done to adjust the temperature.
7. Pour the egg and milk mixture slowly into the hot milk mixture while whipping constantly. Cook until thick and smooth. Remove from the fire.
8. Add vanilla to taste and cool the mixture just slightly.
9. Fill the prebaked pie shells. Place in the refrigerator until the filling sets.
10. Top with a meringue and brown slightly in a hot oven (425°F).

Precautions

- Exercise extreme caution when cooking the butter/sugar mixture to avoid scorching. Use very low heat.
- Add the milk to the butter/sugar mixture while at the same time stirring constantly. The mixture will crystallize if this step is not followed correctly.
- Be sure to add some of the hot milk mixture to the egg mixture. This adjusts the temperature and prevents curdling.
- When preparing cream pies, add the vanilla flavoring after the mixture has been thickened. If added to an extremely hot mixture, a high percentage of its flavor will be lost.

CHIFFON FILLING RECIPES

Pumpkin Chiffon Pie Filling

Beaten egg whites are folded into a chilled pumpkin pie filling to create a mixture that is light and fluffy. The filling is poured into prebaked pie shells and refrigerated.

ten to twelve 8″ pies

Equipment

- Baker's scale
- Double boiler
- Wire whip
- Small stainless steel bowl
- Kitchen spoon
- Mixing machine and wire whip
- Bake pan

Ingredients

4 lb 8 oz canned pumpkin
2 lb 8 oz light brown sugar
1 lb 12 oz egg yolks
¾ oz cinnamon
⅛ oz ginger
⅛ oz mace
½ oz salt
8 oz water
2 oz plain gelatin
1 lb 12 oz powdered egg whites
1 lb 12 oz sugar
¼ oz nutmeg

Preparation

1. Reconstitute the egg whites with water in a stainless steel bowl.
2. Place water in the bottom of the double boiler and bring to a boil.

Procedure

1. Place the pumpkin, brown sugar, egg yolks, spices, and salt in the top of a double boiler. Cook until the mixture becomes thick, stirring constantly with a kitchen spoon.
2. In a small stainless steel bowl, soak the gelatin in the water. Stir in the hot mixture with a kitchen spoon until thoroughly dissolved. Remove the mixture from the heat.
3. Place the pumpkin mixture in a bake pan. Refrigerate until it is cool and starts to thicken. Remove from the refrigerator.
4. Place the reconstituted egg whites in the stainless steel mixing bowl and whip at high speed until a meringue starts to form.
5. Add the sugar slowly while continuing to whip at high speed until stiff peaks form.
6. Using a gentle motion, fold the meringue into the slightly thickened pumpkin mixture with a kitchen spoon.
7. Pour the filling into prebaked pie shells and cool in the refrigerator until the filling sets and becomes firm.

Precautions

- When adding the gelatin to the hot pumpkin mixture, stir continuously until thoroughly dissolved.
- To whip the meringue, the mixer should be running at high speed throughout the operation and the sugar must be added slowly.
- Fold the meringue gently into the thickened pumpkin mixture to preserve as many air cells as possible.

Trade Tip

When mixing pie dough, weight the shortening first and chill it thoroughly before cutting it into the flour. A chilled shortening holds up better during the cutting-in process. The shortening will not break down as quickly from the friction created when mixing by machine, or body heat when mixing by hand.

Strawberry Chiffon Pie Filling

Beaten egg whites are folded into a chilled strawberry pie filling to create a light and very fluffy filling. Chiffon fillings are poured into prebaked pie shells and refrigerated until set.

eight 8" pies

Equipment

- Baker's scale
- 2 gal. sauce pot
- Stainless steel bowl
- Kitchen spoon
- Mixing machine, bowl, and wire whip
- Bake pan
- Skimmer

Ingredients

4 lb frozen strawberries
2 lb strawberry juice and water
¼ oz salt
1 lb sugar
¼ oz lemon juice
1 oz plain gelatin
a few drops of red color, if needed and desired
7 oz cornstarch
12 oz water
1 lb 8 oz powdered egg whites
1 lb 4 oz sugar

Preparation

1. Thaw the frozen berries and drain. Save the juice.
2. Reconstitute the egg whites with water in a stainless steel bowl.

Procedure

1. Place the strawberries, the first amount of liquid, sugar, salt, lemon juice, red color (if used), and gelatin in a sauce pot. Place on the range and bring to a boil.
2. In a stainless steel bowl, dissolve the starch in the second amount of liquid (water). Pour slowly into the boiling strawberry mixture while stirring continuously with a kitchen spoon. Avoid crushing the fruit.
3. Cook slowly until the mixture is thick and clear. Remove from the fire. Pour mixture into a bake pan and refrigerate until the mixture gels slightly.
4. Place the reconstituted egg whites in the stainless steel mixing bowl and whip at high speed until whites start to froth.
5. Add the second amount of sugar slowly, continuing to whip at high speed until meringue becomes fairly stiff (soft peaks).
6. Remove the strawberry mixture from the refrigerator when it is slightly jellied and gently fold in the meringue using a skimmer. Incorporate thoroughly.
7. Pour or spoon the filling into prebaked pie shells and place in the refrigerator until filling sets.

Note: For fresh strawberries, use 4 lb strawberries and 2 lb sugar. For a raspberry chiffon or blackberry chiffon pie, change the fruit.

Precautions

- When adding the dissolved starch to the boiling strawberry mixture, stir rapidly to avoid lumps.
- Stir the thickened strawberry mixture continuously once the starch has been added.
- To whip the meringue, the mixer should be running at high speed throughout the operation and the sugar must be added slowly.

Lemon Chiffon Pie Filling

Beaten egg whites are folded into a lemon cream filling to create a light, fluffy filling. Chiffon fillings are poured into prebaked pie shells and refrigerated until they set.

nine to ten 8" pies

Equipment

- Baker's scale
- Food grater
- Paring knife
- 1 gal. sauce pot
- Wire whip
- Stainless steel bowl
- Kitchen spoon
- Mixing machine and wire whip

Ingredients

3 lb water
2 lb sugar
¾ oz salt
2 oz lemon grating (rind)
1 lb pasturized eggs
1 lb lemon juice
9 oz cornstarch
1½ oz plain gelatin
1 lb hot water
2 lb powdered egg whites
1 lb 8 oz sugar
yellow color as needed

Preparation

1. Grate the rinds of fresh lemons on the medium grid of the food grater.
2. Cut the lemons in half with a paring knife and squeeze out the juice by hand.
3. Reconstitute the egg whites with water in a stainless steel bowl.

Procedure

1. Place the first amount of water and sugar, salt, and lemon grating in a sauce pot. Bring to a boil.
2. In a stainless steel mixing bowl, whip the egg yolks slightly with a wire whip. Add the juice and blend together.
3. Add the cornstarch and stir with a kitchen spoon until dissolved. Pour this mixture slowly into the boiling mixture, whipping vigorously with a wire whip until thickened and smooth. Remove from the heat.
4. In a stainless steel bowl, dissolve the plain gelatin in the hot water and stir into the lemon filling using a kitchen spoon. Add yellow color if desired.
5. Place the egg whites in the stainless steel mixing bowl and whip on the mixing machine at high speed until a meringue starts to form.
6. Add the second amount of sugar slowly, continuing to whip at high speed until stiff peaks form.
7. Using a gentle motion, fold the meringue into the hot lemon filling with a kitchen spoon.

8. Pour the filling into prebaked pie shells and cool in the refrigerator until the filling sets.

Note: This filling may be used for orange chiffon pie by using orange grating in place of the lemon grating and changing the juice ingredient to 14 oz orange juice and 2 oz lemon juice.

Precautions

- When adding the starch mixture to the boiling mixture, whip vigorously to avoid lumps.
- To whip the meringue, the mixer should be running at high speed throughout the operation and the sugar must be added slowly.
- Fold the meringue gently into the lemon filling to preserve as many air cells as possible.

SOFT FILLING RECIPES

Pecan Pie Filling

This filling is simple to prepare because it is mixed in two stages. The pecans are not added to the filling during the mixing stages. They are placed in each individual pie shell before the filling is added so even distribution can be obtained. This pie is very rich and sweet, so small portions are usually served.

seven 8″ pies

Equipment

- Baker's scale
- Mixing machine, bowl, and wire whip
- 1 pt saucepan
- Wire whip
- Stainless steel bowl

Ingredients

9 lb light corn syrup
6 oz pastry flour
6 oz sugar
3 lb 4 oz eggs
½ oz salt
8 oz butter
vanilla to taste
1 lb 12 oz pecans

Preparation

1. Melt the butter in a saucepan.

Procedure

1. Place the flour, sugar, and syrup in the stainless steel mixing bowl. Using the paddle, blend together at slow speed.
2. Place the eggs, salt, melted butter, and vanilla in a stainless steel bowl. Whip with a wire whip until thoroughly blended.
3. Pour the blended egg mixture slowly into the syrup mixture while continuing to mix at slow speed. Mix until all ingredients are thoroughly blended.
4. Place 4 oz pecans in each individual pie shell. Fill the shell with the mixture and bake in a preheated oven at 325°F until the crust is done and the filling is set.

Precautions

- All soft pies are difficult to bake properly. Best results may be obtained by prebaking the shells lightly before adding the pecans and filling.
- Always mix at slow speed. A higher speed will cause splashing.
- Check pies from time to time during the baking period. Do not overbake the filling or underbake the crust.

Trade Tip

Always have the liquid ice cold. The amount of liquid used in pie dough is usually 1 qt to each 4 lb of flour. However, the type of pie dough being made also affects the amount. For exact amounts, follow the recipe being used.

Pumpkin Pie Filling

The highly-seasoned pumpkin mixture is thickened by the reaction of the flour and eggs when heat is applied. This soft pie filling is extremely popular during the fall and winter and has become a traditional favorite at Thanksgiving and Christmas.

eight 8″ pies

Equipment

- Baker's scale
- Mixing machine, bowl, and wire whip
- Stainless steel bowl
- Wire whip

Ingredients

6 lb canned pumpkin
1 lb brown sugar
2 lb sugar
6 lb (3 qt) milk
4 oz cake flour
½ oz cinnamon
¼ oz nutmeg
¼ oz ginger
¼ oz salt
1 lb 6 oz eggs

Preparation

1. Break the eggs into a stainless steel bowl and beat slightly.

Procedure

1. Place the pumpkin, sugar, flour, salt, and spices in the bowl of the electric mixing machine.

2. Mix at slow speed using the paddle until all ingredients are thoroughly blended.
3. Add the beaten eggs alternately with the milk while continuing to mix at slow speed. Mix until all the sugar has been dissolved and all ingredients are thoroughly blended.
4. Remove from the mixing machine and let set approximately 1 hr so all spices incorporate.
5. Pour into unbaked or slightly prebaked pie shells and bake in a preheated oven at 400°F until filling is firm and set. Remove from the oven.
6. Serve topped with whipped cream.

Precautions

- For best results, let the filling stand for at least 1 hr before filling the pie shells. This gives the sugar time to dissolve completely and the spices time to disperse their flavor.
- For best results, prebake the pie shells just slightly before adding the filling. This reduces the chance of overbaking the filling in order to bake the pie shell completely.

Custard Pie Filling

This is one of the most popular soft pies and also one of the most difficult pies to prepare properly. This recipe can also be used for a coconut custard pie.

eight 8″ pies

Equipment

- Baker's scale
- Mixing machine, bowl, and paddle
- Stainless steel bowl
- Kitchen spoon
- 1 pt saucepan

Ingredients

1 lb 6 oz sugar
10 oz dry milk
¼ oz salt
1½ cornstarch
½ oz precooked (pregelatinized) starch
1 lb 6 oz eggs
4 lb water
1 oz vanilla
2 oz butter

Preparation

1. Break the eggs and place them in a stainless steel bowl.
2. Melt the butter in a saucepan.

Procedure

1. Place the sugar, dry milk, salt, cornstarch, and precooked starch in the stainless steel mixing bowl. Mix at slow speed using the paddle until blended.
2. Stir the broken eggs with a kitchen spoon until well blended. Do not whip.
3. Pour the stirred eggs into the blended dry ingredients slowly, while at the same time mixing at slow speed.

4. Add the water and vanilla while continuing to mix at slow speed.
5. Remove from the mixer and let the filling stand for approximately 45 min. Stir in the melted butter.
6. Pour into unbaked or slightly prebaked pie shells and bake in a preheated oven at 425°F until custard is set (slightly jellied or firm in the center). If precooked starch is not used in the filling, only fill the pie shell half full and place in the oven. Allow a 5 min to 6 min baking period before filling the shell to the brim. This two-step fill ensures a better bake and prevents curdling.

Note: For a coconut custard pie, sprinkle the bottom of the pie shell with approximately 1 oz to 1½ oz of unsweetened shredded coconut before adding the filling. Avoid using sweetened coconut because it may darken before the pie is baked. Use a long or medium shred of coconut for a more even distribution throughout the filling.

If a nutmeg flavor is desired, it should be sprinkled very lightly over the top of the filling after the pie shell has been filled to the top. This prevents the spice from settling to the bottom, creating a very undesirable condition.

Precautions

- Overbaking custard results in "weeping." That is, beads of water will form on the surface. To test custard for doneness, insert the blade of a metal knife into the center. If it comes out clean, the custard is done.
- Custard filling continues to cook after leaving the oven. Therefore, the pies should be removed just before they are completely baked.

FRIED PIE FILLING RECIPES

Fried Pie Filling

These fried pie filling recipes are all prepared in the same general manner and are used exclusively in the preparation of fried pies. They differ from conventional pie fillings because more starch is added to develop the thicker consistency necessary when the filling is to be used in a fried product.

Equipment

- 2 gal. sauce pot
- Wire whip
- Kitchen spoon
- Stainless steel bowl
- Colander
- Baker's scale
- Bake pan
- Pint measure
- French knife
- Paring knife

Preparation

1. Drain the fruit, if necessary, in a colander. Save the juice.
2. Chop the fruit, if necessary, with a French knife.
3. Soak and drain the raisins if they are being used.
4. Cut the lemons with a paring knife and squeeze the juice from the lemon by hand.

Apple Filling

100 fillings

Ingredients

7 lb canned apples
1 lb soaked raisins
2 lb 8 oz sugar
1 lb water
½ oz salt
¼ oz cinnamon
1 oz lemon juice
1 lb water
7 oz modified starch or cornstarch

Procedure

1. Place the first seven ingredients in a sauce pot. Bring to a boil and cook until the apples become soft.
2. In a stainless steel bowl, dissolve the starch in the second amount of water. Pour slowly into the boiling mixture, stirring rapidly with a kitchen spoon until thickened and clear.
3. Remove the mixture from the heat, place in a bake pan, let cool, and refrigerate until thoroughly chilled.
4. Remove the apple filling from the refrigerator and prepare the turnovers. Use approximately 2 oz of filling to each turnover.

Cherry Filling

70 fillings

Ingredients

6 lb 8 oz cherries (No. 10 can)
2 lb cherry juice and water
5 oz modified starch or cornstarch
1 lb 8 oz sugar
juice from 1 lemon
red color as needed

Procedure

1. Place the cherry juice and water in a sauce pot, reserving 1 pt for dissolving the starch. Bring to a boil.
2. In a stainless steel container, dissolve the starch in the cherry juice held in reserve. Pour slowly into the boiling juice, whipping vigorously with a wire whip.
3. Bring the mixture back to a boil and cook until thickened and clean.
4. Add the sugar and whip with a wire whip until dissolved.
5. Add the lemon juice and red color. Stir with a kitchen spoon until blended into the thickened juice. Pour over the cherries and blend.
6. Pour into a bake pan, let cool, and refrigerate until thoroughly chilled.
7. Remove the cherry filling from the refrigerator and prepare the turnovers. Use approximately 2 oz of filling for each turnover.

Minor ingredients are added to pie dough for specific purposes. Eggs add richness and coloring to pie dough. Salt adds flavor and has a minor effect in conditioning the pastry. Sugar also adds flavor and produces a crisper crust. Lemon juice and vinegar tenderize the dough by mellowing the effects of gluten.

MERINGUE TOPPING RECIPES

These toppings are prepared with raw egg whites. Food containing raw eggs cannot be served to at-risk populations. Reconstituted powdered egg whites may be substituted.

Meringue Toppings

Equipment

- Baker's scale
- Plastic scraper
- Mixing machine, bowl, and wire whip
- Saucepan
- Kitchen spoon
- Double boiler
- Wire whip

Preparation

1. Separate the egg whites from the yolks by passing the yolk back and forth from one half egg shell to the other until the white has run off. Save the yolks for another preparation by covering them with cold water and placing in the refrigerator.

Hydrogenated vegetable shortening produces the most plasticity and softness in pie dough. One hundred percent shortening pastry produces the most flaky and fragile crusts. Butter produces a rich-tasting, durable, flaky pastry.

Common Meringue

six 8″ pies or four 8″ cakes

Ingredients

1 lb egg whites
2 lb sugar
2 oz tapioca flour

Procedure

1. Place the egg whites in the bowl of the electric mixing machine. Beat at high speed using the wire whip until the whites start to foam.
2. Mix the sugar and tapioca flour together and add slowly to the egg whites while continuing to whip at high speed.
3. Whip until wet or dry peaks are formed. How the meringue is to be used determines the peak needed. A wet peak displays a shiny appearance; a dry peak, a dull appearance. A wet peak is moister and usually spreads better than a dry peak, which is stiffer.

Swiss Meringue

fourto five 8″ cakes

Ingredients

1 lb egg whites
2 lb sugar

Procedure

1. Place the sugar and egg whites in the top of a double boiler. Heat while whipping constantly with a wire whip until sugar is dissolved and mixture warms (120°F).
2. Remove from the heat and place in the bowl of the electric mixing machine. Beat at high speed using the wire whip until wet or dry peaks are formed. How the meringue is to be used determines the peak needed.

Cakes and Icings

Cakes are a very popular dessert item served in food service establishments. The cost of cakes, compared to the cost of other dessert items, is relatively inexpensive. In addition, cakes are easy to prepare in quantity and can be stored successfully for fairly long periods of time.

Cakes can be baked in many different varieties using basic cake recipes and different variations. The introduction of cake mixes has simplified the process of making cakes. Improvement in equipment has also contributed to a saving of time in preparation. However, the best cakes are still made from scratch. The three common cake mixing methods are the creaming, sponge, and two-stage methods.

Icings are used to improve the taste and enhance the appearance of a cake. Icing also seals in the moisture and flavor of the cake. Icing recipes, like cake recipes, can be varied to obtain different flavors and textures.

CAKES AND ICINGS

Cakes are commonly offered as a dessert item on luncheon and dinner menus. A good cake recipe has a proper balance of the various ingredients and has been tested. Most bakers have a recipe file of successful recipes obtained from trade publications, other bakers, or from baking product companies that sell products used in the bakeshop. These companies usually have research or test kitchens that develop new recipes and improve old ones.

If a cake is to be successful, it must have:

- a good recipe
- high-quality ingredients
- careful weighing
- proper mixing
- proper baking

Cake Ingredients

Cake ingredients commonly used include shortening, cake flour, eggs, sugar, baking powder, liquid (milk or water), salt, and flavoring. Some ingredients have one function or act with other ingredients in the recipe. Ingredients may be categorized by their function within the recipe.

Tenderizers: Sugar, shortening, egg yolks, and baking powder

Moisteners: Milk or water, syrups, eggs, and sugar

Tougheners: Flour, dry milk solids, and egg whites

CAKES

SWEET POTATO CHEESECAKE   **CHOCOLATE CAKE**   **CARROT CAKE**

Cakes add eye appeal to dessert offerings.

Driers: Flour, starches, and dry milk solids

Flavorers: Eggs, butter, vanilla or other flavoring liquid, and salt

High-quality ingredients should be used for proper taste, texture, volume, and overall quality of the finished product.

Shortening may be butter, margarine, or any good commercial shortening. If using shortening, a hydrogenated or emulsified shortening will produce the best results. *Hydrogenated shortening* is made from vegetable oils that have been hydrogenated to transform them into a solid white fat that has a flexible melting point. With a flexible melting point, the shortening melts at various temperatures without breaking down and still performs the task of tenderizing. This shortening improves creaming qualities and helps trap and hold a greater amount of air.

Emulsified shortening is made from hydrogenated vegetable oils and has greater emulsifying powers. It is produced for use in cakes with a high sugar content. These cakes are also known as *high ratio cakes*. This shortening blends more readily with the liquid ingredients of the cake batter and produces a cake with greater volume and better keeping qualities.

Cake flour is milled from soft wheat and contains all starch and no gluten. A good-quality cake flour is pure white and has strength, uniform granulation, and high absorption.

Eggs are one of the most important ingredients used in making a cake. Both fresh and frozen eggs produce a quality cake. Fresh eggs deteriorate rapidly, so they should be purchased as necessary for use within about three days. Purchase fresh eggs that have a firm, clear white and a deep yellow yolk. Eggs laid in the spring are the most desirable. Frozen eggs are commonly used in most bakeshops for convenience. Frozen eggs can be purchased as whole eggs, whites, or yolks in 30 lb tins and can be stored easily.

Sugar is extremely important in cake production because cake is a sweet dessert. The sugar used in making cakes must be clear, bright, and pure white.

Baking powder is the leavening agent commonly used to make a batter light and porous. Baking powder may be purchased in three different types: fast-acting, slow-acting, or double-acting. Each type regulates the speed at which the gas is released or generated. Double-acting baking powder is used most often.

Milk used in cakes may be in dry or liquid form. Milk in liquid form should be purchased fresh every day. Dry milk should blend quickly when mixed with water.

Salt is a minor ingredient that brings out the taste and flavor of the cake. The salt used should be pure white and have no bitter taste.

Flavoring is very important and usually quite expensive. Quality flavoring is more expensive than lesser grades or imitations. However, pure flavoring yields more flavoring than imitation flavoring.

Weighing Ingredients

Ingredients used in making a cake are weighed for maximum accuracy. A baker's scale that weighs with weights rather than springs should be used. Take care in weighing each ingredient and use a checklist to verify that each ingredient has been added.

CAKE MIXING

Proper mixing and handling of cake batter is required for good results. The cake batter should always be mixed in accordance with the recipe being used. Each step of the mixing instructions must be followed carefully. The three methods of mixing a cake batter are the *creaming method*, the *sponge* or *whipping method*, and the *two-stage* or *blending method*.

Creaming Method

1. In a stainless steel mixing bowl, cream the sugar, butter or shortening, salt, and spices. During the mixing, small air cells are formed and incorporated into the mix. The volume increases and the mix becomes softer in consistency.

2. Add the eggs gradually and continue to cream at slow speed with the mixing machine. During the mixing, the eggs coat the cells formed during the creaming stage and allow them to expand and hold the liquid, when it is added, without curdling.

3. Add the liquid (milk or water) alternately with the sifted baking powder and flour. Mix until a smooth batter is formed. During this stage, the liquid and the baking powder/flour mixture are added alternately so the batter does not curdle. If all the liquid is added at one time, the cells coated by the eggs will not be able to hold all the moisture and curdling will result.

4. Add the flavoring and blend thoroughly.

Note: Before mixing, all ingredients should be at room temperature. The bowl must be scraped down at intervals throughout the entire mixing process so all ingredients are blended and a smooth batter is obtained. There are some variations to this method of mixing. Follow the mixing instructions of each recipe carefully.

Russell Harrington Cutlery, Inc.

Utensils with different shapes are used to spread icing and serve cakes.

Sponge or Whipping Method

1. Warm the eggs and sugar to about 100°F over hot water. This softens the egg yolks and slightly dissolves the sugar, which allows for quicker whipping and a greater volume. Whip the mixture until the required volume or stiffness is obtained.

2. Slowly add the liquid and flavoring if called for in the recipe.

3. Fold in the sifted flour gently to ensure a smooth and uniform batter.

Note: Before mixing, all the ingredients except the eggs and sugar should be at room temperature. There are some variations to this method of mixing. Follow the mixing instructions of the recipe carefully.

Two-Stage or Blending Method

1. Place all the dry ingredients, shortening, and part of the milk in the mixing bowl. Blend at slow speed for the required period of time.

2. Blend the eggs and the remaining milk and add to the mixture at three different intervals to ensure a smooth, uniform batter.

Note: Before mixing, all ingredients should be at room temperature. At intervals throughout the entire mixing process, the bowl must be scraped down so all ingredients are blended and a smooth batter is obtained.

Round cake pans were used to bake this cake.

CAKE BAKING

Whenever possible, cakes should be placed in the center of the oven where the heat is distributed evenly. If a number of cakes are being baked at one time, the pans should be placed so they will not touch one another or any part of the oven wall. Always allow the heat to circulate freely around each pan. The oven must be preheated to the required temperature and checked periodically with an oven thermometer.

Generally, the larger the cake being baked and the richer the cake batter, the slower it should be heated. However, if the oven heat is too low, the cake will rise and fall, causing a very heavy texture. If the oven heat is too high, the outside of the cake will bake rapidly, forming a crust. When the heat reaches the center, the cake expands, causing the crust to

burst. The baking time of a cake is divided into four stages of development:

1. The cake is placed in the oven and starts to rise. At this stage, use the lowest temperature called for in the baking instructions to prevent quick browning and to keep a crust from forming.

2. The cake continues to rise and the top surface starts to brown. Exercise caution in this stage. Do not open the oven door. The heat may be increased if the recipe suggests it.

3. The rising stops and the surface of the cake continues to brown. The oven door can now be opened if necessary. The heat may be reduced if the cake is browning too fast.

4. The cake starts to shrink, leaving the sides of the pan slightly. It can now be tested for doneness.

Cakes can be tested for doneness by sticking a wire tester or toothpick into the center of the cake. If the tester comes out dry with no batter adhering to it, the cake is done. Another method may be used for heavier cakes, such as fruitcakes. Press the top surface of the cake with a finger. If it feels firm and the impression of the finger does not remain, the cake is done.

When a cake is removed from the oven, it should be placed on wire racks or shelves so that air circulates around the pan. Allow the cake to cool for approximately 5 min. Invert the pan and remove the cake from the pan. If wax or parchment paper is used on the bottom of the pan, remove this also. Place the cake back on the rack and continue to cool thoroughly.

Tables I and II list some baked cake defects and suggested causes and remedies. Table I should be used with cakes made by the creaming or blending methods. Table II should be used with cakes prepared by the sponge method. The recipe used in making a cake must have the ingredients properly balanced so each function to produce a cake has desirable results.

Preparing Cake Pans

Cake pans are prepared by covering the bottom of the pan with parchment paper and greasing the sides lightly with shortening or pan grease (a mixture of one part flour to two parts shortening). Individual cake pans, such as Mary Ann pans or bundt cake pans, are greased thoroughly with pan grease and dusted lightly with sugar to prevent sticking. Butter used to grease the cake pans or used in the preparation of the pan grease improves the flavor of the cake. Pan sprays may also be used.

TABLE I: BATTER CAKES (CREAMING OR BLENDING METHOD)

Defect	Cause	Remedy
Layers uneven	1. Batter spread unevenly 2. Oven racks out of balance 3. Cake tins warped	1. Spread batter evenly 2. Adjust oven racks 3. Do not use damaged tins
Cakes peak in center	1. Insufficient shortening 2. Batter too stiff 3. Too much oven top heat	1. Balance recipe 2. Increase moisture and/or decrease flour content 3. Check drafts and burners
Cakes sag in center, poor symmetry	1. Excessive sugar in recipe 2. Insufficient structure building materials 3. Too much leavening 4. Cold oven 5. Cakes underbaked	1. Balance recipe 2. Increase egg content and/or flour content 3. Balance recipe 4. Correct oven temperature 5. Bake thoroughly
Undersized cakes	1. Unbalanced recipe 2. Oven too hot 3. Oven too cool 4. Improper mixing 5. Cake tins too large for amount of batter	1. Correct recipe balance 2. Check oven temperature 3. Check oven temperature 4. Exercise care in mixing 5. Use proper amount of batter
Dark crust color	1. Oven too hot 2. Too much top heat in oven 3. Too much sugar, too much milk solids	1. Use correct baking temperature 2. Check oven drafts 3. Balance recipe
Light crust color	1. Oven too cool 2. Unbalanced recipe	1. Raise oven temperature 2. Balance recipe
Uneven baking	1. Oven heat not uniform 2. Variations in baking pans	1. Check oven drafts, flues, insulation 2. Use same type tins for entire batch
Tough cakes	1. Insufficient tenderizing 2. Flour content too high 3. Wrong type of flour	1. Increase sugar, shortening, or both 2. Balance recipe 3. Use soft wheat flour
Thick, hard crust	1. Oven too hot 2. Cakes baked too long 3. Slab-type cake tins not insulated	1. Reduce oven temperature 2. Reduce baking time 3. Use insulation around cake molds
Sticky crust	1. Sugar content too high 2. Improper mixing	1. Balance recipe 2. Use care in mixing
Soggy crust	1. Cakes steam during cooling	1. Remove cakes from tins and allow to cool on rack; cool cakes before wrapping
Crust cracks	1. Oven too hot 2. Stiff batter	1. Reduce oven temperature 2. Adjust flour and liquid contents
Poor flavor	1. Inferior materials used 2. Poor flavoring material or wrong combination 3. Materials improperly stored	1. Care in selecting materials 2. Use quality pure flavors; check flavor combinations 3. Material storage space should be free from foreign odors
Lack of flavor	1. Lack of salt 2. Lack of flavoring materials or weak flavoring materials	1. Use correct amount of salt 2. Use sufficient flavoring and correct types
Heavy cakes	1. Too much sugar 2. Too much shortening 3. Liquid content high 4. Insufficient leavening 5. Too much leavening 6. Cakes underbaked	1. Balance recipe 2. Balance recipe 3. Balance recipe 4. Balance recipe 5. Balance recipe 6. Bake out correctly
Cakes too light and crumbly	1. Batter overcreamed 2. Leavening content high 3. Shortening content too high	1. Mix properly 2. Balance recipe 3. Balance recipe
Coarse grain	1. Leavening content high 2. Separation of liquids and fats (curdled characteristic in batter)	1. Balance recipe 2. Add liquids at proper temperatures and only as fast as it will emulsify well
Tough-eating cakes	1. Formula low in tenderizing materials, sugar, and shortening 2. Oven too hot	1. Balance recipe 2. Regulate oven temperature

TABLE II: SPONGE CAKES

Defect	Cause	Remedy
Undersized cakes	1. Overbeating or underbeating 2. Overmixing after flour is added 3. Sugar content is too high 4. Oven too hot 5. Cakes removed from pans too soon after baking 6. Cakes underbaked 7. Greased pans or tins	1. Beat egg whites, sugar, salt, and cream of tartar to a wet peak 2. Fold in just enough to incorporate 3. Balance recipe 4. Regulate oven temperature 5. Allow cakes to cool before removing from tins 6. Bake thoroughly 7. Do not grease tins for angel food cakes
Light crust color	1. Cakes underbaked 2. Cool oven 3. Overbeaten and overmixed batter	1. Bake correctly 2. Regulate oven temperature 3. Mix properly
Dark crust color	1. Oven too hot 2. Cakes overbaked 3. Excessive sugar content causing cake to have sugar crust	1. Regulate oven temperature 2. Give proper bake 3. Balance recipe
Tough crust	1. Oven too hot 2. Sugar content too high 3. Improper mixing	1. Regulate oven temperature 2. Balance recipe 3. Exercise care in assembling batter
Thick and hard crust	1. Overbaking 2. Cold oven	1. Lessen baking time 2. Regulate oven temperature
Strong flavor	1. Off-flavored materials 2. Poor flavoring materials 3. Cakes burned or overbaked	1. Check storage space of materials for foreign odors 2. Use only top-quality flavors 3. Exercise care in baking
Lack of flavor	1. Insufficient salt in recipe 2. Poor flavor combination 3. Poor-quality flavoring materials used	1. Increase salt content 2. Use proper flavor blends 3. Use only top-quality materials
Heavy cakes	1. Overbeaten or underbeaten eggs 2. Overmixing after flour has been added 3. Too much sugar 4. Too high a baking temperature	1. Beat eggs to wet peak 2. Fold flour in just enough to incorporate 3. Balance recipe 4. Regulate oven temperature
Coarse grain	1. Cold oven 2. Overbeaten whites 3. Insufficiently mixed batter	1. Regulate oven temperature 2. Whip to wet peak 3. Fold until smooth
Tough cakes	1. Overmixing ingredients 2. Excessive sugar content 3. Bakes too hot 4. Flour content high or wrong type flour used	1. Mix properly 2. Balance recipe 3. Regulate oven temperature 4. Balance recipe; use soft wheat flour
Dry cakes	1. Low sugar content 2. Overbaking 3. Eggs overbeaten 4. Flour content too high	1. Balance recipe 2. Lessen baking time 3. Whip to wet peak 4. Balance recipe

The amount of cake batter required for common cake varieties varies. See Table III. Most cake batters can be used to produce a variety of shapes and sizes. With the different size pans, the baking temperature will also vary.

TABLE III: CAKE CHART

Cake Type	Scaling Weight	Pan Size	Baking Temperature*
Layer	13 – 14 oz	8″ dia	375 – 385
Bar	5 – 6 oz	2¾″ × 10″	390 – 400
Ring	10 – 14 oz	6½″ dia	390 – 400
Loaf	11 – 24 oz	3¼″ × 7⅛″	385 – 400
Oval Loaf	8 oz	6¼″ long	385 – 400
Sheet	6 – 7 lb	17″ × 25″	375 – 390
Mary Ann	2 oz	3½″ dia	385 – 400

* in °F

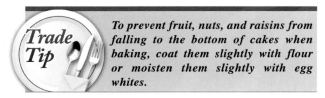

Trade Tip

To prevent fruit, nuts, and raisins from falling to the bottom of cakes when baking, coat them slightly with flour or moisten them slightly with egg whites.

Cake Baking at High Altitude

Most cake recipes are developed in low-altitude locations. The ingredients are balanced so they will produce good results when the cake is baked at or near sea level. If these recipes are to be used in high-altitude areas, adjustments to the recipe are required. This includes reducing the baking powder and sugar and increasing the liquid. The exact amount of ingredients required can be determined by experimenting with the recipe or by consulting with baking product manufacturers.

Rolling Sheet Cakes

Most sheet cakes can be rolled if they are rolled at the proper time and in the proper manner. Let the sheet cake cool to a temperature of approximately 100°F. Turn the cake out of the sheet pan onto a cloth that has been dusted generously with sugar. If the ends of the cake are crusted and slightly crisp, cut slits through them with a knife. The slits should be approximately ½″ in length. Roll the cake inward using the cloth as a support, but do not let the cloth roll into the cake roll. When rolled, let the cake cool thoroughly. When cooled, unroll and spread with the desired filling before re-rolling.

If the cake appears to be too dry to roll, place a wet bath towel on a very hot sheet pan. Cover the towel with a sheet of white freezer paper and turn the sheet cake out of its pan on top of the paper. The steam penetrates the cake and moistens it enough to allow it to be rolled without cracking.

The size of a sheet cake is determined by the size of sheet pan used.

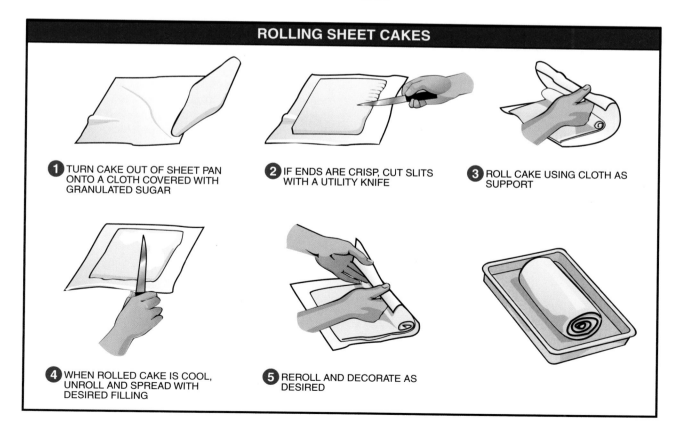

ROLLING SHEET CAKES

1 TURN CAKE OUT OF SHEET PAN ONTO A CLOTH COVERED WITH GRANULATED SUGAR

2 IF ENDS ARE CRISP, CUT SLITS WITH A UTILITY KNIFE

3 ROLL CAKE USING CLOTH AS SUPPORT

4 WHEN ROLLED CAKE IS COOL, UNROLL AND SPREAD WITH DESIRED FILLING

5 REROLL AND DECORATE AS DESIRED

Slicing Layer Cakes

Layer cakes are sliced when filling is to be added. Use a ham or roast beef slicer to produce a more accurate and even slice. The layer cake should be placed on a cake wheel. Cut in the center of the cake. The slicer should be in a horizontal position. Using a forward and backward motion, cut through the center of the cake while rotating the cake wheel.

Russell Harrington Cutlery, Inc.

A roast slicer may be used to slice layer cakes.

CUPCAKES

Most cake batters can be used to make cupcakes. Cupcakes are popular because they are an individual serving that can be iced and set up in a number of attractive and eye-appealing ways. A variation of a cupcake is the butterfly cupcake.

Butterfly Cupcakes

To make butterfly cupcakes:

1. Cut the top off the cupcake where the paper lines end.

2. Cut the top in half.

3. With a pastry bag and star tube, pipe buttercream icing around the edge of the cupcake.

4. Insert each half of the cut top at a 45° angle into the circle of icing to form wings. Fill the center with a spiral of icing, a maraschino cherry, and chopped nuts.

Cupcakes are an easily portable dessert popular for children's parties.

ICINGS

Sugar is the main ingredient in icing forms a protective coating around the item to seal in the moisture and flavor. It also improves the taste, and adds eye appeal. Icing is easy to prepare. Quality icing can be purchased from baker supply companies to save preparation time.

FORMING BUTTERFLY CUPCAKES

1 CUT TOP OFF AT TOP OF PAPER

2 CUT TOP IN HALF

3 PIPE BUTTERCREAM ICING AROUND EDGE OF CUPCAKE WITH PASTRY BAG

4 USE TOPS TO FORM WINGS

Icing Preparation

Icings are usually simple to prepare. The following basic rules should be observed for good results.

- Use the best ingredients, especially shortening if called for in the recipe.
- Use proper combinations of flavoring.
- Color icing in pastel shades for a more attractive appearance.
- Mix most buttercream icings at medium speed. Increase the mixing time to aerate the icing and increase the volume.
- Obtain proper consistency before applying or using the icing. In most cases, the consistency can be controlled by adding or eliminating certain amounts of powdered sugar. The consistency of the icing needed depends upon the use.

Icing Colors

Icing is colored to attract the eye and create a desire to purchase or consume the product. Standard principles regarding color used in icings include:

- Red and yellow create a hungry feeling.
- Pastel colors are more pleasing to the eye.
- Colors in paste and powder forms give better results than in liquid form.
- Certain colors can be blended to create other colors. For example, red and blue create violet, and blue and yellow create green.
- Use two or more color tones whenever possible by placing a layer of each color of icing side by side in a pastry bag.

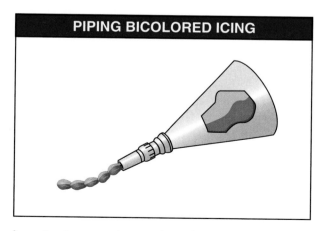

PIPING BICOLORED ICING

A pastry bag can be used to pipe two colors at the same time to create an attractive pattern.

Icing Classification

Icing is classified into six basic types:

Cream icing is one of the most popular types of icings used. It is simple to prepare, easy to keep, and adds eye appeal and taste. It is usually made by creaming shortening or butter, powdered sugar, and in some cases, eggs. Cream icings are light and aerated because more air cells can be retained with this method of mixing. Cream icing colors well. Use pastel shades for best results.

Flat icing is the simplest icing to prepare. It is usually prepared by blending water, powdered sugar, corn syrup, and flavoring. It is heated to approximately 100°F. It is applied by brush or hand to sweet rolls, doughnuts, Danish pastry, and others. Flat icing should be heated in a double boiler because direct heat or overheating causes it to lose its gloss when it cools.

Boiled icing is prepared by combining sugar, glucose, and water. It is boiled to approximately 240°F. The resulting syrup is added to an egg white meringue while still hot. If a heavy syrup is added to the meringue, a heavy icing will result. If a thin syrup is added, the result will be a thin icing. Boiled icing may be colored slightly and must be applied the same day it is prepared. It breaks down if held overnight. This icing is used on cakes and should be applied in generous amounts and worked into peaks.

Fudge icing is a rich, heavy-bodied icing that is usually prepared by adding a hot liquid or syrup to the other ingredients called for in the recipe, while whipping to obtain smoothness. Fudge should be used while still warm. However, if left to cool, it should be reheated in a double boiler before applying. Fudge icing is generally used to ice layer cakes, loaf cakes, and cupcakes. To store, cover and place in the refrigerator.

Fondant icing is a rich, white, cooked icing that hardens when exposed to the air. It is used mainly on small cakes (petit fours) that are picked up with the fingers to be eaten. It is prepared by cooking glucose, sugar, and water to a temperature of 240°F, letting it cool to 150°F, then working it (by mixing) until it is creamy and smooth. Fondant is the most difficult and time-consuming icing to prepare; therefore, most bakers purchase a ready-made fondant or a powdered product known as *drifond* from a baker's supply house. The ready-made product is usually purchased in a 40 lb tin and keeps well if covered with a damp cloth or a small amount of water to keep it from drying out when stored in a cool place. The drifond needs only water and a small amount of glucose added to produce an excellent fondant. When using drifond, one can prepare the amount needed in a very short time.

When needed for use, fondant is heated to about 100°F in a double boiler while stirring constantly. This causes the icing to become thin so it will flow freely over the item to be covered. The secret of covering an item successfully with fondant is the consistency of the icing. If the fondant is too heavy after it is heated, it can be thinned down by using a glaze consisting of one part glucose to two parts water or a regular simple syrup may be used. The fondant may be colored and flavored to suit the need. Exercise caution when heating the fondant. If it is heated over 100°F, it loses its gloss and when it hardens the product will have a dull finish. This icing can also be used as a base for other icings.

Royal icing is simple to prepare. Powdered sugar, egg whites, and cream of tartar are blended to the consistency desired. Royal icing sets up and hardens when exposed to air; therefore, it must be kept covered with a damp towel when not being used. It is used for decorating, flower making, and for dummy cakes used in window displays.

Storing Icing Properly

The type of icing being stored determines how it should be stored. The six basic types of icing and the proper storage method for each are:

- Cream or buttercream icing should be stored in a cool place and covered with plastic wrap or wax paper to avoid crusting. If a cool storage place outside the refrigerator can be found, use it for best results because refrigeration causes the shortening to harden. Considerable mixing would then be required to return the icing to spreading consistency.
- Flat icing should be kept covered with a damp cloth if it is out and not in use. To store, cover with a thin coating of water, plastic wrap, or wax paper. Remember, to reuse, it must be heated to approximately 100°F in a water bath.
- Boiled icing breaks down and loses its volume if stored overnight. Prepare only in amounts needed.
- Fudge icing dries rapidly when stored. It should be covered with plastic wrap and stored in the refrigerator. To reuse, it must be heated slightly in a water bath.

- Fondant icing, like flat icing, must be kept covered with a very thin coating of water, plastic wrap, or wax paper in a cool place. It can be refrigerated, but may lose some gloss when reheated in a double boiler or water bath.
- Royal icing should be stored in a cool place covered with a damp cloth or a very thin film of water to prevent crusting.

Pastry Bags

Filling and using pastry bags properly requires a certain amount of knowledge and skill. A pastry bag made of plastic, canvas, parchment, or parchment paper may be used. Most decorators prefer to make their own pastry bag using parchment or parchment paper. Silicon or parchment paper is used because neither absorbs moisture, which would cause the bag to break. The paper cones are simple to make, easy to handle, and a cone can be set up for each color used. When finished decorating, the paper cones and disposable plastic bags can be discarded whereas the canvas bags must be washed.

When writing with icing, the paper cone is most convenient because a metal tip does not have to be inserted. The tip of the paper cone can be cut to the correct size for the letters to be formed. The canvas and plastic bags as well as the paper cone, when not used for writing, require that a metal tip be inserted in the tip of the bag or cone before filling with icing. Many kinds of metal tips are used to make different designs. There are tips for different kinds of flowers, leaves, and borders. Use the correct tip for each job.

To fill the plastic or canvas bag with icing, the left hand is placed around the middle of the bag using a very delicate grip, and the top half of the bag is drawn over the left hand. The icing is inserted into the bag using a spatula or kitchen spoon. As the utensil is withdrawn, it is gripped by the left hand, which is covered with the top portion of the bag, so all the icing can be removed before the utensil is completely withdrawn. If left-handed, the positions are reversed. When filling the paper cone, only a spatula should be used and the icing deposited in the center of the cone. After the bag or cone is filled, fold over the top several times to prevent the icing from coming out when pressure is applied.

Regardless of which bag or cone is used, only fill about one-half to three-fourths full, deposit the icing down to the tip and away from the top sides, and leave no air pockets in the icing. If this is not done, the decorating job can be ruined if, instead of a smooth flow of icing, a burst of air comes forth.

When decorating, the bag is held with the right hand at the top of the bag and the left hand lightly gripping the lower half. If left-handed, the positions are reversed. The hand at the top of the bag applies all the pressure to cause the icing to flow. The hand on the lower half is used only as a guide. In all decorating tasks, the two most important factors are holding the bag at the correct angle and applying correct pressure to the bag to obtain a smooth, even flow of icing.

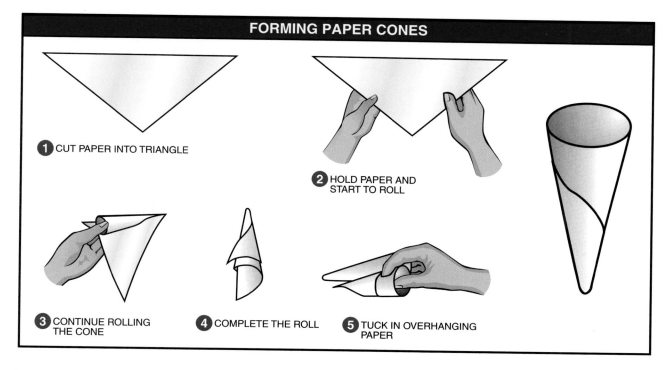

FORMING PAPER CONES

1. CUT PAPER INTO TRIANGLE
2. HOLD PAPER AND START TO ROLL
3. CONTINUE ROLLING THE CONE
4. COMPLETE THE ROLL
5. TUCK IN OVERHANGING PAPER

CAKE: CREAMING RECIPES

Sunny Orange Cake

This is a moist, tender cake with a refreshing orange flavor. This batter can be used for preparing cupcakes, loaves, layer cakes, or ring cakes.

thirteen to eighteen 8″ cakes

Equipment

- Mixing machine and paddle
- Baker's scale
- Plastic scraper
- 8″ cake pans (18)
- Wire whip
- 1 gal. stainless steel container
- Food grinder

Ingredients

2 lb 8 oz cake flour
1 lb 6 oz emulsified vegetable shortening
3 lb 8 oz sugar
1 ½ oz salt
2½ oz baking powder
1 lb skim milk
1 lb 8 oz eggs
1 lb 6 oz skim milk
8 oz ground oranges

Preparation

1. Grind the oranges in a food grinder using the medium chopper plate.

2. Prepare 8″ cake pans. Grease the sides and cover the bottom with parchment paper.
3. Preheat the oven to 375°F.
4. Scale all ingredients carefully using a baker's scale.

Procedure

1. Place the flour and shortening in the stainless steel mixing bowl. Mix at slow speed using the paddle for 3 min to 5 min. Scrape down the bowl and paddle with a plastic scraper at least once in this stage.
2. Add the sugar, salt, baking powder, and skim milk. Mix at slow speed from 3 min to 5 min. Scrape down the bowl.
3. In a stainless steel container, combine the second amount of milk, eggs, and oranges. Beat slightly using a wire whip. Add half of this mixture to the bowl and mix at slow speed until smooth. Scrape down the bowl and mix smooth again.

4. Add the balance of the liquid mixture and continue mixing at slow speed for a total of 3 min to 5 min in this stage, scraping down again to ensure a smooth batter.
5. Scale 12 oz to 14 oz of batter into each prepared 8″ cake pan.
6. Bake in a preheated oven at 375°F until golden brown. Remove from the oven when done.

Precautions

- Scale all ingredients. Double-check all weights.
- All mixing should be done at slow speed.
- Scrape down the bowl at least once during each mixing stage to ensure a smooth batter.
- Check oven temperature with an oven thermometer.

Brown Sugar Cake

This cake is a dark, moist, tender-textured cake with a maple flavor, possessing excellent eating qualities. The batter is mixed using the creaming method.

eight or nine 8″ cakes

Equipment

- Mixing machine and paddle
- Baker's scale
- Plastic scraper
- 8″ cake pans (9)
- Flour sifter
- 1 gal. stainless steel container
- Wire whip

Ingredients

2 lb dark brown sugar
12 oz shortening
¼ oz vanilla
maple flavoring to taste
1 lb eggs
1 lb 12 oz cake flour
½ oz baking powder
½ oz salt
1 lb 8 oz milk

Preparation

1. Prepare 8″ cake pans. Grease the sides and dust lightly with flour. Cover the bottom with parchment paper.
2. Preheat the oven to 375°F.
3. Place the eggs in a stainless steel container and beat slightly with a wire whip.
4. Scale the ingredients carefully using a baker's scale.

Procedure

1. Place the shortening, brown sugar, vanilla, and maple flavoring in the stainless steel mixing bowl. Cream at slow speed using the paddle until light and fluffy.
2. Add the beaten eggs gradually, continuing to cream at slow speed.
3. Sift the flour, baking powder, and salt. Add alternately with the milk to the creamed mixture, mixing at slow speed until the batter is smooth. Scrape down the bowl with a plastic scraper.
4. Scale 12 oz to 14 oz of batter into each prepared 8″ cake pan.
5. Bake in a preheated oven at 375°F until done. Remove from the oven.

Precautions

- Scale all ingredients. Double-check all weights.
- Scrape down the bowl at least once in the final mixing stage to ensure a smooth batter.
- Check the oven temperature with an oven thermometer.

Apple-Nut Cake

This is a rich, spicy, moist, apple-flavored cake that can be formed and baked to produce many different varieties. Apple-nut cake is mixed using the creaming method. It is topped with a pecan mixture before it is baked to produce a sugar/nut topping on the finished product.

ten 8″ cakes

Equipment

- Mixing machine and paddle
- Baker's scale
- Plastic scraper
- 8″ cake pans (10)
- Apple corer
- Paring knife
- French knife
- Flour sifter
- Kitchen spoon

Ingredients

2 lb 8 oz sugar
6 oz shortening
6 oz butter
¼ oz salt
⅛ oz cinnamon
⅛ oz mace
8 oz eggs
1 lb milk
¾ oz baking soda
1 oz baking powder
2 lb 12 oz cake flour
2 lb 12 oz fresh apples

Preparation

1. Core the apples with an apple corer, peel with a paring knife, and chop fine with a French knife.
2. Prepare the topping by blending:
   10 oz chopped pecans
   6 oz butter
   ¼ oz cinnamon
   2 oz sugar
3. Preheat the oven to 375°F.
4. Prepare 8″ cake pans for baking.
5. Scale all ingredients carefully using a baker's scale.

Procedure

1. Place the sugar, shortening, butter, salt, mace, and cinnamon in the stainless steel mixing bowl. Cream at slow speed using the paddle.
2. Add the eggs and continue to cream at slow speed until thoroughly blended.

3. In a stainless steel container, dissolve the baking soda in the milk. Add to the mixture and mix at low speed until thoroughly blended.
4. Combine the cake flour and baking powder. Sift with a flour sifter and add to the mixture. Mix at slow speed to a smooth batter. Scrape down the bowl with a plastic scraper.
5. Fold in the chopped apples with a kitchen spoon until thoroughly blended.

6. Scale 1 lb of batter into each prepared 8″ cake pan. Sprinkle a small amount of the pecan mixture on top of the batter.
7. Bake in a preheated oven at 375°F until golden brown. Remove from the oven.

Precautions

- Scrape down the bowl at intervals during the mixing period to ensure a smooth batter.
- Scale all ingredients. Double-check all weights.

Spice Cake

A high percentage of sugar is used in this cake batter. Spice cake is mixed in three stages using the creaming method. The batter may also be used for cupcakes and loaf and ring cakes.

fifteen 8″ cakes

Equipment

- Mixing machine and paddle
- Baker's scale
- Plastic scraper
- 8″ cake pans (15)
- Stainless steel container

Ingredients

1 lb 14 oz cake flour
1 lb 6 oz emulsified vegetable shortening
3 lb 5 oz sugar
9 oz cake flour
¼ oz salt
¾ oz baking soda
2 oz baking powder
1 oz spice mix
1 lb 4 oz buttermilk
1 lb 10 oz eggs
2 lb 7 oz canned pumpkin

Preparation

1. Prepare the spice mix by combining:
    4 oz cinnamon
    1½ oz mace
    ½ oz allspice
    1½ oz nutmeg
    ½ oz ginger
2. Preheat the oven to 375°F.

3. Scale all ingredients carefully using a baker's scale.

Procedure

1. Place the flour and shortening in a stainless steel mixing bowl. Cream and mix at slow speed using the paddle for 4 min. Scrape down the bowl with a plastic scraper at least once in this stage.
2. Add the sugar, second amount of flour, salt, soda, baking powder, spice mix, and buttermilk. Continue to mix at slow speed for 4 min. Scrape down at least once in this stage.
3. In a stainless steel container, blend the pumpkin and eggs and add approximately half of the mixture to the bowl. Mix at slow speed until smooth. Scrape down and mix until smooth again.
4. Add the balance of the egg and pumpkin mixture and continue mixing at slow speed for 3 min to 5 min. In this stage, scrape down again to ensure a smooth batter.
5. Scale 12 oz to 14 oz of batter into each 8″ cake pan. Bake in a preheated oven at 375°F until golden brown. Remove from the oven.

Precautions

- Scrape down the bowl at intervals during the mixing period to ensure a smooth batter.
- Scale all ingredients. Double-check all weights.

Eggnog Cake

This is a moist, very tender cake with the unusual flavor of eggnog. The batter can be used for preparing cupcakes, sheet cakes, or layer cakes. The batter is mixed using the creaming method.

thirteen to fifteen 8″ cakes

Equipment

- Baker's scale
- Mixing machine and paddle
- Plastic scraper
- 8″ cake pans (15)
- Stainless steel mixing bowl
- Wire whip

Ingredients

2 lb 8 oz cake flour
1 lb 12 oz emulsified vegetable shortening
3 lb 2 oz sugar
1½ oz salt
2½ oz baking powder
4 oz dry milk
12 oz water
2 lb 4 oz eggs
14 oz water
¼ oz nutmeg
rum flavor to taste

Preparation

1. Prepare 8″ cake pans. Grease the sides and cover the bottom with parchment paper.
2. Preheat the oven to 375°F.
3. Scale all ingredients. Double-check all weights.

Procedure

1. Place the flour and shortening in the stainless steel mixing bowl. Mix at slow speed using the paddle for 3 min to 5 min. Scrape down the bowl with a plastic scraper.
2. Add the sugar, salt, baking powder, dry milk, and the first amount of water. Continue to mix at slow speed for 3 min to 5 min. Scrape down the bowl a second time.

3. In a stainless steel bowl, place the eggs, second amount of water, nutmeg, and rum flavor. Blend with a wire whip.
4. Add half of the liquid to the ingredients in the stainless steel mixing bowl. Mix at slow speed until smooth.
5. Add the balance of the liquid mixture and mix for 3 min to 5 min. Scrape down the bowl at least once to ensure a smooth batter.
6. Scale 12 oz to 14 oz of batter into each prepared 8″ cake pan.
7. Bake in a preheated oven at 375°F until golden brown. Remove from the oven when done.

Precautions

- Scale all ingredients. Double-check all weights.
- All mixing should be done at slow speed.
- Scrape down the bowl at least once during each mixing stage to ensure a smooth batter.
- When placing the cakes in the oven, be sure the pans do not touch. If the pans touch, raw spots could occur in the cakes.

Fruitcake

This is the traditional yuletide treat. The dark, moist, heavy-textured cake is full of rich fruit and nuts. The batter is mixed using the creaming method and can be baked in the form of loaves or rings.

fourteen 6″ ring cakes or loaf cakes

Equipment

- Mixing machine and paddle
- Baker's scale
- Baking pans as desired
- Plastic scraper
- 3 gal. stainless steel container
- Kitchen spoon
- 1 qt saucepan
- Colander
- Paring knife
- French knife
- 6″ ring pans (14) or 7⅛″ × 3¼″ × 2¼″ loaf pans (14)

Ingredients

2 lb sugar
1 lb bread flour
1½ oz salt
⅛ oz baking soda
2 lb emulsified vegetable shortening
2 lb eggs
1 lb 8 oz bread flour
8 lb 12 oz fruit mix
1 lb 4 oz water
brandy flavor to taste
4 oz dark molasses
5 lb raisins
mace to taste
cinnamon to taste
1 lb black walnuts
1 lb pecans

1 lb water
Bring to a boil in a saucepan. Use this solution warm as a wash.
4. Preheat the oven to the desired temperature.
5. Wash the fruit thoroughly in a colander and allow it to drain.
6. Cut the washed fruit into medium-sized pieces with a paring knife. Place in a stainless steel container. Add the water, flavoring, raisins, molasses, and spices. Place this mixture in the refrigerator, cover, and let set overnight. This ensures full flavor in the cake.
7. Chop the nuts into medium-sized pieces with a French knife.
8. Scale all ingredients carefully using a baker's scale.

Preparation

1. Prepare pans selected for baking. Grease the sides lightly and cover the bottoms with parchment paper.
2. Prepare the fruit mix using:
    2 lb 8 oz glazed red cherries
    1 lb 8 oz glazed green cherries
    3 lb 4 oz glazed pineapple
    8 oz citron
    8 oz orange peel
    8 oz lemon peel
    Mix all ingredients by hand until thoroughly blended.
3. Prepare a glucose wash using:
    2 lb glucose

Procedure

1. Place the sugar, soda, emulsified vegetable shortening, and first amount of bread flour in the stainless steel mixing bowl. Cream at slow speed using the paddle until light and smooth.
2. Add the eggs slowly and continue to cream at slow speed.
3. Add the second amount of bread flour. Mix for 3 min.
4. Remove the fruit mixture from the refrigerator. Add the chopped nuts and mix thoroughly.
5. Add the fruit/nut mixture to the batter and mix well at slow speed.
6. After the cakes are baked, wash generously with the glucose wash using a pastry brush.

Precautions

- Scale all ingredients. Double-check all weights.
- The cakes should either be covered during baking or baked in an oven containing moisture in order to produce moist cakes and to prevent the tops from becoming too dark.

TABLE IV: FRUITCAKE			
Scaling Weight	**Pan Size**	**Baking Temperature**	**Baking Time**
1 lb 12 oz	6″ ring pan	340°F – 350°F	Approx 1½ hr
1 lb 12 oz	7⅛″ × 3¼″ × 2¼″ loaf pan	340°F – 350°F	Approx 1½ hr

German Chocolate Cake

This is a very rich, chocolate-flavored cake that uses a rich German sweet chocolate in the preparation. A pecan/coconut filling is placed between the layers and spread on top of the cake for added richness and moistness. This cake is prepared using the creaming method.

ten 8″ layer cakes

Equipment

- Baker's scale
- Mixing machine, paddle, and wire whip
- Plastic scraper
- 8″ layer cake pans (10)
- 2 qt sauce pot
- Parchment paper
- Stainless steel mixing bowl

Ingredients

8 oz emulsified vegetable shortening
8 oz butter
1 lb 12 oz sugar
8 oz egg yolks
8 oz German sweet chocolate
8 oz boiling water
½ oz salt
vanilla to taste
¼ oz baking soda
1 lb 2 oz cake flour
1 lb buttermilk
8 oz egg whites

Preparation

1. Place the boiling water and salt in a saucepan. Add the chocolate and let it melt. Cool thoroughly.
2. Separate eggs to acquire yolks and whites.
3. Prepare the 8″ layer cake pans. Cover the bottom with parchment paper and grease the sides lightly.
4. Preheat the oven to 375°F.

Procedure

1. Place the shortening, butter, and sugar in the stainless steel mixing bowl. Cream at slow speed using the paddle.
2. Add the yolks slowly while continuing to mix at slow speed.
3. Add the melted chocolate solution with the vanilla. Mix until blended thoroughly.
4. Sift the flour and baking soda together and add alternately with the buttermilk. Mix until smooth, scrape down the bowl and paddle with a plastic scraper, and mix smooth a second time.
5. Remove the batter from the mixing machine and place in a stainless steel mixing bowl.
6. Wash the mixing bowl, wipe dry, and add egg whites. Using the wire whip, whip at high speed until stiff.
7. Fold the beaten egg whites gently into the chocolate batter.
8. Place 10 oz of batter into each 8″ prepared layer cake pan.
9. Bake in a preheated oven at 375°F until cakes are set and firm.
10. Remove the cakes from the oven. Remove from the pans immediately to reduce shrinkage. Let cool.
11. When setting up the German chocolate cake, place German chocolate cake filling in the center of each of the three layers and spread the filling over the top of the cake.

Note: A white German chocolate cake may be made by substituting white candy coating for German sweet chocolate.

Precautions

- Scale all ingredients. Double-check all weights.
- Scrape down the mixing bowl at least once during the mixing period.
- Do not let cake pans touch during the baking period or raw spots may occur.
- Test cakes for doneness before removing from oven.

CAKE: SPONGE RECIPES

Semi-Sponge Cake

The leavening for this cake is provided by eggs and baking powder. In a true sponge cake, the eggs provide all the leavening. This cake may be used to prepare Boston cream pie, strawberry shortcake, or roll cake.

sixteen 8″ cakes

Equipment

- Mixing machine and wire whip
- Plastic scraper
- Baker's scale
- 8″ cake pans (16)

Ingredients

2 lb 8 oz cake flour
3 lb sugar
1½ oz salt
2½ oz baking powder
¼ oz baking soda
1 lb egg yolks
1 lb eggs
8 oz water
4 oz dry milk
8 oz salad oil
1 lb 4 oz water
vanilla to taste

Preparation

1. Prepare cake pans. Grease the bottom and sides of each pan lightly, cover the bottom with wax paper, and grease the wax paper slightly.
2. Preheat the oven to 375°F.
3. Separate the egg yolks from the whites.
4. Scale all ingredients. Double-check all weights.

Procedure

1. Place the flour, sugar, salt, baking powder, baking soda, egg yolks, eggs, water, and dry milk in the stainless steel mixing bowl. Mix at medium speed using the wire whip for 5 min to 10 min or until mixture becomes lemon-colored. Scrape down the bowl with a plastic scraper.

2. Add the salad oil, second amount of water, and the vanilla. Mix at slow speed for 2 min or 3 min or until blended thoroughly.
3. Scale approximately 10 oz of batter into each prepared cake pan.
4. Bake in a preheated oven at 375°F for 12 min to 15 min. Remove from the oven.

Note: To convert this recipe to a chocolate semi-sponge cake, add 8 oz of cocoa in the first stage and increase the first stage water from 8 oz to 12 oz.

Precautions

- Scale all ingredients. Double-check all weights.
- Check the oven temperature before baking.

Banana Chiffon Cake

This is a light, fluffy, spongy cake with a true banana flavor. The secret of a successful chiffon cake lies in proper mixing and folding the egg white mixture into the batter gently so the air cells will not be broken.

4 cakes

Equipment

- Mixing machine and paddle
- Baker's scale
- Plastic scraper
- 10″ × 4″ center-tube cake pans (4)
- French knife
- Skimmer
- Flour sifter
- Kitchen spoon

Ingredients

1 lb 12 oz cake flour
1 lb 6 oz sugar
1¼ oz baking powder
½ oz salt
14 oz egg yolks
14 oz salad oil
14 oz bananas
8 oz water
banana flavor to taste
1 lb 12 oz egg whites
1 lb sugar
¼ oz cream of tartar

Procedure

1. Sift the flour, the first amount of sugar, baking powder, and salt into the mixing bowl with a flour sifter. Scrape down the bowl with a plastic scraper.
2. Mix at medium speed using the paddle while adding the salad oil, egg yolks, water, and banana flavor in several portions. Mix until smooth. Remove from the mixer.
3. Blend in the chopped bananas with a kitchen spoon.
4. In a separate stainless steel mixing bowl, place the egg whites and cream of tartar. Beat at high speed using a wire whip while adding the second amount of sugar gradually until stiff peaks form.
5. Using a flat skimmer, fold the egg white mixture onto the batter until well-blended.
6. Scale 1 lb 14 oz of batter into each 10″ × 4″ prepared cake pan.
7. Bake in a preheated oven at 350°F until golden brown. Remove from the oven.

Preparation

1. Chop the bananas very fine with a French knife.
2. Separate the eggs by hand.
3. Prepare the 10″ × 4″ center-tube cake pans. Grease lightly and then dust with flour, or grease the sides and cover the bottom with parchment paper.
4. Preheat the oven to 350°F.
5. Scale all ingredients carefully using a baker's scale.

Precautions

- Use a high-grade salad oil.
- Scale all ingredients. Double-check all weights.
- When adding the egg white mixture to the batter, fold only enough to blend the two together. Overworking will break down the air cells.
- Bake immediately after mixing the batter.
- Remove the cakes from the oven and turn upside down immediately to let them cool.

Jelly Roll Sponge Cake

This is a soft, sponge-textured cake that rolls with ease. It can also be rolled with ice cream, lemon filling, vanilla filling, or pineapple filling. It is mixed using the sponge or whipping method.

2 sheet cakes

Equipment

- Mixing machine and whip
- Baker's scale
- Plastic scraper
- 17″ × 24½″ sheet pans (2)
- 1 qt saucepan
- Kitchen spoon
- 1 gal. stainless steel container
- Kitchen spoon

Ingredients

12 oz eggs
8 oz egg yolks
1 lb 10 oz sugar
½ oz salt
vanilla to taste
1½ oz dry milk
12 oz water
4 oz honey
1 lb 6 oz cake flour
½ oz baking powder

Preparation

1. Preheat oven to 375°F.
2. Prepare the sheet pans. Line them with wax or parchment paper.
3. Combine the water and honey. Blend in a saucepan and heat.
4. Scale all ingredients carefully using a baker's scale.

Procedure

1. Place the eggs, egg yolks, sugar, salt, vanilla, and dry milk in the stainless steel mixing bowl. Beat at high speed using the whip for approximately 10 min until mixture becomes lemon-colored.

2. Pour the warm water/honey mixture into the mixture while continuing to mix at medium speed.
3. Combine the cake flour and baking powder. Sift together and fold into the mixture very gently with a kitchen spoon.
4. Pour the batter in two 17″ × 24½″ prepared sheet pans. Scrape the bowl clean with a plastic scraper.
5. Bake in a preheated oven at 375°F until golden brown (approximately 12 min to 15 min).
6. Remove from the oven and turn the cakes out onto white cloths sprinkled with sugar or coconut. Spread with jelly or desired filling and roll.

Note: To prepare a chocolate roll cake, use the same mixture, but change step 3 to:
   1 lb cake flour

6 oz cocoa
½ oz baking soda
½ oz baking powder
Proceed as instructed.

Precautions

- Scale all ingredients. Double-check all weights.
- Bake the cake on the light side.
- Turn the cake out of the pan as soon as it is removed from the oven.
- Spread the batter over the pan evenly before placing it in the oven.

Lemon Chiffon Cake

This is a light, fluffy, spongy cake similar to angel food cake. The egg white mixture must be folded into the batter very gently.

4 cakes

Equipment

- Mixing machine, paddle, and whip
- Baker's scale
- Plastic scraper
- 10″ × 4″ center-tube cake pans (4)
- Box grater
- Skimmer
- Flour sifter
- Wire rack

Ingredients

1 lb 8 oz cake flour
2 lb sugar
1½ oz baking powder
12 oz salad oil
14 oz egg yolks
1 lb cold water
½ oz lemon flavor
1½ oz lemon rind
1 lb 4 oz egg whites
¼ oz cream of tartar
½ oz salt

Procedure

1. Sift the flour, sugar, and baking powder into the stainless steel mixing bowl.
2. Mix at medium speed using the paddle while adding the salad oil, egg yolks, water, and lemon flavor in several portions. Mix until smooth. Scrape down the bowl with a plastic scraper. Do not overmix.
3. Place the cream of tartar, salt, and egg whites in a separate mixing bowl. Beat using the wire whip until they form very stiff peaks. Remove from the mixer.
4. Fold the egg white mixture into the batter using a flat skimmer until well-blended.
5. Scale 1 lb 14 oz of batter into each 10″ × 4″ prepared cake pan.
6. Bake in a preheated oven at 325°F until golden brown. Remove from the oven.

Preparation

1. Grate the lemons on the medium grid of the box grater.
2. Separate the eggs by hand.
3. Prepare the 10″ × 4″ center-tube cake pans. Grease and dust with flour, or grease and dust the sides and cover the bottom with parchment paper.
4. Preheat the oven to 325°F.
5. Scale all ingredients carefully.

Precautions

- Use a high-grade salad oil.
- Scale all ingredients. Double-check all weights.
- When adding the egg white mixture to the batter, fold only enough to blend the two together.
- Bake immediately after mixing the batter.
- When the cakes are removed from the oven, turn upside down immediately on a wire rack to let them cool.

Trade Tip

Dusting cakes with confectioners' sugar is sometimes done to hide mistakes caused in mixing or baking. The idea is to improve appearance when served. The sugar will adhere to the cake better and last longer if the cake is sprinkled with granulated sugar before sifting on the confectioners' sugar.

CAKE: TWO-STAGE RECIPES

White Cake

This is a fine-grain, white, soft-textured cake with excellent eating qualities. It is mixed using the blending or two-stage method. A number of different cakes can be made from this basic recipe.

weight of mix 11lb

Equipment

- Mixing machine and paddle
- Baker's scale
- Plastic scraper
- Cake pans as desired
- 1 gal. stainless steel container

Ingredients

2 lb 8 oz cake flour
1 lb 12 oz emulsified vegetable shortening
3 lb 2 oz sugar
1½ oz salt
2½ oz baking powder
14 oz water
2½ oz nonfat dry milk
10 oz eggs
1 lb egg whites
1 lb water
vanilla to taste

Preparation

1. Preheat the oven to the required temperature.

2. Prepare baking pans selected.
3. Scale all ingredients carefully using a baker's scale.

Procedure

1. Place the first seven ingredients in the stainless steel mixing bowl. Mix at slow speed using the paddle for 5 min. Scrape down the bowl and paddle with a plastic scraper at least once in this stage.
2. Scale eggs, water, and flavor in a stainless steel container and add approximately half of the mixture to the bowl. Mix at slow speed until smooth. Scrape down and mix until smooth again.
3. Add the balance of the liquid ingredients and continue mixing at slow speed for 3 min. Scrape down again.

Precautions

- Scrape down the bowl at intervals during the mixing period to ensure a smooth batter.
- Scale all ingredients. Double-check all weights.

Yellow Cake

This is a fine-grain, yellow, soft-textured cake with excellent eating qualities. It is mixed using the blending or two-stage method. A number of cakes can be made using this basic yellow cake batter.

weight of mix 11 lb

Equipment

- Mixing machine and paddle
- Baker's scale
- Plastic scraper
- Cake pans as desired
- 1 gal. stainless steel container

Ingredients

2 lb 8 oz cake flour
1 lb 6 oz emulsified vegetable shortening
3 lb 2 oz sugar
1 oz salt
1¾ oz baking powder
4 oz nonfat dry milk
1 lb 4 oz water
1 lb 10 oz eggs
12 oz water
vanilla to taste

Preparation

1. Preheat the oven to required temperature.
2. Prepare baking pans selected.

3. Scale all ingredients carefully using a baker's scale.

Procedure

1. Place the first seven ingredients in the stainless steel mixing bowl. Mix at slow speed using the paddle for 5 min. Scrape down the bowl and paddle with a plastic scraper at least once in this stage.
2. Scale eggs, water, and flavor in a stainless steel container and add approximately half of the mixture to the bowl. Mix at slow speed until smooth. Scrape down and mix smooth again.
3. Add the balance of the liquid ingredients and continue mixing at slow speed for 3 min. Scrape down again.

Precautions

- Scrape down the bowl at intervals during the mixing period to ensure a smooth batter.
- Scale all ingredients. Double-check all weights.

Trade Tip

Dry milk in icing preparation gives excellent results and is more economical to use. However, if not treated properly, it may cause lumps in the finished product. To eliminate this possibility, dissolve the dry milk in the liquid ingredients and strain before adding to the other ingredients.

Devil's Food Cake

This is a soft, tender cake with a rich chocolate flavor. Devil's food cake is mixed using the blending or two-stage method. Many different cakes can be made from this batter.

weight of mix 13 lb

Equipment

- Mixing machine and paddle
- Baker's scale
- Plastic scraper
- Cake pans as desired
- 1 gal. stainless steel container

Ingredients

2 lb 8 oz cake flour
1 lb 6 oz emulsified vegetable shortening
3 lb 8 oz sugar
8 oz cocoa
1½ oz salt
¾ oz baking soda
1½ oz baking powder
6 oz nonfat dry milk
1 lb 4 oz water
1 lb 14 oz eggs
1 lb 9 oz water
vanilla to taste

Procter and Gamble Co.

Preparation

1. Preheat the oven to the required temperature.
2. Prepare baking pans selected.
3. Scale all ingredients carefully using a baker's scale.

Procedure

1. Place the first nine ingredients in the stainless steel mixing bowl. Mix for 5 min using the paddle. Scrape down the bowl and paddle with a plastic scraper at least once in this stage.
2. Scale eggs, water, and flavor and add approximately half of the mixture to the bowl. Mix at slow speed until smooth. Scrape down and mix again until smooth.
3. Add the balance of the liquid ingredients and continue mixing at slow speed for 3 min. Scrape down again.

Precautions

- Scrape down the bowl at intervals during the mixing period to incorporate all ingredients and to ensure a smooth batter.
- Scale all ingredients correctly. Double-check all weights.

Fudge Cake

This is a rich, flavorful, chocolate-colored cake. The cake is rich in sugar and eggs and produces a product with above-average eating qualities. It is mixed using the blending or two-stage method and can be used to produce many fudge cake varieties.

weight of mix 11 lb

Equipment

- Mixing machine and paddle
- Baker's scale
- Plastic scraper
- Cake pans as desired
- 1 gal. stainless steel container

Ingredients

2 lb 2 oz cake flour
6 oz cocoa
1 lb 12 oz emulsified vegetable shortening
3 lb 2 oz sugar
1½ oz salt
¾ oz baking soda
1½ oz baking powder
1 lb water
3½ oz nonfat dry milk
2 lb 4 oz eggs
10½ oz water
flavor to taste

Preparation

1. Preheat the oven to the required temperature.
2. Prepare baking pans selected.
3. Scale all ingredients carefully using a baker's scale.

Procedure

1. Place the first nine ingredients in the stainless steel mixing bowl. Mix at slow speed using the paddle for 5 min. Scrape down the bowl and paddle with a plastic scraper at least once in this stage.
2. In a stainless steel container, scale the eggs, water, and flavor and add approximately half of the mixture to the bowl. Mix at slow speed until smooth. Scrape down and mix smooth again.
3. Add the balance of the liquid ingredients and continue mixing at slow speed for 3 min. Scrape down again.

Precautions

- Scrape down the bowl at intervals during the mixing period to ensure a smooth batter.
- Scale all ingredients correctly. Double-check all weights.

Yellow Pound Cake

This is a rich, golden yellow, fine, and smooth-textured cake. The cake is an excellent choice when setting up baked Alaska, eaten by itself, or served topped with fruit. It is mixed using the blending method.

weight of mix 10½ lb

Equipment

- Mixing machine and paddle
- Baker's scale
- Plastic scraper
- Loaf pans as desired
- Stainless steel mixing bowl
- Wire whip

Ingredients

2 lb 8 oz cake flour
1 lb 12 oz emulsified vegetable shortening
3 lb sugar
1½ oz salt
1 lb 4 oz water
2½ oz dry milk
1 lb 12 oz eggs
vanilla to taste

Preparation

1. Preheat oven to either 350°F (for 1 lb loaf cakes) or 330°F (for 3 lb cakes).
2. Prepare the loaf pans selected. The size of the pan is determined by the weight of the cake desired (1 lb or 3 lb). For 1 lb cakes, use 4½″ × 8½″ loaf pans. For 3 lb cakes, use 4½″ × 13″ × 3″ loaf pans.
3. Scale off all ingredients carefully with a baker's scale.

Procedure

1. Place the first six ingredients in the stainless steel mixing bowl. Mix at medium speed using the paddle for 6 min to 8 min. Scrape down the bowl and paddle with a plastic scraper.
2. Place the eggs and vanilla in a stainless steel bowl. Whip slightly with a wire whip. Add half of this mixture to the blended ingredients in the stainless steel mixing bowl. Mix at slow speed until smooth.
3. Add the balance of the egg mixture and continue to mix at slow speed until batter is smooth. Scrape down bowl and mix for 1 min more to ensure a very smooth batter.
4. Bake the 1 lb cakes at 350°F or the 3 lb cakes at 325°F until golden brown. Remove from the oven.

Note: The baking time for a 1 lb cake is approximately 70 min. For the 3 lb cake, it is approximately 2 hr.

Precautions

- Scale all ingredients correctly. Double-check all weights.
- Scrape down the bowl at intervals during the mixing period to ensure a smooth batter.

CAKE ICING RECIPES

New York Buttercream Icing

4 qt

Ingredients

2 lb 8 oz emulsified vegetable shortening
1 lb butter
½ oz salt
12 oz nonfat dry milk
5 lb powdered sugar
1 lb water heated to 110°F
1 oz vanilla

Procedure

1. Place all the ingredients in the stainless steel mixing bowl. Beat at medium speed using the paddle for 5 min, then beat at high speed for 2 min or to desired lightness.

Buttercream Icing for Decorating

1½ qt

Ingredients

1 lb 8 oz shortening
2 lb 8 oz powdered sugar
1 oz egg whites
2 oz cornstarch (variable)
(use cornstarch only in hot weather)

Procedure

1. Place all the ingredients in the stainless steel mixing bowl. Beat at medium speed using the paddle until smooth.

Trade Tip

When adding cocoa to an icing, sift the cocoa with the confectioners' sugar. This will give complete dispersion and prevent the formation of cocoa lumps in the finished product. Once the lumps form in the icing, it is almost impossible to eliminate them.

Flat Icing

1 qt

Ingredients

2 lb 8 oz powdered sugar
4 oz corn syrup
2 oz egg whites
8 oz hot water (variable)

Procedure

1. Place the powdered sugar, corn syrup, and egg whites in the stainless steel mixing bowl. Mix at slow speed using the paddle while adding the hot water.
2. Mix until smooth.

Note: When ready to use, heat the amount of icing needed in a double boiler. Use it on rolls, Danish pastry, coffee cakes, etc.

Roll Icing

1 qt

Ingredients

2 lb 8 oz powdered sugar
4 oz salad oil
6 oz glucose
¼ oz salt
4 oz hot water (variable)
flavoring to taste

Procedure

1. Place the powdered sugar, salad oil, glucose, and salt in the mixing bowl. Mix at slow speed using the paddle while adding the hot water.
2. Mix until smooth.
3. Stir in the flavoring until thoroughly blended.

White Cream Icing

3 qt

Ingredients

1 lb 4 oz emulsified vegetable shortening
½ oz salt
5 oz nonfat dry milk
14 oz water
vanilla to taste
5 lb powdered sugar

Procedure

1. Place all the ingredients in the stainless steel mixing bowl. Mix at slow speed using the paddle for about 5 min.
2. Whip at medium speed for 10 min to 15 min to acquire desired lightness.

Variations

Add the following ingredients to 5 lb of icing.
Nut icing: 8 oz chopped nuts
Raisin icing: 8 oz ground raisins
Cherry icing: 8 oz chopped cherries
Candied fruit icing: 8 oz chopped fruit
Jam or marmalade icing: 8 oz jam or marmalade
Coconut icing: 8 oz macaroon coconut

Procter and Gamble Co.

Fondant icing: 2 lb 8 oz fondant
Peppermint candy icing: 4 oz crushed peppermint candy
Cocoa: 5 oz cocoa plus 5 oz water
Lady Baltimore filling: 1 lb chopped cherries, nuts, and raisins

Boiled Icing

1 gal.

Ingredients

1 lb egg whites
½ oz salt
2 lb sugar
6 oz glucose
8 oz water

Procedure

1. Place the sugar, glucose, and water in a saucepan. Boil to 240°F.
2. Place the egg whites and salt in the stainless steel mixing bowl and beat until soft, wet peaks form.
3. Pour in the hot mixture very slowly while continuing to beat.
4. Beat until desired consistency is obtained.

Chocolate Supreme Fudge Icing

3 qt

Ingredients

1 lb emulsified vegetable shortening
4 oz butter
½ oz salt
12 oz cocoa
5 lb powdered sugar
4 oz honey
14 oz hot water (variable)

Procedure

1. Place the shortening and butter in a saucepan. Melt and place in the stainless steel mixing bowl.
2. Add the cocoa and salt. Mix at slow speed using the paddle until blended.
3. Add the powdered sugar. Mix at slow speed until smooth
4. Mix the honey in the hot water. Add to the mixture slowly to prevent lumping. Continue at slow speed until smooth.

Caramel Fudge Icing

3½ qt

Ingredients

1 lb 12 oz light brown sugar
8 oz butter
½ oz salt
¼ oz cream of tartar
8 oz water
5 lb powdered sugar
2 oz emulsified vegetable shortening
8 oz butter
6 oz milk (variable)
½ oz vanilla

Procedure

1. Place the brown sugar, salt, cream of tartar, first amount of butter, and the water in a saucepan. Boil to 242°F.
2. Place the powdered sugar, shortening, second amount of butter, milk, and vanilla in the stainless steel mixing bowl. Beat at high speed using the paddle until mixture becomes light.
3. Add the hot syrup while mixing at medium speed. Mix until just smooth (approximately 1 min to 2 min). Do not overmix.

Fondant Icing

5 qt

Ingredients

10 lb sugar
1 lb glucose
4 lb water

Procedure

1. Place all the ingredients in a sauce pot. Boil to 240°F. Wash the sides of the kettle carefully, keeping sides of bowl clean by constantly rubbing with a kitchen spoon.
2. Pour the cooked mixture in the stainless steel mixing bowl. Set the bowl in cold water and cool to 150°F.
3. Grain the mixture (smooth to develop the texture) at high speed using the paddle until it becomes stiff and white in color.
4. Place in a container and cover tightly with a damp cloth.

Bittersweet Chocolate Icing

3½ qt

Ingredients

2 lb 8 oz melted bitter chocolate
1 lb 4 oz cocoa
3 lb 12 oz powdered sugar
1 lb 14 oz hot water

Procedure

1. Place the melted chocolate in the stainless steel mixing bowl and add the powdered sugar and cocoa. Blend thoroughly at slow speed using the paddle.
2. Add the hot water and mix until smooth.
3. If adjustment of flavor is desired, salt and vanilla may be added.

Butterscotch Stock

1½ qt

Ingredients

2 lb brown sugar
4 oz glucose
8 oz butter
8 oz water

Procedure

1. Place all the ingredients in a saucepan. Boil to 244°F, stirring occasionally with a kitchen spoon. Cool before using.

Butterscotch Fondant Icing

3 qt

Ingredients

3 lb 4 oz fondant
1 lb 6 oz butterscotch stock
4 oz glucose
1 lb emulsified vegetable shortening
1 oz salt
8 oz evaporated milk (variable)

Procedure

1. Place the fondant, butterscotch stock, and glucose in the stainless steel mixing bowl. Mix slowly using the paddle until smooth.
2. Add the shortening and salt. Mix at slow speed until smooth. Cream 2 min at medium speed.
3. Add evaporated milk, blending in at slow speed. Cream 3 min at medium speed.

Chocolate Fondant Icing

3 qt

Ingredients

3 lb 6 oz fondant
4 oz glucose
4 oz butter
10 oz emulsified vegetable shortening
½ oz salt
1 lb melted bitter chocolate
12 oz evaporated milk

Procedure

1. Place the fondant, glucose, butter, shortening, and salt in the stainless steel mixing bowl. Mix at slow speed using the paddle until smooth. Cream at medium speed for 2 min.
2. Add the melted bitter chocolate and mix at slow speed until smooth.
3. Add the evaporated milk. Mix slowly until smooth. Cream at medium speed for 3 min or 4 min.

Icing Base

2½ qt

Ingredients

1 lb 8 oz emulsified vegetable shortening
3 lb 2 oz powdered sugar
6 oz eggs

Procedure

1. Place the shortening in the stainless steel mixing bowl. Whip at medium speed using the paddle until light.
2. Add the sugar and continue whipping at medium speed.
3. Add the eggs and whip at high speed until light (about 5 min).

Orange or Lemon Fondant Icing

2½ qt

Ingredients

3 lb 12 oz fondant
4 oz ground oranges or 2 oz ground lemons
2 lb 8 oz icing base
citric acid to taste

Procedure

1. Place all the ingredients in the stainless steel mixing bowl. Mix using the paddle for about 10 min to a smooth consistency.

Strawberry Fondant Icing

2½ qt

Ingredients

3 lb 12 oz crushed strawberries
2 lb 8 oz icing base
citric acid to taste

Procedure

1. Place all the ingredients in the stainless steel mixing bowl. Mix at medium speed using the paddle to a smooth creamy consistency (about 5 min).

Royal Icing

1½ qt

Ingredients

4 lb powdered sugar
10 oz egg whites (variable)
½ tsp cream of tartar

Procedure

1. Place the sugar, cream of tartar, and half of the egg whites in the mixing bowl. Mix at slow speed using the paddle while adding the remaining egg whites.
2. Mix until icing is smooth.
3. Keep icing covered with a damp cloth.

CAKE FILLING RECIPES

Fudge Cake Filling

2½ qt

Ingredients

2 lb water
2 lb 8 oz sugar
6 oz cocoa
¼ oz salt
7 oz modified starch
1 lb water
¼ oz vanilla
2 oz emulsified vegetable shortening

Procedure

1. Place the water, sugar, cocoa, and salt in a sauce pot. Bring to a boil.
2. Dissolve the starch in the second amount of water. Pour into the boiling mixture while stirring constantly with a kitchen spoon. Cook until thickened and clear.
3. Remove from the heat and stir in the vanilla and shortening until thoroughly blended.

Orange Cake Filling

3½ qt

Ingredients

3 lb water
1 lb 4 oz frozen orange concentrate
2 lb sugar
½ oz salt
1 lb 8 oz water
10 oz modified starch
8 oz egg yolks
4 oz emulsified vegetable shortening
4 oz butter
4 oz lemon juice

Procedure

1. Place the water, frozen orange concentrate, sugar, and salt in a sauce pot. Bring to a boil.
2. Dissolve the starch in the second amount of water. Add egg yolks and mix with a kitchen spoon. Add to the boiling mixture while stirring constantly with a kitchen spoon. Cool until thickened and clear.
3. Remove from the heat and stir in the shortening, butter, and lemon juice until thoroughly blended.

Strawberry Cake Filling

2½ qt

Ingredients

1 lb water
1 lb 8 oz sugar
¼ oz salt
2 lb fresh or frozen chopped strawberries
6 oz modified starch
1 lb water
½ oz red color
1 oz lemon juice

Procedure

1. Place the first amount of water, sugar, salt, and strawberries in a sauce pot. Bring to a boil.
2. Dissolve the starch in the second amount of water. Add to the boiling mixture while stirring constantly with a kitchen spoon. Cook until thickened.
3. Remove from the heat and add the color and lemon juice.

Butterscotch Cake Filling

3 qt

Ingredients

1 lb 4 oz dark brown sugar
1 lb sugar
2 lb water
½ oz salt
8 oz glucose
10 oz water
8 oz cornstarch
2 oz butter
¼ oz maple flavoring
½ oz vanilla

Procedure

1. Place the brown sugar, sugar, water, salt, and glucose in a sauce pot. Bring to a boil.
2. Dissolve the starch in the second amount of water. Pour into the boiling mixture while stirring constantly with a kitchen spoon. Cook until thickened and clear.
3. Remove from the heat and stir in the butter, maple flavoring, and vanilla.

Specialty Dessert Preparation

Specialty desserts include those dessert items not classified as cookies, pies, or cakes, including puddings and ice creams. Puddings are available as different types, including cream puddings, baked puddings, chilled puddings, soufflé puddings, and steamed puddings. Puddings are easily prepared in quantity for high profit.

Ice creams and sherbets are very popular and can be purchased in a variety of flavors. Ice creams and sherbets can be served plain or with cake, cookies, or other dessert items. Specialty desserts using ice cream combined with fruits, fruit sauces, and liqueurs create eye-appealing parfaits and coupes. Flambés (flaming desserts) such as jubilees and baked Alaska, and crepes Suzette are a la carte items that must be ignited before the customer.

Specialty desserts also include the light, flaky pastries made from puff paste and eclair or choux paste dough. These provide a variety of attractive forms for filling with fruit, ice cream, or other food items.

PUDDINGS

Five types of puddings are commonly used in food service establishments. The common types of puddings in order of importance and popularity are:

- *Cream puddings* or *starch-thickened puddings* are made from hot milk, sugar, starch, vanilla, salt, and eggs. The milk is heated. Save a small amount of cold milk to blend with the sugar and cornstarch. This mixture is then blended into the hot milk. This preparation is usually done in a double boiler. Cream puddings can be served warm or chilled; however, chilled is more popular. Common cream puddings include chocolate pudding, vanilla pudding, coconut pudding, and butterscotch pudding.

- *Baked puddings* or *egg-thickened puddings* include desserts such as rice pudding, bread pudding, and custard. Baked puddings are usually served with a warm sauce. They are usually bound together by a baked custard made of eggs and milk or cream. The preparation is generally baked in a water bath (pans containing water) at a temperature of 325°F to 340°F until the custard has set but not completely cooked. Custard continues to cook after it has been removed from the oven.

Oven temperature is very important when baking an egg-thickened pudding. If the oven temperature is too low, the pudding does not solidify properly. If the oven temperature is too high, the pudding becomes watery. Custard may be tested for doneness by inserting a knife. If the knife comes out clean, the custard is done.

- *Chilled puddings* or *gelatin puddings* are light and fluffy because whipped cream or egg whites are folded into the basic gelatin mixture. Chilled puddings include Bavarian creams, snow puddings, and mousses.

- *Soufflé puddings* or *soufflés* are prepared using different flavors, such as chocolate soufflé and vanilla soufflé. Soufflé puddings can only be made properly when baked to order. The beaten egg whites must be folded into the basic mix gently and baked very carefully to prevent the soufflé from becoming heavy and soggy. This procedure is why soufflés are difficult to prepare.

- *Steamed puddings* or *boiled puddings* are made with a large percentage of fruit, suet (animal fat) as the shortening, flour, eggs and bread crumbs as binders, baking soda if a leavening agent is used, and brown sugar or molasses as the sweetener. The use of the dark-colored sweetener gives this pudding its characteristic dark appearance. Steamed puddings are highly spiced with such spices as ginger, mace, nutmeg, and allspice. Rum, brandy, or both are used to provide aroma and taste.

Florida Department of Citrus

Flan is a custard baked with a caramel glaze.

Steamed puddings can be cooked in large or individual metal containers by steaming in a steam pressure chamber covered with aluminum foil or by baking in water baths in a 350°F oven covered with a damp cloth for approximately 2 hr to 3 hr. Another method of cooking is to place the pudding in a damp muslin cloth that has been dusted with flour, tie the ends of the cloth loosely to allow for expansion, and lower into simmering water. The bag may also be suspended just above the water and cooked by steam vapors. Steamed puddings have a heavy texture and are commonly served hot with a warm sauce that complements the pudding's flavor and color.

CREAM PUDDINGS

This pudding is generally thickened with cornstarch, flour, and eggs. However, if a pudding with a higher sheen is desired, just cornstarch may be used. Chocolate pudding is usually topped with whipped cream or topping and garnished with a cherry or some type of fruit.

Chocolate Pudding

25 servings

Equipment

- Baker's scale
- Wire whip
- Double boiler
- Gallon measure
- Pint measure
- Stainless steel mixing bowl
- Measuring spoons
- Kitchen spoon

Ingredients

1 gal. milk
4 oz cornstarch
3 oz all-purpose flour
1 lb 8 oz sugar
7 oz cocoa
½ tsp salt
1 pt milk
6 eggs
1 ½ oz butter
1 tsp vanilla

Procedure

1. Place the milk and half the sugar in the top of a double boiler, cover, and heat until scalding hot.
2. In the stainless steel bowl containing the beaten eggs, add the pint of milk, remaining sugar, cornstarch, flour, salt, and cocoa. Mix to a smooth paste.
3. Pour the paste mixture into the scalding milk gradually, whipping vigorously with a wire whip.
4. Continue to cook, whipping the mixture at intervals until it becomes smooth and stiff. Remove from the heat.
5. Add the vanilla and butter and stir with a kitchen spoon until thoroughly blended.
6. Pour into champagne or cocktail glasses, chill, serve topped with whipped cream or topping, and garnish with a cherry or some type of fruit.

Precautions

- Work the whip vigorously when adding the egg/starch mixture to the scalding milk so lumps will not form.
- Exercise caution when whipping the hot mixture. Do not splash.

Preparation

1. Break the eggs into a stainless steel bowl and beat slightly with a wire whip.

This pudding is thickened with cornstarch and eggs. Vanilla pudding is similar to French blancmange, but is richer because of the addition of the eggs. Many variations are possible by adding ingredients such as bananas, cocoa, coconut, and pineapple. If convenience is desired, there are excellent pudding mixes on the market.

Vanilla Pudding

25 servings

Equipment

- Baker's scale
- Wire whip
- Double boiler
- Quart measure
- Stainless steel mixing bowl
- Spoon measures
- Kitchen spoon

Ingredients

6 lb (3 qt) milk
12 oz sugar
6 oz cornstarch
8 oz egg yolks
½ tsp salt
3 oz butter
½ tsp vanilla (variable)

Preparation

1. Separate eggs. Hold yolks and store whites in the refrigerator or freezer.

Procedure

1. Place the milk and half of the sugar in the top of a double boiler, cover, and heat until scalding hot. A film will form on the surface of the milk when it is scalding hot.
2. Place the remaining sugar, cornstarch, egg yolks, and salt in a stainless steel bowl. Add some of the scalding milk gradually while whipping steadily until a thin paste is formed.
3. Pour the paste mixture into the scalding milk, whipping briskly with a wire whip.
4. Continue to cook and whip the mixture until it becomes fairly stiff. Remove from the heat.
5. Add the vanilla and butter. Stir with a kitchen spoon until the butter melts and blends into the pudding.
6. Pour into champagne or cocktail glasses, chill, and serve topped with whipped cream.

Variations

Banana pudding: Add 2 lb sliced bananas.
Coconut pudding: Add 8 oz plain or toasted shredded coconut.
Pineapple pudding: Add 12 oz drained crushed pineapple.
Vanilla nut pudding: Add 4 oz chopped nuts (pecans, walnuts, or toasted almonds).
Lemon pudding: Omit the vanilla and add 5 oz lemon juice.

Precautions

- Work the whip constantly when adding the hot milk to the egg/starch mixture.
- Whip vigorously when adding the egg/starch mixture to the scalding milk. Exercise caution when whipping the hot mixture. Do not splash.

This pudding is usually thickened with cornstarch and eggs and prepared in a similar manner as other cream- or starch-thickened puddings. Like chocolate pudding, butterscotch pudding is popular in cafeteria service.

Butterscotch Pudding

25 servings

Equipment

- Baker's scale
- Spoon measures
- Large and small double boilers
- Gallon measure
- Pint measure
- Wire whip
- Stainless steel mixing bowl

Ingredients

1 gal. milk
6 oz cornstarch
1 pt water
1 lb 8 oz dark brown sugar
10 oz butter
6 eggs
1 tsp vanilla
1 tsp maple flavor

Preparation

1. Break the eggs into a stainless steel bowl and beat slightly with a wire whip.
2. Combine the sugar, butter, and salt. Place in the top of the small double boiler and cook until the sugar is melted.

Procedure

1. Place the milk in the top of a double boiler, cover, and heat until scalding hot.
2. In the stainless steel bowl containing the beaten eggs, add the pint of water and cornstarch. Mix until smooth.
3. Pour the egg/starch mixture into the scalding milk gradually while whipping vigorously with a wire whip. Cook until mixture thickens.
4. Add the melted sugar and butter mixture while continuing to whip vigorously. Continue to cook until thick and smooth.
5. Add the vanilla and maple flavor. Blend thoroughly using the wire whip.
6. Pour into champagne or cocktail glasses and chill. Top with whipped cream or topping and garnish with a cherry or some type of fruit.

Note: If desired, ¾ c of chopped walnuts or other nuts may be added as a variation.

Precautions

- Work the whip vigorously when adding the egg/starch mixture to the scalding milk so lumps do not form.
- Exercise caution when whipping the hot mixture. Do not splash.

BAKED PUDDINGS

Bread Pudding

This pudding is popular, economical, and profitable when prepared and served properly. Bread slices are lined up overlapping in a bake pan. Custard is poured over the bread slices until they are thoroughly saturated. The dish is baked in a water bath (one pan sitting in another that contains water) until the custard becomes firm. Bread pudding is commonly served with a sauce.

25 servings

Equipment

- Bake pans (2)
- Wire whip
- Baker's scale
- French knife
- Stainless steel mixing bowl
- Measuring spoons
- Sauce pot

Ingredients

3 qt milk
14 eggs
½ tsp salt
1 tsp vanilla
12 oz sugar
1 lb 8 oz sliced bread
nutmeg to taste
cinnamon to taste

Preparation

1. Trim the crust from the bread slices and cut each slice in half using a French knife. Line the bread slices in a bake pan, slightly overlapping each slice.
2. Break the eggs into a stainless steel mixing bowl and beat slightly with a wire whip.

Procedure

1. Place the milk and sugar in a sauce pot and heat until scalding. Remove from the heat.
2. Add the salt and vanilla to the eggs in the mixing bowl. Pour this mixture into the scalding milk and sugar mixture while whipping vigorously with a wire whip.
3. Pour the custard mixture over the bread slices. Sprinkle lightly with cinnamon and nutmeg.
4. Place on a second bake pan containing water and bake in a preheated oven at 375°F until the custard is just set.
5. Remove from the oven and let set until firm. Portion with an ice cream scoop and serve warm with an appropriate sauce such as brown sugar, cherry, or lemon sauce.

Note: For raisin bread pudding, sprinkle 10 oz raisins over the bread slices before pouring the custard mixture.

Precautions

- Whip vigorously and exercise caution when whipping the eggs into the hot liquid.
- Bake pudding only until custard is just set because custard continues to cook after it leaves the oven.
- Cover all pieces of bread thoroughly with the custard.

Baked Custard

This custard is a mixture of eggs and milk with sweetener and flavoring added. To prepare a successful baked custard, a ratio of 10 oz to 12 oz of eggs to each quart of milk is required. Custard can be baked in a pan or in individual custard cups. Baked custard can be served plain, garnished with a little cinnamon, nutmeg, or both, or with an appropriate sauce.

25 servings

Equipment

- Custard cups (25)
- Bake pans (2)
- Wire whip
- Quart measure
- Spoon measures
- Sauce pot
- Baker's scale
- Kitchen spoon

Ingredients

3 qt milk
2 lb eggs
1½ tsp vanilla
½ tsp salt
1 lb sugar

Preparation

1. Break the eggs into a stainless steel mixing bowl and beat slightly with a wire whip.
2. Butter the inside of each custard cup.

Procedure

1. Place the milk, salt, and vanilla in a sauce pot. Heat until scalding and remove from the heat.
2. Add the sugar to the mixing bowl containing the eggs. Blend thoroughly using a kitchen spoon, but do not whip.
3. Pour some of the hot milk slowly into the egg/sugar mixture while stirring rapidly with a kitchen spoon. Pour this mixture into the remaining hot milk, continuing to stir rapidly.
4. Pour the custard mixture into the buttered custard cups. Place in bake pans approximately two-thirds full of water.
5. Bake in a preheated oven at 375°F until the custard has just set.
6. Remove from the oven and let set until firm and cool.
7. Unmold on a dessert plate and serve topped with cinnamon and nutmeg, whipped cream, or an appropriate sauce such as vanilla, cherry, or lemon.

Variations

Coffee custard: Add ½ c instant coffee to the hot milk.
Caramel custard: Boil 2 lb sugar and 1 lb water until a temperature of 330°F is reached or until the mixture becomes dark. Pour about ¼″ of this syrup into dry custard cups.

Rice pudding is a mixture of cooked rice and custard baked until slightly firm. This dessert should be served warm with cream or an appropriate sauce. It is an excellent choice when serving a large group.

Rice Pudding

25 servings

Equipment

- Bake pans (2)
- Wire whip
- Baker's scale
- Stainless steel mixing bowl
- Double boiler
- Kitchen spoon
- Quart measure

Ingredients

- 1 gal. milk
- 1 lb rice
- 1 lb sugar
- 1 tsp salt
- 10 oz egg yolks
- 1 qt single cream
- 1 tsp vanilla
- nutmeg to taste
- cinnamon to taste

Preparation

1. Separate the egg yolks from the whites. Place the yolks in a stainless steel bowl and beat slightly with a wire whip. Freeze the whites and save for another preparation.
2. Wash the rice in cold water and drain thoroughly.

Procedure

1. Place the milk and salt in the top of a double boiler, cover, and heat until scalding hot.
2. Add the rice and cook, stirring occasionally with a kitchen spoon until the rice is tender. Remove from the heat.
3. Add the sugar, cream, and vanilla to the bowl containing the egg yolks.
4. Pour this mixture slowly into the cooked rice while whipping with a fairly rapid motion.
5. Pour the rice and custard mixture into a bake pan and sprinkle cinnamon and nutmeg lightly over the surface.
6. Place on a second bake pan containing water and bake in a preheated oven at 375°F until mixture has set.
7. Remove from the oven and let set until firm. Portion into a serving dish with an ice cream dipper and serve warm with an appropriate sauce such as brown sugar or cherry sauce.

Variation

Raisin rice pudding: Add 8 oz raisins to the rice after it has been cooked.

Precaution

- When adding the egg mixture to the cooked rice, whip continuously and exercise caution to avoid splashing.

CHILLED PUDDINGS

This is a chilled dessert that has excellent eye appeal because of its two-tone color. The top of this molded dessert contains red gelatin and the bottom contains a creamy rice mixture. Rice imperatrice is served with an appropriate cold sauce.

Rice Imperatrice Eggs

25 servings

Equipment

- Gelatin molds 2½" to 3" deep (25)
- Kitchen spoon
- Skimmer
- Sauce pot
- Stainless steel mixing bowls, medium and small
- Wire whip
- Mixing machine and wire whip
- French knife
- Quart measure
- Baker's scale
- Hotel pan

Ingredients

- 8 oz flavored gelatin (raspberry, cherry, or strawberry)
- 1 pt hot water
- 1 pt cold water
- 1 qt single cream
- 1 tsp vanilla
- 1 lb rice
- 2 oz plain gelatin
- 1 pt cold water
- 10 oz pasturized liquid eggs
- 1 lb sugar
- 1 pt whipping cream
- 5 oz red maraschino cherries or 8 oz candied fruit

Preparation

1. Wash the rice and drain thoroughly.
2. Chop the red maraschino cherries with a French knife, or wash and chop the candied fruit.
3. Heat 1 pt water until scalding hot.
4. Place the plain gelatin in the small stainless steel mixing bowl, add the pint of cold water, stir, and let set.

Procedure

1. Dissolve the flavored gelatin in the pint of scalding water. Stir until thoroughly dissolved. Add the pint of cold water and stir until blended.

2. Pour the dissolved gelatin into the individual molds until about ½″ deep.
3. Place the molds in the refrigerator until the gelatin sets.
4. Place the milk, single cream, and vanilla in a sauce pot and bring to a simmer.
5. Add the washed rice and cook slowly, stirring frequently with a kitchen spoon until the rice is tender. Remove from the heat.
6. Add the dissolved plain gelatin and stir with a kitchen spoon until thoroughly blended.
7. Add the sugar to the egg yolks and whip slightly with a wire whip. Pour slowly into the cooked rice while stirring rapidly with a wire whip.
8. Pour the rice mixture into a hotel pan, place in the refrigerator, and let cool until it starts to set.
9. Place the whipping cream in the bowl of an electric mixer. Whip until stiff.
10. Remove the rice mixture from the refrigerator when it starts to set. Fold in the whipped cream and chopped fruit using a skimmer. Blend thoroughly.

11. Remove the gelatin molds from the refrigerator and fill them with the rice mixture.
12. Return molds to the refrigerator and chill until firm.
13. Unmold by dipping each mold in warm water and serve on a dessert plate, covering each serving with an appropriate sauce such as vanilla or lemon sauce.

Precautions

- Exercise caution when chopping the fruit.
- Dissolve the flavored gelatin in the scalding water thoroughly or the gelatin will not set properly.
- Always soak plain gelatin in cold water before adding it to a hot mixture. It dissolves quicker and does not lump.
- The bowl, whip, and cream should be cold when whipping the cream.
- Use a very gentle motion when folding the whipped cream into the rice mixture.

This is a light, smooth, and fluffy dessert. Whipped cream is folded into a basic gelatin mixture to create a delicate texture that is characteristic of all Bavarian creams. These can be set up and served in individual portions or for group servings. They may be featured in molded form or in a silver cup, cocktail glass, or champagne glass. They may be served with or without a cold sauce.

Vanilla Bavarian Cream Eggs

25 servings

Equipment

- Electric mixer and wire whip
- Quart mixer
- Kitchen spoon
- Baker's scale
- Wire whip
- Double boiler
- Skimmer
- Stainless steel mixing bowls

Ingredients

1½ qt milk
2 oz plain gelatin
1 pt cold water
16 oz pasturized liquid eggs
12 oz sugar
1 qt whipping cream
vanilla to taste

Preparation

1. Place plain gelatin in the stainless steel bowl, add the cold water, and soak to soften the gelatin.

Procedure

1. Place the milk in the top of a double boiler and heat. Remove from the double boiler.
2. Add 8 oz sugar to the egg yolks in the stainless steel bowl, whipping gently until the mixture is stiff and smooth.
3. Pour the hot milk gradually into the sugar and egg yolk mixture while whipping briskly with a wire whip.
4. Add the remaining sugar, vanilla, and the water and gelatin mixture. Stir with a kitchen spoon until the gelatin mixture dissolves and is thoroughly incorporated.

5. Place this mixture in the refrigerator to cool until it begins to set.
6. Place the whipping cream in the bowl of the electric mixer and whip at high speed until stiff. Remove from the mixer.
7. Fold the whipped cream into the cold vanilla mixture using a skimmer or kitchen spoon.
8. Pour the mixture into individual or large-size molds, silver cups, cocktail glasses, or champagne glasses. Serve unmolded with an appropriate cold sauce or leave in a cup or glass and garnish the top with whipped cream and candied fruit.

Variations

Chocolate Bavarian cream: Add 3 oz unsweetened and 12 oz sweet chocolate to the hot milk.
Mocha Bavarian cream: Add 4 tbsp instant coffee to the hot milk.
Walnut Bavarian cream: Add 4 oz to 6 oz finely chopped walnuts to the vanilla Bavarian cream.

Precautions

- When cooling the basic mixture, do not let it become too firm. A jelly-like consistency produces the best results.
- When adding the hot milk to the egg yolk mixture, pour the milk slowly and whip briskly to avoid curdling the egg yolks.
- When whipping cream, have the mixing bowl and whip cold, as well as the cream, for best results. Do not overwhip cream after it has become stiff.
- Fold gently when adding the whipping cream to the vanilla mixture.

This is a light and fluffy dessert. It is similar to a Bavarian cream or a chiffon filling because whipped cream, meringue, or a combination of both is folded into the basic mixture to produce a light, delicate texture. A mousse may contain gelatin to help it set up. This depends on the type of mousse to be made and the ingredients used. Mousse is especially popular in establishments serving French cuisine.

Chocolate Mousse Eggs

25 servings

Equipment

- Electric mixer and wire whip
- Baker's scale
- Pint measure
- Small double boiler
- Medium stainless steel mixing bowls (2)
- Kitchen spoon
- Skimmer

Ingredients

2 lb grated sweet chocolate
12 oz water
12 oz egg yolks or pasturized liquid eggs
12 oz dried egg whites
10 oz sugar
1 pt whipping cream

Preparation

1. Reconstitute the dried egg whites in a medium sized stainless steel bowl with water.

Procedure

1. Place the chocolate and water in the top of a small double boiler. Heat until the chocolate melts and blends with the water. Stir occasionally with a kitchen spoon.
2. Remove the chocolate from the heat and let cool. Stir occasionally to speed cooling.
3. When the chocolate mixture starts to set, add the slightly beaten egg yolks or pasturized eggs while whipping briskly with a wire whip. If the mixture becomes too stiff, add a little milk. Set aside and hold for later use.
4. Place the egg whites in the bowl of the electric mixer and whip at high speed until they start to froth. Add the sugar slowly until a fairly stiff meringue is formed. Place in a stainless steel bowl and hold for later use.
5. Place the whipped cream in the bowl of the electric mixer and whip at high speed until stiff. Remove from the mixer.
6. Fold the whipped cream and meringue alternately into the chocolate mixture using a skimmer or kitchen spoon.
7. Pour the mixture into individual silver cups, cocktail glasses, or champagne glasses and chill in the refrigerator until ready to serve.

Precautions

- The egg yolk and chocolate mixture should be fairly stiff; however, if it is too stiff, problems could develop when folding. The mixture can be made thinner by adding a very small amount of milk.
- For best results in whipping the cream, have the mixing bowl and whip cold, as well as the cream. Whip only until stiff. Do not overwhip.
- When preparing the meringue, whip at high speed and add the sugar gradually after the egg whites start to froth. Continue whipping until soft peaks form.

This pudding, when served in molded form, resembles a mound of snow. A meringue is folded into a whipped lemon gelatin mixture to create this eye-appealing dessert. Serve lemon snow pudding on an appropriate cold sauce with a contrasting color.

Lemon Snow Pudding Eggs

25 servings

Equipment

- Baker's scale
- Quart measure
- Spoon measure
- Stainless steel bowl
- Grater
- Mixing machine and wire whip
- Sauce pot
- Hotel pan
- Kitchen spoon
- Skimmer

Ingredients

1 qt boiling water
8 oz cold water
1¼ oz plain gelatin
1 lb sugar
8 oz lemon juice
2 tbsp lemon rind
12 oz dried egg whites
12 oz sugar

Preparation

1. Reconstitute the dried egg whites in a medium sized stainless steel bowl with water.
2. Place the plain gelatin in a stainless steel bowl, add the cold water, and let soak until gelatin is soft.
3. Grate the lemon rind on the fine grid of the box grater.
4. Squeeze the juice from fresh lemons.

Procedure

1. Place 1 qt of water in a sauce pot, place on the range, and bring to a boil. Remove from the heat.
2. Add the softened gelatin, first amount of sugar, lemon juice, and lemon rind. Stir with a kitchen spoon until the gelatin and sugar are thoroughly dissolved.
3. Pour into a bake pan and place in the refrigerator until the mixture starts to set.
4. Place the egg whites in the bowl of the electric mixer and whip at high speed until the egg whites start to froth. Add the second amount of sugar gradually while continuing to whip at high speed until stiff peaks are formed. Remove from the mixer and hold.
5. Remove the lemon gelatin mixture from the refrigerator, place in the bowl of the electric mixer, and whip at high speed until light and fluffy.
6. Using a skimmer, gradually fold the meringue into the whipped gelatin mixture until thoroughly blended.
7. Pour into large or individual molds and refrigerate until set.
8. Remove from the refrigerator, unmold, and serve on an appropriate cold sauce.

Precautions

- To whip the meringue properly, the mixer should be running at high speed throughout the operation and the sugar should be added slowly.

- Fold the meringue gently into the lemon gelatin mixture to preserve as many air cells as possible.
- Exercise caution when grating the lemon rind.
- Remove the lemon mixture from the refrigerator as soon as it starts to set.

SOUFFLÉ PUDDING

This soufflé produces many variations by adding other ingredients. All soufflés should be prepared to order and served at once with a sauce. Whipping the egg whites, folding them into the basic mix, and proper baking are important steps in producing a successful soufflé.

Basic Vanilla Soufflé

12 servings

Equipment

- Small bake pan
- Electric mixer and wire whip
- Sauce pot
- Wire whip
- Baker's scale
- Pint measure
- Stainless steel bowls (2)
- Soufflé cups or dishes (special casseroles)
- Skimmer

Ingredients

3 oz bread flour
3 oz butter
3 oz sugar
1 pt milk
8 egg yolks
10 egg whites
3 oz sugar
vanilla to taste

Preparation

1. Separate the egg yolks from the whites and place in separate stainless steel bowls.
2. Butter soufflé cups or dish and dredge with sugar.
3. Preheat the oven to 425°F.

Procedure

1. Place the milk and vanilla in a sauce pot. Bring to a boil on the range and remove from the heat.
2. Place the bread flour, butter, and first amount of sugar in a small bake pan. Rub together by hand to form a smooth paste.
3. Add this paste to the hot milk while whipping briskly with a wire whip.
4. Return the mixture to the range and cook gently for approximately 2 min.
5. Place the hot mixture in a round stainless steel bowl and add the egg yolks, one at a time, while whipping with a wire whip after each addition.
6. Place the egg whites in the bowl of the electric mixer and whip at high speed until the egg whites start to froth. Add the second amount of sugar gradually while continuing to whip at high speed until soft peaks form.
7. Fold the meringue immediately into the creamy vanilla mixture using a skimmer. Blend gently.
8. Fill the prepared soufflé cups or dish to within ½″ of the top. Sprinkle heavily with powdered sugar and bake immediately in a preheated oven at 425°F for approximately 25 min (depending on size of cup or dish) until golden brown.
9. Serve immediately with an appropriate sauce.

Variations

Chocolate soufflé: Add 1½ oz melted sweet chocolate to the hot vanilla mixture at the same time the egg yolks are added. Increase the egg yolks to 10 and the egg whites to 12.

Mocha soufflé: Add 2 tbsp instant coffee to the hot vanilla mixture.

Orange soufflé: Add 3 oz Grand Marnier liqueur and 2 tbsp grated orange peel to the hot vanilla mixture.

Cherry soufflé: Add 2 oz Kirschwasser and 4 oz chopped candied cherries to the hot vanilla mixture.

Precautions

- The egg whites should be stiff, not dry, in order to obtain best results.
- Fold the meringue carefully into the creamy vanilla mixture. Do not overmix.
- Never let the oven temperature exceed 425°F. Check the temperature with a thermometer.
- Butter the baking dishes generously with butter and dredge thoroughly with sugar.

Trade Tip

Soufflé means to blow or puff up. Soufflés puff up when they bake. Always serve soufflés immediately after preparation because they can fall very quickly.

STEAMED PUDDING

Plum Pudding

This is one of the most popular steamed puddings. Plum pudding has a very heavy texture because the ratio of fruit to batter is about two to one. It should be served with a hot sauce unless a hard sauce is used. Traditionally, plum pudding is most popular during the Christmas season.

25 servings

Equipment

- Electric mixer and paddle
- Baker's scale
- Steam table pan
- Stainless steel bowl
- Wire whip

Ingredients

8 oz butter
8 oz brown sugar
⅛ oz salt
⅛ oz allspice
½ oz ginger
8 oz eggs
8 oz dark molasses
3 oz rum
3 oz brandy
8 oz bread flour
1 lb raisins
1 lb 8 oz currants
8 oz citrons
8 oz orange peel
4 oz lemon peel
8 oz bread crumbs

Procedure

1. Place the butter, brown sugar, salt, and spices in the bowl of the electric mixer. Cream together at slow speed using the paddle.
2. Add the slightly beaten eggs while continuing to mix at slow speed. Blend thoroughly.
3. Add the molasses, rum, and brandy while continuing to mix at slow speed until mixture is thoroughly blended and smooth.
4. Add the bread flour, raisins, currants, citrons, orange peel, and lemon peel, and mix until blended.
5. Add the bread crumbs and mix until thoroughly incorporated.
6. Pack the pudding in a prepared steam table pan, filling only two-thirds full to allow for expansion in cooking. Cover the pan with aluminum foil, place in the steamer, and steam for 2 hr.
7. Remove from the steamer. For each serving, place a No. 12 scoop of pudding in a dessert bowl or on a dessert plate and serve with a hard sauce or a hot sauce such as brandy, vanilla, or rum.

Preparation

1. Break the eggs and place in a stainless steel bowl. Beat slightly with a wire whip.
2. Grease the steam table pan lightly with butter and dust with flour.

Precautions

- Exercise extreme caution when removing the pudding from the steamer.
- Fill the pan only two-thirds full of pudding because it will expand while in the steamer.

ICE CREAM DESSERT RECIPES

PARFAITS

Parfaits are eye-appealing and elegant ice cream desserts that are prepared by alternating layers of crushed fruit or syrup and various colored and flavored ice creams. They are topped with whipped cream, chopped nuts, and a maraschino cherry. Parfaits may be served immediately or frozen and held for service at a later date. This is helpful when preparing for large group service.

Parfait Crème de Menthe

Alternate layers of crème de menthe and vanilla ice cream. Garnish with whipped cream, chopped nuts, and a maraschino cherry.

Rainbow Parfait

Alternate layers of strawberry sauce and vanilla, strawberry, and chocolate ice cream. Garnish with fresh strawberries and chopped nuts.

Parfait Melba

Alternate layers of melba sauce and vanilla ice cream. Garnish with fresh raspberries and whipped cream.

Pineapple Parfait

Alternate layers of crushed pineapple and vanilla ice cream or lemon sherbet. Garnish with whipped cream, chopped nuts, and a maraschino cherry.

Chocolate Parfait

Alternate layers of chocolate syrup and vanilla or chocolate ice cream. Garnish with chocolate shot, whipped cream, and a maraschino cherry.

Strawberry Parfait

Alternate layers of strawberry sauce and vanilla ice cream. Garnish with whipped cream, chopped nuts, and fresh strawberries.

Butterscotch Parfait

Alternate layers of butterscotch sauce and vanilla ice cream. Garnish with whipped cream, chopped nuts, and a maraschino cherry.

COUPES

Coupes are desserts combining ice cream or sherbet, liqueurs, sauces, fruit, and whipped cream. They are served in champagne glasses or silver cups in a way that will be attractive and eye-appealing. Coupes are economical and quick to prepare and can, in some cases, be partially prepared, frozen, and finished at serving time.

Coupe Melba

Vanilla ice cream covered with a peach half and topped with melba sauce, garnished with whipped cream and a sliced peach.

Coupe Helene

Vanilla ice cream covered with a Bartlett pear half and topped with chocolate sauce, garnished with whipped cream and a maraschino cherry.

Strawberry Coupe

Vanilla ice cream covered with fresh strawberries that have been tossed in curaçao liqueur, garnished with whipped cream and a fresh strawberry.

Pineapple Coupe

Vanilla ice cream covered with diced pineapple flavored with Kirschwasser, garnished with whipped cream and a maraschino cherry.

Coupe Savory

Assorted diced fresh fruit flavored with anisette liqueur, covered with mocha ice cream, and garnished with whipped cream and chopped nuts.

JUBILEES

Jubilees are a combination of ice cream and a flaming liqueur fruit sauce that is poured over the ice cream in front of the guest. It is the most spectacular of desserts and is unique in that the sauce keeps flaming momentarily when it comes in contact with the cold ice cream. The flame is extinguished when all the alcohol has been burned out, and the exquisite flavor remains, blending with the ice cream that is only slightly melted on the surface. The most popular jubilee is the cherry jubilee; however, other fruits such as peaches, strawberries, and oranges may also be used to create a variety of jubilees.

Cherry Jubilee

4 servings

Ingredients

1 pt pitted Bing cherries and juice
¼ c sugar
¼ tsp arrowroot
2 oz Kirschwasser
4 large scoops of vanilla ice cream

Procedure

1. Pour the juice from the pint of cherries into the blazer of a chafing dish. Reserve enough juice to dissolve the arrowroot.

2. Place the blazer pan directly over the flame of the chafing dish and bring the juice to a boil.

3. Dissolve the arrowroot in the reserved cherry juice and pour slowly into the boiling juice while stirring constantly. Cook until the juice is slightly thickened.

4. Add the sugar and reduce the heat to a simmer.

5. Add the cherries, stirring them into the sauce.

6. Heat the Kirschwasser in a separate pan and pour the warm (not hot) liqueur over the cherry mixture.

7. Ignite the liqueur and pour the flaming cherry mixture over each scoop of ice cream. This preparation should be done before the guest.

Strawberry Jubilee

4 servings

Ingredients

1 pt frozen whole strawberries and juice
½ tsp arrowroot
1 oz cointreau
2 oz brandy
4 large scoops vanilla ice cream

Procedure

1. Pour the juice from the thawed berries into the blazer of a chafing dish. Reserve enough juice to dissolve the arrowroot.

2. Place the blazer pan directly over the flame of the chafing dish and bring the juice to a boil.

3. Dissolve the arrowroot in the reserved strawberry juice and pour slowly into the boiling juice while stirring constantly.

4. Add the strawberries and cointreau and stir into the sauce.

5. Heat the brandy in a separate pan and pour the warm (not hot) brandy over the strawberries.

6. Ignite the brandy and pour the flaming strawberry mixture over each scoop of ice cream. This preparation should be done before the guest.

Baked Alaska

Baked Alaska is a combination of cake, ice cream, and meringue. This dessert is unique in that it has an ice cream center, a golden brown outside, and can be set aflame. It can be prepared ahead of service and held in the freezer. The outside meringue covering can then be browned in the oven or with a blowtorch just before serving. **Warning:** Use extreme caution when handling the blowtorch.

Baked Alaska must be served immediately after preparation for the most effective presentation.

Baked Alaska (Standard Recipe)

depends on size set up

Ingredients

ice cream (vanilla, chocolate, strawberry, etc., in any combination)
sponge or pound cake
meringue (as needed)

Procedure

1. Place a layer of sponge cake approximately½" to 1" thick on a silver or stainless steel platter. Cut the cake the shape of the platter, leaving a 3" margin.

2. Cover the cake with ice cream, using the flavor desired or evenly distributed combination of flavors. Mold the ice cream to a height of approximately 4".

3. Cover the ice cream with sheets of sponge cake cut about ¼" thick. Cover with a cloth and place in the freezer for several hours until the ice cream is solid.

4. Remove from the freezer and cover entirely with meringue, using a spatula to produce a smooth surface.

5. Fill a pastry bag with the meringue. Using a star tube in the tip of the bag, decorate the baked Alaska as desired. The meringue may be tinted a pastel color if a two-tone baked Alaska is desired.

6. Dust the entire baked Alaska with powdered sugar and brown very quickly in a hot oven (450°F) or brown with a blowtorch. Using a torch eliminates any chance of melting the ice cream.

7. Serve the baked Alaska immediately with an appropriate sauce, or return it to the freezer for later use. To present a baked Alaska flambé, press half egg shells into the top, hiding the sides of the shells with meringue. Pour a small amount of warm brandy or rum into the shells, ignite, and carry the flaming baked Alaska into the dining room.

Note: The same preparation can be made with either white or yellow pound cake rather than the more usual sponge cake. Pound cake is preferable because it has better eating qualities and is easier to slice.

Individual Baked Alaska

Individual baked Alaska is essentially the same as the standard recipe, differing only in form and certain details of the procedure to be followed.

From a sheet of sponge or pound caked sliced about ½" thick, cut out rounds approximately 3" in diameter and place them on a sheet pan about 3" apart.

Place a No. 12 scoop of ice cream on top of each 3" cake round. Cover with a cloth and place in the freezer until the ice cream is quite solid.

Fill a pastry bag with meringue and, using a star tube in the tip, cover the ice cream entirely, creating a decorative design.

Dust with powdered sugar and brown quickly in a 450°F oven or with a blowtorch.

Hawaiian Baked Alaska

Hawaiian baked Alaska is a variation of the standard recipe. Brandy or rum is used when it is served flambé.

2 servings

Ingredients

1 fresh pineapple
ice cream (vanilla, chocolate, or strawberry, or an assortment of all three)
meringue (as needed)

Procedure

1. Cut the fresh pineapple in half lengthwise.
2. Scoop the meat out of the center of the pineapple, leaving the wall of the pineapple about ½″ thick. Place the scooped-out pineapple half on a platter.
3. Fill the pineapple cavity with the desired type of ice cream and place in the freezer until the ice cream is quite solid.
4. Remove from the freezer. Fill a pastry bag with meringue and, using a star tube in the tip, cover the ice cream entirely, creating a decorative design.
5. Dust with powdered sugar and brown quickly in a 450°F oven or with a blow torch.
6. Serve immediately, plain or flambé. To flambé, pour warm brandy or rum over the Hawaiian baked Alaska, ignite, and serve aflame.

CREPE DESSERTS

Crepe desserts are prepared by using the thin French pancakes known as crepes. Most crepe desserts call for three crepes approximately 5″ in diameter for each order. The crepes are either rolled or folded in four, with or without a filling, and served with a hot, sweet sauce. Crepes may be served to the guest while they are aflame. Brandy, rum, and cognac are the liqueurs used most often because of their alcohol content and flavor.

A liqueur burns best if it is approximately 100 proof and slightly warm. *Note*: 100 proof means that the liqueur is 50% alcohol. Never boil a liqueur because the alcohol will be boiled off and the liqueur cannot be ignited.

Crepes are usually prepared ahead of service and stacked on a sheet pan between layers of silicon or wax paper, covered with a damp towel, and stored in a cool place until ready to use. The most famous and popular crepe dessert is crepes Suzette.

Crepes Suzette

This dessert is a classic in fine restaurants. The final preparation should be done before the guest.

4 servings

Ingredients

12 French pancakes (crepes)
4 oz sugar
1 tbsp orange rind
1 tsp lemon rind
¼ c orange juice
1 tbsp lemon juice
3 oz butter
1 oz Grand Marnier
2 oz cognac

Procedure

1. Sprinkle 3 oz of the sugar in the blazer pan of a chafing dish and melt over low heat while stirring constantly with kitchen spoon.
2. Grate and add the orange and lemon rind, continuing to cook until the sugar is slightly brown.
3. Add the butter, orange, and lemon juice, and stir until thoroughly blended. Do not let the mixture boil.
4. Place the crepes, one at a time, into the sauce. Sprinkle the remaining sugar over the crepes, turn them over at least once, and fold in four, placing them around the sides of the pan.
5. Add warm cognac to the sauce, ignite, and let flame while moving the pan back and forth gently.
6. Serve three crepes to each order on a hot plate and pour on a flaming sauce. This preparation should be done before the guest.

Trade Tip

To improve the taste and increase the rise of puff paste dough, brush the dough with rum before the final folding. Baking puff paste dough with a small amount of steam injected into the oven also improves the rising of the dough.

SPECIALTY PASTRIES

This dough contains no sugar or leavening agent but rises to approximately eight times its original size when heat is applied. Puff paste dough is made by rolling and folding alternate layers of fat and dough a total of five times. Care must be exercised when preparing this type of dough and since a minimum of 15 min to 20 min should be allowed between each of the five rollings and foldings, preparation is time-consuming. Puff paste dough has a texture that is very tender, flaky, and crisp. A wide variety of baked goods can be prepared by using puff paste dough. The dough can be purchased to save prep time.

Puff Paste Dough

determined by the item being prepared

Equipment

- Mixing machine and dough hook
- Baker's scale
- Sheet pans
- Parchment paper
- Rolling pin
- Bench brush
- Pastry wheel
- 3″ and 2″ round cutters (for patty shells)
- Yardstick (for turnovers, cream horns, or lady locks)
- Dough docker or dinner fork

Ingredients

- 5 lb bread flour
- 8 oz eggs
- 8 oz butter
- 1 oz salt
- 2 lb 4 oz cold water (variable)
- 5 lb puff paste shortening

Preparation

1. Scale 5 lb puff paste shortening. Break shortening into fairly small pieces and place on a sheet pan covered with wax paper. Let set at room temperature. This is done so the shortening will have the same consistency as the dough when rolled in. If one is stiffer than the other, a poor dough will result.
2. Prepare the sheet pans. Cover pans with parchment paper if preparing turnovers, patty shells, lady locks, cream horns, or puff paste stars. If preparing Napoleon slices, dampen sheet pan with cold water.
3. Preheat the oven to 375°F.

Procedure

1. Place the bread flour, salt, butter, eggs, and cold water in the stainless steel mixing bowl. Mix at slow speed using the dough hook, then mix at medium speed until a very smooth dough is formed.
2. Remove the dough from the mixer, place on a floured bench, and shape into a smooth ball.
3. Cover the dough with a damp cloth and let rest on the bench for 15 min to 20 min.
4. Roll the dough into a rectangular shape twice as long as it is wide and approximately ¼″ to ½″ thick.
5. Spot the puff paste shortening evenly over two-thirds of the dough's surface. Do not bring the shortening to the edge of the dough. Keep it about ½″ from the edge.
6. Fold the unspotted third of the dough over one-half of the spotted portion. Then fold the remaining spotted one-third of dough over this folded portion, completing a three-fold dough.
7. Brush the dough free of excess flour and give it a half turn so the former length now becomes the width. Roll out the dough into a rectangular shape a second time. This rectangle should be twice as long as it is wide and about ½″ thick. Again, brush off the excess flour.
8. Fold both ends of the dough toward the middle, then double again. This is referred to as a *pocketbook fold*. Place the dough on a sheet pan and cover with a damp cloth. Refrigerate for 15 min to 20 min.
9. Remove the dough from the refrigerator and repeat the rolling and folding process a total of five times, but each time allow the dough to rest in the refrigerator 15 min to 20 min. The width of the previous roll should be rolled into the length each time.
10. After rolling the dough for the fifth time, the dough should be made up into desired units. Allow the units to stand 30 min before baking in a preheated oven at 375°F.

Variations

Patty shells: Roll out a piece of puff paste dough approximately ⅛″ thick. Cut out rounds 3″ in diameter using a plain- or scallop-edged cutter. Place these rounds on a sheet pan covered with parchment paper.

Take a second piece of puff paste dough and roll it to a thickness of ¼″. Using the same size cutter, cut the second piece of dough into 3″ rounds. Then cut the center out of these rounds by using a 2″ cutter.

Wash the first rounds with water and place the rings (cut from the second rounds) on top. Continue this procedure until all of the first rounds are covered with the rings. Wash with egg wash (eggs and milk) and allow to stand at room temperature for 30 min.

Place a piece of greased paper over the top of the patty shells so they will not topple over while they are baking. Place them in the oven and bake in a preheated oven at 375°F until they are crisp and dry. They will rise to approximately 2″ to 2½″ high. Remove from the oven and let cool.

Patty shells can be filled with fruit and topped with whipped cream or, if served as an entree, they can be filled with chicken a la king, shrimp, lobster, seafood Newburg, creamed ham, or cream chicken and mushrooms.

Turnovers: Roll out a piece of puff paste dough to a thickness of approximately ⅛″. Cut the dough into 4″ squares. Dampen the surface of the dough with water.

Spot the center of each square with the desired fruit filling and fold cornerwise to form a triangle. Secure the edges with tines of a dinner fork and puncture the top of the turnover twice with the fork tines.

Place the turnovers on sheet pans covered with parchment paper. Allow them to stand at room temperature for 30 min, then bake in a preheated oven at 375°F.

Turnovers may be made up and baked without filling. In this case, the filling is added after baking by splitting them slightly with a knife and adding the filling with a spoon or pastry bag. Before serving, turnovers may be iced with roll icing or dusted with powdered sugar.

Lady locks or cream horns: Roll the puff pastry dough into a thin rectangle about 48″ long, 15″ wide, and about ⅛″ thick. Cut this sheet of dough into strips approximately 1¼″ wide using a pastry wheel. Wash the surface of the dough slightly with water and roll each strip around a lady lock or cream horn tin, starting at the small end of the tin and overlapping the dough strip slightly as a diagonal direction is followed around the tin. Leave about 1″ of the tin so it will be easier to remove after baking.

Place the lady locks or cream horns on a sheet pan covered with parchment paper and sprinkled with sanding sugar (very coarse sugar). Allow them to rest about 15 min before baking in a preheated oven at 375°F until golden brown.

Remove from the oven, pull the tins out of the horns while they are still hot, and let them cool. After cooling, fill the centers of the baked horns with whipped cream, meringue topping, or cream filling. Dust with powdered sugar and serve.

Puff paste pocketbooks: Roll out the puff paste dough to a thickness of ¼″. Cut the dough into 4″ squares. Turn each of the four corners into the center and secure by pressing the dough with a finger.

Place the pocketbooks on a sheet pan covered with parchment paper, let rest for 30 min, and bake in a preheated oven at 375°F until the dough puffs upward and outward and is golden brown.

Remove from the oven and let cool. Poke a hole into the center of each pocketbook and fill with the desired fruit filling. Dust with powdered sugar and serve.

Napoleon slices: Roll out a piece of the puff paste dough approximately ⅛″ thick. Place the thin sheet of dough on a sheet pan and allow to stand for about 30 min.

Pick the dough with a docker or the tines of a dinner fork all over to prevent blistering while baking.

Bake the dough in a preheated oven at 350°F until dry and crisp, but do not overbrown.

After baking, cut into three equal strips and stack the strips together with a rich cream filling between each layer.

Frost the top with two different colors of fondant icings (for example, white and chocolate). Draw the two colors together before the icing sets by using the edge of a spatula. This creates a marbleized effect. When the fondant sets, cut into bars about 4″ long and 2″ wide.

Cream slices: Roll out the puff paste dough to a thickness of ⅛″. Brush the surface of the dough with water, then sprinkle with shaved almonds. Cut the dough into 2″ wide and 4″ long units.

Place on a sheet pan covered with parchment paper and allow to stand at room temperature for 30 min. Bake in a preheated oven at 375°F until golden brown and crisp. Let cool and split each unit in two. Fill with whipped topping or cream filling, dust with powdered sugar, and serve.

Precautions

- The dough and shortening should always be of the same consistency to prevent the dough walls from rupturing.
- Between each rolling, allow the dough to rest in a cool place for 15 min to 20 min. This permits the gluten in the dough to relax, making the dough easier to roll.
- All excess flour should be brushed off before folding over the dough. Excess flour toughens the dough.
- Care must be taken that the puff paste shortening is evenly distributed through the dough. All corners and ends should be even when the dough is folded.
- Leftover puff paste requires additional rolling and folding before it can be made into units.

Appendix

MATH REVIEW

ADDITION

In addition, numbers or units are combined to find a sum or total. See Addition. The plus sign (+) indicates the operation. Addition problems can be solved horizontally and vertically. To verify the sum, add the addends backward or upward. If both answers are equal, the sum is correct.

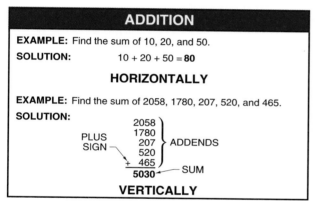

ADDITION
EXAMPLE: Find the sum of 10, 20, and 50.
SOLUTION: 10 + 20 + 50 = **80**
HORIZONTALLY
EXAMPLE: Find the sum of 2058, 1780, 207, 520, and 465.
SOLUTION:
VERTICALLY

SUBTRACTION

Subtraction is the opposite of addition. Subtraction is finding the difference between two numbers. See Subtraction. The minus sign (−) indicates the operation. The number to be subtracted is the *subtrahend*. The number from which the subtrahend is to be subtracted is the *minuend*. The amount left is the *difference*.

Subtraction problems may require borrowing from the units in the tens, hundreds, or thousands column even though the minuend is larger than the subtrahend. When a digit in the minuend is less than the digit that appears beneath it in the subtrahend, a value must be borrowed from the previous digit in the minuend.

Note: The digit 6 in the minuend is smaller than the digit 8 in the subtrahend. Since 8 cannot be subtracted from 6, it is necessary to borrow a value from the digit to the left in the minuend. This increases the value of the 6 to 16 so subtraction can occur, and reduces the 5 to 4.

To verify the difference, add the subtrahend and difference. If the sum equals the value of the minuend, the answer is correct.

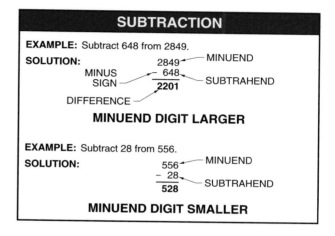

SUBTRACTION

EXAMPLE: Subtract 648 from 2849.

SOLUTION: 2849 — MINUEND / MINUS SIGN — 648 / 2201 — SUBTRAHEND / DIFFERENCE

MINUEND DIGIT LARGER

EXAMPLE: Subtract 28 from 556.

SOLUTION: 556 — MINUEND / − 28 — SUBTRAHEND / 528

MINUEND DIGIT SMALLER

MULTIPLICATION

In multiplication, an amount is increased a specified number of times. See Multiplication. The times sign (×) indicates the operation. The *product* or answer is obtained by multiplying the *multiplicand*, which is the number to be multiplied by another number, which is the *multiplier*.

Two methods may be used to verify that the product is correct. One method is to divide the product by the multiplicand. If the multiplier is obtained, the product is correct. Another method is to divide the product by the multiplier. If the multiplicand is obtained, the product is correct.

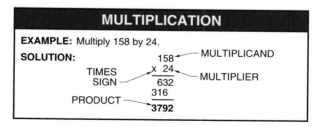

MULTIPLICATION

EXAMPLE: Multiply 158 by 24.

SOLUTION: 158 — MULTIPLICAND / TIMES SIGN — × 24 — MULTIPLIER / 632 / 316 / PRODUCT — **3792**

DIVISION

Division is the opposite of multiplication. Division is finding the number of times a number can be divided into another number. See Division. The division sign (÷) indicates the operation. The *quotient* or answer is obtained by dividing the *dividend* by the *divisor*.

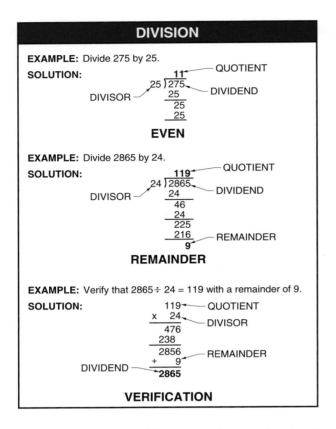

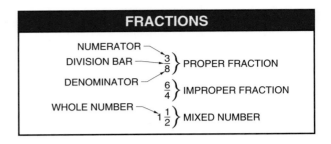

In some division problems, a number remains that is less than the divisor. This number is the *remainder* and may be expressed as a fraction.

The remainder may be expressed as $^9/_{24}$ or $^3/_8$. To verify the solution to a division problem, use multiplication and addition. Multiply the quotient by the divisor and then add the remainder to obtain the dividend. If the dividend is obtained, the quotient is correct.

FRACTIONS

A fraction is a part of a whole and is comprised of three parts: the *numerator*, *denominator*, and *division bar*. See Fractions. The numerator is the number above the division bar. It indicates the number of fractional units. The denominator is the number below the division bar. It represents the number of equal parts that the whole is divided into. For example, if a pie is cut into six equal servings and five servings are used, the fraction representing the used portion is $^5/_6$. A fraction may be *proper* or *improper*. When the numerator is less than the denominator, the fraction is a proper fraction. When the numerator is equal to or greater than the denominator, the fraction is an improper fraction. A *mixed number* is a whole number and a proper fraction.

Reducing Fractions

Reducing a fraction to the lowest terms without changing its value is done by dividing the numerator and denominator by the largest number common to both numbers. See Reducing Fractions.

REDUCING FRACTIONS
EXAMPLE: Reduce $\frac{8}{32}$ to the lowest terms.
SOLUTION: $\frac{8}{32}\ \frac{(\div 8-\text{largest number})}{(\div 8-\text{largest number})} = \frac{1}{4}$

To change an improper fraction to a mixed number, divide the numerator by the denominator and reduce the answer to the lowest terms. To avoid reducing the answer after the conversion, reduce the improper fraction first. See Changing to a Mixed Number.

CHANGING TO A MIXED NUMBER
EXAMPLE: Change $\frac{13}{4}$ to a mixed number.
SOLUTION: $\frac{13}{4} = 13 \div 4 = 3\frac{1}{4}$

To change a mixed number to an improper fraction, multiply the whole number by the denominator and add the sum to the numerator. Place the sum over the denominator. See Changing to an Improper Fraction.

CHANGING TO AN IMPROPER FRACTION
EXAMPLE: Change $3\frac{3}{4}$ to an improper fraction.
SOLUTION: $3\frac{3}{4} = \frac{3 \times 4}{4} + \frac{3}{4} = \frac{12}{4} + \frac{3}{4} = \frac{15}{4}$

Adding Fractions

Fractions may be like fractions, unlike fractions, or mixed numbers. See Adding Fractions. When adding *like fractions* (fractions with the same denominators), add the numerators and place the sum over the common denominator.

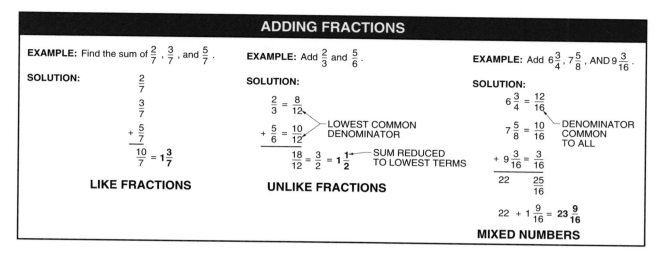

When adding *unlike fractions* (fractions with different denominators), find the lowest common denominator, add the numerators, and reduce the sum to the lowest terms.

When adding mixed numbers, add the fractions first and then add the whole numbers. Place the sum of the fractions next to the sum of the whole numbers.

Subtracting Fractions

Fractions to be subtracted may be like fractions, unlike fractions, or mixed numbers. See Subtracting Fractions. When subtracting like fractions, subtract the numerator of the subtrahend from the numerator of the minuend. The denominator remains the same. Reduce to the lowest terms, if necessary.

When subtracting unlike fractions, find the lowest common denominator and subtract the numerator of the subtrahend from the numerator of the minuend. The denominator remains the same.

When subtracting mixed numbers, find the lowest common denominator. Subtract the proper fractions and the whole numbers. Place the fraction next to the whole number to obtain the difference.

Note: In this problem, borrowing is necessary because the numerator 15 of the subtrahend cannot be subtracted from the numerator 4 of the minuend. A 1 is borrowed from the whole number 7 making it a 6, and the numerator is increased to 22 ($^4/_{18}$ + $^{18}/_{18}$ = $^{22}/_{18}$).

Multiplying Fractions

Fractions may be multiplied by fractions, whole numbers by fractions, and mixed numbers by mixed numbers. See Multiplying Fractions. When multiplying two fractions, multiply the numerators of each fraction and place the product over the product obtained by multiplying the denominators of each fraction. Reduce to the lowest terms, if necessary.

SUBTRACTING FRACTIONS

EXAMPLE: Subtract $\frac{5}{16}$ from $\frac{9}{16}$.

SOLUTION:

$$\frac{9}{16}$$
$$-\frac{5}{16}$$
$$\frac{4}{16} = \frac{1}{4}$$

DIFFERENCE REDUCED TO LOWEST TERMS

LIKE FRACTIONS

EXAMPLE: Subtract $\frac{2}{3}$ from $\frac{7}{8}$.

SOLUTION:

$$\frac{7}{8} = \frac{21}{24}$$
$$-\frac{2}{3} = \frac{16}{24}$$
$$\frac{5}{24}$$

LOWEST COMMON DENOMINATOR

UNLIKE FRACTIONS

EXAMPLE: Subtract $2\frac{1}{2}$ from $4\frac{3}{4}$.

SOLUTION:

$$4\frac{3}{4} = 4\frac{3}{4}$$
$$-2\frac{1}{2} = 2\frac{2}{4}$$
$$2\frac{1}{4}$$

EXAMPLE: Subtract $5\frac{5}{6}$ from $7\frac{2}{9}$.

SOLUTION:

$$7\frac{2}{9} = 7\frac{4}{18} = 6\frac{22}{18}$$
$$-5\frac{5}{6} = 5\frac{15}{18} = 5\frac{15}{18}$$
$$1\frac{7}{18}$$

MIXED NUMBERS

MULTIPLYING FRACTIONS

EXAMPLE: Multiply $\frac{7}{8}$ by $\frac{1}{2}$.

SOLUTION: $\frac{7}{8} \times \frac{1}{2} = \frac{7}{16}$

TWO FRACTIONS

EXAMPLE: Multiply 3 by $\frac{3}{8}$.

SOLUTION: $\frac{3}{1} \times \frac{3}{8} = \frac{9}{8} = 1\frac{1}{8}$

WHOLE NUMBER BY A FRACTION

EXAMPLE: Multiply $4\frac{3}{4}$ by $5\frac{1}{2}$.

SOLUTION: $4\frac{3}{4} \times 5\frac{1}{2} = \frac{19}{4} \times \frac{11}{2} = \frac{209}{8} = 26\frac{1}{8}$

MIXED NUMBERS

When multiplying a whole number by a fraction, convert the whole number into a fraction with a denominator of one. Multiply the numerators and the denominators of each fraction to obtain the product. Reduce to the lowest terms, if necessary.

When multiplying mixed numbers, change the mixed numbers to improper fractions and then multiply the numerators and the denominators of each fraction to obtain the product.

Dividing Fractions

Fractions may be divided by fractions, whole numbers by fractions, and mixed numbers by mixed numbers. See Dividing Fractions. When dividing a fraction by another fraction, *invert* the divisor (place the denominator over the numerator) and proceed as in multiplying fractions.

DIVIDING FRACTIONS

EXAMPLE: Divide $\frac{4}{5}$ by $\frac{2}{3}$.

SOLUTION: $\frac{4}{5} \div \frac{2}{3} = \frac{4}{5} \times \frac{3}{2} = \frac{12}{10} = \frac{6}{5} = 1\frac{1}{5}$

FRACTION BY FRACTION

EXAMPLE: Divide 10 by $\frac{1}{4}$.

SOLUTION: $10 \div \frac{1}{4} = \frac{10}{1} \times \frac{4}{1} = 40$

WHOLE NUMBER BY A FRACTION

EXAMPLE: Divide $7\frac{3}{8}$ by $2\frac{1}{4}$.

SOLUTION: $7\frac{3}{8} \div 2\frac{1}{4} = \frac{59}{8} \times \frac{4}{9} = \frac{236}{72} = 3\frac{20}{72} = 3\frac{5}{18}$

MIXED NUMBERS

When dividing a whole number by a fraction, change the whole number into a fraction with a denominator of one, invert the divisor and proceed as in multiplying fractions.

When dividing mixed numbers, change the mixed numbers to improper fractions, invert the divisor, and proceed as in multiplying fractions.

DECIMALS

The decimal system is a method of counting by tens and in multiples of ten. Decimals may be expressed as fractions with denominators of 10 or in multiples of 10.

$.1 = \frac{1}{10}$ $.01 = \frac{1}{100}$

$.001 = \frac{1}{1000}$ $.0001 = \frac{1}{10,000}$

Decimals are located to the left and right of the decimal point. Whole numbers are located to the left of the decimal point. Their place value starts with the ones column and each column to the left increases in multiples of 10.

1.   10.
100.   1000.

Each column to the right of the decimal point represents a part of a whole number. Their place value starts with one-tenth and decreases in multiples of $\frac{1}{10}$.

$.1, .01, .001, .0001 = \frac{1}{10}, \frac{1}{100}, \frac{1}{1000}, \frac{1}{10,000}$

Changing Decimals to Fractions

To change a decimal to a fraction, eliminate the decimal point and use the number as the numerator. The denominator is obtained by placing a 1 below the division bar and adding a zero for each decimal place. See Changing Decimals to Fractions.

CHANGING DECIMALS TO FRACTIONS

.6	=	$\frac{6}{10}$	SIX TENTHS
.34	=	$\frac{34}{100}$	THIRTY-FOUR HUNDREDTHS
.524	=	$\frac{524}{1000}$	FIVE HUNDRED TWENTY-FOUR THOUSANDTHS
.0425	=	$\frac{425}{10,000}$	FOUR HUNDRED TWENTY-FIVE TEN-THOUSANDTHS

Changing Fractions to Decimals

To change a fraction with a denominator of 10 or multiple of 10 to a decimal, use the numerator as the decimal value. See Changing Fractions to Decimals. Count the number of zeros in the denominator and place the decimal point according to the number of zeros starting from the right and moving to the left. The number of decimal places in the decimal and the number of zeros in the fraction must be equal. If there are fewer zeros than decimal places, zeros must be placed between the decimal point and the first digit in the decimal.

CHANGING FRACTIONS TO DECIMALS

EXAMPLE: Change $\frac{725}{10,000}$ to a decimal number.

SOLUTION: $\frac{725}{10,000}$ = **.0725**

Adding Decimals

When adding decimals, arrange the figures so the decimal points are aligned vertically and proceed as in ordinary addition. If some numbers have fewer digits to the right of the decimal point than other numbers, zeros can be added without changing the value. See Adding Decimals.

ADDING DECIMALS

EXAMPLE: Find the sum of 8.24, 38.464, 42.7, and 9.635.

SOLUTION:

```
    8.240 ――― ZERO ADDED
   38.464
   42.700 ――― TWO ZEROS ADDED
 +  9.635
  ―――――
   99.039
```

Subtracting Decimals

When subtracting decimals, arrange the figures so the decimal points are aligned vertically and proceed as in ordinary subtraction. If some numbers have fewer digits behind the decimal point than other numbers, zeros may be added without changing the value. See Subtracting Decimals.

SUBTRACTING DECIMALS

EXAMPLE: Subtract 24.8 from 68.154.

SOLUTION:

```
   68.154
 - 24.800 ――― TWO ZEROS ADDED
  ―――――
   43.354
```

Multiplying Decimals

Multiplying decimals uses the same method as multiplying whole numbers. See Multiplying Decimals. To multiply decimals, multiply the multiplicand by the multiplier. Count the total number of decimal places to the right of the decimal point in the multiplicand and the multiplier. Mark off this number of decimal places in the product starting from the right and moving to the left.

MULTIPLYING DECIMALS

EXAMPLE: Multiply 8.265 BY 2.9.

SOLUTION:

```
                  8.265 ――― MULTIPLICAND
                x   2.9       (THREE DECIMAL
PRODUCT (DECIMAL ―――――         PLACES)
POINT FOUR        74385
DECIMAL PLACES    16530
FROM RIGHT      ―――――― ――― MULTIPLIER
TO LEFT)         23.9685      (ONE DECIMAL PLACE)
```

Dividing Decimals

When dividing decimals, the method of obtaining the quotient is similar to whole numbers. See Dividing Decimals. The main difference is in placing the decimal point in the quotient. When the divisor is a whole number and the dividend is a decimal, place the decimal point in the quotient directly above the decimal point in the dividend. Divide as with whole numbers.

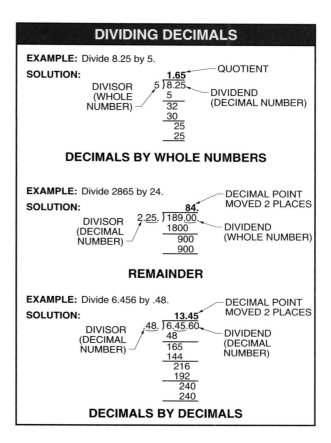

DIVIDING DECIMALS

EXAMPLE: Divide 8.25 by 5.

SOLUTION:

QUOTIENT — **1.65**
DIVISOR (WHOLE NUMBER) — 5)8.25 — DIVIDEND (DECIMAL NUMBER)

DECIMALS BY WHOLE NUMBERS

EXAMPLE: Divide 2865 by 24.

SOLUTION:

DECIMAL POINT MOVED 2 PLACES — **84.**
DIVISOR (DECIMAL NUMBER) — 2.25.)189.00. — DIVIDEND (WHOLE NUMBER)

REMAINDER

EXAMPLE: Divide 6.456 by .48.

SOLUTION:

DECIMAL POINT MOVED 2 PLACES — **13.45**
DIVISOR (DECIMAL NUMBER) — .48.)6.45.60. — DIVIDEND (DECIMAL NUMBER)

DECIMALS BY DECIMALS

When the divisor is a decimal and the dividend is a whole number, move the decimal point in the divisor to the right of the decimal point. Move the decimal point in the dividend the same number of places that the decimal point was moved in the divisor. Place the decimal point in the quotient directly over the decimal point in the dividend. Divide as with whole numbers.

When dividing a decimal by another decimal, move the decimal point in the divisor to the right of the last digit. Move the decimal point in the dividend the same number of places that the decimal point was moved in the divisor. Divide as with whole numbers.

PERCENTAGE

Percentage is a method of expressing a value in hundredths. See Percentage. It is commonly used in making comparisons such as profit and loss margins. A 10% value means 10 out of every 100. Percentages, represented by the percent sign (%), can be expressed as a fractional or decimal equivalent. For example,

$$20\% = {}^{20}\!/_{100} \text{ (or } {}^{1}\!/_{5}) = .20$$

PERCENTAGE

EXAMPLE: A 580 lb side of beef is blocked out and produces a 24 lb rib cut. The rib cut is what percentage of the side of beef?

SOLUTION:

$$\begin{array}{r} .041 = \textbf{4.1\%} \\ 580\overline{)24.000} \\ \underline{2320} \\ 800 \\ \underline{580} \end{array}$$

FINDING % WITH VALUE

EXAMPLE: Change 25% to a proper fraction.

SOLUTION:

$$25\% = \frac{25}{100} = \frac{1}{4}$$

EXPRESSING % AS A PROPER FRACTION

EXAMPLE: Change 45.5% and 75% to decimal numbers.

SOLUTION:

$$45.5\% = \textbf{.455}$$
$$75\% = \textbf{.75}$$

EXPRESSING % AS A DECIMAL

EXAMPLE: Change .246 to a percentage.

SOLUTION:

$$.246 = \textbf{24.6\%}$$

EXPRESSING DECIMAL AS A PERCENTAGE

EXAMPLE: A restaurant takes in $2850 in one week and 38% of that amount is used for expenses. What are the expenses for the week?

SOLUTION:

$$\begin{array}{r} \$\ 2850 \\ \times\ .38 \\ \hline 22800 \\ 8550 \\ \hline \end{array}$$

Expenses = **$1083.00**

FINDING VALUE WITH PERCENT

Percentages express a portion of a whole. If a food service establishment uses $50 for food costs for every $100 received, the establishment has a 50% food cost. To determine percentages, divide the value of the portion by the value of the whole. Move the decimal point three places to the right and multiply by 100 to obtain the tenth of a percent.

To express a percentage as a proper fraction, the percentage is the numerator and 100 is the denominator. Reduce to the lowest terms, if necessary.

To express a percentage as a decimal number, move the decimal point two places to the left and remove the percent sign. (Moving the decimal point two places to the left is the same as multiplying by 100.)

To express a decimal fraction as a percentage, move the decimal point two places to the right and insert a percent sign after the last figure. (Moving the decimal point two places to the right is the same as multiplying by 100.)

To determine an actual value represented by a percentage, multiply the whole value by the decimal equivalent of the percentage.

FOOD PREPARATION MATH

Weights and Measures

Recipes used in commercial kitchens and bakeshops are expressed in weights or measures to ensure that the product will be consistent in quality each time it is prepared. Weights are more accurate than measures because measures may involve more human error resulting from an individual's interpretation of a full measure. Consideration must be made when measuring on how firmly the ingredients are packed into the measuring device or whether the ingredients should be leveled or slightly heaped. These are not concerns when weighing an ingredient.

Common measuring units are teaspoons, tablespoons, cups, pints, quarts, and gallons. These measuring units are usually abbreviated when they appear in a recipe. *(Note: Only abbreviations that form another word are followed by a period.)* The common abbreviations are:

tsp – teaspoon   tbsp – tablespoon

c – cup   pt – pint

qt – quart   gal. – gallon

Common abbreviations used to indicate weight in the commercial kitchen are:

lb – pound (equals 16 oz)

oz – ounce (equals $^{1}\!/_{16}$ lb)

Measuring units and their general equivalents are shown in Table I. Other measuring devices include ladles, kitchens spoons, and scoops. Ladles are used to serve various items when portion control and uniform servings are desired.

The kitchen spoon, which may be solid, slotted, or pierced, holds approximately 3 oz. They may be used to control portion size when serving certain vegetables. The scoop is also for controlling portion size. Ladle sizes and approximate weights are shown in Table II. Scoop and ladle equivalents are shown in Table III. To determine the content of standard size cans by weight or volume, see Table IV.

Weights and measures may be given in the English system or the Metric system. See Table V. Liquid ingredients are measured in volume. Measuring is faster than weighing and accuracy can be maintained if the liquid weight is taken into consideration when measuring.

2 tbsp = 1 fl oz   1 fl c = 8 oz

1 fl pt = 1 lb   1 fl qt = 2 lb

1 fl half gal. = 4 lb   1 fl gal. = 8 lb

Solid ingredients are usually weighed because of the difficulty in measuring them accurately. Some dry ingredients, such as flour and baking powder, may be weighed or measured depending on the amount required in the recipe. When measuring an ingredient, fill the container heaping full, and then level it with a spatula. Do not pack the container. Measurement equivalents may be used when measuring ingredients.

1 pinch = $\frac{1}{8}$ tsp	3 tsp = 1 tbsp
16 tbsp = 1 c	2 c = 1 pt
2 pt = 1 qt	2 qt = $\frac{1}{2}$ gal.
4 qt = 1 gal.	

TABLE I: GENERAL EQUIVALENTS

16 tbsp	1 c
1 c (standard measure)	$\frac{1}{2}$ pt (8 fl oz)
2 c	1 pt
16 oz	1 lb
3 qt (dry)	1 peck
4 pecks	1 bushel
32 oz	1 fl qt
128 oz	1 fl gal.
1 No. 10 can	13 c
1 lb margarine/butter	2 c
1 lb flour	4 c
1 tsp	60 drops
1 pinch (few grains)	$\frac{1}{16}$ tsp
3 tsp	1 tbsp
1 jigger	$1\frac{1}{2}$ oz
2 pt	1 qt
4 qt	1 gal.
16 oz (liquid)	1 lb or 1 pt (liquid)
8 oz (liquid)	1 c (liquid)
1 oz	2 tbsp (approx)

TABLE II: LADLE SIZES AND APPROXIMATE WEIGHTS

Ladle Size	Approximate Weight of Portion
$\frac{1}{4}$ c	2 oz
$\frac{1}{2}$ c	4 oz
$\frac{3}{4}$ c	6 oz
1 c	8 oz

Portion Control

Portion control is the control of the portion size to ensure that the designated amount of an item is served to the guest. It is essential to control portions in all food service opera-

tions if a profit is to be made. The kitchen and serving personnel must be informed of the correct portion size for each item served. This may be accomplished by placing the portion size on recipes, displaying a standardized portion chart in various areas of the kitchen, or listing the portion size of each item on a working menu posted in the production area.

TABLE III: SCOOP AND LADLE EQUIVALENTS

Scoops		
Scoop Size	Equivalent	Approx Scoops per Gallon
No. 6	$\frac{2}{3}$ c	16
No. 8	$\frac{1}{2}$ c	22
No. 10	$\frac{3}{8}$ c	24
No. 12	$\frac{1}{3}$ c	26
No. 16	$\frac{1}{4}$ c	35
No. 20	3+ tbsp	42
No. 24	$2\frac{2}{3}$ tbsp	51
No. 30	2+ tbsp	62
No. 40	$1\frac{1}{2}$ tbsp	70

Ladles	
Ladle Size	Equivalent
1 oz	$\frac{1}{8}$ c
2 oz	$\frac{1}{4}$ c
4 oz	$\frac{1}{2}$ c
6 oz	$\frac{3}{4}$ c
8 oz	1 c

TABLE IV: CONTENT OF STANDARD CANS

Can Number	Weight	Volume*
1	10 oz – 11 oz	$1\frac{1}{4}$
300	15 oz	$1\frac{3}{4}$
303	1 lb	2
1 Tall	1 lb	2
2	1 lb 3 oz	$2\frac{1}{2}$
$2\frac{1}{2}$	1 lb 13 oz	$3\frac{1}{2}$
3	1 qt 13 fl oz	$3\frac{1}{2}$
5	3 lb	$5\frac{3}{4}$
10	6 lb 8 oz	12 – 13

* in c

TABLE V: WEIGHTS AND MEASURES

Unit	Fluid Dram	Tsp	Tbsp	Fl Oz	¼ c	Gill (½ c)	C	Liquid Pt	Liquid Qt	Milliliter	Liter
Fluid Dram	1	¾	¼	⅛	¹⁄₁₆	¹⁄₃₂	¹⁄₆₄	¹⁄₁₂₈	¹⁄₂₅₆	3.7	0.004
Tsp	¹⁄₁₃	1	⅓	⅙	¹⁄₁₂	¹⁄₂₄	¹⁄₄₈	¹⁄₉₆	¹⁄₁₉₂	4.9	0.005
Tbsp	4	3	1	½	¼	⅛	¹⁄₁₆	¹⁄₃₂	¹⁄₆₄	15	0.015
Fl Oz	8	6	2	1	½	¼	⅛	¹⁄₁₆	¹⁄₃₂	30	0.030
¼ c	16	12	4	2	1	½	¼	⅛	¹⁄₁₆	59	0.059
Gill (½ c)	32	24	8	4	2	1	½	¼	⅛	118	0.118
C	64	48	16	8	4	2	1	½	¼	237	0.237
Liquid Pt	128	96	32	16	8	4	2	1	½	473	0.473
Liquid Qt	256	192	64	32	16	8	4	2	1	946	0.946
Milliliter	0.27	0.20	0.068	0.034	0.017	0.0084	0.0042	0.0021	0.0011	1	¹⁄₁₀₀₀
Liter	270	203	67.6	33.8	16.9	8.45	4.23	2.11	1.06	1000	1

Purchasing is another function that must be considered for a successful portion control program. A buyer must purchase foods that portion well and cut waste to a minimum. For example, purchase link sausage by the count per pound that best suits the portion amount. Purchase hams that are easily cut into steaks of the amount and size desired with little or no waste.

Controlling portion size may be the responsibility of the cook, pantry person, baker, pastry chef, meat cutter, or service personnel serving the food in a cafeteria operation. Methods such as count, weight, equal portion, volume, and controlled fill are used to control portion size. See Table VI.

TABLE VI: PORTION CONTROL METHODS AND EXAMPLES

Count	6 meat balls per order 3 pieces of chicken per order 6 fried shrimp per order
Weight	2½ oz cooked beef per order 12 oz sirloin steak 2½ oz of cooked ham
Equal Portion	cake cut into 12 equal wedges pie cut into 8 equal wedges gelatin cut into 12 equal squares
Volume	2 oz portion of sauce over meat 3 oz kitchen spoon of peas or carrots No. 12 scoop of bread dressing
Controlled Fill	5 oz glass of orange juice 4 oz cup of custard 8 oz casserole of beef stew

Portion control assists in purchasing, food production, determining portion cost, and setting menu prices. It also reduces food costs, resulting in a lower food cost percentage. Portion size must be known to determine the amount to order, an accurate amount to prepare, and the approximate number of servings obtained from a given amount.

The Mil

When dealing with monetary figures, the first two digits behind the decimal point are cents. The third place, referred to as a *mil*, represents a thousandth of a dollar or one-tenth of one cent.

When a mil appears in the answer of a monetary value, it is usually rounded to the nearest cent. If the mil is 4 or less, it is dropped. If it is 5 or more, it is rounded up and another cent is added to the digit that appears before it.

Example:

$75.142 = **$75.14** $10.546 = **$10.55**

$22.683 = **$22.68** $43.728 = **$43.73**

Determining Portion Cost

To determine the portion cost (cost per serving), convert the total weight of the item into ounces. Divide the total cost of the item by its total weight to obtain the cost of one ounce. Multiply the cost per ounce by the portion size. Carry the division out to three places to determine the exact cost per ounce in mils.

Example: A 2½ lb box of frozen peas costs $2.60. Determine the cost of a 3 oz serving.

Solution

1. Convert pounds to ounces.

   $oz = lb \times 16$

   $oz = 2.5 \times 16$

   $oz = $ **40 oz**

2. Determine cost per ounce.

   $cost\ per\ oz = total\ cost \div oz$

   $cost\ per\ oz = 2.60 \div 40$

   $cost\ per\ oz = $ **$0.065 (6.5¢)**

3. Determine cost per serving.

cost per serving = cost per oz × portion size

cost per serving = .065 × 3

cost per serving = **$0.195 = $0.20 (20¢)**

If cost per pound is listed rather than total cost, divide the cost per pound by 16 to obtain the cost per ounce. Multiply the cost per ounce by the portion size to obtain the cost per serving.

Example: A 5 lb box of frozen lima beans costs $.88/lb. Determine the cost of a 3 oz serving.

Solution

1. Determine cost per ounce.

cost per oz = cost per lb ÷ 16

cost per oz = .88 ÷ 16

cost per oz = **$0.055 (5.5¢)**

2. Determine cost per serving.

cost per serving = cost per oz × portion size

cost per serving = .055 × 3

cost per serving = **$0.165 (16.5¢) = $0.17 (17¢)**

If the item must be trimmed or boned, or if shrinkage is expected, the amount lost must be subtracted from the original weight before the cost per ounce is determined. The exact cost per ounce is determined by dividing the total cost by the actual usable amount.

Example: A 22 lb rib of beef costs $1.85/lb. Four pounds are lost through trimming and 3 lb are lost through shrinkage when it is roasted. Determine the cost of a 6 oz serving.

Solution

1. Determine the ounces purchased.

oz pur = lb × 16

oz pur = 22 × 16

oz pur = **352 oz**

2. Determine the ounces lost.

oz lost = lb × 16

oz lost = (4 + 3) × 16

oz lost = 7 × 16

oz lost = **112 oz**

3. Determine the usable amount of item.

usable amount = oz purchased − oz lost

usable amount = 352 − 112

usable amount = **240 oz**

4. Determine total cost of purchased product.

total cost = lb × cost/lb

total cost = 22 × 1.85

total cost = **$40.70**

5. Determine cost per ounce.

cost/oz = total cost ÷ oz

cost/oz = 40.70 ÷ 240

cost/oz = **$0.169 (16.9¢)**

6. Determine cost per serving.

cost per serving = cost per oz × portion size

cost per serving = .169 × 6

cost per serving = **1.014 = $1.01**

Determining Amount to Prepare

The amount of prepared food items must be controlled or shortages or leftovers may occur. Leftovers for certain foods may be utilized, but it is recommended to keep the amount of leftovers to a minimum.

To determine the approximate amount of an item to be prepared, multiply the number of servings required by the suggested portion size. Convert the weight of the contents in a container (box, can, or package) to ounces. The number of ounces required is divided by the weight of the contents of the container to determine the number of containers required to serve the suggested number of servings. If a remainder results when dividing the amount required by the contents of the container, an additional container will need to be purchased.

Example: A 4 oz portion of frozen lima beans is served to each of 180 people. Determine the number of 2½ lb boxes of frozen lima beans to be cooked.

Solution

1. Determine the total ounces required.

total oz req'd = # of servings × suggested portion

total oz req'd = 180 × 4

total oz req'd = **720 oz**

2. Determine the ounces/container.

oz/container = lb × 16

oz/container = 2½ × 16

oz/container = **40 oz**

3. Determine the number of containers required.

# of containers = total oz req'd ÷ oz/container

# of containers = 720 ÷ 40

# of containers = **18**

Determining Number of Servings

To determine the approximate number of servings that can be obtained from a given amount of food, the actual usable amount must first be established. Subtract the amount lost due to trimming, boning, or other preparation from the original weight. Divide the usable amount by the serving size to determine the number of servings.

Example: An 18 lb sirloin of beef is purchased and 2 lb 10 oz is lost through trimming and boning. Determine the number of 12 oz sirloin steaks that can be cut from this sirloin.

Solution

1. Determine the original weight.

original weight = lb × 16

original weight = 18 × 16

original weight = **288 oz**

2. Determine the nonusable weight.

nonusable weight = nonusable product × 16

nonusable weight = (2 × 16) + 10

nonusable weight = 32 + 10

nonusable weight = **42 oz**

3. Determine the usable weight.

usable weight = original weight − nonusable weight

usable weight = 288 − 42

usable weight = **246 oz**

4. Determine the number of servings.

# of servings = usable weight ÷ serving size

# of servings = 246 ÷ 12

# of servings = 20.5 = **20**

Determining Amount to Purchase

Determining the approximate amount of food to purchase for a specific number of people is accomplished by multiplying the single serving amount by the number of people being served. Convert the common purchase quantity (pound, quarts, pints, etc.) into ounces and divide this amount into the number of ounces required. The amount lost through trimming, boning, and other preparation must also be given consideration.

Example: Chop steak is the entree selected for a party of 128 people with each person receiving a 5 oz portion. Determine the amount of ground beef that must be purchased.

Solution

1. Determine the total ounces required.

total oz req'd = servings × portion

total oz req'd = 128 × 5

total oz req'd = **640 oz**

2. Determine the amount to be purchased.

purchase amount = total oz req'd ÷ lb

purchase amount = 640 ÷ 16

purchase amount = **40 lb**

Note: In this example, 16 oz is used as the common purchase quantity because ground beef is purchased in pound increments (16 oz = 1 lb).

FOOD SERVICE MANAGEMENT MATH

Determining Food Cost Percentage

The food cost percentage indicates what part of every dollar received through register sales is used for the cost of food. A percentage is a method of expressing a rate in terms of hundredths. The food cost percentage that must be main-

tained is usually determined by the budget. The monthly food cost percentage is calculated once a month to determine whether the menu price and the costs for each item are within the budgeted amount. When calculating the percentage, carry the division out to three places.

To determine the monthly food cost percentage, use the following formula:

monthly food cost % = (beginning inventory + monthly purchases) − final inventory ÷ total monthly sales

The *monthly food cost %* is the portion of money received that is spent on food served during the month.

The *beginning inventory* for the month is the final inventory from the previous month. An inventory is a detailed list of food on hand or in storage and its estimated value. It may also be called the physical inventory because the amounts are determined by counting the supplies.

The *monthly purchases* are a total cost of all food purchased during the month. The figure is determined by adding the totals of all invoices for food purchased.

The *final inventory* is an estimate of all food still on hand or in storage. The inventory is usually taken at the end of each month.

The *total monthly sales* are determined by totaling the daily register tapes for the month. Quite often, a weekly summary is used to determine the total monthly sales. In this case, the weekly summaries are added to obtain to obtain the total monthly sales.

Example: Determine the cost of food sold and the monthly food cost percentage.

Monthly sales $20,000.00

Beginning inventory 1250.00

Monthly purchases 7750.00

Final inventory 1400.00

Solution

1. Determine the cost of food sold.

Cost of food sold = (beginning inventory + monthly purchases) − final inventory

Cost of food sold = (1250 + 7750) − 1400

Cost of food sold = 9000 − 1400

Cost of food sold = **$7600**

2. Determine the monthly food cost percentage.

Monthly food cost % = (beginning inven + monthly purchases) − final inven ÷ total monthly sales

Monthly food cost % = (1250 + 7750) − 1400 ÷ 20,000

Monthly food cost % = 9000 − 1400 ÷ 20,000

Monthly food cost % = 7600 ÷ 20,000

Monthly food cost % = **38%**

Note: In this example, it is not necessary to carry the division out to three places. 38% is an excellent monthly food cost percentage and may represent a profit for the month.

Pricing Menus

Pricing the menu can be a difficult task. Consideration must be given to raw food cost, rent, equipment cost, taxes, and other operating variables when determining a profitable menu price. The menu must be priced so that the establishment can attract customers and make a profit.

The use of computers has made it possible for a food service operator to determine an accurate overall operational cost. Menu pricing in the past was not standardized due to inconsistent record keeping and because most food service operations were small or family-owned businesses.

One standardized practice is that the raw food cost of an item must be determined before a *mark-up* (amount added to the raw food cost) can be determined to acquire a menu price. The mark-up rate may be indicated as a whole number, fraction, or percentage. Percentages are the easiest and most accurate means of calculating mark-ups.

The mark-up rate varies depending on the type of food service establishment. A gourmet-type restaurant may add a mark-up of two to three times the raw food cost. A cafeteria may add a mark-up of 50%, while another operation may add a mark-up of 75%. In general, menu prices may be determined by using two methods. In one method, the menu price equals the raw food cost plus the markup. In another method, the menu price equals the raw food cost divided by the monthly food cost percentage.

Example: The raw food cost is $3.56 and the mark-up rate is ¾. Determine the menu price using the mark-up rate.

Solution

1. Determine the menu price.

    menu price = raw food cost + (raw food cost × mark-up rate)

    menu price = 3.56 + (3.56 × ¾)

    menu price = 3.56 + 2.67

    menu price = **$6.23**

Example: The raw food cost is $3.89 and the monthly food cost percentage is 43%. Determine the menu price using the monthly food cost percentage.

Solution

1. Determine the menu price.

    menu price = raw food cost ÷ monthly food cost %

    menu price = 3.89 ÷ .43

    menu price = $5.441 = **$5.44**

It is customary in the food service industry to terminate menu prices in multiples of $0.25 ($0.25, $0.50, $0.75, $1.00) to facilitate change handling and check totaling. If a menu price is determined to be $6.23, the adjusted menu price is $6.25. If a menu price is determined to be $5.44, the adjusted menu price is $5.50. Other examples are:

Determined Menu Price	Adjusted Menu Price
$2.10	$2.25
$4.40	$4.50
$6.63	$6.75
$8.85	$9.00

Determining Approximate Recipe Yield

The yield of a recipe is the number of servings it produces. Approximate recipe yield is determined by adding the weight of all ingredients used in the preparation and dividing the total weight by the weight of one portion.

Example: What is the approximate recipe yield for the yellow cake recipe if each cake requires 12 oz of batter?

Yellow Cake

Ingredients

2 lb 8 oz cake flour
1 lb 6 oz shortening
3 lb 2 oz granulated sugar
1 oz salt
1¾ oz baking powder
4 oz dry milk
1 lb 4 oz water
1 lb 10 oz whole eggs
12 oz water

Solution

1. Determine the total weight of the ingredients.

    total weight = sum of all ingredients

    total weight = 2 lb 8 oz + 1 lb 6 oz + 3 lb 2 oz + 1 oz + 1¾ oz + 4 oz + 1 lb 4 oz + 1 lb 10 oz + 12 oz

    total weight = 40 oz + 22 oz + 50 oz + 1 oz + 1.75 oz + 4 oz + 20 oz + 26 oz + 12 oz

    total weight = 176.75 = **177 oz**

2. Determine the approximate yield.

    approx yield = total weight ÷ portion size

    approx yield = 177 ÷ 12

    approx yield = 14.75 = **15**

Determining Standard Recipe Cost

The raw food cost of a preparation must be calculated before a selling or menu price is determined. Standardized recipe forms are commonly used to record pertinent information. The forms, which differ from one food service operation to another, are generally kept on file in the kitchen office. A typical standardized recipe form contains the following information:

- Name of preparation
- Approximate yield
- List of ingredients used in the preparation

- Amount of each ingredient used
- Market or unit price of each ingredient
- Extension cost of each ingredient
- Total cost of the preparation
- Cost per serving or unit

To determine the standard recipe cost, the cost of each ingredient must first be determined. The ingredient cost is then totaled and divided by the approximate yield to determine a unit cost. A unit cost is the cost of one serving of the recipe.

In a standardized recipe form, the market prices are listed in quantities frequently associated with the item or quoted by the purveyor. The extension cost is determined by multiplying the quantity by the market or unit price. When determining the extension cost, the quantity value and market or unit price must represent the same quantity: ounces must be multiplied by cost per ounce, pints multiplied by cost per pint, etc. Extension cost, total cost, and cost per serving are indicated to the mil to determine an accurate cost. While the cost per serving may be indicated to the mil, it is often expressed in dollars and cents for convenience.

Braised Swiss Steak

Ingredient	Quantity	Market or Unit Price	Extension Cost
6 oz round steaks	50	$2.88 per lb	$54.00
onions, minced	12 oz	$0.64 per lb	0.48
garlic, minced	1 oz	$0.95 per lb	0.059
brown stock	1 pt	$0.90 per gal.	0.113
tomato puree	6 qt	$3.44 per gal.	5.16
salad oil	3 c	$2.96 per qt	2.22
flour	12 oz	$0.44 per lb	0.33
salt	½ oz	$0.50 per lb	0.016
pepper	¼ oz	$3.90 per lb	0.061
		Total cost = $62.439	
		Cost per serving = **$1.25**	

Glossary

A

additive: Substance combined with certain foods to extend keeping qualities and shelf life.

agar: A seaweed product with gelatinous properties used as a thickening agent.

aging: Holding meat at a temperature of 34°F to 36°F to improve its tenderness.

agneau: French term for lamb.

a la: With, in the manner or fashion of.

a la Americaine: In the American style or fashion.

a la Anglaise: In the English style or fashion.

a la bouquetiere: Served with a variety of vegetables in season. Usually associated with broiled meat or fish surrounded with a variety of colorful vegetables.

a la bourgeoisie: Plain, family-style. Meats garnished with assorted vegetables that are cut into large sections.

a la brioche: Cooked on a skewer.

a la carte: Food ordered separately. Generally, the food is prepared to order.

a la goldenrod: Hard-cooked egg whites that are coarsely chopped and placed in a cream sauce, served on toast, and garnished with hard-cooked yolks.

a la Holstein: A fried veal cutlet served on tomato sauce with a fried egg on top, and garnished with lemon, capers, and anchovy.

a la Italienne: In the Italian style or fashion.

a la king: Foods served in a white sauce including mushrooms, green pepper, and pimientos. Commonly flavored with sherry.

a la maison: Specialty of the house.

a la Marengo: Sautéed chicken simmered in a brown sauce consisting of wine, tomatoes, mushrooms, and ripe and green olive slices.

a la Maryland: Disjointed chicken breaded with bread crumbs and deep fat fried. Served with cream sauce, crisp bacon, and corn fritters.

a la mode: 1. Ice cream served on top of pie or cake. **2.** Beef prepared and served in a special way, such as beef a la mode.

a la Newburg: Cream sauce colored with a small amount of paprika and flavored with sherry. Usually associated with seafood.

a la provencale: With garlic and oil.

a la reine: To the queen's taste. Commonly applied to soup to indicate the presence of finely chopped chicken or turkey white meat.

a la Russe: In the Russian style or fashion.

a la Suisse: In the Swiss style or fashion.

al dente: Slightly chewy or firm to the bite. Usually associated with cooked vegetables and pasta.

allemande: White sauce that includes egg yolks.

almond paste: Paste made from blanched and ground almonds, sugar, and egg whites. Used in making marzipan.

amandine: Prepared or served with almonds.

ambrosia: Dessert of assorted fruits and shredded coconut.

anchois: French term for anchovy.

anchovy: A small fish in the herring family. Anchovies are salted and packed in oil when canned.

anisette: A cordial liquor flavored with anise seed.

antipasto: An Italian salad used as an appetizer.

apple strudel: A thin, rolled-out dough made from water, flour, and fat or oil. Filled with thin apple slices, flavored, and rolled.

apprentice: In the culinary field, an individual working with an experienced cook or baker for the purpose of learning to be a cook or baker.

arrowroot: A starch extracted from the roots of a West Indian plant. Used as a thickening agent in certain soups and sauces and also brings forth a high sheen.

artichoke: Vegetable consisting of compact thistle-like leaves and a "choke" or center portion that is discarded.

arugula: Rich green, intensely flavored, spicy leaves used in salads.

aspic: A clear meat, fish, or poultry jelly.

au (aux): With, in the manner or fashion of.

au four: Baked in an oven.

au gratin: Foods covered with a sauce, sprinkled with cheese and/or bread crumbs, and baked to a golden brown.

au jus: With natural juices.

au lait: With milk.

aux croutons: With croutons.

avocado: Pear-shaped tropical fruit with a thick skin and green buttery flesh. Also known as an alligator pear.

B

baba au rhum: Small rum-flavored cake usually served with a topping of whipped cream.

bagels: Crisp, hard rolls in the shape of a ring.

bain-marie: 1. Pan or container of hot water into which other pans, containing food, are placed to keep hot. **2.** A stainless steel food storage container.

bake: To cook foods by dry heat.

baked Alaska: Dessert consisting of ice cream on cake covered with meringue and delicately browned in a quick oven.

baking sheet: Large pan (approximately 18″ × 26″) with shallow sides to give maximum exposure to oven heat. Also referred to as a bun pan.

banquet: Elaborate and often ceremonious meal attended by many people.

barbecue: 1. To cook over the embers of an open fire. **2.** A highly seasoned tomato-based sauce.

barde: To cover poultry or game with thin slices of bacon or salt pork when roasting to inject flavor and juice.

baste: To ladle drippings over food while cooking to prevent drying out.

batch: Quantity of material prepared at one time.

batter: A semi-fluid mixture of flour, sugar, eggs, milk, etc.

Bavarian: A dessert of gelatin and whipped cream that is folded together before setting.

bearnaise: Hollandaise sauce combined with a tarragon vinegar mixture.

beat: To mix to inject air and create a smooth mixture.

bechamél: A white sauce consisting of milk or cream thickened with roux.

beef a la stroganoff: Sautéed thin slices of beef tenderloin poached in a sour cream sauce.

beignet: French term meaning fritters.

Belgian endive: Salad greens of tapering white, compact leaves with a slight greenish tinge.

bercy: A brown sauce consisting of shallots, lemon juice, and white wine. Usually served with meat or fish.

Bibb lettuce: Cup shaped, delicate, sweet flavored tender green leaves used for salads.

bigarade: A sweet-sour brown sauce flavored with orange peel and juice. Usually served with roast duck.

bind: To hold or stick together.

biscuit: Small round quickbread made light with baking powder.

biscuit tortoni: A mousse frozen in individual paper cups and sprinkled with macaroon crumbs that have been soaked in sherry or rum.

bisque: A thick cream soup usually made from shellfish.

blanc: French term meaning white.

blanch: To partially cook an item under boiling water for a short period of time.

blancmange: Molded white pudding made of milk, sugar, and cornstarch.

blanquette: A stew of chicken, veal, or lamb in a white sauce.

bleeding: Dough that has been cut and left unsealed, permitting air and gas to escape.

blend: To thoroughly mix two or more ingredients.

blintz: Russian pancakes, usually served with caviar.

blue points: Small oysters served raw on the half shell.

boeuf: French term meaning beef.

boil: To cook foods in liquid at approximately 212°F.

bombe: A molded dessert consisting of two or more ice creams or sherbets.

bordelaise: A brown sauce flavored with red wine, usually served with beef entrees.

bordure: A ring of vegetables, usually duchess potatoes, surrounding a food item.

borscht: A Russian soup consisting of beef stock, tomatoes, lemon juice, and beets, usually topped with sour cream.

Boston cream pie: A two-layer sponge cake filled with cream filling and topped with icing or fruit.

Boston lettuce: Soft, tender, pale green leaves with a delicate flavor used for salads.

bouchee: Petite patty shells usually filled with a savory paste of meat or fish.

bouillabaisse: Thick soup or stew made with five or six different fish or shellfish, flavored with white wine, and seasoned with saffron.

bouillon: A rich liquid, similar to a stock, usually made of beef.

bouquet garni: A combination of herbs tied together in a small cheesecloth bag for seasoning which is cooked with the food and then removed when the proper flavor has been obtained.

bourguignonne: With Burgundy wine.

braise: To brown an item and then cook slowly in a covered pan.

brandy: Alcoholic liquor distilled from wine or fermented fruit juice.

braten: German term meaning roast.

bread: To coat an item with flour, egg wash, and bread crumbs.

break: To separate ingredients bound by an emulsification, such as hollandaise and bearnaise sauces.

breton: Items that contain or are garnished with beans.

brew: To extract color and flavor by infusion, such as in coffee and tea.

brine: A liquid of water or vinegar and salt used for pickling.

brioche: 1. A type of rich roll with more eggs and butter than the average roll. 2. Traditional French breakfast cake.

brochette: Meat or other foods broiled or roasted on a skewer.

broil: To cook by direct heat from above.

broth: Liquid in which meat, fish, poultry, or vegetables have been simmered.

brown betty: A pudding with apples, bread or cake crumbs, spices, and sugar. Usually served with a vanilla or lemon sauce.

brunoise: Assorted vegetables cut into small squares, used to garnish soups and consommés.

buffet: A display of ready-to-eat hot and cold foods. It is generally self-service.

buttercream: An icing made mainly of butter and/or shortening and confectioner's sugar.

butterflied: Cut partially through, spread open, and flattened to increase its surface area. Pork chops are often butterflied.

C

cacciatore: Sautéed chicken baked in a seasoned (basil and oregano) tomato sauce with diced mushrooms and chives.

cafe: French term meaning coffee.

cafe au lait: Beverage consisting of equal parts coffee and hot milk.

cafe noir: Black coffee.

calorie: 1. The amount of heat required to raise 1 g of water 1°C. 2. A measure of food energy.

Camembert: A soft, full-flavored cheese made in the region of Camembert, France. Commonly served as a dessert.

Canadian bacon: Smoked loin of pork that has been trimmed and pressed.

canape: Small open-faced appetizer consisting of a toasted bread or cracker covered with a savory paste.

canard: French term meaning duck.

candy: To cook certain fruits or vegetables in a heavy sweetened syrup.

cannelloni: An Italian pasta shaped like a tube with ends cut at an angle. Frequently stuffed with meat or cheese and served as an entree.

caper: Marinated flower bud used for seasoning or garnish.

capon: Castrated male chicken noted for its flavorful, textured meat.

carafe: Glass container with a narrow neck and spherical body commonly used to serve wine.

caramel: Heavily browned sugar used for coloring and flavoring.

carbohydrate: Compound, such as sugar or starch, composed of carbon, hydrogen, and oxygen. Carbohydrates supply energy to the body.

carte du jour: Menu of the day.

carve: To cut meat or poultry into slices or pieces for serving.

casaba melon: Large oval-shaped melon with yellow skin and white flesh.

casserole: Earthenware dish in which certain food items are baked and served.

cater: To provide food and service for a social or group affair.

caviar: Salted roe (eggs) of a sturgeon.

Chablis: A white, good-bodied wine, sometimes referred to as white Burgundy.

champignon: French term meaning mushroom.

chantilly sauce: Hollandaise sauce with unsweetened whipped cream folded in.

charlotte: A mold lined with ladyfingers and filled with fruit and whipped cream or custard.

charlotte russe: A mold lined with ladyfingers and filled with a Bavarian cream.

chasseur: French term meaning hunter style. A sauce consisting of equal parts of brown sauce and tomato sauce with mushrooms, onions, and lemon juice.

chateaubriand: A thick beef tenderloin steak weighing approximately 1 lb.

chaud-froid: Jellied white sauce used to cover cold decorated meats.

cheese: A dairy food consisting of coagulated, compressed, and usually ripened curd of milk separated from the whey.

chef: Person in charge of the kitchen or a department of the kitchen.

cherries jubilee: Dark, sweet cherries in slightly thickened syrup with Kircshwasser added. The preparation is served aflame with ice cream.

chicory: Lacy, fringed leaves with a somewhat bitter taste used for salads. The darker, outer leaves are stronger in flavor than the bleached inner leaves.

chiffonade: Shredded or chopped vegetables used in soups or salad dressing.

chili: **1.** Hot red, yellow, or green pepper. **2.** A sauce made with browned ground meat and onions simmered in a liquid with chili powder.

chili con carne: Chili preparation containing beans.

Chinese celery cabbage: Tapered stalks with crispy, crinkly-edged, celery-colored, white ribbed leaves used for salads.

chive: Long, slender onion-like sprouts with a mild flavor, used mainly in sauces and salads.

chlorophyll: Green pigment in fruits and vegetables.

chop: To cut into irregular pieces using a knife or other type of sharp tool.

choux paste: A paste of eggs, water, salt, shortening, and flour used in making eclairs and cream puffs.

chowder: A thick soup of fish, shellfish, and/or vegetables with milk and diced potatoes added.

chutney: A relish of fruits and spices usually served with curry dishes.

cilantro: Spanish name for coriander. The fresh leaves impart a lemony, pleasingly pungent flavor.

citron: A large lemon-like fruit with thicker skin and less acid.

clarify: To make clear or transparent and free from impurities. Butter is often clarified.

coagulate: The process of changing a liquid substance to a thickened mass.

coat: To cover the surface of a food. Meats are often coated with flour before frying.

cobbler: A deep-dish fruit pie.

cocoa: Finely ground and processed cacao bean.

cocotte: French term meaning small earthen cooking ware.

coddle: To cook or simmer slowly just below the boiling point. Eggs may be coddled.

colbert sauce: Sauce of brown sauce, shallots, claret wine, butter, and lemon juice.

compote: Fruits stewed in a syrup.

condiment: A seasoning for food, such as a spicy or pungent relish.

consommé: Clear soup made from well-seasoned stock.

convection oven: Oven in which heated air is circulated by a fan.

coq au vin: French term for chicken in wine.

core: To remove the center of a fruit or vegetable.

corned beef: Beef cured in a brine solution.

cottage pudding: Cake served with a warm sweet sauce.

coupe: **1.** Shallow dessert dish. **2.** Popular dessert consisting of diced fruit topped with whipped cream.

course: Part of a meal served at one time.

court bouillon: A liquid of water, vinegar, or wine and seasoning in which fish is poached.

couverture: Chocolate used for filling and finishing cakes, pastries, and cookies.

cranshaw melon: Oval melon with a mottled green and yellow skin and a sweet orange flesh.

cream: **1.** Working of one or more foods until soft and creamy. **2.** Yellowish part of milk containing 18% to 40% butterfat.

crecy: French term meaning items composed of, or garnished with, carrots.

crème: French term meaning cream.

crepe: French term meaning pancake.

crepes suzette: Thin pancakes folded and served aflame with a rich brandy sauce.

croissant: Crescent-shaped roll.

croquette: Ground food product bound together with a thick cream sauce and eggs. It is formed into balls or cones, breaded, and deep-fat fried.

crouton: Toasted or deep-fat fried bread cubes.

cruller: Long, twisted baking powder doughnut.

crustacean: Class of aquatic sea life with a shell and no backbone or spinal column. Common crustaceans include lobster, shrimp, crab, etc.

cube: To cut into small blocks with each side forming a square.

cuisine: A style of cooking.

curd: The thick casein-rich part of coagulated milk.

cure: To preserve by pickling, salting, or drying.

curry: East Indian stew or dish containing curry powder.

custard: A baked or boiled mixture of eggs, milk, sugar, and flavoring commonly served as a dessert.

cut in: To blend one part of a mixture into another.

cutlet: A small, flattened, boneless piece of meat.

D

dandelion greens: Sharply indented, tender, sightly bitter salad leaves.

deglaze: To add water to a pan in which meats have been sautéed or roasted in order to dissolve dried juices on bottom and sides of the pan.

demi: French term meaning half.

demiglacé: A rich brown stock reduced until it is half of the original amount.

demitasse: A small cup of black coffee.

devil: To flavor an item with a hot condiment such as pepper, mustard, or hot red pepper sauce.

dice: To cut into small cubes.

Dijon mustard: A prepared mustard, originally made in Dijon, France, which can be mild or highly seasoned.

dissolve: To cause a dry substance to be absorbed into a liquid.

divider: Device used to cut dough into equal portions.

dot: To place small particles of butter intermittently over the surface of an item.

dough: A thick, soft, uncooked mass of moistened flour associated with bread, cookies, and rolls.

drawn butter: Melted butter.

dredge: To coat an item with dry ingredients, usually flour.

dress: To trim and clean. Commonly associated with poultry and fish.

drippings: Fat and natural juices extruded from roasting meats.

drumstick: Poultry leg.

dry: 1. Without moisture. **2.** In spirits or wine, a low amount or absence of sugar content.

duchess potatoes: Boiled potatoes whipped with egg yolks and pressed through a pastry tube.

dugleré: White sauce of white wine, shallots, and crushed tomatoes.

du jour: French term meaning of the day.

dumpling: Starch product of simmered or steamed dough.

dust: To sprinkle an item with flour or sugar.

d'uxelles: Stuffing of mushrooms, shallots, and seasoning with a base of tomatoes or brown sauce.

E

eclair: Thin, oblong shell of choux paste filled with cream filling and iced.

eggplant: Large, dark purple, pear-shaped vegetable.

eggs benedict: Poached eggs and broiled ham on a toasted English muffin covered with hollandaise sauce.

emince: French term meaning to cut fine.

emulsion: Uniform mixture of one liquid suspended in another.

en: French term meaning in or on, such as in en coquille.

en brochette: To cook on a skewer.

en chemise: With skin on. Commonly associated with potatoes.

enchiladas: Mexican dish of tortillas filled with meat or cheese and rolled. Usually topped with melted cheese.

en coquille: To serve in a shell.

English muffin: Round bread baked on a griddle.

enriched: Resupplied with vitamins and minerals lost during processing.

en tasse: To serve in a cup.

entree: Main course.

entremets: French term meaning desserts.

epigramme: Small cutlet of tender meat.

escallop: 1. To cut into thin slices. **2.** To bake in a white sauce with a topping of crumbs.

escargot: French term meaning snail.

escarole: Wide, flat, slightly bitter salad leaves, curly at the edges with green outer leaves and yellow or cream colored inner leaves.

Escoffier: 1. Famous French chef (1846 – 1935). **2.** Trade name for a bottled table sauce.

espagnole: Sauce of brown stock thickened with roux.

essence: Extract of meat flavors.

etuver: To cook or steam an item in its own juices.

extract: To pull or draw out.

F

farce: French term meaning stuffing.

farci: French term meaning stuffed.

feather bones: Small bones joining vertebrae in ribs of beef.

fennel: Feathery leaves on the stalks of the fennel bulb that are used for salads. It has an anise flavor.

fermentation: Chemical breakdown of an organic compound caused by the action of living organisms.

fermiere: Served with small slices of carrots, turnips, onions, potatoes, cabbage, and celery.

filet mignon: Thick slice of beef tenderloin, usually free of all fat.

fines herbes: Mixture of three or four finely chopped herbs.

Finnan haddie: Lightly smoked haddock.

flake: To separate into small scale-like particles.

flambé: Served aflame.

flambeau: To serve on a flaming torch.

flan: An open tart containing fruit. It is traditionally baked in a metal flan ring.

Florentine: With spinach.

flute: To twist the edges of a pie shell.

foie gras: Liver of a fattened goose.

fold: To add ingredients carefully so as not to lose air bubbles.

fondant: An icing or candy made by boiling sugar and water to the crystallization point and whipping to a creamy mass.

fond lie: Sauce of a rich brown stock with cornstarch or similar starch to thicken. Used as a base for brown sauces and variations.

fondue: A warm cheese entree into which cubes of bread, meat, or fruit are dipped.

forcemeat: Finely ground and highly seasoned meat or fish bound together and used for stuffing.

forestiere: Garnished with mushrooms.

fortified: Supplied with more vitamins and minerals than were present in the natural state.

Francaise: In the French style or fashion.

frappe: Partially or completely frozen.

Frenched: To scrape meat and fat from the flank end or the ribs.

French toast: Bread dipped in a batter of milk and eggs and fried on both sides until golden brown.

fricassee: Pieces of chicken, lamb, or veal stewed in a liquid and served in a sauce of the same liquid. No browning occurs before stewing.

frit: French term meaning fried.

fritter: Food dipped or coated with a batter and deep-fat fried to a golden brown.

froid: French term meaning cold.

fromage: French term meaning cheese.

fry: To cook in oil or fat in a skillet.

fumet: A stock of fish, meat, or game reduced with wine until concentrated.

G

garbanzo: Chickpeas.

garde manger: French term meaning guardian of the cold meats. Usually refers to the cold meat department or person in charge of it.

garnish: To decorate a dish with an edible item to improve its appearance.

garniture: French term for garnish.

gateau: French term meaning cake.

gazpacho: A cold vegetable soup of Spanish origin.

gefilte fish: Jewish entree consisting of fish fillets stuffed with a ground fish mixture and poached.

gelée: Jelly or jellied.

gherkin: A small pickled cucumber.

giblets: Gizzard, heart, and liver of poultry.

glacé: To cover with a glazed coating.

glacé de viande: Reduced meat glaze.

glaze: To coat an item with a glossy coating.

gnocchi: Italian potato dumpling.

goulash: A rich Hungarian stew seasoned with paprika.

grate: To wear into small particles by rubbing on a rough surface.

green meat: Meat that has not had enough time to develop flavor and tenderness after slaughter.

griddle: A large heavy cooking plate with heat applied from the bottom.

grill: To cook on a griddle.

gristle: Hard elastic tissue found in meat.

grits: Coarsely ground hominy.

Gruyère: Type of cheese resembling Swiss cheese. Made in France and Switzerland.

guacamole: Paste made of mashed avocado and seasoned with condiments. Commonly served as an appetizer dip.

guava: Tropical pear-shaped fruit used for jelly and jam.

gumbo: A rich Creole dish of chicken broth, onion, celery, green peppers, okra, tomatoes, and rice.

gum paste: White paste of confectionery sugar, gelatin or soaked gum tragacanth, and water. Used for molding decorative pieces.

gum tragacanth: A gum obtained from various Asian or Eastern European plants used to give firmness.

H

hacher: French term meaning to hash or mince.

hard sauce: Dessert sauce or thick cream of butter and sugar with flavoring as desired.

haute cuisine: Classical French cuisine.

head cheese: A spiced and jellied meat of the hog's head and other edible parts.

hearth: Heated floor of an oven.

heifer: A young cow that has not borne a calf.

herbs: Savory leaves such as tarragon, sage, basil, and parsley used for seasoning.

homard: French term meaning lobster.

hominy: Hulled corn, coarsely ground or broken, usually boiled.

homogenize: To break fat globules into very small particles.

honeydew melon: Pale-skinned melon with sweet green flesh.

hors d'oeuvres: Appetizers served as the first course of a meal.

humidity: Amount of moisture in the air.

hush puppies: A southern food of deep-fried cornmeal shaped into small balls.

I

iceberg lettuce: Crisp, broad, green leaves on the outside and pale leaves on the inside of the head are used for salads. It has a sweet, mild flavor.

Indian pudding: Dessert of yellow cornmeal, eggs, brown sugar, milk, raisins, and seasoning.

Indienne: Generally refers to dishes flavored with curry powder.

Irish stew: Stew of lamb, carrots, turnips, potatoes, onions, dumplings, and seasoning.

Italienne: In the Italian style or fashion. Usually contains pasta.

jambalaya: Creole dish of meat or seafood, rice, tomatoes, onions, and seasonings.

jambon: French term meaning ham.

jardiniere: Garnished with fresh garden vegetables including carrots, celery, and turnips cut approximately $\frac{1}{4}''$ thick by $1''$ long.

julienne: To cut into long thin strips.

jus: Natural meat juice.

K

kartoffel klosse: German potato dumpling.

kebab: Small cubes of meat and/or vegetables arranged on a skewer.

kirschwasser: A liqueur made from cherries. It is frequently used to flame certain dishes.

knead: To press, fold, and stretch the air out of dough.

kohlrabi: Vegetable of the cabbage family with an enlarged edible turnip-shaped stem.

kosher: Food processed according to Hebrew religious customs.

kuchen: German cake made with sweet yeast dough.

kummel: Liqueur flavored with caraway seed.

kumquat: Small, orange-colored citrus fruit used primarily in making preserves.

lait: French term meaning milk.

larding: To insert strips of salt pork into meat to add flavor and prevent drying.

leaf lettuce: Broad, tender, loose, smooth leaves that grow from a central stem are used for salads. It has a delicate flavor. The leaves are curly and may be green-tipped or red-tipped.

leek: Member of the green onion family. Its green stems are used to season foods.

legumes: Vegetables consisting of beans, lentils, and split peas.

lentil: Flat edible seed of the pea family used in soup.

liaison: A thickening agent for sauces consisting of cream and egg yolks.

limburger cheese: Soft, pungent cheese originally made in Belgium.

London broil: Boneless cut of beef, usually a flank steak, sliced on the bias and commonly served with a rich mushroom or Bordelaise sauce.

lox: Smoked salmon that has been cured in brine.

lyonnaise: To prepare and serve with onions.

macaroon: Small cookie made of sugar, egg whites, and almond or kernel paste.

Madeira: Wine, similar to sherry, produced in the Madeira Islands.

madrilene: A consommé flavored with tomato.

maître d'hotel: Person in charge of the dining room service.

malt vinegar: Vinegar with a deep brown color made from fermented barley.

maraschino: A cherry preserved in an imitation maraschino liqueur.

marbled: Visible streaks of fat within meat indicating the quality of a steak or roast.

marinade: Brine or pickling solution in which meat is soaked before cooking to change or enrich the flavor.

marinate: To soak an item in a marinade.

marmite: An earthenware pot in which soup is heated and served.

marrow: Soft tissue from the center bones.

Marsala: A semi-dry Italian sherry.

masking: To cover an item completely, usually with a sauce.

matzo: Thin pieces of unleavened bread eaten by people of Jewish descent during Passover.

matzo balls: Small dumplings prepared with matzo flour and cooked in broth.

mayonnaise: A rich salad dressing emulsified by combining eggs, oil, and vinegar.

medallion: Small round or oval serving of food, commonly a meat fillet.

melba: Dessert consisting of fruit on ice cream covered with a melba sauce.

melba toast: Thin toasted slices of white rolls or bread.

menthe: French term meaning mint.

menu: 1. Bill of fare. **2.** Order in which foods are served at a meal.

meringue: Egg whites and sugar beaten together to form a white frothy mass. Used to top pies and cakes.

meunière: Rolled lightly in flour and sautéed in butter.

mignon: Petite or small piece, usually of beef tenderloin.

milanaise: Style of preparation developed in Milan, Italy which uses pasta, such as minestrone milanaise.

mince: To cut into very small pieces.

mincemeat: A blended mixture of finely chopped cooked beef, currants, apples, suet, and spices.

minestrone: A thick Italian vegetable soup usually served with Parmesan cheese.

minute steak: Small, thin, boneless sirloin steak.

mirepoix: Diced vegetable mixture of carrots, onions, and celery.

mix: To combine two or more ingredients.

mix grill: A combination of any four broiled or grilled items. Usually lamb chop, bacon, sausage, and tomato slices.

mocha: Coffee flavoring commonly used for icings.

mold: 1. Metal form in which foods may be shaped. **2.** To obtain a desired shape by placing in a mold.

mollusk: Aquatic sea life with soft bodies covered by a hard shell, such as oysters, clams, and scallops.

mornay sauce: A rich cream sauce with eggs and Parmesan cheese.

mousse: 1. Chilled dessert made mainly of whipped cream, sweetening, and flavoring. **2.** Gelatin entree of ground poultry, meat, or fish lightened by the addition of whipped cream.

Mozzarella: A soft, unripened cheese with a rubbery texture. Commonly used in pizza.

mulligatawny: A thick East Indian soup consisting of chicken stock, rice, vegetables, and highly seasoned with curry powder.

mussel: An aquatic bivalve with a thin black shell closely related to the oyster and clam.

mutton: The flesh of mature sheep.

N

Napoleon: French pastry made by separating layers of puff paste with a cream filling and topping with fondant icing.

navarin: A rich, brown mutton stew garnished with carrots and turnips.

nesselrode: Mixture of diced fruits in a rum sauce used in puddings, pies, and ice cream.

noir: French term meaning black.

noisette: Small pieces of loin of lamb or pork without the bone and fat.

nougat: Confection of pastry consisting of sugar, almonds, and pistachio nuts.

O

O'Brien: With diced green pepper and pimiento.

oeuf: French term meaning egg.

omelet: Beaten eggs seasoned and fried in butter or grease. When the eggs begin to puff, they are rolled or folded over.

P

panache: Of mixed colors. Two or more kinds of one item in a dish.

papaya: A long, subtropical fruit used as a thirst quencher and a tenderizer.

papillote: Cooked and served in paper.

parboil: To partially cook or boil.

Parisienne: Cut into small round shapes.

parmentier: Served with potatoes.

Parmesan: A hard Italian cheese usually grated and used for seasoning.

parmigiana: Made or covered with Parmesan cheese.

parsley: Herb used to garnish foods.

pastry bag: Conical bag with a metal tip at the small end used to decorate foods.

pâte: Ground meat or liver paste.

paysanne: In the peasant style. A dish with diced or shredded vegetables.

perigord: Served with truffles.

persillade: Garnished with parsley.

petit: French term meaning small.

petite marmite: A strong consommé and chicken broth blended together and served with diamond-cut cooked vegetables, beef, and chicken.

petits fours: Small, decorated, fondant-iced cakes in various shapes.

pilaf: Rice cooked in chicken stock with minced onions and seasoning.

pimiento: Sweet red pepper.

pipe: To force dough or icing through a pastry tube producing a narrow stream.

piquant: A sharp, tart item.

pistachio nut: Small, thin-shelled, light green tropical nut.

plank: To serve meat or fish on a board or hardwood plank. Usually garnished with duchess potatoes and vegetables.

poach: To cook gently in slightly boiling water.

pocket: Cavity in a cut of meat to be stuffed or to hold forcemeat.

pois: French term meaning peas.

poisson: French term meaning fish.

poissonier: Fish cook.

polenta: Italian term meaning cornmeal.

polonaise: Polish garnish consisting of bread crumbs freshly browned and mixed with chopped parsley and hard boiled eggs.

pomme: French term meaning apple.

pommes de terre: 1. French term meaning apples of the earth. **2.** Potatoes.

popover: A puffed-up quickbread of milk, sugar, eggs, and flour.

portabella: Type of mushroom with a very large cap.

porterhouse steak: A cut of beef with the T-bone left in the loin. Contains sirloin and a large amount of tenderloin.

potage: Thick soup.

pot pie: Meat and vegetables in a rich sauce and covered with a pie crust.

poularde: French term meaning roasting chicken. The chicken weight is between 3½ lb to 5 lb.

poulet: French term meaning young fowl.

prawn: A large shrimp.

printaniere: Served with an assortment of spring vegetables cut into small pieces.

profiterole: Small round pastry filled with savory fillings and covered with a sauce.

proof: To allow yeast dough to rise by setting it in a warm, moist place.

prosciutto: Italian spicy ham that has been salted, pressed, and dried. It is generally served in very thin slices.

pullman bread: Rectangular loaf of bread.

pumpernickel: A dark, coarse bread made from coarsely ground unbolted rye. Known for its distinctive, slightly acidic flavor.

puree: 1. Food cooked and strained to produce a pulp. **2.** A thick soup.

Q

quahog: A large, thick-shelled, Atlantic Coast clam.

quartz lamp: Heating device to keep food hot.

quenelle: A poached oval dumpling, usually made of chicken or veal.

quiche: A custard cheese pie.

R

radicchio: Ruby red, slightly bitter leaves used with other salad greens.

ragout: Thick, savory brown stew.

raisin: Dried grape.

ramekin: Small, shallow baking dish in which foods are baked and served.

rasher: Three slices of bacon.

ravigote: A tart cold sauce of a mayonnaise base, chopped green herbs, and tarragon vinegar.

ravioli: Small, square pieces of noodle dough filled with seasoned ground meat and spinach and served with a meat sauce.

reconstitute: To restore a food product, such as frozen juice, to its original consistency.

reduction: Liquid that has been concentrated by cooking until part of the water has evaporated.

remoulade sauce: A seasoned cold sauce with mustard and ground pepper. Similar to tartar sauce.

render: To extract grease from animal fat.

risotto: A rice dish baked with minced onions, Parmesan cheese, and meat stock.

rissole: French term meaning browned.

roe: Fish eggs.

romaine: Firm, crisp, nutty-flavored, dark green salad leaves.

Roquefort: French blue-veined cheese.

roti: French term meaning roast.

rotisseur: Cook who prepares broiled, roasted, and braised meats.

rouge: French term meaning red.

roulade: Rolled meat.

roux: Mixture of equal parts of flour and fat used to thicken liquids.

royale: Mixture of cream and eggs baked into a custard and used as a garnish for consommé and broth.

S

saccharin: Coal tar product used as a substitute for sugar. It has no food value.

sachet bag: Small cloth bag of selected herbs used to season stock or soup.

saffron: Seasoning made of the dried orange stigmas of a purple crocus.

salamander: A small broiler used to brown and glaze servings of certain preparations.

salami: A highly-seasoned dried sausage of pork and beef.

salmonella: Food-borne disease spread by poor sanitation and improper food handling.

saturated fat: Fat that is normally solid at room temperature.

saucier: Cook who prepares the sauces, stews, and sautéed foods.

sauerbraten: A sour beef dish in which the meat is marinated in a vinegar solution and served with a sour sauce.

sauté: To cook quickly in shallow grease.

sauterne: A French white wine of the Bordeaux region.

sautoir: A heavy, flat, copper saucepan.

scald: 1. To heat to a point just below the boiling point. **2.** To dip an item into very hot or boiling water to facilitate removal of outer skin.

scallion: A green onion with a long, thick stem and a small bulb.

scallop: 1. The muscle of a mollusk that operates the opening and closing of the two shells. **2.** To bake in a sauce topped with bread or cracker crumbs.

scampi: 1. Shellfish similar to a large shrimp. **2.** Large shrimp that has been broiled and brushed with garlic butter.

scone: Scottish hot bread or cake in a triangular shape that is cooked on a griddle or baked in the oven.

score: To make superficial cuts to improve appearance or increase tenderness.

scrod: Young cod or haddock.

sear: To brown the surface of meat by intense heat.

shad: A saltwater fish with markings similar to whitefish. It is valuable because of its highly prized roe (eggs).

shallot: A small onion-like vegetable with a strong flavor.

sherbet: Frozen dessert of fruit juices, milk, and sugar.

shirred egg: Egg baked in a shallow casserole with butter.

shred: To cut into thin strips.

shuck: To remove the meat of bivalves such as oysters and clams.

sift: To pass dry ingredients through a fine screen or sieve.

simmer: To cook in liquid that is just below the boiling point (185°F to 200°F).

slurry: A mixture of cold liquid and raw starch used as a thickening agent.

smorgasbord: A buffet of light appetizers displayed for self-service as a first course.

smother: To cook in a covered container until tender.

sole: A white meat flatfish from the Atlantic and Pacific Oceans.

sorrel: Sour, pungent-flavored, long, arrow-shaped, green leaves are combined sparingly with milder greens for salads.

soufflé: A light food item made by folding beaten egg whites into a basic batter.

spaetzel: A heavy Austrian noodle prepared by running a heavy batter through a large-hole colander and into boiling stock.

spinach: Dark green, crisp, flat or crinkled leaves served as a vegetable or used in salads.

spit: Pointed metal rod for roasting meats over an open fire.

spoon bread: Type of southern corn bread baked in a casserole and served with a spoon.

spumoni: A rich Italian ice cream containing fruit and nuts.

squab: A young pigeon that has never flown.

steak tartare: Raw, highly-seasoned ground steak. Usually served with a raw egg yolk and onions.

steam: To cook by using direct steam.

steep: To soak in a hot liquid to extract flavor and color.

steer: A young, male beef animal that has been castrated.

stew: 1. To simmer or boil in a small amount of liquid. **2.** A meat and vegetable dish in gravy.

stock: Liquid in which meat, poultry, fish, or vegetables have been cooked.

stroganoff: Sautéed pieces of beef tenderloin cooked gently in a sour cream sauce.

strudel: Viennese dessert consisting of a thin pastry dough filled with apples, cherries, and other fruits.

suet: Hard fat around the kidney and loins of mutton and beef animals used in making tallow.

sweetbread: Thymus gland of a calf or lamb.

Swiss chard: The leaves of a variety of beets. They are used as a vegetable and for salad.

Swiss steak: Tough beef cut into steaks and browned. It is then baked in its own juice until tender.

T

table d'hote: French term referring to a type of menu in which the price listed includes all the courses offered for the complete meal.

tamale: Mexican dish of crushed corn and ground meat seasoned with red pepper. It is dipped in oil and steamed in a corn husk wrapper.

tapioca: Starch prepared from the cassava plant. Used in puddings and for thickening soups.

tarragon: European herb used in cooking and to flavor vinegar.

tart: Small pies without a top crust filled with fruit or fruit and cream.

tasse: French term meaning cup.

tenderloin: A strip of tender meat taken from inside the loin cavity of beef, pork, lamb, and veal.

torte: A delicate, filled layer cake with fancy decoration.

tortilla: A flat, unleavened Mexican corn cake baked on a heated stone or iron.

toss: To mix with a rising and falling action.

tripe: The edible lining of a beef stomach.

truffle: A black fungus similar to the mushroom, grown mainly in France. They are used for seasoning and garnishing.

truss: To bind poultry with string or skewers for roasting.

turbot: A delicately flavored white meat fish.

tureen: A large deep vessel in which soup is served.

U

unsaturated fat: A type of fat that is liquid at room temperature.

V

velouté: 1. French term meaning velvety. **2.** A smooth creamy white sauce made by combining stock and roux.

vermicelli: Long fine strings of pasta, similar in appearance to spaghetti but thinner.

vert: French term meaning green.

viande: French term meaning meat.

vichyssoise: Cream of potato soup served cold.

vin: French term meaning wine.

vinaigrette: Cold sauce of vinegar, salad oil, chopped pickles, shallots, fine herbs, chopped hard-boiled eggs, chopped pimientos, salt, and pepper.

W

watercress: Crisp, refreshing, peppery-tasting sprigs with small, deep green leaves that are often combined with other greens for salads.

wellington: Beef tenderloin baked in a rich dough until tenderloin is slightly rare and crust is crisp and golden brown.

Welsh rarebit: Melted Cheddar cheese flavored with beer, mustard, and Worcestershire sauce and served hot over toast.

whey: The watery part of milk.

whip: To beat rapidly to increase volume and incorporate air.

whisk: To beat rapidly with a wire whip or a piano whip.

whitewash: Thickening agent consisting of equal parts of flour and cornstarch diluted in cold water.

wiener schnitzel: A breaded veal cutlet that is fried and served with a slice of lemon and anchovy.

wild rice: The brown seed of a tall northern water grass that is usually served with wild game.

Y

Yorkshire pudding: A batter of flour, milk, salt, and eggs that is baked with roast rib of beef.

Z

zest: A rind of lemon or orange.

zucchini: Italian squash resembling the cucumber.

Index